PETROLOGY OF LAMPROITES

PETROLOGY OF LAMPROITES

Roger H. Mitchell
Lakehead University
Thunder Bay, Ontario
Canada

and

Steven C. Bergman
ARCO Oil and Gas Company
Plano, Texas

PLENUM PRESS • NEW YORK AND LONDON

Library of Congress Cataloging-in-Publication Data

Mitchell, Roger H.
  Petrology of lamproites / Roger H. Mitchell and Steven C. Bergman.
     p. cm.
  Includes bibliographical references and index.
  ISBN 0-306-43556-X
  1. Lamproite. 2. Petrology. 3. Geochemistry. 4. Rocks, Igneous.
I. Bergman, Steven C. II. Title.
QE462.L35M57 1991
552'.3--dc20   90-25226
                                                                  CIP

ISBN 0-306-43556-X

© 1991 Plenum Press, New York
A Division of Plenum Publishing Corporation
233 Spring Street, New York, N.Y. 10013

All rights reserved

No part of this book may be reproduced, stored in a retrieval system, or transmitted in any form or by any means, electronic, mechanical, photocopying, microfilming, recording, or otherwise, without written permission from the Publisher

Printed in the United States of America

This work is dedicated to

REX T. PRIDER

Emeritus Professor of Geology
University of Western Australia

in recognition of his pioneering studies of
the mineralogy and petrology of lamproites.

Preface

In this book, the first dedicated entirely to the petrology of lamproites and their relationships to other potassium-rich rocks, the objective of the authors is to provide a comprehensive critical review of the occurrence, mineralogy, geochemistry, and petrogenesis of the clan. Although lamproites represent one of the rarest of all rock types, they are both economically and scientifically important and we believe the time is ripe for a review of the advances made in their petrology over the past two decades.

Many of these advances stem from the recognition of diamond-bearing lamproites in Western Australia and the reclassification of several anomalous diamond-bearing kimberlites as lamproites. Consequently lamproites, previously of interest only to a small number of mineralogists specializing in exotica outside the mainstream of igneous petrology, have become prime targets for diamond exploration on a worldwide basis.

Contemporaneously with these developments, petrologists realized that lamproites possess isotopic signatures complementary to those of midoceanic ridge basalts, alkali basalts, kimberlites, and other mantle-derived melts. These isotopic studies provided new insights into the long-term development of the mantle by suggesting that the source regions of lamproites were metasomatically enriched in light rare earth and other incompatible elements up to 1–2 Ga prior to the melting events leading to generation of the magma.

We believe that an understanding of the nature of lamproites is essential for geologists concerned with exploration for diamond, and petrologists and geochemists interested in continental alkaline magmatism and mantle evolution. In the book we present new interpretations of various aspects of lamproite petrology. Hence, the work is not simply a summary of existing information and accepted concepts; rather it proposes hypotheses that it is hoped, will serve to stimulate further studies of these unusual rocks.

The book is the work of two authors, and although in overall agreement regarding the content and philosophy of the text, each assumes primary responsibility for any errors or omissions to be found in his particular contributions. In addition to the style and overall editing of the work, Roger Mitchell was responsible for Chapters 1, 2, 6–8, and 10 and Steven Bergman was responsible for Chapters 3–5 and 9.

Alan Edgar and Henry Meyer are thanked for reviewing initial drafts of many of the chapters. Valerie Dennison is especially thanked for proofreading the entire work.

Many people have contributed to the production of this book. We are particularly appreciative of preprints, thin sections, and rock samples provided by and discussions with

Lynton Jaques, Barbara Scott Smith, Alan Edgar, Danielle Velde, Henry Meyer, Bram Janse, Leendert Krol, Peter Berendsen, John Lewis, Steve Foley, Gino Venturelli, Mike Roden, Peter Nixon, Carter Hearn, Ted Scamboz, Howard Coopersmith, Hugh O'Brien, Tony Irving, Ken Collerson, Tony Erlank, Nick Rock, and Dave Nelson. Special thanks go to Henry Meyer for providing many hours of microprobe time at Purdue University, Barbara Scott Smith for photographic work, and Ken Foland for geochronological data. J. Toney, D. J. Henry, and L. Liang assisted with some of the microprobe analyses in Dallas.

We thank P. A. Sheahan, K. Weissman, P. A. Carper, L. Batzle, and J. M. Prokesh for literature searches and the acquisition of citations. Technical assistance at ARCO in Dallas was provided by J. Talbot, V. Mount, S. R. Yang, F. Stiff, and E. Kinsel, and at Lakehead University by A. MacKenzie, M. Downey, A. Hammond, R. Viitala, W. Bons, and S. Millar. Sam Spivak drafted all of the original diagrams.

Mitchell's work on lamproites is supported by the Natural Sciences and Engineering Research Council of Canada and Lakehead University. Much of Bergman's contribution was supported by the Anaconda Minerals Company (Industrial Minerals Group) and the ARCO Oil and Gas Company Exploration and Production Research Laboratory, in conjunction with their (presently expired) diamond exploration and minerals research programs.

Roger Mitchell wishes to thank Valerie Dennison for her continued encouragement of his studies of lamproites and maintenance of an environment favorable to the preparation of this book.

Steve Bergman acknowledges the support and consideration provided by Mary, Allison, Everett, Louise, and Maggie Bergman.

Roger Mitchell
Thunder Bay
Steven Bergman
Dallas

Contents

Chapter 1.	**The Lamproite Clan—Etymology and Historical Perspective**	1
	1.1. Introduction	1
	1.2. Initial Discoveries—1870–1906	1
	1.3. Etymology of Lamproite	2
	1.4. Western Australian Discoveries—The Legitimization of Lamproites	3
	1.5. Johannsen and Modal Classifications of Rocks	4
	1.6. Further Developments, New Occurrences, and Diamond-Bearing Lamproites	5
	1.7. Reclassification of Anomalous Kimberlites	6
	1.8. Recent Developments	7
Chapter 2.	**Potassic Rocks and the Lamproite Clan**	9
	2.1. Alkalinity, Sodic, Potassic, and Ultrapotassic Rocks	9
	2.2. Alkali–Alumina Relationships	10
	2.3. Potassic Rocks—General Petrographic Characteristics and Terminology	11
	2.3.1. Lamproites	11
	2.3.2. Roman Province Leucitites	12
	2.3.3. Ultrapotassic Leucitites	13
	2.3.4. Kalsilite-Bearing Lavas (Kamafugitic Rocks)	14
	2.3.5. Potassic Lamprophyres	15
	2.3.6. Kimberlite and Micaceous Kimberlites	15
	2.3.7. The Shoshonite Association	16
	2.3.8. Potassic Intrusive Rocks	17
	2.4. Petrochemical Classifications of Potassic Rocks	17
	2.4.1. Niggli Parameters	18
	2.4.2. K_2O–Na_2O Diagrams	19
	2.4.3. Total Alkali–Silica Classifications	19
	2.4.4. Sahama (1974)	21
	2.4.5. Barton (1979)	23

		2.4.6. Bogatikov *et al.* (1985)	26
		2.4.7. Foley *et al.* (1987)	29
	2.5.	Petrographic and Mineralogical Classifications	31
		2.5.1. Petrographic Classifications	31
		2.5.2. Mineralogical Classifications	32
	2.6.	The Lamproite Clan	35
		2.6.1. Geochemical Criteria for Lamproite Recognition	37
		2.6.2. Mineralogical Criteria for Lamproite Recognition	37

Chapter 3. Description of Lamproite Occurrences—Distribution, Age, Characteristics, and Geological Framework — 39

- 3.1. Introduction — 39
- 3.2. North American and Greenland Lamproites — 40
  - 3.2.1. Group Name: Prairie Creek — 43
  - 3.2.2. Group Name: Leucite Hills — 46
  - 3.2.3. Group Name: Smoky Butte — 50
  - 3.2.4. Group Name: Francis — 52
  - 3.2.5. Group Name: Hills Pond — 55
  - 3.2.6. Group Name: Sisimiut — 57
  - 3.2.7. Group Name: Yellow Water Butte — 59
  - 3.2.8. Group Name: Froze-to-Death Butte — 61
- 3.3. European Lamproites — 63
  - 3.3.1. Group Name: Murcia-Almeria — 65
  - 3.3.2. Group Name: Sisco — 69
  - 3.3.3. Group Name: Sesia-Lanzo and Combin — 70
- 3.4. African Lamproites — 72
  - 3.4.1. Group Name: Kapamba — 73
  - 3.4.2. Group Name: Pneil, Postmasburg, Swartruggens — 75
  - 3.4.3. Group Name: Bobi — 76
- 3.5. Australian Lamproites — 79
  - 3.5.1. Group Name: Argyle — 80
  - 3.5.2. Group Name: West Kimberley — 83
- 3.6. Antarctic Lamproites — 86
  - 3.6.1. Group Name: Gaussberg — 86
  - 3.6.2. Group Names: Mount Bayliss and Priestley Peak — 87
- 3.7. Asian Lamproites — 89
  - 3.7.1. Group Name: Chelima — 90
  - 3.7.2. Group Name: Majhgawan — 93
  - 3.7.3. Group Name: Barakar — 96
  - 3.7.4. Group Name: Murun — 97
- 3.8. South American Lamproites — 98
  - 3.8.1. Group Name: Coromandel — 98
- 3.9. Conclusions — 101

CONTENTS

Chapter 4. Tectonic Framework of Lamproite Genesis — 103
- 4.1. Age and Temporal Relations of Lamproite Magmatism — 104
- 4.2. Regional Geological and Tectonic Setting Generalizations — 107
  - 4.2.1. Lamproites and Plate Tectonics — 108
  - 4.2.2. Lamproites and Contemporaneous Subduction Zones — 108
  - 4.2.3. Fracture Zones, Transform Faults, and Continental Lineaments — 110
  - 4.2.4. Mantle Plumes, Hot Spots, and Hot Regions — 111
  - 4.2.5. Continental Rift Zones and Aulacogens — 114
  - 4.2.6. Orogeny and Postorogenic Relaxation — 115
- 4.3. Lithospheric History of Lamproite Settings — 116
  - 4.3.1. Regional Structure — 116
  - 4.3.2. Basement Age, Composition, and Evolution — 117
  - 4.3.3. Contemporaneous, Previous, and Subsequent Magmatism — 117
  - 4.3.4. Importance of Paleosubduction and Fossil Benioff Zones — 119
  - 4.3.5. Mantle Metasomatism—Another Necessary Condition — 120
  - 4.3.6. Comparison with Kimberlites, Alkali Basalts, Lamprophyres, and Potassium-Rich Rocks — 121
- 4.4. Tectonic Framework of Four Mesozoic-to-Cenozoic Lamproite Type-Locality Magmatic Fields — 122
  - 4.4.1. Leucite Hills — 122
  - 4.4.2. West Kimberley — 122
  - 4.4.3. Murcia-Almeria — 123
  - 4.4.4. Prairie Creek — 123
- 4.5. Conclusions and Preferred Model — 123

Chapter 5. Petrological Facies and Igneous Forms of the Lamproite Clan — 125
- 5.1. Introduction — 125
  - 5.1.1. Historical Development — 126
  - 5.1.2. Lamproite Igneous Forms — 126
  - 5.1.3. Lamproite Facies Classification — 127
- 5.2. Lava Flow Facies — 128
  - 5.2.1. Examples of Lava Flow Facies Lamproites — 130
  - 5.2.2. Comparisons with Other Mafic and Intermediate Lavas — 136
  - 5.2.3. Summary — 137
- 5.3. Crater and Pyroclastic Facies — 137
  - 5.3.1. Vent Morphology — 138
  - 5.3.2. Pyroclastic Fall Deposits — 139
  - 5.3.3. Base Surge and Pyroclastic Flow Deposits — 142
  - 5.3.4. Epiclastic Deposits — 146
  - 5.3.5. Examples of Crater and Pyroclastic Facies Lamproites — 147

	5.4.	Hypabyssal Facies	157
		5.4.1. Examples of Hypabyssal Facies Lamproites	158
	5.5.	Plutonic Facies	161
	5.6.	Generalized Model of Eruptive Sequences	162
	5.7.	Comparison with Kimberlite Diatremes	165
	5.8.	Summary	166

Chapter 6. Mineralogy of Lamproites — 169

- 6.1. Phlogopite — 169
  - 6.1.1. Paragenesis — 169
  - 6.1.2. Leucite Hills, Wyoming — 172
  - 6.1.3. West Kimberley, West Australia — 179
  - 6.1.4. Argyle, West Australia — 190
  - 6.1.5. Prairie Creek, Arkansas — 191
  - 6.1.6. Murcia-Almeria, Spain — 192
  - 6.1.7. Smoky Butte, Montana — 195
  - 6.1.8. Kapamba, Zambia — 195
  - 6.1.9. Sisimiut, Greenland — 197
  - 6.1.10. Hills Pond—Rose Dome, Kansas — 198
  - 6.1.11. Bobi, Ivory Coast — 198
  - 6.1.12. Francis, Utah — 199
  - 6.1.13. Yellow Water Butte, Montana — 200
  - 6.1.14. Sisco, Corsica — 201
  - 6.1.15. Majhgawan, India — 201
  - 6.1.16. Gaussberg, Antarctica — 203
  - 6.1.17. Probable Lamproites—Murun, Presidente Oligario, and Chelima — 203
  - 6.1.18. Minor and Trace Elements — 205
  - 6.1.19. Summary of Mica Compositional Variations — 206
  - 6.1.20. Solid Solutions in Lamproite Mica — 207
  - 6.1.21. Comparisons with Mica in Other Potassic Rocks — 211
  - 6.1.22. Conditions of Crystallization — 215
- 6.2. Amphibole — 217
  - 6.2.1. Classification — 217
  - 6.2.2. Paragenesis — 218
  - 6.2.3. Composition — 219
  - 6.2.4. Minor and Trace Elements — 229
  - 6.2.5. Comparisons with Amphiboles in Other Potassic Rocks — 230
  - 6.2.6. Conditions of Crystallization — 232
- 6.3. Clinopyroxene — 233
  - 6.3.1. Paragenesis — 233
  - 6.3.2. Composition of Phenocrystal and Groundmass Pryoxenes — 234
  - 6.3.3. Compositions of Pyroxenes from Parageneses 2 and 3 — 239

	6.3.4. Comparison with Pyroxenes in Other Potassic Rocks	240
6.4.	Orthopyroxene	242
	6.4.1. Paragenesis	242
	6.4.2. Composition	242
6.5.	Olivine	244
	6.5.1. Paragenesis	244
	6.5.2. Composition	245
	6.5.3. Summary	249
6.6.	Leucite	249
	6.6.1. Paragenesis	249
	6.6.2. Composition	251
6.7.	Analcite	253
6.8.	Sanidine	254
	6.8.1. Paragenesis	254
	6.8.2. Composition	255
6.9.	Spinel	257
	6.9.1. Paragenesis—Primary Spinels	257
	6.9.2. Composition—Primary Spinels	258
	6.9.3. Comparison with Primary Spinels from Kimberlite and Lamprophyre	264
	6.9.4. Secondary Aluminous Spinels	266
6.10.	Priderite	268
	6.10.1. Paragenesis	268
	6.10.2. Composition	268
	6.10.3. Ba-Titanates Related to Priderite	272
	6.10.4. Comparison with Hollandites in Other Potassic Rocks	272
6.11.	Jeppeite	273
6.12.	Iron Titanium Oxides	274
	6.12.1. Armalcolite	274
	6.12.2. Ilmenite	276
	6.12.3. Titanium Dioxides	278
6.13.	Potassium Zirconium Silicates	279
	6.13.1. Wadeite	279
	6.13.2. Dalyite	281
6.14.	Apatite	282
	6.14.1. Paragenesis	282
	6.14.2. Composition	283
6.15.	Perovskite	286
	6.15.1. Paragenesis	286
	6.15.2. Composition	286
6.16.	Titanosilicates	287
	6.16.1. Shcherbakovite	287
	6.16.2. Davanite	289
	6.16.3. Unnamed K-Ba-Titanosilicate	290
6.17.	Minor Accessory and Secondary Minerals	290

	6.17.1. Zeolites	291
	6.17.2. Carbonates	291
	6.17.3. Roedderitelike Phases	291
	6.17.4. Cerium-Bearing Minerals	292
	6.17.5. Silicon Dioxide	293
	6.17.6. Barite	294
	6.17.7. Other Minerals	294
	6.17.8. Secondary Phases	294
Chapter 7.	**The Geochemistry of Lamproites**	**295**
7.1.	Major Element Geochemistry	295
	7.1.1. General Characteristics	295
	7.1.2. Compositional Relationships to Other Potassic Lavas	298
	7.1.3. Intraprovincial Characteristics	301
7.2.	Compatible Trace Elements	314
	7.2.1. First Period Transition Elements	314
	7.2.2. Platinum Group Elements	318
7.3.	Incompatible Trace Elements—1: Ba–Sr, Zr–Hf, Nb–Ta, Th–U	319
	7.3.1. Barium and Strontium	319
	7.3.2. Zirconium and Hafnium	323
	7.3.3. Niobium and Tantalum	325
	7.3.4. Thorium and Uranium	326
7.4.	Incompatible Trace Elements—2: Rare Earth Elements and Yttrium	327
	7.4.1. Rare Earth Elements	327
	7.4.2. Yttrium	337
7.5.	Incompatible Trace Elements—3: Alkali Elements	337
	7.5.1. Lithium	337
	7.5.2. Rubidium	338
	7.5.3. Cesium	240
7.6.	Volatile Trace Elements: Fluorine, Sulfur, and Chlorine	341
7.7.	Other Trace Elements	341
7.8.	Interelement Relationships	342
7.9.	Isotopic Composition	343
	7.9.1. Strontium and Neodymium	343
	7.9.2. Lead	348
	7.9.3. Oxygen	350
7.10.	Summary	351
Chapter 8.	**Experimental Studies Relevant to the Formation and Crystallization of Lamproites**	**353**
8.1.	Low-Pressure Studies of Lamproites	353
	8.1.1. Anhydrous Melting Relationships	353

	8.1.2.	Water-Saturated Melting Relationships	354
8.2.	High-Pressure Phase Relationships of Natural Lamproites		356
	8.2.1.	Phlogopite Lamproites	357
	8.2.2.	Olivine and Madupitic Lamproites	360
8.3.	Synthetic Systems		362
8.4.	The Oxidation State of Lamproite Magmas		367
8.5.	Summary		370

Chapter 9. Diamonds, Xenoliths, and Exploration Techniques — 371

9.1.	Diamonds and Xenoliths: Alien, Yet Beneficial, Companions of Lamproites		371
9.2.	Diamonds		371
	9.2.1.	Type, Morphology, Color, and Size	372
	9.2.2.	Mineral Inclusion Suite	372
	9.2.3.	Isotopic Composition	374
	9.2.4.	Comparison with Kimberlite Diamonds	374
	9.2.5.	Discussion	374
9.3.	Xenoliths and Xenocrysts		375
	9.3.1.	Xenoliths in Olivine Lamproites	375
	9.3.2.	Xenoliths in Leucite and Phlogopite Lamproites	377
	9.3.3.	Discussion	377
9.4.	Exploration Techniques for Diamondiferous Lamproite		378
	9.4.1.	Remote Sensing Techniques	379
	9.4.2.	Geophysical Techniques	380
	9.4.3.	Geochemical Surveys	382
	9.4.4.	Indicator/Heavy Mineral Sampling	383
	9.4.5.	Summary	384

Chapter 10. Petrogenesis of Lamproites — 385

10.1.	Introduction		385
10.2.	Previous Petrogenetic Models		385
	10.2.1.	Fractionation of Peridotite and Kimberlite	385
	10.2.2.	Contamination of Kimberlite	387
	10.2.3.	Partial Melting of Garnet Lherzolite	387
	10.2.4.	Incongruent Melting of Phlogopite	389
	10.2.5.	Fractional Fusion and Diapiric Uprise	390
	10.2.6.	Partial Melting of Enriched Mantle	392
	10.2.7.	Partial Melting of Harzburgitic Sources	393
	10.2.8.	Subduction-Related Models	395
10.3.	Genesis of the Lamproite Clan		396
	10.3.1.	Character of the Source	396
	10.3.2.	Melting of the Source	398
	10.3.3.	Relationships between Olivine and Phlogopite Lamproites and the Origins of the Isotopic Signatures of the West Kimberley Lamproites	399

	10.3.4.	Relationship of Madupitic Lamproites to Phlogopite Lamproites and the Origins of Isotopic Signatures of the Leucite Hills Lamproites	400
	10.3.5.	The Anomalous Murcia-Almeria Lamproites	401
10.4.	Relationships to Kimberlites		401
10.5.	Relationships to MARID-Suite Xenoliths		403
10.6.	Relationships to Other Potassic Rocks and Lamprophyres		404
10.7.	Summary		406

Postscript   407

References   409

Index   443

The trouble is not with classification but with nature which did not make things right.
  Albert Johannsen (1931)

The Lamproite Clan
Etymology and Historical Perspective

1.1. INTRODUCTION

Igneous rock classifications have never been graced with the elegance and simplicity of terminology employed in other branches of science. Consequently rock names present us with a bewildering plethora of unrelated, noninformative, and commonly unpronounceable terms. This unfortunate and unsatisfactory situation resulted from the empirical approach to rock nomenclature developed in the nineteenth century coupled with active resistance by petrologists to objective classifications. Petrographers, using the then recently invented polarizing microscope, coined a new name for each newly recognized assemblage of minerals. Different names were also proposed for rocks that differed only in the modal proportions of a particular mineral assemblage. The name was usually derived from the geographic locality at which the rock was discovered. Commonly, as a consequence of poor communication, different names were applied to rocks consisting of the same mineral assemblages.

In the case of alkaline rocks in general, and potassic rocks in particular, this type-locality approach to nomenclature resulted in the creation of a terminology that is daunting even to the specialist and incomprehensible to the neophyte. The method has burdened the literature with many unnecessary and cacophonous terms that may only be translated into mineral assemblages by the use of an appropriate lexicon. In addition, many of the type-locality definitions are so vague that they do not allow for precise description of the rock.

The terminology devised for the rocks now known as lamproites illustrates particularly well the confusion resulting from the legacy of the early years of igneous rock classification. To this day, this archaic nomenclature has obscured petrological relationships within this group and between other groups of potassic rocks.

1.2. INITIAL DISCOVERIES—1870–1906

In 1871 while undertaking a survey of the 40th parallel in the western United States, S. F. Emmons identified the first occurrence of leucite-bearing rocks on the North American continent. In recognition of this discovery, the series of buttes and mesas composed of these rocks in southwestern Wyoming became known as the Leucite Hills (Emmons 1877). Subsequent study of the occurrence by Whitman Cross (1897) resulted in the definition of three new rock types; *wyomingite* (after the state of Wyoming), *orendite* (after Orenda Mesa—etymology unknown), and *madupite* (from the Shoshone word "madupa" meaning "sweetwater" and the name of the county in which the Leucite Hills are located).

As a corollary to the introduction of the names, Cross (1897) noted that they could also be classified as leucite trachytes or leucite phonolites using classifications devised by Zirkel (1893) and Rosenbusch (1877), respectively. These names reflect the presence of modal leucite and sanidine but ignore the characteristic phlogopite. Designation as trachyte or phonolite implies the presence of plagioclase or nepheline, respectively, both of which are absent from the Leucite Hills rocks. The widespread acceptance of the Zirkel–Rosenbusch classifications and their derivatives has resulted in the perception, to this day, that the Leucite Hills rocks are members of the leucitite family (see 2.3.2). This misconception is exemplified by the extreme view of Rittman (1973), who regards madupite as synonymous with hauyne melilite leucitite, despite the obvious mineralogical differences, i.e., absence of hauyne or melilite. Clearly, problems of nomenclature surrounded these rocks from the time of their discovery, problems unfortunately compounded by further discoveries of similar rocks in southeastern Spain.

In 1889 Alfred Osann described a glassy volcanic rock, which he named *verite* (after the village of Vera) from Cabo de Gata in the Province of Almeria, Spain. Subsequently R. A. de Yarza (1895) described from Fortuna, in the province of Murcia, a rock termed *fortunite* associated with a "trachyte." Osann (1906) later determined that de Yarza's fortunite was identical to the Cabo de Gata verite and that the "trachyte" was a crystalline variety of verite. Thus, Osann (1906) proposed that the term verite be reserved for the glassy rocks and that the crystalline type be termed fortunite. Alas, Osann did not see fit to term the verites "hyalo-fortunites" and eliminate some of the terminological confusion.

Osann (1906) also proposed that a holocrystalline rock, found near the town of Jumilla, be called *jumillite*. Although Osann's (1906) paper demonstrates his familiarity with the work of Cross (1897), he declined to equate jumillite with amphibole-rich orendite and failed to recognize its affinities with madupite.

The above studies demonstrate the failure of many petrologists at the turn of the century to recognize modal variants of a petrographic clan of rocks and to deal rationally with the problems of heteromorphy. Recognition of the deficiencies of modal classifications led to attempts to classify rocks objectively and solely in terms of their chemical composition. Whereas this approach is clearly advantageous for glassy volcanic rocks, it is less appropriate for holocrystalline plutonic and hypabyssal rocks, whose compositions may not even remotely resemble those of their parental magmas. It was these attempts at objective classification that ultimately led to the introduction of the term *lamproite*.

1.3. ETYMOLOGY OF LAMPROITE

Johannsen (1939) has provided a comprehensive summary of the numerous chemical classifications proposed during the earlier part of this century. Most of these have been relegated to history, for although many are logical and quantitative, the results of the classification were found, to all but their progenitors, to be as incomprehensible as the previous type-locality names. The much admired classification scheme proposed by Cross *et al.* (1903) provides an excellent example. This classification method mirrors the hierarchical principle used in the biological sciences. In this scheme orendite, for example, becomes Class III (salfemic), Order 5 (perfelic), Rang 1 (peralkalic), Subrang 1 (perpotassic)! The original name at least has the merit of brevity, and although the latter classification is apparently chemically precise, it does not provide the slightest clue as to the modal mineralogy.

In Europe, Paul Niggli (1920) devised a detailed chemical system of classification based upon calculated parameters termed Niggli values (see Section 2.4.1 in Chapter 2) and found that the rocks described by Cross (1897) and Osann (1906) were similar with respect to their Niggli k and mg values, both being in excess of 0.80. No other rock types known at that time exhibited such extremes of k and mg, which resulted from the unusual compositional feature of high K_2O contents coupled with high MgO contents. To identify the magma type possessing these characteristics, Niggli introduced the term *lamproite* (Niggli 1923, pp. 182–185). The name is derived from the Greek root, λᾱμπρος, meaning "glistening" and refers to the characteristic presence in these rocks of shiny phenocrysts of phlogopite. No genetic relationship with *lamprophyres*, i.e., the glistening porphyries of von Gumbel (1874), was implied by this definition. Phlogopite phenocrysts are the only feature that these disparate rock types have in common. However, Tröger (1935, p. 322) redefined lamproites as the extrusive equivalents of certain lamprophyres rich in K_2O and MgO. This usage led to ambiguity, as the term lamproite was used strictly in a petrochemical sense by Niggli, and is therefore not synonymous with a modal macroscopic descriptive term. Nevertheless, this hybrid meaning ultimately gained acceptance owing to the stature of Tröger as a teacher and petrographer in Europe. The name was subsequently firmly entrenched in the pantheon of potassic rock terminology by Wade and Prider (1940) in their description of leucite-bearing rocks from Western Australia (see below). *It should be noted that the current usage of lamproite is not synonymous with either Niggli's or Tröger's usage of the term (see Section 2.6).*

1.4. WESTERN AUSTRALIAN DISCOVERIES— THE LEGITIMIZATION OF LAMPROITES

The first recognition of leucite-bearing rocks in Western Australia was by Fitzgerald (1907), who described a "mica leucitite" from the Lennard River area. Subsequently, during the 1920s, exploration for petroleum in the Fitzroy River Basin resulted in the description of "lamprophyres" together with leucite-bearing lavas, tuffs, and agglomerates (Farquharson 1922, Wade 1924, Simpson 1925). Systematic study of these occurrences by Rex Prider later formed the basis of a Ph.D. thesis at Cambridge University. The results of this benchmark study were communicated at a meeting of the Geological Society in London in 1939 and were subsequently published as Wade and Prider (1940). This study demonstrated the existence of a third major lamproite province and was particularly noteworthy in that four new type-locality names were introduced. This was unusual as petrographers had by this time abandoned this approach to terminology. The new terms were: *cedricite*, from Mount Cedric (named after Wade's son); *fitzroyite*, after the Fitzroy River (Lord Fitzroy was Governor of Australia in the nineteenth century); *wolgidite*, from the Walgidee (formerly Wolgidee) Hills (unknown aboriginal derivation), and *mamilite*, from Mamilu Hill (named after Wade's wife). The names were criticized by Cecil Tilley (Prider's thesis supervisor) and others for their lack of euphony (Wade and Prider 1940, p. 98) but were not objected to on petrological grounds.

The paper by Wade and Prider (1940) was significant in that, for the first time, related rocks of diverse modal assemblages were described collectively as lamproites (*sensu* Tröger 1935). The work unfortunately imposed a further taxonomic burden that actually served to obscure rather than clarify petrological relationships between the major lamproite provinces. Certainly a long-lasting impression was created that each province was unique with respect to its mineralogy and petrology.

Wade and Prider (1940) introduced fitzroyite (leucite phlogopite lamproite) and cedricite (diopside leucite lamproite) on the grounds that these rocks contained only one major mafic phase and were thus not wyomingites (leucite diopside phlogopite lamproite). However, in defining these terms Wade and Prider (1940) were selective and did not put sufficient emphasis on transitional rock types, e.g., the wyomingites occurring at P Hill, and the presence of accessory mafic phases, which clearly demonstrate that there is a complete spectrum of modal variation between fitzroyite, cedricite, and wyomingite.

Wade and Prider (1940) further stated that primary sanidine did not occur in the Western Australian rocks. This claim was their justification for the introduction of the term wolgidite. Unfortunately, Wade and Prider (1940) were unable to determine that the brownish matrix of wolgidite contains abundant primary sanidine (Mitchell 1985, Jaques *et al.* 1986). Recognition of such sanidine by Wade and Prider (1940) would have enabled them to term the rocks jumillites or mela-orendites. Mamilite is an amphibole-rich variant of wolgidite with petrographic similarities to amphibole-rich orendites from the Leucite Hills and therefore does not require a separate name. Clearly, terminology existing in the 1930s was adequate to describe the Western Australian lamproites.

1.5. JOHANNSEN AND MODAL CLASSIFICATIONS OF ROCKS

Importantly not all petrologists were as keen as Wade and Prider (1940) to follow the Niggli-Tröger terminology. Albert Johannsen in particular eschewed chemical classifications, and especially that of Niggli as evidenced by his response (Johannsen 1932) to Niggli's attempt (Niggli 1931) to modify his own modal classification scheme. Consequently, Johannsen's (1939) epic study *A Descriptive Petrography of the Igneous Rocks* does not describe any rocks as lamproites. In terms of Johannsen's quantitative modal classification, orendite becomes rock type 2121E, wyomingite, 2125E, and madupite, 3125! In common with the chemical classifications, these designations are no improvement on the original type-locality name. Johannsen did, however, recognize that verite and fortunite were heteromorphs and assigned them to the same family (219E). Even Johannsen did not entirely discard descriptive terminology; orendites, for example, are also referred to as phlogopite leucite trachytes. This designation stems from the fact that the Johannsen classification is a derivative of the earlier Zirkel–Rosenbusch petrographic schemes.

Johannsen's influence upon petrography in the first third of this century was profound and has had unfortunate consequences for potassic rock petrology. For example, his classification places lamproites in petrographic families devised without regard to genetic relationships. Thus jumillites (also termed mela-olivine orendite) and orendite are placed in the naujaite-fasinite family (No. 21) along with nephelinite-phonolite and blairmorite!

Similarly, wyomingite is grouped in the feldspar-free family (No. 25) with nephelinite, melilitite, and leucitite. Clearly, classifications based upon petrographic criteria alone can result in strange bedfellows.

Johannsen's groupings were unfortunate in that some lamproites were further identified as merely modal variants of leucite-bearing rocks described in the nineteenth century from Germany and Italy, and not as the products of crystallization of a distinct magma type. Many influential texts have perpetuated this supposition. Thus, Turner and Verhoogen (1960) discuss all leucite-bearing rocks collectively and suggest or imply a genetic relationship. Even recent works, e.g., McBirney (1984), Hall (1986), and Yoder (1986), are not exempt from this belief. A consequence of this approach is the multiplicity of complex hypotheses proposed in catholic models of leucite-bearing rock petrogenesis.

1.6. FURTHER DEVELOPMENTS, NEW OCCURRENCES, AND DIAMOND–BEARING LAMPROITES

The discovery of the Western Australian lamproites did not herald an upsurge of interest in the study of lamproites, despite the fact that Wade and Prider (1940) suggested a genetic link between these rocks and diamondiferous kimberlites. Petrologists at that time were more concerned with the origins of granites than with exotic mafic alkaline "trivia." Consequently, the discoveries of Wade and Prider remained an isolated event outside the mainstream of petrology, and lamproites in general were largely ignored for the next two decades.

Interest in the 1960s was concerned principally with reinvestigation of the Murcia-Almeria province (summarized by Fuster *et al*. 1967) and a compilation of the geology of the Western Australian province (Prider 1960). Carmichael's (1967) study of the Leucite Hills lamproites, based upon a suite of specimens collected by Stuart Agrell (Cambridge University) was the first modern study of a major lamproite province using the electron microprobe. This work has been widely quoted and until recently remained one of the few detailed investigations of lamproite mineralogy. Velde (1965, 1968) also used the electron microprobe in a study of the Sisco lamproite, which at that time was termed a minette.

From about 1970 lamproite studies followed three main directions: description of new lamproite occurrences; further investigation of known provinces, which resulted in a reevaluation of their character; and reclassification of rocks previously considered to be kimberlites or lamprophyres, as lamproites.

Foremost of the new occurrences were lamproites described from Smoky Butte, Montana (Matson 1960, Velde 1975); Francis (formerly Moon Canyon or Kamas, Best *et al*. 1968, Henage 1972); Sisimiut (formerly Holsteinsborg), West Greenland (Scott 1979, 1981); and Kapamba, Zambia (Scott Smith *et al*. 1989). Fortunately, new type-locality names were not coined even though some of the mineral assemblages identified were unlike any previously described. Hyalo-olivine armalcolite phlogopite lamproites from Montana could justifiably have been termed smokybutteite!

Of special significance was the reevaluation of the Western Australian province. An exploration program, based upon the hypothesis that lamproites are differentiates of kimberlites (Wade and Prider 1940, Prider 1960), led to the discovery of diamonds in the Ellendale region of the Fitzroy River basin in 1976/77. Alluvial diamonds in Smoke Creek in the East Kimberleys were traced back in the same period to the Argyle lamproite vent, thus also defining a new lamproite province. This exploration case history is interesting in that indicator mineral techniques devised for southern African kimberlites were, not surprisingly, unsuccessful in locating an entirely different type of diamond-bearing rock. The exploration philosophy was based upon an incorrect petrogenetic hypothesis and it was not until diamond and chromite were used as indicator minerals that the Ellendale and Argyle vents were discovered. The moral of the story is that if one wishes to find diamonds one should look for diamonds! A comprehensive history of the prospecting activities in Western Australia is given by Jaques *et al.* (1986).

From a petrological viewpoint the discoveries were of tremendous significance. First, it was realized that kimberlites are not the only primary source of diamonds. Second, it was shown that lamproite tuffs may form champagne-glass-like volcanic vents with little or no topographic expression. This latter observation prompted a radical revision of ideas concerning the style of lamproite volcanism.

In economic terms, the occurrences were monumental in that diamond-bearing tuffs were shown to be far richer in diamonds (e.g., Argyle lamproite 500 carats/100 tons) than kimberlites (20–80 carats/100 tons). This discovery alone catapulted lamproites from petrological limbo to being one of the most desirable exploration targets. An immediate consequence was an upsurge of interest in the petrology of lamproites and several worldwide exploration programs were initiated by major multinational mining organizations.

Unfortunately, reports of the discoveries of diamond-bearing lamproites were initially clouded with confusion. Atkinson *et al.* (1982) and Jaques *et al.* (1982) presented the first descriptions of the rocks at the Third International Kimberlite Conference in Clermont-Ferrand, France, and, together with an important isotopic study (McCulloch *et al.* 1983), referred to them as kimberlites, kimberlitoids, and even ultrapotassic kimberlites. These appellations were given either for proprietary reasons or simply because the rocks contained diamonds, and despite clear petrographic, mineralogical, and geochemical evidence that the rocks were related to the lamproites described by Wade and Prider (1940). Fortunately, before this incorrect designation became entrenched, Scott Smith and Skinner (1984a,b) demonstrated succinctly that the Ellendale rocks were not kimberlites but were simply olivine-rich lamproites. In later publications, Atkinson *et al.* (1984) and Jaques *et al.* (1984), despite revision of their terminology in accordance with the recommendations of Scott Smith and Skinner (1984a,b), still referred to some of the rocks as "kimberlitic." This terminological confusion has, one would hope, been laid to rest with the publication of a comprehensive review of the Western Australian lamproite provinces (Jaques *et al.* 1986) in conjunction with the Fourth International Kimberlite Conference held in Perth, Western Australia in 1986. Unfortunately, the legacy of the early papers (especially McCulloch *et al.* 1983) lingers among geoscientists not familiar with the complexities of alkaline rock terminology.

1.7. RECLASSIFICATION OF ANOMALOUS KIMBERLITES

Recognition of diamond-bearing lamproites prompted a reexamination of some kimberlites that were anomalous, in terms of their petrography and style of occurrence, relative to archetypal South African kimberlites. The most instructive example in this regard is the reappraisal of the Prairie Creek (Arkansas) "kimberlite" as a lamproite.

Igneous rocks were known from this occurrence as early as 1840 (Powell 1842) and initially described as peridotites (Branner and Brackett 1889). Diamonds were discovered in the rocks in 1906 and the affinities with kimberlite were speculated upon by Miser and Ross (1923), who nevertheless regarded them as micaceous peridotite. This designation was ultimately transmogrified into kimberlite in spite of the anomalous petrographic character of the rock and, perhaps, not insignificantly because of the presence of diamond.

On the basis of an observation by Mitchell (Scott Smith and Skinner 1984b, p. 256) that the Prairie Creek peridotites are petrographically similar to madupite from the Leucite Hills, Scott Smith and Skinner (1982, 1984b) undertook a detailed mineralogical examination of these rocks. This study, in conjunction with the observations of Scott Smith and Skinner (1984a) on the Western Australian discoveries, conclusively demonstrated their lamproitic character. Mitchell and Lewis (1983) provided confirmation of these conclusions with the recognition of priderite-bearing potassium richterite diopside lamproite inclusions in a madupitic lamproite host. Scott Smith and Skinner (1984a,b) also showed that tuffaceous rocks at Prairie Creek are lamproite tuffs analogous to those found in the Ellendale lamproite vents.

These observations demonstrated that Prairie Creek is a diamond-bearing lamproite, and not a kimberlite, a conclusion that explained why geophysical studies (Bolivar and Brookins 1979, Meyer *et al.* 1977) revealed the intrusion to have a champagne-glass-like structure and not that of a typical carrot-shaped kimberlite diatreme. The reclassification also served to explain the absence of typical kimberlite indicator minerals at Prairie Creek.

It is only with hindsight that we can note that all the evidence required to describe Prairie Creek as a lamproite vent was available before the Western Australian discoveries prompted reinvestigation of this intrusion.

Recent petrographic and mineralogical studies of the Majhgawan (India) diamond-bearing vent by Scott Smith (1989) have demonstrated that this "kimberlite" is actually a hyalo-olivine lamproite lapilli tuff. This intrusion was known to be diamond-bearing as early as 1827 (Halder and Ghose 1974), long before the discovery of kimberlite in South Africa in 1869. Curiously, Majhgawan may now be regarded as one of the first lamproites to be discovered, yet one of the last to be recognized as such.

Other "kimberlites" reclassified as lamproites at this time were those found at Hills Pond, Kansas (Mitchell 1985), Bobi, Ivory Coast (Mitchell 1985), and Kapamba, Zambia (Scott Smith *et al.* 1989). Priderite-bearing rocks from Sisco, Corsica (Velde 1965), originally described as minettes, are now regarded as lamproites (Bergman 1987). Leucite-bearing lavas from the Gaussberg volcano, Antarctica (Sheraton and Cundari 1980) have been shown to have the mineralogical and geochemical characteristics of lamproites (Bergman 1987, Foley *et al.* 1987) and not leucitites.

1.8. RECENT DEVELOPMENTS

Classifications of potassium-rich rocks that include lamproites with leucite-bearing rocks (*sensu lato*), coupled with the confusing archaic terminology, has obscured petrological relations between diverse groups of rocks. The initial steps towards resolution of these problems was provided in seminal papers by Sahama (1974) and Barton (1979). These studies were the first to suggest that lamproites could be regarded as a distinct magma type. The three groups of potassic rocks recognized by Barton (1979) were supported and redefined by Mitchell (1985) and Foley *et al.* (1987).

Lamproite nomenclature was revised by Scott Smith and Skinner (1982, 1984a,b), who suggested that the rocks be named on the basis of the modal abundance of the principal primary minerals. The method was based upon an analogous scheme devised by Skinner and Clement (1979) for kimberlites. Their proposal was adopted by Jaques *et al.* (1984) and modified by Mitchell (1985). As a consequence of these revisions the old lamproite terminology was abandoned in favor of compound names describing the mineralogy of the rock. Summaries of the occurrence, mineralogy, and geochemistry of lamproites by Bergman (1987), Mitchell (1985), and Foley *et al.* (1987), respectively, suggested the existence of lamproites as a distinct clan of potassic rocks.

Contemporaneous with these developments, petrologists realized that, despite their small volume, lamproites were important in that they are mantle-derived and possess strontium and neodymium isotopic signatures complementary to those of midoceanic ridge basalts and other mantle-derived alkaline rocks. The unusual enrichment in the incompatible and compatible trace elements was interpreted to be indicative of derivation from an ancient metasomatized lithospheric source (Fraser *et al.* 1985, McCulloch *et al.* 1983). A knowledge of lamproite petrogenesis is thus a prerequisite in contributing to a clearer understanding of the evolution of the subcontinental lithospheric mantle.

In summary, the last decade has witnessed a revolution in lamproite studies. These rocks have been elevated from the status of mere petrological curiosities to objects of immense economic and petrological significance. These advances in lamproite petrology form the basis of this book.

Many protests have been made against the use of long compound names but unless they become ridiculously cumbrous they are thoroughly justified in the interests of clearness.

Arthur Holmes (1920)

Potassic Rocks and the Lamproite Clan

2.1. ALKALINITY, SODIC, POTASSIC, AND ULTRAPOTASSIC ROCKS

Sørenson (1974, p.7), in a review of the varied definitions proposed for alkaline rocks, concludes that the term has been used by so many petrologists in so many different ways and usually so vaguely that "it is hard to know what is covered by the term." Alkalinity implies enrichment in the alkali elements, sodium and potassium, but this in turn can be related to either the silica or alumina content of the rock. There is no general agreement among petrologists as to the amounts of Na_2O and K_2O that need to be present for a rock be termed alkaline, and commonly the question is carefully avoided by many petrologists in discussions of alkaline rocks.

Petrologists usually agree in designating feldspathoid-bearing, silica-undersaturated rocks as alkaline. Silica-oversaturated rocks are generally not termed alkaline unless sodic pyroxenes, sodic-calcic amphiboles, and/or alkali amphiboles are present. A high content of alkali feldspar (typical of many lamproites) does not in itself provide sufficient grounds for identifying a rock as "alkaline." Such rocks, which may be silica-saturated, are thus alkaline by virtue of the presence of minerals deficient in alumina. This relationship of total alkalies to either silica or alumina content (or both) was recognized by Shand (1922), who proposed that an alkaline rock is one in which the alkalies are in excess of the alkali feldspar molecular ratio [$(Na_2O + K_2O):Al_2O_3:SiO_2$] of 1:1:6, with either Al_2O_3 or SiO_2 being deficient.

The precise definition of "alkaline" is beyond the scope of this work, and following Sørenson (1974), we accept Shand's (1922) definition as a useful working concept rather than choosing some arbitrary Na_2O and K_2O content as delineator of alkalinity. Most lamproites by this definition are alkaline. Their alkalinity does not always result from silica undersaturation, and the presence of a feldspathoid cannot be used to define them as such. A glass-bearing leucite diopside phlogopite lamproite (wyomingite) analyzed by Carmichael (1967) provides an informative illustration of the concept of alkalinity. This rock (Carmichael 1967, Table 12, analysis LH7) is silica oversaturated with 6.2% q in the CIPW-norm, yet contains leucite phenocrysts. The alkalinity index is 1.0:0.64:6.19, and the rock is clearly alkaline by virtue of the alumina deficiency and not on the basis of the presence of leucite.

All alkaline rocks may be regarded as belonging to a *sodic* or a *potassic* series. Conventionally, potassic rocks are loosely defined as those in which K_2O (wt % or molar) exceeds Na_2O (wt % or molar). Thus a rock is sodic if the K_2O/Na_2O (wt % or molar) ratio is less than unity and potassic if it is greater than unity. Recently, Le Bas *et al.* (1986) have suggested that the terms sodic and potassic may be applied when $(Na_2O - 4) > K_2O$ and $Na_2O < K_2O$, respectively. Using any of these criteria, lamproites are potassic rocks. Potassic rocks are further subdivided into *potassic* and *ultrapotassic* varieties.

Johannsen (1939) defined potassic rocks as those with molar K_2O/Na_2O ratios between 1 and 3. This group includes such relatively common rocks as granites, trachytes, latites, leucite tephrites, leucite basanites, and minettes together with the shonkinite suite.

Currently, there is no consensus as to the constitution of ultrapotassic rocks as defined on a chemical basis. Rocks with molar K_2O/Na_2O ratios in excess of 3 have conventionally been deemed ultrapotassic (Johannsen 1939, Carmichael *et al.* 1974, Bergman 1987). Bogatikov *et al.* (1985), however, suggest that the term be limited to those rocks in which the K_2O/Na_2O (wt %) ratio exceeds 10. *High potassic* rocks are defined as those with K_2O/Na_2O (wt %) between 5 and 10, and those with this ratio between 1 and 5 are simply potassic. Recently, Foley *et al.* (1987) revised the definition in a manner opposite to the extreme viewpoint of Bogatikov *et al.* (1985) and suggested that a rock is ultrapotassic when $K_2O > 3$ wt %, $K_2O/Na_2O > 2$, and $MgO > 3$ wt %. The lower alkali ratio consequently permits more rocks to be designated as ultrapotassic; however, this effect is offset by the magnesia criterion, which excludes many salic rocks. By any of the existing definitions lamproites are ultrapotassic rocks, as are some minettes, leucite phonolites, juvites, kimberlites, and mafurites.

2.2. ALKALI–ALUMINA RELATIONSHIPS

The relative amounts of the alkalies to alumina are defined by the molecular $(K_2O + Na_2O)/Al_2O_3$ ratio. This ratio is known as the *peralkalinity index* (Shand 1943). If this index is greater than unity the rock is termed *peralkaline*. This alkali–alumina ratio has also been termed the *agpaitic coefficient* (Ussing 1912) and rocks are described as being *agpaitic* or *miascitic* depending upon whether the ratio is greater or less than unity, respectively. This latter terminology is particularly favored by Soviet petrologists, e.g., Bogatikov *et al.* (1985), who believe that any magma typically exhibits both agpaitic and miascitic differentiates. Peralkaline rocks are characterized in their CIPW-norms by normative acmite and sodium or potassium metasilicate (ns—Na_2SiO_3 and ks—K_2SiO_3, respectively). The mineralogical expression of peralkalinity is the common occurrence of sodic pyroxenes and amphiboles, sodic or potassic titanosilicates, and the presence of ferric iron in tetrahedral lattice sites in aluminosilicates, where it proxies for aluminum to accommodate the inherent aluminum deficiency. Peralkaline rock types include some granites, some nepheline syenites, commendites, pantellerites, minettes, lamproites, and some kimberlites.

Most lamproites, potassium-rich kimberlites, i.e., the group 2 varieties of Smith *et al.* (1985), and some minettes are unusual in that they are so enriched in K_2O that it is present in excess of molar Al_2O_3. Such rocks are described as *perpotassic* (Johannsen 1939). The characteristic of being simultaneously ultrapotassic, perpotassic, and peralkaline serves to distinguish most lamproites from most other rock types and in particular from leucite phonolites and leucitites.

2.3. POTASSIC ROCKS—GENERAL PETROGRAPHIC CHARACTERISTICS AND TERMINOLOGY

Potassic rocks include the following groups:

1. Leucite and sanidine-bearing volcanic rocks termed lamproites (Niggli 1923, Tröger 1935, Wade and Prider 1940) or orendites (Sahama 1974) as found in the Leucite Hills, Wyoming, Murcia-Almeria, Spain, and the West Kimberley region of Australia.
2. Leucite, plagioclase, and alkali feldspar-bearing volcanic rocks as initially described from the Roman comagmatic province (Washington 1906) of western central Italy.
3. Leucitites and leucite basalts such as those found in central New South Wales (Cundari 1973) and the central Sierra Nevada (Van Kooten 1980).
4. Mafic kalsilite, melilite, and leucite-bearing lavas as found in the western branch of the East African Rift (Holmes 1950).
5. Potassic lamprophyres, e.g., minette, alnöite, alvikite, and aillikite (Rock 1987, 1986).
6. Kimberlite and micaceous kimberlite (Mitchell 1986).
7. The shoshonite association (Joplin 1968, Morrison 1980).
8. Potassium-rich intrusive rocks represented by shonkinite, missourite, malignite, and yakutite.

Not included in this section are potassic variants of common rock types, e.g., potassic granites and trachytes. Many leucite-bearing rocks are also omitted as the mere presence of leucite is not considered as sufficient to confer a potassic character to a rock. For example, the Laacher See volcanics of the Eifel, West Germany are only mildly potassic (K_2O/Na_2O = 1.0–1.5), but more significantly are related to sodic basaltic and nephelinitic magma types (Wimmenauer 1974).

2.3.1. Lamproites

Lamproites exhibit a mineralogy that reflects their peralkaline ultrapotassic nature. As a group they are characterized by the presence of titanian phlogopite, titanian potassium richterite, titanian tetraferriphlogopite, sodium- and aluminum-deficient leucite, iron-rich sanidine, aluminum-poor diopside, potassian barian titanates (priderite, jeppeite), and potassian zirconian or titanian silicates (wadeite, davenite, shcherbakovite). Minerals that are characteristically *absent* from lamproites described from the best-studied localities (Leucite Hills, West Kimberley) include: nepheline, melilite, kalsilite, sodium-rich alkali feldspar, plagioclase, monticellite, titanium- and zirconium-bearing garnets and aluminum-rich augite (Mitchell 1985, Bergman 1987).

As noted in Chapter 1, lamproite terminology has historically been one of the most complex and uninformative of any group of igneous rocks. Table 2.1 summarizes the nomenclature that was developed on the basis of the petrography of individual lamproite provinces. Although this terminology is now considered to be obsolete (Scott Smith and Skinner 1984b, Mitchell 1985, Bergman 1987, Jaques *et al.* 1986), a working knowledge is nevertheless necessary for comprehension of the earlier literature. The revised terminology is described in Section 2.5.

Table 2.1. Otiose Terminology of Lamproites

Rock	Principal minerals[a]							Occurrence	Originator
	opx	ol	amp	cpx	phl	leu	san		
Wyomingite				×	×	×		Leucite Hills	Cross (1897)
Orendite				×	×		×	Leucite Hills	Cross (1897)
Madupite				×	×			Leucite Hills	Cross (1897)
Cedricite				×		×		West Kimberley	Wade and Prider (1940)
Mamilite			×			×		West Kimberley	Wade and Prider (1940)
Wolgidite			×	×	×	×		West Kimberley	Wade and Prider (1940)
Fitzroyite				×	×	×		West Kimberley	Wade and Prider (1940)
Verite		×		×	×			Murcia-Almeria	Osann (1906)
Jumillite		×	×	×	×	×	×	Murcia-Almeria	Osann (1906)
Fortunite	×				×		×	Murcia-Almeria	De Yarza (1895)
Cancalite	×		×		×		×	Murcia-Almeria	Parga Pondal (1935)
Cocite[b]		×		×	×	×	×	North Vietnam	Lacroix (1933)
Kajanite[b]		×			×			Borneo	Lacroix (1926)
Gaussbergite		×		×		×		Antarctica	Lacroix (1926)

[a]opx, orthopyroxene; ol, olivine; amp, amphibole; cpx, clinopyroxene; phl, phlogopite; leu, leucite; san, sanidine. N.B. All varieties may have a glassy groundmass.
[b]No longer considered to be a lamproite.

2.3.2. Roman Province Leucitites

The leucite-bearing rocks of the Roman comagmatic province (Washington 1906) were among the first potassic lavas to attract scientific attention. The province consists of eight major Pliocene–Pleistocene volcanoes situated to the west of the Appenines. Although this was originally considered to be a unique petrographic province, similar rocks are now known from other areas, e.g., Indonesia (Stolz *et al.* 1988), and the Philippines (Knittel 1987).

Studies of the petrography of the Roman province rocks in the late nineteenth century led to the introduction of the terminology listed in Table 2.2. In common with other alkaline rocks, there is no agreement on the precise meaning of some of these names. For example leucitophyre is considered to be nepheline-free or nepheline-bearing by Hatch *et al.* (1961) and Sørenson (1974), respectively. Discussion of the complexities of the terminology of these rocks is beyond the scope of this work. The summary given in Table 2.2 is presented to facilitate discussion of the question as to why lamproites should not be regarded as members of this suite or described as phlogopite leucitophyre (orendite) or phlogopite leucitite (wyomingite).

Minerals that are characteristically absent as major constituents of the Roman comagmatic province rocks include melilite, titanian potassium richterite, sanidine, and potassian titanian oxides and silicates.

Petrochemical investigation of the Roman comagmatic province has resulted in the recognition of two rock series at each volcanic center. A high-potassium (or ultrapotassic) silica undersaturated series and a low-potassium (potassic) silica saturated (hy and Q normative) series. The former are represented by leucitites, leucite phonolites, tephritic leucitites, phonolitic leucitites, etc., and the latter by leucite basanites, olivine basalts, trachybasalts, and rocks variously described as latites or shoshonitic basaltic andesites (Appleton 1972, Civetta *et al.* 1981, Holm *et al.* 1982).

Table 2.2. Terminology of Some Leucite-Bearing Rocks

Rock	cpx	leu	plg	ol	afd	nep
Leucitite[b]	×	×		×		
Leucite tephrite	×	×	×			
Leucite basanite	×	×	×	×		
Leucitophyre	×	×			×	
Leucite phonolite	×	×			×	
Phonolite	×				×	×
Italite		×				

[a]cpx, clinopyroxene (typically Al-Ti-augite and aegirine augite); leu, leucite; plg, plagioclase; ol, olivine; afd, alkali feldspar; nep, nepheline.
[b]Synonymous with leumafite (Hatch et al. 1961) and leucite basalt (Zirkel 1893).

2.3.3. Ultrapotassic Leucitites

Ultrapotassic leucite-bearing rocks, associated with contemporaneous but not necessarily coeval, potassic and alkaline olivine basalt magmatism, occur in central New South Wales, Australia (Cundari 1973) and the Sierra Nevada, U.S.A. (Van Kooten 1980) and also possibly within the Baikal-Aldan belt of potassic rocks (Kostyk 1983). These rocks, unlike those of the Roman province, are not easily related to subduction zones.

Van Kooten (1980) has described ultrapotassic (K_2O/Na_2O wt % = 3.5–4.2) "basanites" characterized by phenocrystal and groundmass phlogopite, poikilitic sanidine, and groundmass leucite, together with diopside, magnetite, pseudobrookite, apatite, and minor olivine. As the rocks lack plagioclase they are not strictly basanites and are best termed leucitites. Associated with these rocks are potassic "olivine basalts" (actually basanites) consisting of phenocrysts of olivine, diopside, and minor phlogopite set in a groundmass of sanidine, plagioclase, leucite, apatite, pseudobrookite, and magnetite. The ultrapotassic rocks bear a mineralogical resemblance to lamproites; however, they are not peralkaline (see Section 2.4.5) or even ultrapotassic on the basis of their low molar K_2O/Na_2O (< 3) ratios (see Section 2.6.1).

Leucite-bearing rocks from New South Wales (Cundari 1973) are potassic to ultrapotassic (molar K_2O/Na_2O up to 5.5) in composition. They consist of phenocrysts of phlogopite, diopside, sanidine, leucite, and olivine set in a groundmass of phlogopite, pyroxene, nepheline, alkali feldspar, and titanium- and potassium-bearing richterite. The presence of nepheline and sodium-bearing alkali feldspar suggests that these rocks are not lamproites. Although some of the essential minerals of lamproites are present in both the Sierra Nevada and New South Wales suites, they have compositions that are typically more sodic and aluminous than those found in lamproites (see Chapter 6). The Sr–Nd isotopic composition of the New South Wales leucitites is unlike that of lamproites and indicates derivation from a bulk earthlike source (Nelson et al. 1986). The lead isotopic composition of the Sierra Nevada "ultrapotassic rocks" (Van Kooten 1981) indicates derivation from a source that, like the source of lamproites, has undergone long-term depletion in uranium. In contrast, the associated potassic "basalts" have lead isotopic compositions similar to Roman province-type lavas (see Section 7.9.1).

Although these suites of potassic leucitites are not considered to be lamproites *sensu stricto* they are closely allied to the lamproite clan and warrant much further investigation.

2.3.4. Kalsilite-Bearing Lavas (Kamafugitic Rocks)

Mafic kalsilite-bearing lavas represent the rarest of extrusive rocks and are known only from two petrographic provinces. Holmes (1937, 1950) described two lavas, *katungite* and *mafurite* (Table 2.3), from the Toro-Ankole field of the West African Rift Valley, Uganda (Holmes and Harwood 1932) which represent the extreme of silica undersaturation in the province. With a slight increase in silica activity katungite grades into *ugandite* (Table 2.3), a variety containing leucite and augite in place of kalsilite and melilite, respectively. Importantly, these lavas are associated with other volcanoes which erupt lavas of less extreme composition and which include leucitites, leucite basanites, shoshonitic rocks, and nepheline melilitites.

Sabatini (1899) and Rosenbusch (1899) simultaneously designated phlogopite kalsilite-bearing lavas found at San Venanzo, central Italy, as *venanzite* and *euktolith*, respectively (Table 2.3). The presence of phlogopite and leucite distinguishes the rocks from katungites. Interestingly, Rosenbusch (1899) considered euktolith, and rocks of the Leucite Hills and the Murcia-Almeria suites, as examples of extrusive lamprophyres.

Subsequently, Sabatini (1903) described an olivine- and leucite-free variety of venanzite, termed *coppaelite*, from Coppaeli di Sotto (now Cupaello), central Italy. Both these rocks and venanzite/euktolith were erupted to the east of the main Roman comagmatic province activity. Their relationship to this activity remains unclear (Peccerillo *et al.* 1988).

The above kalsilite-bearing and related rocks have also been termed *kamafugites* (*ka*tungite-*ma*furite-*ug*andite) by Sahama (1974), *kalamafite-leumafites* (*kal*silite or *leu*cite plus *maf*ics) by Hatch *et al.* (1961), and *Toro-Ankole type* lavas by Barton (1979). Of these terms, kamafugite has been accepted as a group name by many petrologists (e.g., Foley *et al.* 1987, Gallo *et al.* 1984) and will be used in this work in the sense of Sahama (1974).

Kamafugites clearly differ from lamproites with respect to the degree of silica saturation (see Section 2.4.4) as expressed petrographically by the presence of kalsilite and melilite and the absence of sanidine.

Potassic dike rocks occurring in the Navajo volcanic field have been referred to as katungites (Laughlin *et al.* 1989). The rocks have compositions similar to those of katungite but lack modal kalsilite. Their affinities to kamafugites are uncertain and they may be best described as phlogopite olivine melilitites.

Table 2.3. Terminology of Kamafugitic Rocks

Rock	Principal minerals[a]							Originator
	phl	cpx	leu	kal	mel	ol	gls	
Ugandite		×	×			×	×	Holmes (1950)
Mafurite[b]		×		×		×	×	Holmes (1950)
Katungite				×	×	×	×	Holmes (1950)
Venanzite[c]	×	×	×	×	×	×		Sabatini (1899)
Coppaelite	×	×		×	×			Sabatini (1903)

[a]phl, phlogopite; cpx, clinopyroxene; leu, leucite; kal, kalsilite; mel, melilite; ol, olivine; gls, glass.
[b]Synonymous with mela-olivine kalsilitite and kalamafite (Hatch *et al.* 1961).
[c]Synonymous with euktolith (Rosenbusch 1899).

2.3.5. Potassic Lamprophyres

Lamprophyres or "glistening porphyries" were so named by von Gumbel (1874) when describing the "mica-traps" of the Fichtelgebirge, Germany, with reference to the lustrous character conferred by the presence of abundant biotite phenocrysts. Subsequently, the term has been applied to any melanocratic dike rock regardless of the nature of the ferromagnesian phenocrysts (Rosenbusch 1877, Hatch *et al.* 1961, Rock 1977).

It should be realized "lamprophyre" is essentially a *field term* and the grouping of rocks of diverse character and origin under this banner serves no rational petrogenetic purpose. Unfortunately such associations commonly suggest genetic relationships where none exist. Rock (1989, 1987, 1977) has even extended the original Rosenbusch classification to include "leucite lamproites" and kimberlites, apparently on the basis of their macroscopic character, while recognizing that their origins are probably quite distinct from other lamprophyre subgroups. We believe that such extensions of the lamprophyre group are regressive and that petrologists, instead of creating more lamprophyres, should be working towards the complete elimination of the term. Discussion of lamprophyre classification is beyond the scope of this work and the reader is referred to Rock (1977, 1984, 1987) for summaries of their petrology.

Of the many varieties of lamprophyre, only *minette* bears a resemblance to some lamproites. These mildly potassic rocks (K_2O/Na_2O molar = 1.0–1.5) consist of phenocrysts of biotite or phlogopite together with clinopyroxene set in a sanidine/orthoclase matrix. Studies of the petrology of minettes suggest that two varieties may be distinguished, those associated with granite intrusions, e.g., Spanish Peaks, Colorado (Knopf 1936), and those that are apparently "primitive" mantle-derived and lacking associated magmas, e.g., the Navajo Buttes, Arizona (Rogers *et al.* 1982, Roden 1981). Most minettes can be distinguished from lamproites on the basis of their mineralogy, especially the evolutionary trend of phlogopite composition (Section 6.1.19.1), and geochemistry (Chapter 5). Some minettes are peralkaline, e.g., Pendennis Point (Hall 1982), and are very difficult to distinguish from lamproites.

Ultramafic lamprophyres (Rock 1986, 1987) containing phenocrystal phlogopite, such as alnöite, aillikite, and alvikite, are mineralogically unlike lamproite in that they are characterized by the presence of melilite, monticellite, nepheline, melanite-grossular-andradite garnet, and carbonates, in addition to being sanidine- and leucite-free.

2.3.6. Kimberlite and Micaceous Kimberlites

Currently two groups of kimberlites are recognized:

(1) Kimberlites (*sensu* Mitchell 1986) or group 1 kimberlites (Smith *et al.* 1985, Skinner 1989) characterized by the presence of macrocrysts and phenocrysts (two generations) of olivine, phlogopite, monticellite, spinel, perovskite, calcite, and apparently primary serpentine. These kimberlites contain the typical upper mantle-derived megacryst suite of titanian pyrope magnesian ilmenite, subcalcic diopside, enstatite, and zircon. Diopside is not a characteristic mineral of the hypabyssal variants of this group and when present, particularly in diatreme facies rocks, can be related to contamination of the magma by crustal xenoliths (Mitchell 1986). Some of these kimberlites have a superficial resemblance to olivine lamproites, but can be easily distinguished from the latter on the basis of their different groundmass mineralogy.

(2) Micaceous kimberlites (Wagner 1914, Mitchell 1986) or group 2 kimberlites (Smith *et al*. 1985, Skinner 1989) are characterized by the occurrence of phlogopite as a phenocrystal and groundmass phase. Primary diopside is common and monticellite is typically absent. Group 2 kimberlites contain macrocrystal olivine but are poor in megacrysts relative to group 1 kimberlites. Magnesian ilmenite is characteristically absent (Skinner 1989). The high modal content of phlogopite imparts an extreme ultrapotassic (K_2O/Na_2O molar ratios up to 21) character to these rocks. Micaceous kimberlites exhibit mineralogical and isotopic similarities to some phlogopite-rich lamproites. Mitchell (1986, 1989a) has suggested that the mineralogical, isotopic, and compositional differences between the two groups of kimberlites are so distinctive that a good case can be made for considering each group to be a distinct magma type, unrelated to each other or to lamproites. These relationships are discussed in Chapter 10.

2.3.7. The Shoshonite Association

The term *shoshonite* was coined by Iddings (1895) for orthoclase-bearing alkaline rocks from Wyoming. These are essentially basaltic rocks composed of olivine and augite phenocrysts set in a groundmass of orthoclase-rimmed labradorite, olivine, and leucite. With increasing modal amounts of olivine, they are gradational into *absarokite* and, with increasing silica into *banakite*. Joplin (1968) extended the usage of the term by referring to a shoshonitic association of magma series for a suite of basaltic to trachytic rocks that she considered as the potassium-rich equivalent of the alkali basalt series. Both silica oversaturated and silica undersaturated members of the series were recognized. Subsequently, classification of volcanic rock associations from island arcs and continental margins, using K_2O versus SiO_2 bivariate diagrams, e.g., Morrison (1980), has led to a proliferation of chemically-defined shoshonites (*sensu* Joplin 1968). Even lamprophyres have been included in the association (Joplin 1968, Rock 1977). Many lamproites could unrealistically be termed shoshonitic using this simple classification scheme.

Included in the association are subalkaline rocks of the continental interior which are more potassic than the common calk-alkaline series of the continental margins. These rocks define a high-potassium series consisting of trachybasalts, trachyandesites, and latites, distinguished by the presence of modal sanidine or orthoclase. In all respects these magmas follow normal calc-alkaline differentiation trends. Barker (1983) and Morrison (1980), in particular, note that their designation as shoshonitic obscures their relationship to less potassic calc-alkaline rocks and recommend that they are better termed potassium-rich latites.

In a review of the shoshonite association, Morrison (1980) notes that they are identified by their chemical characteristics, being hypersthene olivine normative *basalts* with high Na_2O plus K_2O (> 5 wt %) and high K_2O/Na_2O ratios (> 1.0 at 55 wt % SiO_2), together with high Al_2O_3 (14–19 wt %) contents, high Fe_2O_3/FeO ratios (> 0.5), and low TiO_2 (< 1.3 wt %) contents. Their major petrographic feature is the common occurrence of plagioclase phenocrysts rimmed by alkali feldspar and the coexistence of plagioclase and sanidine in the groundmass.

Although shoshonites and lamproites cannot be confused on a petrographic basis, they may be confused using simple petrochemical classifications of volcanic rocks (see Section 2.4.3).

Studies of the tectonic setting of shoshonites associated with subduction zones in island arc and continental margin settings have demonstrated that they are always the youngest subduction-related magmas and occur above the deeper parts of the Benioff zone (Morrison 1980). The increasingly potassic character with increasing depth of subduction-related magmatism could ultimately lead to lamproite magma, and it is in this context that the relationships, if any, between shoshonites and lamproites are important. The relationship of lamproites to paleo-Benioff zones is discussed in Chapter 4.

2.3.8. Potassic Intrusive Rocks

There exists a diverse suite of potassic to ultrapotassic intrusive rocks whose origins and relationships to potassic lavas are as yet unexplained. The principal varieties are as follows:

(1) *Shonkinite*, a potassic mela-syenite, consisting of sanidine, augite, olivine, biotite, and pseudoleucite (nepheline plus orthoclase).

(2) *Fergusite*, a leucocratic leucite/pseudoleucite aegirine augite rock.

(3) *Missourite*, a mafic olivine, augite leucite/pseudoleucite rock.

All gradations between these "end members" can be found. The principal occurrence of the association is in the Highwood Mountains potassic petrographic province of Montana (Larsen 1940). The rocks are mildly potassic with molar K_2O/Na_2O ratios of approximately 1.5.

(4) *Malignite*, a nepheline augite orthoclase ($> Or_{90}$) rock commonly containing fingerprintlike intergrowths of nepheline and orthoclase or kalsilite and orthoclase. These intergrowths are believed to be formed by the breakdown of leucite (Gittins et al. 1980). Some malignites from the Selawik Hills, Alaska are extremely ultrapotassic (K_2O/Na_2O molar = 13) and contain significant quantities of kalsilite (Miller 1972). Malignites are prone to subsolidus reequilibration leading to the formation of garnet-biotite-orthoclase rocks such as ledmorite and borolanite. Malignites, as defined by Mitchell and Platt (1979), are not equivalent to mela-foyaites and represent a distinct lineage of potassic syenites. The principal feature distinguishing malignites from foyaites and litchfieldites is the occurrence of a single potassium-rich orthoclase in the former and two feldspars (alkali feldspar plus albite) in the latter.

(5) *Yakutites* are kalsilite-aegirine-orthoclase rocks. Their modes vary widely with some examples containing up to 40 vol. % kalsilite. Nepheline is absent. Yakutites are known only from the Murun potassic complex located in eastern Siberia (Smyslov 1986, Orlova 1987). This complex is considered by some to contain lamproites (Vladikin 1985, Bogatikov et al. 1986).

2.4. PETROCHEMICAL CLASSIFICATIONS OF POTASSIC ROCKS

A fundamental problem of igneous petrology is the determination of the relationships (if any) between the varied potassic rocks described in Section 2.3. Despite many years of study these relationships remain obscure. This is in part a consequence of the tendency, until recently, to consider all leucite-bearing rocks as originating from a common source, e.g., Gupta and Yagi (1980). This assumption stems principally from the use of petrochemical classification schemes that fail to discriminate between mineralogically different potassic rocks. Recently, three major groups of potassic rocks—lamproites, Roman province-type lavas, and kamafugitic lavas—have been recognized as possessing chemical and mineralogical characteristics that do not permit their derivation from a single magma type (Barton 1979, Mitchell 1985). Petrochemical classifications of potassic rocks are described in the following sections.

2.4.1. Niggli Parameters

Niggli (1923) considered lamproites to be rocks whose Niggli k [molar $K_2O/(K_2O + Na_2O)$] and mg [molar $MgO/(MgO + FeO + Fe_2O_3 + MnO)$] values are in excess of 0.80. Figure 2.1 illustrates the composition of several suites of potassic rocks in terms of these parameters and demonstrates that this type of plot does not discriminate between lamproites, group 2 kimberlites, minettes, and the mafurite variety of kamafugite. The majority of these rocks have Niggli k greater than 0.8 and all could be termed lamproite using this criterion alone. Discrimination between lamproites and Roman province-type lavas (RPT-lavas) is possible as the latter typically have Niggli $k < 0.8$ and $mg < 0.7$.

Particularly important, with regard to the original limits defined for lamproites, is the wide variation of Niggli mg as revealed by recent studies of lamproites—e.g., the Noonkanbah field of Western Australia (Jaques et al. 1986) with mg ranging from 0.49 to 0.84. Figure 2.1 also shows that olivine lamproites from the Ellendale field of Western Australia (Jaques et al. 1986) have for the most part higher mg and lower k than the Noonkanbah, Leucite Hills, and Smoky Butte lamproites. This is a reflection of their high olivine and low titanian potassium richterite contents.

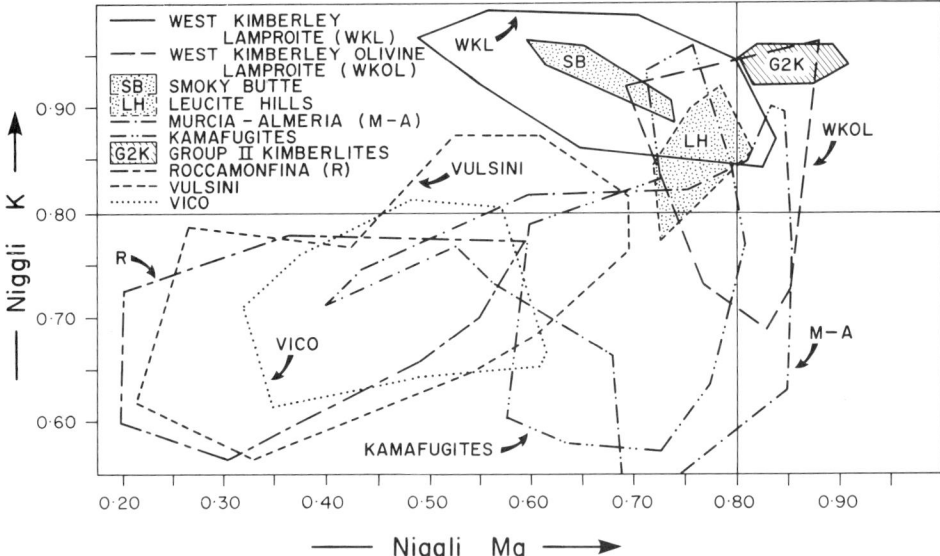

Figure 2.1. Niggli parameter diagram for diverse potassic rocks. Lamproite data from West Kimberley (Jaques et al. 1986), Smoky Butte (Mitchell et al. 1987, this work), Leucite Hills (Carmichael 1967, Ogden 1979, Kuehner 1980), Murcia-Almeria (Fuster et al. 1967, Venturelli et al. 1984). Data for kamafugites from El-Hinnawi (1965), Gallo et al. (1984), Peccerillo et al. (1988); group 2 kimberlites from Smith et al. (1985); Roman province high-potassium leucitites (Appleton 1970, Cundari and Mattias 1974, Holm et al. 1982).

Typically, lamproites from Murcia-Almeria have lower Niggli k values than other lamproite suites and their compositional field overlaps that of the RPT-lavas. Only the diopside sanidine phlogopite (cancalite) lamproites commonly have k and mg greater than 0.8. The exceptionally wide range in k and mg for these lamproites is a reflection of their anomalous geochemical character as compared to other lamproite suites (see Sections 7.8 and 7.9).

The low k values exhibited by some Murcia-Almeria and Smoky Butte lamproites are a consequence of the alteration of glass by ground waters. The effect is to replace potassium by sodium, resulting in a trend of decreasing k at essentially constant mg.

2.4.2. K_2O–Na_2O Diagrams

Middlemost (1975), in an attempt to revise the nomenclature of volcanic rocks, suggested that those containing between 44 and 53.5 wt % SiO_2 should be considered members of the "basalt clan." Most lamproites, by virtue of their silica content, are included by Middlemost (1975) in this clan. Alkaline rocks in this strictly compositionally based classification are defined as those containing greater than arbitrarily defined contents of K_2O and Na_2O at a given silica content. Three divisions of alkaline rocks are defined on the basis of K_2O/Na_2O ratios (wt %): a high potash series, a potash series, and a soda series. Figure 2.2 illustrates the compositions of some potassic rocks in terms of Middlemost's (1975) scheme and shows that the majority belong to the high potash or leucitite series. Some hyalo-lamproites from Murcia-Almeria and Smoky Butte plot in the potash and soda fields as a result of secondary alteration of glass.

Middlemost's (1975) classification serves only to highlight the potassium-rich character of the rocks and does not discriminate between the various suites. The principal disadvantage of such simple bivariate chemically based classifications is that rocks of diverse mineralogy are considered as members of a single all-embracing petrochemical group. Such an association unfortunately implies genetic relationships where none necessarily exist. Designation of practically all potassic rocks as leucitites (Middlemost 1975, p. 350) presents the further disadvantage of applying a petrographic term to a chemically defined group. Mineralogical differences are obscured and there is the tacit suggestion that all members of the group contain leucite and clinopyroxene. Inclusion of lamproite in the basalt clan is particularly inappropriate since plagioclase is absent from all *bona-fide* lamproites.

2.4.3. Total Alkali–Silica Classifications

The IUGS Subcommission on the Systematics of Igneous Rocks has recommended that *all* volcanic rocks should be classified on a *nongenetic* basis using the total alkali-silica (TAS) diagram (Le Bas *et al*. 1986). Figure 2.3 illustrates the compositional fields defined by Western Australian lamproites (Jaques *et al*. 1986), in terms of this scheme, and shows that the rocks plot in nine of the fourteen TAS classification fields. Members of the suite may therefore be referred to by any one of nine "root names" ranging from picrite to phonolite (Figure 2.3). Importantly, none of the root names provides any information regarding the modal mineralogy. More significantly, the terminology suppresses the consanguinity of the suite. Lamproites from the Leucite Hills plot as two distinct groups covering three TAS

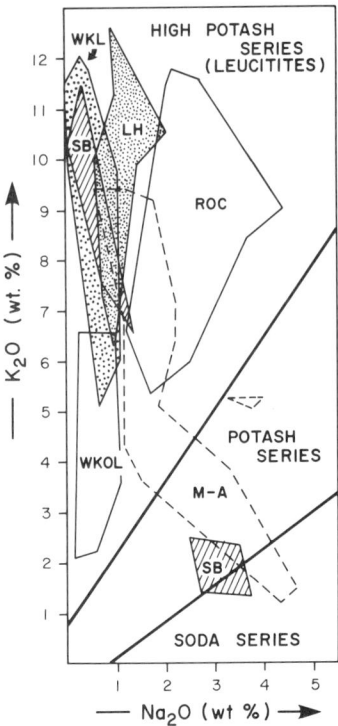

Figure 2.2. K$_2$O versus Na$_2$O diagram afater Middlemost (1975). WKL and WKOL, West Kimberley phlogopite and olivine lamproites, respectively (Jaques *et al.* 1986); SB, Smoky Butte (Mitchell *et al.* 1987, this work); LH, Leucite Hills (Carmichael 1967, Ogden 1979, Kuehner 1980); M-A, Murcia-Almeria (Fuster *et al.* 1967, Venturelli *et al.* 1984); ROC, Roccamonfina (Appleton 1970).

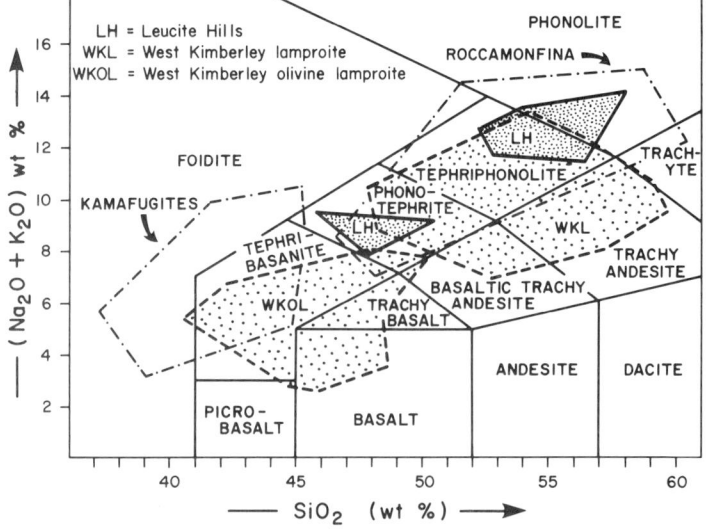

Figure 2.3. Total alkali–silica diagram after Le Bas *et al.* (1986). Data sources given in Figures 2.1. and 2.2.

fields. Madupitic lamproites are classified as phonotephrites, while phlogopite lamproites may be either tephriphonolites or phonolites, depending upon their TAS parameters. This latter division clearly serves no petrogenetic purpose.

The compositional fields of kamafugites and RPT-lavas overlap those of lamproites (Figure 2.3) and demonstrate that the TAS diagram does not discriminate between modally different suites of potassic rocks. Group 2 kimberlites and minettes, although not extrusive, may be considered as subvolcanic and are plotted in Figure 2.3 for comparative purposes.

Although the TAS classification appears to be satisfactory for common volcanic rocks, including RPT-lavas, we suggest that it is unsuitable for lamproites and kamafugites. Its use in practice may hinder rather than assist in determining petrogenetic relationships between potassic rocks.

The principal drawbacks to the TAS classification are as follows:

(1) Consanguineous rock suites are divided into arbitrary subgroups. Such compositional "pigeonholing" obscures genetic relationships.

(2) The classification is based upon a bivariate diagram which ignores significant compositional differences (e.g., TiO_2 or Al_2O_3 contents) between rocks possessing similar TAS parameters.

(3) Root names are suggestive of a particular modal mineralogy that is totally inappropriate for lamproites or kamafugites. Notwithstanding the cautions of the originators that the TAS classification is nongenetic, it is inevitable that users will forget this particular injunction. Thus, designation of a diopside leucite phlogopite lamproite (wyomingite) as a phonolite will suggest a relationship or comparison with *bona fide* (i.e., petrographically named) phonolite belonging to the nephelinite-phonolite series serving no petrogenetic purpose. Adherence to the classification may also hinder the recognition of new lamproite provinces if they are described in the literature as consisting of leucitites or phonolites.

In summary, although the TAS system permits uniform characterization of common volcanic rocks we see no benefits stemming from its application to lamproites, because it obscures petrogenetic relationships.

2.4.4. Sahama (1974)

The initial step in the process leading to the recognition of lamproites as a distinct clan of rocks was made by Sahama (1974), who classified potassic rocks into two groups: orendites (synonymous with lamproite) and kamafugites (Section 2.3.3) on the basis of their major element composition. Sahama (1974) recognized that kamafugitic rocks represent alkaline ultramafic rocks that are strongly to extremely undersaturated with respect to silica, and it is this feature that sets them apart from the near to oversaturated orendites (Figure 2.4). Additional chemical differences further distinguish the two groups: the orendites being commonly peralkaline, poorer in CaO (Figure 2.5) and containing less FeO_T (total iron expressed as FeO) for a given MgO content than kamafugites. Although Sahama (1974) did not specifically compare RPT-lavas with orendites and kamafugites, it is clear that they were regarded as a distinct compositional group (Sahama 1974, p. 106).

Sahama (1974) stressed that the differing degrees of silica saturation were reflected in the modal mineralogy of the suites, kamafugites being characterized by kalsilite and melilite and orendites by sanidine, with leucite being common to both suites. The compositional and mineralogical differences together indicate that the rocks are derived from two magma types that are not related by a simple petrogenetic process.

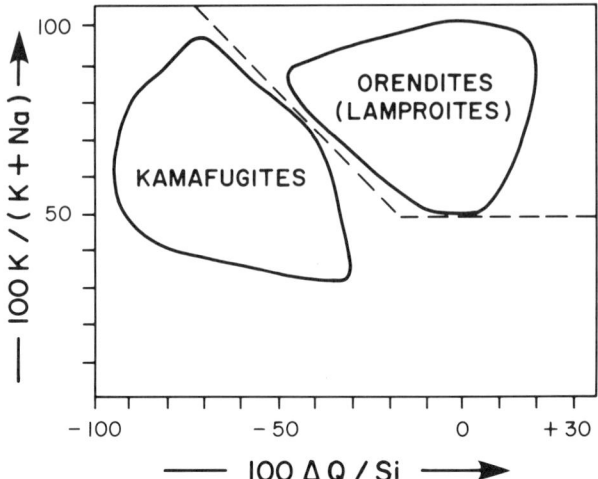

Figure 2.4. Alkali ratio versus degree of silica undersaturation after Sahama (1974). Compositional field for orendites has been revised to exclude altered lamproites from Spain. Q is the excess or deficiency in silica in the saturated norm. K, Na, and Si are atomic proportions of cations.

Sahama's (1974) observations may be quantified by reference to the silica activity of the magmas involved. Figure 2.6 illustrates the extreme silica undersaturation of kalsilite-bearing rocks relative to the silica activity of sanidine-bearing rocks, and justifies their separation into two distinct suites of potassic rocks. It is very unlikely that groups of rocks having such disparate silica activities are genetically related.

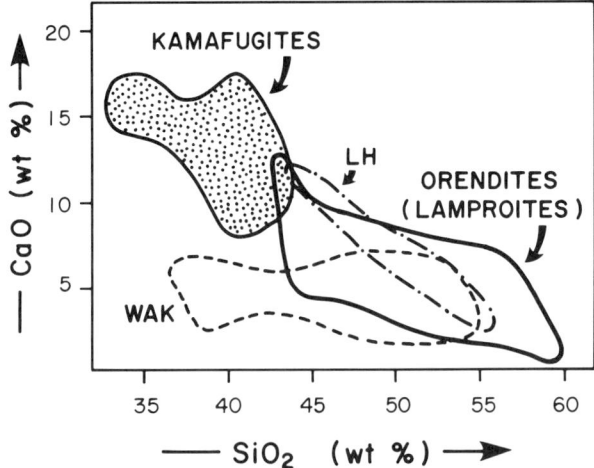

Figure 2.5. CaO versus SiO_2 diagram for lamproites and kamafugites after Sahama (1974). WAK, West Kimberley (Jaques *et al.* 1986) and LH, Leucite Hills (Carmichael 1967, Ogden 1979, Kuehner 1980).

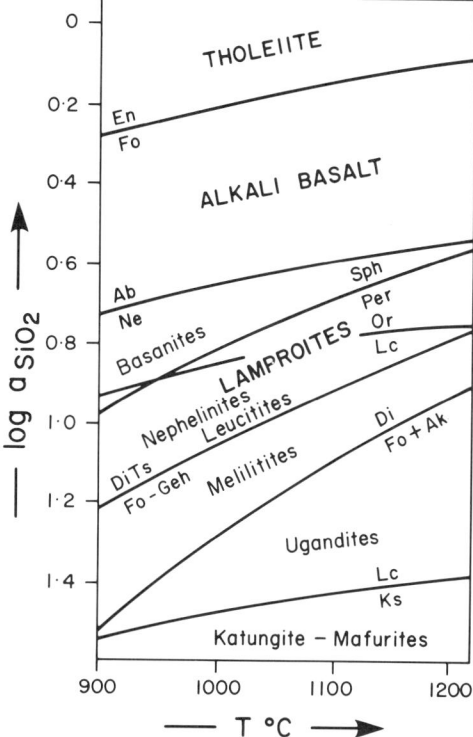

Figure 2.6. Activities of silica versus temperature for various silica buffers (after Hughes 1982). En, enstatite; Fo, forsterite; Ab, albite; Ne, nepheline; Sph, sphene, Per, perovskite; Or, potassium feldspar; Lc, leucite; DiTs, Ca-Tschermak's pyroxene; Geh, gehlenite; Di, diopside; Ak, akermanite; Ks, kalsilite.

2.4.5. Barton (1979)

Barton (1979) proposed a threefold classification of potassium-rich rocks on the basis of their petrography, mineralogy, and composition. This work provided the framework upon which current classifications of ultrapotassic leucite-bearing rocks are based. *The study was the first to recognize that compositional classifications alone are inadequate in characterizing potassic rocks.* The three groups recognized are as follows:

1. The Leucite Hills type (LHT): peralkaline lavas relatively rich in silica ($SiO_2 > 42$ wt %) characteristically containing aluminum-poor pyroxenes, micas, and amphiboles together with alkali-bearing accessory phases (priderite, wadeite) and ferric iron-rich leucite and sanidine.
2. The Roman province type (RPT): metaluminous (miascitic) lavas relatively rich in silica ($SiO_2 > 42$ wt %) and typically containing aluminum-rich pyroxenes, iron-poor leucite, and sanidine together with plagioclase.
3. The Toro-Ankole type (TAT): metaluminous (miascitic to peralkaline) silica-poor ($SiO_2 < 44$ wt %) lavas containing silica-deficient or stoichiometric leucites together with kalsilite, melilite, and perovskite.

The Leucite Hills and Toro-Ankole types correspond to the orendite (lamproite) and kamafugitic groups of Sahama (1974), respectively.

Barton (1979) suggested that a plot of molar $(K_2O + Na_2O)/(Al_2O_3)$ against SiO_2 (wt %) may be used to classify potassium-rich rocks into one of the above groups. The fields defining the group boundaries, as shown in Figure 2.7, are based upon compositional data from Vico volcano, Italy (RPT), the Leucite Hills (LHT), and the Toro-Ankole lavas (TAT). Note that the overlap of the TAT and LHT fields results from the presence of the relatively silica-poor madupitic lamproites in the LHT suite.

Figure 2.8 demonstrates that phlogopite and leucite lamproites from the Noonkanbah and Ellendale fields of the West Kimberley lamproite province plot in the LHT-field, while consanguineous olivine lamproites from Ellendale extend into the TAT-field. Figure 2.9 presents further data for RPT-lavas. All fall within Barton's originally defined field and confirm the usefulness of this plot in the recognition of the RPT-suite. Figure 2.10 illustrates data for Murcia-Almeria lamproites and New South Wales potassium richterite-bearing leucitites. Both suites straddle the boundary between the RPT and LHT-fields. The diagram emphasizes the unusual chemistry of the Spanish lamproites relative to other lamproite suites. The data may be interpreted as suggesting either that the lamproite field overlaps the RPT-field, or that the Spanish lamproites have been contaminated during emplacement. We favor the latter interpretation (see Chapter 10). The New South Wales leucitites cannot be assigned to any of Barton's groups and their affinities must be assessed using mineralogical criteria.

Figure 2.11 shows that most minettes plot in the RPT-field with only peralkaline minettes from St. Helier (Jersey) and Pendennis (Cornwall) falling within the lamproite field. Potassic rocks from Monticatini and Orciatico (Italy) similarly plot within the RPT-field and are unlikely, as suggested by Peccerillo *et al.* (1988), to be lamproites. Kamafugites

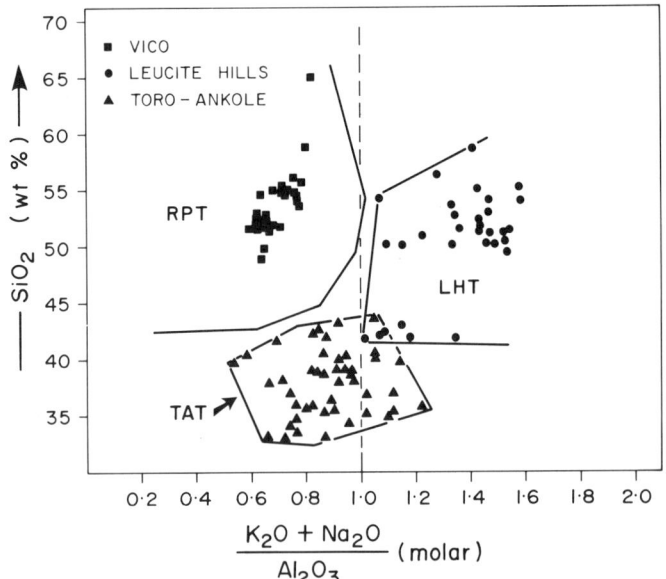

Figure 2.7. Plot of SiO_2 (wt %) versus peralkalinity index for the three groups of potassic lavas recognized by Barton (1979). RPT, Roman province type lavas; LHT, Leucite Hills type lavas; TAT, Toro-Ankole type lavas.

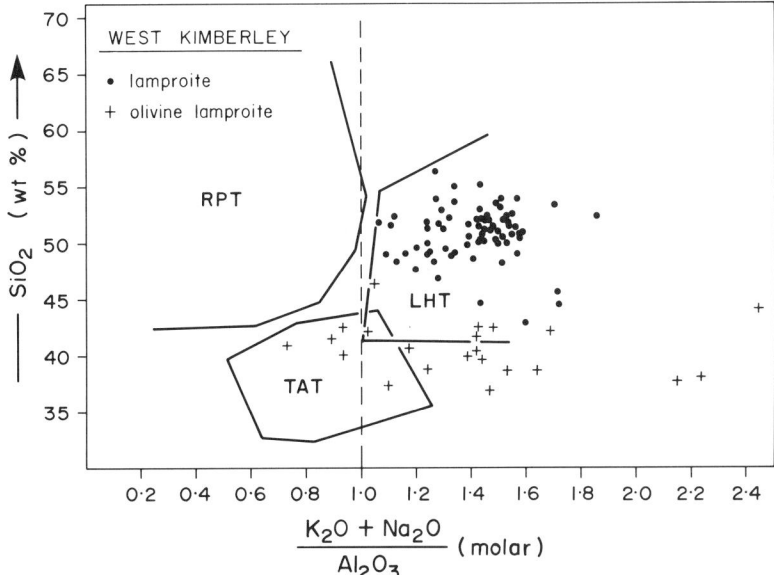

Figure 2.8. Plot of SiO_2 (wt %) versus peralkalinity index for West Kimberley lamproites (Jaques *et al.* 1986). For field definitions see Figure 2.7.

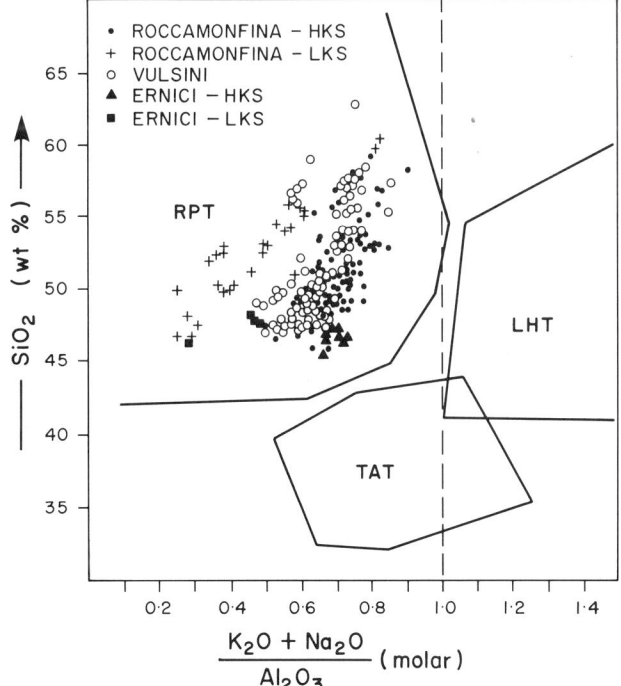

Figure 2.9. Plot of SiO_2 (wt %) versus peralkalinity index for Roman province leucitites. Data sources: Roccamonfina (Appleton 1970), Vulsini (Holm *et al.* 1982), Ernici (Civetta *et al.* 1981). For field definitions see Figure 2.7.

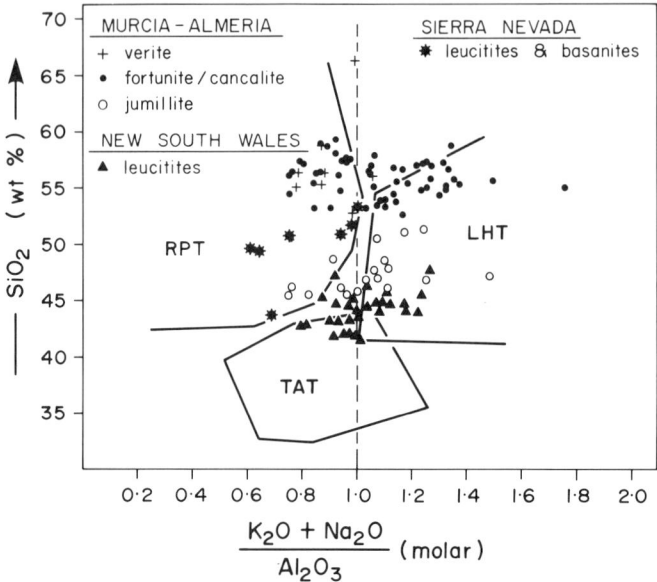

Figure 2.10. Plot of SiO$_2$ (wt %) versus peralkalinity index for lamproites from Murcia-Almeria (Fuster *et al.* 1967, Venturelli *et al.* 1984) and leucitites from New South Wales (Cundari 1973) and the Sierra Nevada (Van Kooten 1980). For field definitions see Figure 2.7.

from San Venanzo fall as expected within the TAT-field, but those from Cupaello plot in the LHT-field. The latter, however, are easily distinguished from lamproites on a mineralogical basis (Sections 2.3.3 and 2.6.2). N.B. The overlap in composition on this bivariate plot does not imply that these rocks and lamproites are heteromorphs as they possess other compositionally distinguishing characteristics. Bona fide lamproites from Smoky Butte and Sisco, plotted in Figure 2.11, lie partially within Barton's LHT-field. Importantly, they do not plot within the RPT-field and their position suggests that Barton's LHT-field may be enlarged slightly. Group 2 kimberlites, olivine lamproites, and TAT-lavas cannot be distinguished on this diagram and their characterization requires mineralogical and petrographic data.

In summary, Barton's (1979) diagram is useful in that it serves to emphasize the peralkaline character of many lamproites and the miascitic nature of many minettes and RPT-lavas. The diagram enables one to distinguish RPT-lavas from lamproites but is based upon an insufficient number of compositional variables to be useful in categorizing all potassic rocks.

2.4.6. Bogatikov *et al.* (1985)

Subsequent to the description of rocks purported to be lamproites from the USSR (Vladikin 1985, Bogatikov *et al.* 1986, 1987, Vishnevskii *et al.* 1986), it has become imperative to understand exactly what the term implies to Soviet geologists. Although different groups of petrologists apparently utilize different classification schemes, all follow

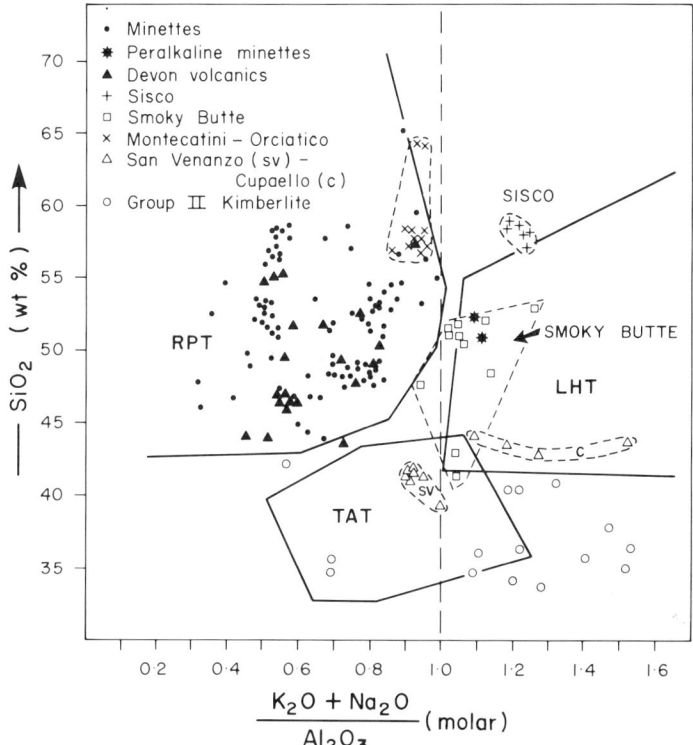

Figure 2.11. Plot of SiO$_2$ versus peralkalinity index for lamproites from Smoky Butte (Mitchell *et al.* 1987, this work) and Sisco (Peccerillo *et al.* 1988), diverse Italian potassic rocks (Peccerillo *et al.* 1988, Gallo *et al.* 1984), minettes (Bergman *et al.* 1988, Van Bergen *et al.* 1983, Rogers *et al.* 1982, Roden 1981, Turpin *et al.* 1988, Luhr and Carmichael 1981, Allen and Carmichael 1984, Ehrenberg 1978, McDonald *et al.* 1985, Hall 1982, Cosgrove 1972), and group 2 kimberlites (Smith *et al.* 1985).

a common approach with the belief that lamproites can be identified on the basis of their chemical composition.

The most important classification scheme used by Soviet petrologists is that devised by Bogatikov *et al.* (1985). Lamproites are considered to be members of a group of high-magnesium rocks that includes boninites and komatiites. They are distinguished from the latter by virtue of their potassic nature. Potassium-rich rocks are defined as *potassic* when K$_2$O/Na$_2$O = 1–5, *high potassic* when K$_2$O/Na$_2$O = 5–10, and *ultrapotassic* when K$_2$O/Na$_2$O > 10.

The initial premise of the classification scheme is that *potassic high-magnesium* rocks, i.e., lamproites, can be distinguished from *potassic normal magnesium* rocks on the basis of their K$_{mg}$ and SiO$_2$ contents [K$_{mg}$ = 100 MgO/(MgO + FeO) molar with Fe$_2$O$_3$/FeO = 0.15]. According to this scheme *any* rock plotting in the high-magnesium field on the K$_{mg}$–SiO$_2$ diagram (Figure 2.12) belongs to the *lamproite series*. It should be noted that there is a discrepancy between the position of the field boundary, as illustrated by Bogatikov *et al.* (1985, Figure 2, p. 5), and the coordinates specified by these authors in the text of their paper (Bogatikov *et al.* 1985, p. 4).

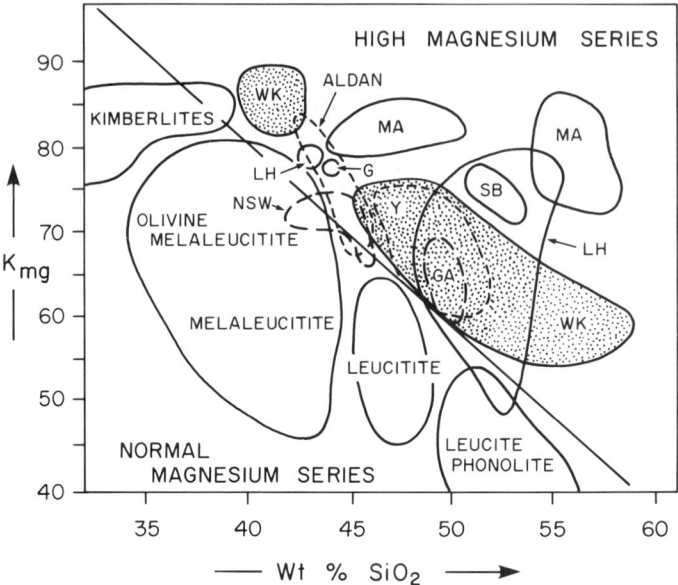

Figure 2.12. K_{mg} versus SiO_2 diagram for rocks of the normal (leucitites, leucite phonolites, kimberlites) and high (lamproites) magnesium series after Bogatikov *et al.* (1985). LH, Leucite Hills; MA, Murcia-Almeria; WK, West Kimberley; SB, Smoky Butte; G, Sisimiut; GA, Gaussberg; Y, Yugoslavia; NSW, New South Wales.

Members of the lamproite series are stated to be potassic high magnesium, slightly silica undersaturated rocks characterized by a wide range in SiO_2 (38–59 wt %) and total alkali (2–15 wt %) content coupled with low Al_2O_3 and CaO contents. Nine rock "species" are recognized as belonging to the lamproites series (Table 2.4). Based on the SiO_2 content it is possible to recognize ultrabasic, basic, and intermediate families within the series. One important aspect of the classification is that lamproites are not identified on the basis of their alkalinity. Hence Bogatikov *et al.* (1985) recognize *alkaline* and *subalkaline lamproites*. The former are considered to be typically phlogopite-bearing while the latter are characterized by the presence of olivine and diopside and are typically poor in, or lacking, phlogopite. Subalkaline lamproites include rocks previously defined as batukites (augite-olivine-leucite rocks from Batuku, Celebes), verite, and fortunite. Thus, verite in this classification scheme is termed a subalkaline phlogopite trachyte of the lamproite series.

Bogatikov *et al.* (1985) restrict usage of the term lamproite to rocks containing 38–45 wt % SiO_2, 2.5–5.9 wt % total alkalies, 1.6–5.1 wt % K_2O, 19.0–29.2 wt % MgO with K_{mg} = 78–89, i.e., ultrabasic members of the alkaline picrite family (Table 2.4). This very restricted use of the term drastically limits the number of rocks that may be called lamproites. None of the rocks from the Leucite Hills, Murcia-Almeria, or the majority of the Western Australian occurrences are lamproites in terms of this scheme. Thus, diopside leucite phlogopite lamproite (wyomingite) is a phlogopite leucite melanophonolite of the lamproite series. In fact, only olivine lamproites from Ellendale, Western Australia, and Prairie Creek, Arkansas qualify as lamproites (Table 2.4).

Further divisions within the groups are based upon alkali-alumina relationships, hence

Table 2.4. Classification and Characteristics of Volcanic Rocks of the Lamproite Series (Bogatikov *et al.* 1985)

Rock species	Rock family	SiO$_2$ (wt. %)	K$_2$O (wt. %)	K$_{mg}$	Type-locality name
Lamproite	Ultrabasic alkaline picrite	38.8–44.0	1.6–5.1	78–89	Olivine lamproite
Phlogopite melaleucitite	Alkaline ultrabasic-foidite	41.9–43.6	5.1–8.0	70–80	Madupite
Amphibole phlogopite leucitite	Alkaline basic foidite	43.5–50.5	8.2–11.6	73–77	Wolgidite
Phlogopite feldspar leucitite	Alkaline basic foidite	44.4–49.0	2.8–7.4	66–85	Jumillite
Phlogopite leucite melanophonolite	Alkaline basic phonolite	48.9–51.7	4.1–11.0	60–83	Wyomingite, orendite, cocite
Phlogopite leucoleucitite	Alkaline phonolite	51.0–59.0	8.6–12.6	55–80	Fitzroyite, cedricite
Phlogopite leucite phonolite	Alkaline phonolite	52.0–55.5	7.6–12.0	56–84	Orendite
Phlogopite alkaline trachyte	Alkaline trachyte	51.8–57.4	7.7–10.0	67–85	Cancalite
Phlogopite trachyte	Subalkaline-trachyte	53.0–59.0	3.4–9.1	65–80	Verite, fortunite

agpaitic and miascitic varieties are recognized. The presence of titanian potassium richterite is considered to be a characteristic mineralogical feature of agpaitic lamproites.

Bearing the above in mind, particular scrutiny must be given to Soviet publications describing lamproites. Bogatikov *et al.* (1986) have described lamproites from the Aldan Shield. The rocks are named on the basis of their K$_{mg}$–SiO$_2$ parameters and are miascitic lamproites (ultrabasic alkaline picrites) belonging to the lamproite series. In terms of their mineralogy they range from phlogopite pyroxenites to shonkinite–missourite-like rocks lacking titanian potassium richterite. They would not be regarded as lamproites using the chemical and mineralogical criteria proposed by Scott Smith and Skinner (1984b), Mitchell (1985), or Bergman (1987). In contrast, other dike rocks also from the Aldan region, described by Vladikin (1985), appear to be *bona-fide* lamproites but would be described by Bogatikov *et al.* (1985) as agpaitic phlogopite melaleucitites of the lamproite series.

In addition to the arbitrary delineation of the lamproite series field on the K$_{mg}$–SiO$_2$ diagram, the Bogatikov *et al.* (1985) classification suffers from all the disadvantages noted above with regard to the IUGS-TAS classification of volcanic rocks (Section 2.4.3). The recognition of a lamproite series is useful, but further subdivision into "species," bearing names that are not informative of the mode and which have petrogenetic/petrographic implications is not desirable. A particular limitation is the restriction of lamproite as a rock name to only the ultrabasic members of the series. Other Soviet petrologists e.g., Vishnevskii *et al.* (1986), do not use the Bogatikov *et al.* (1985) classification and instead follow Sahama (1974).

2.4.7. Foley *et al.* (1987)

Foley *et al.* (1987), eschewing all mineralogical classifications of potassic rocks as undesirable, have attempted to devise a "resemblance classification by which all rocks are grouped on the basis of similarity to standard members and dissimilarity to standard nonmembers" (Foley *et al.* 1987, p. 83). Thus, potassic rocks are considered to have compositions that are transitional between end members rather than being partitioned into a smaller number of mineralogically distinct groups. This approach is believed by the authors to avoid nomenclature problems resulting from heteromorphism.

Foley *et al.* (1987) define ultrapotassic rocks as those having $K_2O/Na_2O > 2$, $K_2O > 3$ wt %, and $MgO > 3$ wt %, and, using this compositional screen, recognize three end member compositional groups of ultrapotassic rocks, namely:

Group I: Rocks with 36–60 wt % SiO_2, < 14 wt % Al_2O_3, < 10 wt % CaO, $K_2O/Al_2O_3 > 0.6$, that are commonly perpotassic ($K_2O/Al_2O_3 > 1$). Standard members are lamproites from Western Australia and the Gaussberg volcano, Antarctica. Group I corresponds to the Leucite Hills type and the orendite group of Barton (1979) and Sahama (1974), respectively. Foley *et al.* (1987) use lamproite as a name for this group.

Group II: Rocks characterized by low SiO_2 (< 46.0 wt %) content, < 14 wt % Al_2O_3, > 10 wt% CaO and $K_2O/Al_2O_3 < 0.9$. Standard end members are the Toro-Ankole lavas and the group is equivalent to Barton's (1979) Toro-Ankole type of potassic rocks. Following Sahama (1974), Foley *et al.* (1987) refer to the group as kamafugites.

Group III: Rocks exhibiting a wide range of SiO_2 content (42–63 wt %), that are enriched in Al_2O_3 (> 11 wt %) relative to group I and II rocks and have $K_2O/Al_2O_3 < 0.5$. The standard end member is represented by lavas from the Roman comagmatic province. The group is thus synonymous with Barton's (1979) Roman province type magmas.

Foley *et al.* (1987) conclude that rocks of unknown affinity may be placed in the above groups by reference to bivariate plots of $CaO–Al_2O_3$, $(K_2O/Al_2O_3)–SiO_2$, $CaO–SiO_2$, $CaO–MgO$, Al_2O_3/Na_2O, and (K_2O/Al_2O_3)-*mg* number. The $CaO–Al_2O_3$ diagram that effects the maximum separation between the groups is illustrated in Figure 2.13. The field boundaries shown include the majority of compositions within each defined group. It should be noted that the boundaries are quite arbitrary and that Foley *et al.* (1987) present no justification for the locations shown. Foley *et al.* (1987) contend that the boundaries are not intended as strict boundaries but only as reference lines.

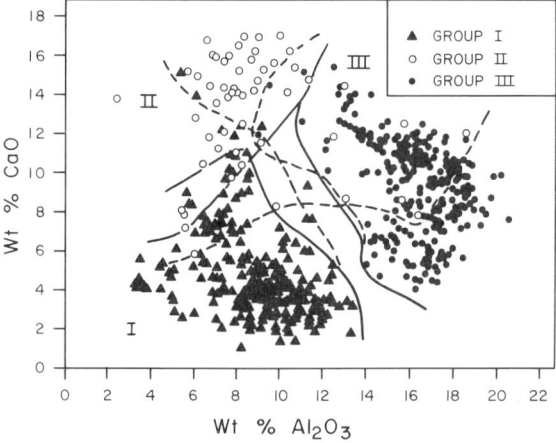

Figure 2.13. CaO versus Al_2O_3 variation in ultrapotassic rocks. Solid lines defining field boundaries are the approximate delimiters of the groups defined by Foley *et al.* (1987). Dashed lines are group limits using all data assigned to a particular group (this work).

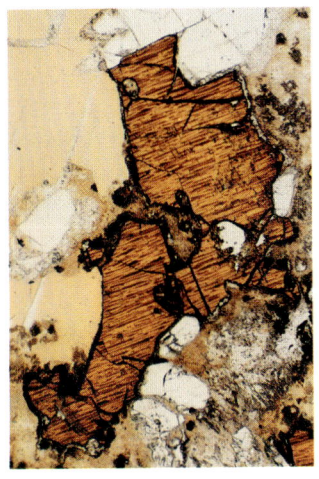

Figure 2.14. Characteristic yellow-to-pinkish orange pleochroism of titanian phlogopite. Note the polysynthetic twinning visible in this plane polarized light photomicrograph of a resorbed phlogopite phenocryst from Fishery Hill, West Kimberley. Field of view approximately 1 mm.

Figure 2.15. Characteristic reddish brown-to-yellow pleochroism of titanian potassium richterite. This composite photomicrograph illustrates the appearance of a strongly zoned poikilitic plate of amphibole from Mt. North, West Kimberley in two orientations. Note the characteristic "fingerprintlike" fluid inclusions. Field of view 1.5 mm.

Figure 2.16. Coarse-grained priderite from the Walgidee Hills pluton, West Kimberley. Note the characteristic striped appearance. Field of view 1.6 mm.

Figure 2.17. Characteristic appearance and habit of wadeite (a) in plane polarized light and (b) with crossed polars. The wadeite is enclosed by phlogopite and is associated with diopside (colorless and twinned), perovskite (deep yellow and isotropic grains), and striped brown priderite. Walgidee Hills pluton, West Kimberley. Field of view approximately 1 mm.

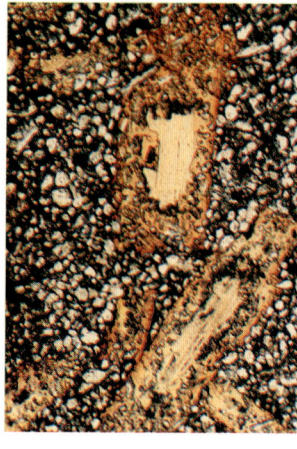

Figure 2.18. Olivine lamproite, Ellendale 11, West Kimberley. (a) Plane polarized light (b) crossed polars. Note the presence of the polycrystalline "dog-tooth" macrocryst showing undulose extinction together with euhedral phenocrystal olivines. The serrate margins of the large macrocryst probably result from resorption. Field of view 4 mm.

Figure 2.19. Olivine lamproite, Kimberlite Mine, Arkansas. Note the presence of resorbed macrocrysts of "dog-tooth" olivine (colorless) with serrate margins together with smaller euhedral-to-subhedral phenocrystal olivines. The unresolved groundmass is rich in phlogopite, pyroxene, spinel, and perovskite. Field of view 10 mm.

Figure 2.22. Transitional madupitic lamproite, Middle Table Mountain, Leucite Hills, Wyoming. Inclusion-free, early-formed titanian phlogopite phenocrysts are mantled by groundmass inclusion-bearing phlogopites that are richer in titanium and iron. Colorless microphenocrysts occurring as inclusions and throughout the groundmass are altered leucite, diopside, and apatite. Field of view approximately 2 mm.

Figure 2.20. Madupitic olivine lamproite, Prairie Creek, Arkansas. This photomicrograph illustrates the characteristic appearance of groundmass poikilitic phlogopite in olivine lamproites. Details of the groundmass are shown in Figure 2.21. Field of view 4 mm.

Figure 2.21. Groundmass poikilitic or madupitic titanian phlogopite. Madupitic olivine lamproite, Prairie Creek, Arkansas. Colorless prismatic crystals are diopside. Opaque euhedral crystals are spinel and perovskite. Field of view 1.6 mm.

Figure 2.23. Hyalo-olivine leucite lamproite. Oscar, West Kimberley. A large euhedral resorbed phenocryst of olivine is set together with microphenocrystal fresh rounded leucite (colorless) and opaque prismatic ilmenite in a groundmass consisting of red-brown titanian phlogopite and pale yellowish-brown glass. Note the presence of inclusions of glass in the leucite. Field of view approximately 2 mm.

Figure 2.24. Hyalo-armalcolite diopside phlogopite lamproite. Smoky Butte, Montana. Phenocrysts of resorbed olivine (colorless) and titanian phlogopite together with microphenocrysts of armalcolite (opaque) and diopside (pale green prisms) are set in pale brown glass matrix. Field of view approximately 2 mm.

Figure 2.25. Leucite phlogopite lamproite, Emmons Mesa, Leucite Hills, Wyoming. Phenocrysts of resorbed titanian phlogopite set in matrix of microphenocrystal colorless rounded leucite crystals. Also present are small prismatic crystals of diopside (colorless). Field of view 1.6 mm.

Figure 2.26. Leucite richterite lamproite, Endlich Hill, Leucite Hills, Wyoming. Resorbed fresh phenocrysts of leucite (colorless) are poikilitically enclosed by titanian potassium richterite showing yellow-to-brown pleochroism. Field of view 0.6 mm.

Figure 2.27. Phlogopite sanidine lamproite, Smoky Butte, Montana. Euhedral laths of zoned titanian phlogopite enclosed by subhedral prisms of sanidine (colorless). Interstices between the sanidine prisms are filled by brown titanian potassium richterite and/or glass. Opaque phase is armalcolite. Field of view 1.6 mm.

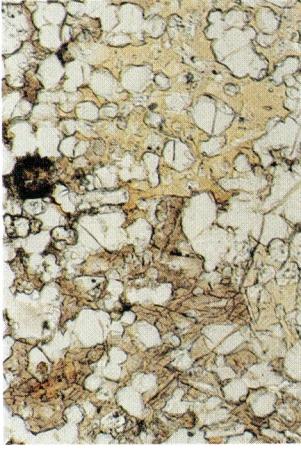

Figure 2.28. Richterite sanidine lamproite, Cerro Negro, Cancarix, Murcia-Almeria Province. Strongly zoned titanian potassium richterites are intergrown with colorless sanidine. Field of view approximately 2 mm.

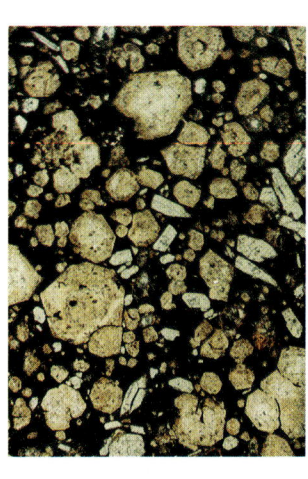

Figure 2.29. Diopside leucite lamproite, Machell's Pyramid, West Kimberley. Subhedral altered phenocrysts of leucite and colorless diopside are set in an opaque unresolved matrix. Field of view approximately 4 mm.

Figure 2.30. Olivine lamproite lapilli tuffs. Irregular clasts of strongly altered olivine lamproite set in a finely comminuted matrix of lamproite and anhedral crystals of quartz (colorless) derived from the local country rocks. (a) Argyle, East Kimberley (b) Calwynyardah, West Kimberley. Field of view 10 mm.

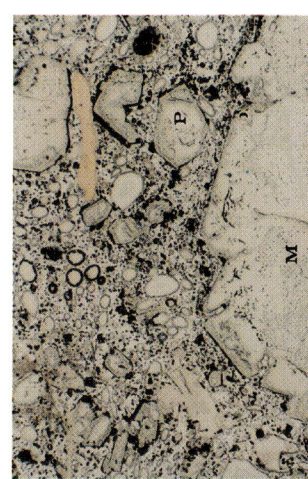

Figure 2.31. Hyalo-olivine lamproite lapilli tuff, Prairie Creek, Arkansas. Portion of a lapillus containing an olivine macrocryst (M) and a primary phenocrystal olivine (P) set in a light brown vesicular glass. Field of view 1.6 mm.

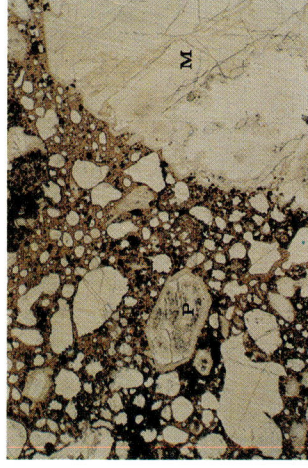

Figure 2.32. Hyalo-olivine lamproite lapilli tuff, Majhgawan, Madhya Pradesh. Portion of a lapillus containing a resorbed phlogopite phenocryst together with pseudomorphs after macrocrystal (M) and phenocrystal (P) olivines. The groundmass contains numerous vesicles set in a clear glass matrix. Field of view 1.6 mm.

The Foley et al. (1987) classification has the benefit that potassic rocks are recognized as a group of consanguineous rocks that exhibit wide compositional variation. However, the disadvantages of the scheme far outweigh this singular virtue. Basically the conclusions simply echo Barton's (1979) threefold classification of potassic rocks while ignoring useful mineralogical discriminants. Consequently it does not represent any petrogenetic advance, despite the pleas of Foley et al. (1987) as to its heuristic aspects.

The most serious criticism must be levelled at the lack of rigor associated with the classification. There is no discussion by Foley et al. (1987) of the meaning of similarity or dissimilarity with regard to end-member compositions. No attempt is made to use multivariate statistical methods to define discriminant functions that would address this question. End members are chosen subjectively with the *a priori* assumption that they are compositionally different. This hypothesis is never actually tested. Curiously these end members appear to have been chosen from suites that were originally defined as being distinct on the basis of the mineralogical classifications that Foley et al. (1987) profess to deplore. The work would stand on firmer ground if all the compositions meeting the ultrapotassic criteria had been subjected to divisive cluster analysis in order to establish an objective set of end-member compositions.

Foley et al. (1987) themselves demonstrate the failure of the classification by finding it necessary to establish a fourth group of transitional rocks. Many of these rocks have compositions similar to one or other of the three end members. They are assigned to group IV rather than apportioned, for example between groups I and III, simply on the grounds that the suite of rocks is from the same locality. Further a large part of group IV is made up of rocks which are not usually considered at all in discussions of ultrapotassic rocks, e.g., many minettes and shoshonites.

2.5. PETROGRAPHIC AND MINERALOGICAL CLASSIFICATIONS

Classifications described in this section are of two types:

(1) Petrographic or Modal Classifications. In such classifications a rock is named on the basis of the dominant (volume percentages) minerals present. The compositions of the minerals are not taken into consideration. This approach to nomenclature has no genetic significance.

(2) Mineralogical Classifications. Modal mineralogical information is used in conjunction with compositional data for some or all of the phases present. Classification is based upon assemblages of typomorphic minerals, some of which would be relegated to accessory status in purely modal classifications. This approach has genetic implications in that rocks are assigned to a petrological clan or suite of consanguineous rocks of widely varying texture and modal mineralogy. The mineral compositional data allow discrimination between rocks which, on a petrographic basis, would be given the same name but which are genetically different, having crystallized from different magma types.

2.5.1. Petrographic Classifications

Most of the rocks now collectively termed lamproites were named according to simple petrographic classifications. Their modal and textural diversity led to the confusing and redundant type-locality nomenclature given in Table 2.1. Unfortunately, recent petrographic classifications of volcanic rocks have not resulted in any improvements in the classification and description of potassic rocks beyond those employed in the nineteenth century, as the following examples with respect to the widely used IUGS classification of volcanic rocks (Streckeisen 1978) illustrate.

In this system volcanic rocks are classified on the basis of the relative abundance of quartz (Q), alkali feldspar (A = sanidine, anorthoclase, alkali feldspar, albite), plagioclase (P = >5% anorthite), and feldspathoids (F = leucite, nepheline, kalsilite, etc.). Rock names are assigned by completely ignoring the mafic mineralogy and plotting the mode of the felsic phases, recalculated to 100%, in the QAPF double triangle. Consider now how the following assemblages would be named according to this scheme:

Rock 1 = sanidine (30%) + phlogopite (20%) + leucite (20%) + diopside (10%) + priderite (5%) + glass (15%)
Rock 2 = alkali feldspar (50%) + leucite (30%) + nepheline (10%) + aegirine-augite (5%) + nosean (5%)
Rock 3 = sanidine (55%) + nepheline (35%) + riebeckite (10%)

All of the above are classified as phonolites, despite their obvious mineralogical differences. Rock 1 is equivalent to diopside leucite phlogopite lamproite (wyomingite) and description as phonolite does not convey its lamproitic character relative to a sodic phonolite (rock 2) or an agpaitic sodi-potassic phonolite (rock 3).

The practice of naming feldspathoid-bearing volcanic rocks on the basis of their felsic mineralogy is also responsible for the confusion between lamproitic rocks and modally similar sodic rocks. Use of the term phonolite in the IUGS system introduces further ambiguity, as there is no agreement among petrographers as to the meaning of this term. Thus, assemblages of alkali feldspar (Afd), nepheline (Ne), and leucite (Lc) are variously described as follows:

Phonolite = (Afd + Ne), or (Afd + Lc), or (Afd + Ne + Lc)   (Streckeisen 1978)
Phonolite = (Afd + Ne)   (Zirkel 1893)
Leucite phonolite = (Afd + Lc)   (Rosenbusch 1877)
Leucite phonolite = (Afd + Ne + Lc)   (Hatch *et al.* 1961)
Leucitophyre = (Afd + Lc)   (Hatch *et al.* 1961)
Leucite trachyte = (Afd + Lc)   (Zirkel 1893)

Thus, diopside leucite phlogopite lamproite (wyomingite) might be described as phonolite, leucite phonolite, leucitophyre, or leucite trachyte. Clearly such simple petrographic classifications are totally inappropriate for lamproites and kamafugitic rocks. Those based upon felsic mineralogy lead to ambiguity and those based upon felsic plus mafic minerals lead to a proliferation of uninformative type-locality names.

2.5.2. Mineralogical Classifications

Significant advances in the study of potassic rocks were made with the realization that lamproites possess mineralogical characteristics that set them apart from leucitites and kamafugites. The initial observations by Sahama (1974) and Barton (1979) have led to the current concept of a lamproite clan of rocks (Mitchell 1985, 1988, Bergman 1987) and to complete revision of their anachronistic nomenclature.

2.5.2.1. Sahama (1974)

Compositional differences between potassic rocks enabled Sahama to recognize the existence of an orendite (lamproite) and kamafugite series of rocks (Section 2.4.4). Sahama (1974) further noted that each of these series exhibits characteristic mineralogical features. Thus, orendites are typified by the presence of nonstoichiometric Fe-rich leucite, Fe-rich sanidine, titanian potassium richterite, priderite, and wadeite and by the absence of kalsilite and melilite. Kamafugites in contrast contain essential kalsilite and melilite together with titanian salite, stoichiometric Fe-poor leucite and Fe-poor sanidine. The most important conclusion of this study was that leucites and sanidines found in each magma series were compositionally distinct. Sahama's (1974) work permitted the recognition of lamproites as a group of rocks of similar mineralogy, an important step in the establishment of a genetic classification of potassic rocks. However, revisions to the nomenclature of the rocks comprising the orendite series were not made.

2.5.2.2. Barton (1979)

Barton extended and supported Sahama's (1974) original conclusions by recognizing three groups of potassic rocks on the basis of whole rock compositions (Section 2.4.4). Barton further noted that compositional differences between similar minerals in the different suites were a direct reflection of variations in magma bulk compositions. Thus Leucite Hills type lavas are peralkaline and contain iron-rich leucite and sanidine together with aluminum-poor pyroxene and phlogopite. Plagioclase is absent and richterite and alkali-bearing, alumina-free accessories such as priderite and wadeite are commonly present. Roman province type lavas in contrast are metaluminous and contain iron-poor leucite and sanidine, aluminum-rich pyroxene and plagioclase. Toro-Ankole type lavas in addition to being metaluminous are strongly silica-undersaturated, consequently pyroxenes are typically aluminous. These lavas are characterized by the presence of kalsilite, melilite, and perovskite. Feldspars are absent.

The most important conclusion stemming from Barton's (1979) study is the recognition of three potassic magma types and that a rock may be characterized as belonging to one of these types from a combination of petrographic, geochemical, and mineral compositional data. For the first time it was possible to go beyond simple petrographic classifications and demonstrate that not all "leucitites" or "phonolites" are equivalent in a genetic sense. Barton (1979) did not attempt to revise the nomenclature of the Leucite Hills type rocks.

2.5.2.3. Scott Smith and Skinner (1982, 1984a,b)

Scott Smith and Skinner, recognizing the problems of existing potassic rock terminology, attempted to redress the prevailing confusion by classifying lamproites using a scheme similar to that devised by Skinner and Clement (1979) for kimberlites. Scott Smith and Skinner (1982) proposed that six subdivisions of the lamproite clan may be recognized according to the modal predominance of leucite, amphibole, phlogopite, clinopyroxene, olivine, sanidine, and glass. Initially, leucite was not included in the list of classification phases as it was believed to be ubiquitous (Scott Smith and Skinner 1982, 1984b). Subsequently, Scott Smith and Skinner (1984a) realized that leucite is not an essential mineral in the definition of lamproite and therefore should not be excluded from the above list. The basic names are compounded by the addition of a prefix reflecting the relative abundance of other essential phases. Specifically these should equal or exceed two thirds of the volume abundance of the dominant classification mineral. Prefixes are given in order of increasing abundance of essential phases. Thus, wyomingite, using this scheme becomes clinopyroxene leucite phlogopite lamproite.

Scott Smith and Skinner's (1982, 1984a,b) scheme heralded a major advance in the *descriptive petrography* of lamproitic rocks. However, it must be clearly appreciated that *the terminology may only be used once a given rock has been identified as belonging to the lamproite clan*. To aid in such recognition Scott Smith and Skinner (1982, 1984a,b) presented a definition of lamproite based upon compositional and mineralogical criteria (see Section 2.6.1).

The Scott Smith and Skinner terminology has been widely adopted with some modifications (Section 2.5.2.4) by most petrologists currently studying lamproites (Mitchell 1983, 1985, 1988, Jaques *et al.* 1984, 1986, Bergman 1987). The principal advantage of the nomenclature is that it enables comparisons to be made between lamproites from diverse localities, and for similar rock types, to be recognized immediately.

Interestingly, this significant advance in lamproite petrology arose from an investigation of an anomalous "kimberlite" (Prairie Creek) and not from study of lamproites occurring at the type localities.

2.5.2.4. Mitchell (1983, 1985)

Mitchell (1983, 1985) suggested the following modifications to the Scott Smith and Skinner (1982, 1984a,b) classification scheme:

(1) As glass is not a mineral, use of this essentially textural term in nomenclature is undesirable. Thus, glass is not considered to be a part of the petrological name but is useful as a modifier describing textural varieties of lamproite. Thus, verite may be described as a hyalo-phlogopite lamproite.

(2) The classification does not reflect the differing habits of phlogopite which have a petrogenetic significance. Groundmass micas are typically aluminum-poor and iron-rich, i.e., more evolved, relative to phenocrystal micas (Section 6.1.16). It was recommended that rocks with phenocrystal mica be termed *phlogopite lamproites* and those with poikilitic groundmass phlogopite and tetraferriphlogopite be termed *madupitic lamproite* to reflect these important textural and compositional differences. Use of the term "madupitic" may be objected to on the grounds that it is derived from the archaic nomenclature and that some knowledge of the petrography of madupite from Pilot Butte, Wyoming is required to appreciate its meaning. While recognizing these disadvantages, we believe that the term is useful in that it conveys more information that "poikilitic" and it serves to indicate the *composition and texture* of these characteristic groundmass micas. In some lamproites, phenocrystal micas are mantled by discrete overgrowths of mica whose composition is similar to that of groundmass phlogopite. Commonly such mantles are transitional into groundmass poikilitic plates, and are clearly formed at the same time (see Section 6.2.2). Such lamproites are referred to as transitional madupitic lamproites (this work).

(3) A definition of lamproite was presented that emphasized the assemblage and compositional range of the minerals occurring in the lamproite suite. The intent of this definition was to permit the identification of lamproite by means of typomorphic assemblages and mineral compositions without recourse to whole rock compositional data. In

particular the general terms clinopyroxene and amphibole in Scott Smith and Skinner (1984a,b) were replaced by diopside and richterite, respectively.

The archaic type-locality nomenclature recast into the new descriptive mineralogical nomenclature is listed in Table 2.5. Representative examples of the principal varieties of lamproite are illustrated in Figures 2.14–2.32.

2.6. THE LAMPROITE CLAN

In common with kimberlites and alnöites it is not possible to devise a definition of lamproites based solely upon their modal mineralogy, for the following reasons:

(1) Simple modal definitions ignore mineral compositional data and the presence of typomorphic accessory phases. Rocks belonging to genetically different clans may be composed of similar essential modal phases, e.g., leucite plus alkali feldspar, which are compositionally quite distinct and occur in association with different accessory minerals. Barton (1979) and Mitchell (1985, 1988, 1989a) have shown that these compositional differences are a direct reflection of the composition of the magma from which they crystallized and hence are of genetic significance. Such rocks would be described by the same name using modal classifications. Such an appellation would be misleading and obscure genetic differences.

(2) Lamproites exhibit an extremely wide range in modal mineralogy. This is a consequence of the potentially large number of liquidus phases which may crystallize from compositionally diverse lamproitic magmas, coupled with the mineralogical diversity resulting from subsequent differentiation. Consequently, olivine-rich lamproites bear little petrographic resemblance to leucite richterite lamproites, even though both belong to the same clan.

(3) Glassy and fine-grained rocks are common.

Studies of lamproites, until recently, did not result in the formulation of a useful definition of the clan beyond that of Niggli (1923), as the type-locality nomenclature and

Table 2.5. Revised Nomenclature of Lamproites[a]

Historical name	Revised name
Wyomingite	Diopside leucite phlogopite lamproite
Orendite	Diopside sanidine phlogopite lamproite
Madupite	Diopside madupitic lamproite
Cedricite	Diopside leucite lamproite
Mamilite	Leucite richterite lamproite
Wolgidite	Diopside leucite richterite madupitic lamproite
Fitzroyite	Leucite phlogopite lamproite
Verite	Hyalo-olivine diopside phlogopite lamproite
Jumillite	Olivine diopside richterite madupitic lamproite
Fortunite	Hyalo-enstatite phlogopite lamproite
Cancalite	Enstatite sanidine phlogopite lamproite

[a]Rocks containing phenocrystal and groundmass phlogopite may be referred to as transitional madupitic lamproite.

paucity of mineral compositional data precluded recognition of common characteristics. The first modern approach to defining lamproites was taken by Scott Smith and Skinner (1982, 1984a,b), who state that:

> Lamproite is an ultrapotassic magnesian igneous rock. It is characterized by high K_2O/Na_2O ratios, typically greater than five. Trace element concentrations are extreme with high concentrations of Cr and Ni, typical of ultrabasic rocks, as well as those more typical of highly fractionated or acid rocks, e.g., Rb, Sr, Zr, and Ba. Generally CO_2 appears to be absent.
>
> Lamproite contains, as primary phenocrystal and/or groundmass constituents, variable amounts of leucite and/or glass and usually one or more of the following minerals are prominent: phlogopite (typically titaniferous), clinopyroxene (typically diopside), amphibole (typically titaniferous potassic richterite), olivine, and sanidine. Other primary minerals may include priderite, perovskite, apatite, wadeite, spinel, and nepheline. Other minerals such as carbonate, chlorite, and zeolite, if present, may not be primary.
>
> Upper-mantle-derived xenocrysts and xenoliths may or may not be present. (Scott Smith and Skinner 1984b, pp. 279–280)

Note that the above definition relies substantially upon chemical criteria for the identification of lamproite and that nepheline and xenolithic materials are included as characteristic phases. The approach taken by Scott Smith and Skinner (1984a,b) is to recognize lamproite on a compositional basis and to describe the rocks as outlined in Section 2.5.2.2.

Jaques *et al.* (1984, 1986) *following* Scott Smith and Skinner (1982, 1984a,b) provided a nearly identical definition of lamproite which in addition referred to the rocks as being lamprophyric.

Mitchell (1985) formulated a mineralogical–genetic definition of the lamproite clan.

> A group of ultrapotassic mafic rocks characterized by the presence of widely varying modal amounts of titanian (2–10 wt % TiO_2), Al_2O_3-poor (12–5 wt %) phlogopite, titanian (5–10 wt % TiO_2) tetraferriphlogopite, potassian (ca. 5 wt % K_2O) titanian (3–5 wt % TiO_2) richterite, forsteritic olivine, diopside, sanidine, and leucite as the major phases. Minor and accessory phases include enstatite, priderite, apatite, wadeite, magnesiochromite, titanian magnesiochromite, ilmenite, shcherbakovite, armalcolite, perovskite, and jeppeite. Analcime is common as a secondary mineral replacing leucite and/or sanidine. Other secondary phases include carbonate, chlorite, zeolites, and barite. (Mitchell 1985, p. 413)

The principal differences between the above definition and that of Scott Smith and Skinner (1984b) is the emphasis given to the composition of the minerals present and in particular to their titanian nature. Note that nepheline is not considered to be a typomorphic phase of lamproites.

Subsequent definitions of lamproites by Bergman (1987) and Mitchell (1988, 1989a) have elaborated upon these original definitions (see below). *Currently the lamproite clan is defined on the basis of geochemical and mineralogical criteria and the rocks are described petrographically using Mitchell's (1985) modification of the Scott Smith and Skinner (1982, 1984a,b) mineralogical classification scheme.*

Although petrographers and petrochemists alike may deplore this hybrid approach, we believe that it is the only means by which positive identification of these complex rocks can be achieved. We further suggest that petrological science has advanced beyond the merely descriptive phase and that it is now possible to devise modern mineralogical–genetic classifications for rocks that are similar in their gross petrographic mode but distinct in their detailed mineralogy, this being a reflection of their derivation from genetically unrelated magmas.

2.6.1. Geochemical Criteria for Lamproite Recognition

A rock may be included in the lamproite clan if it possesses the following compositional characteristics:

- Molar $K_2O/Na_2O > 3$; i.e., ultrapotassic.
- Molar $K_2O/Al_2O_3 > 0.8$ and commonly > 1.0; i.e., perpotassic.
- Molar $(K_2O + Na_2O)/Al_2O_3 > 0.7$ and typically > 1.0; i.e., peralkaline.
- Niggli $mg = 45–85$.
- Niggli $k > 70$ (lower values may result from alteration).
- Typically FeO_T and CaO < 10 wt %.
- High Ba contents (typically > 2000 and commonly > 5000 ppm) coupled with high TiO_2 (1–7 wt %).
- Typically Zr > 500 ppm, Sr > 1000 ppm, La > 200 ppm.
- The majority of olivine/madupitic lamproites fall in the compositional range 40–51 wt % SiO_2, 1–5 wt % TiO_2, 3–9 wt % Al_2O_3, 5–9 wt % FeO (total), 4–13 wt % CaO, 12–28 wt % MgO, < 2 wt % Na_2O, 3–9 wt % K_2O, 1–3 wt % P_2O_5, 2–8 wt % H_2O (composition recalculated to 100% on a CO_2-free basis with total iron expressed as FeO).
- The majority of phlogopite lamproites fall in the compositional range 50–60 wt % SiO_2, 1–7 wt % TiO_2, 7–12 wt % Al_2O_3, 3–7 wt % FeO (total), 3–12 wt % MgO, 2–7 wt % CaO, < 2 wt % Na_2O, 7–13 wt % K_2O, 1–3 wt % P_2O_5, 1–3 wt % H_2O (composition recalculated to 100% on a CO_2-free basis with total iron expressed as FeO).
- Fluorine contents that typically range from 0.2 to 0.8 wt %.
- Nd and Pb isotopic compositions that indicate derivation of the magma from mantle sources that have undergone long-term enrichment in the light rare elements and depletion in uranium.

2.6.2. Mineralogical Criteria for Lamproite Recognition

(1) Lamproites are characterized by the presence of widely varying amounts (5–90 vol. %) of the following primary phases:

- Titanian (2–10 wt % TiO_2), Al_2O_3-poor (5–12 wt %) phenocrystal phlogopite.
- Titanian (5–10 wt % TiO_2), groundmass poikilitic tetraferriphlogopite.
- Titanian (3–5 wt % TiO_2) potassium (4–6 wt % K_2O) richterite.
- Forsteritic olivine.
- Al_2O_3-poor (< 1 wt %), Na_2O-poor (< 1 wt %) diopside.
- Nonstoichiometric Fe-rich (1–4 wt % Fe_2O_3) leucite.
- Fe-rich sanidine (typically 1–5 wt % Fe_2O_3).

It should be particularly noted that the presence of all of the above phases in a rock is *not* required in order that it be classified as a lamproite. Any one mineral may be modally dominant and in association with two or three others determine the petrographic name.

(2) Minor and common accessory phases include priderite $[(K,Ba)(Ti,Fe^{3+})_8O_{16}]$, wadeite ($K_4Zr_2Si_6O_{18}$), apatite, perovskite, magnesiochromite, titanian magnesiochromite, magnesian titaniferous magnetite. Less common, but nevertheless characteristic, accesso-

ries include jeppeite [$(K,Ba)_2(Ti,Fe^{3+})_6O_{13}$], armalcolite [$(Mg,Fe)Ti_2O_5$], shcherbakovite [$(Ba,K)(K,Na)Na(Ti,Fe,Nb,Zr)_2Si_4O_{14}$], ilmenite, and enstatite.

(3) Alteration or other secondary phases comprise analcime, barite, quartz, TiO_2 polymorphs, and zeolites (commonly barium-rich).

(4) If a rock carries any of the following it falls outside of the lamproite group as defined above:

a. Primary plagioclase (common in Roman province leucitites).
b. Melilite and/or monticellite. Their presence indicates alnöites, kamafugites, or kimberlites.
c. Kalsilite. Its presence indicates kamafugitic affinities.
d. Nepheline, sodium-rich alkali feldspar, sodalite, nosean, hauyne. These sodic phases are not compatible with an ultrapotassic assemblage.
e. Melanitic, schorlomitic, or kimzeyitic garnets (common in alnöites and other lamprophyres and some group 2 kimberlites).

Lamproites as defined in this work refer to a clan of rocks and not a specific rock variety. Individual rocks belonging to this clan might not possess all of the diagnostic features of the clan but are named lamproites on the basis of their containing one or more typomorphic minerals, having the requisite geochemical character, and occurring with other *bona fide* members of the suite.

These are a very unusual lot of rocks, Prider.
Cecil E. Tilley

Description of Lamproite Occurrences
Distribution, Age, Characteristics, and Geological Framework

3.1. INTRODUCTION

Lamproites have been recognized in over 25 occurrences (provinces or fields) on all continents (Figure 3.1). In common with kimberlites, they are restricted to continental environments. As noted in Chapter 4, the formation of both kimberlite and lamproite magmas has occurred from the Precambrian to the Cenozoic, although the Mesozoic Era was the time of most frequent kimberlite emplacement, and the Cenozoic Era the most frequent lamproite formation (see Table 4.1). Although both kimberlites and lamproites exhibit intrusive and extrusive forms, lamproites are dominantly characterized by extrusive forms, whereas kimberlites mostly occur as intrusive diatremes, hypabyssal root zones, and dikes. The overall volume and number of individual intrusive or extrusive bodies of lamproite are dwarfed by those of kimberlite and other members of the mantle-derived alkaline mafic rock clan. We estimate that the total volume of known kimberlite intrusive rocks is on the order of over 5000 km^3, whereas that of known lamproites is less than 100 km^3. The large volume of extrusive lamproites, compared with intrusive varieties, may explain the paucity of recognized lamproites in the geological record. Once erupted, lamproite volcanic rocks are subject to rapid erosion and are rarely preserved for times in excess of 5–30 Ma. The generally younger age and shallower erosion level of lamproites compared with kimberlites can be linked to large-scale geodynamic processes: any petrogenetic model must therefore take into account these intrinsic differences in age and igneous form. A discussion of the tectonic controls on lamproite emplacement and their igneous forms can be found in Chapters 4 and 5, respectively.

Kimberlites form fields and provinces above cratonized regions of continental lithosphere (Mitchell 1986, pp. 105ff). Examples are now known from most cratons. In contrast, many cratons possess only one or several lamproite suites of limited volumetric importance, and there are many cratonic regions throughout the world without any recognized lamproites. It is possible that this disparity between kimberlites and lamproites is due to the long-term economic interest in kimberlites and that future exploration will discover

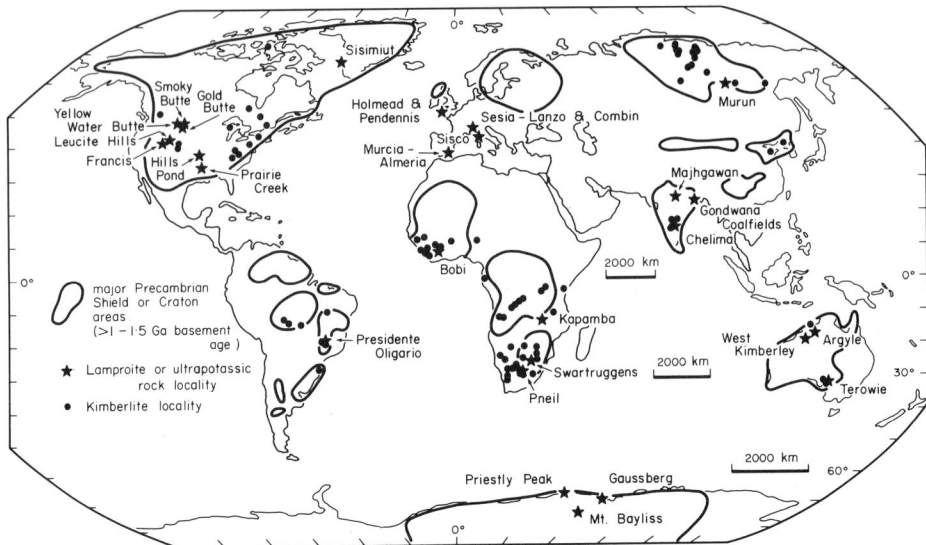

Figure 3.1. World map (Robinson projection) showing the distribution of lamproites relative to kimberlite fields and Precambrian shields (after Condie 1989, and Hoffman 1988).

lamproite fields and provinces in all major cratonic regions. It is perhaps more probable that lamproite source characteristics and the tectonic environment required for lamproite magma generation are much rarer than those of kimberlite.

Our classification of lamproite occurrences is based on a scale that differs from that used for kimberlites. A lamproite province is defined as a group of cogenetic lamproite occurrences of similar age that formed in a limited geographical area (<500 km² area). Lamproite provinces can be classified on the basis of the relative volume or number of individual component intrusive/extrusive bodies as follows:

Group 1 lamproites (or provinces) are major provinces composed of many individual intrusive/extrusive bodies (>10–100). Occurring over a wide area (>50–100 km²), they are of large aggregate volume (>10 km³), and can be subdivided into separate fields.

Group 2 lamproites (or fields) comprise several (2–20) intrusive/extrusive bodies of intermediate aggregate volume (on the order of 1 km³).

Group 3 lamproites (or localities) are isolated occurrences of minor volume and extent (typically <0.05 km³).

This chapter systematically summarizes the major characteristics of all lamproite provinces, fields, and localities.

3.2. NORTH AMERICAN AND GREENLAND LAMPROITES

Lamproites occur in eight fields or provinces in North America and Greenland (Figures 3.2–3.4): Prairie Creek (group 2); Leucite Hills (group 1); Smoky Butte (group 2); Francis (group 2); Hills Pond (group 2); Sisimiut (group 2); Yellow Water Butte (group 2); and Froze-to-Death Butte (group 2). All of these except the two latter groups, are generally accepted as *bona fide* lamproites (Mitchell 1985, Bergman 1987). Rocks from the Yellow

DESCRIPTION OF LAMPROITE OCCURRENCES

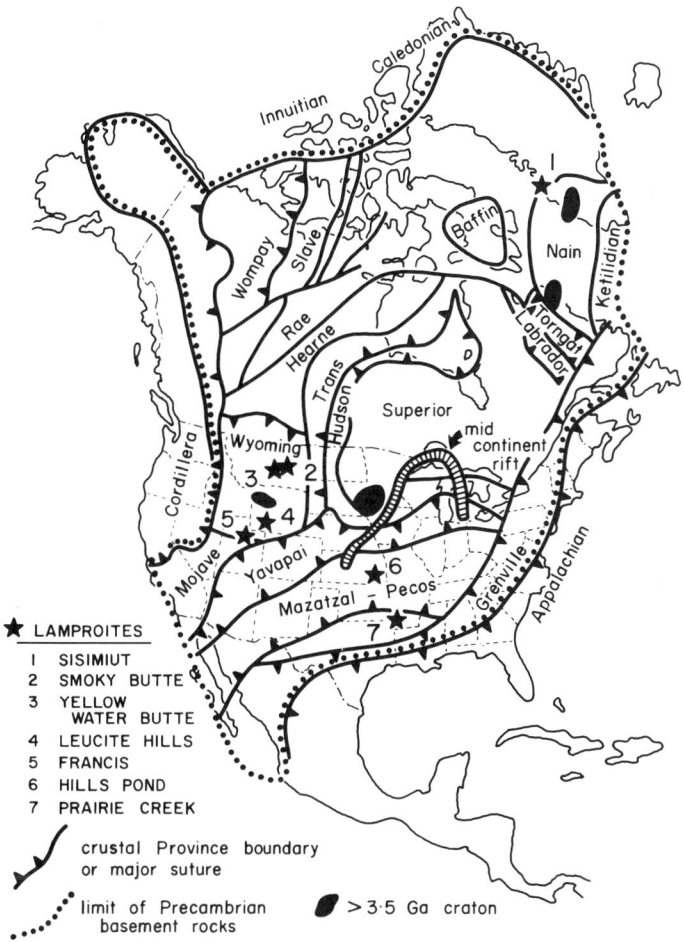

Figure 3.2. Map of North American lamproite provinces and fields relative to Archean and Proterozoic cratonic domains (after Hoffman 1988) and several other major tectonic features.

Water Butte and Froze-to-Death Butte fields have many of the mineralogical characteristics of lamproites but have only recently been recognized as such (this work). Bergman (1987, p. 109) suggested that the Late Proterozoic-to-Early Paleozoic vermiculite bodies of the Enoree district, South Carolina were lamproites. Detailed mineralogical studies are lacking as the rocks are intensely metamorphosed. Their primary mineralogy has clearly been modified, and it is not possible to determine unambiguously whether or not these rocks are lamproites. Therefore, pending further detailed investigations, the Enoree vermiculites will be considered as possible lamproites.

There are many potassic-to-ultrapotassic rock suites in North America occurring in close proximity to lamproites: however, they do not fulfill the mineralogical and geochemical requirements to classify them as such. These suites include: the 3–4 Ma ultrapotassic basanitic lavas and plugs from the Central and Eastern Sierra Nevada, California; the 45–55 Ma potassic rocks from the Highwood Mountains subprovince; members of the Montana alkaline petrographic province (Absaroka Mountains, Sweet Grass Hills, Bearpaw Moun-

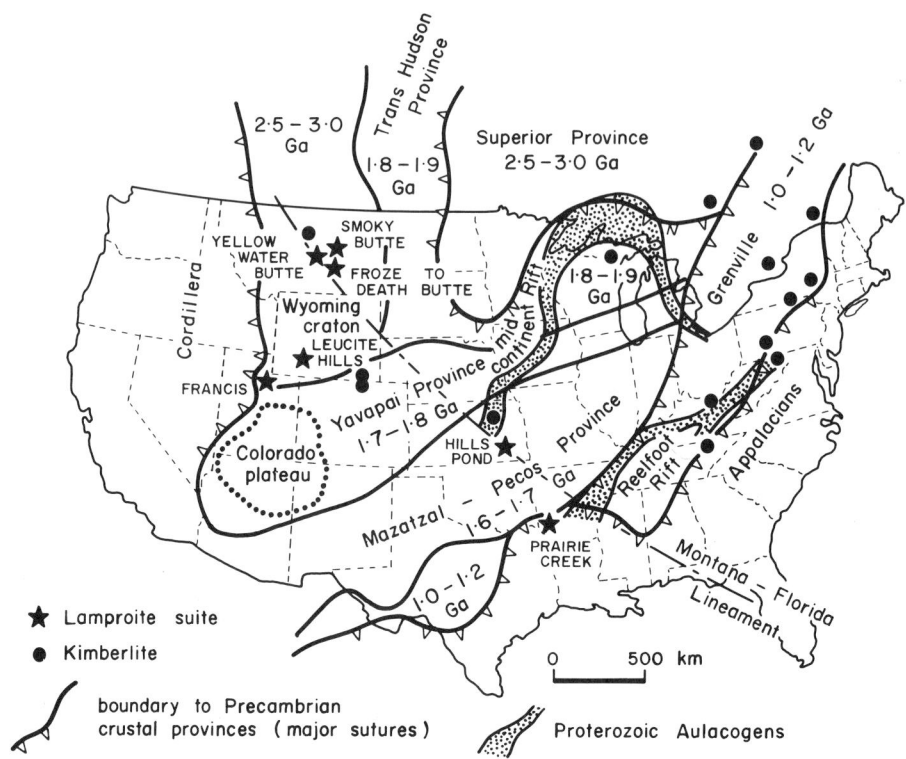

Figure 3.3. Map of the United States and southern Canada showing the distribution of lamproites relative to kimberlites and Precambrian basement domains, major Proterozoic aulacogens, and several important structural elements (after Condie 1989, and Kinsland 1985).

tains, Missouri Breaks diatremes, etc.); the 19–34 Ma potassic rocks from the Navajo Reservation, Arizona; the Fortification dike, Colorado (termed verite by Ross 1926); Two Buttes, Colorado (originally included in Niggli's lamproite clan); and the priderite(?)-bearing dikes from Aillik Bay, Newfoundland (K. Collerson, personal communication 1984). The reader is referred to Bergman (1987, pp. 127–128) for a tabulation of these localities, their petrologic characteristics and tectonic framework, and the reasons for their not being classified as lamproites.

All lamproites from North America are located within 100–200 km of the margins of areas with Archean-to-Proterozoic cratonic basement, in contrast with the occurrence of kimberlites and carbonatites both in the interior and near the margins of Precambrian cratons (Bergman 1987, p. 109). The basement rocks of all North American lamproites are regions that have experienced multiple phases of Proterozoic-to-Mesozoic compressional orogeny and/or (usually aborted) ancient rifting. These regions are underlain by ancient zones of basement weakness. None of the lamproites are contemporaneous with nearby active or passive rifting events, subduction events, or wrench tectonic events. However, these processes are apparently important in preparing the mantle source regions for the eventual lamproite magmatism (cf. Chapter 4). All North American lamproites occur in areas of negative regional magnetic anomalies (ca. -100 to -200 nT). The Arkansas and Kansas lamproites occur in slight regional Bouger gravity lows (ca. -50 milligal), whereas, the

DESCRIPTION OF LAMPROITE OCCURRENCES

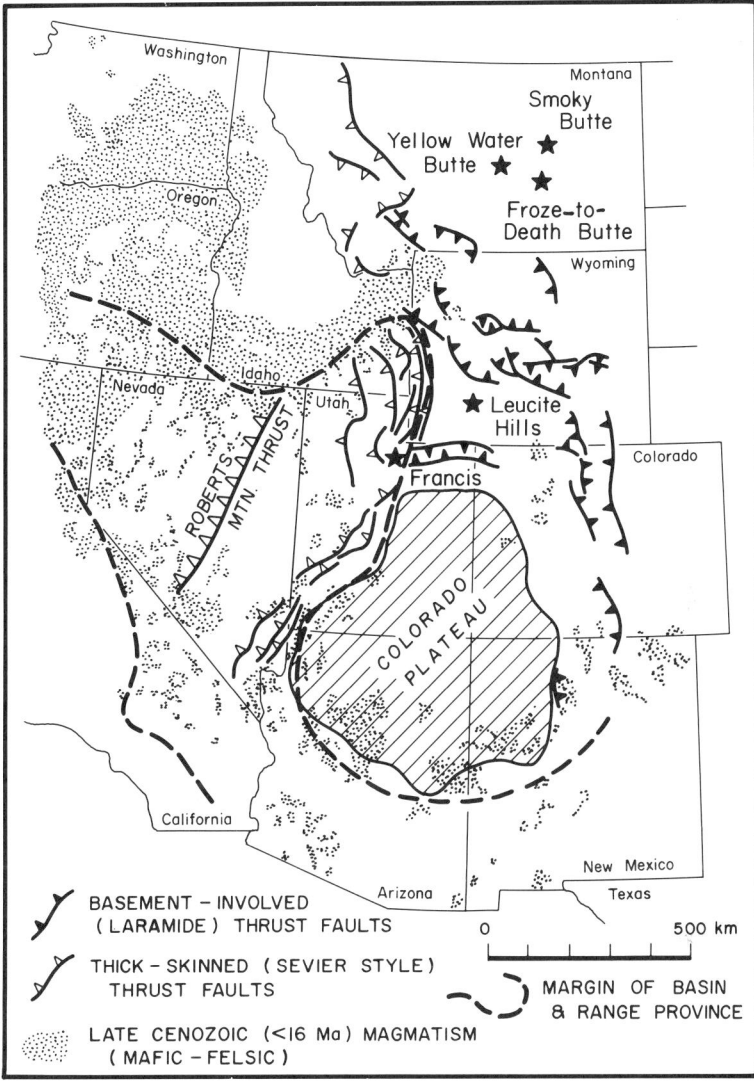

Figure 3.4. Map of lamproites in the Rocky Mountain region of the western United States showing important structural elements and Cenozoic volcanic domains (after Cross 1986 and Leudke and Smith 1983).

Utah, Montana, and Wyoming lamproites occur in areas with an intense gravity low (-100 to -150 milligal), relative to average North American continental crust.

3.2.1. Group Name: Prairie Creek

Location: [34.033 N, 93.667 W]. The Prairie Creek lamproites occur in a northeast-trending 2×4-km zone, 150 km southwest of Little Rock, 1–3 km southeast and east of Murfreesboro and Kimberley, Pike County, southwestern Arkansas, U.S.A. (Figure 3.5).

Synonyms: Crater of Diamonds State Park, Pike County intrusives, Murfreesboro.

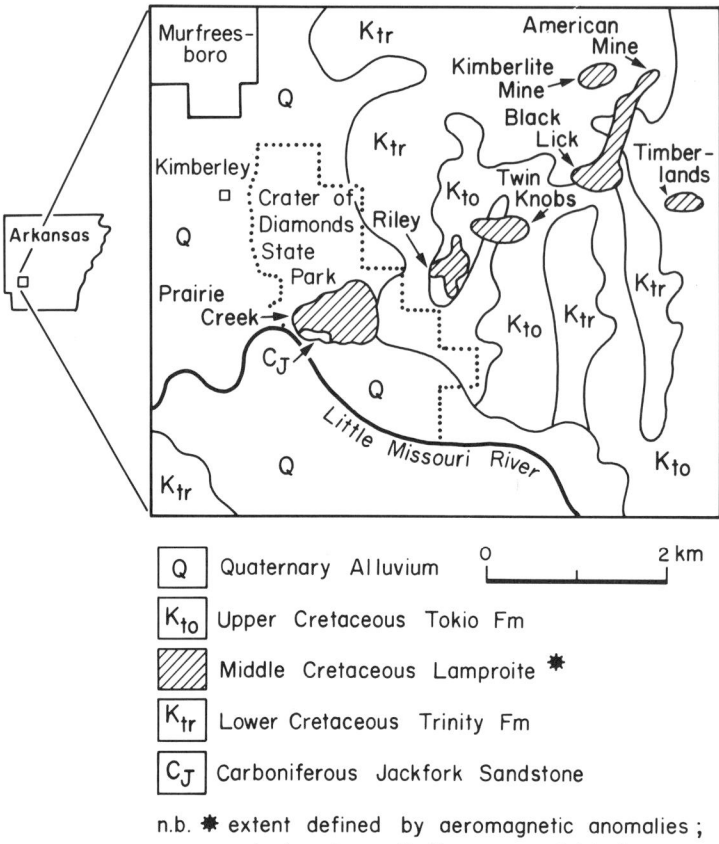

Figure 3.5. Geological map of the Prairie Creek lamproite field (after Miser and Ross 1923, Bolivar 1977, 1984, Waldman et al.1987, D. Dunn, personal communication, 1988).

Age: 97–106 Ma. [Mid-Cretaceous] (Zartman 1977, Gogineni et al. 1978).

Nature of Country Rocks: The lamproites occur in a Carboniferous-to-Cretaceous clastic sedimentary sequence, including the Early Cretaceous Trinity Formation, and the Carboniferous Jackfork sandstone.

Local Structural Setting: The lamproites intrude a nearly flat-lying (gently southward dipping) sedimentary section. No major structures have been recognized, probably because of the young sedimentary cover. The lamproites undoubtedly ascended along deeply penetrating fractures, perhaps along the northwestern marginal fault of the Reelfoot Rift.

Regional Tectonic Setting: Prairie Creek lamproites occur in a region that has experienced repeated major tectonic disturbances in the Proterozoic (extensional), Late Paleozoic (compressional), Late Mesozoic (extensional), and present-day (transcurrent). Prairie Creek is located on the southeast margin of the Mississippian–Pennsylvanian Ouachita orogenic belt, along the western edge of the Cretaceous–Tertiary Mississippi embayment (Burke and Dewey 1973, a Mesozoic failed rift), and just northwest of the projected underlying Proterozoic–Cambrian Reelfoot Rift (Ervin and McGinnis 1975, Van Schmus et al. 1987, a

Proterozoic failed rift or aulacogen). The area is also characterized by recent seismicity associated with the New Madrid Seismic zone, with focal mechanisms indicating dextral fault displacements (Braile *et al.* 1986). The lamproites are located less than 100 km from the edge of the Proterozoic North American Craton margin.

The Prairie Creek lamproites occur near the confluence of the Reelfoot Rift, Ouachita orogen and the Montana–Florida lineament. The latter has been interpreted as a major Proterozoic (to early Paleozoic) basement strike slip (transform) fault along which 800 km of right lateral offset has been proposed (Thomas 1978, Kinsland 1985). Interestingly, when the 800 km of offset is reconstructed, the Reelfoot Rift realigns with southeastern Kansas, which places the lithospheric basement of the Prairie Creek lamproites in close proximity to that of the Hills Pond lamproite intrusives. At the time of lamproite magmatism, the regional tectonics were characterized by the onset of extension associated with initiation of the Mississippi embayment downwarping.

Nature of Basement Rocks: The lamproites are underlain by deformed Early-to-Middle Paleozoic sedimentary, metamorphic, and igneous rocks of the Mississippian–Pennsylvanian Ouachita orogen as well as Proterozoic (1.3–1.8 Ga) metamorphic rocks. The local crustal thickness is about 46 km (Braile *et al.* 1986), and regionally elevated due to thickening associated with the Ouachita orogen. U–Pb ages of four (basement-derived) zircon xenocrysts range from 442 to 1852 Ma (Reichenbach and Parrish 1988).

Intrusive/Extrusive Forms: Vent complexes and dikes of massive hypabyssal rocks and intrusive breccias; extrusive sheets of pyroclastic tuffs, lapilli tuffs, and tuff breccias.

Geomorphology: The vents are intensely dissected. Hypabyssal phases are exposed. Marginal intrusive breccias and extrusive tuffs are preserved and only partially eroded.

Area/Volume: Prairie Creek intrusive/extrusive—0.6 km^2; Twin Knobs—0.1 km^2; Black Lick and American Mine—0.3 km^2; Kimberlite Mine—0.04 km^2; Riley—0.1 km^2; Timberlands—0.1 km^2, total near surface volume—0.2 to 0.5 km^3.

Component Localities: American Mine and Black Lick (two distinct surface exposures of the same intrusive), Kimberlite Mine, Twin Knobs, Riley, Timberlands, Prairie Creek (West Hill phlogopite lamproite tuff, Middle Hill massive olivine lamproite).

Nearby Contemporaneous Igneous Rocks: The Prairie Creek lamproites occur at the southwest limit of the northeastern-trending Arkansas alkaline province. About 80–150 km to the northeast is a 15-km-wide, northeast-trending belt of 86–101 Ma alkaline rocks along the northern edge of the Mississippi embayment, including quartz syenite, ijolite, nepheline syenite, sannaite, phonolite, trachyte, lamprophyres, and carbonatite (Mullen 1987). The post-Pennsylvanian Blue Ball kimberlite (probably Cretaceous) occurs in Scott County, Arkansas 100 km north of Prairie Creek (Salpas *et al.* 1985). Nepheline syenites, of presumed Cretaceous age, were encountered in wells in western-most Tennessee, about 400 km northeast of Prairie Creek along the trend of the Reelfoot Rift (Kidwell 1951). Late Cretaceous (79–91 Ma) alkaline intrusive and extrusive rocks have been encountered in over 100 wells in the Monroe Uplift, 200–300 km southeast of Prairie Creek in the tri-state area of Arkansas, Louisiana, and Mississippi (Sundeen and Cook 1979). The Woodson County, Kansas lamproites (88–91 Ma) occur 450 km north-northeast of Prairie Creek. About 500–750 km to the southwest and south of Prairie Creek, Mid-to-Late Cretaceous (66–100 Ma) alkali basalt volcanic centers and shallow intrusive rocks are abundant along the Gulf Coast, where they were extruded onto the carbonate shelf of the shallow epicontinental sea (Ewing and Caran 1982, Barker *et al.* 1987).

Rock Types: Massive, hypabyssal madupitic olivine lamproite; phlogopite lamproite tuff; lapilli tuff and tuff breccia.

Mineralogical Comments: Heavy mineral concentrates of the Prairie Creek intrusives include rare pyrope-almandine garnets and common chromite. Priderite and wadeite are rare in these rocks.

References: Powell (1842); Branner and Bracket (1889); Williams (1891a,b); Kunz (1907); Kunz and Washington (1908); Miser (1914); Miser and Ross (1922, 1923a,b,c); Miser and Purdue (1929); Ross *et al.* (1929); Moody (1949); Thoenen *et al.* (1949); Stone and Sterling (1964); Gregory (1969); Gregory and Tooms (1969); Melton *et al.* (1972); Langford (1974); Giardini *et al.* (1974); Giardini and Melton (1975a,b,c); Melton and Giardini (1975, 1976, 1980); Bolivar *et al.* (1976); Meyer (1976); Lewis *et al.* 1976); Lewis and Meyer (1977); Bolivar (1977, 1982, 1984); Lewis (1977); Meyer *et al.* (1977); Zartman (1977); Newton *et al.* (1977); Gogineni *et al.* (1978); Bolivar and Brookins (1979); Brookins *et al.* (1979); Pantaleo *et al.* (1979); Watson (1979); Steele and Wagner (1979); Scott Smith and Skinner (1982, 1984a,b,c); Mitchell and Lewis (1983); Mitchell (1985); Waldman *et al.* (1987); Morris (1987); Bergman (1987); Fraser (1987); Alibert and Alberede (1988); and other references cited in the bibliochrony by Janse and Sheahan (1987).

Comments: Prairie Creek is the only known commercial-grade (?) diamondiferous lamproite field in North America: It was originally mined in the early twentieth century (Ozark, Mauney, and Arkansas mines). During 15–20 years of intermittent operation over 10,000 carats of diamonds are thought to have been produced from the breccia phases of the intrusives. During the 1950s, Prairie Creek was opened as a private tourist attraction and subsequently developed into The Crater of Diamonds State Park.

3.2.2. Group Name: Leucite Hills

Location: [41.833 N, 109.000 W]. The 22 occurrences forming the Leucite Hills occur in a 30×50-km zone near Superior, 25–50 km northeast of Rock Springs, Sweetwater County, Wyoming, U.S.A. (Figures 3.6, 3.7).

Synonym: Rock Springs Volcanics.

Age: 1.1–1.25 Ma [Quaternary] (Bradley 1964, Mcdowell 1966, 1971); 1.4–3.1 Ma (range of K–Ar ages of phlogopite from five volcanic centers: Steamboat Mountain—2.4 ± 0.1 Ma; Spring Butte—2.2 ± 0.1 Ma; Zirkel Mesa—1.6 ± 0.1 Ma; Emmons Mesa—1.4 ± 0.1 Ma; Boars Tusk—3.1 ± 0.2 Ma, this work).

Nature of Country Rocks: The lamproites are hosted by a sequence of Lower Tertiary (Eocene and Paleocene: Fort Union Formation), to Upper Cretaceous (Lance, Lewis, Almond, Ericson, and Rock Springs Formations) clastic sedimentary rocks, including sandstones, siltstones, coals, shales, and mudstones.

Local Structural Setting: The lamproites were intruded through, and extruded over, a nearly flat-lying Upper Cretaceous-to-Eocene sedimentary sequence (gentle northeast dip, northwest strike) at the northern nose of the Rock Springs Uplift, a doubly plunging Laramide basement-involved uplift or anticline in the center of the greater Green River Basin in the Wyoming Basin. There is a preferred northwest orientation to the lamproite vents which parallels the Farson Lineament (Blackstone 1972) and other Laramide structures, such as a buried Maastrichtian east-dipping thrust fault just west of Rock Springs. Johnston (1959) noted that the 12 igneous bodies between Zirkel Mesa and North Table

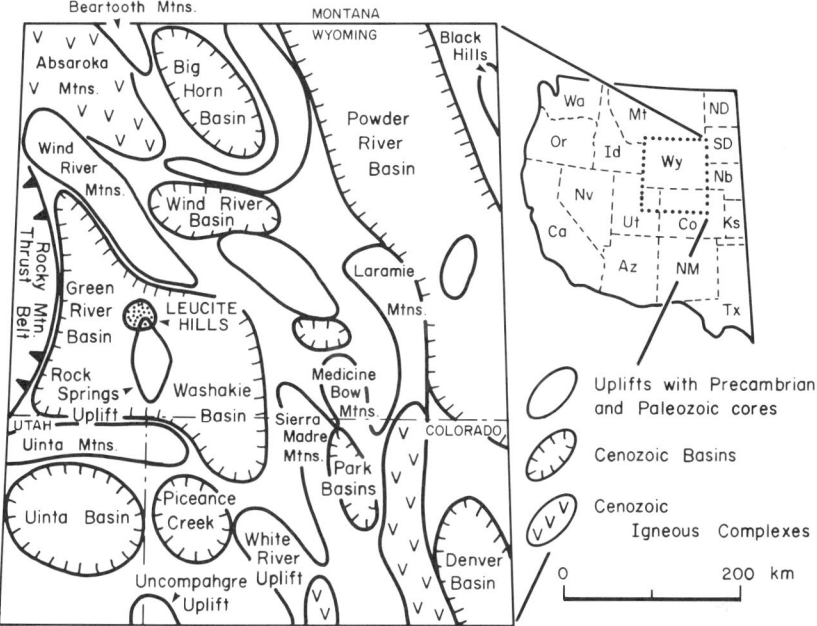

Figure 3.6. Regional map of the position of the Leucite Hills lamproite field relative to major Cenozoic regional structural or tectonic elements (after Kuehner 1980).

Mountain are aligned about N30–35°W. This NW trend is also exhibited by several dike segments on the east flank of Middle Table Mountain. Many northeast-trending normal faults cut across the Rock Springs Uplift in the vicinity of the Leucite Hills, however, there is no obvious relationship between these and the lamproites.

Regional Tectonic Setting: The Leucite Hills occur in the northern Colorado Plateau area, midway between the Uinta Mountain uplift (to the southwest) and the Wind River Range (to the north). These two Proterozoic-to-Archean complexes were structurally inverted by thrusting in the Late Cretaceous-to-Paleogene. Leucite Hills lamproites occur less than 200 km north of the southern limit of the Archean Wyoming Craton. This limit is defined by the Cheyenne Belt, a major shear zone separating Proterozoic rocks of the Colorado Province from the Archean Wyoming craton (Houston *et al.* 1979). The Cheyenne shear zone has been described by Hills and Houston (1979) as a Proterozoic suture developed at a south-dipping paleo-Benioff zone. At the time of lamproite genesis, the area was relatively stable, having experienced regional uplift, and was relatively insulated from Basin and Range extensional faulting, being located over 200 km east of its eastern limit. Zoback and Zoback (1980) suggest a present-day strike-slip to compressive state of stress for the Rock Springs Uplift and Colorado Plateau region, although regional extension characterizes the nearby Northern Rocky Mountains and Southern Great Plains.

Nature of Basement Rocks: The Leucite Hills are underlain by the Archean Wyoming Craton with crystallization ages of > 2.6–3.2 Ga (Peterman 1979, Ernst 1988, and references therein).

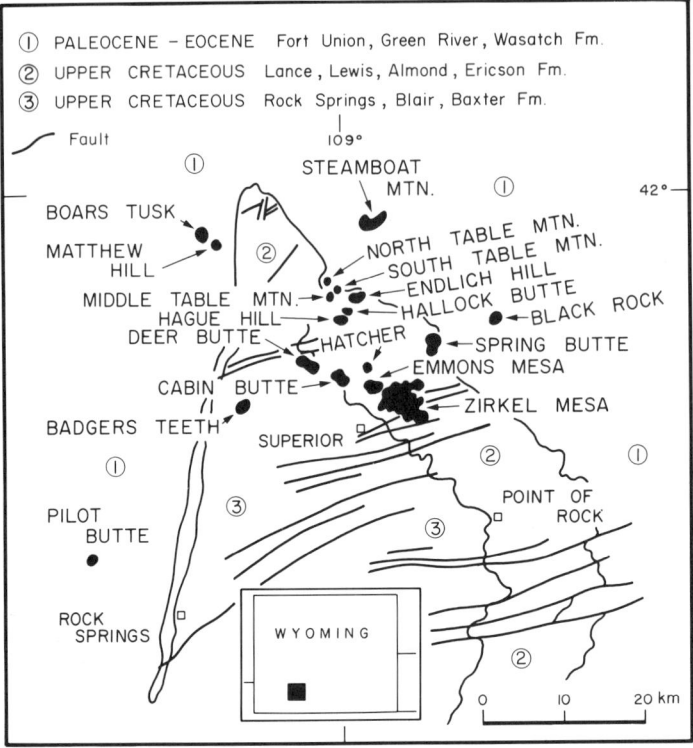

Figure 3.7. Geological map of the Leucite Hills lamproite field (after Kemp and Knight 1903, Love *et al.* 1955, Ogden 1979).

Intrusive/Extrusive Forms: Because of their recent age, the lamproites are dominated by volcanic forms, e.g., cinder cones, lava flows, plugs, volcanic necks, dikes, intrusive sheets, and rare pyroclastic deposits.

Geomorphology: The bulk of the province consists of lavas in the form of extensive mesas. Slight to moderate erosion of their upper flow surfaces has occurred. Cinder cones developed on the surface of these flows represent the last eruptive events. The cones are typically slightly denuded (Emmons Cone, Steamboat Mountain Cone), but have been entirely removed in some instances (e.g., North Table Mountain). Some necks (e.g., Boars Tusk) are intensely eroded. Note that Cross (1897) suggested that much of the Leucite Hills once formed a coalesced lava shield but have since suffered sufficient denudation to produce isolated volcanic centers. However, as recognized by Ogden (1979), there is abundant evidence that indicates the contrary.

Area/Volume: Collectively, 2000–2500 km^2, 25–50 km^3.

Component Localities (older terminology in parentheses): Pilot Butte, Boars Tusk, Matthew Hill, Steamboat Mountain, Black Rock, Spring Butte (Orenda Mesa), Badgers Teeth (Twin Rocks), Zirkel Mesa, Emmons Mesa, Hatcher Mesa, Hague Hill, Endlich Hill, North Table Mountain (North Table), Middle Table Mountain (South Table Butte), South Table Mountain (North Table Butte, Black Butte, Table Mountain, North Pilot Mesa),

Deer Butte (Cross Mesa), Cabin Butte (Osborn Mesa), Wortman Dike, Iddings-Weed-Hallock Buttes.

Nearby Contemporaneous Igneous Rocks: Early-to-Late Tertiary alkaline rocks are abundant 200 km to the north in the Absaroka Range. Pliocene (ca. 3–4 Ma) basaltic rocks occur 150 km north-northwest of the Leucite Hills in the southern Absaroka Mountains. Miocene (8–11 Ma) basaltic and rhyolitic igneous rocks occur 150 km southeast of the Leucite Hills near Steamboat Lake in northern Routt County, Colorado (Leudke and Smith 1978). The nearest significant occurrences of Miocene-to-Quaternary igneous rocks are the hot-spot-related magmatic rocks at Yellowstone (dominantly rhyolitic) and the Snake River Plain (dominantly basaltic), 250–350 km to the northwest (Leudke and Smith 1983). The post-Eocene Fortification dike is located 125 km southeast of the Leucite Hills. Eocene alkaline rocks occur in the Rattlesnake Hills, 200 km northeast of the Leucite Hills in central Wyoming. McCandless (1982, 1984) reported mantle-derived detrital (magmatic host unknown) Cr-pyrope, Cr-diopside, and enstatite in ant hills, 50–150 km to the west of the Leucite Hills, in the Green River Basin.

Rock Types: The lamproites are dominantly vesicular lavas, scoria, autolithic intrusive breccias, lapilli tuffs, tuff breccias, and agglomerates of diopside leucite phlogopite lamproite, diopside sanidine-phlogopite lamproite, and diopside madupitic lamproite.

Mineralogical Comments: Madupitic rocks contain a REE and strontium-rich perovskite. Shcherbakovite is common in vesicles in nearly all the lavas (this work). Fresh leucite and very iron-rich sanidine are characteristic of this province.

References: Zirkel (1876); Emmons (1877); Cross (1897); Kemp (1897); Kemp and Knight (1903); Schultz and Cross (1912); Carey (1955); Johnston (1959); Smithson (1959); Bradley (1964); MacDowell (1966); Yagi and Matsumoto (1966); Carmichael (1967); Kay and Gast (1973); MacKenzie *et al.* (1974); Barton (1975, 1976, 1979, 1988); Sobolev *et al.* (1975); Ogden (1978, 1979); Ogden *et al.* (1977, 1978, 1980); Kay *et al.* (1978); Barton and Hamilton (1978, 1979, 1982); George (1979); Sheriff *et al.* (1979); Kuehner (1980); Sheriff and Shive (1980); Barton and van Bergen (1981); Kuehner *et al.* (1981); Vollmer *et al.* (1984); Gunter *et al.* (1983); Lange *et al.* (1984); Mitchell (1985); Salters and Barton (1985); Bergman (1987); Eggler (1987); Wagner and Velde (1986a); Fraser (1987); Nelson and McCulloch (1989).

Comments: S.F. Emmons was the first to discover leucite in North America and the Leucite Hills were named by him during the 40th Parallel survey in 1867–1873. The Leucite Hills Field represents perhaps the best type locality for lamproites in the world and additionally possesses a wide variety of igneous forms. The Leucite Hill's young age and consequent lack of erosion permits a clear understanding of the volcanological evolution and emplacement history. The volcanic features of the Leucite Hills are similar to those of alkali basalt volcanic fields from all over the world. The relatively short duration of magmatism exhibited by Leucite Hills, coupled with its large volume, requires extremely high rates of magma emplacement and effusion. A relatively large phlogopite-rich source mantle lithosphere, which has been subjected to a rapid melting event, is also required. The Leucite Hills were considered a commercial potash resource for fertilizer for some time (200 million tons K_2O: Schultz and Cross 1912) and the Liberty Potash Co. mined the lamproites from the Zirkel Mesa quarry until 1920. Pumice and cinders from Zirkel Mesa cones have intermittently been mined commercially for use as abrasives, road metal, building aggregate, decorative stone, and cinder radiants for barbecues.

3.2.3. Group Name: Smoky Butte

Location: [47.317 N, 107.050 W]. This 3-km-long, intrusive and extrusive complex occurs 200 km northeast of Billings and 13 km west of Jordan, in Garfield County, northeastern Montana, U.S.A. (Figures 3.8 and 3.9).

Age: 27 Ma. [Miocene], K–Ar phlogopite (Marvin *et al.* 1980).

Nature of Country Rocks: The lamproites occur in a nearly flat-lying Early Tertiary-to-Late Cretaceous clastic sedimentary sequence (i.e., the Paleocene Fort Union Formation, the Late Cretaceous Hell Creek and Fox Hill Sandstones, Bearpaw Shale, and the Judith River Formation) which form a very broad, gently dipping syncline.

Local Structural Setting: Smoky Butte forms a N30°E-trending dike system, along, but oblique to, the axis of the broad (100-km-wide) east-trending Blood Creek syncline, about 30 km northwest of the Paleocene Freedom Dome.

Regional Tectonic Setting: Smoky Butte occurs 350 km east of the Rocky Mountain front, in the central foreland basin position of the Laramide orogen. The lamproites intruded through and extruded over stable platform clastic sedimentary deposits of the Cretaceous epicontinental seaway, which were deposited in the reactivated basement Big Snowy Trough (Maughan 1983). They are located 300 km west of the boundary between the Archean Wyoming Craton and the 1.8–2.5-Ga Trans-Hudson orogen of the western Dakotas (delineated by the North American conductivity anomaly, Bickford *et al.* 1986). Smoky Butte is located 200 km east of the northwest-trending Montana–Florida lineament, interpreted by Kinsland (1985 and references therein) as a major Proterozoic basement strike-slip fault along which >800 km of dextral slip has occurred. Smoky Butte is localized at the north edge of a Proterozoic-to-Mississippian aulacogen, the Proterozoic Belt embayment and the Montana trough. Over 500 m of Upper Mississippian clastic sedimentary and evaporite rocks were deposited in the 59×300-km Montana trough.

Nature of Basement Rocks: These lamproites are underlain by Archean Wyoming Craton with crystallization ages of >2.6 Ga (Ernst 1988 and references therein).

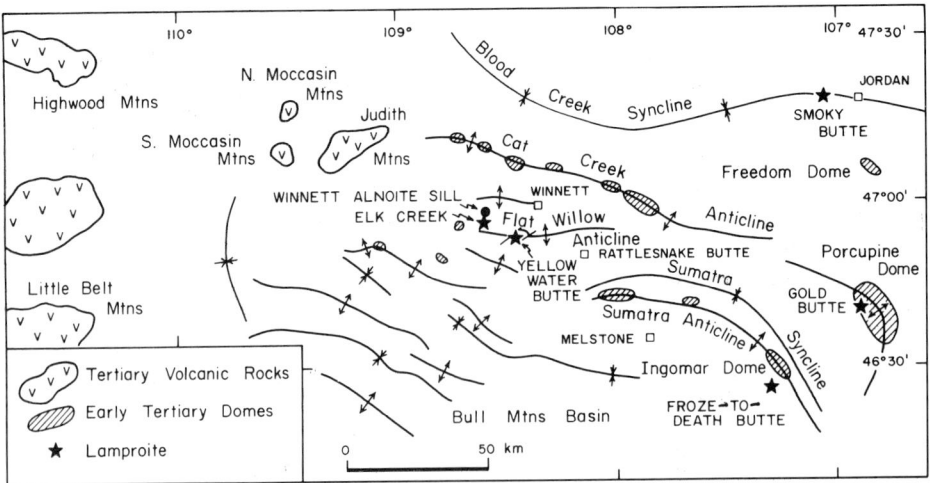

Figure 3.8. Regional map showing the Tertiary structure and volcanic setting of Smoky Butte and nearby lamproite fields (after Bowen 1915, Heald 1926, Ross 1926a, Larsen 1940, Johnson and Smith 1964).

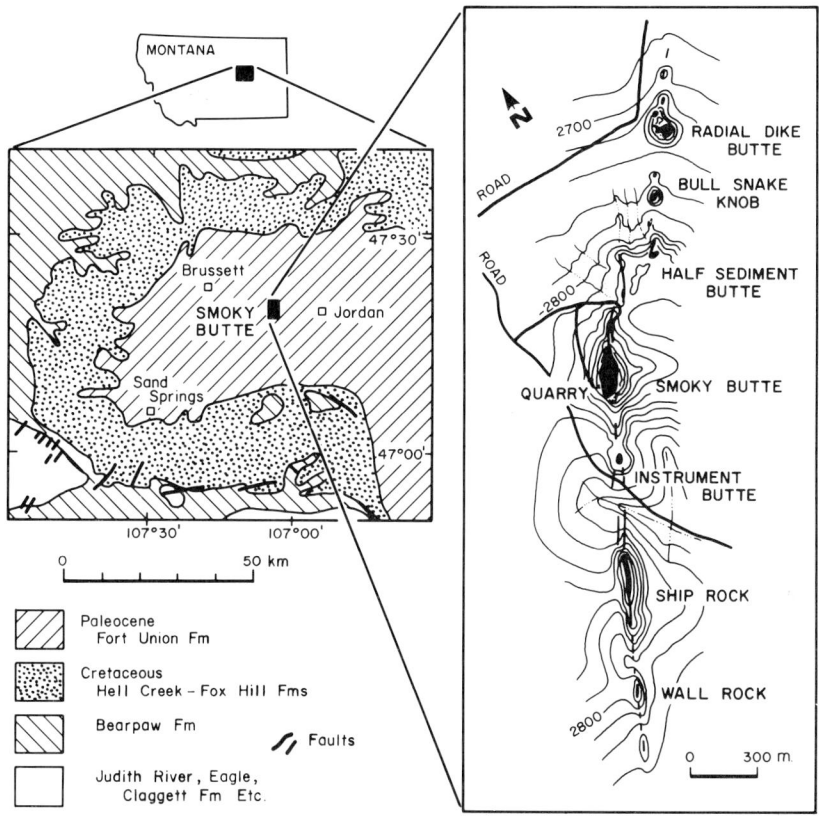

Figure 3.9. Geological map showing the nature and setting of the Smoky Butte lamproite field (after Matson 1960, Ross *et al.* 1955, Mitchell *et al.* 1987). Contours show elevation in feet (above sea level).

Intrusive/Extrusive Forms: The occurrence consists principally of multiply intruded hyalo-phlogopite lamproite dike segments (1–10 m wide). Smoky Butte is a plug/vent area, where the dike swells up to a width of 40 m. Most of the rocks are vesicular, massive, or glassy. Also present are minor agglutinate tuffs, pyroclastic rocks, and intrusive breccias.

Geomorphology: These moderately dissected dikes form erosional highs. Extrusive rocks are preserved at the highest structural levels of the plug and in several of the northern most dike exposures.

Area/Volume: 0.1 km², >0.5 km³.

Component Localities: Radial Dike Butte, Bull Snake Knob, Half Sediment Butte, Smoky Butte, Ship Rock, Wall Rock, Instrument Butte.

Nearby Contemporaneous Igneous Rocks: The Smoky Butte lamproites are among the easternmost and youngest components of the Montana alkaline province (mostly 45–55 Ma). An isolated suite of 30–55 Ma alkaline rocks occurs 420 km to the southeast, in the Devils Tower and Black Hills area, near the northern Wyoming–South Dakota border (Lisenbee *et al.* 1981). Slightly older olivine lamproites occur 75–100 km south of Smoky Butte at Froze-to-Death Butte and Gold Butte, on the Ingomar and Porcupine Domes (Heald

1926), respectively, and 125 km west of Smoky Butte in the vicinity of Yellow Water Reservoir and Elk Creek (Johnson and Smith 1964).

Rock Types: Glassy, massive, vesicular hypabyssal intrusive breccias, contact breccias, agglomerates, piperno, and tuffs of sanidine diopside richterite phlogopite lamproite.

Mineralogical Comments: Smoky Butte is the principal occurrence of terrestrial armalcolite-bearing rocks. Analcite has replaced all preexisting leucite at this locality.

References: Matson (1960); Velde (1975); Marvin *et al.* (1980); Mitchell and Hawkesworth (1984); Fraser *et al.* (1985); Wagner and Velde (1986a,b); Mitchell *et al.* (1987); Rådde (1987); Bergman (1987); Fraser (1987).

Comments: The glassy lamproite has been mined from a small quarry at the main intrusion for use as road metal.

3.2.4. Group Name: Francis

Location: [40.667 N, 111.200 W]: Several lamproite intrusions occur in a 15×25-km zone, 70 km east of Salt Lake City, 10–20 km north and east of Kamas, Summit County, northeastern Utah, U.S.A. (Figures 3.10–3.12).

Synonyms: Kamas, Moon Canyon.

Age: 13–40 Ma [Eocene-Miocene] (Best *et al.* 1968). Phlogopites from four different sills in and south of Moon Canyon give Oligocene $^{40}Ar/^{39}Ar$ plateau ages: 35.21 ± 0.10 Ma; 34.93 ± 0.12 Ma; 34.58 ± 0.11 Ma; 34.10 ± 0.35 Ma (K.A. Foland analyst, this work). Phlogopite from the White's Creek dike gives a $^{40}Ar/^{39}Ar$ plateau age 13.93 ± 0.09 Ma (this work).

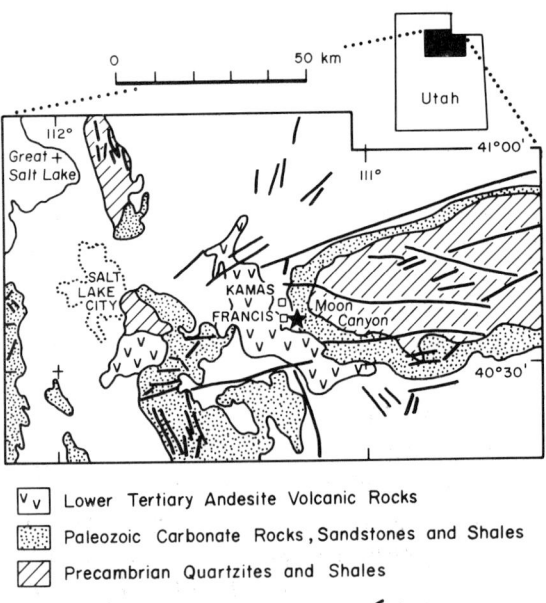

Figure 3.10. Geological map showing the setting of the Francis lamproite field (after Hintze 1980, Bergman 1987, this work).

DESCRIPTION OF LAMPROITE OCCURRENCES

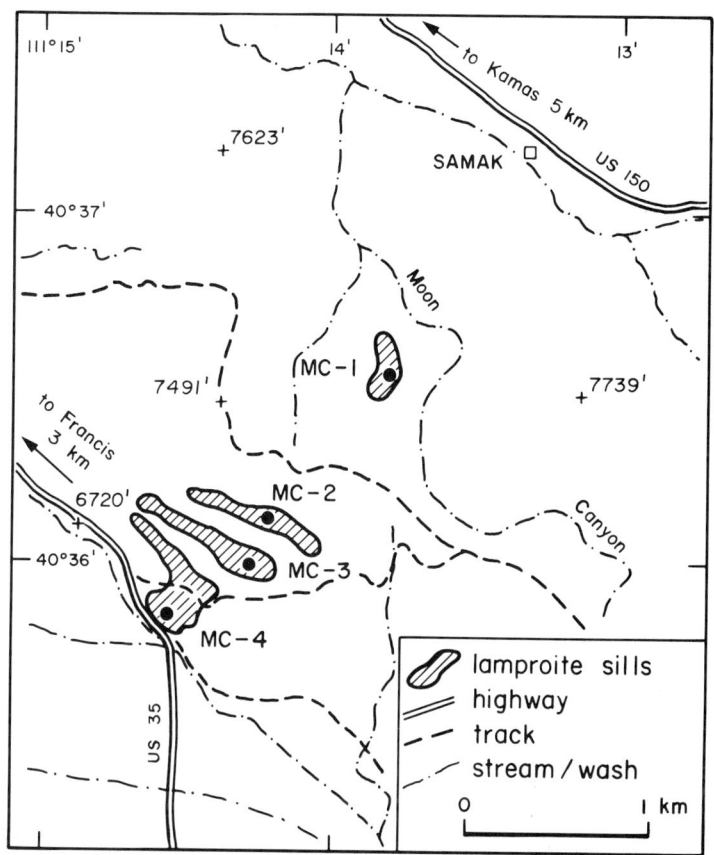

Figure 3.11. Location map showing the extent of selected lamproite sills in the Francis field (this work).

Nature of Country Rocks: The Francis lamproites intrude a nearly flat-lying Tertiary (Eocene) andesite pyroclastic and sedimentary sequence, which overlies Paleozoic carbonate rocks, shales and sandstones, and Proterozoic sedimentary and metasedimentary rocks. White's Creek dike intrudes Cretaceous shales and sandstones of the Frontier and Wanship Formations.

Local Structural Setting: The Francis and related lamproites occur at the intersection of the western margin of the east-trending Uinta arch and the north-trending Wasatch Mountains. Prominent east-trending high-angle faults were undoubtedly important in their emplacement.

Regional Tectonic Setting: The lamproites occur just west of the Colorado Plateau near the eastern margin of the Basin and Range Province, between the Uinta Uplift and the Toole Arch. A 3–4-km-thick sequence of Proterozoic metaclastic rocks in the Uinta arch comprise an inverted Proterozoic failed rift basin. Francis is located between two nearby major tectonic lines; Kay's Wasatch Line and Welsh's Las Vegas Line (Stokes 1976), extensions of the right-lateral Garlock Fault in California and Nevada. The Wasatch line separates low-velocity (7.4–7.6-km/sec), 30-km-thick crust with high heat flow (>2 HFU) to the west, from high-velocity (7.8–7.9-km/sec), 40-km-thick crust with low heat flow (<2 HFU) to

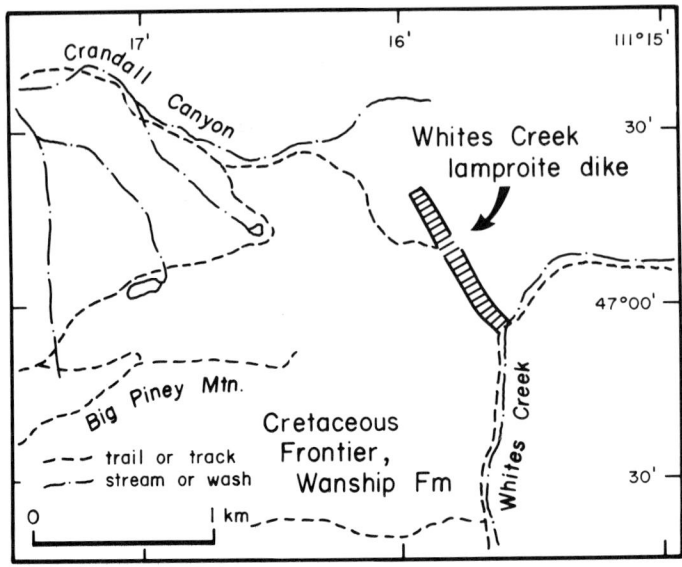

Figure 3.12. Location map showing the outcrop pattern of White's Creek dike (this work).

the east (Stokes 1976). The Wasatch Line forms the eastern margins of the Great Basin, the Laramide thrust belt and a Proterozoic rift with a 1.4-Ga age of inception, and therefore represents a major ancient zone of weakness in the lithosphere. The lamproites are located within 100 km of the southern margin of the Archean Wyoming Craton. The southern limit of the Wyoming Craton is defined by the Cheyenne Belt, a major shear zone which separates Proterozoic rocks of the Colorado Province from the Archean Wyoming craton (Houston et al. 1979). The Cheyenne shear zone has been described by Hills and Houston (1979) as a Proterozoic suture developed at a south-dipping paleo-Benioff zone. If the 800 km of post-Proterozoic dextral motion is reversed along the Montana–Florida strike slip fault of Kinsland (1985), then the Francis lamproites fall along a southwestern extension of the Mid-Continent Rift system. Just prior to lamproite intrusion, the area was in a foreland basin position, associated with the Sevier and Laramide orogenic activity to the west (Late Cretaceous–Paleocene). A major phase of uplift of the Uinta Mountains occurred in Paleogene times (Hintze 1988). At the time of lamproite formation, the western Utah region was experiencing widespread explosive caldera and stratovolcano activity (15–37 Ma), followed by Basin and Range extensional faulting (15 Ma to present); however, the Uinta Mountain region was in a position of tectonic quiescence.

Nature of Basement Rocks: In the Uinta and Wasatch Mountain Ranges of northeastern Utah, there are deformed sequences of Late Proterozoic and Archean metasedimentary rocks (0.8–1.1 Ga Uinta Mountain Group; 2.4–2.6 Ga Red Creek Quartzite) and older Archean metaigneous and metasedimentary rocks (2.6–3.0 Ga Farmington Canyon Complex; 2.7–2.8 Ga Owiyukuts Complex) of the Archean Wyoming Craton (Hintze 1988, Bryant and Nichols 1988 and references therein). Therefore, Mid-to-Late Archean rocks probably form the basement of the Francis and related lamproites.

Intrusive/Extrusive Forms: The lamproites occur as hypabyssal, shallow dikes (1–30 m wide), and vesicular sills (5–30 m thick).

Geomorphology: The lamproites form ridges and erosional remnants on mesa faces. Some dikes and sills (Moon Canyon) are poorly exposed and covered with alluvium.

Area/Volume: >0.5 km², >1.2 km³.

Component Localities: Francis Garage site sills; Moon Canyon sills; Park City Mine dikes; White's Creek dike.

Nearby Contemporaneous Igneous Rocks: The host andesite pyroclastic rocks are Paleocene-to-Eocene and mildly shoshonitic in character. Their relationship to the lamproites is unclear, but the pyroclastic succession most probably predates the lamproite intrusive activity by 10–20 Ma. A post-Paleocene, 300 m × 2 km phlogopite-rich dike (14.8 Ma phlogopite $^{40}Ar/^{39}Ar$ plateau age: this work) at White's Creek near Crandall Canyon was reported by Morris (1953) to contain nepheline. The melilite-kalsilite-bearing Smith–Morehouse dike (Larsen 1954, Wagner *et al.* 1987) occurs along the Erickson Basin fault plane, a vertical fault juxtaposing two Proterozoic quartzite units. The closest major volume of contemporaneous rocks are the Miocene basaltic and rhyolitic rocks, north and west of the Great Salt Lake, 200 km west and northwest of Moon Canyon (Leudke and Smith 1978). Miocene shoshonitic lavas occur 250 km to the southwest in eastern Juab County, Utah (Hogg 1972), Tertiary dikes and sills of alkaline rocks occur in the San Rafael Swell area, about 200 km south-southeast of Kamas (Gilluly 1927).

Rock Types: Vesicular, massive, hypabyssal sanidine diopside richterite phlogopite lamproite.

Mineralogical comments: The Francis lamproites contain a roedderitelike phase and pseudobrookite (Wagner and Velde 1986a).

References: Boutwell (1912); Morris (1953); Larsen (1954); Crittenden and Kistler (1966); Henage and Best (1968); Best *et al.* (1968); Henage (1972); Wagner and Velde (1986a); Bergman (1987); this work.

3.2.5. Group Name: Hills Pond

Location: [37.667 N, 95.750 W]. Two localities (Hills Pond and Rose Dome) occur in a 2×8-km area, 135 km east of Wichita, 7–15 km southwest of Rose, Woodson County, southeastern Kansas, U.S.A. (Figure 3.13).

Synonyms: Silver City lamproite, southeastern Kansas intrusives

Age: 88–91 Ma [Late Cretaceous]. (Zartman *et al.* 1967).

Nature of Country Rocks: The Hills Pond and nearby Rose Dome lamproites intrude nearly flat-lying Late Pennsylvanian shales and limestones of the Douglas and Lansing Groups, respectively. The Hills Pond lamproite intrudes the Ireland Sandstone and the Vinland Shale. The Rose Dome lamproite intrudes the Weston Shale and the Stanton Limestone.

Local Structural Setting: Both lamproites intrude gently domed portions of a relatively undeformed homoclinal sequence of Late Paleozoic platform sedimentary rocks.

Regional Tectonic Setting: The lamproites intrude broadly domed Late Paleozoic stable platform sedimentary rocks overlying stable North American craton. The Hills Pond and Rose Dome lamproites are located about 200 km southeast of the southwestern extent of

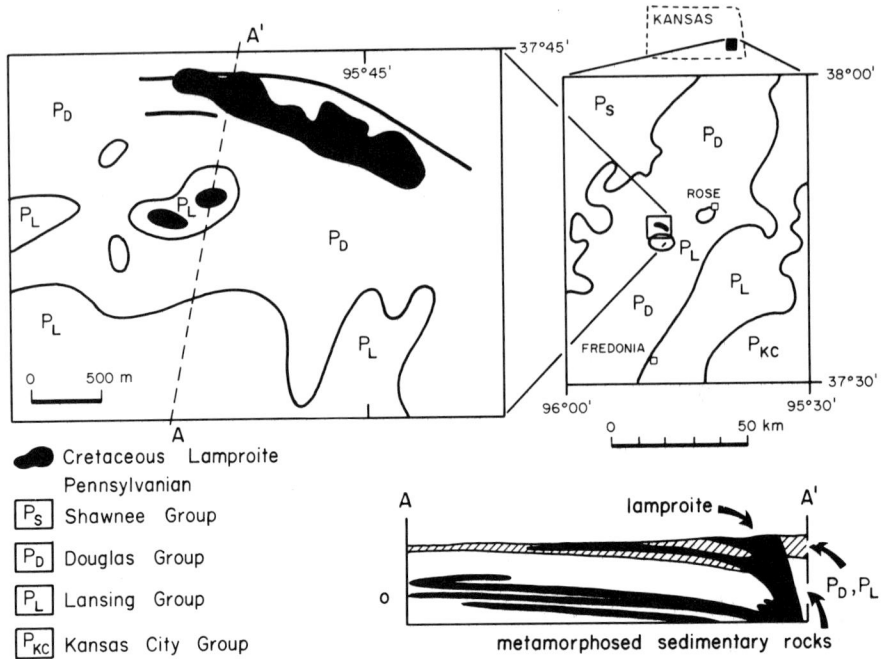

Figure 3.13. Geological map and cross section showing the extent of the Hills Pond lamproite (after Wagner 1954, Jewett 1964, Bickford *et al.* 1986).

the 1.1-Ga Midcontinent Rift System (Van Schmus and Hinze 1985). They occur near the southeast extension of a northwest-trending basement high-angle fault zone, the Fall River tectonic zone or the Wilson–Burns structural element (Berendsen and Blair 1987). Hills Pond occurs along the northwest-trending major Proterozoic basement strike slip fault defined by the Montana–Florida lineament, along which 800 km of dextral movement has been inferred by Kinsland (1985). Interestingly, when the 800 km of offset is reconstructed, southeastern Kansas realigns with the Reelfoot Rift, which places the lithospheric basement of the Hills Pond lamproites in close proximity to that of the Prairie Creek lamproite intrusives. The lamproites also occur just east of the north-trending Nemaha uplift, which is generally interpreted as a Late Paleozoic structure associated with Ouachita orogenic activity. The location of these lamproites is coincident with the northern limit of the contiguous 1.3–1.4 Ga Western granite-rhyolite province of Bickford *et al.* (1986). The Hills Pond lamproites occur along an extension of the axis of tectonic and igneous activity in the eastern and central U.S.A. along the 38th Parallel (Snyder and Gerdemann 1965, Zartmann 1877). At the time of lamproite intrusion, southeastern Kansas occupied a position of relative tectonic stability, situated in the eastern foreland basin position of the Rocky Mountain fold and thrust belt.

Nature of Basement Rocks: These lamproites are underlain by Proterozoic North American Craton with crystallization ages of 1.3–2.0 Ga (Bickford *et al.* 1986, Ernst 1988 and references therein). Basement rocks encountered in hundreds of petroleum wells in southeastern Kansas include 1.4-Ga epizonal granitic intrusive rocks and related rhyolitic-to-dacitic volcanic rocks and 1.4–1.7-Ga sheared mesozonal granitic-to-quartz monzonitic intrusive rocks (Bickford *et al*. 1979).

Intrusive/Extrusive Forms: The Hills Pond lamproites consist predominantly of sills and vents.

Geomorphology: The surface exposures of the Hills Pond rocks are intensely weathered to vermiculite and form topographic depressions. Relatively fresh lamproite has been encountered in shallow drilling. The Rose Dome rocks do not crop out at the surface and were discovered in shallow auger and drill holes.

Area/Volume: Hills Pond: Surface = 0.8 km^2; near surface = 1–3 km^3, Rose Dome: scattered subcrop over a 2 km^2 area; near surface = 1–3 km^3.

Component Localities: Hills Pond, Rose Dome, several unnamed lamproite vents in the vicinity (Coopersmith, personal communication, 1985).

Nearby Contemporaneous Igneous Rocks: Slightly older (Early Cretaceous—115–135 Ma) kimberlite diatremes occur 225 km northwest of the Hills Pond sills in Riley County. The kimberlites intrude Permian strata along the crest of the Nemaha Uplift (Brookins 1970) and are localized along the southeast margin of the northeastern-trending 1.1–1.2-Ga Midcontinent Rift System (Van Schmus and Hinze 1985). The Post-Pennsylvanian (Cretaceous?) Elk Creek carbonatite intrudes Pennsylvanian strata, also along the crest of the Nemaha uplift, in southeast Nebraska (Brookins *et al*. 1975, Burchett *et al*. (1983).

Rock Types: Hills Pond: coarse-grained, porphyritic olivine phlogopite richterite diopside madupitic lamproite. Rose Dome: extremely altered and carbonated olivine phlogopite (richterite diopside?) lamproite.

Mineralogical Comments: The Hills Pond lamproite contains well-developed euhedral K-richterite grains.

References: Hay (1883); Twenhofel and Edwards (1921); Twenhofel and Bremer (1928); Knight and Landes (1932); Wagner (1954); Franks (1959, 1966); Zartman *et al*. (1967); Franks *et al*. (1971); Bickford *et al*. (1971); Merrill *et al*. (1977); Berendsen *et al*. (1985); Cullers *et al*. (1985); Mitchell (1985); Bergman (1987); Coopersmith and Mitchell (1989).

Comments: Bulk testing of the Hills Pond and Rose Dome lamproites by Cominco (Coopersmith and Mitchell 1989) failed to detect any diamonds. The Hills Pond lamproites represent one of the few well-developed olivine lamproite sill bodies. The current commercial use of the Hills Pond lamproite (vermiculite at the surface) is as a diluent in the local cattle feed.

3.2.6. Group Name: Sisimiut

Location: [66.750 N, 53.000 W]. The Sisimiut lamproites occur over a broad 40×120-km zone, 50–75 km north and 40–50 km east to southeast of Sisimiut (formerly Holsteinsborg), central West Greenland (Figure 3.14).

Synonym: Holsteinsborg.

Age: 1206–1240 Ma [Proterozoic], K–Ar phlogopite (Winter 1974, Brooks *et al*. 1978); 1227 ± 12 Ma (Rb/Sr 7 mineral concentrate isochron, Scott 1981); 1200 ± 10 Ma (Rb–Sr whole rock, Stecher *et al*. 1987).

Nature of Country Rocks: The lamproite dikes intrude Late Archean (2.8 Ga)-to-Early Proterozoic (1650–1740 Ma) medium-to-high grade metamorphic rocks (gneisses, meta-

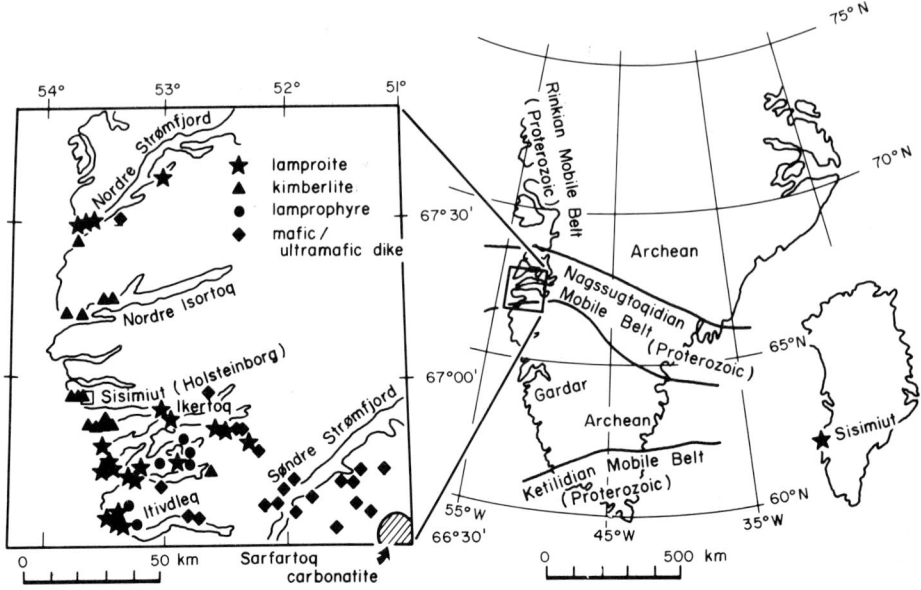

Figure 3.14. Location map of the Sisimiut lamproites and kimberlites and a regional map showing the major regional Precambrian belts (after Noe-Nygaard and Rambert 1961, Brooks *et al.* 1978, Scott, 1977, 1981, Larsen 1980, Thy *et al.*1987).

sedimentary rocks, and marbles) in the Ikertoq and Norde Strømfjord shear zones of the Isortoq (granulite facies) and Ikertoq (amphibolite facies) metamorphic complexes of the Nagssugtoqidian mobile belt.

Local Structural Setting: The lamproites are all posttectonic, largely undeformed dike swarms, localized along two shear zones near the southern limit of the intensely deformed Archean-to-Proterozoic Nagssugtoqidian mobile belt within 100 km of the undeformed (2.9–3.0 Ga) Archean craton (Escher *et al.* 1976).

Regional Tectonic Setting: The lamproites occur in a part of the stable North American craton which has remained relatively undeformed since the Early Proterozoic, except for the formation of the Danian rift between Baffin Island and Greenland. At the time of lamproite magmatism, the region had not experienced major orogenic deformation for 400–500 Ma, although 1200 Ma was a time of widespread rifting in North America and Western Europe. Elsewhere in Greenland, at about 1.0–1.3 Ga, widespread activity associated with the Caroliniadian orogeny took place and alkaline plutons were emplaced at the time of the eruption of basaltic lavas. The latter are intercalated with sandstones in fault-bounded blocks (Escher and Watt 1976). Many workers have suggested an extensive (Canada–Greenland–Europe) Proterozoic rifting event in the North Atlantic region, analogous to the East African Rift System (Kumarapeli and Saull 1966, Åberg 1988, see below).

Nature of Basement Rocks: The lamproites and host Proterozoic rocks are probably underlain by Archean craton with zircon crystallization ages of 2.9–3.0 Ga (Escher *et al.* 1976).

Intrusive/Extrusive Forms: Closely spaced veins and dike swarms or isolated dikes predominate (from a few centimeters to 7 m wide, averaging 1–2 m), with several rare sills or irregular intrusive bodies.

Geomorphology: The area is a deeply eroded midcrustal section with epizonal (6–12-km depth) lamproite intrusive rocks exposed along coastal outcrops.

Area/Volume: Over several hundred dikes and veins comprise a surface area of about 0.1 km^2 and a near surface volume of 0.2 km^3.

Component Localities: Many dikes along Norde Strømfjord, Ikertoq, and an unnamed fjord between Ikertoq and Itivdleq fjords.

Nearby Contemporaneous Igneous Rocks: Much younger (Late Proterozoic, 585 Ma) posttectonic kimberlites occur in the same area. These tend to be localized within unsheared blocks between the shear zones containing the lamproite dikes. The lamprophyre-bearing Sarfartoq carbonatite complex (Larsen 1980) formed at the same time (about 600 Ma) at the transition zone between the mobile belt and Archean basement, about 150 km south of Sisimiut. However, the Sarfartoq lamprophyres are apparently much older (1786–1974 Ma, Larsen *et al.* 1983) than the Sisimiut lamproites. Jurassic and significantly older (2.62–2.69 Ga) carbonatite complexes occur at Qaqarssuk and Tupertalik (Larsen *et al.* 1983). The 1100–1300-Ma Gardar alkaline igneous province, which contains ultramafic (lamprophyric-to-carbonatitic) dikes (Stewart 1970, Emeleus and Upton 1976, Blaxland *et al.* 1978, Upton *et al.* 1985), is located 700–800 km to the south of Sisimiut. Mid-Proterozoic alkaline anorogenic magmatism characterizes the Protogine zone of southern Sweden (Åberg 1988); many alkaline igneous complexes of broadly the same age occur to the west on Baffin Island in the Northwest Territories and in Ontario, eastern Canada (Gittins *et al.* 1967, Currie *et al.* 1975). The 1100–1250-Ma magmatic event also comprised widespread dolerite intrusions in Canada, S. Greenland, and Fennoscandia (Patchett 1978) and is generally regarded as the product of a major rifting event in the Proterozoic North Atlantic region (Åberg 1988 and references therein). This region in western Greenland is exceptional in that it contains over a dozen lithospheric mantle-derived alkaline rock suites (lamprophyres, carbonatites, and kimberlites) ranging in age from 120 Ma to 2.65 Ga: this indicates mantle melting events recurring in the same region for more than half of Earth history (Larsen *et al.* 1983). The formation of the Sisimiut lamproites occurred at about the halfway point in the 2.5-Ga evolution of these alkaline mantle melting episodes.

Rock Types: Massive, hypabyssal, sometimes vesicular olivine richterite diopside madupitic lamproite.

Mineralogical Comments: The Sisimiut lamproites contain rare priderite and pyrope and are characterized by the presence of abundant calcite.

References: Noe-Nygaard and Ramberg (1961); Winter (1971, 1974); Escher and Watterson (1973); Scott (1977, 1979, 1981); Brooks *et al.* (1978); Korstgård (1979); Larsen (1980); Getreuer and Rehkopff (1980); Thy (1982, 1985); Stecher and Thy (1982); Sörensen (1983); Stecher *et al.* (1987); Thy *et al.* (1987); Nelson (1989).

Comments: Diamonds have been found in heavy mineral suites in the Fiskenaesset area along the west Greenland Coast, about 500–700 km south of Sisimiut (Nilsen 1976). Diamonds have also been recovered from kimberlites in the Ivigtut–Frederikshåb area (Andrews and Emeleus 1971).

3.2.7. Group Name: Yellow Water Butte

Location: [46.967 N, 108.617 W to 46.917 N, 108.467 W]. The lamproites occur in a 10×20-km zone, 120 km north of Billings, 6 km southwest to 10 km south-southeast of Winnett, Petroleum and Fergus Counties, east-central Montana (Figures 3.8, 3.15).

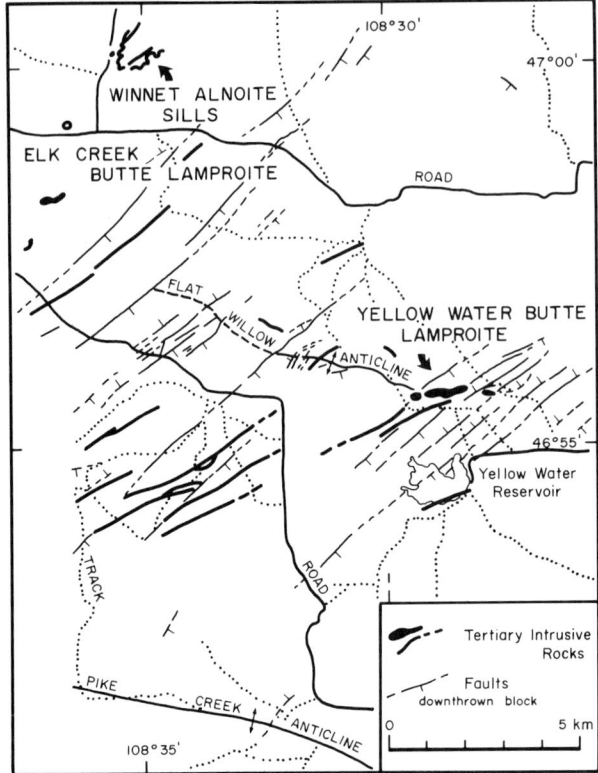

Figure 3.15. Location and structural map of the Yellow Water Butte and nearby lamproites and alnöites (after Johnson and Smith 1964, this work).

Synonym: Winnett area intrusives.

Age: 50–51 Ma (Paleocene); Yellow Water Butte—51.41 ± 0.13 Ma); Elk Creek massive phase—50.34 ± 0.16 Ma (phlogopite $^{40}Ar/^{39}Ar$ plateau ages, K.A. Foland analyst, this work).

Nature of Country Rocks: The lamproites occur in a nearly flat-lying, faulted, and broadly folded Early-to-Late Cretaceous shales and sandstones of the Colorado Shale Group, Telegraph Creek Formation, Eagle Sandstone, Claggett Shale, Judith River Formation and Bearpaw Shale.

Local Structural Setting: These lamproites occur in close proximity to the eastern-trending Flat Willow anticline, between the Button Butte and Porcupine Domes. The lamproite dikes and vents are elongate with east-trending orientations in a complicated series of northeast-trending, high-angle local fault blocks.

Regional Tectonic Setting: The lamproites occur in the core of the Big Snowy Anticlinorium or the Central Montana Uplift, in the Central Montana Trough and Proterozoic Belt Embayment. They lie 120 km to the east of the Late Proterozoic–Lower Paleozoic uplift complex of the Little Belt Mountains. The field occurs just east (within 50 km) of the northwest-trending Montana–Florida lineament, interpreted by Kinsland (1985) as a major Proterozoic basement strike slip fault along which >800 km of dextral displacement has occurred.

DESCRIPTION OF LAMPROITE OCCURRENCES

Nature of Basement Rocks: The lamproites intrude Archean Wyoming Craton with crystallization ages >2.6–3.2 Ga (Ernst 1988, references therein).

Intrusive/Extrusive Forms: The Yellow Water Butte intrusion is an east-trending elongate dikelike body of variable width (5–30 m), with a few subsidiary vent complexes, dikes, sills, and irregular intrusive bodies. The Teigen area intrusives are isolated plugs and dikes (2–100 m across). The Elk Creek body is an east-trending elongate vent complex.

Geomorphology: The lamproites are erosional highs, forming knobs and ridges.

Area/Volume: Collectively, 2 km², near-surface volume = 3 km³.

Component Localities: Yellow Water Butte, numerous dikes and plugs south of Teigen, Elk Creek dike/vent complex.

Nearby Contemporaneous Igneous Rocks: These lamproites occur on the eastern part of the belt forming the Missouri River Breaks–Winnett province of ultramafic lamprophyre (mostly alnöite) diatremes and dikes. The ages of most rocks in this province are 45–60 Ma, probably predating the lamproite magmatism. The Tertiary magmatism throughout central Montana includes the Montana alkaline province (Larsen 1940), in which the 50–54-Ma Highwood Mountains complex comes closest to, yet is distinct from, the lamproite clan (O'Brien *et al.* 1989). Probably contemporaneous lamproites at Gold Butte and Froze-to-Death Butte, Montana, occur on the Porcupine and Ingomar Domes, 60 km southeast and 80 km east-southeast, respectively, of the Yellow Water Butte area.

Rock Types: The Yellow Water Butte lamproites are extremely altered and carbonated massive-to-brecciated olivine phlogopite diopside lamproite and massive hypabyssal olivine lamproite. Intrusive breccias, lapilli tuffs, and lapillistones are present at the Elk Creek dike/vent complex.

Mineralogical Comments: The Yellow Water Butte lamproites are intensely altered with abundant secondary calcite and barite. Phlogopites and spinels in hypabyssal phases have compositions characteristic of lamproites.

References: Johnson and Smith (1964); Scambos (1987); this work.

Comments: The Yellow Water Butte lamproites have only recently been recognized as lamproites (this work), and their relationship to the central Montana alkaline province is not yet understood. The rocks were mapped as ultramafic dikes and sills and felsic volcanic rocks by Johnson and Smith (1964). The lamproites have been used commercially as road metal; several quarries occur along the length of Yellow Water Butte. Further studies of their mineralogy are required to confirm their lamproitic character.

3.2.8. Group Name: Froze-to-Death Butte

Location: [46.433 N, 107.3 W]. The 100 m-diameter pipe/vent complex of Froze-to-Death Butte occurs 120 km northeast of Billings, 15 km north-northwest of Hysham, in Treasure County, northeastern Montana, U.S.A. Gold Butte occurs about 40 km to the northeast of Froze-to-Death Butte at 46.672 N, 106.925 W, about 1 km southeast of the locality named Gold Butte on the 7.5′ Flat Bottom Coulee, NW, topographic map (Figures 3.8, 3.16, 3.17).

Age: Probably 25–55 [Paleogene], K–Ar phlogopite (Marvin *et al.*1980). A ^{40}Ar–^{39}Ar step-heating analysis of phlogopite attempted in this work was indeterminate (91.6 total gas age, no plateau, no isochron).

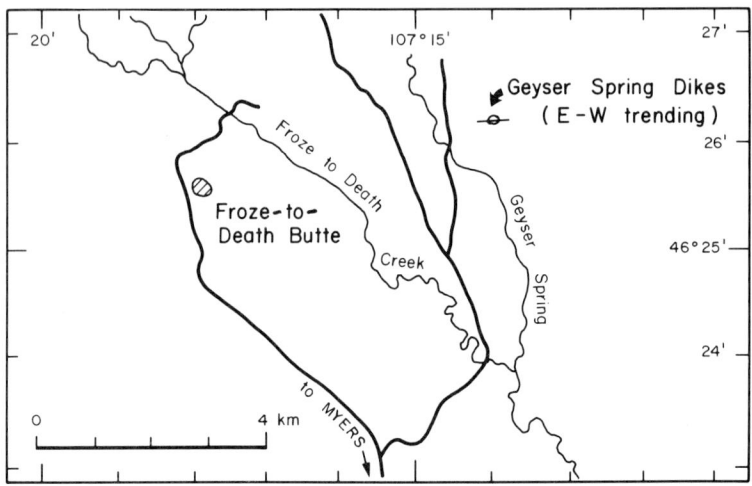

Figure 3.16. Location map of Froze-to-Death Butte and nearby lamproites (after Heald 1926, this work).

Nature of Country Rocks: The Froze-to-Death Butte lamproites occur in a nearly flat-lying Early Tertiary-to-Late Cretaceous clastic sedimentary sequence, which includes the Claggett Shale, the Bearpaw Shale, and the Judith River Formation.

Local Structural Setting: Froze-to-Death Butte forms a subcircular 80–100 m-diameter pipe, occurring just south of the axis of the small (10–20 km wide × 30–50 km long) northwest-trending Ingomar Dome, about 30 km southwest of the Paleocene Porcupine Dome, and 60 km northeast of the center of the Bull Mountain Syncline. Small (1–2 m wide) discontinuous dikes, about 5 km east of Froze-to-Death Butte, appear subvertical and trend east–west. The Ingomar Dome is cut by northeast-trending high-angle faults. The regional Cretaceous–Paleocene deposits have been deformed into broad arches and folds with northwest- and west-trending axes and have been cut by northeast-trending high-angle normal faults.

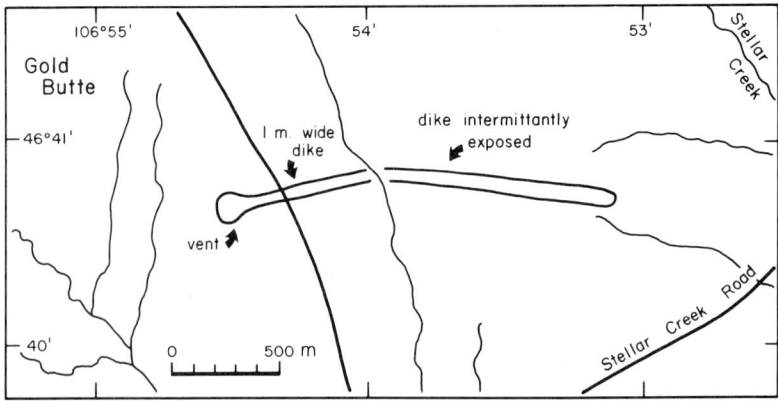

Figure 3.17. Location map of Gold Butte (after Johnson and Smith 1964, this work).

Regional Tectonic Setting: Froze-to-Death Butte occurs about 300 km east of the Rocky Mountain front in the central foreland basin position of the Laramide orogen. The lamproites intruded through and extruded over stable platform clastic sedimentary deposits of the Cretaceous epicontinental seaway, which were deposited in the Big Snowy Trough, a reactivated basement trough (Maughan 1983). These lamproites are located 300 km west of the boundary between the Archean Wyoming Craton and the 1.8–2.5-Ga Trans-Hudson orogen of the western Dakotas (delineated by the North American conductivity anomaly, Bickford *et al.* 1986). Froze-to-Death Butte is located 200 km east of the northwest-trending Montana–Florida lineament, interpreted by Kinsland (1985) as a major Proterozoic basement strike-slip fault along which >800 km of dextral slip has occurred: it is localized over the north edge of a Proterozoic-to-Mississippian aulacogen, the Proterozoic Belt embayment and Montana trough.

Nature of Basement Rocks: These lamproites are underlain by Archean Wyoming Craton with crystallization ages of >2.6 Ga (Ernst 1988 and references therein).

Intrusive/Extrusive Forms: The Froze-to-Death Butte lamproite is composed of a multiple pipe/vent complex dominated by massive rocks with varying proportions of foreign fragmental material.

Geomorphology: This moderately dissected pipe/vent forms an erosional high, protruding about 30 m above the incised rolling flatlands. However, nearby (about 4–5 km to the east), possibly consanguineous micaceous dike rocks have no topographic expression.

Area/Volume: 0.05 km^2, >0.2 km^3.

Component Localities: Froze-to-Death Butte; Gold Butte; Geyser Spring dikes.

Nearby Contemporaneous Igneous Rocks: Lacking a reliable radiometric age we will assume that the age of the Froze-to-Death intrusion is bracketed by that of Smoky Butte and the remainder of the Montana alkaline province rocks (i.e., 27–55 Ma). The Froze-to-Death lamproites are among the easternmost components of the Eocene and younger Montana Alkaline province (mostly 45–55 Ma). An isolated suite of 30–55 Ma alkaline rocks occurs 420 km to the southeast in the Devils Tower and Black Hills area near the northern Wyoming–South Dakota border (Lisenbee *et al.* 1981). Possibly younger (?) lamproites occur 75–100 km north-northeast at Smoky Butte. Approximately coeval (?) lamproites occur 100 km northwest of Froze-to-Death Butte in the vicinity of Yellow Water Reservoir and Elk Creek.

Rock Types: Massive hypabyssal intrusive breccias, and contact breccias of altered olivine phlogopite diopside lamproite.

Mineralogical Comments: Froze-to-Death Butte rocks contain phlogopites and spinels typical of lamproites. The rocks are strongly altered.

References: Heald (1926); Marvin *et al.* (1980); Scambos (1987); this work.

Comments: The locality requires an age determination by means other than K–Ar or ^{40}Ar-^{39}Ar, as both methods have proven to be inadequate. Further studies of the mineralogy are required to confirm their lamproitic character.

3.3. EUROPEAN LAMPROITES

Three lamproite provinces, in a variety of tectonic settings, have been recognized in Europe: Murcia-Almeria (group 1); Sisco (group 3); and Sesia-Lanzo/Combin (group 2)

(Figure 3.18). Other potassic to ultrapotassic rock suites occur in Europe (cf. Bergman 1987, p. 135, and references therein), some of which have been designated as lamproites by other workers. However, we do not include them in the lamproite clan because the mineralogical or compositional criteria for lamproites, outlined in Chapter 2, are not met. These suites include the 4.1-Ma selagitic minettes (termed orenditic lamproites by Peccerillo and Manetti 1985) from Montecatini Val di Cecina and Orciatico of the Roman comagmatic province in the Northern Appenines, Italy (Borsi *et al.* 1967, Barberi and Innocenti 1967, Wagner and Velde 1986a); the 0.2–0.4-Ma kamafugitic monticellite-bearing melilitites (termed madupites by Peccerillo and Manetti 1985) from San Venanzo and Cupaello in the Central Appenines, Italy (Holm and Munksgaard 1982, Gallo *et al.* 1984); a diverse suite of minette dikes from the Bohemian Massif, Czechoslovakia (Němec 1972, 1973, 1974, 1975, 1985, 1987, 1988, Schulze *et al.* 1985), some of which contain priderite (Němec, 1985); apatite-rich, biotite- katophorite-bearing rocks from Svidnya, Bulgaria (Stefanova 1966, Stefanova and Boyadzhieva 1974, 1975, Stefanova *et al.* 1974). The Pendennis/Holmeade

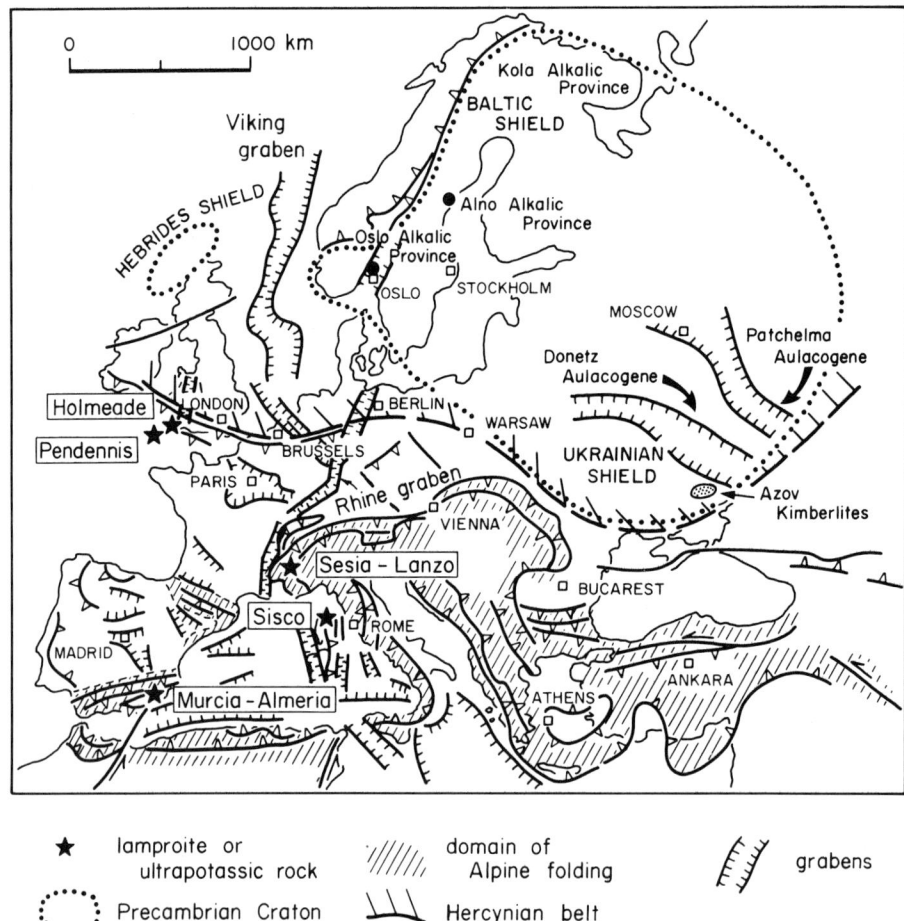

Figure 3.18. Map of European lamproite provinces, field and localities showing the main tectonic elements and Precambrian shields (after Aubouin 1980).

Farm field of Permian dikes and lavas were considered by Bergman (1987) to be lamproites; however, we now prefer to group them with ultrapotassic rocks such as minettes, pending more detailed mineral/chemical work on these altered rocks.

Perhaps the most interesting characteristic of European lamproites along the Mediterranean rim (as well as potassic rocks in this area in general) is their close association in space and time with Cenozoic subduction or continental collision, the rocks being emplaced within 5–30 m.y. of the cessation of subduction-related magmatism.

All the Mediterranean lamproites were apparently formed during an extensional rebound phase immediately following a phase of subduction and orogeny. The lamproites are adjacent to present-day extensional basins which formed by collapse of the originally thickened orogen (cf. Dewey 1988 for a discussion of the model of extensional collapse of orogens). These lamproites are located in the direction of extensional collapse adjacent to the extensional basins. In addition to this apparent subduction affinity, all of the European lamproites occur in terrains that have experienced several major episodes of compressional orogeny and extensional tectonics in the Paleozoic and Mesozoic. Another key feature is the presence of lamproites in parts of the Alpine orogen where the lithosphere is exceptionally thick (ca. 200 km) relative to typical Alpine lithosphere (ca. 70–140 km, Suhadolc and Panza 1988).

European lamproites also possess some compositional features characteristic of subduction-related melts, despite the fact that they postdate known subduction events. The major and trace element compositions of all European lamproites are typically depleted in titanium and niobium (see Section 7.8) relative to lamproites from other continents. These and other compositional features are shared by European lamproites and subduction-related intermediate-to-mafic volcanic rocks from the Mediterranean and elsewhere (Ewart and LeMaitre 1980, Gill 1981, Briqueu et al. 1984). Whereas Cenozoic subduction may have played a role in determining the composition of the Mediterranean rim lamproites, we view the nature and evolution of the lithospheric source as a more fundamental parameter in the formation of these melts. Therefore, Cenozoic subduction and postsubduction extensional relaxation may have provided the tectonic framework that led to melting of ancient modified lithosphere.

3.3.1. Group Name: Murcia-Almeria

Location: [37.500 N, 1.700 W]. These lamproites form a 90 × 240-km northeast-trending belt within 100 km of the southeast Mediterranean coast, about 360 km southeast of Madrid, between the cities of Almeria, Alicante, and Valencia, in the provinces of Almeria and Murcia, Spain (Figures 3.19, 3.20).

Synonym: Mazarron basin lamproites.

Age: 5.7–10.8 Ma [Late Miocene] (Nobel *et al.* 1981, Bellon 1976, 1981, Bellon and Brouse 1977, Montenat, 1973).

Nature of Country Rocks: The lamproites are hosted by Mesozoic-to-Tertiary sedimentary rocks. Volcanic rocks are interbedded with Late Miocene lacustrine and fluvial calcareous and marly sedimentary rocks of the Vera and Mazarron basins (Calvo *et al.* 1978). Intrusive rocks cut Mid-to-Late Miocene arenaceous, bioclastic marine sedimentary rocks, as well as older rocks. These Late Tertiary sediments unconformably overlie isoclinally folded Lower Cretaceous clastic sedimentary rocks onto which Jurassic carbonate rocks have been thrust from the northwest.

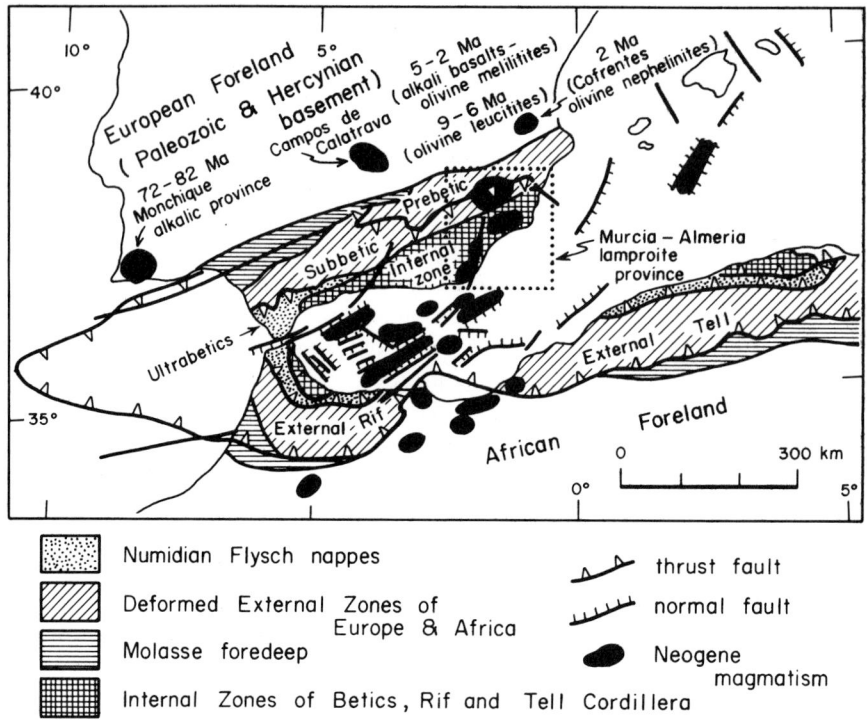

Figure 3.19. Regional tectonic element map of the western Mediterranean showing the position of the Murcia-Almeria lamproite province (after Horvath and Berckhemer 1982).

Local Structural Setting: All Murcia-Almeria lamproites are posttectonic, with many of the vents and dikes localized along thrust faults and strike slip faults occurring in the northeast-trending shear zone of basement weakness.

Regional Tectonic Setting: The lamproites occur in several zones of the Alpine orogenic belt in southeast Spain: on the southeast edge of the prebetic zone (rocks not metamorphosed by Alpine events); in the Subbetic zone (folded and thrusted Tertiary and Mesozoic rocks with only low-grade Alpine metamorphism); Betic zone (nappe complexes). The lamproites form two subbelts in distinct structural provinces separated by a major LANDSAT lineament (e.g., Rahman and Wright 1976). The northern belt is localized along a northeast-trending basement hinge zone, which is primarily in the Prebetic zone and possesses southeast vergence marking intense Tertiary deformation. This contrasts with dominant northward vergence elsewhere in the Prebetic zone. The southern belt occurs mainly in the Internal zone (the nappe complex). Dewey (1988) presented a model of extensional collapse to explain the formation of 15-km-thick crust in the Alboran Sea following Paleogene crustal thickening (ca. 60 km) in the Betic-Rif orogen. The Murcia-Almeria lamproites were erupted in a belt paralleling the orogen about 30–40 m.y. following the orogenic thickening event. At about the time of lamproite magmatism, the region was experiencing a north–south-directed convergent stress field (Ott d'Estevou and Montenat 1985). Since the time of lamproite emplacement the magmatism has been more alkali basaltic in character. Suhadolc

DESCRIPTION OF LAMPROITE OCCURRENCES

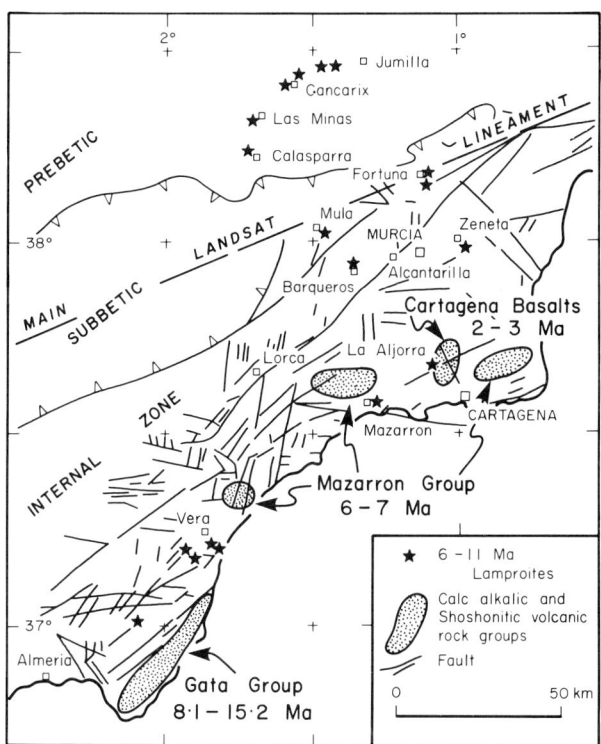

Figure 3.20. Map of the local structure and geology of the eastern Betic Cordillera showing the distribution of Murcia-Almeria lamproites and other Neogene magmatism (after Fuster *et al.* 1967, Bellon and Brouse 1977, Lopez Ruiz and Rodriguez Badiola 1980, Nobel *et al.* 1981, Bellon *et al.* 1983, Venturelli *et al.* 1984a, Hernandez *et al.* 1987).

and Panza (1988) noted the presence of deep (about 200 km) lithospheric roots in southern Spain, in contrast to normal lithospheric thicknesses of about <100 km in the rest of Spain.

Nature of Basement Rocks: The lamproites are underlain by Paleozoic (pre-Silurian?) continental crust (graphitic quartz–mica schist) metamorphosed in the Hercynian orogeny (270–340 Ma). Alpine greenschist facies metamorphism has overprinted rocks in the nappe complexes.

Intrusive/Extrusive Forms: Breccia-mantled pipes or vents (0.4–1 km diameter), sills, dikes, flows, plugs, small plutons, and pyroclastic rocks including hyaloclastite tuffs, lapilli tuffs, tuff breccias, and tuffs.

Geomorphology: The Murcia-Almeria lamproites are moderately-to-intensely dissected, forming isolated erosional highs.

Area/Volume: 100 km^2, 10 km^3.

Component Localities: Vera, Fortuna, Cancarix, Jumilla, Las Cabras, La Celia, Puebla de Mula, Las Minas, Barqueros, Cerro de Monagrillo (Cerro de Salmeron), Mazarron, Calasparra, Cabezo Maria, Cabezo Negro, Zaneta.

Nearby Contemporaneous Igneous Rocks: The lamproites are minor components of a regional volcanic belt on the southern coast of Spain and France of Mid-to-Late Tertiary

(<30 Ma) shoshonitic, calc-alkaline, anatectic rhyolites and alkali basaltic rocks (DiBattistini et al. 1987, Bordet 1985, Bordet and De Larouzière 1983). The rhyolites (18–22 Ma) and shoshonites (16–8 Ma) are generally older and the alkali olivine basalts much younger (2–3 Ma) than the 5–10 Ma lamproites (Nobel et al. 1981, Bellon et al. 1983). Miocene calc-alkaline magmatism also characterizes the Alboran Sea, to the south of the Murcia-Almeria lamproites. The calc-alkaline andesites have been interpreted as subduction-related melting products of a northward-dipping slab below southern Spain. The shoshonitic rocks are generally regarded as late-stage subduction-related melts (Hortvath and Berckhemer 1982). These workers also interpret the lamproites as eruption products formed during the senile stages of subduction. Olivine leucitite volcanic rocks of the same age (6.4–8.6 Ma) as Murcia-Almeria lamproites occur to the northwest at Campos de Calatrava, south of Ciudad Real in south Central Spain (Ancochea 1982). Contemporaneous (2–16 Ma) shoshonitic, calc-alkaline, rhyolitic, and alkaline basaltic magmatism characterizes much of the North African margin, in the area between Tunisia and Morocco (Bellon and Letouzey 1977, Bellon and Brousse 1977, Megartsi 1985). High-potassium rocks and shoshonites (5–10 Ma) are present in Morocco from Guilliz to Gourougou (Hernandez 1975) and at Azzaba, Algeria (Villa et al. 1974). All of these potassium-rich rocks occur within the Alpine orogen, yet are clearly post-tectonic. This Mid-to-Late Tertiary alkaline magmatism (including lamproites) clearly postdates, and appears unrelated to, the 70–80-Ma Iberian alkaline igneous province, which has been interpreted as a manifestation of an opening phase of the North Atlantic (cf. Rock 1982 and references therein).

Rock Types: Glassy, vesicular-to-massive diopside richterite sanidine (hypersthene-olivine) phlogopite lamproites.

Mineralogical Comments: Several of the Murcia-Almeria lamproites contain hypersthene. Priderite and wadeite have not yet been reported from this suite. Amphiboles and feldspars are unusually sodic relative to those found in other lamproites.

References: Osann (1889, 1906); de Yarza (1895); Meseguer Pardo (1924); Jeremine and Fallot (1928); Fallot and Jeremine (1932); Hernandez-Pacheco (1935, 1965); Parga Pondal (1935); San Miguel (1935, 1936); Burri and Parga Pondal (1937); San Miguel and de Pedro (1945); San Miguel et al. (1951); Fuster and de Pedro (1953); Fuster (1956); Fuster et al. (1954, 1967); Fuster and Gastezi (1964); Coello and Castanon (1965); Marinelli and Mittempergher (1966); Fermoso (1967a,b); Borley (1967); Velde (1969b); Powell and Bell (1970); Fernandez and Hernandez-Pacheco (1972); Pellicer (1973); Sahama (1974); Caraballo (1975); Montenat (1975, 1977); Bellon (1976, 1981); Bellon and Brousse (1977); Bellon and Letouzey (1977); Bijou-Duval et al. (1977); Lopez Ruiz and Rodriguez Badiola (1980); Molin (1980); Bellon et al. (1981, 1983); Nobel et al. (1981); Ancochea (1982); Bordet and Larouziere (1983); Larouzière and Bordet (1983); Nixon et al. (1984); Venturelli et al. (1984a, 1988); Larouzière (1985); Hertogen et al. (1985); Wagner and Velde (1986a); Ancochea and Nixon (1987).

Comments: The Murcia-Almeria lamproites were used by Niggli (1923) as the type locality for lamproites. Ironically, most of the Murcia-Almeria lithologies are unlike the other well-characterized lamproites (e.g., Leucite Hills, West Kimberley) in terms of mineral and rock major and trace element composition. The Murcia-Almeria olivine diopside richterite madupitic lamproites (jumillite) are closest to typical olivine madupitic lamproites from other suites.

Many of the lamproites might have been contaminated by crustal material.

3.3.2. Group Name: Sisco

Location: [42.833 N, 9.417 E]. The Sisco sill occurs just west of the hamlet of Sisco, 15 km north of Bastia, northern Corsica, France (Figure 3.21).

Age: 13.5–15.4 Ma [Miocene] (Bellon 1981).

Nature of Country Rocks: The sill intrudes the Mesozoic "schistes lustres" unit, composed mostly of metabasaltic glaucophane schists and ultramafic rocks. Potassium feldspar is developed in a contact metamorphic aureole (Velde 1967).

Local Structural Setting: The 1.5–4-m-thick sill intrudes an igneous sequence, metamorphosed during the Alpine orogeny.

Regional Tectonic Setting: Northern Corsica occurs on the margin of the Alpine nappes. The area was subjected to anatectic plutonism in Permo-Triassic times. In Early Miocene times, Corsica and Sardinia were almost joined to southern France and Spain. Since then, drift and counterclockwise rotation has led to their present-day configuration. The lithospheric thickness in the region is about 80 km (Suhadolc and Panza 1988). The Sisco lamproite lies just west of the Tyrrhenian Basin, a Miocene-to-Recent extensional basin,

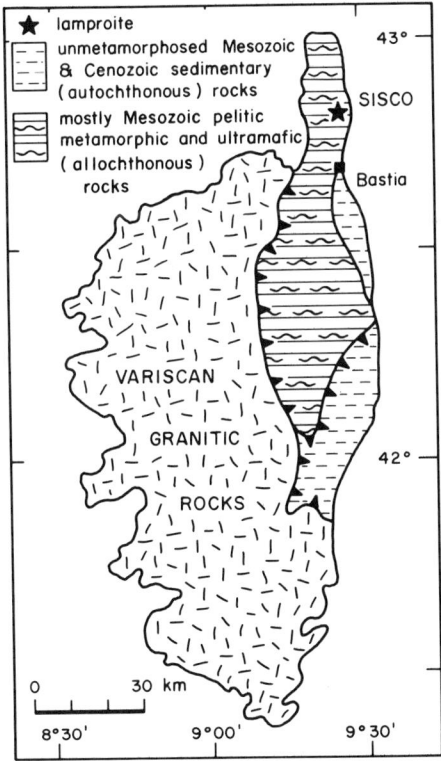

Figure 3.21. Map of the geology of Corsica showing the location of the Sisco lamproite (after Zwart and Dornslepen 1978)

which formed on an Early Cenozoic orogenic belt as a result of collapse following crustal thickening (Dewey 1988, Wezel 1982).

Nature of Basement Rocks: The area is probably underlain by Paleozoic granitic and rhyolitic-to-dacitic volcanic rocks.

Intrusive/Extrusive Forms: The lamproite forms a small hypabyssal sill (1.5–4 m thick).

Geomorphology: The lamproite occurs in a mid-to-lower crustal section recently exposed at the surface by uplift and erosion.

Area/Volume: >0.001 km^2, >0.00001 km^3 near surface.

Component Localities: The Sisco sill is the only known lamproite in the area.

Nearby Contemporaneous Igneous Rocks: Oligocene-to-Miocene (30–25 Ma) calc-alkaline magmatic rocks occur within 300 km of Sisco in southeastern France, southern Corsica, and Sardinia. Miocene dike rocks of glassy potassic olivine trachyte occur 750 km to the southwest along trend in the Alpine orogen at Azzaba, Algeria (Villa *et al.* 1974).

Rock Types: Porphyritic-to-equigranular, massive, diopside sanidine olivine phlogopite lamproite.

References: Primal (1963); Velde (1967, 1968); Bellon and Brousse (1977); Feraud *et al.* (1977); Civetta *et al.* (1978); Bellon (1981); Wagner and Velde (1986a).

3.3.3. Group Name: Sesia-Lanzo and Combin

Location: [45.583 N, 7.750 E]. The Sesia-Lanzo and Combin lamproite field is located about 120 km west of Milano, 15–40 km northwest of Ivrea, in the states of Piemonte and Valle-D'Aosta, NW Italy (Figure 3.22).

Synonyms: Northwestern Alps lamproites.

Age: 29–33 Ma [Late Oligocene] (Dal Piaz *et al.* 1973, Hunziker 1974, Scheuring *et al.* 1974, Zingg *et al.* 1976).

Nature of Country Rocks: The lamproites intrude intensely deformed allochthonous rocks of the Sesia-Lanzo and Combin units including: the Piemonte ophiolite nappe; the Gneiss Minuti complex; the Micascisti Eclogitic complex.

Local Structural Setting: These northwest-trending posttectonic dikes occur north and west of the Canavese tectonic line, in the western sector of the Peri-Adriatic (Insubric) lineament.

Regional Tectonic Setting: The region has experienced Hercynian compressional metamorphism, Triassic–Jurassic extension, and early Alpine (Maastrichtian–Paleocene) and Lepontine (Eocene–Early Oligocene) compression associated with continent–continent collision. Post-Middle Oligocene extension occurred near the boundary between high-temperature metamorphic rocks of the nappes, and the more stable rocks of the southern Alps (Dal Piaz and Ernst 1978 and references therein). The dikes were emplaced in the Internal zone of the NW Alpine orogen, 4–9 m.y. after the thermal peak of the Lepontine metamorphism (38 Ma), during a phase of regional uplift that began in the Middle Oligocene times and has continued to the present. The lithosphere (about 200 km) characterizing this part of the Alps is anomalously thick, compared with 90–140 km in the central Alps. In common with Sisco and Murcia-Almeria lamproites, the Northwestern Alps lamproites occur adjacent to a lithospheric region that has undergone extensional collapse (both the

DESCRIPTION OF LAMPROITE OCCURRENCES

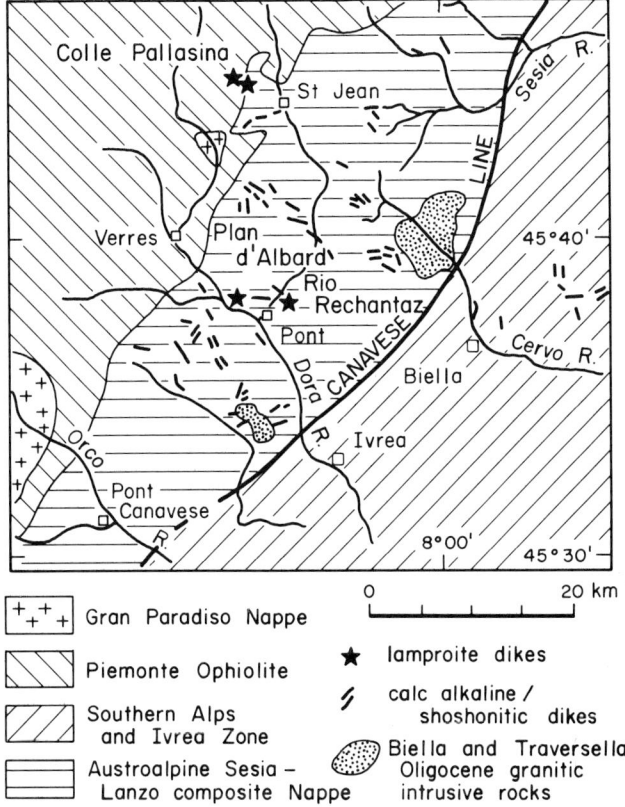

Figure 3.22. Geological map of NW Italy showing the distribution of the Sesia-Lanzo and Combin lamproites and other Neogene magmatism (after Dal Piaz et al. 1979, Venturelli et al. 1984b).

northern Alps and the Tyrrhenian basin to the south) following crustal thickening associated with the Alpine orogeny (cf. Dewey 1988).

Nature of Basement Rocks: The bulk of the basement in the region is thought to consist of Mid-to-Late Paleozoic crystalline rocks and sedimentary deposits. All were subjected to Hercynian and Alpine metamorphism (Debelmas and Lemoine 1970).

Intrusive/Extrusive Forms: Mostly dikes and rare flows.

Geomorphology: The region is one of high relief, and intensely dissected by ongoing rapid uplift and denudation.

Area/Volume: <0.001 km^2, <0.0001 km^3.

Component Localities: (1) Champoluc (Ayas valley); (2) East of Colle Pallasina; (3) Rio Rechantaz (lower Gressoney valley); (4) Plan d'Albard, Bard (Aosta valley).

Nearby Contemporaneous Igneous Rocks: The lamproites are temporally and spatially associated with shoshonitic calc-alkaline volcanic and intrusive rocks (Dal Piaz et al. 1979).

Rock Types: Massive diopside sanidine alkali amphibole lamproite.

Mineralogical Comments: The lamproites contain alkali amphiboles close to riebeckite-arfvedsonite in composition.

References: De Marco (1959); Krummenacher and Evernden (1960); Carraro and Ferrara (1968); Dal Piaz *et al*. (1973, 1979); Hunziker (1974); Scheuring *et al*. (1974); Zingg *et al*. (1979); Venturelli *et al*. (1984b).

3.4. AFRICAN LAMPROITES

African lamproites occur in three main regions: Swartruggens, Pneil, and Postmasborg, South Africa; Kapamba in the Luangwa Graben, Zambia, East Africa; Bobi, Ivory Coast, West Africa (Figure 3.23). Ages range from Proterozoic to Mesozoic and all either contain diamonds or are temporally and spatially associated with diamondiferous kimberlites. Unfortunately, little has been published on any of these lamproite suites. The African lamproites all occur near the outer margins of Archean-to-Lower Proterozoic cratons, adjacent to Proterozoic-to-Early Paleozoic mobile zones.

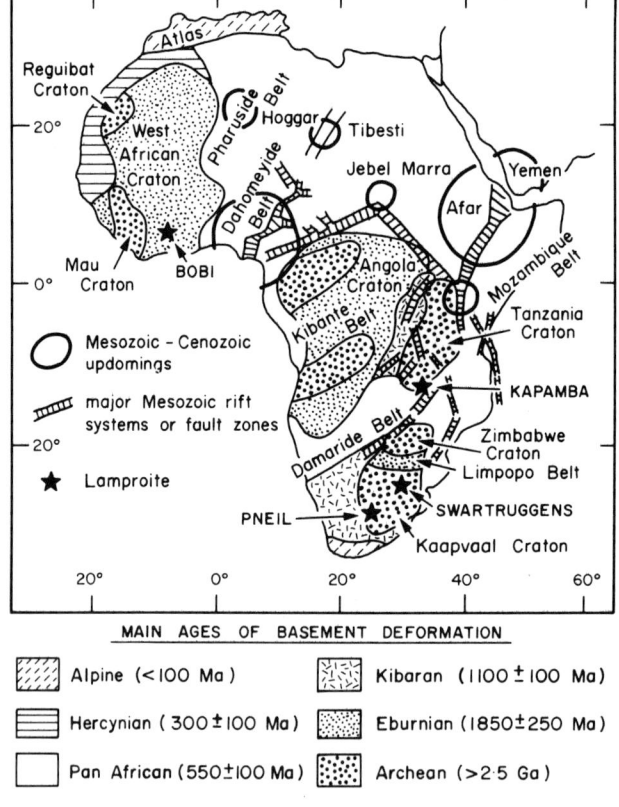

Figure 3.23. Map of Africa showing Precambrian domains, tectonic elements, Mesozoic rift systems, Mesozoic-Cenozoic upwellings and lamproite and kimberlite fields and provinces (after Dawson 1980, Janse 1985, Cahen *et al*. 1984, Wright 1985).

DESCRIPTION OF LAMPROITE OCCURRENCES

Reid and Barton (1983), Barton (1983), and Hunter and Reid (1987) have noted the presence of phlogopite-, amphibole-, potassium-feldspar-, sphene-bearing dikes (lamproite ?) in the northeast-trending 1.04 Ga Middelplatt dike swarm of the Spektakel intrusive suite, western South Africa. Some workers (e.g., Vollmer and Norry 1983) have incorrectly included rocks from the kamafugitic Toro-Ankole volcanic province, Uganda, within the lamproite clan.

3.4.1. Group Name: Kapamba

Location: [13.00 S, 31.34 E]. The Kapamba lamproite field forms a belt 25 km in length along the Kapamba River, a tributary to the Luangwa River, about 150 km WNW of Chipata in Eastern Zambia (Figures 3.24, 3.25).

Synonym: Luangwa Graben lamproites.

Age: About 220 Ma based on the modal age of phlogopite concentrates, or <250 Ma, probably Cretaceous or Jurassic, based on geological relationships (Scott Smith *et al.* 1989).

Nature of Country Rocks: The lamproites intrude an Upper Karoo (Jurassic) sedimentary sequence of mudstones, siltstones, sandstones, and grits. The host Karoo sedimentary rocks overlie, and occur just east of, exposed Proterozoic basement rocks of the Irumide mobile belt (graphitic schists, marble, granitic gneisses, and granulites).

Local Structural Setting: A group of 14 small vents and associated dikes form a 25-km-long northwest-trending belt, which is orientated orthogonally to the down-faulted, northeast-trending Luangwa Graben and the prominent north-northeast structural grain of the Proterozoic basement. Tertiary faulting has occurred in the area, but the intrusive rocks are

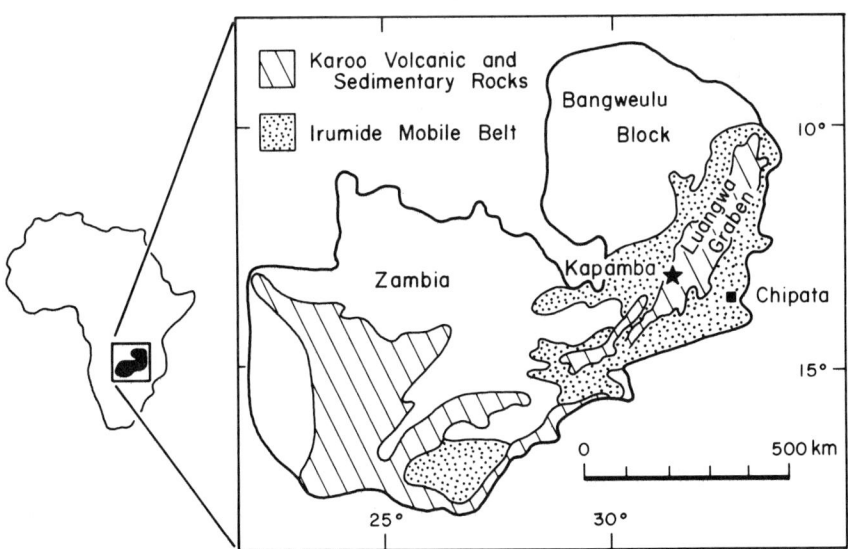

Figure 3.24. Regional map of Zambia showing the tectonic situation of the Kapamba lamproite field (after Cahen *et al.* 1984).

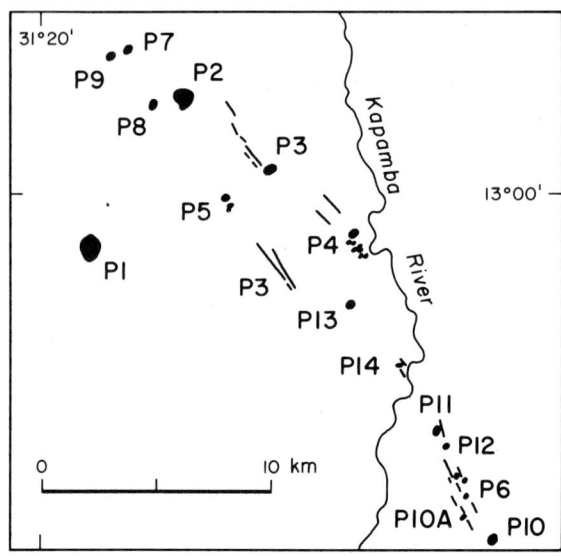

Figure 3.25. Location map showing the distribution of pipes and dikes comprising the Kapamba lamproite field (after Scott Smith *et al.* 1989).

not associated with these faults. The lamproites occur near the northwestern margin of the Luangwa Graben.

Regional Tectonic Setting: The Luangwa Graben is a tributary graben to the Western Branch of the East African Rift System (Chapman and Pollack 1977). The Luangwa Graben consists of a northeast-trending Mesozoic rift basin filled with Karoo and some younger sedimentary deposits. Unlike other segments of the East African Rift System, the Luangwa Graben is not characterized by abundant Cenozoic-to-Recent alkaline magmatism. Recent seismicity and graben sedimentation occur along the trough. The graben occurs in the Proterozoic (ca. 1350 Ma) Irumide and Mozambique Tectonic Belts (Cahen *et al.* 1984). The northern part of the basin represents a symmetrical trough bounded by two major fault scarps, whereas the southern portion is asymmetrical, with the southwest margin containing fewer normal faults. The northwest trend of the dikes parallels the structural grain of the nearby Ubendia–Ukinga (1.8–2.2 Ga) and the Dodoman (2.5–3.0 Ga) orogens of East Africa (Ackerman 1962). The Ubendia–Ukinga orogen represents an Early Proterozoic mobile belt developed on the Archean Tanganyika Cratonic nucleus.

Nature of Basement Rocks: The lamproites are underlain by Middle Proterozoic (1.8–2.1 Ga) metasedimentary crystalline rocks of the Kibaran Formation. These were deformed in the Proterozoic (1350 Ma) and during the widespread early Paleozoic (ca. 500 Ma) Pan-African event.

Intrusive/Extrusive Forms: The Kapamba lamproites occur as vents and dikes with crater-facies and massive hypabyssal rocks. Country rock-vent contacts are characteristically silicified.

Geomorphology: The lamproites are well exposed and most vents form topographic highs. One oval vent (P7) forms a depression.

Area/Volume: >10 km², composite near-surface > 5 km³.

Component Localities: vents P1–P14, dikes D1–D3.

Nearby Contemporaneous Igneous Rocks: Cretaceous syenites occur 100 km east of Kapamba in the central Luangwa graben. Post-Karoo kimberlite intrusions occur at Panela, North Luangwa, and Isoka, all within 150–300 km of Kapamba (Scott Smith *et al.* 1989 and references therein). Cretaceous carbonatite and syenite complexes occur in the Chilwa Province, Malawi (Snelling 1965). Tertiary (<45 Ma) rift-related magmatism in the East African Rift system (mostly north on Kapamba) was dominated by tholeiites and silica-undersaturated alkaline mafic rocks and carbonatites. The Kapamba lamproites appear to be related to Karoo rifting; however, they most likely postdate active extension and rift basin sedimentation by 10–100 m.y.

Rock Types: Pyroclastic rocks; lapilli tuffs; massive lavas of madupitic diopside olivine phlogopite lamproite; diopside–leucite lamproite; diopside lamproite; phlogopite leucite lamproite; glassy olivine lamproite.

Mineralogical Comments: The Kapamba lamproites contain a lamproite mineral assemblage, although the amphiboles and feldspars are relatively sodic. Priderite and wadeite have not been reported. Secondary nepheline has been observed to replace leucite (Scott Smith *et al.* 1989).

References: L.G. Murray, personal communication to Dawson (1970); Mossman (1976); Dawson (1980); Scott Smith *et al.* (1989).

Comments: These intrusions were originally designated kimberlites (Dawson 1980) as diamond occurs in subeconomic proportions in some of the vents (Scott Smith *et al.* 1989). Heavy mineral concentrates contain a variety of mantle and crustal phases, including Cr-pyrope, Ti-pyrope, pyrope-almandine, almandine, spinel, Mg-chromite, and Cr-diopside.

3.4.2. Group Name: Pneil, Postmasburg, Swartruggens

Location: Pneil and Postmasburg (27.9 S, 22.8 E), Swartruggens (25.7 S, 27.0 E). Pneil and Postmasburg occur in the Kimberley area, Cape Province, South Africa. Swartruggens is located 100 km west of Pretoria, in the western Transvaal, South Africa.

Synonyms: Kimberley lamproites; Hellam Mine (Swartruggens).

Age: Swartruggens, 142–156 Ma [Jurassic] (Allsopp and Kramers 1977, Allsopp and Barrett 1975, McIntyre and Dawson 1976, Smith *et al.* 1985); Kimberley area ca. 114–120 Ma [Early Cretaceous] (Dawson 1980, Smith *et al.* 1985).

Nature of Country Rocks: The Pneil and Postmasburg lamproites intrude a Lower Mesozoic sequence (Lower Griquatown/Pretoria Group, Ghaaplateau dolomite) overlying Precambrian crystalline rocks of the Kapvaal craton. The Swartruggens dikes intrude Lower Proterozoic Transvaal system quartzites and shales.

Local Structural Setting: The Swartruggens dikes generally trend east–west and therefore parallel to the regional strike of the host quartzites and shales of the Transvaal system. The Pneil and Postmasburg dikes strike east–west and do not follow the strike of the host rocks.

Regional Tectonic Setting: The Kimberley area lamproites occur near the western margin of the Archean Transvaal Craton, adjacent to the Late Proterozoic (1.0–1.3 Ga) Orange River, or Namaqua, mobile belt. The Swartruggens lamproites occur within 100 km of the northern margin of the Kapvaal Craton, adjacent to the Archean-to-Proterozoic Limpopo (3.2–2.0 Ga) mobile belt.

Nature of Basement Rocks: The area is underlain by Archean craton with crystallization ages up to >3.5 Ga, which has not experienced major orogeny since 1.5 Ga (Hunter and Pretorius 1981).

Intrusive/Extrusive Forms: The lamproites occur as dikes.

Geomorphology: The lamproites are dissected and poorly exposed.

Area/Volume: <3 km^2, <1 km^3.

Component Localities: Pneil, Postmasburg, and Male fissure (Swartruggens).

Nearby Contemporaneous Igneous Rocks: The Kimberley area lamproites occur in Barkly West group 2 kimberlite field of Mitchell (1986) and are part of the widespread South African Jurassic–Cretaceous kimberlite province. The Swartruggens Male fissure lamproites occur in association with group 2 kimberlite dikes (Main, Changehouse, and Normal fissures).

Rock Types: Hypabyssal massive olivine richterite phlogopite lamproite; olivine leucite phlogopite lamproite.

Mineralogical Comments: The Swartruggens rocks consist essentially of variable proportions of altered olivine; diopside; phlogopite; altered leucite (?); nepheline (?); sanidine; and secondary zeolite. Priderite and wadeite appear to be absent. The Pniel rocks contain abundant groundmass poikilitic titanian potassian richterite.

References: Wagner (1914); Williams (1932); Fourie (1958); Erlank (1973); Skinner and Scott (1979); Hargraves and Onstott (1980); Smith (1983); Smith *et al.* (1985); Skinner (1989); Kirkley (1987); Kirkley *et al.* (1989).

Comments: The Swartruggens lamproites are unusual in their association with apparently consanguineous group 2 kimberlites. It is possible that other lamproites, or rocks transitional between lamproites and group 2 kimberlites, occur in South Africa (see Section 10.4). Further studies are required to define their relationships to group 2 kimberlites.

3.4.3. Group Name: Bobi

Location: [8.167 N, 6.567 W]. A 2-km-long discontinuous lamproite dike is located 1 km west of Watson, 3 km northwest of Bobi, 25 km north-northeast of Seguela, Ivory Coast, West Africa (Figures 3.26, 3.27).

Synonym: Seguela.

Age: 1410–1455 ± 60 Ma [Proterozoic] (Lemarchand and Papon 1969); 1150 Ma (Bardet and Vachette 1966).

Nature of Country Rocks: The region is composed of Archean-to-Early Proterozoic granitic crystalline rocks, including Eburnian (1.95–2.10 Ga metamorphism) and Pre-Birrimian (>1.9 Ga) basement rocks.

Local Structural Setting: The Bobi dike strikes north–south in a region characterized by north-trending faults and aeromagnetic anomalies. The dike is discontinuous and has been segmented by east-trending faults.

Regional Tectonic Setting: The Bobi dike occurs in the Leo Shield (also called the Guinea Rise and "Dorsal de Man"), of the southern West African Craton. The lamproite

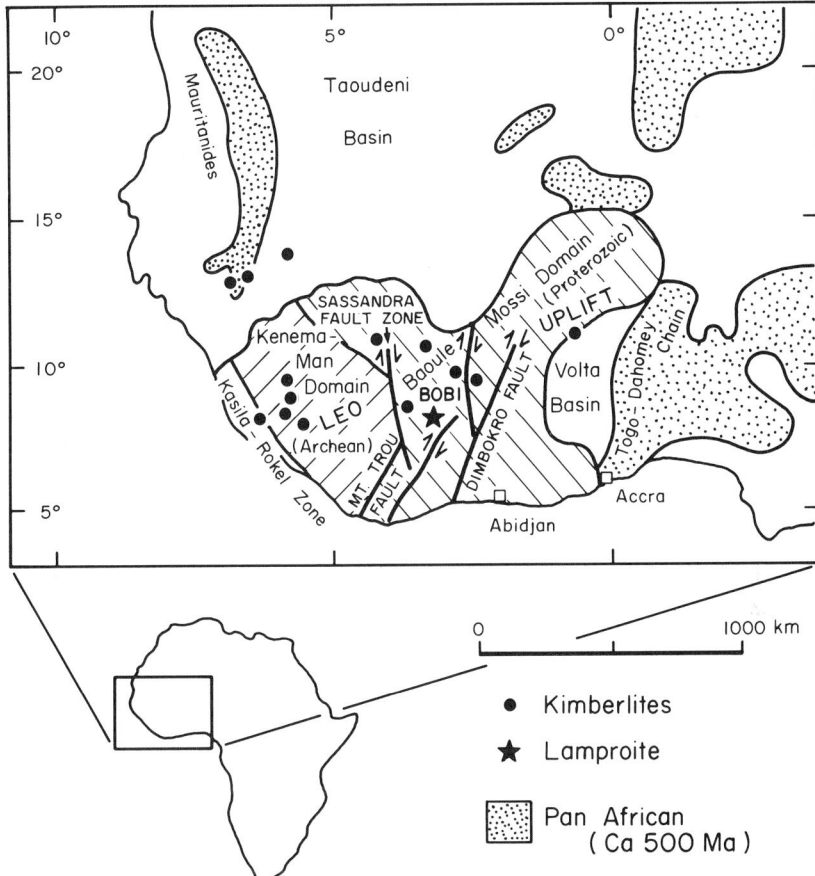

Figure 3.26. Regional map of southern West Africa showing the tectonic setting of the Bobi lamproite field and important kimberlite localities (after Wright 1985, and Cahen et al. 1984).

occurs in the western Baoule–Mossi domain, 100 km east of the Sassandra (mylonitic) fault zone which forms the eastern boundary of the Kenema–Man domain, or the Archean (Liberian) nucleus, of the West African Craton (Cahen et al. 1984). The Baoule–Mossi domain consists of Lower Proterozoic Birimian supracrustal rocks, Eburnian granitoids and older basement (from 2.0 to > 2.5 Ga) rocks. The main tectonic/orogenic events that modified the West African Craton include: the east-trending 2.96 Ga Leonean event; the north-trending 2.75 Ga Liberian event; the 1.95–2.1 Ga Eburnean event; and the 0.6–1.1 Ga Pan African fold belts, which border the east and west margins of the southern West African Craton (Cahen et al. 1984). The lamproites were intruded at a time of relative tectonic quiescence, approximately midway between the latter two deformational events.

Nature of Basement Rocks: In the vicinity of the Bobi lamproites, the Archean–Lower Proterozoic Leo Shield of the West African Craton was affected by the Lower Proterozoic (ca. 2.0 Ga) Eburnean orogeny and the region has been stable since 1.6–1.8 Ga. The basement therefore probably includes Archean rocks as old as 2.5–3.0 Ga.

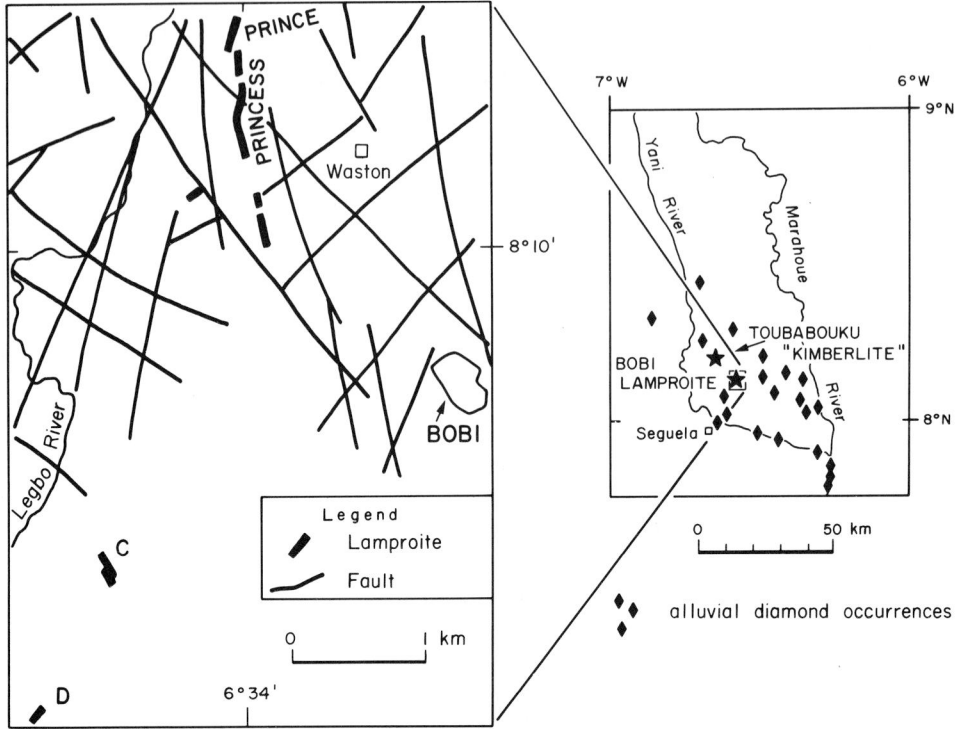

Figure 3.27. Local maps of the Seguela area showing the distribution of alluvial diamond deposits, the Toubabouku kimberlite and Bobi lamproite field (after Knopf 1970).

Intrusive/Extrusive Forms: Massive lamproite forms a narrow (<5–10 m wide), 2 km long, nearly vertical dike.

Geomorphology: The dike forms a mild topographic depression in an area of subdued relief.

Area/Volume: <0.1 km², <0.1 km³.

Component Localities: Prince dike, Intermediate dike, and Princess dike are all segments of the Bobi dike. Phlogopite lamproite float also occurs nearby in the Kohu River.

Nearby Contemporaneous Igneous Rocks: Proterozoic (ca. 1.4 Ga) kimberlite and alnöite dikes (at Toubabouko, in the Haut Nzi kimberlite field) occur within 20–40 km of Bobi (Knopf 1970). Much older and unrelated ultrabasic volcanic and plutonic rocks occur in the Eburnean basement. Proterozoic kimberlites from Mali (1.1–1.25 Ga) and Liberia (1.4 Ga) (Haggerty 1982, Tompkins and Haggerty 1984, Janse 1985) occur 400–800 km to the west of the Bobi lamproites. The much younger West African kimberlite fields in Guinea and Sierra Leone (Bardet 1974), and the alkaline igneous complexes at Songo, Freetown, and Babge (Culver and Williams 1979) are all apparently Cretaceous and presumably associated with the breakup of Gondwanaland.

Rock Types: Highly altered carbonated olivine madupitic lamproite breccia (Bobi). Fresh phlogopite lamproite (Kohu River).

DESCRIPTION OF LAMPROITE OCCURRENCES

Mineralogical Comments: The Bobi rock is intensely altered and contains abundant anatase.

References: Knopf (1970); Bardet and Vachette (1966); Bardet (1974); Dawson (1980); Mitchell (1985, 1986); Bergman (1987).

Comments: The Bobi lamproites are one of the few suites in which diamond is present in economic proportions. In addition alluvial diamonds have also been found downstream of the dikes (Bardet 1974). The Bobi lamproites were previously regarded as metakimberlites (Bardet 1974).

3.5. AUSTRALIAN LAMPROITES

Diamondiferous and barren lamproites occur on the southwest and southeast margins of the Kimberley Craton of northwest Australia in the Fitzroy Trough (group 1) and at Argyle (group 2) (see Figure 3.28). The Australian lamproites are post-tectonic and occur in Proterozoic mobile belts which mantle the Archean Kimberley Craton.

There are several other ultrapotassic mafic rock suites in Australia that approach lamproites in composition but are not included because of significant mineralogical differences. The most important suites are the Lake Cargelligo area leucitites of New South Wales (Wellman *et al.* 1970, Cundari 1973) and the Terowie, South Australia "fitzroyite" dikes (Colchester 1972, 1982, 1983). Lamproites may occur at Terowie (Jaques *et al.* 1985); however, the compositional data required to verify their identity is not available. Olivine nephelinite dikes have been recognized at Terowie (Carr and Olliver 1980, Bergman 1987) in the vicinity of the Orroroo and Calcutteroo kimberlites (Colchester 1972, 1982, 1983, Stracke *et al.* 1979, Scott Smith *et al.* 1984, Jaques *et al.* 1985).

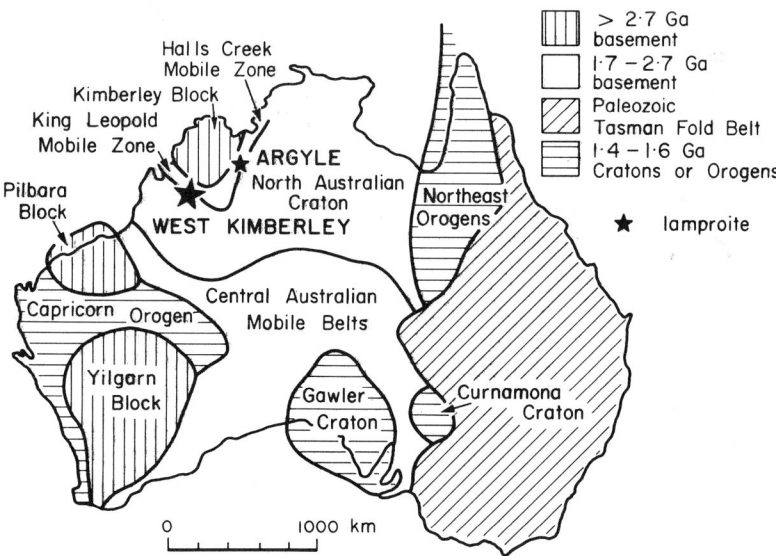

Figure 3.28. Map of Australian lamproite localities showing the main Precambrian tectonic elements (after Plumb 1979, Veevers 1984, Jaques *et al.* 1986).

3.5.1. Group Name: Argyle

Location: [16.233 S, 128.383 E]. The Argyle lamproite vent is located about 120 km south-southwest of Kununurra, West Australia, Australia (Figures 3.29, 3.30). Probably cogenetic lamproites occur 10 km to the southwest at Lissadell Road, where narrow dikes discontinuously extend for >1 km (Atkinson *et al.* 1984).

Synonyms: AK1 pipe, Argyle pipe, Argyle mine, East Kimberley field.

Age: Mid-Proterozoic: 1126 ± 9 Ma (Rb–Sr model age, Skinner *et al.* 1985); 1178 ± 47 Ma (K–Ar on phlogopite and Rb–Sr on phlogopite-apatite or whole rock pairs, Sun *et al.* 1986; Pidgeon *et al.* 1989).

Nature of Country Rocks: The Argyle lamproite intrudes and overlies Early Proterozoic igneous, metamorphic, and sedimentary rocks of the Lamboo complex (Tickalara Metamorphics and Bow River Granite), the Revolver Creek Formation and Upper Proterozoic sedimentary deposits (Carr Boyd Group–Hensman sandstone, Golden Gate siltstone and Lissadell Formation). The Argyle lamproite is postdated by Lower Paleozoic rocks (the Early Cambrian (?) Antrim Plateau Volcanics and the Devonian Ragged Range conglomerate), which outcrop as close as several hundred meters to the north and west of the

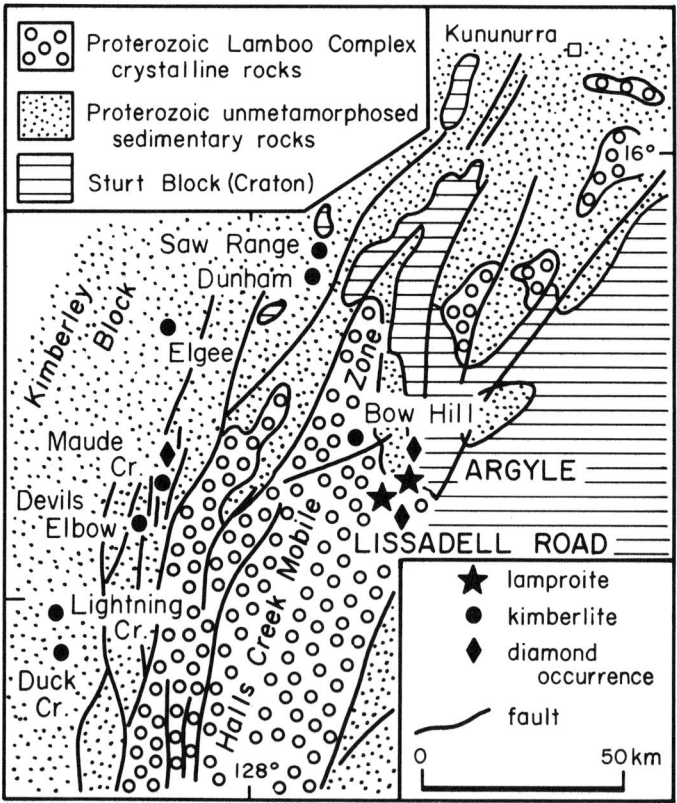

Figure 3.29. Regional geologic map of part of the East Kimberley area, showing the structural and geological framework of the Argyle lamproite field (after Jaques *et al.* 1986).

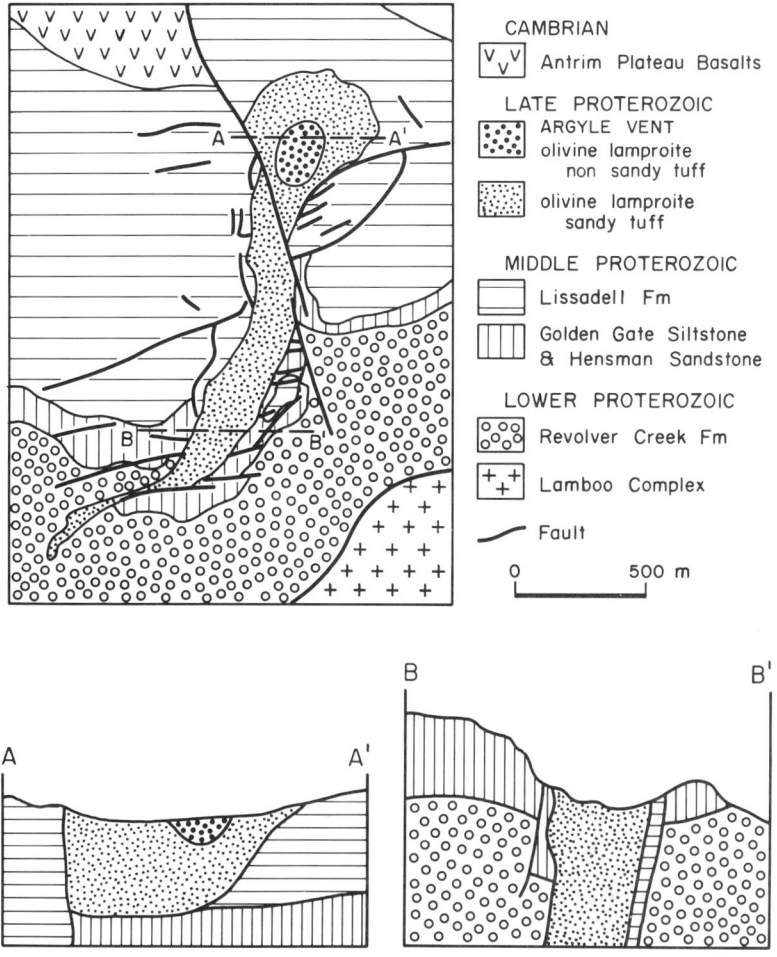

Figure 3.30. Geological map and cross sections of the Argyle lamproite pipe (after Jaques *et al.* 1986, and Boxer *et al.* 1989).

pipe. The Lissadell Road lamproites intrude the Proterozoic Bow River Granite (1.85–1.90 Ga) of the Lamboo Complex (Plumb *et al.* 1981).

Local Structural Setting: The Argyle lamproite is a 2-km-long, north-northeast-trending, elongated, tilted vent which was intruded along several well-established, intersecting faults. The vent has been cut by several north-northwest and east-trending, high-angle faults with minor offset. The vent is located 5 km west of the Halls Creek Fault, a major north-northeast-trending high-angle fault near the southern margin of the Halls Creek Mobile Zone. The Lissadell Road lamproites consist of more than a dozen discontinuous, en-echelon, northwest-striking dikes and stringers arranged in a northeast trend (Jaques *et al.* 1986). Their attitude is suggestive of a north-northeast-trending sinistral shear couple, which is also supported by regional structural analysis (Katz 1979, Plumb and Gemuts 1976).

Regional Tectonic Setting: The Argyle and Lissadell Road lamproites occur at the northeastern extent of the northeast-trending Early Proterozoic (ca. 1.9–2.0 Ga deformation, metamorphism, plutonism) Halls Creek Mobile Zone (Plumb and Gemuts 1976, Hancock and Rutland 1984), near the point where the Mobile Zone is overlain by Lower Paleozoic sedimentary rocks. The Halls Creek Mobile Zone consists of Lower Proterozoic sedimentary rocks which were: deposited in a trough built on extended Archean crust adjacent to the Pine Creek geosyncline; metamorphosed at about 1.85–1.86 Ga (Page and Hancock 1988); and subsequently intruded by late-tectonic anatectic felsic magmatic rocks. The Halls Creek Mobile Zone marks the eastern boundary of the Archean Kimberley Craton. Associated north-northeast-trending faults resulted from southeast-vergent oblique regional convergence (Hancock and Rutland 1984). Many of these faults have experienced over 100 km of sinistral transcurrent offset (Plumb and Gemuts 1976). The Argyle vent is situated midway between the two major Mobile Zone-bounding faults, the Halls Creek and Greenvale faults, in a zone of regional positive Bouguer anomaly (Wellman 1988). The Argyle lamproite was emplaced 700 m.y. after the main orogenic activity in the Halls Creek Mobile Zone, at about the same time that syn-depositional faulting was initiated during the deposition of the Carr Boyd Group.

Nature of Basement Rocks: The Argyle vent is underlain by Early Proterozoic (ca. 2.2 Ga) metasedimentary rocks, presumably deposited on Archean basement. Sun *et al.* (1986) reported Nd isotope data (ϵ_{Nd} = 0 to −2) for the Hart dolerite and Lamboo Complex magmatic rocks (1.9–2.0 Ga) of the Halls Creek Mobile Zone which supports the presence of Archean basement.

Intrusive/Extrusive Forms: The Argyle vent is composed primarily of a wide variety of pyroclastic rocks (lapilli tuffs, hyaloclastites etc.). Hypabyssal olivine lamproite dikes are a minor late stage phase of the activity.

Geomorphology: The vent occurs in a dissected steep terrain and was originally concealed beneath a thin layer of soil.

Area/Volume: 0.5 km^2, 0.3 km^3.

Component Localities: Argyle vent, Lissadell Road dikes.

Nearby Contemporaneous Igneous Rocks: A Late Proterozoic (803–826 Ma; Pidgeon *et al.* 1989) north-northeast-striking swarm of micaceous ultramafic-to-mafic lamprophyre dikes occur at Bow Hill, about 20–25 km west to northwest of Argyle (Sun *et al.* 1986, Fielding and Jaques 1989). Five occurrences of Late Proterozoic (802–810 ± 10 Ma) kimberlite pipes and dikes are located 250–300 km north-northwest of Argyle in the Kimberley Basin at Pteropus Creek and Skerring (Atkinson *et al.* 1984a, Jaques *et al.* 1986, Pidgeon *et al.* 1989). Eight northeast-trending kimberlite dikes occur 50–100 km west to northwest of Argyle (Atkinson *et al.* 1984a, Jaques *et al.* 1984a, 1985). The 905–1012-Ma (Rb–Sr and U–Pb zircon ages, respectively) Cummins Range carbonatite–pyroxenite complex occurs 350 km south-southwest of Argyle, at the intersection of the Halls Creek and King Leopold Mobile Zones (Jaques *et al.* 1985, Sun *et al.* 1986). The 732 ± 5 Ma Mud Tank Carbonatite occurs in the Arunta Block (Black and Gulson 1978), and a suite of alnöite-melilitite-nephelinite dikes (849 ± 2 Ma) occur in the southwest Yilgarn Block (Robey *et al.* 1986). Ultrapotassic felsic volcanic rocks form the 1064 ± 23 Ma Tollu Volcanics of the Musgrove Block (Giles 1981). Dolerite dikes of the northern Kimberley Basin are thought to be 900–1100 Ma (Wyborn *et al.* 1987). Peralkaline dikes of the Mudginberri complex (1316 ± 40 Ma) occur in the Pine Creek inlier. Ultrapotassic *ne*-normative rocks (1210 ±

DESCRIPTION OF LAMPROITE OCCURRENCES

90 Ma) occur in the Mordor complex, Arunta block, central Australia (Langworthy and Black 1978).

Rock Types: Massive, brecciated, glassy and tuffaceous olivine-phlogopite lamproite; lapilli tuffs; tuffs; tuff breccias; hyaloclastites; hyalotuffs; and accretionary lapilli tuffs.

Mineralogical Comments: The Argyle lamproite tuffs characteristically contain abundant quartz xenocrysts and autolithic clasts with altered leucite.

References: Atkinson *et al.* (1984a, 1984b, 1984c); Jaques *et al.* (1984, 1985, 1986, 1989a, 1989b, 1989c); Ajarzo (1985); Hall and Smith (1985); Harris and Collins (1985); O'Neill *et al.* (1986); Sun *et al.* (1986); Wyborn *et al.* (1987); Nixon *et al.* (1987); Griffin *et al.* (1988); Boxer *et al.* (1989); and Pidgeon *et al.* (1989).

Comments: The Argyle pipe is the largest and highest grade diamond pipe in the world (about 50 hectares, 100 million tonnes, 600 carats/100 tonne). Because of the high proportion (90%) of industrial diamonds in the deposit the value per tonne is low.

3.5.2. Group Name: West Kimberley

Location: [18.0 S, 124.75 E]. The West Kimberley lamproite province contains over 100 magmatic bodies. These occur in a broad 150 × 200-km belt, centered 50 km northwest of Fitzroy Crossing, 150 km southeast of Derby, West Kimberley, West Australia (Figure 3.31). The lamproites form four main fields: Ellendale, Calwynyardah, Noonkanbah, and Eastern Lennard Shelf. Seven isolated intrusions occur in the Fitzroy Valley.

Synonyms: Fitzroy Basin lamproites, Fitzroy volcanics, Fitzroy lamproites.

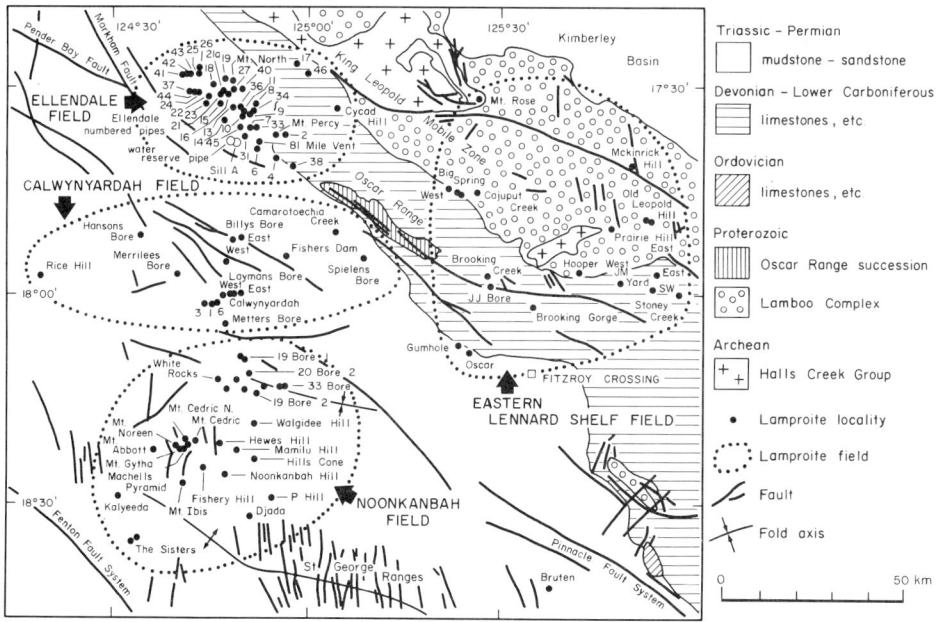

Figure 3.31. Regional geological map of Western Australia showing the four lamproite fields comprising the West Kimberley lamproite province (after Jaques *et al.* 1986, Smith 1984).

Age: [Early Miocene]. The K–Ar and Rb–Sr ages of 19 intrusions ranges from 17–24 Ma (Wellman 1973, Nixon *et al.* 1982, Jaques *et al.* 1984a, Allsopp *et al.* 1985). These ages supercede earlier work by Prider (1960) and Kaplan *et al.* (1967) which indicated Jurassic ages. The Ellendale field appears to be slightly older (20–22 Ma) than the Noonkanbah field (18–20 Ma).

Nature of Country Rocks: The Eastern Lennard Shelf lamproites intrude Archean and Proterozoic crystalline and metasedimentary rocks of the King Leopold Mobile Zone (Halls Creek Group, Lamboo Complex and Oscar Range Succession). These and lamproites of the other three fields intrude: Devonian limestones and clastic sedimentary rocks; Permian sandstones, mudstones and limestones; and Triassic sandstones.

Local Structural Control: Many of the intrusions are structurally controlled by prominent northwest-trending normal faults along the northern margin of the Fitzroy Trough, and north- and east-trending faults in the interior of the Trough. The lamproites of the Noonkanbah Field are located on the southern margin of the west-northwest-trending Quanbun syncline and the northern margin of the St. Georges Range anticline. The Ellendale and Calwynyardah Field intrusions are aligned, or elongated along, east-trending basement structures.

Regional Tectonic Setting: The lamproites occur at the northeastern margin of the Canning Basin and at the southern margin of the Proterozoic King Leopold Mobile Zone. The Canning Basin is a composite basin formed by at least four tectonostratigraphic sequences, deposited in an Early Paleozoic aulacogen and during subsequent phases of rifting. The Fitzroy Graben consists of two subbasins, the northwestern Fitzroy Trough and the southeastern Gregory subbasin. The Canning Basin forms the northwestern limit of the west-to-northwest-trending Amadeus Transverse Zone (Rutland 1973), which bisects the Australian continent. Veevers (1984) suggests that the structural evolution is similar to that of the Southern Oklahoma aulacogen (see Prairie Creek, Section 3.2.1). The Eastern Lennard Shelf intrusions occur in Proterozoic granitic basement and Paleozoic carbonate rocks at the northern margin of the Lennard Shelf. The Ellendale intrusions occur over the Markham Ridge, a regional basement high. Other Fitzroy lamproites occur in the middle of the Fitzroy Trough, which contains 8–10 km of Phanerozoic sedimentary rocks. There are many regionally extensive northwest-trending structural elements (Jaques *et al.* 1986) including: the Sandy Creek Shear (a Proterozoic sinistral strike-slip fault zone at the easternmost limit of the province), the Markham Ridge; the Markham and Pinnacle–Harvey Faults (southwest-dipping normal faults); several fold systems (Quanbun syncline and St. Georges anticline); and many regional gravity and magnetic trends (Wellman 1988).

Nature of Basement Rocks: The underlying basement rocks consist of the Proterozoic King Leopold Mobile Zone, including the Halls Creek Group and Lamboo complex, which have metamorphic ages of 1.8–2.2 Ga (Plumb and Gemuts 1976, Page 1988). These metasedimentary and metaigneous rocks represent Proterozoic accretion to the Archean Kimberley Craton to the north. Drummond (1988) reviewed geophysical data relevant to the nature of the northwest Australian lithosphere, and presented evidence for a thick cold tectosphere underlying much of the region, with 45–50-km-thick crust forming the Proterozoic mobile belts.

Intrusive/Extrusive Forms: Both intrusive and extrusive lithologies have been recognized in the West Kimberley lamproites. Intrusive rocks predominate because the overlying volcanic portions have largely been removed by denudation. Intrusive forms include necks, pipes, dikes, sills, and small plutons. Extrusive-to-subvolcanic forms are principally vents containing a wide variety of pyroclastic deposits (pyroclastic flow and surge deposits, lapilli tuffs, tuff breccias, and agglutinate tuffs). Epiclastic deposits formed in crater lakes are rarely found.

Geomorphology: The regional topography is subdued and many of the lamproite intrusions are partially-to-fully covered by Quaternary alluvium. Some intrusive bodies have erosion-resistant silicified margins and contact zones and therefore have positive relief. Many of these form isolated hills in a region of low relief. Some of the vents, especially those consisting of olivine lamproite, are erosional lows and completely covered by alluvium. An exception is Mt. Abbot, which contains silicified rocks.

Area/Volume: The West Kimberley lamproites collectively comprise an area of 25–35 km^2 and a volume of <30 km^3.

Component Localities: Ellendale Field—Mount North, Mount Percy, 81 Mile Vent, Water Reserve Pipe, Sill A, and 40 intrusions named Ellendale 2–46 (noninclusive) by CRA (referred to in Jaques *et al.* 1986). *Eastern Lennard Shelf Field*—Mount Rose, McKinrick Hill, Big Spring, Big Spring West, Cajaput Creek, Old Leopold Hill, Old Leopold West, Prairie Hill East, Hooper West, Brooking Creek, J J Bore, J K Yard (3 intrusions), Stoney Creek, Brooking Gorge, Gumhole, Oscar. *Calwynyardah Field*—Camarotoechia Creek, Spielers Bore, Fishers Dam, Billys Bore East, Billys Bore West, Hansons Bore, Merrilees Bore, Rice Hill, Calwynyardah, Laymans Bore (two intrusions), Laymans Bore East, Metters Bore (Nos. 1, 3, 6), Seltrust Pipes 1, 2, and 3. *Noonkanbah Field*—19 Bore (No. 1, 2), 20 Bore, 33 Bore (two intrusions), White Rocks, Walgidee Hill, Mount Cedric North, Mount Cedric, Mount Cedric South, Howes Hill, Mount Noreen, Machels Pyramid, Mount Gytha, Mount Abbott, Mamilu Hill, Hills Cone, Fishery Hill, Noonkanbah Hill, Mount Ibis, Nerrima (four sills), Kalyeeda, P Hill, Djada Hill, The Sisters (two intrusions); and an isolated intrusion, Bruten, to the southeast of these four fields.

Nearby Contemporaneous Igneous Rocks: The West Kimberley lamproites are Cenozoic alkaline magmatic rocks isolated in both time and space. The nearest late Tertiary igneous rocks occur over 1000 km to the north in Timor and New Guinea. Collision of the Australian plate with the New Guinea arc to the north occurred in the Miocene. Apart from the widespread Cambrian plateau basalts, 500–800 km to the east, the only significant volume of Phanerozoic volcanic rocks in northwest Australia consists of the scattered basalts produced by Early Mesozoic seafloor spreading when India separated from Australia (Prider 1969, Veevers 1984).

Rock Types: Massive-to-fragmental and vesicular, leucite phlogopite lamproite; leucite richterite lamproite; leucite diopside lamproite; diopside leucite richterite madupitic lamproite; massive olivine lamproite.

Mineralogical Comments: The West Kimberley lamproites are the type locality for priderite, wadeite, and jeppeite. Sanidine is relatively rare in this province.

References: Hardman (1884); Fitzgerald (1907); Wade (1924, 1937); Farquharson (1920, 1922); Skeats and Richards (1926); Blatchford (1927); Prider (1939, 1960, 1965, 1969, 1982); Wade and Prider (1940); Prider and Cole (1942); Casey (1958); Thomas (1958); Veevers (1958); Kaplan *et al.* (1967); Bell and Powell (1970); Derrick and Gellatly (1972); Derrick and Playford (1973); Wellman (1973); Sahama (1974); Bardet (1977); Mason (1977); Crowe and Towner (1981); Mitchell (1981, 1985); Jaques *et al.* (1982, 1983, 1984a,b, 1985, 1986, 1989a,b,c); Atkinson *et al.* (1982, 1983, 1984a,b); McCulloch *et al.* (1983); Nixon *et al.* (1982, 1984, 1987); Smith (1984); Hall and Smith (1985); Nicoll (1981); Jaques and Foley (1985); Foley *et al.* (1987); Smith and Lorenz (1989).

Comments: Subeconomic diamond deposits were discovered in the Ellendale field during the 1970s as a result of a concerted exploration effort by the Ashton Joint Venture (Jaques *et al*. 1986). This late recognition of diamond was principally because traditional, kimberlite-based, indicator mineral techniques are not successful in locating lamproite intrusions (see Chapter 9).

3.6. ANTARCTIC LAMPROITES

Lamproites have been recognized at three isolated localities in Eastern Antarctica: Gaussberg (Quaternary); Priestly Peak; Mount Bayliss (Paleozoic). Each of these suites is located on or within 800 km of the coast (Figure 3.32). Many other lamproite suites are undoubtedly present in the region, but are covered by the Antarctic ice cap.

3.6.1. Group Name: Gaussberg

Location: [66.8 S, 89.317 E]. Gaussberg is located about 250 km west of Mirny, and east of Davis, on the coast of Kaiser Wilhelm II Land, Australian Territory, Antarctica (Figure 3.32).

Age: 0.056 ± 0.005 Ma. K–Ar on leucite (Tingey *et al*. 1983). This age supercedes earlier suggested ages of 9–20 Ma (Ravich and Krylov 1964, Soloviev 1972). Note that

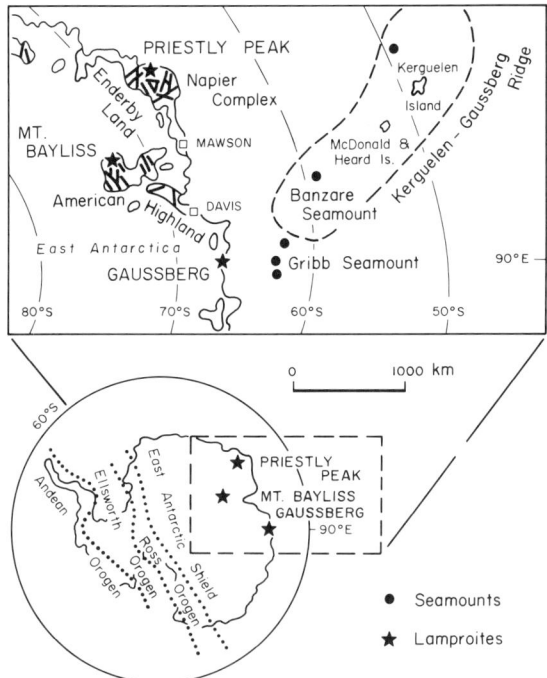

Figure 3.32. Maps of Antarctica and East Antarctica showing the Gaussberg, Priestly Peak and Mt. Bayliss lamproites relative to the main orogenic domains and regional tectonic elements (after Dort 1972, Sheraton *et al*. 1987a,b)

Gleadow (in Tingey *et al.* 1983) determined an apatite fission track age of 0.025 ± 0.012 Ma, and Dort (1972) suggested a Recent-to-Late Pleistocene age based on geomorphology.

Nature of Country Rocks: On the basis of the rocks observed in nearby moraine (Nockolds 1940), the Gaussberg volcano is considered to be built on high-grade Archean and Proterozoic gneisses similar to those of the Vestfold Block. Collerson and McCulloch (1983) reported Proterozoic model ages (Sr: 0.7–1.09 Ga, Nd: 1.97–2.15 Ga) on basement-derived inclusions of quartzofeldspathic gneiss, and nonfoliated rapakivi granite and metabasite in the lavas.

Local Structural Setting: Nothing is known about the structural geology in the immediate vicinity because of the permanent ice cover.

Regional Tectonic Setting: Gaussberg is situated on a passive margin, at the Antarctic continental extension of the Kerguelan Plateau. The regional structural fabric of nearby Archean crystalline rocks trends from northeast to east.

Nature of Basement Rocks: The region consists of high-grade Archean metaigneous rocks, which suffered Proterozoic metamorphism similar to that seen in the Vestfold Block and the more distant Napier and Rayner Complexes of the East Antarctic Shield.

Intrusive/Extrusive Forms: Pillow lavas; flow breccias; cinder deposits; tuffs; lapilli tuffs.

Geomorphology: The Gaussberg cone is a relatively undissected, extremely fresh volcano.

Area/Volume: About 10 km^2, 1 km^3.

Component Localities: Gaussberg volcano. Subice dikes possibly occur in the vicinity.

Nearby Contemporaneous Igneous Rocks: On the other side of the East Antarctic Shield (over 2500 km away), Late Cenozoic volcanic rocks (alkali basalts) are widespread in the Andean, Ross, and Borchgrevink Orogens in west Antarctica (Dort 1972). Cenozoic (mostly 5–27 Ma) volcanic rocks occur within 500–1000 km along the Kerguelen–Gaussberg Rise at the Gribb and Banzare Seamounts and on Kerguelen Island.

Rock Types: Glassy, vesicular, nearly-aphyric leucite diopside olivine lamproite.

Mineralogical Comments: Despite the glass-rich nature of the Gaussberg lavas, a typical lamproite assemblage, consisting of olivine, diopside, leucite, and phlogopite, is observed.

References: Drygalski (1912); Philippi (1912); Reinisch (1912); Lacroix (1926); Nockolds (1940); Vyalov and Sobolev (1959); Ravich and Krylov (1964); Dort (1972); Soloviev (1972); Sheraton and Cundari (1980); Tingey (1981); Collerson and McCulloch (1983); Tingey *et al.* (1983); Collerson *et al.* (1988).

Comments: Gaussberg was discovered in March, 1902 by the 1901–1903 German Antarctic Expedition and was named while the ship, the Gauss, was immobilized for nearly a year because of sea ice entrapment (Sheraton and Cundari 1980). The volcano is the youngest lamproite known and deserves further examination.

3.6.2. Group Names: Mount Bayliss and Priestley Peak

Location: Mount Bayliss [73.433 S, 62.833 E]; Priestley Peak [67.183 S, 50.366 E]. Mount Bayliss is located about 650 km south of Mawson, in the southern Prince Charles Mountains, MacRobertson Land. Priestly Peak is located in the northern Scott Mountains, about 500 km west of Mawson, near the coast of Enderby Land, East Antarctica. Both localities are in the western Australian Antarctic Territory (Figure 3.32).

Age: Mount Bayliss (413–430 ± 12 Ma, Silurian), range of 3 K–Ar ages on K-richterite and K-arfvedsonite concentrates (Sheraton and England 1980); Priestley Peak (482 ± 3 Ma, Ordovician), whole rock and mineral Rb–Sr isochron (Black and James 1983).

Nature of Country Rocks: The Mount Bayliss dike intrudes Proterozoic-to-Archean crystalline rocks of the Prince Charles Mountains. The Priestley Peak dike intrudes a fresh Amundsen dolerite dike (1.19 Ga) which is emplaced in Proterozoic-to-Archean (1.5–3.9 Ga) crystalline rocks of the Napier Complex and adjacent domains.

Local Structural Setting: The Priestley Peak dike consists of a single, undeformed, discordant, lensoid, <1–5 m wide subvertical dike which crops out for >1 km. The dike represents the youngest known igneous event in the region. The Mount Bayliss dike is 5 m wide, undeformed and subvertical.

Regional Tectonic Setting: Both dikes occur in regions of Paleozoic-to-Recent structural stability, near the passive margin of the East Antarctic Shield. The lamproite magmatism represents the youngest magmatic episode in the Napier complex, and the Southern Prince Charles Mountains. The tectonic history of the regions includes major continental crust-forming events at 3.9 and 2.0 Ga; granulite to amphibolite facies metamorphic episodes at about 3.0 and 1.0 Ga (with intervening intense deformation); intrusion of mafic dike swarms during several Late Archean-to-Late Proterozoic episodes. The formation of lamproites in the Napier Complex and southern Prince Charles Mountains was coeval with the intrusion of A-type granitic rocks in the Prydz Bay area. These diverse magmatic types are thought to have been associated with the initial fracturing of Gondwanaland prior to the eventual rifting which separated East Antarctica from India (Sheraton *et al.* 1987b).

Nature of Basement Rocks: Both regions of lamproite activity are underlain by Archean-to-Early Proterozoic (3.8–3.9 Ga) basement rocks.

Intrusive/Extrusive Forms: The rocks occur as massive, medium-to-finegrained 1–5-m-wide dikes.

Geomorphology: Both regions represent intensely dissected, lower crustal sections, which have been uplifted, eroded, and planed-off by recent glacial activity.

Area/Volume: Collectively, <0.1 km^2, <0.01 km^3.

Component Localities: Mount Bayliss and Priestly Peak.

Nearby Contemporaneous Igneous Rocks: Early Paleozoic lamproite magmatism at Priestly Peak and Mount Bayliss was coeval with alkali basalt magmatism (420–500 Ma) in the Northern Prince Charles and MacRobertson Land Coast (Sheraton *et al.* 1987b). Chemically similar, but more fractionated dikes occur at Hydrographer Island (Sandiford and Wilson 1983). These magmatic episodes of small volume postdate extensive Pan-African age (500–700 Ma) A-type granitic batholiths in the region. Paleozoic (258–455 Ma, Rex 1972) dolerite dikes occur in northeast Heimefrontfjella and Mannefallknausane, western Drønning Maud Land (Juckes 1972, Sheraton *et al.* 1987a). LIL element-enriched dolerite dikes of similar age (297–457 Ma, Rex 1972) occur in the Shackleton Range, Coats Land (Clarkson 1981, Sheraton *et al.* 1987a). The 110-Ma Radok Lake alnöite sills occur about 300 km to the north-northeast of Mount Bayliss (Trail 1963, Walker and Mond 1971, Sheraton 1983). Also located in the Northern Prince Charles and MacRobertson Land Coast areas are 50-Ma tristanite lava flows (Sheraton *et al.* 1987b).

DESCRIPTION OF LAMPROITE OCCURRENCES

Rock Types: Medium-to-fine grained phlogopite K-feldspar richterite lamproites.

Mineralogical Comments: The amphibole in these rocks exhibits a very wide range in composition from K–Ti-richterite to K–Ti-arfvedsonite. Anatase is common.

References: Trail (1963); Sheraton and England (1980); Sheraton *et al.* (1980, 1987a, 1987b); Tingey *et al.* (1981); Sheraton (1983); Nelson and McCulloch (1989).

3.7. ASIAN LAMPROITES

Lamproites have been recognized in four fields in Asia: Chelima and Majhgawan, India (both Proterozoic); Barakar, India (Cretaceous); and possibly within the Murun complex, USSR. The two Proterozoic Indian fields occur in Proterozoic mobile belts which mantle Archean cratonic nuclei (Figure 3.33). The Cretaceous lamproites of the Barakar region were intruded about 200 m.y. after the formation of the host Gondwana Damador Rift basin. The Cretaceous Murun complex formed in a paleorift setting, the Baikal Rift, which was initiated in Late Archean times. The Murun complex contains apparently consanguineous alkaline igneous rocks of nonlamproite composition.

Lacroix (1926, 1933) included the cocites from Coc Pia, North Vietnam in Niggli's lamproite clan. However, Wagner and Velde (1986c) have presented modern mineral and rock chemical data that demonstrate the available cocite specimens are not true lamproites.

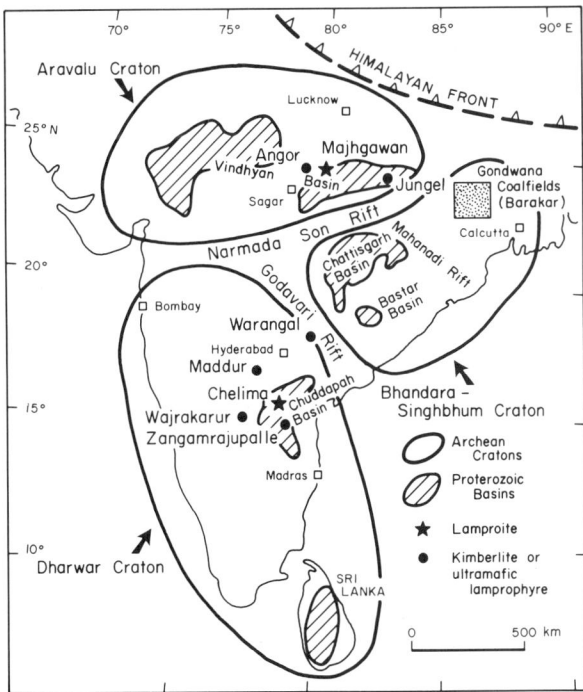

Figure 3.33. Map of India showing the main tectonic elements and kimberlite and lamproite fields (after Naqvi and Rogers 1987).

Phlogopite and diopside compositions are excessively aluminous for lamproites. It is nevertheless possible that lamproites occur in the Coc Pia and Sin Cao regions, North Vietnam (Bergman 1987).

Reddy (1987) and Nayak *et al.* (1988) have assigned incorrectly kimberlitic rocks from the Maddur area (near Wajrakrarur, India) to the olivine lamproite clan. Leucite-bearing minette dikes from the Gondwana Coalfields, eastern India, may also be lamproites. Modern mineralogical studies of these rocks have not been undertaken. The only *bona-fide* lamproites yet recognized are those of the Barakar area (Middlemost *et al.* 1988).

Potassic-to-ultrapotassic rocks occur in the Baikal-Aldan Belt in the Soviet Union and in the Pamir Mountains, Monogolia; however, none of these are lamproites (Bergman 1987, Mitchell 1988). Several alkaline rock complexes (e.g., Talakhtakh diatreme) in the USSR have been described as lamproites by Vishnevskii *et al.* (1986) and Bogatikov *et al.* (1986); however, these are not lamproites as defined in this work. They are probably potassium-rich trachytes and shonkinite/missouritelike rocks, respectively (Mitchell 1988). Volynets *et al.* (1987) described Miocene plagioclase-bearing potassium-rich alkali basalts from Western Kamchatka as lamproites. The presence of plagioclase indicates that the rocks are not lamproites. Satian and Khazatyana (1987) have described rocks from Armenia as lamproites, using the compositional criteria of Bogatikov *et al.* (1985). The rocks are poor in K_2O (1–7 wt %) and highly aluminous (12–7 wt % Al_2O_3) and unlikely to be *bona fide* lamproites.

The potassium-rich dike rocks (kajanites) from the Kajan River, Central Kalimantan, Indonesia (Brouwer 1909, Lacroix 1926), consist of olivine, diopside, phenocrystal and groundmass phlogopite, leucite, and richterite. They are not currently regarded as lamproites, as they contain nepheline, and the mineral assemblage present is more sodic and aluminous than that typical of lamproites (Wagner 1986).

3.7.1. Group Name: Chelima

Location: [15.45 S, 78.717 E]. The Chelima dikes ($n > 30$) occur over a 7×10-km area within about 6 km of the Chelima Railway Station, about 85 km southeast of Kurnool, Andrha Pradesh, India (Figures 3.34, 3.35, 3.36).

Age: 1319–1371 ± 20 Ma, K–Ar on phlogopite (Murty *et al.* 1987, Bhattacharji 1987); 1140–1225 Ma, Rb/Sr (Crawford 1969, Crawford and Compston 1973).

Nature of Country Rocks: The dikes intrude weakly-metamorphosed, folded Proterozoic (1350–1490 Ma) shales, slates, and phyllites of the Cumbum Subgroup (Giddalur and Pullampet Formations), Nallamalai Supergroup. The younger Kurnool Group deposits overlie the Nallamalai strata just to the west of Chelima. The dikes were emplaced prior to the deposition of the Banganapalle Formation (Meijerink *et al.* 1984).

Local Structural Setting: Over thirty individual <1–5-m-wide dike segments strike east-southeast to southeast, dip steeply to the north, and form a 10-km-long en-echelon swarm of probably 5–10 discrete dike bodies. These trends are orthogonal to the structural fabric in the area, as defined by fold axes that trend dominantly north. However, the dikes approximately parallel a major lineament pattern and structural fabric present throughout the Cuddapah Basin and surrounding Archean basement (Venkatakrishnan and Dotiwalla 1987). The dikes postdate folding, which presumably accompanied the 1360-Ma metamorphism in the region.

Regional Tectonic Setting: The Chelima lamproites occur in the center of the Cuddapah

DESCRIPTION OF LAMPROITE OCCURRENCES

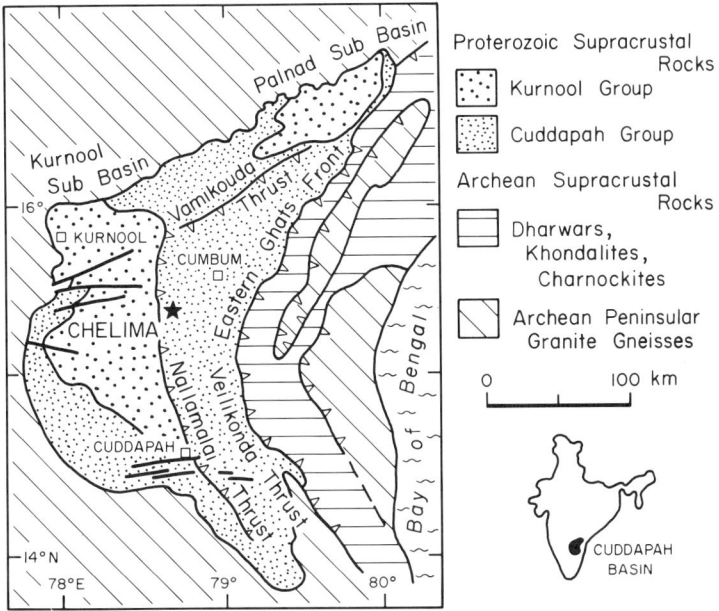

Figure 3.34. Map of the Cuddapah Basin showing the tectonic position of the Chelima lamproites and nearby kimberlites (after Naqvi and Rogers 1987, Bergman 1987).

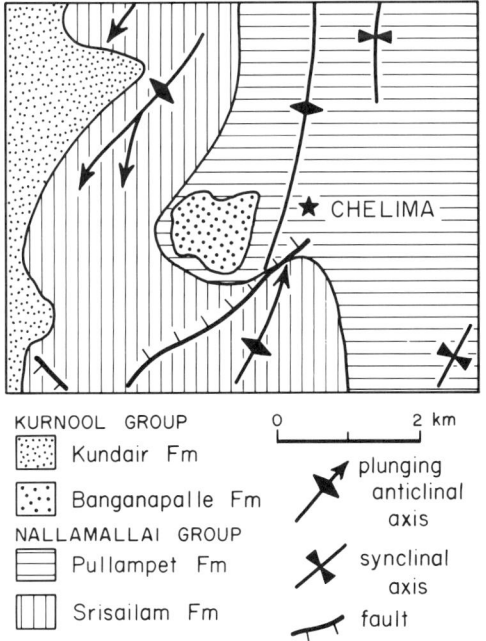

Figure 3.35. Geological map of the Chelima area (after Venkatakrishnan and Dotiwalla 1987).

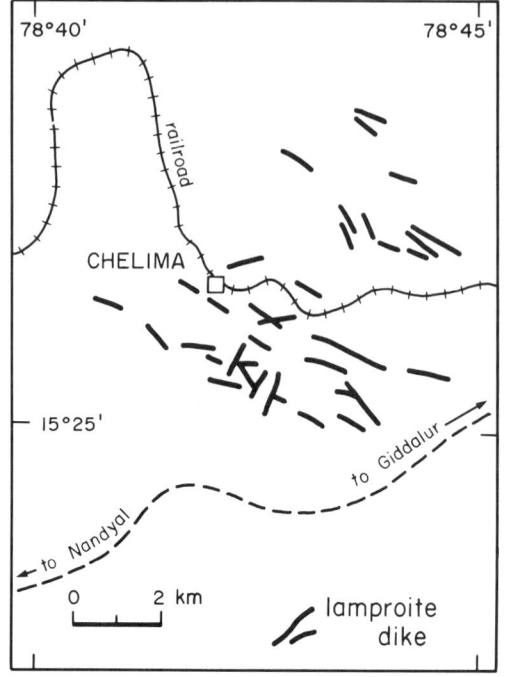

Figure 3.36. Locality map of the Chelima area showing the distribution of lamproite dikes (after Bergman 1987).

Basin, in the Archean Dharwar Craton. The Cuddapah Basin overlies the Early Proterozoic Mobile Belt surrounding the Karnataka Nucleus (Radnakrishna and Naqvi 1986). The basin formed in Middle Proterozoic times as the result of lithospheric flexure around the craton margins (Radnakrishna and Naqvi 1986) and is truncated to the east by the East Ghats front, an east-dipping major thrust fault. The Chelima lamproites occur between the Racherla and Rudravaram Lines, which bound a zone of low-amplitude harmonic folding (Meijerink *et al.* 1984). Seismic data for the region demonstrate the presence of deep crustal faults in the area, some even off-setting the Moho (Kaila *et al.* 1979, 1987). Leelanandam (1981) suggested that the Chelima–Zangamrajupalle–Mangampet lineament (Kaila *et al.* 1979) was a deep crustal fracture related to an upwarp which localized the distribution of alkaline magmas. The lamproites were emplaced about 100–400 m.y. after the basin formed, prior to the emplacement of widespread dolerite dikes in the region (870–1090 Ma, Crawford and Compston 1973).

Nature of Basement Rocks: Archean crystalline rocks (Mafic Schist Belt and meta-granites of the Peninsular Gneiss group) of the Dharwar Craton with crystallization ages of 3.0–3.3 Ga and metamorphic ages in the range 2.4–3.0 Ga (Naqvi and Rogers 1987) form the basement of the Chelima dikes. The craton has been stable since ~2.0 Ga.

Intrusive/Extrusive Forms: Massive, metamorphosed, medium-grained, carbonated dikes.

Geomorphology: The dikes are poorly-exposed in a hilly forested area of moderate relief.

Area/Volume: 0.03 km², <0.02 km³.

Component Localities: Individual dike segments are not named.

Nearby Contemporaneous Igneous Rocks: The Chelima dikes occur on the western margin of a Proterozoic alkaline rock and carbonatite province in the Eastern Ghats orogenic belt, centered to the east of the Cuddapah Basin (Leelanandum 1981, Nag 1983). A micaceous kimberlite dike rock (phlogopite-calcite) was reported to intrude metamorphosed and folded Cumbum carbonate and clastic rocks 3 km west-southwest of Zangamrajupalle (14.767 S, 78.883 E), only 80 km southeast of apparently identical dikes at Chelima (Bhaskara Rao 1976). Eight kimberlite pipes (840–1023 Ma) from the Wajrakarur, Lattavaram, Muligiripalle, and Venktampalli areas occur about 150 km west-southwest of Chelima (Scott Smith 1989). Note that Reddy (1987) suggested that two of the Wajrakarur pipes were lamproites. However, Scott Smith (1989) argues that all eight pipes are typical monticellite-bearing kimberlites, a conclusion with which we agree.

Rock Types: Altered, metamorphosed, and carbonated olivine phlogopite lamproite.

Mineralogical Comments: The rocks are apatite-, chlorite- and phlogopite-rich as a consequence of metamorphism and alteration. Olivine and clinopyroxene pseudomorphs are observed.

References: Vemban (1946); Appavadhanulu (1966); Sen and Narasimha Rao (1967, 1970, 1971); Crawford (1969); Crawford and Compston (1973); Sarma (1983); Meijerling *et al.* (1984); Bergman and Baker (1984); Bhattacharji (1987); Murthy *et al.* (1987); Naqvi and Rogers (1987); Scott Smith (1989).

Comments: The Chelima dikes have generally been regarded as minettes or rocks intermediate between carbonatites and kimberlites. However, Bergman and Baker (1984) and Bergman (1987) demonstrated that they contain tetraferriphlogopites, which are similar in composition to those found in other well-documented lamproites. Although alluvial diamonds have been found nearby, the Chelima dikes have not been tested for diamonds using modern techniques. Bergman and Baker (1984) nevertheless suggested that they are diamondiferous lamproites. Early workers (Newbold 1846, King 1872) noted alluvial diamond workings within 8 km of Chelima at Busavapuram (Sen and Narasimha Rao 1971). Sakuntala and Krishna Brahmam (1984) reviewed the widespread distribution of alluvial diamonds in the Kurnool district.

3.7.2. Group Name: Majhgawan

Location: [24.45 N, 80. E]. The Majhgawan and nearby (3 km to the northwest) Hinota pipes occur about 20 km southwest of Panna, about 175 km northeast of Sagar, in north-central Madhya Pradesh, India (Figures 3.37, 3.38).

Age: Majhgawan pipe, 840–1140 Ma (Rb–Sr), the preferred age is 1140 ± 12 Ma (Grantham 1969, Crawford 1969, Crawford and Compston 1970, 1973, Paul *et al.* 1975). Hinota, 1170 Ma (K–Ar, Paul *et al.* 1975); 974–1120 ± 45 Ma (K/Ar phlogopite ages, Paul *et al.* 1975).

Nature of Country Rocks: The Majhgawan and Hinota intrusives cut relatively undisturbed flat-lying quartz arenites of the Middle-to-Upper Proterozoic Kaimur Group, Vindhyan Supergroup, just south of the 2.5–2.6 Ga Bundelkhand metabatholithic complex. The Vindhyan Supergroup strata are generally regarded as correlative and similar to the Cuddapah Basin strata (Naqvi and Rogers 1987). The presence of alluvial diamonds in the Upper

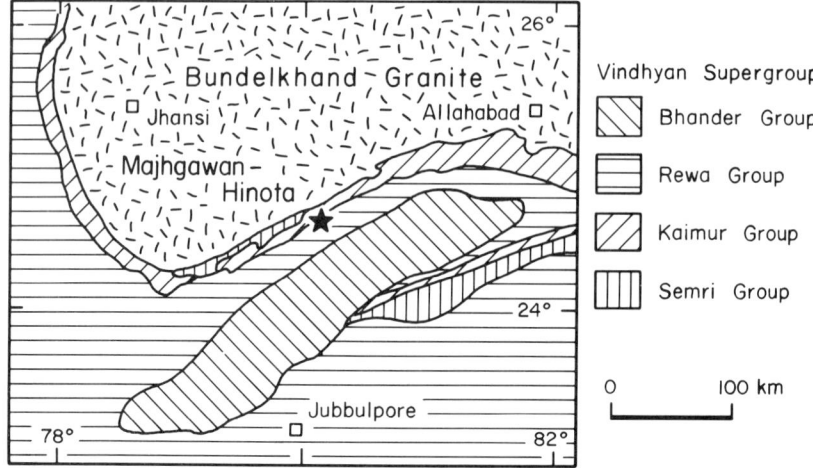

Figure 3.37. Regional geological map of a portion of the Aravalli craton showing the position of the Majhgawan lamproite field (after Naqvi and Rogers 1987).

Proterozoic lower Rewa, basal Jhiri, and upper Rewa conglomerates around Panna (Mathur 1981) suggests that the pipes were emplaced prior to Upper Vindhyan sedimentation.

Local Structural Setting: The two lamproites form circular intrusions in a relatively undisturbed region of flat-lying unmetamorphosed strata. The Majhgawan pipe possesses a small pointed bulge on its western margin. The local structural grain trends east-northeast. Haldar and Ghosh (1974) found no fundamental fractures in the area that would have controlled emplacement.

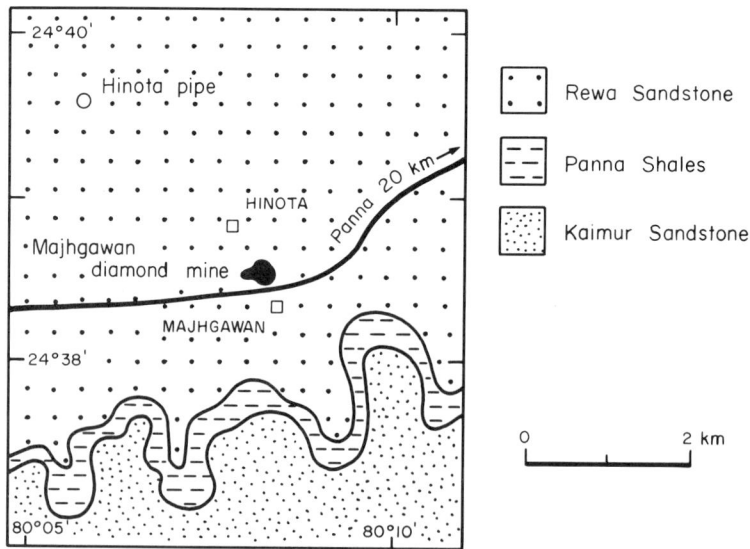

Figure 3.38. Local geological map of the Majhgawan area (after Mehr 1952).

Regional Tectonic Setting: The Majhgawan and Hinota pipes occur to the east of the Great Boundary Fault within 50 km of the southeast margin of the Aravali Craton (as defined by Naqvi and Rogers 1987). The host Vindhyan strata form an extensive sedimentary basin which is truncated to the south by the Son Valley Thrust, a major south-dipping thrust fault. A broad northeast-trending crest of a wide, southwest-plunging anticlinorium in the region might have controlled pipe emplacement (Haldar and Ghosh 1974). The Majhgawan lamproites were emplaced relatively soon after basin formation and deposition of the host strata (within 100–200 m.y.).

Nature of Basement Rocks: The host Proterozoic sedimentary strata overlie 2.5–2.6 Ga metagranitic rocks of the Bundelkand Complex and metasedimentary and metaigneous rocks of the 2.9–3.5-Ga Banded Gneissic Complex (Naqvi and Rogers 1987).

Intrusive/Extrusive Forms: The Majhgawan and Hinota rocks apparently consist entirely of crater facies rocks in the form of a vent. Although the rocks are altered, no massive lithologies have been recognized and only fragmental rocks have been found.

Geomorphology: The rocks in this low relief area are poorly exposed.

Area/Volume: 0.15 km^2, 0.1 km^3.

Component Localities: Majhgawan and Hinota pipes.

Nearby Contemporaneous Igneous Rocks: Archean and Proterozoic alkaline intrusive complexes are widespread throughout the Aravalli Craton (Naqvi and Rogers 1987). The Angor "kimberlite pipe" (Nane 1971, Puri 1972, Mathur 1981, 1986), located 100 km west of Majhgawan, has been reclassified as a layered peridotite–pyroxenite–gabbro complex (Scott Smith 1989). The 919-Ma Jungel kimberlites, located 250 km to the east of Majhgawan (Chattopadhyay and Venkataraman 1977, Balasubrahmanyan *et al.* 1978), have been reclassified as metavolcanic rocks of nonlamproite and nonkimberlite affinity by Scott Smith (1989). The 959-Ma Newania carbonatite, near Udaipur (Deans and Powell 1968), is located 600 km west of Majhgawan.

Rock Types: Fragmental volcanic rocks, including tuff breccias, lapilli tuffs, and tuffs of phlogopite olivine lamproite.

Mineralogical Comments: The rocks contain a typical lamproite assemblage. Anatase, Mn-ilmenite, perovskite, rutile, and titanomagnetite are also present.

References: Vredenburg (1906); Sinor (1930); Dubey and Mehr (1949); Mehr (1952); Mathur (1951, 1953, 1955, 1962, 1981, 1986); Venkataraman (1960); Mathur and Singh (1963, 1971); Grantham (1964, 1969); Crawford (1969); Kumar (1971); Singh (1971); Gupta and Phukan (1971); Das and Lakshmanan (1971); Crawford and Compston (1973); Halder and Ghosh (1974); Paul *et al.* (1975, 1977, 1979); Kresten and Paul (1976); Paul (1979, 1980); Murthy *et al.* (1980); Murthy (1980); Deshpande (1980); Banerjee and Agarawal (1980); Middlemost and Paul (1984); Gupta *et al.* (1986); Scott Smith (1989).

Comments: The Majhgawan and Hinota intrusives have conventionally been termed kimberlites, although Grantham (1964, 1969) noted some significant contrasts with archetypal kimberlites. The rocks have recently been reclassified as lamproites by Scott Smith (1989). The diamondiferous Majhgawan lamproite has operated as India's only significant primary diamond mine (with intense mining for 27 years from 1937 to 1964), with a grade of about 10 carats/100 tonnes. N.B. Some workers have rejected a volcanic origin for the tuff and preferred to view the rock as a greywacke (Ahmad 1956).

3.7.3. Group Name: Barakar

Location: [23.755 N, 86.850 E]. The Barakar dikes occur 3 km north of Kulti, 16 km northwest of Asansol, 200 km northwest of Calcutta, in West Bengal, India (Figure 3.39). The rocks are a part of an extensive group of micaceous mafic dikes occurring in the Raniganj Coalfield, between the Barakar and Ajay Rivers, and the nearby Bokaro and Jharia Coalfields.

Synonyms: Ramnagar Colliery, Asansol, Raniganj.

Age: Although age determinations for these dikes are not available, they are probably contemporaneous with 105–121 Ma (K–Ar biotite) lamprophyre dikes, in the Jhariak, Raniganj and Darjeeling Coalfields (Sarkar *et al.* 1980).

Nature of Country Rocks: The Barakar dikes intrude the Permian coal-bearing Barakar Formation. Other micaceous mafic dikes in the Raniganj Coalfield intrude Upper Carboniferous-to-Permian coal-bearing and associated sedimentary sequences.

Local Structural Setting: The Barakar lamproites occur in a mildly deformed (faulted, tilted, and folded) Late Paleozoic rift basin, which formed during the breakup of Gondwanaland. Local faults trend northwest.

Regional Tectonic Setting: The Raniganj Coalfield is part of the 20–50 × 400 km, east-trending Damodar Valley, a Gondwana rift basin situated in the middle of the Chotanagpur Terrain. The Chotanagpur Belt (metasedimentary rocks, granitoids, mafic/ultramafic schists, anorthosites) is adjacent to the Proterozoic Mobile belt which forms the northern boundary at the Singhbhum Craton. The dominant structural grain of the Chotanagpur platform trends east–west, and the belt is thought to range from Archean-to-Late Proterozoic in age (Naqvi and Rogers 1987).

Nature of Basement Rocks: Crystalline basement rocks of Archean-to-Proterozoic age underlie the Late Paleozoic basin fill. Granitic orthogneiss as old as 2.8 Ga is thought to be underlain by 3.8-Ga crystalline rocks (Naqvi and Rogers 1987).

Intrusive/Extrusive Forms: The lamproites form thin (1–2 m wide) dikes and sills of massive fine-to-medium-grained hypabyssal facies rocks.

Geomorphology: The region possesses subdued topographic relief. The dikes and sills have been encountered during exploration drilling and open pit and subsurface coal mining.

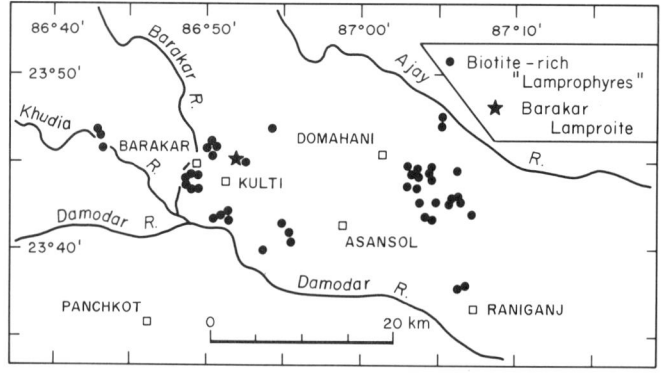

Figure 3.39. Location map of the Barakar lamproites, Raniganj coalfield.

DESCRIPTION OF LAMPROITE OCCURRENCES

Area/Volume: <0.05 km², <0.001 km³.

Component Localities: Individual dikes are not named.

Nearby Contemporaneous Igneous Rocks: The closest coeval igneous rocks in the Raniganj and Bokaro coalfield comprise a diverse suite of lamprophyres including: biotite peridotites, olivine minettes, leucite minettes, kersantites, microsyenites, and monchiquites (Banerjee 1953, Chatterjee 1974).

Rock Types: Altered and carbonated, porphyritic phlogopite lamproite.

Mineralogical Comments: The Barakar lamproites contain titanium-rich phlogopite (9 wt % TiO_2 10 wt % Al_2O_3) and priderite (4–9 wt % BaO) (Middlemost *et al.* 1988).

References: Hughes (1870); Holland and Saise (1895); Fox (1930); Gee (1932); N.N. Chatterjee (1937); Ghosh (1949); Banerjee (1953); Mitra (1953); Mukherjee (1961); Sanyal (1964); S.C. Chatterjee (1974); Sathe and Oka (1975); Sarkar *et al.* (1980); Paul and Potts (1981); Paul and Sarkar (1986); Middlemost *et al* (1988).

Comments: The Barakar intrusives, known for more than a century, have only recently been reclassified as lamproite (Middlemost *et al*. 1988). N.B. Middlemost *et al*. (1988) refer to the presence of only one lamproite dike in the Barakar field. Other dikes, strictly on a petrographic basis, are considered to be lamprophyres (minettes). However, the composition of the phlogopites and feldspars in these rocks is typical of a lamproite paragenesis. Thus, we regard all of the sanidine-bearing dikes as lamproites and do not accept the minette–lamproite consanguinity proposed by Middlemost *et al*. (1988). It is possible that many other lamproites occur in the Gondwana coalfields.

3.7.4. Group Name: Murun

Location: [58.4 N, 119.3 E]. The Murun complex is located near the Chara River, about 170 km north-northeast of Chara, 700 km southeast of Yakutsk, in the Severo-Baykalskove and Patomskoye Nagorye districts of the Soviet Union (Figure 3.40).

Age: 115–143 Ma (Kostyuk 1983).

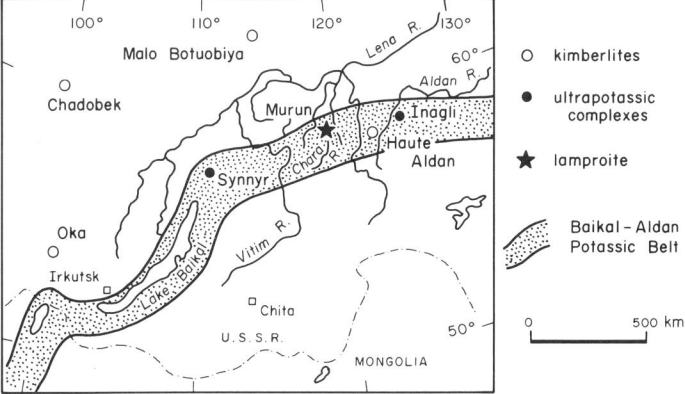

Figure 3.40. Map of eastern USSR showing the location of Murun in the Baikal-Aldan potassic magmatic belt (after Kostyuk 1983).

Nature of Country Rocks: The dikes intrude the Murun pluton, a potassic magmatic complex, composed of nepheline syenite, shonkinite, tinguaites, and kalsilite-orthoclase rocks (yakutites).

Local Structural Setting: Unknown.

Regional Tectonic Setting: The Murun complex is one of many Mesozoic-to-Cenozoic potassium-rich alkaline igneous complexes in the Baikal-Aldan Belt (Kostyuk 1983). The Baikal Rift was initiated in the Late Archean and is located on the margin of the Archean Shield.

Nature of Basement Rocks: The complex is underlain by Proterozoic-to-Archean crystalline basement.

Intrusive/Extrusive Forms: Hypabyssal dikes.

Area/Volume: Unknown.

Nearby Contemporaneous Igneous Rocks: The Murun complex is located within 400 km of several other potassium-rich alkaline igneous complexes, e.g., the Synnyr and Inagli complexes, which contain shonkinite, fergusite, missourite etc.

Rock Types: Sanidine phlogopite lamproite.

Mineralogical Comments: These rocks contain a typical lamproite assemblage of phlogopite, sanidine, diopside, apatite, and spinel.

References: Arkhangelskaya (1974); Kostyuk (1983); Smyslov (1985); Orlova (1987, 1988); Vladikin (1985); Bogatikov *et al*. (1986, 1987); Mitchell (1988).

Comments: These rocks require detailed mineralogical and petrological study to assess their relationship to the potassic rocks of the Murun complex.

3.8. SOUTH AMERICAN LAMPROITES

Poor access and exposure have hindered the discovery of kimberlites and lamproites in South America, and lamproites have only recently been recognized near Coromandel, in Minas Gerais, Brazil (Figure 3.41). Kimberlite pipes and dikes and alluvial diamonds are well known in this part of Brazil (Svisero *et al*. 1979a, 1979b, 1984, Ulbrich and Gomes 1981).

Nixon *et al*. (1989) have identified intensely weathered Proterozoic micaceous rocks carrying diamonds, yimengite, and Cr-garnets, in the Guaniamo district of Bolivar Province, Venezuela. These rocks may have lamproite affinities. Unfortunately detailed mineralogical studies are not possible because of the extreme alteration.

3.8.1. Group Name: Coromandel

Location: [18.6 S, 46.6 W]. Four lamproite localities occur over a 100×10-km east-trending belt. The Presidente Oligario and Pantanão lamproites are located near the Paranaiba River, approximately 75 and 45 km east of Coromandel. Limeira and Indaia are 25 km west of Coromandel (Figures 3.42, 3.43).

Synonyms: Presidente Oligario

Age: 80–87 Ma (Cretaceous), zircon U-Pb (Davis 1977, Meyer *et al*. 1988).

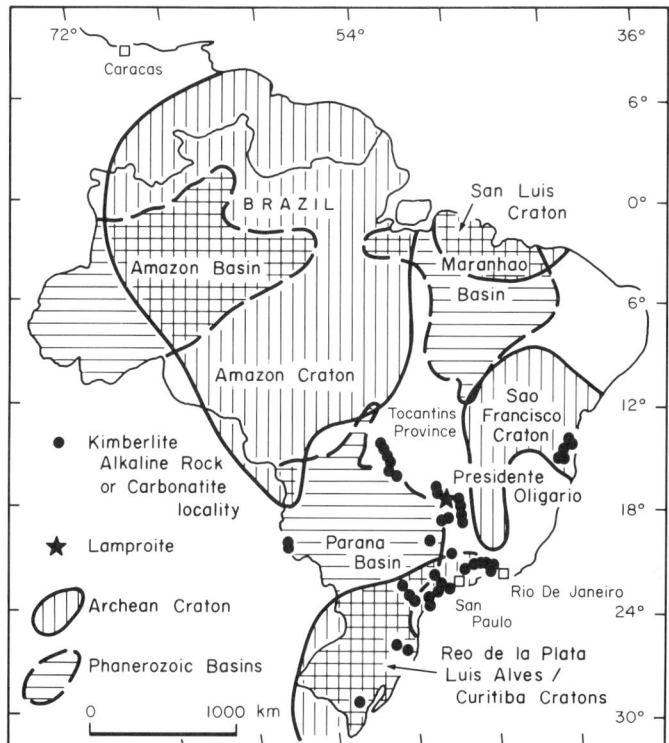

Figure 3.41. Regional map of Brazil showing the structural and alkaline magmatic context of the Coromandel lamproite field (after Cordani *et al.* 1988, Herz 1977, Ulbrich and Gomez 1981).

Nature of Country Rocks: The lamproites intrude Jurassic and Cretaceous sedimentary rocks, Mid-to-Late Proterozoic supracrustal metasedimentary rocks of the Bambui, Canastra, and Araxa Groups and Archean granitic metaigneous rocks.

Local Structural Setting: Unknown.

Regional Tectonic Setting: The Coromandel lamproites occur south of the Pirineus Megaflexure in the Alto Paranaiba Uplift, a regional gravity high, between the Parana and Alto Sanfranciscana Intracratonic Basins. The lamproites occur in the paired Uruacu and Brasilia Proterozoic Mobile belts which mantle the western part of the Archean São Francisco Craton. The mobile belt experienced metamorphism in Late Proterozoic times (1.0–1.2 Ga) of basaltic rocks and sedimentary deposits derived from Lower Proterozoic (2.0–2.2 Ga) crystalline rocks. The region is dominated by a northwestern-trending structural grain, defined by major folds and faults, lineaments, and aeromagnetic anomalies. The Late Proterozoic northwest-trending folds verge to the east. The lamproites appear to be contemporaneous with kimberlite, carbonatite, lamprophyre, and other alkaline intrusive rocks of the Alto Paranaiba kimberlite province (Svisero *et al.* 1984, Meyer *et al.* 1988). These rocks were emplaced during a phase of regional uplift about 20–60 m.y. after the eruption of immense volumes of the Parana and Alto Sanfranciscana flood basalts in the region (Svisero *et al.* 1984), which was temporally associated with continental breakup and

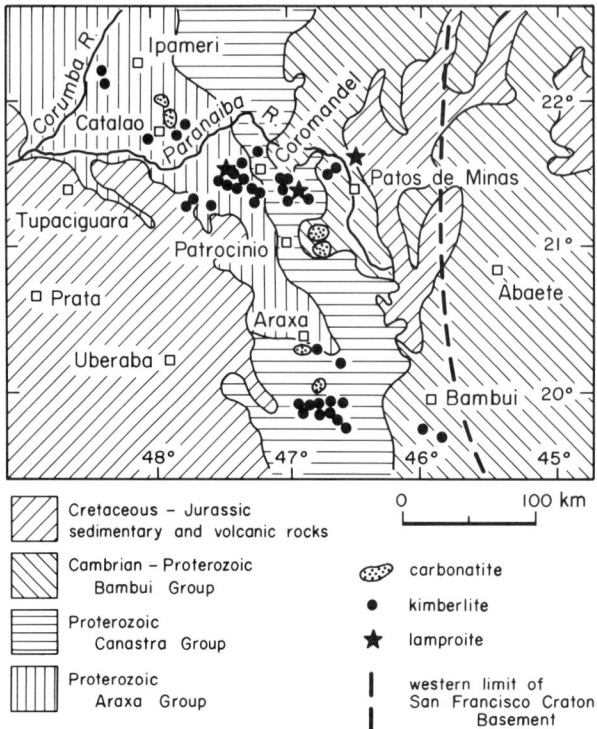

Figure 3.42. Geological map of a portion of eastern Brazil showing the structural context of the Coromandel area alkaline intrusive rocks (after Bardet 1974, Meyer et al. 1988, Ulbrich and Gomez 1981, Almeida et al.1971, Svisero et al. 1984).

drift. Herz (1977) suggested that the 130-Ma Jacupiranga alkaline intrusive rocks (located 700 km south of Coromandel) formed as the result of a diffuse hot spot near a triple junction.

Nature of Basement Rocks: The lamproites are underlain by Proterozoic rocks and probably by Archean basement with crystallization ages of 2.6–3.1 Ga (Cordani et al. 1988).

Intrusive/Extrusive Forms: The lamproites form dikes and vents. Hypabyssal and fragmental rocks (tuffs and breccias) have been recognized.

Area/Volume: <5 km^2, <2 km^3.

Component Localities: Presidente Oligario, Pantanão, Limeira II, Indaia II.

Nearby Contemporaneous Igneous Rocks: Indaia II and Limeira II occur adjacent to small intrusions of hypabyssal kimberlite (Meyer et al. 1988). The lamproites form part of the 80–87-Ma Alto Paranaiba kimberlite–carbonatite–lamprophyre province. Also in the area are 45–85-Ma alkaline intrusions (syenite, shonkinite, etc.) of the northwest-trending Goias–Minas Gerais trend, and an orthogonal 120–144-Ma belt of alkaline intrusive rocks (Herz 1977, Marsh 1973, Svisero et al. 1984). A small (500-m^2) volcanic neck of 78-Ma alkali olivine basalt occurs at Pantanão (Hasui and Cordoni 1968, Barbosa et al. 1970, Ulbrich and Gomes 1981). At Catalão, northwest of Coromandel, 83-Ma phlogopite glim-

DESCRIPTION OF LAMPROITE OCCURRENCES

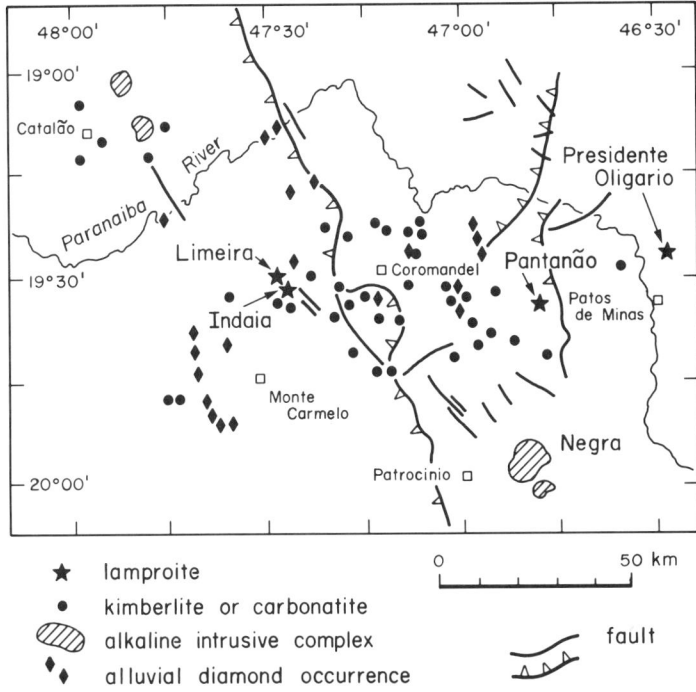

Figure 3.43. Location map of the Coromandel area showing the distribution of alkaline intrusive rocks (after Bardet 1974, Meyer *et al.* 1988, Ulbrich and Gomez 1981, Almeida *et al.* 1970, Svisero *et al.* 1984).

merites, pyroxenite, peridotite syenite, and carbonatite form a 27-km² circular stock (ibid). About 150 km south of Coromandel is the 43-Ma (?) Sacramento suite consisting of "ugandite" lavas, breccias, and tuffs (Murta 1965, Barbosa *et al.* 1970, Ulbrich and Gomes 1981).

Rock Types: Massive to fragmental olivine phlogopite diopside madupitic lamproites.

References: Barbosa *et al.* (1970); Fragomeni (1976); Davis (1977); Berbert *et al.* (1981); Ulbrich and Gomes (1981); Svisero *et al.* (1979a,b, 1984); Leonardos and Ulbrich (1987); Meyer *et al.* (1988).

Comments: The Coromandel lamproites are among the few occurrences of apparently contemporaneous lamproites, kimberlites and other alkaline igneous rocks.

3.9. CONCLUSIONS

Compared with other mantle-derived alkaline rock suites, lamproites form relatively small aggregate volumes (ca. 100 km³ worldwide), in relatively few occurrences (24 recognized worldwide). These occurrences range from volumetrically small dikes or sills (group 3), to large-volume lamproite provinces (group 1). Lamproites range in age from Quaternary (0.05 Ma) to Proterozoic (1.4 Ga). The majority of lamproites are of Cenozoic age. All lamproites overlie continental lithospheric basement that was deformed and metamorphosed in the Proterozoic. Nearly all lamproites occur in regions that experienced multiple phases of Proterozoic-to-Mesozoic compressional orogeny and/or (usually aborted) ancient rifting. Lamproites generally occur external to, or on, the margins of Archean cratons. In contrast, kimberlites occur both on the margins and in the central parts of Archean cratons. Many lamproites occur on the margins of orogens in regions that have suffered extensional collapse following crustal thickening events. Lamproites are located mainly in regions of excessive lithospheric thickness compared with those of non-lamproite-bearing areas. Lamproites are post-tectonic (anorogenic) igneous rocks, and are not contemporaneous with local (within 100 km) active or passive rifting events, subduction events, or wrench tectonic events. Magmatism may occur 5–2500 m.y. after these events. At the time of continental breakup, several lamproites were emplaced about 500 km away from the eventual passive margin. The processes of subduction and rifting are apparently important in preparing the mantle source regions for the eventual lamproite magmatism (see Chapters 4 and 10). Many lamproite suites are approximately spatially and temporally associated with other mantle-derived alkaline magmas. However, there are many lamproites that formed in relatively isolated regions without any nearby (i.e., within 200–2000 km) contemporaneous alkaline or other igneous activity.

Nature goes her own way, and all that to us seems an exception is really according to order.
   Goethe

Tectonic Framework of Lamproite Genesis

Lamproites are temporally and spatially widespread; however, they form only volumetrically minor igneous occurrences relative to most, if not all, other mantle-derived alkaline rocks. In fact, all presently known lamproite bodies comprise an aggregate volume of merely 75 km^3, corresponding to a sphere with a radius of under 3 km or a 20-m-thick layer spread over Long Island, NY! In common with other alkaline magmas, lamproite melts have formed in the mantle and ascended to the surface over at least the last 1.4 Ga.

Perhaps the most intriguing aspect of lamproites is the compositional, mineralogical, and textural uniformity of lamproite fields that were erupted on opposite sides of the globe. This petrological consistency is surprising given their exotic nature relative to many other mafic alkaline rocks. The uniformity suggests that common features exist in their mantle sources, petrogenesis, and the tectonic framework that controls their emplacement. Hence, despite their small volumes and petrological peculiarities, lamproites can potentially provide much intrinsic information on the nature and evolution of their mantle source regions. In addition study of their genesis provides some interesting tectonic insights into the nature of postrift and postcollisional mantle lithospheric magmatism.

Several fundamental questions are raised from consideration of the spatial and temporal distribution of lamproites and compositionally allied magmatic rocks:

1. Are the parental magmas of lamproites different from those of other approximately contemporaneous alkaline mantle melts occurring in the same region?
2. Are the mantle source regions of lamproite melts different from those of kimberlite, minette, alkali basalt, or related melts?
3. Are there unique tectonic framework or regional geological conditions leading to the formation of lamproites on the one hand, and kimberlites, alkali basalts, or other mantle-derived alkaline magmatic rocks on the other?

We contend that the answer to all of these questions is yes. The first two questions are addressed in subsequent chapters (6, 7, and 8), where it is shown that there are unique characteristics to the mantle source and parental magma compositions of lamproites, compared with compositionally allied rocks.

This chapter attempts to clarify the unique tectonic framework and regional geological history shared by lamproite fields, provinces, and localities in order to place constraints upon hypotheses concerning their petrogenesis and emplacement. The tectonic framework data taken mainly from Chapter 3 are integrated with isotopic, geochemical, and mineralogical information from Chapters 6 and 7. Unfortunately, only 24 lamproite provinces, fields, or localities are known, and thus any genetic model is limited by this paucity of occurrences. Fortunately, about half of all lamproite fields or occurrences are Cenozoic in age, and because of their youth it is possible to infer or deduce the plate tectonic settings prevailing at the time of their emplacement. However, it is much more difficult to deduce the ambient plate tectonic settings of the Proterozoic to Mesozoic lamproites. Assuming uniformitarianism, many of the tectonic framework conclusions reached concerning the emplacement of these lamproites are based upon those inferred for Cenozoic occurrences. Undoubtedly, as further lamproite provinces and localities are recognized and all occurrences are better understood, the conclusions stated below will require revision.

4.1. AGE AND TEMPORAL RELATIONS OF LAMPROITE MAGMATISM

Lamproites have been erupted during at least a third of the Earth's history. Known lamproite magmatism ranges in age from Mid-Proterozoic (1.2–1.4 Ga: Sisimiut, Bobi, Chelima, and Argyle) to Plio-Pleistocene (0.05–3 Ma: Gaussberg and Leucite Hills) (Table 4.1). In comparison, the oldest well-documented kimberlites are the 1.6–1.7-Ga Kuruman Province (South Africa), 1.4-Ga Gabon and Liberia pipes (West Africa), 1.2-Ga Premier and National pipes (South Africa), Toubabouko (Ivory Coast), and Wajrakarur (India) pipes, whereas the youngest known kimberlites are Early Tertiary (Eocene) to Late Cretaceous (Nzega, Mwadui, Bushmanland, South Cape, Namibia, Kimberley, and Lesotho) (Janse

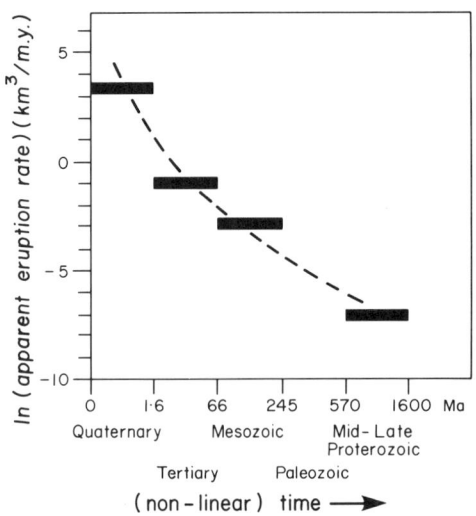

Figure 4.1. Histogram plot (logarithmic scales) of the total preserved erupted volume of lamproites as a function of age of eruption.

Table 4.1. Compilation of the Age and Volume of Lamproite Provinces, Fields, and Localities

	Locality/field/province[a]	Age (Ma)	Volume (km³)	
1.	Gaussberg	0.05	1	Quaternary
2.	Leucite Hills	1.6	30	
3.	Murcia-Almeria	8	10	Tertiary
4.	Sisco	14	0.0001	
5.	West Kimberley	20	20	
6.	Smoky Butte	27	0.5	
7.	Sesia-Lanzo and Combin	31	0.001	
8.	Francis	35	1	
9.	Yellow Water Butte	52	3	
10.	Froze-to-Death Butte	52?	0.3	
11.	Gold Butte	55?	0.3	
12.	Coromandel	85	1	Cretaceous
13.	Hills Pond	90	2	
14.	Prairie Creek	100	0.5	
15.	Murun	130	0.5?	
16.	Swartruggens	150	0.01	Jurassic
17.	Kapamba	220	5	
18.	Mount Bayliss	425	0.001	Paleozoic
19.	Priestly Peak	482	0.001	
20.	Majhgawan	1140	0.1	Proterozoic
21.	Argyle	1150	0.3	
22.	Sisimiut	1227	0.2	
23.	Chelima	1345	0.02	
24.	Bobi	1430	0.1	

[a]Data sources are given in Chapter 3. The age represents a weighted average for the given province/field. Volumetric values are estimates.

1984, Skinner et al. 1985, Bristow et al. 1989). Of the 24 lamproite provinces, fields, or localities recognized in this work, 5 are Proterozoic, 2 are Paleozoic, 6 are Mesozoic, and 11 are Cenozoic, suggesting therefore, that the degree of preservation of lamproites apparently decreases with age.

The Mid-Proterozoic Helikian period (ca. 900–1600 Ma) was a time of worldwide mantle-derived alkaline magma formation. In addition to the 5 lamproite and 7 kimberlite occurrences noted above, the Mid-Proterozoic witnessed the formation of numerous alkaline intrusive complexes, alnöites, and other mafic or ultramafic lamprophyres in northwestern Europe and North America (e.g., Woolley 1987, Åberg 1988). The abundant kimberlite magmatism that occurred in the Paleozoic and Mesozoic is not contemporaneous with lamproite formation.

Lamproites (e.g., Leucite Hills, Gaussberg) have erupted during the Quaternary, although there have been no observations of the volcanism. Cenozoic lamproites volumetrically account for the largest proportion of *preserved* lamproites (Figure 4.1). If erosion alone is responsible for this feature, one might expect preserved Precambrian lamproites to consist mainly of deeper-seated feeder systems (dike swarms, etc.), and Cenozoic lamproites to be dominated by only slightly denuded lamproite vents, pipes, or volcanoes. This is the case for Cenozoic lamproites, but only three of the five Proterozoic lamproites consist of dikes or dike swarms (Sisimiut, Bobi, and Chelima). Whereas erosion may account for most of this imbalance in the aggregate volume of Proterozoic versus Cenozoic lamproites, selective preservation has played a role, since extrusive sequences are preserved in two of the five Proterozoic lamproites.

Assuming that lamproites are essentially preserved, the time-averaged lamproite eruption rate systematically decreases with age, from about 31 km^3/m.y. for the Quaternary, through 0.4 km^3/m.y. for the Tertiary, 0.05 km^3/m.y. for the Mesozoic, to 0.0008 km^3/m.y. for the Proterozoic (Figure 4.2). Whereas there is clearly a high probability for the selective denudation of lamproite volcanic fields with time, it is difficult to resolve this apparent 7 orders of magnitude increase in the time-averaged volume of lamproite from the Proterozoic to the Quaternary.

The abundance of Cenozoic lamproites compared with those of pre-Cenozoic age is a feature shared with mantle-derived alkali basaltic rocks in general. The number of Late Cenozoic (<5 Ma) mantle xenolith-bearing alkali basaltic rocks is more than three times the number emplaced during Early to Mid-Cenozoic times (5–65 Ma). The latter period spans over 12 times the duration of the former and accounts for more than a 30-fold increase in apparent eruption rate during recent times (Nixon and Davies 1987, pp. 742–743). However, kimberlites and lamprophyres do not seem to increase in abundance from the Early Phanerozoic to more recent times. Paleozoic and Mesozoic kimberlites and lamprophyres volumetrically account for the largest proportion of both lithologies. These age relations led Bergman (1987, p. 146) to suggest that alkali basalt cinder cones might be underlain by consanguineous alkaline lamprophyre feeder dikes. This feature undoubtedly also led Tröger (1935) to suggest that lamproites represent the extrusive equivalents of lamprophyres enriched in potassium and magnesium.

Whereas Nixon and Davies (1987) suggested that the Late Cenozoic was an exceptional time of extensional tectonics (with a more brittle and more metasomatized lithosphere) relative to earlier periods in Earth history, we prefer a more uniformitarian view. We consider the abundance of relatively recent lamproites (and xenolith-bearing alkali basalts) to reflect simply the greater probability of denudation of volcanic rocks exposed on the Earth's

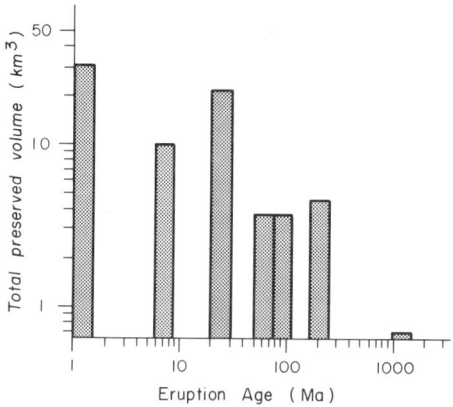

Figure 4.2. Plot of the logarithm of apparent eruption rate (based only on preserved volumes) as a function of geological time, divided into four major periods of dissimilar lengths.

surface, and lesser probability of preservation in the geologic record of earlier formed rocks. Clearly, much less than 0.1%–10% of the volume of these types of mafic volcanic rocks erupted on the surface are expected to form the underlying near-surface feeder systems of these volcanic outpourings.

Some continents contain lamproite fields or localities which are both ancient and modern (e.g., Australia, North America, Africa, Antarctica), whereas others (e.g., Europe, South America) contain lamproites of a more limited age range. Although Proterozoic lamproites are relatively rare and volumetrically insignificant, they are apparently the most diamondiferous, compared with their Phanerozoic counterparts. Lamproite magmas have ascended from the lithospheric mantle, from depths at which diamond is stable (>140 km), in a manner in which diamond is preserved, from the Mid-Proterozoic (Argyle, Majhgawan, Bobi, Chelima?) to the Miocene (West Kimberley).

In common with continental alkali basaltic volcanic fields, lamproite fields are formed during relatively short magmatic episodes, compared with the time taken for alkaline intrusive complex emplacement. Although detailed geochronological studies ($n > 5$–10 bodies) have been undertaken on only two lamproite provinces (West Kimberley, Murcia-Almeria), these data indicate that the time span represented by lamproite magmatism within a given field or province is of the order of 5 m.y. Lower limits include the Leucite Hills field, which was apparently erupted in less than 2 m.y., and Gaussberg volcano, which was formed in less than 0.02 m.y. If the West Kimberley and Murcia-Almeria durations are representative of lamproites as a group, we would expect recurrent future lamproite magmatism in the Leucite Hills and at Gaussberg.

Eruption of lamproites over the last 1.4 Ga indicates that the tectonic conditions required for lamproite melt formation and emplacement, near or on the surface, have operated for a time period of more than one-third the age of the oldest preserved crustal rocks (ca. 3.8 Ga). The source mantle having the requisite composition for lamproite magma generation must have existed for at least the last 1.4 Ga. As noted in the isotope geochemistry section below (Section 7.9), the lead and neodymium model ages of Proterozoic to Recent lamproite source regions is on the order of 1–3 Ga. The considerable difference between the lamproite eruption ages and their source model ages suggests that, either a significant time passes randomly before the metasomatized peridotite source is subjected to the tectonic environment needed for melting and ascent, or the lithospheric mantle source must "age" or ripen before it is capable of producing a lamproite melt. The latter may be the principal cause for the apparent progressive increase in lamproite eruption rate as a function of time (Figure 4.1).

4.2. REGIONAL GEOLOGICAL AND TECTONIC SETTING GENERALIZATIONS

There are two important features concerning the geological and tectonic setting of lamproite occurrences. These are: the regional geological characteristics preceding the eruption of lamproites that were instrumental in "preparing" the mantle source regions for lamproite magma generation; and the setting attending the eruption of lamproites, which was presumably instrumental in forming the lamproite melts and guiding their ascent to the surface. This section synthesizes the latter whereas the subsequent section (Section 4.3) addresses the former.

4.2.1. Lamproites and Plate Tectonics

This section attempts to place lamproite eruption in a general plate tectonic context. Wilson (1963, 1968, 1973), Burke and Dewey (1973), Morgan (1968, 1972), and Miyashiro *et al.* (1982) have proposed a variety of related models of plate tectonic and associated orogenic activity (summarized by Seyfert 1987). These models form the overall tectonic framework in which to place lamproite magmatism. The general plate tectonic settings for terrestrial magmatism include oceanic and continental rift zones; subduction-related arcs; back-arcs; continent–continent collision zones; transform faults (wrench zones); and other regions of mantle upwelling (hot spots). The tectonic framework of most mafic and intermediate mantle-derived magmas can be placed neatly into these environments. Mafic and intermediate magmatic rocks erupted from the mantle in these domains exceed 30 km^3 per year in eruption rate (Nakamura 1974, Fisher and Schmincke 1984). Plate tectonic environments can also be grouped into plate boundary and intraplate domains, and oceanic and continental lithosphere domains. The vast majority of all recent magmas form at plate boundary settings, with intraoceanic and intracontinental plate settings accounting for only 10% of the total present-day to recent magma production rates. The eruption rates of these magmas exceed lamproite eruption rates by over eight orders of magnitude, emphasizing the exotic and exceptional nature of lamproite magmatism, even when viewed in the context of intraplate continental magmatism.

Lamproites are essentially continental intraplate alkaline magmas (Figure 4.3). In common with alkaline rocks in general, lamproites were erupted through a variety of continental lithosphere types in diverse tectonic environments. Lamproites do not apparently erupt in close proximity to contemporaneous plate boundaries (i.e., within <200 km). Cenozoic lamproites bear no spatial relationship to coeval convergent, divergent, or transverse plate margins, and there is no obvious evidence for the emplacement of Mesozoic to Proterozoic lamproites near plate margins. The closest lamproites to contemporaneous plate boundaries are located in Europe and include the Murcia-Almeria province, Sesio Lanzo/Combin, and Sisco, each located less than 200 km from an ambient plate boundary. The remaining lamproites are located from 400 to 5000 km from inferred ambient or present-day plate boundaries.

Lamproites have not been encountered in magmatic suites which have penetrated oceanic crust. "Lamproitic" assemblages (glimmerite or MARID lithologies) have not been found to occur in mantle-derived xenolith suites from oceanic alkali basaltic rocks. Leucite-bearing rocks which are much less potassic than lamproites rarely occur in oceanic areas (Gupta and Yagi 1980). The only known leucite-bearing rocks intruding *bona fide* oceanic crust occur at Tristan da Cunha (Baker *et al.* 1964). Leucite-bearing rocks are known from other oceanic island complexes (Cape Verde, Kerguelen, and Marquesas); however, these islands are thought to have formed over drowned attenuated continental crust (Gupta and Yagi 1980). Note that many leucite-bearing, subduction-related arc complexes occur in island arc settings, where oceanic crust is being subducted beneath continental crust, or in the distal portions of magmatic arcs, where the island arc is well established (>75 Ma).

4.2.2. Lamproites and Contemporaneous Subduction Zones

Although some workers have proposed a link between subduction and kimberlites (Sharp 1974, Helmstaedt *et al.* 1979), the hypothesis has been rejected by Mitchell (1986, pp. 132–133) on a variety of petrological, tectonic, and geochemical grounds. With the

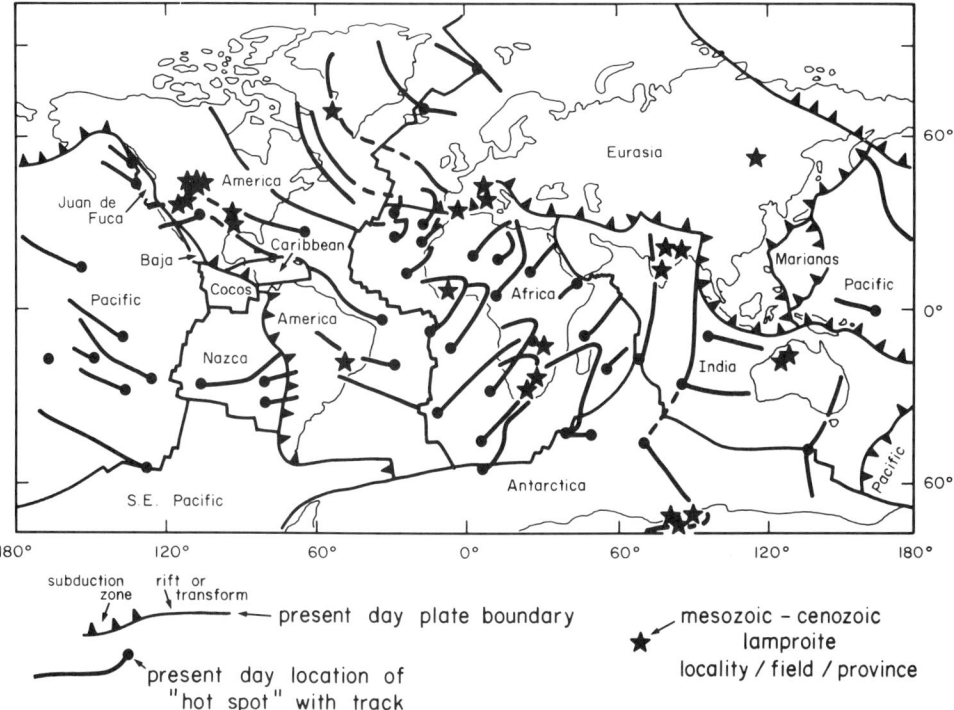

Figure 4.3. World map showing the position of lamproites relative to present-day major plate boundaries and hot spots, together with their predicted 50–100-Ma tracks (after Morgan 1981, Vink *et al.* 1986).

possible exception of the European occurrences, lamproites are not a part of the magmatic activity directly associated with subduction-related magmatic arcs (i.e., overlying active Benioff zones). Nevertheless, lamproites are compositionally allied to potassic calc-alkaline magmas or shoshonites (Joplin 1965, 1968, Jakes and White 1972), which tend to be localized over the deeper parts of Benioff zones, such as in the Indonesian arc (Morrison 1980, Foden 1983, Foden and Varne 1980). This similarity has led some workers to suggest that lamproite parental melts occur in the Phillipine magmatic arc. As summarized in Chapter 2, shoshonites are distinct from lamproites and the presence of the latter has not been verified in subduction-related arc magmatic assemblages and their postulated or observed parental melts. Morrison (1980), from consideration of the temporal and tectonic framework of shoshonite petrogenesis, concluded that these magmas are a widespread but relatively low volume proportion of subduction-related suites. He found that shoshonites form mostly in association with block faulting and uplift accompanying the waning stages of subduction, especially when the convergence angle of the downgoing slab becomes too oblique and deformation by strike-slip faulting increases in the arc region. Note that recent work on the Mariana island arc by Stern *et al.* (1988) demonstrates the occurrence of shoshonites along the magmatic front of a primitive, intraoceanic arc. These workers suggested that shoshonites need not characterize mature arcs, but may also form in the initial states of arc construction after an episode of back-arc rifting.

4.2.3. Fracture Zones, Transform Faults, and Continental Lineaments

Mantle-derived magmatism associated with oceanic transform plate boundaries is dominated by olivine- and hypersthene-normative or nepheline-normative basaltic rocks (Barker 1983). Where oceanic transforms intersect continental lithosphere inland of convergent boundaries (such as the Pacific rim), basaltic volcanism (including both tholeiitic and alkali types) is generally observed in relatively minor volumes compared with nearby convergent and divergent plate boundaries. In cases where oceanic crust converges obliquely upon continental plates it is conceivable that the continental extensions of oceanic transform faults could sweep laterally beneath the continent. The overlying crust would be subjected to transient thermal effects due to the presence of variably aged and thickened crust on both sides of the transform fault. This effect has been used to explain the time–space relations exhibited by western North American Neogene mafic to felsic magmatism, which was possibly influenced by the passage of the Mendicino Fracture Zone (Johnson and O'Neil 1984). Small volumes of alkaline rocks are occasionally observed along the passive margin continental extensions of oceanic transforms (e.g., West Africa, Brazil, Eastern North America, Southeastern Australia). Many workers have suggested that these purported preexisting zones of crustal weakness are capable of localizing or focusing the near-surface emplacement of kimberlites and other mafic alkaline rocks (Marsh 1973, Williams and Williams 1977, Sykes 1978, Stracke *et al.* 1979, Haggerty 1982, Taylor 1984). Mitchell (1986) argued that whereas the transform fault extension hypothesis may explain several kimberlite fields, it is certainly incapable of explaining most kimberlite occurrences in a given province containing many kimberlite fields.

In general, lamproite fields are located too far inland to be affected by the continental extensions of oceanic transform faults and the vast majority of major oceanic transforms possess no lamproites along their continental extensions. For example, lamproites in the midcontinent of North America (Hills Pond and Prairie Creek) are over 1000 km away from the continental extension of the Blake Fracture Zone of the Atlantic Ocean. The majority of lamproite fields do not occur along the extensions of major oceanic transforms (all those from Antarctica, Australia, South America, and Asia and most from North America). Nevertheless, several lamproite fields are located along the continental extensions of oceanic transform faults, e.g., the Bobi lamproites along the Sierra Leone Fracture Zone, and the Leucite Hills lamproites, which lie temporarily along the extension of the rapidly moving Mendicino Fracture Zone.

Some major continent-scale lineaments are thought to be long-lived zones of structural weakness (formed by "resurgent tectonics": Wells 1956, Umbgrove 1947, Watson 1980, White *et al.* 1986), and it is possible (if not probable) that lamproites could be localized along these deeply penetrating lithospheric features. Several lamproites are found along continent-scale dislocations or lineaments. For example: the Kapamba lamproites (Zambia) are located along the northeast-trending Schliesen–Mwembeshi–Chimaliro lineament of Daly (1986); the West Kimberley province lamproites and Argyle lamproites are located along the No. 3 and No. 5 lineaments of O'Driscoll (1986); and the Montana–Florida lineament ("North American transcontinental transform" of Kinsland 1983) passes within 100–200 km of the Montana, Kansas, and Arkansas lamproites (Figure 4.4). However, many lamproites are not related to any recognized major lineaments, e.g., the Wyoming and Utah lamproites are over 500–1000 km from the North American transcontinental lineament.

The near-surface emplacement of lamproites is undoubtedly controlled by significant

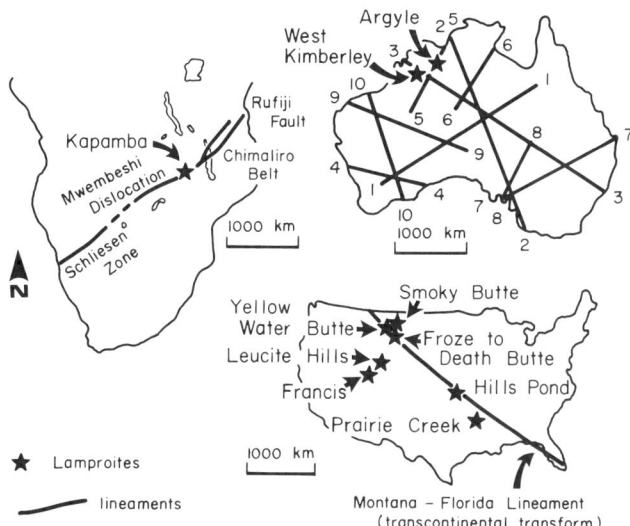

Figure 4.4. Relationship between selected lamproites and major lineaments or continent-scale structures for south central Africa (after Daly 1986), Australia (after O'Driscoll 1986), and the United States (after Kinsland 1985).

zones of crustal weakness. Unfortunately, geoscientists are only now beginning to unravel the complex relationship between well-established zones of crustal weakness and topographic or geophysical lineaments. Consequently, for many areas, where literally hundreds of obliquely intersecting lineaments have been proposed to exist, the important few that are deep-seated weak zones are generally poorly understood (Lathram and Raynolds 1977, Kutina and Carter 1977). The Wyoming Craton lamproites effectively illustrate the problem. The relationship between these lamproites and Phanerozoic lineaments is shown in Figure 4.5. Several lamproites are coincident with suggested lineaments or the margins of Mesozoic-Cenozoic uplifts, but the abundance of orthogonal lineaments, each closer than 50–100 km from adjacent lineaments, reduces the coincidence to a possibly random phenomenon.

4.2.4. Mantle Plumes, Hot Spots, and Hot Regions

Mantle plumes are narrow columns (100–200 km diameter) of hot, upwelling material that rises from depths within the mantle in excess of 700 km (Wilson 1963, Morgan 1972). Hot spots are the surface manifestations of inferred underlying mantle plumes characterized by apparent point-sources generating unusually large volumes of mafic to felsic volcanic rocks. Mantle plumes generally produce 1000–2000 km diameter domal uplifts, whose centers are uplifted 1–2 km above the surrounding terrain (Kinsman 1975, Crough 1979, 1981). Plumes are considered to be fixed, relative to each other, and are thought to move relative to the overlying plates, producing hot spot tracks (a.k.a. aseismic ridges, Morgan 1972). Plumes are most abundant in ocean basins within oceanic plates or along divergent plate margins: there are only several present-day or post-Mesozoic continental hot spots. Based on hot spot tracks, the typical life span of a plume is on the order of 100 Ma (Vink *et al.*

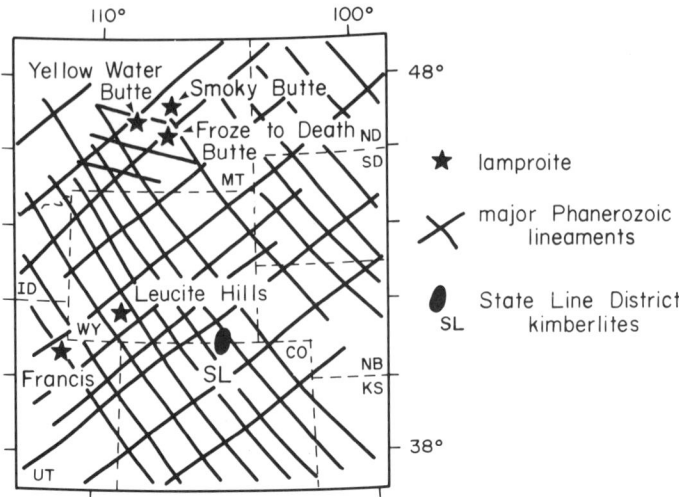

Figure 4.5. Relationship between lamproites of the Wyoming Craton and Phanerozoic lineaments proposed by Maugham (1983).

1985). Hot spots have been suggested by some workers as the dominant tectonic feature of the Archean (Lambert 1984).

Miyashiro (1986) identified hot regions as another type of mantle upwelling also producing intraplate magmatism. In this case, magmatism is caused by randomly migrating, shallower, and larger-scale mantle upwelling than that believed to be associated with deep mantle plumes. He suggested that hot regions beneath continents produce widespread alkaline (with common to rare potassic) mafic magmatism at rates about $\frac{1}{50}$th of that of deep mantle plumes in areas such as the eastern Australian highlands, Borneo/Southeast Asia, northern China and Mongolia.

Volcanic provinces associated with deep mantle plumes are dominantly subalkalic to tholeiitic basalt and rhyolite (with minor alkaline magmas) on continents, and tholeiite and alkali basalt in oceans (Vink et al. 1985). There are many instances of continental intraplate alkaline (even potassic) magmatism attributed to hot spots (Rhodes 1971, Burke and Wilson 1972, Wellman and McDougall 1974, Theissen et al. 1979, Onuoha 1985). For example, Wellman and McDougall (1974) and Wellman (1983) suggested that Cenozoic, and some Mesozoic alkaline-to-subalkaline igneous complexes in southeastern Australia and New Zealand, displayed temporal migration patterns consistent with a hot spot origin. In these areas, the regional duration of this apparent hot spot magmatism is on the order of 30–100 m.y. The Australian plate is thought to have moved over the mantle plume at a rate of 66 mm/yr. Although closely related ultrapotassic leucitites from New South Wales fall along Wellman's (1983) "hot spot track," lamproites *sensu stricto* do not comprise the apparent hot spot magmatism in this part of Australia. Jaques et al. (1986) suggested that the West Kimberley lamproite province possessed a temporal migration pattern (southward younging at a rate of 30 mm/yr) that is consistent with the direction of plate motion over a mantle plume. There are problems with such a hot spot model for the West Kimberley field in that (1) the apparent rate is half that suggested by Wellman (1983) for SE Australia; (2) there is no

TECTONIC FRAMEWORK OF LAMPROITE GENESIS 113

apparent present-day manifestation of any hot spot in northwestern or western Australia; and (3) the duration of West Kimberley Miocene magmatism is only 5 m.y.

Crough *et al.* (1980), Crough (1981), and recently Nickolaysen (1985) and le Roex (1986) suggested that the paleotracks of plumes presently found in ocean basins can be traced beneath adjacent continents. Such plumes have apparently produced a track of kimberlite and/or alkaline magmatism along the calculated plume trace. This hot spot model is capable of explaining only a portion of the kimberlites in eastern North America and South Africa, the two continents for which good continental hot spot paleotracks are available. The main arguments against this model are that kimberlite magmatism of the right age is absent from the vast majority of continental hot spot tracks, and present day to recent hot spots beneath continents (e.g., Yellowstone) have failed to produce kimberlitic eruptive rocks. Mitchell (1986, pp. 131–132) summarized the main objections to the hot spot model as applied to kimberlite magmatism, to which the reader is referred for more details. In the case of eastern North America, the hot spot proposition has also been questioned by Phipps (1988), who suggested that deep rifts were important in the formation of much of the alkaline mafic magmatism in eastern North America.

The location of Mesozoic–Cenozoic lamproites, present-day hot spot positions and paleo plume traces are illustrated in Figure 4.3. Figure 4.6 shows a reconstructed map of <200 Ma lamproites relative to the positions of the 32 primary plumes postulated to be associated with the fragmentation of Pangea in Triassic times. In nearly all cases, lamproites

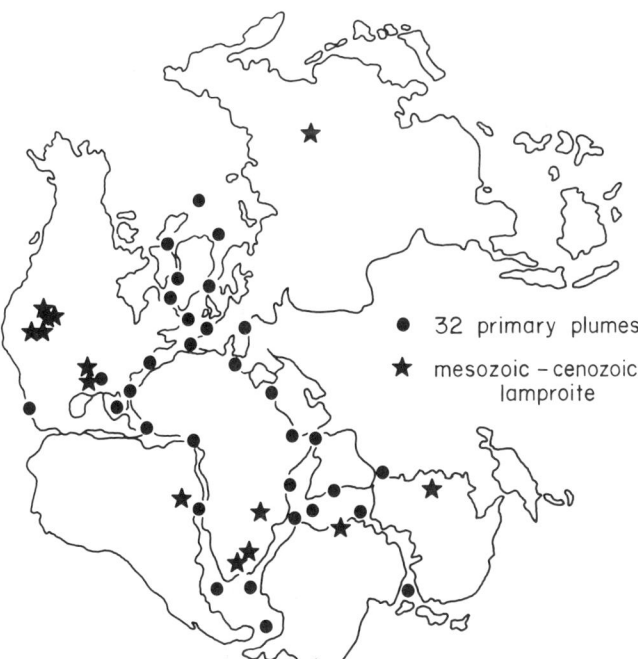

Figure 4.6. Distribution of Mesozoic and Cenozoic lamproites on a Pangea reconstruction (ca. 200 Ma) relative to proposed primary mantle plume locations (after Seyfert 1987).

are located on the fringes of the lithosphere affected by mantle plumes, yet they do not comprise the large volume basaltic magmatism associated with present-day continental hot spots such as the Yellowstone–Snake River Plain track. This spatial coincidence is because of the immense amount of plume-related lithospheric and asthenospheric melting and extensive associated domal uplifts. There is a close temporal coincidence of the 1-Ma age of the Leucite Hills lamproite field with the time that the fringe (200–300 km off axis) of the Yellowstone hot spot passed by Rock Springs. Many Mesozoic–Cenozoic lamproites from North and South America, Africa, and Antarctica fall along paleo-plume tracks and it is possible that past mantle upwelling modified the lithosphere in such a way as to produce the parent rocks of lamproite magmas. Hot regions or mantle plumes offer an effective mechanism for heating large portions of the subcontinental mantle lithosphere to temperatures above the solidus. Based on the coincidence of plume tracks and lamproite localities, close proximity of the Leucite Hills field to the Yellowstone hot spot, and the predicted effect of mantle plumes on the lithosphere, we consider mantle plumes as important precursors to the petrogenesis of lamproites.

The combination of thermal anomalies induced by mantle plumes acting on lithospheric regions underlying aulacogen or paleo-orogenic zones offers an even more attractive mechanism for lamproite melt formation (see below). Since lamproites are apparently small-volume and limited partial melts of ancient enriched mantle lithosphere, we propose that if mantle plumes or hot spots are related to lamproites, lamproite melts would be expected to form on the fringes of the asthenospheric plume where the requisite mantle lithosphere was available to melt and where the thermal anomalies were not excessive. The immense portions of mantle suspected to have melted to relatively large degrees due to mantle plume activity (e.g., East Greenland, White *et al.* 1987) would tend to dilute and obliterate the metasomatized lithospheric mantle source of lamproites.

4.2.5. Continental Rift Zones and Aulacogens

Several recent workers have suggested a genetic relationship between continental rift zones, or aulacogens, and mantle-derived alkaline magmatism, including kimberlites, lamproites, ultramafic lamprophyres, and related magmas (LeBas 1971, Bailey 1974, 1977, A.H.G. Mitchell and Garson 1981, Seyfert 1987, Phipps 1988). Continental rift zones are elongate domains in which the entire lithosphere has fractured due to extension (Burke 1977). Aulacogens are paleorift zones that have developed on ancient platforms and may have experienced multiple episodes of compression and extension (Milanovsky 1981). Rift zones are typically associated with broad domal uplifts, mafic alkaline volcanism and plutonism, and high regional heat flow, although the precise relationships between these processes is a subject of continuing controversy (LeBas 1987). Continental rifts have been grouped into active and passive types, depending on whether the rifting was due to upwelling of the underlying asthenosphere or the regional plate tectonic stress field, respectively (Ramberg and Morgan 1984, and references therein). Stated another way, mantle upwelling is the cause of rifting in the active type and the result in the passive type. Different rifts possess variable types and quantities of magmatic rocks. Despite this diversity, magmatic products are predominantly alkaline in the early stages and progress to tholeiitic compositions with time. Bimodal magma compositions are common in rifts. Magma volumes and eruption rates vary widely, from 5,000 km^3 and 100–200 km^3/m.y. in the Baikal Rift, to 500,000 km^3 and 25,000 km^3/m.y. in the eastern branch of the East African Rift System (Ramberg and Morgan 1984). Mitchell (1986) has shown that kim-

TECTONIC FRAMEWORK OF LAMPROITE GENESIS

berlites do not belong to the magmatic suites of rift zones, although there may be some distant, indirect relationship between some kimberlites and paleorift zones (Phipps 1988).

Lamproites are not generally found among the magmatic products formed in rift troughs during active rifting. The alkaline portion of magmatism in most rift zones consists of sodic types (Barberi *et al.* 1982). However, there are exceptions such as the kamafugitic lavas which occur in the western branch of the East African Rift System; the potassic alkaline rocks of the Baikal Rift; and the leucite-bearing lavas from the Eifel Province of the Rhine Graben. The Kapamba lamproite field in the Luangwa Graben is arguably perhaps the only lamproite field that was erupted during active rifting (Scott Smith *et al.* 1989), although we lack the chronological control to verify this hypothesis (cf. Section 3.4.1.2). Nevertheless, lamproites are commonly found in zones that have experienced rifting sometime in their history (Figure 4.7). Unfortunately, the full extent of paleorift zones is not presently known, and the compilation by Burke *et al.* (1978) is clearly only a partial one. It is therefore possible that the mantle lithospheric source regions of many lamproite fields may have been affected by metasomatism or plutonism associated with rifting. In addition, Cenozoic lamproites of western North America and western Europe occur in the back-arc positions of Cenozoic subduction zones where regional extension and mantle upwelling are indicated. In these cases, lamproite melts are not directly associated with typical subduction processes, but are indirectly associated with the tensional stress field associated with mantle upwelling processes. We therefore suggest that there are indirect relationships between lamproite magmatism and paleoextensional tectonic processes, mainly paleorifting.

4.2.6. Orogeny and Postorogenic Relaxation

Although lamproites are postorogenic magmatic products, orogenic zones appear to be fundamentally related to their formation. Nearly all lamproites occur in regions that have been subjected to Precambrian and/or more recent compressional orogeny. Many lamproites

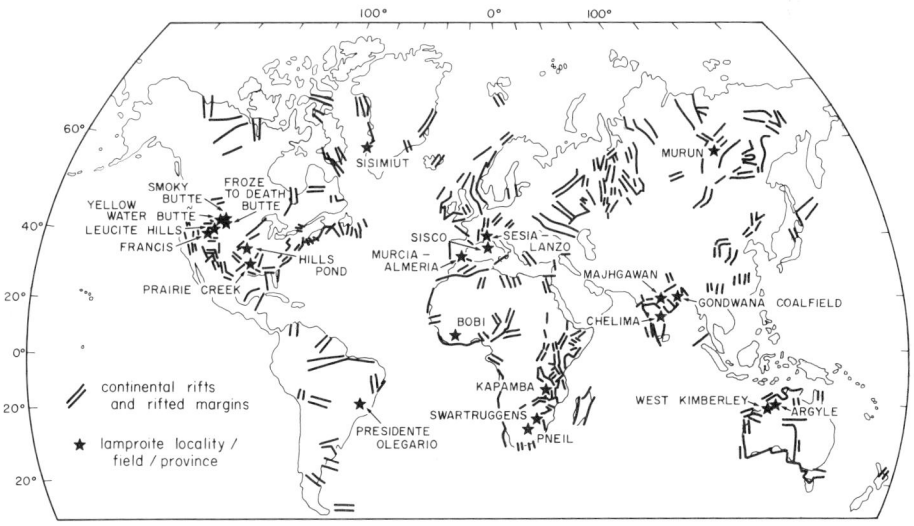

Figure 4.7. World map showing the simplified distribution of continental rift zones (Precambrian–present) relative to lamproite occurrences (after Burke *et al.*1978).

were emplaced through crust that has experienced several stages of compressional deformation. The amount of time separating active orogenic deformation and lamproite formation is highly variable, ranging from over 1–2 g.y. to 10–50 m.y.

Some lamproites were apparently emplaced during a phase of postorogenic relaxation, e.g., the European lamproites at Murcia-Almeria, Sisco, and Sesia-Lanzo and Combin, and some of the Wyoming Province lamproites (Smoky Butte, Francis, Leucite Hills). In each of these areas, lamproites were emplaced about 5–30 m.y. after a major orogenic event, which typically was associated with overthickening of the crust and/or lithosphere. The orogenic deformation, isostatic rebound, and collapse of the overthickened lithosphere, together create a tectonic environment in which ancient lithospheric mantle could partially melt to form lamproite magmas. These magmas would be induced to ascend to the surface under the tensional conditions associated with extensional collapse.

There are many features supporting a relationship between lamproites and ancient orogenic zones, including: the proximity of Phanerozoic lamproites to the margins of cratons where orogenic zones were accreted throughout the long-term evolution of continents; the occurrence of Proterozoic lamproites within much older (>200–1000 m.y.) mobile belts; the similarity between lamproite model Nd–Sr ages and the age of major crust-forming (orogenic) events; and the geochemical character of lamproites (e.g., Nb, Hf, Zr enrichments), indicating a complementary signature to typical subduction-related melts, which tend to be depleted in these elements, and suggesting that lamproite melts may represent partial melts of old Benioff Zones.

4.3. LITHOSPHERIC HISTORY OF LAMPROITE SETTINGS

Several features that preceded the formation and eruption of lamproite melts are considered relevant and will be discussed in this section. These include: the regional structural configuration, such as near-surface regional fault and fold patterns and basement structural fabric; the nature, evolution, and age of crystalline crustal basement rocks; the nature and occurrence of contemporaneous, preexisting, and subsequent nonlamproite magmatism in the region; the importance of paleosubduction and fossil Benioff zones; and the metasomatic evolution of the mantle source regions. Clearly many of these features are related and each has been separated only to aid discussion.

4.3.1. Regional Structure

Since many lamproite melts have ascended rapidly (if not explosively) to the surface from the upper mantle, one might expect that the surficial regional structural fabric would play at least a partial role in the localization and distribution of lamproite magmatic bodies within fields or provinces. Interestingly, the near-surface distribution of lamproites is only infrequently controlled by regional structural fabrics defined by fault and fold patterns, e.g., the Leucite Hills, Yellow Water Butte, Murcia-Almeria, West Kimberley, Argyle, and Bobi. In most cases the surficial structural fabric is not significantly reflected in the morphology or distribution of lamproite bodies. The majority of lamproites exhibit no obvious surface structural control.

Although lamproite emplacement does not appear to be related to near-surface structural elements, deeply penetrating structures are most important in guiding the ascent of magma from the upper mantle. Lamproites tend to occur near either regional structural boundaries, i.e., well-established zones of crustal weakness, or the intersections of prominent lineaments, which perhaps reflect deep-seated lithospheric structural discontinuities. In several lamproite fields, e.g., Kapamba, Prairie Creek, the structural control exhibited by the intrusive/extrusive bodies reflects the structural grain of the underlying basement rocks instead of the near-surface, commonly orthogonal, structural grain.

4.3.2. Basement Age, Composition, and Evolution

The vast majority of lamproites were intruded through, or extruded over, relatively undeformed flat-lying platform sedimentary rocks. These sedimentary platforms overlie crystalline basement rocks that generally are 0.5–2.5 g.y. older than the platform deposits. These shallow sedimentary sequences are considered to play a role only in near-surface intrusive and volcanological processes involved with the eruption of lamproites (see Chapter 5). The underlying crystalline basement, however, bears a much more fundamental relationship to the genesis of lamproites. The continental crust through which lamproites intrude is generally 35–55 km thick and the underlying ancient lithospheric mantle is generally thicker than 50–100 km. With the exception of the European occurrences, all lamproites are underlain by Archean ($n = 16$) or Proterozoic ($n = 5$) continental basement rocks, and presumably similarly ancient subcontinental mantle lithosphere. The Archean Wyoming craton of western North America contains the largest lamproite volumes and number of lamproite fields of all the continents. There is a wide range in the difference in age of the continental basement and the age of lamproite eruption, ranging from 1–1.5 Ga for the Proterozoic lamproites (Bobi, Chelima, Sisimiut, Argyle, Majhgawan) to 1.3–3.0 Ga for the Cenozoic, Mesozoic, and Paleozoic lamproites of North and South America, Africa, Australia, and Antarctica.

The composition of the crustal basement underlying lamproite provinces generally is typical of continental domains and characterized by granitic-to-intermediate compositions. The basement history of this continental crust is nevertheless complex. In the Wyoming craton, extensive crust-forming events have been documented in the Mid- to Late Archean (3.6–2.7 Ga), and mantle metasomatic events are thought to have occurred in Early Proterozoic times. At the same time (1.9–1.6 Ga) a major regional metamorphic event occurred in the overlying crust (Mueller *et al.* 1982, 1983, Wooden and Mueller 1988, Eggler *et al.* 1989). These times overlap the model lead and neodymium ages exhibited by the Smoky Butte and Leucite Hills lamproites (Fraser *et al.* 1985, Fraser 1987) suggesting a link between orogenic events involving continental crust separation from, and enrichment events in, the mantle lithospheric source regions of these two lamproite occurrences. Therefore, the evolution of mantle source regions of lamproites is closely related to the general evolution of the crust/mantle lithospheric system.

Although lamproites are posttectonic magmatic rocks, they occur in regions that have experienced intense lithospheric deformation or orogeny. Nearly all are either underlain by collisional orogenic basement and/or sutures, or aulacogens or failed rifts. In all known occurrences, lamproite magmatism postdates regional orogenic activity by 30 to >1000 m.y.

4.3.3. Contemporaneous, Previous, and Subsequent Magmatism

Lamproites typically occur in regions characterized by a wide variety of magmatic rocks erupted prior to, or subsequent to, lamproite emplacement. Compared with mafic alkaline magmas in general, most lamproites erupt in relatively isolated regions with the nearest contemporaneous magmatism being more than 500–1000 km away.

The Wyoming Craton and adjacent Colorado Plateau have been the sites of diverse mantle-derived alkaline magmatism including: Devonian kimberlites of the Colorado–Wyoming State Line District (McCallum et al. 1975, Smith 1977); Cretaceous shoshonites of the Eastern Absaroka Belt (Meen and Eggler 1987); Tertiary (mainly Eocene) alkalic magmatic rocks of the Montana petrographic province (ultramafic lamprophyres, mafic phonolites, syenites, alkali basalts, and related rocks, Larsen 1940, Hearn 1968, Marvin et al. 1980); and Eocene-to-Quaternary lamproites. The Wyoming Craton lamproites possess eruption ages of 1–55 Ma, most being younger than the widespread Eocene magmatism which was temporally associated with the Laramide orogeny in western North America. The oldest lamproites in the Wyoming province (Yellow Water Butte) overlap in age with widespread and voluminous Eocene back-arc magmatism of the Montana petrographic province, which also comprises ultramafic lamprophyres (alnöites, etc. of the Missouri Breaks and Haystack Butte areas) and kimberlites (Williams Pipe). Further to the west, extensive Eocene subduction-related calc-alkalic magmatism has been well documented (Lipman et al. 1981). However, most of the Wyoming craton lamproites (Smoky Butte, Francis, Leucite Hills) erupted long after termination the Laramide magmatism, contemporaneously with the eruption of immense volumes of the Basin and Range felsic volcanic piles to the south, and the flood basalts and felsic caldera eruptions of the Snake River Plain and Yellowstone area. These mafic and felsic magmas are generally regarded as lithospheric mantle melts and crustal melts, associated with an extensional ("back-arc") stress field, in the case of the Basin and Range melts, and a superimposed hot spot in the case of the Snake River Plain–Yellowstone melts. It is noteworthy that the Leucite Hills lamproites are the youngest, large-volume alkaline volcanic field to have erupted in the Wyoming Craton. The Yellowstone mantle plume is considered to play a fundamental role in the formation of the Leucite Hills lamproites (Section 4.2.4). On the basis of the unique trace element and isotope geochemistry of the Wyoming Province lamproites, relative to the Laramide and Post-Laramide nonlamproite magmatism, we infer distinct source regions for these contrasting melts. The tectonic framework of the lamproites may have been influenced by Laramide subduction; however, we view this influence as indirect at best. Presumably, the postsubduction extensional stress field, in combination with elevated thermal conditions needed to induce lithospheric mantle melting were both necessary conditions leading to the formation of the Wyoming Craton lamproites. These conditions were superimposed on a portion of the Wyoming Craton in which Archean-to-Proterozoic melting and orogenic events, and Phanerozoic rifting events, were extensive.

The Presidente Oligario lamproites are contemporaneous with a widespread and voluminous alkaline mafic magmatic province in Minas Gerais, and are one of the few *bona fide* occurrences closely associated in time and space with kimberlites and carbonatites. The Bobi lamproites are apparently contemporaneous with nearby kimberlite and alnöite dikes at Toubabouko. The Prairie Creek lamproites occur at the distal end of a 150-km linear belt of Late Cretaceous syenitic intrusives, lamprophyres, and carbonatites of the central Arkansas alkalic province. Although the rocks of this province are contemporaneous, they are not consanguineous and are undoubtedly derived from different mantle sources. The Kapamba lamproites occur in south-central East Africa, 200–500 km away from areas where Mesozoic kimberlites are abundant. Mesozoic syenites occur within 100 km of Kapamba. The apparent association of these lamproite fields with kimberlites and carbonatites suggests that differing mantle source regions are required to produce kimberlites, carbonatites, or lamproites. Low degrees of melting of large parts of the mantle are indicated by the voluminous and widespread nature of these alkaline provinces.

The European lamproites are exceptional in that the Murcia-Almeria and Sesia-Lanzo/Combin lamproite fields both occur as members of nearly contemporaneous shoshonitic and calc-alkaline magmatic suites. These magmatic provinces are formed dominantly from subduction-related and post-subduction-related melts, thus suggesting that the European lamproites may be intimately related to subduction processes. We prefer to view the European lamproites as exceptions, not only on the basis of tectonic framework, but also on the basis of their chemistry and mineralogy (see Chapters 6 and 7).

At the other end of the spectrum of tectonic environments are isolated lamproite bodies, fields, or provinces not associated with any nearby contemporaneous magmatism, e.g., Hills Pond/Rose Dome, West Kimberley, Argyle, Gaussberg, Chelima and Majhgawan. These lamproites were undoubtedly formed from unique mantle sources which were evidently not capable of producing other alkaline magmas. This was presumably due to the limited volume of lamproite source mantle available for melting, and the adjacent mantle being incapable of partial melting under the ambient thermal regime. Alternatively, highly focused thermal perturbations in the underlying mantle lithosphere may be responsible for these lamproites.

Some lamproite fields form distinct parts of complicated sequences of alkaline magma lineages spanning several billion years. The Sisimiut lamproites of West Greenland are apparently the only alkaline magmatic products emplaced in a 400-km^2 area at about 1.2 Ga. Prior to the eruption of lamproites, 2.7 Ga carbonatites were emplaced nearby at Qaqarssuk and Tupertalik and 1.8 Ga lamprophyres were emplaced at Sarfartoq. The same general vertical section of mantle lithosphere was subsequently capable of producing 0.6 Ga kimberlites and carbonatites and 0.1–0.2 Ga lamprophyres and carbonatites. These features support the view that distinct portions of the subcontinental lithospheric mantle, each with subtle differences in composition and/or metasomatic evolution, were subjected to melting events at distinct periods.

4.3.4. Importance of Paleosubduction and Fossil Benioff Zones

The location and compositional traits of lamproites suggest that they may represent partial melts of fossil Benioff Zones. Lamproites are located along the margins of Archean cratons in regions that have experienced lithospheric accretionary events or were most likely to have developed an Andean-type margin at some time in their most recent 1–3 g.y. of evolution. Many lamproites occur within Proterozoic or Early Phanerozoic mobile belts which were undoubtedly associated with paleosubduction or other collisional events.

Subduction-related arc rocks become progressively enriched in potassium with increasing distance from the plate boundary. This relationship has led many workers (Dickinson 1975, Beswick 1976) to speculate that a potassic phase such as phlogopite is preserved to great depths in the mantle and may persist as a residual phase. Paleosubduction zones or fossil Benioff zones (the mantle wedge overlying the downgoing slab) that have been accreted on the margins of cratons would be expected to be enriched in phlogopite (Wyllie and Sekine 1982), and exhibit depletions in incompatible major elements such as Na, Ca, and Al. These elements would have been extracted during previous partial melting events at a ridge crest or as the result of subduction. Importantly, present-day subduction-related arc rocks are depleted in Nb, Hf, Zr and Ti, elements that are particularly abundant in lamproites. These elements are generally thought to be held in residual phases during partial melting of the mantle wedge in an active Benioff zone (Green 1981, Green and Pearson 1986). Later partial melting of these titanates and niobates, perhaps during metasomatic modification of these regions, could lead to the formation of Ti-, Nb-, Hf-, and Zr-enriched melts. The predicted composition of fossil Benioff zones fulfills two important compositional features required of a lamproite source: enrichment in K, Nb, Ti, Zr and Hf; and depletion in Na, Ca and Al. Experimental studies have shown that phlogopite can persist to upper mantle depths of >100–150 km in reasonable silicate and volatile phase assemblages (Wyllie 1988, Eggler 1987). The predicted locations of paleosubduction zones are most likely to occur at the margins of cratons and in the mobile belts that weld cratons together. Therefore, our understanding of the source regions of lamproites is consistent with the inferred nature of fossil Benioff zones, and accounts for their occurrence at craton margins.

4.3.5. Mantle Metasomatism—Another Necessary Condition

During the last 15 years, mantle metasomatism has received considerable attention for explaining the geochemical and isotopic peculiarities of alkaline mafic rocks (Menzies and Hawksworth 1987). The process is supported by the recognition of mantle-derived xenoliths containing mineral assemblages that clearly have been superimposed upon preexisting lherzolite (Erlank *et al.* 1986). Many petrologists favor the hypothesis despite the general lack of a viable mechanism describing the origin and transport of material in the metasomatic fluid. Of all the mantle-derived alkaline rocks, lamproites are perhaps the most exotic with regard to their distinctive extreme enrichment in incompatible large ion lithophile and high field strength elements such as K, Ba, Sr, Rb, Y, Nb, Zr, P, etc. (see Chapter 7). The ancient Nd–Sr–Pb isotope model ages of most lamproites suggest that metasomatic events occurred in their mantle sources long ago.

Depletions in several incompatible elements are also important in understanding the metasomatic evolution of the source material of lamproites. Several elements depleted in lamproites are distinct from those in alkali basalts and many other alkalic magmas. Three elements, incompatible in a peridotitic mantle/partial melt system are Na, Ca and Al. They are generally enriched in alkali basaltic rocks, yet depleted in lamproites by more than a factor of 2. These elements may have been concentrated in residual phases or were not present in large quantities in the mantle source regions. We consider that the Na, Ca and Al depletions in lamproites reflect source depletions in these elements, and infer that the source was poor in clinopyroxene. To account for the strong enrichment in K, Rb, Sr and Ba in lamproites relative to alkali basalts, kimberlites, lamprophyres, and related rocks, it is necessary to postulate that these elements were added by metasomatism to a silicate phase assemblage with relatively low Na and Ca contents. Barring major mineralogical compositional control (i.e., K enrichment and Na depletion in phlogopite), in a geochemical sense Na should behave in a manner similar to K and Rb, whereas Ca should follow Sr and Ba. The apparent fractionation of K and Rb from Na, and Sr and Ba from Ca in lamproites, relative to most other alkalic melts, indicates that Na and Ca were not added in substantial quantities during metasomatism and were present in only minor amounts in the silicates comprising the source mantle.

The metasomatic alteration of lamproite mantle source regions could have originated from several processes or events. We infer that the metasomatism was episodic rather than continuous, because of the Nd and Sr isotope systematics summarized in Section 7.9. As the lithospheric mantle source is inferred to be situated at cratonic margins and within domains that have been subjected to several major tectonic events, we suggest that the metasomatism may have been associated with, or strongly affected by, rifting (failed or active), and/or collisional or subduction (orogenic) events. Schreyer *et al.* (1987) recently proposed an elegant model for the formation of ultrapotassic rocks (including lamproites) resulting from the interaction of deeply subducted crustal rocks with mantle materials. K feldspar and trioctahedral micas (phlogopite, biotite), which characterize typical continental crust assemblages (granites, felsic gneisses), are expected to develop a K- and Mg-rich hydrous fluid resulting from subduction to 50–70-km depths at temperatures of 450–650°C. Since these fluids are expected to be highly reactive and contain Ba, Ti, and other large-ion-lithophile elements, they would easily metasomatize any mantle peridotite they contacted. The resulting metasomatites would form an ideal source for lamproites, provided the rocks were subsequently heated to suprasolidus temperatures (800–1100°C), as the result of a thermal event (hot spot, etc.) or a change in geothermal gradient.

4.3.6. Comparison with Kimberlites, Alkali Basalts, Lamprophyres, and Potassium-Rich Rocks

Simple unifying models explaining the worldwide occurrence or tectonic framework of kimberlites, alkali basalts, lamprophyres, and related alkaline rocks do not exist. The formation of these magmas is recognized to be related to a complicated interplay of parameters including: the composition of the silicates and fluids in the mantle source and the long-term evolution of this material; the type and degree of partial melting and subsequent fractionation and contamination history of the melt; and the style of magma ascent, eruption, and emplacement history.

Lamproites form a unique group of alkaline rocks easily distinguished from all others, yet other rocks occur that are transitional between lamproites and alkali basalts, i.e. ultrapotassic and K-rich basanites from the eastern Sierra Nevada, California, minettes (Pendennis Point and Holmeade Farm, England) and group 2 kimberlites from South Africa. These transitional groups may be derived from sources intermediate in composition between those of particular "end-member" magmatic suites. These sources, however, appear to be generated and/or sampled infrequently and may represent metasomites of unique character (see Chapter 10). There does not appear to be a complete continuum of source compositions in the mantle.

In summary, there are several characteristics shared by all mantle-derived alkaline magmas, together with some distinctive features of lamproites, which are relevant to discussion of their tectonic settings:

1. All are melts of lithospheric or asthenospheric mantle that has been subjected to some type of metasomatic event. Lamproites are unique in their extreme enrichment in elements added by metasomatic alteration, their extreme fractionation of K and Rb relative to geochemically similar Na (and Sr and Ba relative to Ca), and their extreme radiogenic isotope (Nd–Sr) ratios reflecting variable time-integrated enrichments with Rb/Sr > bulk earth and Nd/Sm < bulk earth.

2. Lamproites are exclusively mantle lithospheric melts, in contrast to all the others which may be either lithospheric or asthenospheric melts.

3. All alkaline rock groups are apparently intimately related in their genesis to some

type of tensional lithospheric stress regime, ranging from the broad lithospheric swells of some kimberlite-bearing cratons, to the elongate trends of Cenozoic alkali basalt fields delineating back-arc environments around the Pacific rim.

4. Kimberlites and lamproites are the only major primary sources of diamond, indicating that perhaps only the members of these two groups are capable of ascending from >140-km depths to near-surface regimes in a manner capable of preserving diamond. Several alkali basalts and lamprophyres are reported to contain diamonds (Kaminskii 1984, Rock 1988, Haggerty and Nagieb 1989). Some of these occurrences are not well substantiated, but they do suggest that rocks other than lamproites and kimberlites may be sources of diamond.

4.4. TECTONIC FRAMEWORK OF FOUR MESOZOIC-TO-CENOZOIC LAMPROITE TYPE-LOCALITY MAGMATIC FIELDS

This section compares and contrasts the tectonic settings and histories of four of the most important "type-locality" lamproite fields. A more detailed discussion of the tectonic framework of each field is presented in Chapter 3. Each type locality is seen to possess distinct peculiarities with respect to age and tectonic framework. These localities illustrate the difficulties in deriving a general model of the tectonic controls on lamproite emplacement.

4.4.1. Leucite Hills

The Leucite Hills lamproite field consists of over 20 volcanic cones, lava flows, and hypabyssal complexes which were erupted through Upper Cretaceous-to-Eocene sedimentary deposits lying upon Archean basement crystalline rocks. It is located over 1000 km east of the nearest plate boundary, the dextral strike slip margin separating the Pacific and North American Plates. Prior to 30 Ma the margin was a convergent one. The Leucite Hills field is located at the northern nose of the Late Cretaceous–Paleocene Rock Springs Uplift, near the southern boundary of the Wyoming Craton just north of the Colorado Plateau, and east of the Basin and Range Province. The lamproite field is located 200–300 km south-southeast of the 1 Ma–present position of the Yellowstone hot spot, which has erupted several thousand cubic kilometers of tholeiitic basaltic lava over the last 8 m.y. Lamproite eruption ages are tightly clustered around 1–3 Ma and historic volcanic activity has not been recognized. Cenozoic magmatism in western North America includes: extensive 50–80 Ma calc-alkaline to alkaline subduction and back-arc magmatism related to Laramide subduction and orogeny; 20–40 Ma dominantly silicic rocks associated with a Mid-Tertiary orogeny; and Late Tertiary basaltic-to-felsic magmatism associated with Basin and Range extension and hot spots. The lamproites are postorogenic and are located in a continental back-arc tectonic position.

4.4.2. West Kimberley

Over a hundred subvolcanic and intrusive complexes form four lamproite fields comprising the West Kimberley lamproite province. The rocks are emplaced in a variety of lithologies ranging from Permo-Triassic sedimentary deposits to Archean-Proterozoic crystalline rocks. Lamproites were emplaced from 17 to 24 Ma and represent the only Mid- to

Late-Tertiary magmatism within a radius of over 1000 km. The West Kimberley Province is located > 1000 km from the nearest plate boundary. The lamproites intruded the Paleozoic Fitzroy Graben near its northern margin, overlapping and just south of the Proterozoic King Leopold Mobile Belt. This province is the type example of lamproite intrusion into a mobile belt adjacent to a craton.

4.4.3. Murcia-Almeria

More than ten Late Miocene lamproite volcanic centers, occurring in the Alpine orogenic belt of southeastern Spain, were erupted about 8 Ma as a posttectonic magmatic phase of 5 m.y. duration. The lamproites were emplaced in Cenozoic and Mesozoic sedimentary rocks, overlying Paleozoic crystalline rocks affected by Hercynian metamorphism. Well-developed structural dislocations and lineaments, orthogonal to the Alpine belt, occur in the area. Typical subduction-related rocks (calc-alkaline and shoshonitic) are widespread in the region; however, these magmas predate the lamproites by 5–25 m.y. Plio-Pleistocene alkali basalts form the youngest vestiges of volcanism in the region. The lamproites were erupted 30–40 m.y. after a major lithospheric thickening event, at about the time a north–south directed convergent stress field changed to one dominated by extension, resulting from the collapse of an overly thickened orogen. The time of lamproite eruption coincided with widespread regional extension and normal faulting, including such events as the extensional uplift, which transported mantle core complexes (e.g., the Rhonda peridotite massif) into the upper crust.

4.4.4. Prairie Creek

At least six Mid-Cretaceous lamproite intrusive/extrusive complexes of the Prairie Creek field form the southwestern limit of the Arkansas alkaline province (Mid-Cretaceous nepheline syenites, carbonatites, lamprophyres, etc.). The lamproites were emplaced in Cretaceous to Carboniferous sedimentary rocks overlying Proterozoic crystalline basement. The lamproites are located approximately at the intersection of the Reelfoot Rift and the Ouachita orogenic belt in a region which has experienced multiple convergent and divergent tectonic events from the Proterozoic to the present.

4.5. CONCLUSIONS AND PREFERRED MODEL

The emplacement of lamproites in a wide variety of tectonic settings precludes the development of a universal model explaining their temporal, geologic, and tectonic position. Nevertheless, there are several features shared by most, if not all, lamproites. These are as follows:

1. Lamproites are extremely rare products of melting of a geochemically exceptional lithospheric mantle source in a continental intraplate setting.
2. Lamproite fields and provinces are formed during a relatively short-lived magmatic episode (<3–10 m.y. in duration) and are posttectonic magmatic products.
3. Lamproites have erupted during the last 1.4 Ga of earth history. The vast majority are Cenozoic, demonstrating the selective preservation of young volcanoplutonic complexes in the geologic record.

4. Although most lamproites are not directly related to active subduction or rifting events, both processes have generally affected the crustal and mantle basement regions through which they have erupted.
5. Lamproites occur along the margins of cratons or in accreted ancient mobile belts in regions of thick crust (>40–55 km) and thick lithosphere (>150–200 km). This lithosphere typically records multiple episodes of resurgent tectonic events (both extensional and compressional), some of which possess metamorphic ages coincident with lamproite Nd–Sr model ages.
6. Many lamproites occur along continent-scale dislocations or lineaments. Ancient zones of structural weakness have been important in localizing their near-surface emplacement or subaerial eruption.

The model of continental extension of oceanic transform faults is not considered to be an especially important tectonic control on the distribution of lamproite fields, although several may be broadly related to such extensions. Mantle plumes are considered important in providing the thermal conditions necessary for mantle melting events, and for the chemical modification of the mantle lithospheric source of lamproites. Hence, lamproites, or their sources, generated in the peripheral regions of hot spots, may be considered as an indirect form of hot spot magmatism. Paleo-Benioff zones are considered excellent candidates for representing the nature of the mantle lithospheric source of lamproites. The extensional collapse of several Cenozoic orogenic zones is considered to be a fundamental process in lamproite genesis, ascent, and emplacement at or near the surface. Mantle metasomatism of a peculiar type (mainly ancient, >1–2 Ga) has affected the lithospheric source regions of lamproites. These above generalities are clearly oversimplifications and will require revision in the light of further studies of the nature of lamproite emplacement and genesis.

On me dit que là-bas les plages sont noires
De la lave allée à la mer
Et se déroulement au pied d'un immense pic fumant de
  neige
                    André Breton

Petrological Facies and Igneous Forms of the Lamproite Clan

5.1. INTRODUCTION

The diversity of form and facies displayed by lamproites is virtually unparalleled by other mafic to intermediate igneous rocks. Lamproites occur in nearly all major groups of intrusive and extrusive igneous forms and in a wide spectrum of facies. The latter are dominated by volcanic and near-surface facies, but include several deeper-seated facies. Volcanic eruptive mechanisms are similarly variable and include: relatively quiescent magmatic eruptions of lava flows; dominantly phreatomagmatic eruptions characterized by pyroclastic fall, flow, and surge deposits; and rare phreatic eruptions such as those found at Mt. Abbott (West Kimberley).

Lamproite magmas begin their rapid ascent from the upper mantle at great depths (as deep as >150 km) and are emplaced or erupted as essentially undifferentiated volatile-rich magmas. It is the deep-seated origin and volatile-rich nature of lamproite magmas that results in a variety of eruptive mechanisms and textural types.

As expected, the geology of the most recently erupted lamproites provides us with an opportunity for understanding their eruptive mechanisms. Recent (< 2 Ma) lamproites are known to form a subglacial cinder cone composed of pillow lavas and bedded scoriaceous deposits at Gaussberg, and scoriaceous cinder cone fields with associated lava flows and pipelike agglomeratic vents in the Leucite Hills. Fortunately, in many of the Tertiary and older volcanic complexes, shallow and surficial pyroclastics are preserved, e.g., Murcia-Almeria and Argyle. The older lamproites allow characterization of the feeder systems of lamproite volcanoes, including the economically important diamondiferous vents. Uplift and erosion has exposed the lower parts of lamproite lava flow complexes, e.g., the lava lakes found at Mt. North and Cancarix, or the deeper parts of crater facies deposits, e.g., Argyle and Majhgawan.

Lamproites display a wide variety of igneous textures, ranging from holocrystalline to vitric, equigranular to porphyritic, vesicular to dense, and massive to fragmental. Magmatic (intrusive) and volcanic breccias are ubiquitous. Glass-rich lamproites are common and have been recognized in eight lamproite provinces or fields. As lamproite magmas are enriched in volatiles (H_2O, F), vesicular magmatic rocks are characteristic of all but the deepest hypabyssal intrusives.

The purposes of this chapter are to describe and summarize the variety of igneous forms and lithological facies exhibited by lamproites; to summarize important petrographic and textural features of the most common lithologies; to outline several important features related to lamproite eruption mechanisms; and to propose a genetic facies classification scheme consistent with our present understanding of the eruption dynamics of lamproites.

5.1.1. Historical Development

Rocks now known to be lamproites have been placed historically into one of five major petrologic groups: lamproites; leucitites or leucite basalts; lamprophyres; Roman province type volcanic suites; and mica peridotites. Since each of these compositional groups possesses a set of distinct petrologic facies, the development of a unified lamproite facies model was not possible until recent revisions to nomenclature permitted recognition of a distinct lamproite clan.

Tröger (1935) first suggested that lamproites represented the extrusive equivalent of lamprophyres enriched in potassium and magnesium (essentially the minette group), and thus gave the clan a facies and compositional restriction. Lamproites may slightly overlap some minettes in their mineralogy or mineral and bulk rock chemistry (see Chapter 6 and 7). However, the two groups are clearly not heteromorphs, as demonstrated by the absence of volcanic–hypabyssal complexes containing both varieties of rock. Those lamproites and minettes that are similar in their gross petrographic mode differ in detail with regard to their assemblage of accessory minerals and in the compositional trends exhibited by major phases (see Chapters 2 and 6). Many lamproite complexes are dominated by intrusive rocks and, according to Tröger's (1935) concept, should not be termed lamproite. Several extrusive minette lavas are known (e.g., Buell Park and Colima Graben) that are distinct from lamproites in mineral and bulk rock chemistry. These occurrences clearly show that lamproites and minettes are not derived from the same parental magma. Thus, Tröger (1935) was correct only in recognizing the importance of extrusive processes in lamproite genesis. N.B. Rock (1987, 1989) persists in including lamproites in the lamprophyre clan. Justification for separating lamproites from lamprophyres is presented in Chapter 2 and discussed further in Chapter 10.

Recently, Middlemost (1975, 1985) has included lamproites in the basalt clan, because of their broadly mafic compositions and similarities in igneous forms—e.g., cinder cones in the Leucite Hills are identical to typical basaltic cinder cones in morphology. Common volcanological features are, however, insufficient petrological grounds for grouping lamproites with the basalt clan. More importantly, lamproites differ from basalts with respect to their significant compositional (mineralogical and chemical) distinctions; difference in lithospheric mantle source; and differences in primary melt compositions.

5.1.2. Lamproite Igneous Forms

Nearly all intrusive bodies of lamproite now exposed at the surface were emplaced in an epizonal environment, where sharp, discordant contacts with country rocks are observed. Mesozonal or catazonal rocks have not been recognized, although it is possible that

MARID-suite (Dawson and Smith 1977), richterite-phlogopite-bearing metasomatized lherzolite and glimmerite (Jones *et al.* 1982, Erlank *et al.* 1986) xenoliths found in south African kimberlites represent deep-seated lamproite melt/fluid products which crystallized in the upper mantle (Bergman 1987, p. 170; Waters 1987, see Chapter 10). Lamproite intrusions occur dominantly as dikes, plugs, necks, or sills, and rarely as small plutons. Owing to their typically small volumes, they do not form batholiths, lacoliths, lopoliths, or large stocks. Dike swarms are commonly composed of subparallel dikes, suggesting that their emplacement was controlled by a regional stress field or the structural fabric of the country rocks. Cone sheets and ring dikes of the type found associated with nepheline syenite–carbonatite complexes have not been recognized. Their absence is probably due to a lack of shallow, large-volume evolving magma chambers under the volcanic vents.

The near-surface emplacement and subaerial eruption of lamproite magmas produces a wide variety of extrusive or volcanic forms. Extrusive complexes built by relatively quiescent eruptions, include lava flows, lava lakes, lava domes, cinder cones, lava cones, and cinder sheets. Vents resulting from more explosive eruptions contain a variety of pyroclastic rocks. These include tuffs, lapilli tuffs, and tuff breccias, which may result from pyroclastic fall and flow or base surge processes. Intrusive (contact) tuff breccias and autolithic breccias are common. Some lamproites contain epiclastic deposits resulting from the reworking of the above lithologies by fluvial, lacustrine, or other surface processes.

Lamproites do not form stratovolcanoes or calderas due to the relatively small volumes of magma involved in their extrusion compared with more common eruptive rocks. Consequently, it is expected that lamproite ash clouds will be much smaller than other ash clouds. In addition, lamproite eruptive sequences are expected to exhibit more complexity over short distances owing to their small volumes. The presence of diamond-bearing lithic tuffs and base surge deposits indicates that many lamproite eruptions are undoubtedly highly explosive. Diamonds are derived from depths greater than 150 km, and are unstable in an oxidizing environment. Their presence attests to a rapid ascent from the upper mantle, culminating in a violently explosive surface eruption. Unfortunately, since lamproite eruptions are so rare in the geological record and have never been observed, it not possible to describe completely their eruption history. Characterization of their eruptive style is based on the evidence preserved in eroded sequences.

5.1.3. Lamproite Facies Classification

The textural and morphological variety of lamproite igneous bodies and the processes responsible for their formation necessitate a rather complex facies classification. Whereas kimberlites can be neatly placed in crater, diatreme, and hypabyssal facies (Clement and Skinner 1979, Mitchell 1986), and continental basalts into shield volcano, cinder cone, maar and tuff ring, and flood or valley-fill lava facies (Cas and Wright 1987), lamproites form facies that may be placed in categories belonging to both of these groups.

We propose dividing lamproites into four main facies groups: lavas; crater and pyroclastic; hypabyssal; and plutonic. These facies are schematically illustrated in Figure 5.1. Highly idealized facies diagrams for kimberlites and alkali basalts are shown in Figures 5.2 and 5.3, respectively, for comparison. The kimberlite crater facies includes lavas, pyroclastic rocks, and epiclastic rocks (Mitchell 1986). We suggest separating lamproite lavas from the lamproite crater facies because of the importance or dominance of these two facies in lamproites relative to kimberlites.

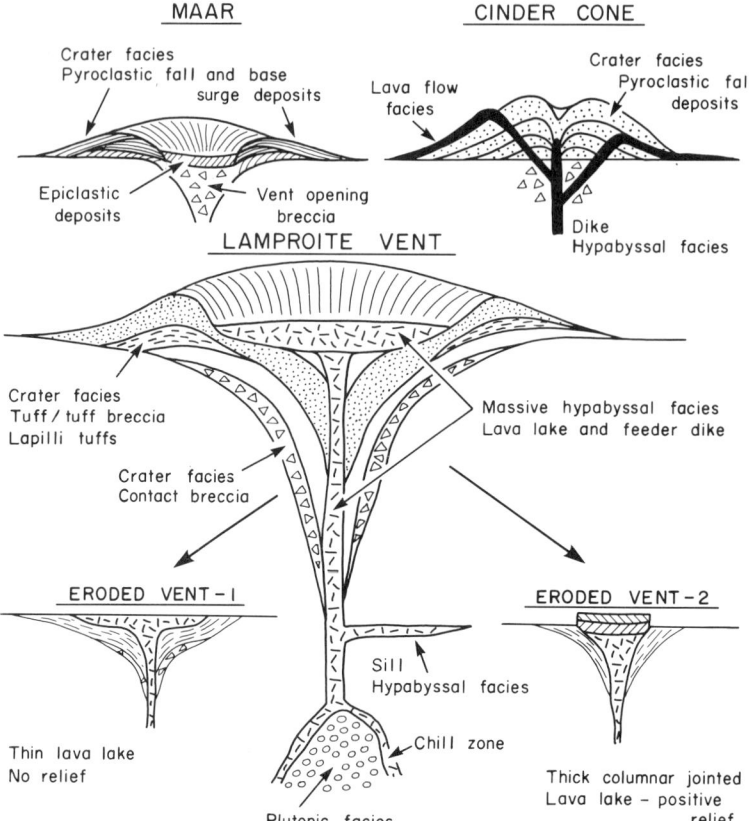

Figure 5.1. Schematic diagram (not to scale) illustrating the idealized morphological relationships of lava, crater or pyroclastic, hypabyssal and plutonic facies lamproites in several idealized vent complexes. Also illustrated are the apperances of lamproite vents after erosion of the maar crater. Depth of a typical lamproite vent is on the order of 0.5–1.0 km.

5.2. LAVA FLOW FACIES

Many well-developed lavas flows are preserved in Cenozoic lamproite provinces, thus permitting comparison with other mafic lavas. Lavas form where eruption rates are high and fragmentation rates low. In common with basaltic eruptions, a given lamproite eruptive center may experience an early stage of fragmental (pyroclastic) eruption forming dominantly tuffaceous rocks, followed by the later effusion of massive, typically vesicular, lavas. Depending on vent geometry, it is also possible to form lava domes, lava ponds, and/or lava lakes, where effusion rates are sufficiently high and topographic closure is developed.

Lamproite lava flows are found at the Leucite Hills (Kemp and Knight 1903, Ogden 1979), Gaussberg (Vyalov and Sobolev 1959, Tingey *et al*. 1983), and rarely (Barqueros) in the Murcia-Almeria province (Fuster *et al*. 1967). Lava domes and/or ponds/lakes are thought to occur at Mt. North (West Kimberley, Jaques *et al*. 1986) and at Cancarix (Murcia-Almeria, this work). Flow types are variable, precluding development of a simple model of the morphology of lamproite lava flows. Massive, vesicular flows with autobreccias, and

PETROLOGICAL FACIES AND IGNEOUS FORMS OF THE LAMPROITE CLAN 129

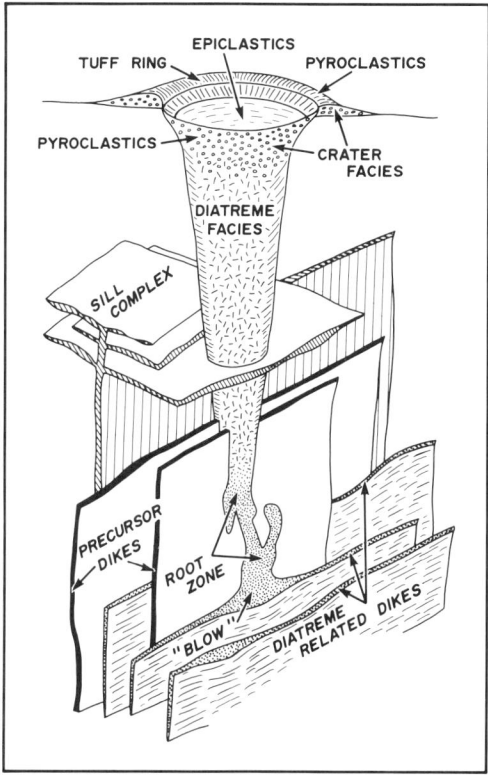

Figure 5.2. Schematic diagram showing the facies relationships of kimberlites (after Mitchell 1986).

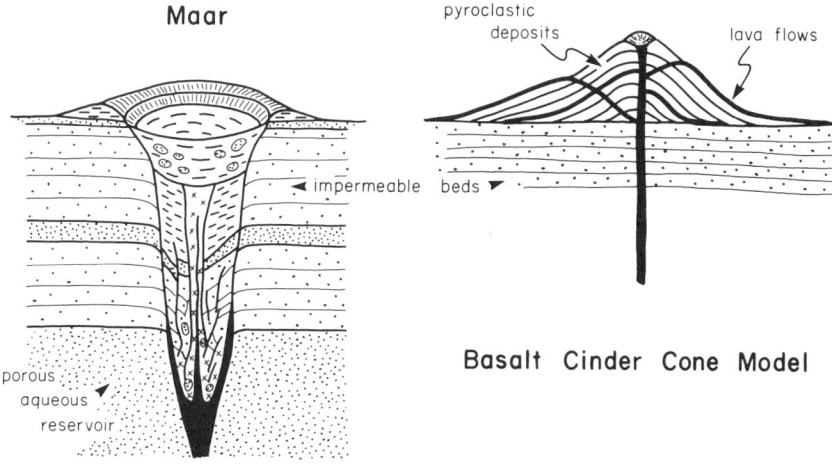

Figure 5.3. Schematic diagram illustrating the facies relationships of alkali basalt vent complexes.

flow-induced rubble and flow breccias are commonly present. Each of the four principal lava flow surface types—pahoehoe, aa, block, and pillow (Williams and McBirney 1979)—are developed to varying degrees in lamproites.

Many lamproite lavas are quenched upon eruption and form glassy-to-aphanitic (yet typically porphyritic) rocks. More voluminous lava lakes or domes cool slowly enough to form coarser-grained, holocrystalline lithologies. Lamproite lavas are characteristically vesicular. A wide range of vesicle morphologies is exhibited by different lavas, ranging from equidimensional and spherical, to varidimensional and deformed. Flow alignment of platy phlogopite phenocrysts is a characteristic feature of massive and vesicular lavas.

Individual lava flows include simple and compound types (terminology of Williams and McBirney 1979). The flows are typically limited in area (about 0.005–5 km^2 each), and relatively thin (about 1–20 m) compared to most mafic lavas (Ogden 1979). The lavas display a wide range of fluidity and degree of supercooling. All are thought to be products of subaerial (or subglacial) eruption. Lamproite hyaloclastites have not been recognized, with the possible exception of Gaussberg. Lamproite lavas are dominantly vesicular (15%–40%) and flow banded. They possess flat flow fronts with abundant flow-surface and flow-front autobreccias, basal rubble zones, flow levees, and squeeze-up spines. Lava tubes are rarely observed in the Leucite Hills (Ogden 1979). Compound lavas are typically 5 m thick and comprise several (commonly 3–5) massive vesicular flow units with interflow rubble zones. Red oxidized flow tops are observed in the lavas forming Steamboat Mountain (Leucite Hills). Figures 5.4–5.7 illustrate the diversity of lamproite lavas.

5.2.1. Examples of Lava Flow Facies Lamproites

5.2.1.1. Leucite Hills

The best preserved lamproite lava flows are found in the Leucite Hills province (Kemp and Knight 1903, Ogden 1979). They are the most voluminous of several associated volcanic forms present in this province. As much as 50% of the originally erupted flows and cones have been eroded in the last 1–2 Ma (Ogden 1979). Between 30–40 lava flow units were erupted from at least 11 vents. Lavas flows, cones, plug domes, and rings form multiple occurrences. Leucite Hills lava flows are limited in areal extent compared with other mafic to intermediate flows. The flows include both simple (dominant) and complex (minor) types. They range from 20 cm to 40 m in thickness (mostly 1–25 m), 0.005–5 km^2 in area (average 0.7 km^2), and 0.00003–0.05 km^3 in volume (average 0.01 km^3).

The best example of a compound lava flow (Steamboat Mountain) consists of a 27-m section composed of 6 individual 3–6 m-thick flow units sandwiched by 2–3 m-thick rubble zones without interflow soil development (Ogden 1979). Elsewhere in the Leucite Hills, typical individual simple flows contain a basal rubble zone of angular vesicular to massive blocks forming an autobreccia, a thin platy intermediate zone with flow foliated phlogopites, and a massive middle zone with moderate vesiculation and orthogonal cooling joints. Flow tops, where preserved, are scoriaceous, and squeeze-up spines protrude less than 1 m above the preserved flow surfaces (Ogden 1979). Concentric flow ridges, similar to those found in alkali basalt flows, occur at some localities (e.g., North Table Mountain). The lava flows are thought to have advanced continuously as broad fronts, and rarely in tubelike conduits (Ogden 1979). Lavas also form shieldlike symmetrical lava cones (with subordinate pyroclastic debris) at four vents at Zirkel Mesa and Steamboat Mountain. These lava cones

Figure 5.4. Zirkel Mesa, Leucite Hills. View, looking east, from the west end of the mesa, showing blocky lava flow and cinder/lava cones of phlogopite lamproite.

are 0.7–1.6 km in diameter and 75–100 m in height. Lava rings form 200-m-diameter doughnut-shaped lobes or levees, in the center of which is a depression thought to have been formed by the draining back of lava into the vent (Ogden 1979).

One important feature of the Leucite Hills lava flows that has yet to be satisfactorily explained in detail is the presence of alternating centimeter scale bands of vesicle-rich sanidine phlogopite lamproite and vesicle-poor leucite phlogopite lamproite. Examples of this phenomenon are well displayed at North Table Mountain. Carmichael (1967) and Gunter *et al.* (1983) proposed a shear mechanism associated with flowage to explain the contrasting mineralogy of these chemically identical rocks. A slight difference in volatile (water) content would explain the contrasting stabilities of sanidine and leucite at pressures approaching one atmosphere. Differences in dissolved water would produce contrasting viscosities and therefore contrasting flow behavior for the two bands.

5.2.1.2. Gaussberg

The Gaussberg nunatak is composed of glass-rich (50%–65%) olivine diopside leucite phlogopite lamproite lavas forming a 373 m-high, 1.5 km-diameter, symmetrical cone. Subglacial fragmented lavas, composed of scoriaceous lapilli, blocks, and bombs, and massive pillowed pahoehoe lavas, form short, stubby, vesicular flows with palagonite rims (Tingey *et al.* 1983). The bulk of the Gaussberg cone is constructed of monotonous pillow-

Figure 5.5. Lava flows forming South (S), Middle (M) and North (N) Table Mountains, Leucite Hills. View, looking northwest, from Hague View, looking northwest, from Hague Hill. The dike that outcrops as Iddings (I) and Weed (W) Buttes is visible at lower right.

Figure 5.6. Columnar jointed transitional madupitic lamproite overlain by blockly phlogopite lamproite. Middle Table Mountain, Leucite Hills.

Figure 5.7a and b. Gaussberg volcano, Antarctica (a) subglacial pillows (b) photomicrograph of hyalo-diopside leucite lamproite (field of view = 4.0 cm).

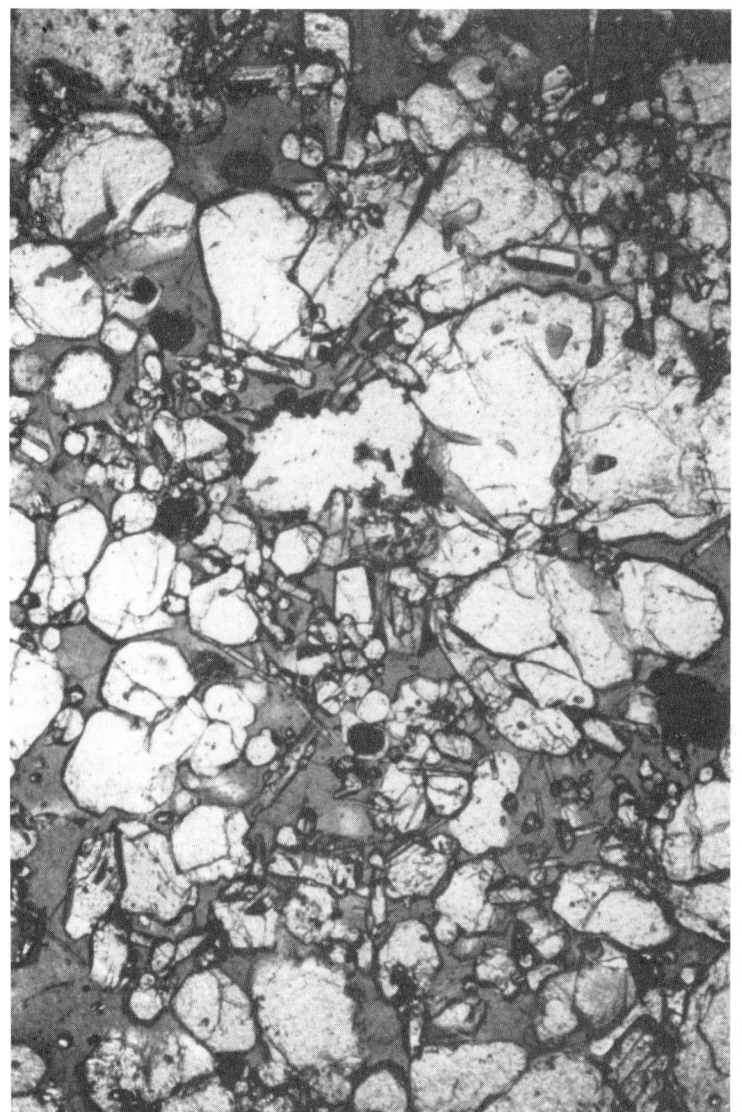

Figure 5.7a and b. (Continued)

shaped tongues of vesicular pahoehoe lavas (Figure 5.7), which led Tingey *et al.* (1983) to suggest that it was formed by a single eruptive episode. Pillow lavas, blocks, and bombs possess a crust with irregular shrinkage cracks. Individual pillows are typically 0.5–2.0 m across and have a thin (3–5 cm) glassy crust with a ropey surface (Sheraton and Cundari 1980). Tuffs and cinder beds rarely occur. Although many workers restrict pillow lavas to submarine environments, the uniformity of vesicle shapes, sizes, and abundances, from the top of the cone to its base, supports an origin by a nonsubmarine subaqueous eruption (Tingey *et al.* 1983). Native sulfur is abundant in veins, vesicles, and inclusions (Vyalov and Sobolev 1958), suggesting a S-bearing hydrous volatile phase was exsolving and contributing to the vesicles during eruption.

5.2.1.3. West Kimberley

As denudation of the West Kimberley province has been rather extensive (in excess of 50–100 m) since the Early Miocene, only remnants of lavas are preserved in the central cores of vent complexes. Massive lamproite lava domes or lakes are thought to occur at Mount North (Jaques *et al.* 1986). Here, two massive 20–30-m-thick, columnar-jointed lavas, occur composed of nonvesicular, medium-grained phlogopite lamproite with flow-foliated phlogopite phenocrysts. Columns are vertical and range from 0.3 to 2 m in diameter, indicating varying degrees of slow cooling for the distinct units. These columnar-jointed flows are continuously exposed on the cliffs surrounding Mount North and intruded by thin sills and dikes.

5.2.1.4. Smoky Butte

The >3-km-long Smoky Butte "dikelike" vent complex (typically 1–10 m wide) develops into a 300-m-long by 100-m-wide "swell" at the Smoky Butte Quarry. Here, a shallow "intrusive" is comprised of vitric and fine-grained columnar-jointed lamproite capped by coarser-grained phlogopite lamproite. It is possible that these units formed a lava blister or the lower portion of a lava dome or pond. The columnar joints (20–50 cm in diameter) are curved, vertical in the core and nearly horizontal at the margins. These massive facies are bounded on the margin by an earlier-formed pyroclastic sequence of intrusive breccias, agglutinate tuffs, and tuff breccias. The shallow level of the preserved rocks at Smoky Butte and the relatively large volumes of lamproite within the central dome suggest that lavas were erupted from this vent. These have now been eroded and the occurrence of rare blocks of pumice is the only testament to their former presence.

5.2.2. Comparisons with other Mafic and Intermediate Lavas

Lamproite lavas are morphologically identical to those found in many alkali basaltic lava flow complexes, suggesting that the main dynamic processes of lamproite eruption are broadly similar to those of alkali basalts. As the eruption dynamics are controlled by variables such as viscosity, eruption rate, volatile exsolution rate, and vent geometry, we infer that these variables are similar for both varieties of lava flow. Unfortunately, few of these variables have been characterized for lamproites because of the absence of any current eruptions. The viscosity of lamproitic silicate melts has yet to be determined experimentally. It is predicted that lamproites will possess lower melt viscosities than alkali basalt melts. This hypothesis is based on the fact that lamproite melts will have lower contents of network-forming cations or molecules (CO_2) and total tetrahedral cations (Si + Al) and much higher contents of network-breaking anions, cations, or molecules (K, H_2O, F, Cl). It is possible to estimate lamproite melt viscosity using empirical relationships established between composition and viscosity, assuming reasonable primary volatile contents. On the basis of relationships developed by Shaw (1972) and McBirney and Murase (1984), we infer that supraliquidus viscosities for madupitic to leucite lamproite compositions will be in the range 10^{0-3} poise for a crystal- and vesicle-free melt with 2–5 wt % dissolved water. The effect of increasing the dissolved water is counterbalanced by the effect of adding crystals. The addition of small amounts of vesicles would be expected to decrease melt viscosity, therefore the natural (vesicular) supraliquidus melt viscosities of lamproites are probably lower than those of alkali basalts. This lower viscosity has a variety of implications regarding eruption and lava flow dynamics. For example, lamproite lavas should be thinner (for a given length) than alkali basalt lavas erupted from vents with similar geometries and effusion rates. Field observations at the Leucite Hills show that thin long flows are typical of lamproites. Theoretically, lamproites are expected to be less capable than alkali basalts in transporting dense mantle xenoliths to the surface. As predicted, mantle xenoliths in lamproites are generally small and rare (see Chapter 9) compared to their ubiquity in alkali basalts (Nixon 1987).

5.2.3. Summary

Lamproite lavas are essentially similar to other mafic-to-intermediate lavas in morphology and variety. Lamproite lava flows exist in all of the major flow morphologies (pahoehoe, aa, pillow, and block), and lava tubes have been recognized. Subglacial lamproite pillow lavas occur. Most lavas are vesicular with both spherical (indicating postemplacement boiling) and elongate/deformed (indicating pre- or synemplacement boiling) vesicles. Nevertheless, much remains to be understood about the dynamics of these lavas as systematic studies are lacking. Basic data as to their viscosity as a function of temperature, pressure, crystallinity, and vesicularity are required. Until the effects of the solubility of H_2O, HF, SO_2, and CO_2 in lamproite melts, and the pre- and posteruption volatile compositions of lamproite magmas are determined, the volcanological dynamics of lamproite lavas will remain enigmatic.

5.3. CRATER AND PYROCLASTIC FACIES

Crater and pyroclastic facies lamproites are the most important and texturally variable of all lamproite facies. Diamondiferous lamproite vents belong to the crater facies. In such vents, pyroclastic or fragmental rocks have typically the highest diamond grade relative to associated magmatic rocks. Thus, a thorough understanding of crater facies lamproites is of economic significance (see Chapter 9). The abundance of pyroclastic lithologies attests to an explosive eruption stage in the evolution of lamproite vent complexes. We interpret these eruptions as resulting from phreatomagmatic and phreatic activity involving two major processes: the late-stage, near-surface exsolution of a water-dominated volatile phase; and the interaction of magma with near-surface meteoric water. Both processes are capable of pulverizing and fragmenting not only the erupting magma, but also large volumes of country rocks and unconsolidated surface sediments.

Crater facies lamproites comprise several subtypes: (1) intrusive contact breccias and autolithic breccias; (2) lapilli tuffs and crystal-lithic tuffs; (3) agglutinate tuffs, lapillistones, and cinder deposits; (4) epiclastic deposits. Facies (1–3) are thought to have originated by base surge, pyroclastic flow and pyroclastic fall processes.

Crater facies lithologies are present in the majority of lamproite occurrences, fields, or provinces. The best examples can be found at Prairie Creek, West Kimberley, Argyle, Kapamba, Majhgawan, Smoky Butte, Yellow Water Butte, Leucite Hills, and Murcia-Almeria. An idealized model, showing the geometry of a lamproite vent and the distribution of pyroclastic (crater facies) lithologies relative to more massive lithologies, is shown in Figure 5.1. Crater facies lithologies generally form early in the eruptive evolution of lamproite vents. Hence, they are typically marginally distributed and intruded by massive magmatic (hypabyssal) phases. In the vent complex proper, some proximal crater facies lithologies typically form bedded deposits which dip shallowly to moderately (most about 20–40°, depending on the degree of erosion) toward the interior of the vent. Distal pyroclastic lithologies (exterior to the vent proper) and proximal epiclastic deposits form horizontal bedded units.

5.3.1. Vent Morphology

Lamproite vents commonly have the shape of a champagne glass or funnel, with 0.5–1 km of vertical flaring (Jaques *et al.* 1986, Bergman 1987, p. 142). This structure (Figure 5.1) is in marked contrast to the carrot shape of typical kimberlite diatremes with 2–3 km or more of vertical flaring (Mitchell 1986). Lamproite vents possess relatively gentle marginal dips averaging 30° (typical range = 25–60°), compared with steep margins of kimberlite diatremes which average 82° (Mitchell 1986). The uniform shape of lamproite vents indicates that a violent and explosive phase of eruption takes place at relatively shallow depths. The difference in shape between kimberlite diatremes and lamproite vents is not controlled by differences in denudation or age, because even old or denuded lamproite vents exhibit a champagne glass shape.

The plan sections of lamproite vents (Figure 5.8) are dominantly elongate-to-subcircular in shape, with average length/width ratios of 1–2. In this respect they are similar to cross sections of kimberlite diatremes, although some lamproite vents (e.g., Argyle, Ellendale 9, Black Lick/American Mine) are extremely elongate and irregular in plan view (with length/width ratios of 2–5). The elongate morphology of such vents indicates significant near-surface structural control of emplacement by strongly anisotropic country rocks. The size of lamproite vents varies widely from 0.1 to 2 km (average 0.5 km) in diameter or maximum dimension. On average, lamproite vents possess larger cross-sectional areas than kimberlites. However, the area of plan sections of lamproite vents is a more important function of depth than that of kimberlite diatremes, because of the shallower marginal dips of lamproite vents. In terms of area, the largest lamproite vents are from the Ellendale field (Ellendale pipes 4 and 6, Calwynyardah), Argyle, P1 and P2 from the Kapamba field, and the Crater of Diamonds and Black Lick/American Mine from the Prairie Creek field. The greatest plan section areas of lamproite vents are on the order of 2–3 km^2.

The marginal and internal contacts of lamproite vents characteristically consist of heterolithic breccias composed of lamproite pyroclasts and autoclasts, together with country rock xenoliths, set in a contaminated, altered tuffaceous groundmass enriched in lamproite magmatic components. Autolithic breccias are poor in country rock clasts, and are dominated by lamproite pyroclasts and autoclasts set in an altered tuffaceous matrix. These

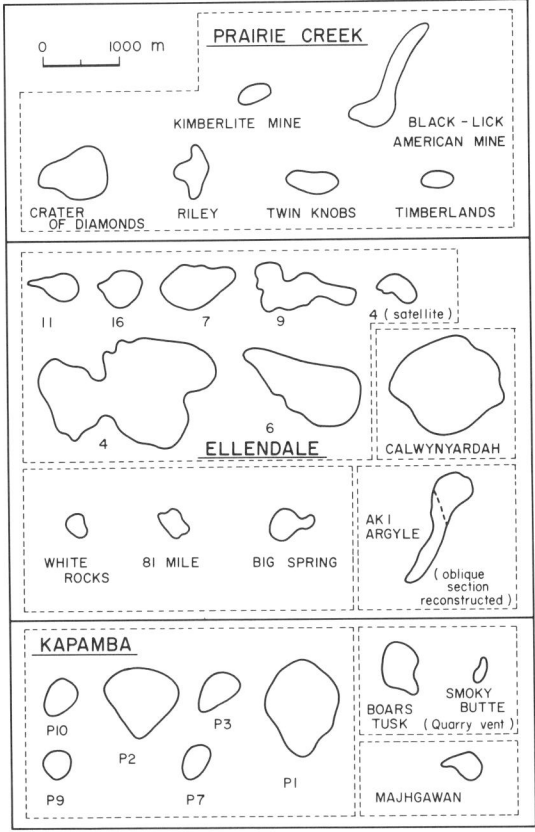

Figure 5.8. Plan cross sections of some typical lamproite vents.

fragmental lithologies result from either the explosive intrusion and extrusion of volatile-supersaturated lamproite magmas which induces autofragmentation, and/or the phreatic fragmentation of magmatic debris. Examples of contact breccias and autolithic breccias are shown in Figure 5.9.

5.3.2. Pyroclastic Fall Deposits

5.3.2.1. Agglutinate Tuffs, Lapillistones, and Cinder Deposits

Proximal agglutinate tuffs (piperno) and associated lithologies have been recognized from Smoky Butte, the Leucite Hills, West Kimberley, and Murcia-Almeria. Agglutinate tuffs and lapillistones form in close proximity to lamproite vents from the fallout of lamproite magma spatter from fountaining eruptions. Following Fisher and Schmincke (1984), lapillistones contain dominantly 2–64 mm juvenile pyroclasts with as much as 25% ash size (<2 mm) particles. Agglutinate tuffs refer to those deposits containing more ash and perhaps a wider range in pyroclast sizes. Lamproite lapilli and agglutinate magma droplets include vesicular and dense magma fragments which are deformed (elongate) by

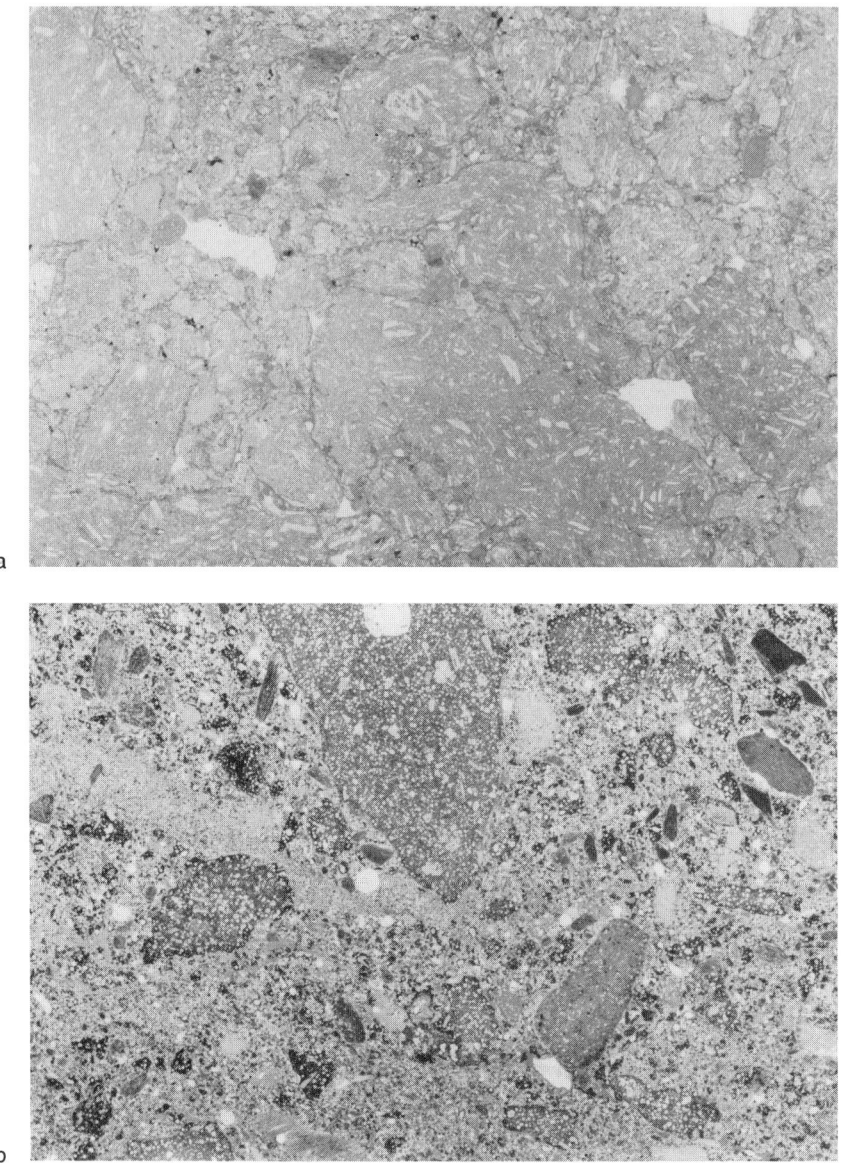

Figure 5.9. Photomicrographs of contact and autolithic lamproite breccias from (a) Matthew Hill, Leucite Hills Field, and (b) Calwynyardah, West Kimberley (field of view = 4.0 cm).

the welding of hot, plastic pyroclasts. These pyroclasts form the bulk of the rock, although as much as 10–50 vol % may be interstitial tuffaceous material. These deposits are typically bedded and form 10–50 cm-thick beds in sequences which may be as thick as 10–50 m. Agglutinate tuffs and lapillistones are dominated by juvenile fragments, although cognate (accessory) and accidental xenoliths can also occur in subordinate proportions. Representative photomicrographs of agglutinate tuffs are illustrated in Figure 5.10.

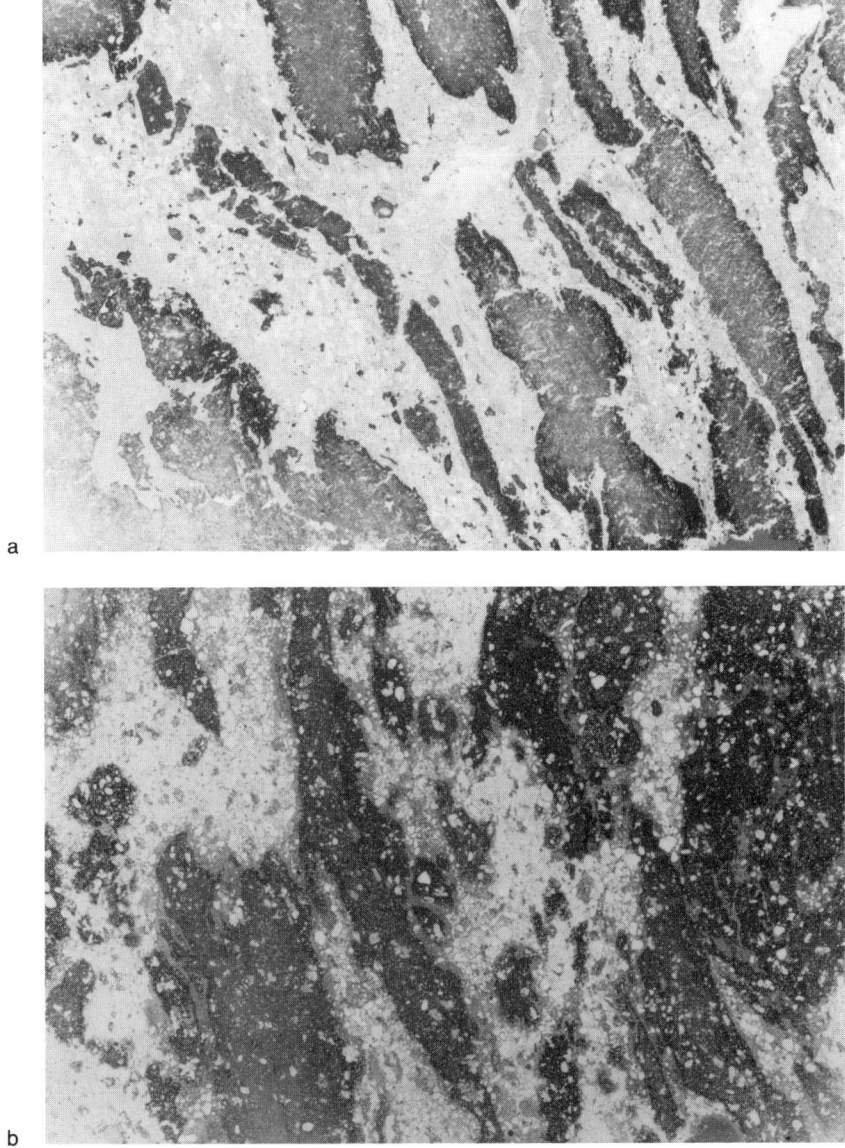

Figure 5.10. Photomicrographs of lamproite agglutinate tuffs and piperno breccias from (a) Smoky Butte (quarry), and (b) Mount Percy, West Kimberley Province (field of view = 4 cm).

Lamproite cinder deposits, very similar to those observed in basaltic cinder cones and sheets, occur in several vent complexes in the Leucite Hills. Cinder cones and associated bedded tephra deposits composed of scoria blocks, bombs, and ash occur at Emmons Mesa, Deer Butte, Spring Butte, Zirkel Mesa, and Steamboat Mountain. Cinder deposits are typically welded in the proximal vent positions and nonwelded in the more distal situations. On the basis of the nature of the best preserved pyroclastic facies found in the Leucite Hills, it is suggested that pyroclastic fall deposits probably mimic those formed by Hawaiian- or Strombolian-type eruption columns. However, more explosive eruption columns, such as those associated with Surtseyan- or Vulcanian-type eruptions (terminology of Walker 1973), were undoubtedly involved with the more explosive vent-forming eruptions in the West Kimberley Province. Because of the limited volume of lamproite eruptions, it is unlikely that Plinian, Ultraplinian, and Phreatoplinian eruptions ever occur.

5.3.2.2. Lapilli Tuffs and Crystal–Lithic Tuffs

Lapilli tuffs and crystal–lithic tuffs occur in lamproite vents from the Argyle, West Kimberley, Smoky Butte, Prairie Creek, Kapamba, Majhgawan, Yellow Water Butte, and Elk Creek fields or provinces. Both clast-supported and matrix-supported lithologies are found. Lamproite lapilli are typically smaller than 2–4 cm (average 5–10 mm) and exhibit a wide spectrum of shapes, ranging from well-rounded and spherical, through amoeboid or angular, to highly irregular and deformed. Both welded and nonwelded tuffs occur. Most lapilli are similar in composition to the juvenile portion of enclosing tuff matrix, and differ in degree of alteration. The lapilli are typically vesicular and glassy with variable proportions of phenocrysts of olivine, phlogopite, diopside, etc. Armored or concentrically zoned lapilli have been recognized in several lapilli tuffs. Lapilli tuffs and crystal–lithic tuffs are dominated by juvenile pyroclasts, but may contain large proportions of cognate or xenolithic fragments. Lapilli and crystal–lithic tuffs are commonly contaminated with ash-sized (<2 mm) matrix xenocrystal quartz, feldspar, and other lithic and crystal components derived from disaggregated country rocks. Lamproite crystal–lithic tuffs are better sorted than lapilli tuffs but are broadly similar in bulk composition to these units. Tuffs and lapilli tuffs are typically laminated and form coherent bedded sequences exhibiting a variety of "sedimentary" textures, including grading and cross lamination. Representative photomicrographs of lapilli tuffs and crystal–lithic tuffs are shown in Figure 5.11.

5.3.3. Base Surge and Pyroclastic Flow Deposits

Base surge deposition results from the turbulent explosive expansion and transport of a hot gas cloud, which is capable of supporting, by fluidization, a wide range in particle sizes (Williams and McBirney 1979). A variety of bed forms can develop in a base surge deposit (Fisher and Schmincke 1984): plane parallel, massive, and sandwave (composed of dunes, ripples, and internal cross-stratified units). Lamproites from many vents in the Ellendale field (e.g., 81 Mile Vent), Argyle, Kapamba, Majhgawan, Prairie Creek (?), and Barqueros (Murcia-Almeria) (Figure 5.12) possess probable base surge deposits. These occurrences exhibit dominantly plane parallel and sandwave bed types with common internal cross stratification. Typical plane parallel tuff beds, thought to result from base surge deposition, are 5–10 cm thick (total range from 1 to >100 cm). The most characteristic feature of lamproite base surge deposits is the abundance of xenocrystal quartz or other accidental crystal–lithic components (typically >50%, and up to 75%–95%) in the bedded deposits. These deposits are associated with the most violent eruptive phases, in which country rocks are extensively pulverized, mixed, and entrained with juvenile constituents. The relative role of phreatic steam (ground or meteoric water) versus primary magmatic steam, in this phase of the eruptive sequence, is unknown. As virtually identical deposits are observed in several provinces it is possible that magmatic steam is most important in their formation.

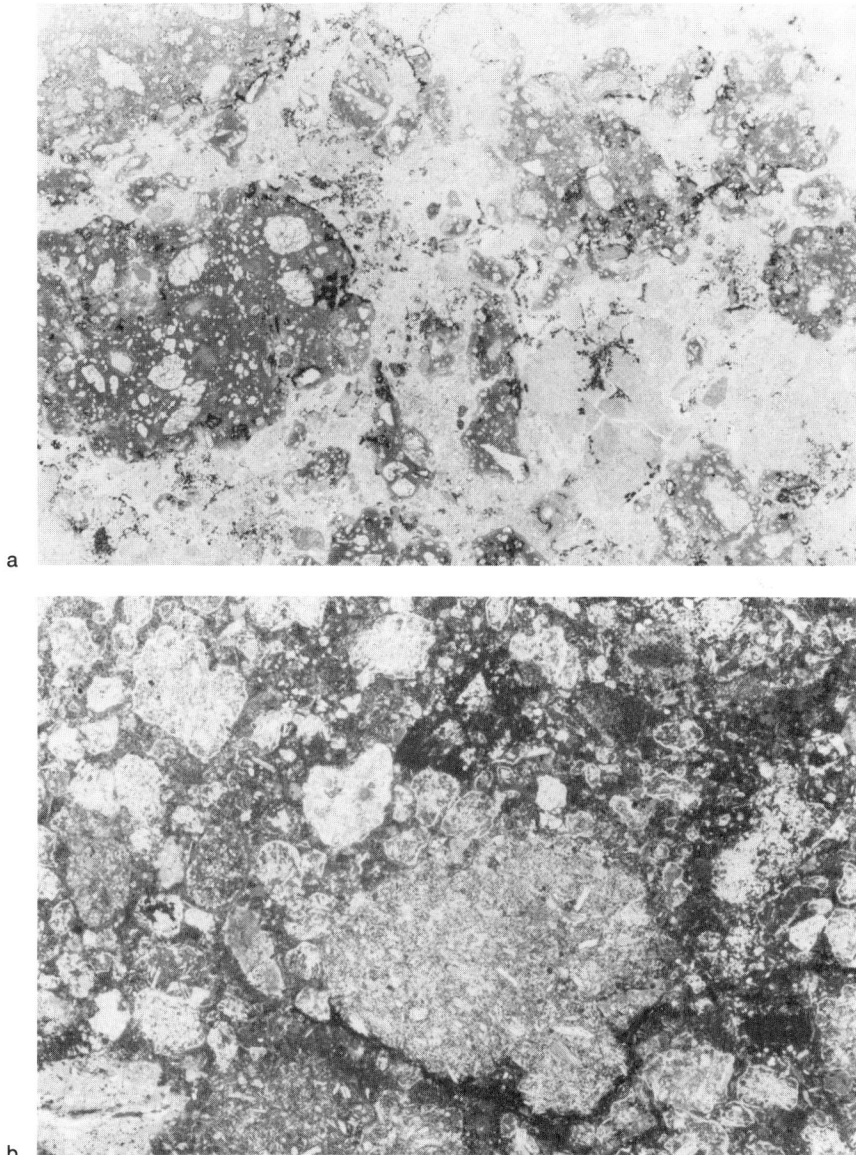

Figure 5.11. Photomicrographs of lamproite lithic lapilli tuffs and crystal-lithic tuffs from (a) Twin Knobs, Prairie Creek field (b) Crater of Diamonds, Prairie Creek field, (c) fine tuff (base surge), and (d) lapilli tuff both from 81 Mile Vent, West Kimberley, (e) lapilli tuff from Argyle, and (f) lapilli tuff from Smoky Butte quarry (field of view = 4.0 cm).

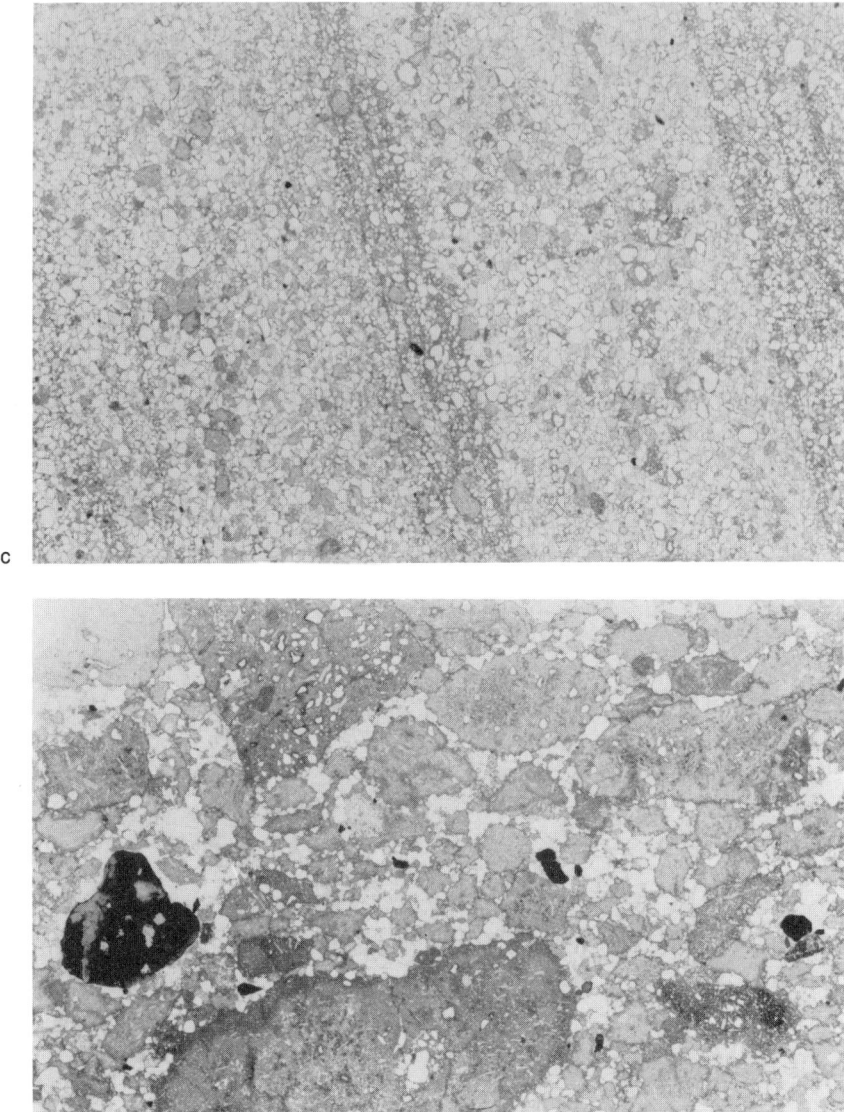

Figure 5.11. (*Continued*)

Significantly, units described as originating by base surge processes occur in highly permeable sandstone. Further volcanological study of these deposits is required.

Pyroclastic flow deposits probably occur in many vent complexes, including Prairie Creek and the West Kimberley province. Unfortunately continuous exposures of lamproitic pyroclastic rocks are lacking in nearly all provinces, hence a detailed description of this facies is not possible at present. Observations of recent examples of pyroclastic flows formed

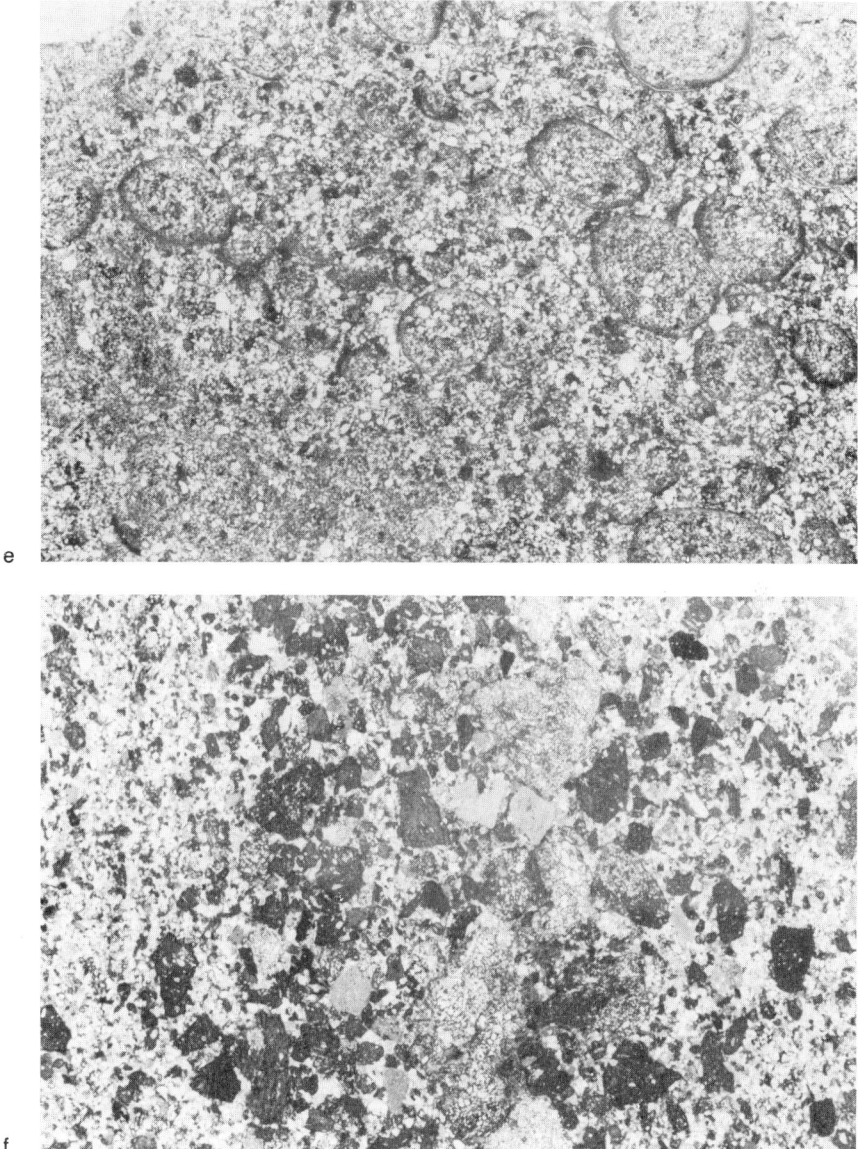

Figure 5.11. (*Continued*)

from other magmas show that they commonly avalanche down the flanks of several-kilometer-high calc-alkaline stratovolcanoes as a result of column collapse. Because of the difference in the scale of the activity, the transportation and deposition mechanisms of lamproite pyroclastic flows are expected to be quite different. We suggest that most lamproite pyroclastic flows probably originate by column collapse, although lava dome or lava flow collapse may play a minor role in their genesis.

Figure 5.12. Base surge deposit (35 cm thick, light-colored fine-grained unit) interbedded with crystal-lithic tuffs. Barqueros, Murcia-Almeria.

5.3.4. Epiclastic Deposits

Epiclastic deposits are formed by the reworking and redeposition of previously formed primary pyroclastic deposits by surface sedimentary processes (Fisher and Schmincke 1984). We do not include simply weathered pyroclastic lithologies in this facies. Because epiclastic deposits form at the surface, they are the first to be eroded and are therefore relatively rarely preserved. They are mostly restricted to those vents in which a substantial crater was buried and preserved. Epiclastic deposits have been recognized in the Calwynyardah vent of the West Kimberley Province (see Section 5.3.5.2), and possibly at Kapamba, Majhgawan, and Prairie Creek.

5.3.5. Examples of Crater and Pyroclastic Facies Lamproites

5.3.5.1. Prairie Creek

Two major groups of crater facies lithologies form significant exposures at the Crater of Diamonds and associated vents in the Prairie Creek field: lithic tuffs, tuff breccias, and lapilli tuffs; and olivine phlogopite lamproite autolithic breccias (Scott Smith and Skinner 1984b, Bolivar 1977, 1984, Miser and Ross 1923). Fragmental lithologies apparently formed early in the eruptive history and both the breccias and tuffs are intruded by massive, dense olivine phlogopite lamproite (Figure 5.13). Many tuff exposures form isolated knobs within the massive hypabyssal facies rocks and presumably have been rafted into place by later eruptive activity. Schematic cross sections illustrating the predicted facies relations are shown in Figure 5.14. Lithic tuffs and tuff breccias are composed of fragments of country rock and juvenile lamproite lapilli, with variable amounts of phlogopite and olivine, set in a distinctive blue-green chlorite-rich altered matrix containing abundant quartz grains. The tuffs, lapilli tuffs, and tuff breccias are typically bedded, possessing well-sorted and graded intervals. The abundance of quartz indicates primary eruptive disaggregation of country rocks, such as the Jackfork Sandstone, at depth. The presence of large rafted blocks of hydrothermally altered Jackfork Sandstone within the breccia phase of the lamproite supports the hypothesis. This interpretation is in contrast to that proposed by Scott Smith and Skinner (1984b) and Bolivar (1984), which suggested an epiclastic detrital origin to account for the presence of abundant quartz.

Lamproite breccias are dominantly altered (serpentine-rich) and contain juvenile olivine phlogopite lamproite lapilli set in a vesicular glassy matrix. In common with the tuffs, they are typically bedded, and may exhibit graded bedding. The latter bedform is possibly indicative of pyroclastic flow or base surge deposition.

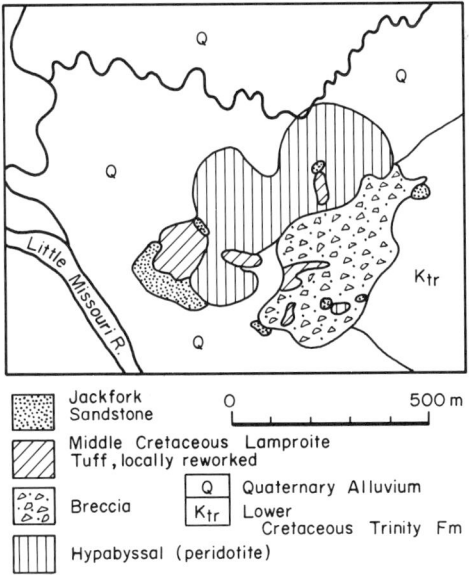

Figure 5.13. Geological sketch map of the Crater of Diamonds vent complex, Prairie Creek, Arkansas (after Bolivar 1977, 1985).

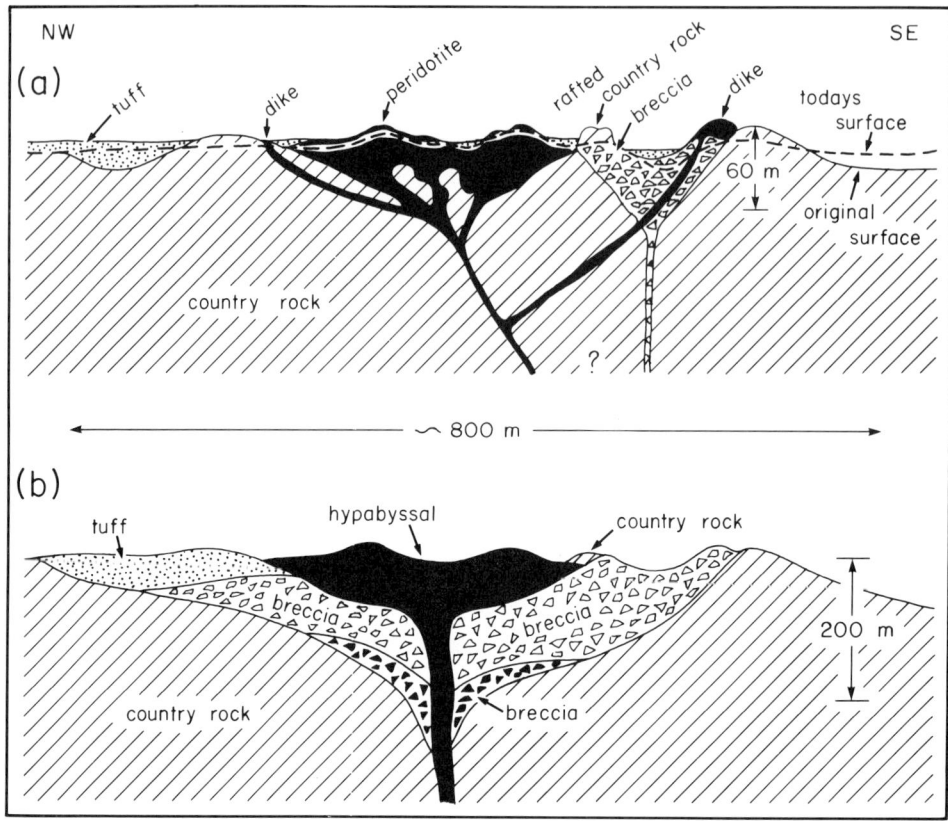

Figure 5.14. Schematic intepretive cross section through the center of the Crater of Diamonds vent complex, Prairie Creek, Arkansas, showing the relationships bewteen various lamproite facies (a) after Bolivar (1977, 1984), (b) this work.

5.3.5.2. Ellendale Field, West Kimberley Province

Crater facies lamproites occur in the majority of the over 100 intrusive/extrusive complexes in the four fields in the West Kimberley Province. Only representative examples from the best preserved vents in the Ellendale and Calwynyardah fields are highlighted here. The reader is referred to Jaques *et al.* (1986) for detailed descriptions. The volcanic centers range from single vents (81 Mile vent) to complex multiple vents (Ellendale 4).

Eighty-One Mile vent (Figure 5.15) was formed by an initial explosive phase that produced a variety of pyroclastic rocks, which were subsequently intruded by sills. Activity ceased after the intrusion of the magmatic core. The pyroclastic rocks consist of laminated and well-bedded lapilli tuffs exhibiting a variety of "sedimentary" structures. Fine-grained, laminated, quartz-rich layers showing low-angle cross-bedding are interpreted as having resulted from base surge deposition (Jaques *et al.* 1986). The laminated tuffs consist of fractured quartz grains (70%), ash-sized lamproite clasts, phlogopite flakes, and interstitial very finely comminuted lamproite ash. The proportions of juvenile to lithic fragments vary widely between laminae. The majority of the bedded lapilli tuffs appear to be airfall deposits similar to those associated with maar-type volcanoes (Jaques *et al.* 1986). They contain only

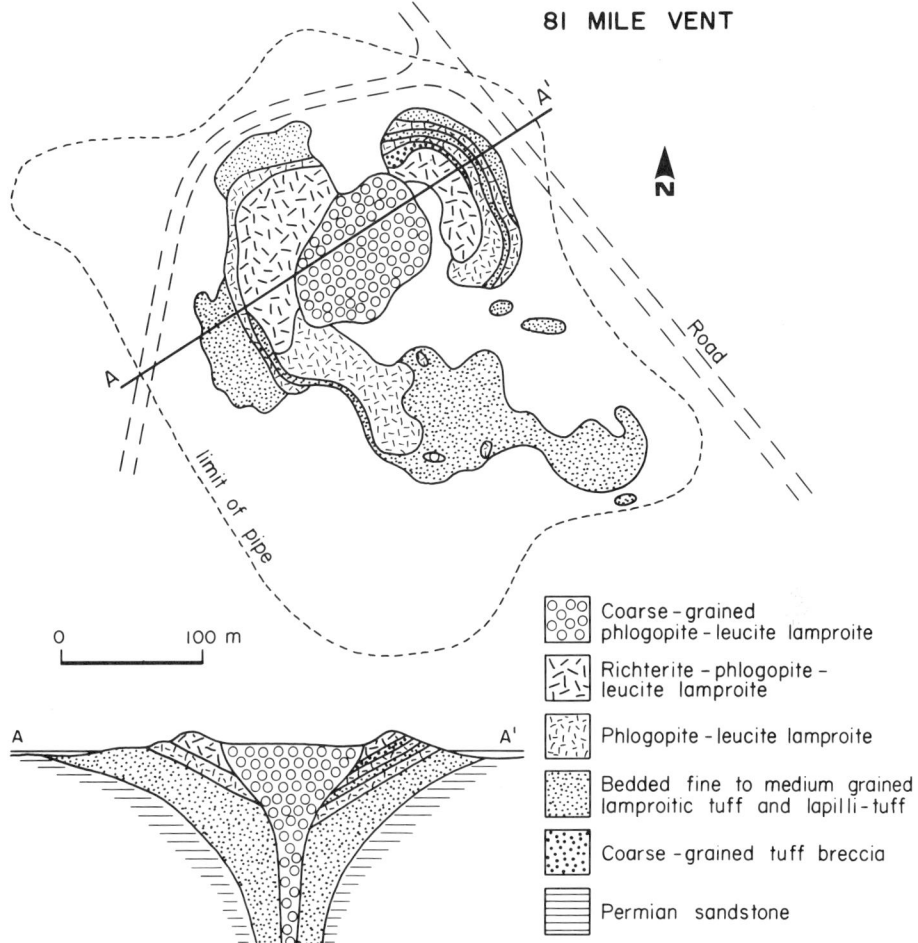

Figure 5.15. Geological sketch map and cross section of the 81 Mile Vent, West Kimberley Province, Western Australia (after Jaques *et al.* 1986).

about 30% rounded quartz grains together with clasts of shale and lamproite. Many of the lamproite lapilli appear to have been incorporated into the tuff while still plastic. Lapilli tuffs near the center of the vent are quartz-poor.

The sills consist of very fine-grained phlogopite lamproite. The basal sill shows autobrecciation and mixing with the underlying tuff. This results in the formation of rocks with the appearance of coarse lapilli tuff. The rocks of the central core are massive coarse-grained phlogopite lamproites.

Ellendale 9 is a vent of greater complexity than 81 Mile vent and is composed of two lobes (Figure 5.16). Jaques *et al.* (1986) interpret this center to have been formed by explosive volcanism occurring at a central vent in the west and a fissure in the east. The initial products of the eruption were a thick pyroclastic sequence of quartz-rich lapilli tuffs. The earliest tuffs are richest in quartz derived by the disaggregation of country rock sandstone. The quartz content of subsequent tuffs decreased as the eruption progressed,

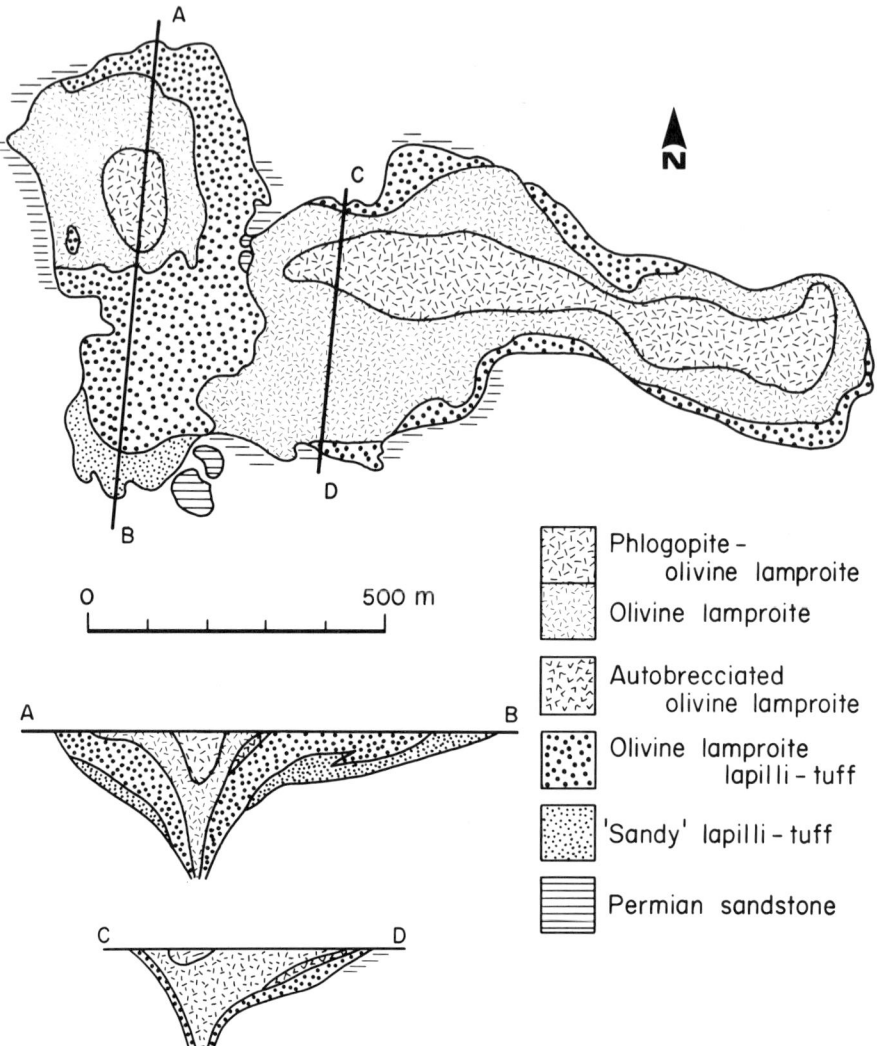

Figure 5.16. Geological sketch map and cross section of the Ellendale 9 vent, West Kimberley Province, Western Australia (after Jaques *et al.* 1986).

forming lamproite-rich sandy tuffs and lamproite lapilli tuffs. Many of the well-bedded pyroclastic rocks are interpreted to represent base surge deposits. Initially, activity is believed to have proceeded simultaneously in both lobes, but was confined to the western vent in the closing stages. The final stage of activity was the upwelling of olivine lamproite magma to form two separate lava lakes.

Ellendale 4 (Figure 5.17) illustrates the complexity that may develop in lamproite vents as a consequence of multiple eruptions. This vent results from the coalescence of two major complex eruptive centers. Much of the surface area of the pipe is covered by magmatic olivine lamproite, with pyroclastic rocks confined to the margins. However, drilling

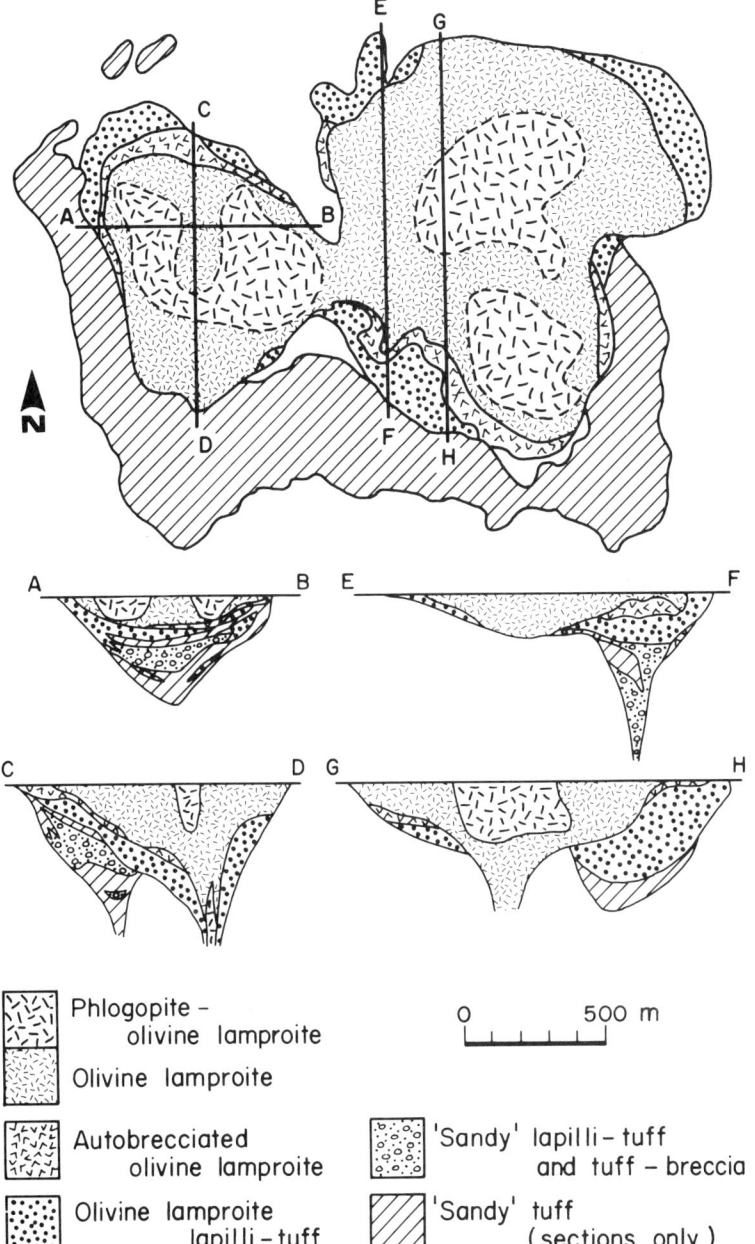

Figure 5.17. Geological sketch map and cross section of the Ellendale 4 vent, West Kimberley Province, Western Australia (after Jaques *et al.* 1986).

shows that the surficial geology is not an accurate guide to the subsurface structure. The magmatic material is in fact a thin cover over the pyroclastic rocks which comprise the bulk of the intrusion.

Extensive drilling of this vent has shown that each lobe consists of multiple nested vents, together with several conduits which fed the magmatic core. The western lobe consists of three vents arranged in a north–south direction along the axis of the lobe. Jaques *et al.* (1986) suggest that the activity migrated from north to south as deposits from the southern vent overlie the tuffs in the central vent. The final stage of activity was marked by the upwelling of magmatic lamproites from the southern vent. The eastern lobe also contains three vents. The two southern vents form an elongate trough along the southern margin of the pipe. The northern largest vent in this lobe is responsible for the overall shape. This vent produced pyroclastic deposits which blanket the two southern vents.

The pyroclastic rocks range from quartz-rich tuffs dominated by xenocrystal material, to lamproite lapilli tuffs, containing few country rock components. These different tuffs are commonly interbedded, suggesting the occurrence of a more complex volcanic history than one eruptive episode of simple vent clearing. The interfingering of the different pyroclastic facies may result from slumping into the vent of earlier-formed deposits from surrounding tuff rings. In other cases the presence of well-developed bedding suggests that base surge deposits, produced from sequentially erupting closely spaced vents, were intermixed with air-fall deposits of lapilli tuff.

The Calwynyardah vent is distinctive in being largely composed of pyroclastic rocks together with a shallow cover (up to 150 m thick) of lacustrine epiclastic rocks. The vent narrows rapidly with depth from a broad surface feature to a relatively narrow conduit. The pyroclastic rocks filling the vent include tuffs, tuff breccias, and lapilli tuffs with varying degrees of xenocrystal quartz and country rock contamination. Both juvenile and cognate lapilli of olivine phlogopite lamproite form irregular-to-ovoid lapilli in an altered and carbonated lamproitic matrix. Only a small amount of massive hypabyssal facies lamproite is present as thin sheets and a small plug.

The epiclastic rocks at Calwynyardah are fine-grained silty tuffs and mudstones interbedded with siltstones and lignites. They fill a depression within the pyroclastic deposits, and are believed to have derived from a surrounding tuff ring. Deposition took place in a crater lake (Jaques *et al.* 1986). The mudstones and siltstones are dominated by accidental components and only a minor assemblage of juvenile material (rare phlogopite and olivine pseudomorphs) has been recognized. The bulk of the reworked tuffs are composed of chlorite, clay, muscovite, quartz, and feldspar (Jaques *et al.* 1986).

5.3.5.3. Argyle

Pyroclastic rocks form the bulk of the 50 ha diamondiferous Argyle vent (Figure 3.30), which has been faulted and tilted 30° to the north (Jaques *et al.* 1986). The two dominant crater facies rocks are quartz-rich polygenetic lamproite ("sandy") tuff and quartz-lacking monogenetic lamproite ("nonsandy") tuff. Individual lithologies include lapilli tuff, coarse tuff, fine tuff, and rare tuff breccia. Possible accretionary lapilli tuffs and hyaloclastites have been reported by Jaques *et al.* (1986). Juvenile lapilli range from irregularly shaped vesicular vitric clasts, to dense blocky porphyritic microcrystalline pyroclasts. Juvenile clasts form >30%–50% of the "nonsandy" tuffs and <5%–30% of the "sandy" tuffs. "Sandy" and "nonsandy" tuff beds exhibit contrasting forms. "Sandy" tuff beds are massive-to-planar (0.01–1-m thick). They are typically poorly sorted but may possess normal or reverse grading, and low-angle (10°) cross-bedding. Soft sediment deformation (slump and water-escape) structures are present in some "sandy" tuffs. The "nonsandy" tuffs form massive to poorly bedded units of hyalotuffs, hyaloclastite, and lamproite autobreccias. Representative examples are illustrated in Figure 5.18.

Jaques *et al.* (1986) determined that the "sandy" tuff was apparently the earliest eruptive phase, followed by the "nonsandy" tuff. The relative proportions of accidental and

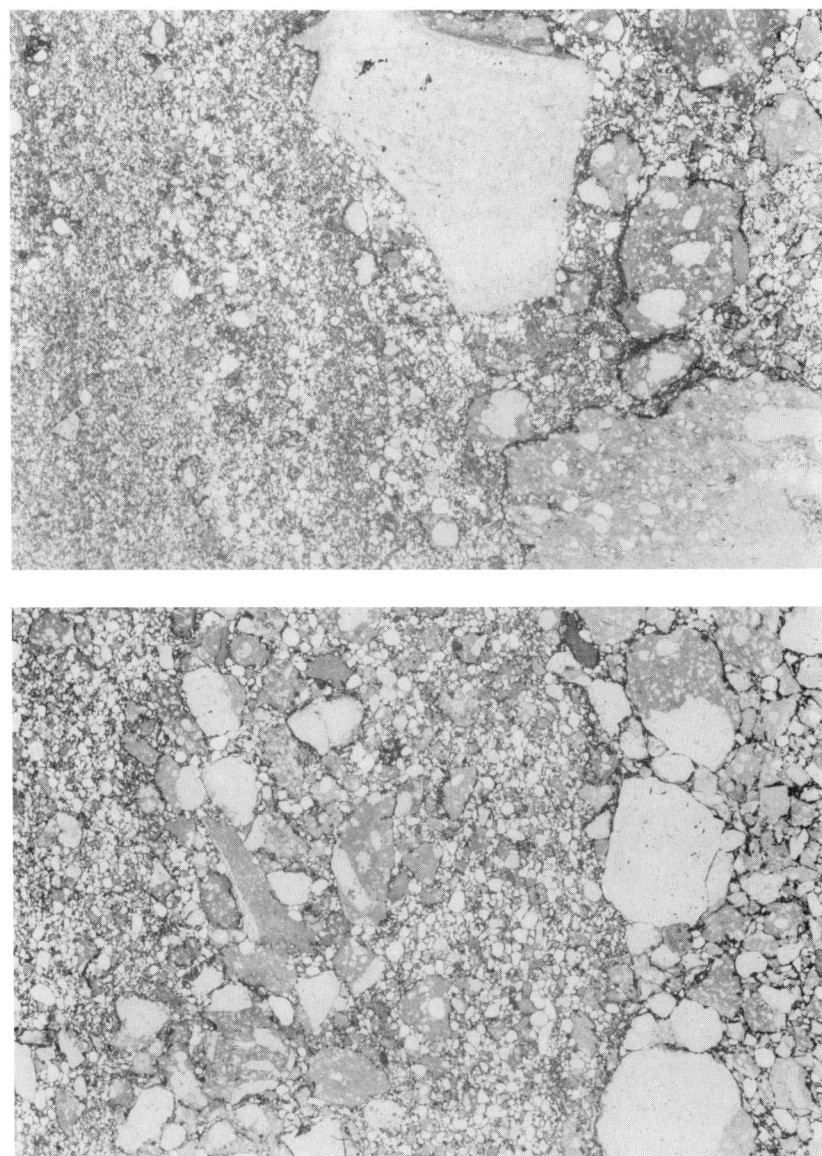

Figure 5.18. Photomicrographs (a and b) of "sandy" tuffs from the Argyle vent, Western Australia (field of view = 4.0 cm).

juvenile components changed as the eruption progressed, the accidental clasts being replaced by juvenile pyroclasts (Jaques *et al.* 1986).

The bulk of the Argyle vent is composed of "sandy" tuff, the "nonsandy" tuff being restricted to a small, 200 m-wide depression at the northern terminus of the vent. Minor late-stage phlogopite olivine lamproite dikes, less than 1 m wide, intrude the "sandy" tuff.

5.3.5.4. Leucite Hills

Cinder cones, volcanic necks, and vents comprise the crater facies rocks of the Leucite Hills. More than 14 cinder cones occur at Zirkel Mesa, Spring Butte, Steamboat Mountain, Emmons Mesa, and Deer Butte. The Leucite Hills cinder cones are smaller than typical alkali basalt cinder cones. They range from 30 to 115 m in height and approach 1 km in diameter, with average present-day (denuded) slope angles in the range 15° to 20°. In common with alkali basalt cones, their morphology is principally semicircular; however, several are slightly asymmetrical and many are breached. The cinder cones are composed of highly vesicular cinders (lapilli), ribbon bombs, spindle bombs, and blocks (the latter 0.5–13 m in size) of altered pale-tan lamproite scoria which are welded to varying degrees. Most cinder cones indicate pyroclastic activity which postdated the eruption of massive (yet vesicular) lava flows. Several cinder cones (at Emmons Mesa, Deer Butte, and Spring Butte) formed contemporaneously with lava flows, although some have been breached or overridden by later lavas (Ogden 1979). Ogden (1979) described several rootless vents at Steamboat Mountain and near Black Rock where cylindrical reentrants are exposed (25 m high) on the walls of eroded mesas. Pyroclastic debris (cinders and ribbon bombs) line the concentric reentrants.

Other crater facies lithologies include tuff breccias, lapilli tuffs, and related agglomeratic rocks from Matthew Hill, Boars Tusk, Badgers Teeth, and Iddings, Weed, and Hallock Buttes. The fragmental rocks contain a wide range of xenolithic components, and some are composed almost entirely of cognate, comagmatic fragmental clasts (pyroclasts) in a tuffaceous lamproite matrix. At Iddings, Weed, and Hallock Buttes, massive (dense) nonvesicular phlogopite leucite diopside lamproite intrudes marginal tuffaceous rocks. These consist of autobrecciated vesicular pyroclasts, country rock xenoliths, and scoria set in a lamproite tuff matrix. The Badgers Teeth exposures consist mostly of autobrecciated lamproite with minor amounts of massive dense lava restricted to the core of a given exposure. At Boars Tusk, vesicular-to-dense autobrecciated lavas form portions of xenolithic tuff breccias (pyroclasts as large as 15 cm) and lapilli tuffs that are virtually identical to similar lithologies from the West Hill tuff at the Crater of Diamonds (Prairie Creek), Calwynyardah (West Kimberley), and Cerro de Monagrillo (Murcia-Almeria). However, the Leucite Hills tuffaceous rocks seem to lack the abundant quartz grains derived from disaggregated country rocks observed in many West Kimberley, Argyle, and Prairie Creek crater facies lithologies.

5.3.5.5. Cerro Negro, Cancarix, Murcia-Almeria

Cerro Negro is a prominent hill composed of columnar jointed, massive, coarse-grained phlogopite richterite lamproite. These rocks overlie a complex sequence of interbedded lavas and pyroclastic rocks (Figure 5.19). The sequence is approximately 50 m in thickness. Individual pyroclastic units are poorly to well-bedded rocks, up to 1 m in

Figure 5.19. Photograph of columnar jointed massive lamproites overlying interbedded lavas (dark) and pyroclastic rocks (light) at Cerro Negro, Cancarix, Murcia-Almeria, Spain.

thickness, consisting of pyroclastic breccias, lapillistones, and coarse tuffs. The presence of "cow dung" bombs and drop-stones suggests that they are predominantly airfall deposits. Minor thin units, close to the contact with the country rock limestones, exhibit graded bedding, suggesting the existence of some base surge activity. In the southeastern and eastern parts of the complex the pyroclastic beds dip towards the center of the vent at about 30–45°. The poor exposure in the western parts of the vent does not permit determination of the attitude or thickness of the pyroclastic unit. The interbedded lavas are fine-grained, flow banded to massive, weakly vesicular rocks. The flows appear to have been extruded over the pyroclastic beds. Contacts with the pyroclastics range from planar to irregular, the latter representing autobrecciation but not mixing with the underlying tuff.

Cerro Negro has many geological similarities with Mt. North and other vents in the Ellendale field. The style of activity is similar in that an initial sequence of pyroclastic activity was followed by the quiescent emplacement of lamproite magma. The columnar jointed magmatic lamproites are thus interpreted to be lava lakes occupying the core of the pyroclastic facies vent complex (Scott Smith personal communication 1986, and this work). The relative paucity of base surge deposits may be related to the impermeable nature of the country rock.

5.3.5.6. Kapamba

Pyroclastic rocks compose the bulk of all of the vents in the Kapamba Field (Scott Smith *et al.* 1989). Lapilli tuffs of variable composition are the dominant lithology, although crystal tuffs, coarse crystal–lithic tuffs, and tuff breccias also occur. The lapilli tuffs are composed of up to 70% juvenile lamproite lapilli with accidental xenoliths and angular grains of quartz and feldspar set in an altered fine-grained matrix. Juvenile lapilli are typically <2 cm, porphyritic, vesicular (phlogopite-diopside-leucite) olivine lamproite vitroclasts.

The vents exhibit a pseudoconcentric arrangement of distinct lapilli tuff units, decreasing in their accidental component content from the vent margin to the core. The crater facies lithologies have been intruded by a massive hypabyssal olivine lamproite (Figure 5.20).

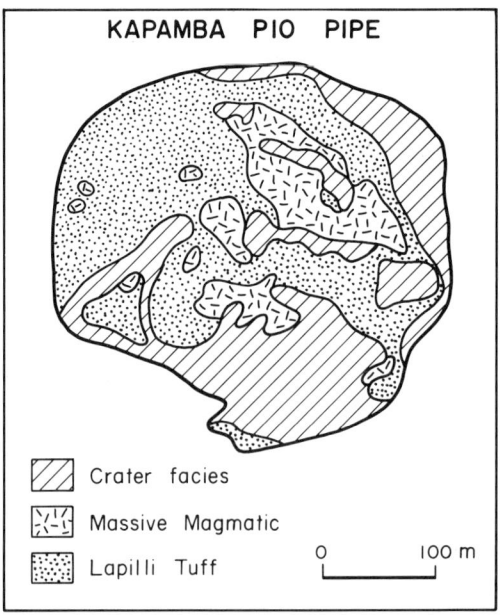

Figure 5.20. Geological sketch map of the Kapamba 10 vent, Luangwa Graben, Zambia (after Scott Smith *et al.* 1989).

5.3.5.7. Majhgawan

Although the agglomeratic tuff at the Majhgawan vent has been known for many years (Sinor 1930), its geology is not yet well characterized. Scott Smith (1989) has summarized new and previous work on the geology and petrology of the vent. On the basis of shallow (<120 m) drilling, the vent appears to have the shape of an inverted cone with 70–75° marginal dips. Sixteen samples examined by Scott Smith (1989) suggest that the bulk of the pit is composed of altered vitric lapilli tuffs, with vesicular pyroclasts of phlogopite olivine hyalolamproite set in an altered vitric matrix. Clearly, modern volcanological studies of this important diamondiferous lamproite are warranted.

5.4. HYPABYSSAL FACIES

Hypabyssal facies lamproites are dominantly massive, dense holocrystalline rocks formed by the *in situ* crystallization of intrusive bodies of lamproite magma. Fragmental hypabyssal facies lithologies may occur in the marginal contact zones of more massive intrusive bodies, and hypabyssal facies breccias may form by the incorporation of country rock xenoliths in lamproite magmas. Hypabyssal facies lamproites occur within the interior zones of many lamproite vents and may also form sills and dikes. Hypabyssal facies lamproites can be considered the intrusive equivalent of lamproite lava facies. Clearly they are very similar and many hypabyssal lithologies are petrographically indistinguishable from lavas.

Hypabyssal facies lamproites are widely distributed and occur in nearly every field or province. Most lamproite vents are composed of pyroclastic facies lamproites intruded by late-stage hypabyssal facies lithologies in the form of irregular intrusive bodies, plugs, dikes, or sills. In addition, lamproite dikes may intrude country rocks far from an associated vent. The root zones of lamproite vents are principally composed of hypabyssal facies rocks. Where extensive denudation has occurred, only hypabyssal facies lamproites may be preserved as dikes or sills.

The morphology and size of hypabyssal lamproite bodies is governed by that of associated vents (see Section 5.3). There is a wide range in the proportion of pyroclastic to hypabyssal rocks in vents containing both facies. This range is not simply related to the extent of erosion; thus, some vents are composed mainly of hypabyssal lithologies (e.g., many in the West Kimberley Province), whereas others are apparently poor in these lithologies (e.g., Argyle).

Volcanic necks are the denuded feeder systems or filled conduits of now-eroded volcanic vents. They are the most abundant form of hypabyssal lamproites. We infer that they were originally overlain by volcanic edifices, on the basis of the occurrence of marginal fragmental lithologies or associated pyroclastic facies rocks. Necks or irregular bodies occur in many fields and provinces, including many in the West Kimberley Province, Prairie Creek, Kapamba, Leucite Hills, Yellow Water Butte, Murcia-Almeria, Froze-to-Death Butte, and Presidente Oligario. In common with typical basalt or andesite necks, those of lamproites are typically 50–100 m in diameter and subcircular to irregular in cross section. Lamproite necks do not usually form positive topographic features, although they may be preserved in some cases by silicification of the lamproites and country rocks, e.g., Machells Pyramid (Noonkanbah Field).

Lamproite dikes or dike swarms occur at Sisimuit, Chelima, Leucite Hills, Smoky

Butte, Sesia-Lanzo and Combin, Bobi, Kapamba, Mount Bayliss, Priestly Peak, Argyle, West Kimberley, and Murcia-Almeria. Occurrences range from long (3 km), thin (1–10 m) single dikes (e.g., Smoky Butte) to dike swarms composed of many subparallel, 1–5-m-wide dikes (e.g., Sisimuit, Chelima). Some dikes are associated with pyroclastic or crater facies rocks (Leucite Hills, Smoky Butte, Kapamba, West Kimberley, Argyle, Murcia-Almeria), whereas others lack any associated pyroclastic lithologies (Sesia-Lanzo and Combin, Sisimuit, Chelima, Mount Bayliss, and Priestly Peak). Dikes exhibit varying degrees of contact metamorphism of country rocks and mainly produce silicified contact zones. Most dikes display chilled fine-grained margins or marginal fragmental zones containing country rock xenoliths.

Lamproite sills occur at Hills Pond, Francis, in several vents in the West Kimberley Province, at Cerro Negro (Cancarix), and Sisco. Sills range from thin (1–3 m) isolated bodies (West Kimberley and Sisco) to multiple, stacked sills (each >2–30 m thick) at Francis and Hills Pond.

5.4.1. Examples of Hypabyssal Facies Lamproites

Only a few representative examples of hypabyssal lamproites will be highlighted, as they occur in nearly all fields and provinces. Generalizations concerning this facies are, however, made on the basis of the geology of all occurrences. Detailed descriptions of hypabyssal facies rocks are provided in many of the references cited in Chapter 3. Hypabyssal lamproites are dominantly holocrystalline and porphyritic; however, they can range from vitric to coarse grained, and equigranular varieties exist. Vesicular rocks also occur, in which the vesicles are typically subspherical and undeformed. Most hypabyssal lithologies are composed dominantly of magmatic material.

5.4.1.1. Leucite Hills

Dikes are widespread in the Leucite Hills field and display a northwest-trending preferred orientation. The dikes are typically less than 50 m long and the longest (Wortman Dike) is 215 m long. Some dikes are offset by about 1 m. The majority of dikes lack any marginal breccia zone and are less than a meter wide (0.3–0.6 m). Ogden (1979) found that the widest dikes (8–12 m wide) tend to possess a breccia zone composed of rounded fragments of lava and sedimentary rock fragments in a volcanic matrix. A dike emanating from Pilot Butte contains globular segregations of fine-grained phlogopite lamproite (this work).

Volcanic necks occur at Pilot Butte, Boars Tusk, and Badgers Teeth. At Pilot Butte, vertical flow foliation of phlogopite is observed near the base of the 30 m-diameter neck, which feeds a 20-m-thick lava flow with horizontally flow foliated phlogopites (Ogden 1979). Boars Tusk is a 500 m-diameter neck (Figure 5.21) consisting principally of tuff breccia (Figure 5.22), cored by a 100 m-diameter irregular zone of massive nonvesicular phlogopite lamproite. The bulk of Badgers Teeth is composed of tuff breccia, with the core of the neck being composed of dense phlogopite lamproite.

5.4.1.2. Sisimiut and Chelima

More than a hundred thin (mostly <1–2 m wide), subvertical dikes, over a kilometer in length, form an east-trending en-echelon swarm within a 20 × 30 km area near Sisimuit

Figure 5.21. Boars Tusk volcanic neck, Leucite Hills, Wyoming.

Figure 5.22. Boars Tusk breccia consisting of xenoliths of country rock (light-colored large clasts) and autobrecciated lava.

(Brooks *et al.* 1978, Scott 1977, 1979, 1981, Thy *et al.* 1987). The dikes are massive, nonvesicular porphyritic phlogopite olivine lamproites. Flow differentiation has apparently concentrated phenocrysts near the center of the dikes (Scott 1977, 1981).

More than thirty 1–5 m-wide subvertical lamproite dikes form a 10 km-long by 6 km wide en-echelon swarm near Chelima. The dikes exhibit chilled fine-grained margins and are composed of altered and weakly metamorphosed nonvesicular porphyritic phlogopite lamproite.

5.4.1.3. Smoky Butte

A 3 km-long, 1–20 m-wide subvertical, discontinuous dike swells into several 50–100 m-wide plugs or vents at the Smoky Butte Quarry and several other Buttes (Matson 1960, Mitchell *et al.* 1987). The dike exhibits chilled margins and there is evidence of multiple injections of as many as 5 units along the length of the exposed dike. The dike pinches and swells and is on average 1–5 m wide. The dike and plugs are composed of massive hyalophlogopite lamproite with variable degrees of vesicularity.

5.4.1.4. Rice Hill, Francis and Hills Pond

Lamproite sills are well developed at Rice Hill, Francis, and Hills Pond. At Rice Hill, a 45 m-thick sill dips 10–14° northeast and is composed of medium-to-coarse grained richterite phlogopite leucite lamproite with altered olivine and a devitrified glass base (Jaques *et al.* 1986). The sill is underlain by a thin tuffaceous lamproite but is otherwise isolated from other lamproite occurrences. The Francis locality consists of at least five subhorizontal sills (5–30 m thick) of vesicular phlogopite sanidine lamproite (this work). The margins exhibit a fine-grained chilled zone. The Hills Pond (Silver City) sill complex contains multiple stacked sills (Figure 3.13), which are 2–20 m thick and about 2 km long. They are composed of medium to coarse-grained madupitic richterite diopside phlogopite olivine lamproite (Cullers *et al.* 1985). The sills produced hornblende hornfels facies contact metamorphism, up to 300 m away from their margin.

5.5. PLUTONIC FACIES

Plutonic facies lamproites result from the relatively quiescent intrusion and crystallization of large volumes of magma ($>$1–5 km^3) in a near-surface environment ($<$2–5 km depths). Fragmental lithologies are limited to the outer margins of the intrusive body. Because of the rarity of these facies, a generalized model is not appropriate. Included within the lamproite plutonic facies are large ($>$0.5–3 km long dimension) magmatic intrusions. In addition to the size contrast of plutonic versus hypabyssal igneous bodies, plutonic facies lamproites are coarser grained than hypabyssal rocks and are generally holocrystalline medium-to-coarse grained ($>$1–10 mm grain size) rocks.

Lamproite plutons are extremely rare. Nevertheless, we predict that lamproite fields and provinces are underlain by plutonic lamproites in stagnated intrusive bodies in the middle to lower crust and upper mantle. This conclusion is based on the observation that the intrusive equivalents of mafic igneous volcanic rocks roughly outnumber surface eruptives by a factor of 10 or more (Spera 1979).

The Walgidee Hills intrusion in the Noonkanbah field is the only known plutonic lamproite complex. This intrusion is unfortunately very poorly exposed, being mainly covered by alluvium. As it has not been extensively drilled or trenched, knowledge of its structure is based on the very few surface exposures. The body is about 2.7×2.5 km in size and has a semicircular cross section, suggesting a volume of at least 5–10 km^3. Structural deformation due to intrusion is not evident. The intrusion is rimmed by brecciated and tuffaceous lamproite. The margins are steep-sided and juxtapose tuff breccia, brecciated lamproite, and tuff (containing abundant country rock xenoliths) with silicified country rocks. Prider (1960) interpreted the intrusive body to be a laccolith; however, the near vertical contacts on its margins (Jaques *et al*. 1986) indicate otherwise. Shallow drilling on the east side of the body failed to delineate the base of the intrusive but constrains its depth to being greater than 60 m (Jaques *et al*. 1986). The magmatic rocks are zoned with respect to their grain size, and very coarse pegmatitic rocks of unusual mineralogy make up the core.

5.6. GENERALIZED MODEL OF ERUPTIVE SEQUENCES

Conclusions derived from the study of the geology of individual centers from over 10 lamproite fields and provinces permit the formulation of a general model of the eruptive style of lamproite volcanism and of vent formation. Obviously, lacking direct observations of lamproite volcanism, the model is not well constrained with respect to the details of the pyroclastic activity. Apart from some exceptions, the following general model satisfies the vast majority of geologic constraints. It is notable that lamproite eruptions follow a similar sequence to those exhibited by many common mafic to intermediate magmas. However, the morphology and nature of lamproite vent complexes contrast markedly with those of the common mafic-to-intermediate rocks. The formation of lamproite vents progresses through three major distinct episodes: (1) crater formation or vent clearing; (2) pyroclastic sequence formation within and surrounding the vent; (3) eruption of lava, development of lava lakes, and formation of hypabyssal rocks in the throat of the vent. Any of these episodes may be repeated at any time in the history of the vent.

The development of an eruption model is hindered as detailed geological data are available only for vents that have undergone 50–500 m of erosion, e.g., West Kimberley province, Argyle, Prairie Creek. Although many pyroclastic lithologies were undoubtedly deposited subaerially, they were quickly buried by subsequent deposits and/or eroded. Such surface assemblages have not usually been preserved.

It is suggested here, by Jaques *et al*. (1986) and Smith and Lorenz (1989), that basaltic tuff cones and rings are possible analogs of subaerial lamproite volcanism. These volcanic landforms have been extensively studied, and detailed eruption models based in part upon observed volcanic events are available (Ollier 1967, Lorenz *et al*. 1970, Wohletz and Sheridan 1983, Lorenz 1985, 1986). Basaltic tuff cones and rings are small (<5 km diameter), circular, monogenic volcanoes composed of tuffs resulting from hydromagmatic explosions (Wohletz and Sheridan 1983). Their form and facies relations are schematically illustrated in Figure 5.23. Wohletz and Sheridan (1983) studied 11 basaltic volcanoes associated with hydromagmatic activity and developed a model of eruption sequences for basaltic magmas intruded into near-surface aquifers or standing bodies of water and found that cinder cones result from relatively dry lava fountaining under conditions of low water/magma ratios. Abundant ground water leads to the formation of tuff rings, and shallow

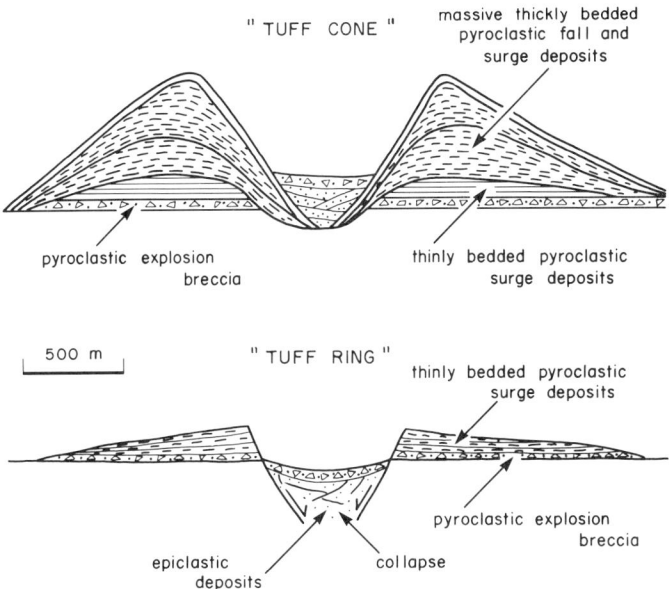

Figure 5.23. Schematic diagram illustrating the various facies components of tuff rings and cones (after Wohletz and Sheridan 1983).

standing water to the development of tuff cones. Wohletz and Sheridan (1983) showed that the vent-filling material is dominated by explosion breccias, envisioned as being poorly stratified to nonstratified. Tuff beds are thinly to thickly bedded and dip primarily away from the vent in distal positions outside of the vent margin. Within lamproite vents, pyroclastic lithologies include explosion breccias and tuffs which may form well-stratified units typically dipping toward the center of the vent. Outside of the vent proper stratified pyroclastic rocks typically dip away from the vent center.

The model lamproite eruptive sequence begins with an explosion, forming a broad crater, typically 500–2000 m in diameter and 200–500 m deep. This explosion is envisioned as one of great energy and violence, far in excess of those associated with many common volcanic explosions (except typical Plinian, Vulcanian, and caldera-forming eruptions). We propose that this explosion is due to phreatomagmatic activity, and involves a complex interplay between expansion of aqueous fluids that exsolved due to decompression from a lamproite melt, and those that expanded due to superheating of meteoric or surface water by interaction with a lamproite melt.

We view primary magmatic volatiles as an important controlling factor in the violence of a lamproite eruption. The vent shape and the ratio of pyroclastic to lava flow facies deposits is governed by several variables, including (1) the predicted shallow (<1 km) depth of vesiculation of an H_2O-rich magma; (2) abundance of meteoric or surface water; (3) the strength, permeability, and porosity of near-surface country rocks and sediments. Most of the fragmented country rocks and juvenile debris are pulverized and deposited in both proximal and distal positions relative to the vent center. Some country rocks and surface sediments are intensely fragmented, pulverized, mixed with juvenile magmatic debris, and redeposited outside the vent margins or collapse into the vent. Such deposits contain a wide range in abundance of juvenile material, but are typically poor in these components. Fragmentation of poorly consolidated or friable country rocks around the vent, coupled with mixing with juvenile material, may be followed by redeposition external to the vent proper. The deposits formed by this process are essentially reworked country rocks in which juvenile components are limited to grain coatings or minor lapilli. In addition, vent margin "intrusion breccias" are characteristically developed.

The initial explosion undoubtedly produces a large ash cloud, followed by deposition of material in the immediate vicinity of the vent by pyroclastic fall and flow/surge processes. Pyroclastic surge deposits are initially formed by a phreatomagmatic explosion or the collapse of an eruption column and are preserved within and surrounding the vent proper.

Following the vent-forming event and pyroclastic flow/fall phase of deposition, hypabyssal facies lamproite breccias, with 20%–80% foreign clasts, typically intrude the crater facies sequences. The proportion of hypabyssal facies lamproite breccias in vents is variable and sometimes negligible; however, most vents contain a substantial proportion (20%–40%) of these lithologies. Massive hypabyssal facies rocks are then formed by the later, relatively quiescent, intrusion of xenolith-poor magma. These magmas may form lava lakes and/or lava flows, sills, dikes, and small plutons.

The characteristic champagne glass shape of lamproite vents on a worldwide basis is suggestive of uniform trends in the process of vent formation. The flared shape of lamproite vents is most probably a consequence of the explosive activity occurring at relatively shallow levels. Jaques *et al.* (1986) envision vent enlargement as being caused partly by exfoliation from the wall rocks, but principally by the inward slumping of blocks of country rock along low-angled faults. Such blocks are subsequently disaggregated and incorporated into the pyroclastic deposits forming the quartz-rich and lithic-dominated lapilli tuffs common in the lower portions of the pyroclastic sequence. Further eruptions result in the formation of tuffs dominated by juvenile lapilli and the reworking of the earlier-formed tuffs.

The role of groundwater in the formation of the vents is uncertain. Jaques *et al.* (1986) note that the vents are commonly located on topographic highs. Consequently hydrovolcanic processes such as those advocated by Lorenz (1984), relating maar formation to valleys, are unlikely to have occurred. N.B. Selective preservation of the vents due to silicification could easily produce these topographic features. However, Jaques *et al.* (1986) note that the best examples of bedded tuffs and base surge deposits in the Ellendale field are found where lamproite magma has intruded groundwater aquifers. In the adjacent Noonkanbah field, where bedded tuffs are less well-developed, the lamproites intrude compact sandstones, siltstones, and mudstones. The presence of tuffs and a champagne glass shape for the Prairie Creek vent, which intrudes into the permeable Jackfork sandstone, is in agreement with the above model. Unfortunately other localities (Kapamba, Majhgawan) have been inadequately studied from a volcanological viewpoint, but it is significant that the Cerro Negro (Murcia-Almeria) vent has a low pyroclast/magmatic rock ratio and is emplaced in impermeable limestones.

The change from pyroclastic to magmatic activity must be caused by changes in the near-surface hydrological and/or magmatic volatile regime. Studies of maar volcanism have demonstrated that if the supply of water to the magma is interrupted, phreatomagmatic eruptions cease. Subsequent activity results in lava formation, fire-fountaining, and intrusion of magma as dikes, sills, and plugs. Clearly, this style of volcanism is analogous to that found in evolved lamproite vents.

During magma ascent and decompression, the least dense and most volatile-rich portion of the melt is likely to be the first to erupt or ascend. Thus, the initial phase of eruption is probably the most explosive. Exsolution of water from the melt will allow separation of a low-density, low-viscosity, gas phase at near-surface conditions. Primary magmatic volatiles thus may contribute significantly to the initial pyroclastic phase of lamproite eruptions. Later eruptions would result from the quiescent effusion of degassed magma.

It is also possible that the formation of the pyroclastic sequence leads to the isolation of the vent from ground water, leading to the final, relatively quiescent, stages of eruption (Boxer *et al.* 1989). In these late stages the magma simply degasses, and because of the low viscosity, pyroclastic deposits are not formed. The activity at this point probably resembles fluid Hawaiian basaltic volcanism as is found in the Kilauea lava lakes.

The Leucite Hills field is dominated by lava flows, cinder cones, and several breccia-containing volcanic necks. There is a conspicuous lack of lamproite vents of the type found in the Ellendale field, which is curious, as the rocks in both provinces have similar chemical compositions. Only Matthew Hill is composed of pyroclastic rocks that are similar in character to those of the Ellendale vents. This vent is partially covered by aeolian sands and has very low (1 m) positive topographic relief. Nothing is known concerning its structure.

The apparent absence of vents may simply be related to their burial by recent unconsolidated sediments. However, as throughout much of the region such sediments are thin or absent, we do not favor this hypothesis. A further possibility is that vents will eventually be formed. This hypothesis assumes that the Leucite Hills volcanic field is still active, and is based upon the duration of magmatism (3–5 m.y.) in the West Kimberley province.

Differences between the Leucite Hills and Ellendale activity may reflect differences in the near-surface hydrological regime. If vent formation is related to magma–groundwater interactions, then it follows that the Leucite Hills magmas must have intruded dry rocks—i.e., the Eocene-to-Cretaceous sandstones of the Rock Springs Uplift were not aquifers 2 Ma ago. There is evidence (e.g., vesicles, whole rock H_2O contents, etc.) that water was present in the Leucite Hills magmas. However, the expulsion of this presumably juvenile water from the magma was not of an explosive nature. The reasons for the relatively quiet degassing are not understood, but may be related to the very low viscosity and high eruption rates of these water-rich magmas.

5.7. COMPARISON WITH KIMBERLITE DIATREMES

Kimberlite diatremes (Figure 5.2) are steeply inclined, cone or carrot-shaped bodies (Mitchell 1986, Clement and Reid 1989). Their marginal dips are steep, ranging from 75° to 85° and independent of the mechanical or hydrological properties of the host rocks. The typically constant dip and downward tapering of the diatreme causes the cross-sectional area to decrease regularly with depth. Approximately circular or elliptical outcrop plans are characteristic of the diatreme proper. The diatremes are filled with several petrographically distinct varieties of tuffisitic kimberlite breccia. These differ in the size, shape, and types of xenoliths and cognate clasts. The xenolith suite contains angular clasts derived from the local country rocks, together with fragments of formations derived from higher stratigraphic levels. Clasts derived from these now eroded formations have in some instances been estimated to have descended up to 1 km within the diatreme. Other features of note are rare sunken blocks of epiclastic kimberlite, derived from the crater zone, and the absence of metamorphic effects on lithic clasts.

Diatremes taper downwards until they are about 100–200 m in diameter. With increasing depth they are gradational into the irregularly shaped bodies of hypabyssal facies kimberlites which comprise the root zone. Diatremes appear to represent low-temperature nonviolent emplacement events. Important features include the extensive mixing of autolithic clasts (juvenile lapilli) and pelletal lapilli with angular unmetamorphosed country rocks fragments and mantle-derived xenoliths and xenocrysts.

Importantly, air-fall tuffs, base surge deposits, and other pyroclastic rocks typical of lamproite vents are not present in kimberlite diatremes. Although the diatreme filling is termed "tuffisitic," it is important to realize that these rocks are not tuffs or tuffisites. Mitchell (1986) has suggested that they are best termed volcaniclastic breccias to eliminate this terminological confusion. Diatremes do not contain central intrusions of hypabyssal kimberlite and there are no equivalents of the lava lakes found in lamproite vents.

Comparison of the above features with lamproite vents demonstrates clearly that such vents have little in common with kimberlite diatremes. This difference is a consequence of different mechanisms of formation. In the case of kimberlite diatremes there is no agreement between those who favor hydrovolcanic processes (Lorenz 1984), fluidization (Dawson 1980), or various forms of embryonic pipe modification (Clement and Reid 1989, Mitchell 1986). For a detailed discussion of these hypotheses the reader is referred to Mitchell (1986).

The difference in volcanic style may result from the different volatile contents of the two magmas. Lamproite volatile budgets are dominated by H_2O and F, and are lacking in CO_2. These former volatiles are highly soluble in silicate melts, and although their solubility has not been experimentally determined in melts of lamproite composition, it is predicted that their combined solubility in supraliquidus lamproitic melts is on the order of 3–10 wt % or more, even at low pressures (0.5–1 kb). Thus, during the ascent of volatile-rich lamproite magmas, it is not until shallow depths (0.5–1 km) that significant exsolution of an aqueous fluid phase will occur. Expansion of this fluid at these low pressures is thought to provide much of the energy required to form the funnel-shaped lamproite vents.

In contrast, CO_2 is the dominant volatile constituent of kimberlite magmas. It is much less soluble than water in silicate melts, and the exsolution of a CO_2-rich fluid phase is expected to occur at relatively greater depths than those suggested for a H_2O-dominated fluid. Thus, kimberlite magmas can be expected to exsolve volatiles at greater depths than lamproites. Such deeper explosions may result in the formation of the root zone and embryonic kimberlite pipe at depths 1–2 km below those at which lamproite vent formation takes place.

5.8. SUMMARY

Lamproites are petrologically diverse and complex. They exist in a variety of igneous forms and facies, including many of those found in common igneous rock bodies, such as dike swarms, sill complexes, necks, pipes, vents, cinder cones, shallow subvolcanic bodies, lava lakes, lava flows, and a variety of complex pyroclastic flow and fall deposits. Lamproites are classified into four distinct facies: lava flow, crater or pyroclastic, hypabyssal, and plutonic. Lamproite textures vary from massive to fragmental, vesicular to dense, porphyritic and aphanitic to phaneritic and equigranular, and vitreous to holocrystalline. Quenched glass is characteristic of lamproites and is present in a majority of lamproite provinces or fields.

Lamproites ascend rapidly from the mantle as fluid H_2O-rich magmas. These are emplaced in the upper crust as dike swarms acting as feeders to volcanic centers. These centers exhibit a characteristic style of eruption in which an initial pyroclastic phase of activity is followed by quiescent magmatic activity. Lamproite vents are not diatremes and are distinct in structure from kimberlite diatremes.

Our understanding of the eruptive styles of lamproite volcanoes is limited, not only because of the absence of any current lamproite volcanism, but also because of insufficient study of known vents and lack of experimental data. Such basic data as the solubility of H_2O, CO_2, CO, CH_4, F, S, Cl and other volatiles, and physical properties (viscosity, density, surface tension, etc.) have yet to be measured in lamproite liquids as a function of pressure, temperature, and composition. Once these basic data are available, modeling of lamproite ascent and eruption will be possible.

Minerals are the archives of the rocks
O.F. Tuttle

Mineralogy of Lamproites

The mineralogy of the lamproite clan is perhaps one of the most exotic of all alkaline rocks. The perpotassic and peralkaline traits of the magmas lead to a distinctive mineral assemblage that is virtually unparalleled. The paragenesis and composition of the minerals comprising lamproites are detailed in this chapter. The approach followed is to describe the minerals approximately in order of their modal importance in the sequence: Mafic, felsic, and accessory minerals.

Of the major minerals occurring in lamproites, only phlogopite exhibits extensive compositional variation. Amphiboles, feldspars, spinels, and priderite show limited variation, whereas pyroxenes, leucites, and olivines are essentially of constant composition. The compositions of lamproite minerals are compared and contrasted, where appropriate, with those of the same phases in other types of potassic rocks.

6.1. PHLOGOPITE

Titanian phlogopite, although not ubiquitous, is one of the characteristic minerals of lamproite. It is also one of the few phases that exhibits extensive compositional variation and is therefore of importance in assessing petrogenetic relationships between different varieties of lamproite. Because of this, the compositional variation of phlogopite with respect to the major elements FeO_T (total Fe expressed as FeO), TiO_2, and Al_2O_3, within individual lamproite provinces, is described in detail, and then compared and contrasted with that of phlogopite in minettes, kimberlites, and other ultrapotassic rocks.

6.1.1. Paragenesis

Phlogopite occurs in the following parageneses:

1. Phenocrysts;
2. Groundmass poikilitic plates;
3. Mantled micas;
4. Phlogopite pyroxenite and phlogopitite inclusions; and
5. Coronas around olivine and rarely chromite.

Phenocrystal micas are strongly pleochroic from colorless to pale yellow and from pinkish-hued yellow to reddish brown (Figure 2.14). The intensity of pleochroism (normal) increases with increasing titanium content. Increasing intensity of pleochroism from core to margin is rarely observed and reflects continuous zonation towards marginal iron and titanium enrichment. Optical absorption spectra are not yet available. Kink banding, undulose extinction, and extensive chloritization are not usually evident. The majority of the phenocrysts are inclusion-free, although rarely euhedral inclusions of chromite or armalcolite may be present.

Many phlogopites have corroded and embayed margins, indicative of resorption in the magmas that formed their host rocks. Very thin (<1 μm) rims of tetraferriphlogopite are found on some phenocrysts. The phenocrysts in some of the Leucite Hills lamproites may be completely resorbed, their former presence being indicated by sagenitic grids of priderite (Kuehner 1980). Olivine has also been reported by Kuehner (1980) as replacing phlogopite along cleavage planes.

Groundmass micas are poikilitic subhedral plates, up to 1 cm in maximum dimension, enclosing earlier crystallizing phases such as diopside, leucite, priderite, apatite, and spinel. They are strongly pleochroic in shades of yellow to reddish-brown, increasing intensity of pleochroism being correlated with increasing total iron and titanium contents. The plates commonly exhibit a single twin plane. Zonation towards margins of tetraferriphlogopite showing a dark red to dark brown normal or reversed pleochroism is common (Figures 2.21, 2.22).

In many lamproites, phlogopite appears not to have been a liquidus phase throughout the entire crystallization interval of the magma, and the mineral may occur exclusively either as phenocrysts or groundmass plates. This observation is the basis for the twofold petrographic division of lamproites (Section 2.5.2.4). A number of occurrences are known in which resorbed phenocrysts are mantled by discrete overgrowths of mica whose composition is similar to that of groundmass phlogopite. Commonly such mantles are transitional into groundmass plates and are clearly contemporaneous with groundmass mica crystallization. The mantles are euhedral to subhedral plates that are crowded with euhedral inclusions of priderite, wadeite, and apatite which stand in marked contrast to the inclusion-free irregularly resorbed cores (Figure 6.1). Polysynthetic twinning is common and has developed after the growth of the mantles, as the twins cross the core–mantle boundaries with optical continuity. The mantles always exhibit stronger pleochroism than the cores.

Rocks containing mantled phlogopites of this type are referred to in this work as *transitional madupitic lamproites* to emphasize their petrographic characteristics. They are particularly important in establishing the compositional evolution of lamproite micas. Examples are known from Mount North (West Kimberley), Middle Table Mountain, Badger's Teeth, and Iddings Butte (Leucite Hills). The Mount North occurrence is important in that consanguineous lamproites contain compositionally similar phenocrystal phlogopite. However, mantles are only developed in the more slowly cooled batches of magma.

Phenocrysts consisting of inclusion-free cores, which exhibit stronger pleochroism, and are richer in Fe than their margins, are not synonymous with micas in transitional madupitic lamproites. In examples from the Leucite Hills the mantles are of the same composition as phenocrystal micas such cores. Commonly, the core–mantle interface is not well defined and represents a region of strong continuous zoning rather than a discrete overgrowth. These micas probably result from magma mixing processes in which crystals of relatively more evolved phlogopite are incorporated into relatively less evolved batches

MINERALOGY OF LAMPROITES

Figure 6.1. Back-scattered electron image of a mantled phlogopite phenocryst in a transitional madupitic lamproite from Mount North (West Kimberley). Note that inclusions of priderite (white), and apatite (grey) occur only in the mantle.

of magma. Such reversely zoned/mantled phenocrysts are rare in lamproites but are common in lamprophyres and kimberlites.

Phlogopite occurs intergrown with olivine and pyroxene in pyroxenite inclusions and as monomineralic aggregates. The inclusions are coarse grained and may represent cognate accumulations from lamproite magmas, crystallizing at relatively high pressures, which have been incorporated into subsequent batches of magma.

Single crystals of anhedral olivine, and rarely spinel, in the Leucite Hills lamproites are commonly characterized by coronas consisting of interlocking laths of phlogopite (Figure 6.2). Commonly the laths are arranged tangentially around the nucleus. Phlogopite in these coronas appears to have nucleated preferentially upon olivine rather than being the result of a reaction between olivine and liquid (Edgar *et al.* 1976), as olivines with and without coronas may coexist in the same lamproite. The coronas probably developed when olivine crystals

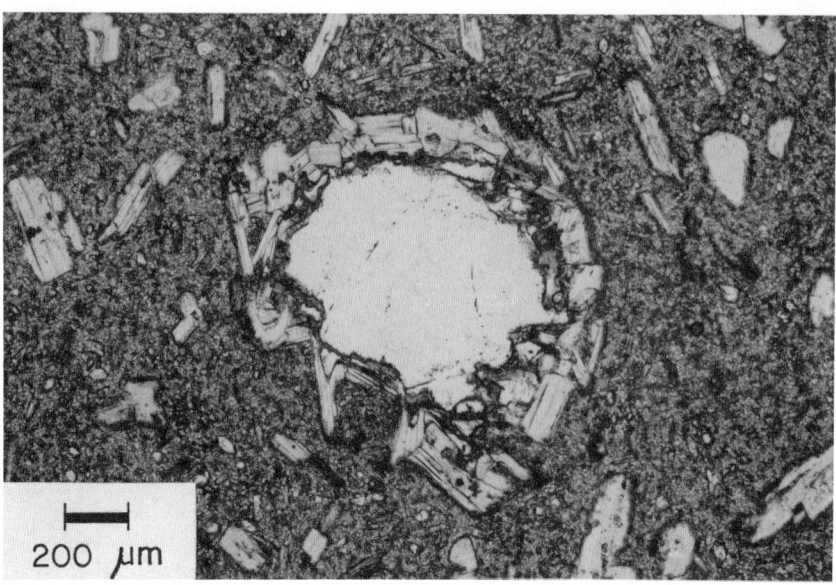

Figure 6.2. Olivine macrocryst mantled by phlogopite in a diopside leucite phlogopite lamproite from South Table Mountain (Leucite Hills). Plane polarized light.

were added to magma saturated in phlogopite, their addition simply promoting further crystallization of the current near-liquidus phase.

6.1.2. Leucite Hills, Wyoming

Mica occurs in four parageneses, that correspond to three compositionally distinct groups:

1. Titanian phlogopite phenocrysts;
2. Reaction mantles of titanian phlogopite around olivine;
3. Groundmass poikilitic plates, or mantles, upon phenocrysts of paragenesis 1 and composed of Ti, Fe-rich phlogopite, and tetraferriphlogopite; and
4. Aluminous phlogopites occurring in mica-rich inclusions, as macrocrysts and as the cores of some phenocrysts.

Discussion of the compositional variation exhibited by the Leucite Hills micas is based primarily upon data presented by Mitchell (1989b), supplemented by the few compositions determined by Carmichael (1967), Kuehner (1980), and Barton and van Bergen (1981).

6.1.2.1. Phenocrysts

Phenocrystal micas in sanidine-bearing or sanidine-free lamproites exhibit a very restricted range in composition and are low-Ti phlogopites containing 1.5–2.8 wt % TiO_2, 2–4 wt % FeO_T, 11–13 wt % Al_2O_3, 0.1–1.5 wt % Cr_2O_3, 0.05–0.4 wt % NiO, 0.2–0.7 wt % BaO, and less than 0.3 wt % Na_2O, with *mg* ranging from 0.91 to 0.95. Representative compositions are given in Table 6.1.

MINERALOGY OF LAMPROITES

Table 6.1. Representative Compositions (wt %) of Leucite Hills Phenocrystal Phlogopite[a]

	1	2	3	4	5	6	7	8	9	10
SiO_2	41.47	41.37	41.23	42.31	41.81	41.75	41.11	41.77	41.65	41.73
TiO_2	2.34	2.36	2.02	2.01	1.99	2.43	1.91	1.94	2.17	2.39
Al_2O_3	11.89	12.03	11.89	11.79	11.91	12.55	12.66	12.06	12.05	12.49
Cr_2O_3	1.05	0.20	0.53	0.83	0.25	1.40	0.48	0.44	0.54	0.94
FeO[b]	2.84	3.30	2.53	2.80	2.77	3.26	3.05	2.64	2.58	2.76
MnO	n.d.	0.01	0.03	0.04	0.01	0.02	0.02	n.d.	0.09	0.04
MgO	25.20	25.39	25.57	24.10	25.24	23.91	24.94	25.02	25.23	24.52
CaO	n.d.	0.20	n.d.	n.d.	n.d.	n.d.	0.01	n.d.	n.d.	n.d.
Na_2O	—	0.08	0.16	0.29	0.24	0.08	0.05	0.32	0.10	0.08
K_2O	10.44	10.82	10.59	10.73	10.66	10.79	10.58	10.52	10.91	10.38
NiO	0.16	0.22	0.37	0.21	0.23	0.29	0.17	0.29	0.27	0.21
	95.39	95.71	94.92	95.11	95.11	96.48	94.98	95.00	95.59	95.54
Structural formulas based on 22 oxygens:										
Si	5.869	5.851	5.867	6.003	5.929	5.866	5.845	5.926	5.888	5.885
Al	1.983	2.005	1.994	1.971	1.991	2.078	2.121	2.017	2.008	2.076
Ti	0.249	0.251	0.216	0.214	0.212	0.257	0.204	0.207	0.231	0.253
Cr	0.117	0.022	0.060	0.093	0.028	0.156	0.054	0.049	0.060	0.105
Fe	0.336	0.358	0.301	0.332	0.329	0.383	0.363	0.313	0.305	0.325
Mn	—	0.001	0.004	0.005	0.001	0.002	0.002	—	0.011	0.005
Mg	5.316	5.352	5.423	5.096	5.335	5.007	5.285	5.291	5.316	5.154
Ca	—	0.030	—	—	—	—	—	—	—	—
Na	—	0.022	0.044	0.080	0.066	0.022	0.014	0.088	0.027	0.022
K	1.885	1.952	1.922	1.942	1.928	1.934	1.919	1.904	1.967	1.867
Ni	0.018	0.025	0.042	0.024	0.026	0.033	0.019	0.033	0.031	0.024
CAT	15.774	15.871	15.873	15.761	15.846	15.738	15.829	15.829	15.844	15.716

[a]1, Black Butte; 2, Deer Butte; 3, Cabin Butte; 4, Spring Butte; 5, South Table Mountain; 6, Badger's Teeth; 7, Middle Table Mountain; 8, Iddings Butte; 9, Hallock Butte; 10, Emmons Mesa. All data from Mitchell (1989b). CAT, cation sum; n.d., not detected.
[b]Total Fe calculated as FeO.

Figures 6.3 and 6.4 show that phenocrysts in the nine lavas and cinder cones occurring at Spring Butte are identical in composition with respect to their Ti, Al, and Fe contents. Systematic differences in minor element contents are not found. The data demonstrate the absence of a compositional trend at a single evolving volcanic center.

Figures 6.5 and 6.6 demonstrate that phenocrysts, and the cores of mantled phenocrysts, are similar and that the compositional variation found at any one locality is similar to that exhibited by the province as a whole. Micas from the Boar's Tusk, Middle Table Mountain, and Hatcher Mesa are on average slightly richer in Al than micas from the other centers. The principal compositional variation is a weak zonation with respect to Cr and Al. Typically, phenocrystal cores are richer in Al, Cr (Figure 6.7), and Ni than their margins. Slight increases in FeO_T, which rarely exceed 4.0 wt %, at essentially constant Ti contents accompany the Cr–Al depletion.

6.1.2.2. Coronas Around Olivines

Phlogopites forming coronas around olivines are also titanian phlogopites. Their composition does not differ from that of associated phenocrystal phlogopite, as illustrated in Figures 6.5 and 6.6 for examples from South Table Mountain and Endlich Hill. This

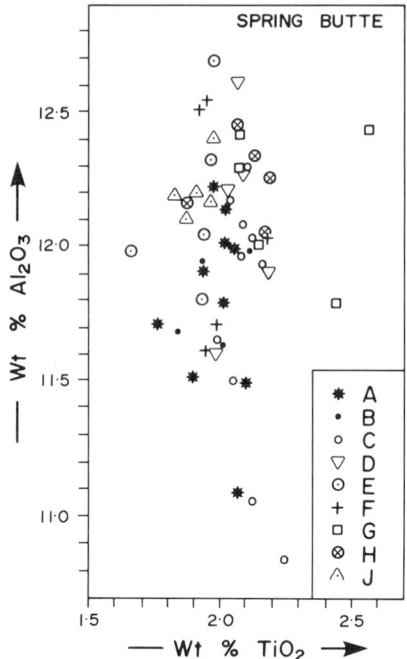

Figure 6.3. Al$_2$O$_3$ versus TiO$_2$ for phenocrystal phlogopites in the Spring Butte (Leucite Hills) lamproites (Mitchell 1989b). Lava eruption sequence from A to J is from Ogden (1979).

compositional similarity supports the contention that these micas are merely near-liquidus phases and not reaction products (see Section 6.1.1).

6.1.2.3. Groundmass Micas

Groundmass micas are Al-deficient and Ti-, Fe-rich, relative to phenocrystal micas, and exhibit an extremely wide range in composition. Representative examples given in Table 6.2 demonstrate that they are titanian biotite–phlogopite solid solutions and titanian tetraferriphlogopites. The overall compositional range is 3–12 wt % TiO$_2$, 4–11 wt % FeO$_T$, 5–11 wt % Al$_2$O$_3$ with <0.2 wt % NiO and <0.1 wt % Cr$_2$O$_3$. The sodium contents lie primarily between 0.2 and 0.6 wt % Na$_2$O with the exception of the Badger's Teeth intrusion where madupitic micas contain 1.0–1.5 wt % Na$_2$O. Barium contents range from 0.7 to 2.4 wt % BaO in the Badger's Teeth and Pilot Butte micas. Increasing Ti contents are accompanied by increasing Fe and decreasing Al. Weakly developed zonation follows this trend. Figure 6.8 shows that each volcanic center contains a groundmass or mantling mica of distinct composition with respect to TiO$_2$ content. Figure 6.9 shows that the trend of Al depletion and Fe increase is similar at all centers, with the exception of Hatcher Mesa where the degree of Al depletion is not as pronounced.

6.1.2.4. Aluminous Phlogopite

Some Al-rich phlogopites occur in cryptogenic mica-rich inclusions in association with clinopyroxene and olivine. These inclusions may be either cognate or xenolithic.

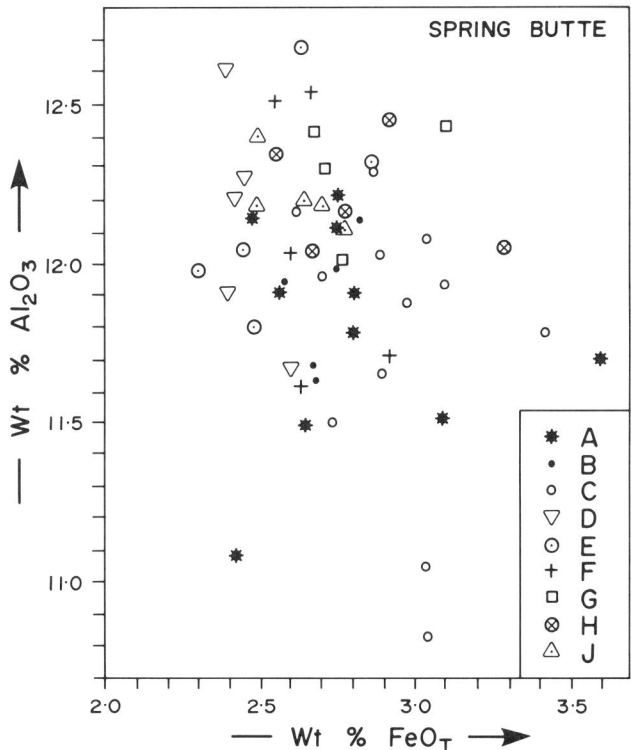

Figure 6.4. Al_2O_3 versus FeO_T for phenocrystal phlogopites in the Spring Butte (Leucite Hills) lamproites (Mitchell 1989b). Lava eruption sequence from A to J is from Ogden (1979).

An apatite-phlogopite pyroxenite from the Emmons Mesa cinder cone contains mica whose compositional range is 1.5–1.6 wt % TiO_2, 14.6–15.0 wt % Al_2O_3, 11.5–12.4 wt % FeO_T, 0.02–0.06 wt % Cr_2O_3, 0.06–0.36 wt % Na_2O with *mg* from 0.71 to 0.72 (Table 6.3 analysis 1). An olivine phlogopite pyroxenite from the same locality contains mica whose compositional range is 1.1–1.2 wt % TiO_2, 15.5–16.2 wt % Al_2O_3, 9.2–10.9 wt % FeO_T, 0.3–0.6 wt % Cr_2O_3, 0.3–0.7 wt % Na_2O with *mg* of 0.77 (Table 6.3 analysis 2).

Figures 6.8 and 6.9 show that each inclusion contains mica of distinct composition and that these micas are richer in Fe and Al and poorer in Ti than phenocrystal micas. Although the Fe contents are similar to those of the groundmass micas they may be distinguished from the latter on the basis of their higher Al and lower Ti contents.

Phlogopite of significantly different composition from the above has been described by Barton and van Bergen (1981) from mica-rich inclusions in the Hatcher Mesa lavas. These micas are richer in Al_2O_3 (>15 wt %), slightly richer in TiO_2, and poorer in FeO_T than the Emmons Mesa inclusion micas (Table 6.3 analysis 3). Megacrysts (1–2 mm) and phenocrystal cores in the same lava are unzoned but exhibit wide intergrain variations in Al at essentially constant Ti and Fe (Table 6.3 analyses 4 and 5, Figures 6.8 and 6.9). The composition of some of these crystals is similar to phlogopites in the mica-rich inclusions. All of these Al-rich micas are mantled by phlogopite identical in composition to the phenocrysts found in this lava.

At Hallock Butte and South Table Mountain, aluminous micas form aggregates of

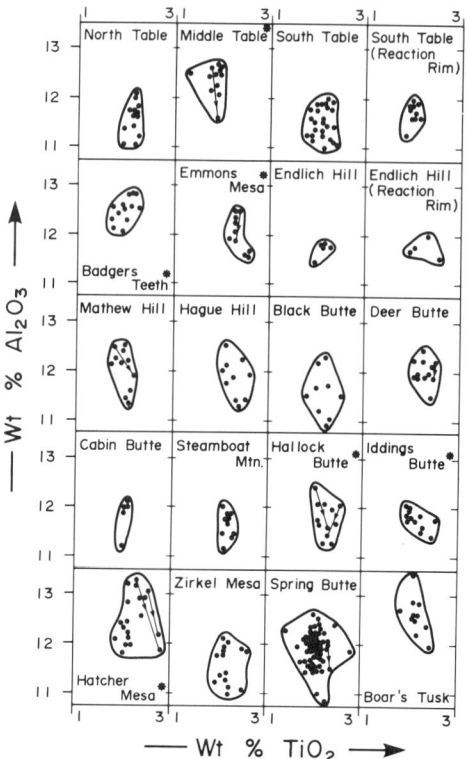

Figure 6.5. Al$_2$O$_3$ versus TiO$_2$ for phenocrystal phlogopites from the volcanic centers of the Leucite Hills (Mitchell 1989b). Localities that also contain groundmass micas indicated by (∗).

intergrown crystals (microinclusions). Commonly these are mantled by phenocryst-type micas (Figures 6.8 and 6.9). The micas differ from those found at Hatcher Mesa in that they are enriched in Cr$_2$O$_3$ (2.1–3.9 wt %) and exhibit wide variations in Ti contents (Table 6.3 analyses 6 and 7).

Micas consisting of discrete reddish to brown pleochroic cores, mantled by pale yellow phenocryst-type micas, occur in the Emmons Mesa and South Table Mountain lavas (Kuehner *et al.* 1981, this work). The micas vary widely in composition and are richer in Fe, Ti, and Al than phenocrystal micas (Table 6.3 analyses 8–10; Figures 6.8 and 6.9).

6.1.2.5. Summary

The principal conclusions to be drawn from the Leucite Hills phlogopite compositional variation are as follows:

1. Phenocrystal micas throughout the whole province are essentially of similar composition. This observation suggests that all of the lamproites in this region were derived from a common magma, which upon eruption contained titanian phlogopite phenocrysts.

2. At some volcanic centers cooling was slow enough to allow the nucleation of Fe- and Ti-rich, Al-poor mica mantles upon phlogopite phenocrysts. Individual centers are characterized by different trends of composition with respect to the Ti content of the mantling

MINERALOGY OF LAMPROITES

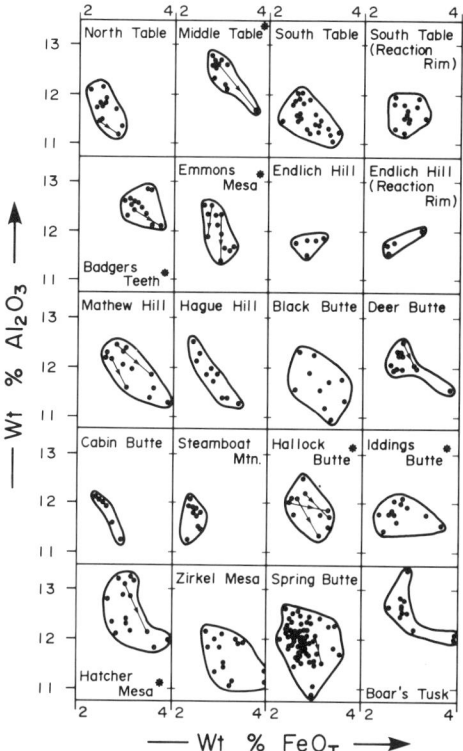

Figure 6.6. Al$_2$O$_3$ versus FeO$_T$ for phenocrystal phlogopites from the volcanic centers of the Leucite Hills (Mitchell 1989b). Localities that also contain groundmass micas indicated by (∗).

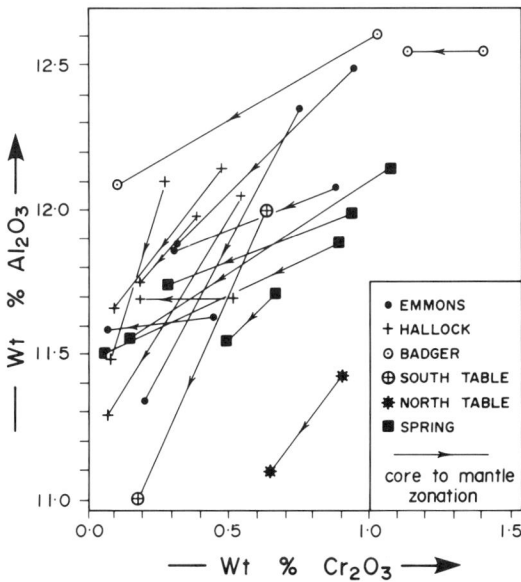

Figure 6.7. Representative zonation trends with respect to the Al$_2$O$_3$ and Cr$_2$O$_3$ content of phenocrystal phlogopites from the Leucite Hills.

Table 6.2. Representative Compositions of Leucite Hills Mantling and Groundmass Phlogopite[a]

	1	2	3	4	5	6	7	8	9	10
SiO_2	40.28	41.49	39.80	40.71	38.18	38.45	41.29	41.81	40.37	43.59
TiO_2	5.64	6.79	7.74	8.81	8.75	10.63	5.13	4.65	3.95	3.62
Al_2O_3	9.13	6.92	9.66	8.47	9.79	9.53	5.27	4.48	9.41	6.50
Cr_2O_3	0.06	0.02	0.19	0.03	0.31	0.17	n.d.	n.d.	0.04	0.04
FeO[b]	5.92	9.57	5.75	7.09	10.09	8.99	8.91	9.14	5.39	7.14
MnO	0.11	0.09	0.09	0.04	n.d.	0.10	0.08	0.13	n.d.	0.08
MgO	21.68	17.96	20.78	19.68	16.28	16.70	20.71	21.53	24.29	23.50
CaO	0.03	0.08	n.d.	n.d.	n.d.	0.00	0.13	0.03	n.d.	n.d.
Na_2O	0.14	0.80	0.41	0.40	0.35	0.46	1.35	1.24	0.07	1.02
K_2O	10.01	9.84	9.91	9.70	9.34	8.76	9.62	9.78	10.09	10.29
BaO	1.17	1.40	1.80	1.46	2.28	3.55	2.46	2.70	2.27	1.27
NiO	0.13	0.04	0.12	0.12	0.07	n.d.	0.05	0.07	0.06	n.d.
	94.30	95.00	96.25	96.51	95.44	97.34	95.00	95.56	95.94	97.05
Structural formulas based on 22 oxygens:										
Si	5.918	6.167	5.763	5.884	5.708	5.645	6.197	6.259	5.862	6.256
Al	1.581	1.212	1.649	1.443	1.725	1.649	0.932	0.790	1.610	1.100
Ti	0.623	0.759	0.843	0.958	0.984	1.174	0.579	0.523	0.431	0.391
Cr	0.007	0.002	0.022	0.003	0.037	0.020	—	—	0.005	0.005
Fe	0.727	1.190	0.696	0.857	1.262	1.104	1.118	1.144	0.654	0.857
Mn	0.014	0.011	0.011	0.005	—	0.012	0.010	0.016	—	0.010
Mg	4.748	3.979	4.485	4.239	3.628	3.654	4.633	4.804	5.257	5.027
Ca	0.005	0.013	—	—	—	—	0.021	0.005	—	—
Na	0.040	0.231	0.115	0.112	0.101	0.131	0.393	0.360	0.020	0.284
K	1.876	1.866	1.831	1.788	1.781	1.641	1.842	1.868	1.869	1.884
Ba	0.067	0.082	0.102	0.083	0.134	0.204	0.145	0.158	0.129	0.071
Ni	0.015	0.005	0.014	0.014	0.008	—	0.006	0.008	0.007	—
CAT	15.622	15.515	15.531	15.386	15.368	15.233	15.875	15.936	15.844	15.885

[a] 1–2, Middle Table Mountain; 3–4, Iddings Butte; 5–6, Hatcher Mesa; 7–8 Badger's Teeth; 9–10 Pilot Butte. Data for 1–9 from Mitchell (1989b); 10 from Kuehner (1980). Analyses 1–6 are mantling micas; 7–10 are groundmass micas. CAT, cation sum; n.d., not detected.
[b] Total Fe calculated at FeO.

mica, presumably as a consequence of the local conditions of crystallization. With respect to Fe, all of the centers, except Hatcher Mesa, follow a similar trend of Fe enrichment and Al depletion, leading to the formation of micas rich in the tetraferriphlogopite component. The similar trends reflect similar redox conditions during crystallization.

3. Groundmass poikilitic micas in the madupitic lamproite at Pilot Butte are similar to mantling micas and therefore must be considered to be evolved relative to phenocrystal compositions. Pilot Butte lamproites lack phenocrysts, and the magma that formed these rocks must have been subject to some differentiation process that removed phenocrysts and caused Fe-enrichment. These observations imply that madupitic lamproites are not representative of primary magma compositions.

4. Aluminous micas are not in equilibrium with their host lavas, as evidenced by the development around them of mantles of phenocrystal composition. This observation suggests that they were present in the magma prior to formation of the phenocrystal-type micas. They may be high-pressure cognate phases, or xenocrysts in a magma already crystallizing phlogopite. This suite of micas is inadequately characterized and it is important to determine their origins. If they are cognate, study of their compositional variation will provide evidence regarding the crystallization of lamproite magma at high pressures. In particular, a

MINERALOGY OF LAMPROITES

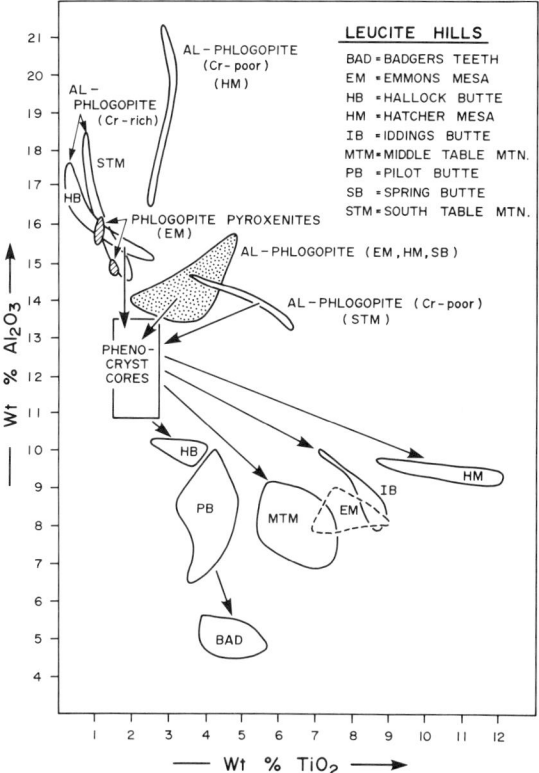

Figure 6.8. Al_2O_3 versus TiO_2 for groundmass micas, aluminous phlogopite macrocrysts (Al-phlogopite) and micas in phlogopite pyroxenite inclusions from the Leucite Hills (Mitchell 1989b).

period of aluminous phlogopite crystallization may lead to the development of peralkalinity in derivative magmas.

6.1.3. West Kimberley, West Australia

The compositional variation exhibited by mica in this province is extensive, and is described separately for each of the three principal varieties of lamproite.

6.1.3.1. Olivine Lamproites

Micas occurring in the Ellendale olivine lamproites have been described by Jaques *et al.* (1984, 1986). Representative compositions are given in Table 6.4.

Phlogopite occurs as a microphenocryst (up to 2 mm) and as a groundmass phase. The microphenocrysts are zoned from weakly pleochroic cores to more strongly pleochroic inclusion-bearing margins. Groundmass micas occur either as relatively coarse (up to 1 mm) poikilitic plates or small (<100 μm) flakes.

Microphenocrystal cores contain 3.5–5.0 wt % TiO_2, 9–13 wt % Al_2O_3 with *mg* ratios of 0.88–0.92 (Figure 6.10). They are poor in Na_2O (<0.5 wt %) with Cr_2O_3 and BaO contents ranging up to 1.3 wt % and 2.0 wt %, respectively. The cores are zoned to Fe-rich,

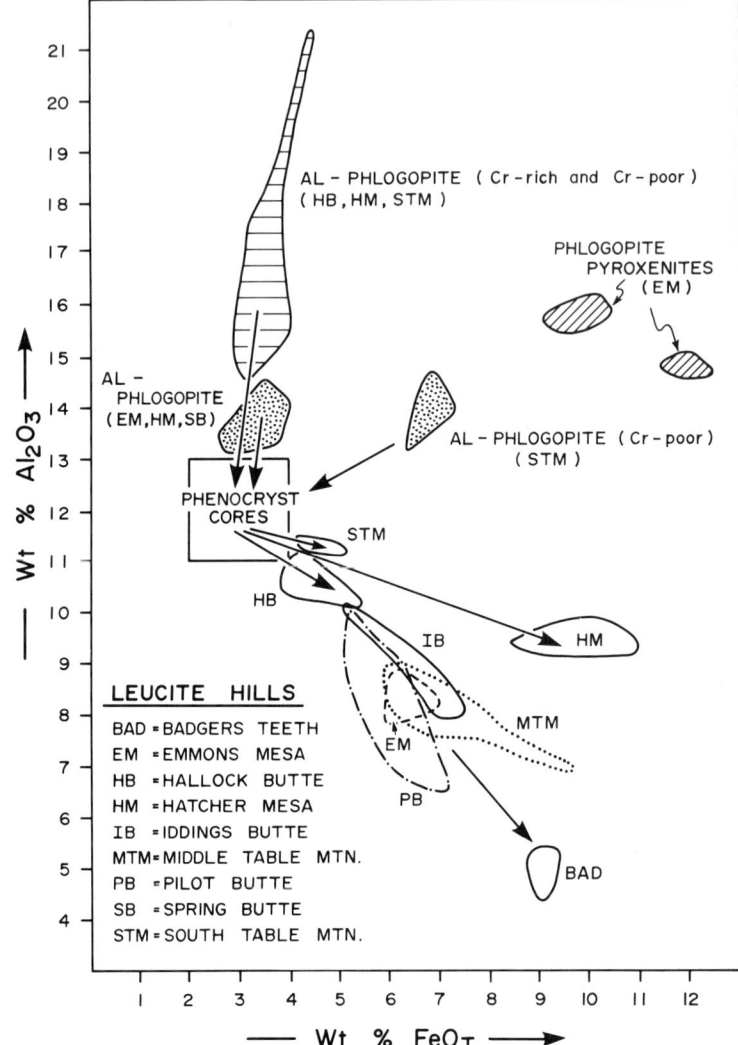

Figure 6.9. Al$_2$O$_3$ versus FeO$_T$ for groundmass micas, aluminous phlogopite macrocrysts (Al-phlogopite) and micas in phlogopite pyroxenite inclusions from the Leucite Hills (Mitchell 1989b).

Al-poor margins with low Cr contents that are compositionally similar to the groundmass micas. Fluorine contents vary from 1.6 to 5.5 wt % and increase with decreasing *mg* (see Figure 6.17).

Poikilitic groundmass plates contain 4–8 wt % TiO$_2$, 5–10 wt % Al$_2$O$_3$, <0.2 wt % Cr$_2$O$_3$, and up to 0.65 wt % Na$_2$O and 2.0 wt % BaO with *mg* ratios of 0.90–0.63. Zonation is towards increasing Fe, Ti, and Na at the expense of Mg and Al. In contrast to the microphenocrysts, fluorine contents (0–7.2 wt %) decrease as *mg* decreases.

Small phlogopite flakes exhibit a very wide range in Fe, Ti, and Al content (Figure 6.10). Relatively Al-rich varieties (4–9 wt % Al$_2$O$_3$) with *mg* ratios of 0.90–0.80 exhibit

MINERALOGY OF LAMPROITES

Table 6.3. Representative Compositions of Leucite Hills Aluminous Phlogopite[a]

	1	2	3	4	5	6	7	8	9	10
SiO_2	38.23	40.20	35.6	35.4	40.01	38.83	37.38	36.59	39.72	38.93
TiO_2	1.52	1.15	3.15	2.85	3.27	2.55	0.48	4.81	6.47	3.60
Al_2O_3	14.81	16.17	20.0	21.3	14.65	15.08	17.31	15.75	13.24	14.61
Cr_2O_3	0.02	0.31	0.40	0.24	0.73	2.13	2.48	0.86	0.07	n.d.
FeO[b]	11.54	10.11	4.19	4.49	3.48	3.31	3.40	8.70	6.43	6.87
MnO	0.16	0.09	n.d.	n.d.	n.d.	0.10	0.09	0.06	0.10	0.05
MgO	17.04	16.29	21.5	21.7	22.73	21.66	22.51	18.52	19.85	20.64
CaO	n.d.	n.d.	n.d.	n.d.	n.d.	n.d.	n.d.	n.d.	n.d.	n.d.
Na_2O	0.24	0.50	0.43	0.47	0.15	0.20	n.d.	0.10	0.10	0.05
K_2O	9.49	10.03	9.45	9.58	10.34	10.50	10.91	10.38	10.35	10.74
BaO	2.18	0.46	n.d.	n.d.	0.39	n.d.	n.d.	n.d.	1.50	0.52
NiO	0.00	0.15	n.d.	n.d.	0.15	0.16	0.13	0.21	n.d.	n.d.
	95.23	95.46	94.72	96.03	95.9	94.51	94.69	95.98	97.83	96.01
Structural formulas based on 22 oxygens:										
Si	5.711	5.854	5.094	5.002	5.660	5.589	5.390	5.328	5.638	5.607
Al	2.607	2.775	3.373	3.547	2.443	2.559	2.942	2.703	2.215	2.480
Ti	0.171	0.126	0.339	0.303	0.348	0.276	0.052	0.527	0.691	0.390
Cr	0.002	0.036	0.045	0.027	0.082	0.242	0.284	0.099	0.008	—
Fe	1.442	1.231	0.501	0.531	0.412	0.399	0.410	1.059	0.763	0.827
Mn	0.020	0.011	—	—	—	0.012	0.011	0.007	0.012	0.006
Mg	3.794	3.536	4.585	4.570	4.793	4.648	4.838	4.019	4.200	4.431
Ca	—	—	—	—	—	—	—	—	—	—
Na	0.070	0.141	0.119	0.129	0.041	0.056	—	0.028	0.028	0.014
K	1.808	1.863	1.725	1.727	1.866	1.928	2.007	1.928	1.874	1.973
Ba	0.128	0.018	—	—	0.022	—	—	—	0.083	0.029
Ni	—	0.018	—	—	0.017	0.018	0.015	0.024	—	—
CAT	15.753	15.617	15.781	15.836	15.683	15.727	15.948	15.723	15.511	15.757

[a]1,2, phlogopite pyroxenites, Emmons Mesa (this work); 3,4, phlogopite in mica-rich inclusions and as megacryst respectively, Hatcher Mesa (Barton and van Bergen 1981); 5, discrete core in phenocryst, Hatcher Mesa (this work); 6–7, phlogopite microxenolith, Hallock Butte (this work); 8, discrete core, Emmons Mesa; 9–10, discrete cores, South Table Mountain (Kuehner 1980). CAT, cation sum; n.d., not determined.
[b]Total Fe calculated as FeO.

normal pleochroism. Al-poor types (<4 wt % Al_2O_3) are reversely pleochroic tetraferriphlogopites, containing up to 0.9 wt % Na_2O.

6.1.3.2. Coarse-Grained Richterite Madupitic Lamproites

Relatively slowly cooled batches of lamproitic magma formed coarse-grained richterite madupitic lamproites at Mount North, Rice Hill, and the Walgidee Hills. Phlogopites within these rocks exhibit the greatest compositional range of all lamproite micas so far examined and have been described by Mitchell (1981) and Jaques et al. (1984, 1986).

The leucite diopside richterite transitional madupitic lamproite (wolgidite) from Mount North contains mica phenocrysts in which discrete cores and mantles are evident, together with groundmass poikilitic plates. These latter micas are similar in optical character and composition to the outermost margins of the phenocryst mantles.

The phenocryst cores are of *uniform* composition (Table 6.5) and characterized by low TiO_2 (5–7 wt %), low FeO_T (3–5 wt %), high Al_2O_3 (10–12 wt %), low Na_2O (<0.1 wt %),

Table 6.4. Representative Compositions of Phlogopites from Olivine Lamproites from the West Kimberley Province[a]

	1	2	3	4	5
SiO_2	38.65	40.31	38.88	39.98	39.94
TiO_2	4.11	3.70	4.53	5.45	8.31
Al_2O_3	11.95	7.20	11.18	9.04	6.87
Cr_2O_3	1.25	0.01	0.78	0.05	0.01
FeO[b]	3.67	8.71	4.16	5.43	9.25
MnO	0.03	0.06	0.08	0.04	0.03
MgO	23.23	22.92	23.80	22.46	18.75
CaO	0.03	—	—	0.05	0.06
Na_2O	0.05	0.23	0.11	0.48	0.65
K_2O	10.49	10.05	10.80	9.45	9.20
BaO	0.90	1.55	0.88	1.62	1.97
NiO	0.05	—	—	0.07	—
F	1.71	4.83	3.21	5.63	2.13
	96.12	99.57	98.41	99.75	97.17
Structural formulas based on 22 oxygens:					
Si	5.624	5.995	5.632	5.896	5.946
Al	2.051	1.263	1.909	1.567	1.206
Ti	0.450	0.414	0.494	0.603	0.930
Cr	0.144	0.001	0.089	0.006	0.007
Fe	0.445	1.084	0.504	0.668	1.152
Mn	0.004	0.008	0.009	0.005	0.004
Mg	5.043	5.085	5.143	4.924	4.162
Ca	0.005	—	—	0.008	0.009
Na	0.014	0.066	0.031	0.139	0.188
K	1.949	1.908	1.997	1.773	1.748
Ba	0.051	0.091	0.050	0.093	0.115
Ni	0.003	—	—	0.005	—
CAT	15.783	15.915	15.858	15.687	15.461

[a]1,2, microphenocryst and groundmass, Ellendale 11; 3, poikilitic plate, Ellendale 4; 4,5, core and rim of poikilitic plate, Ellendale 9. All data from Jaques et al. (1986). CAT, cation sum.
[b]Total iron calculated as FeO.

and high Cr_2O_3 (0.1–1.5 wt %) relative to the mantling phlogopites. The mantles have higher FeO_T (5.5–17.5 wt %), higher TiO_2 (7–9 wt %), and lower Al_2O_3 (5–9 wt %) and no detectable Cr_2O_3 (Table 6.5). Zonation trends from inner to outer mantle are of increasing TiO_2, FeO_T, BaO (0.5–1.1 wt %), and Na_2O (0.15–0.7 wt %) coupled with moderately decreasing Al_2O_3 (Figures 6.11 and 6.12). Figure 6.13 illustrates that the trend of Ba enrichment is reversed as Na (and Fe) increase. Groundmass micas are similar in composition to the outermost portions of the mantles.

The richterite madupitic lamproites from Rice Hill contain only zoned groundmass phlogopites. These are similar in composition (Table 6.5) to the Mount North mantle mica, but on average are richer in TiO_2 (8–10 wt %) and FeO_T (6.5–29.0 wt %). Zonation trends (Figures 6.11 and 6.12) are similar to the Mount North micas but proceed to a greater degree of Fe enrichment. The ultimate composition of the groundmass micas is that of titanian tetraferriphlogopite (4–8 wt % TiO_2, <2 wt % Al_2O_3). The trend of Na enrichment is coupled with Ba depletion (Figure 6.13).

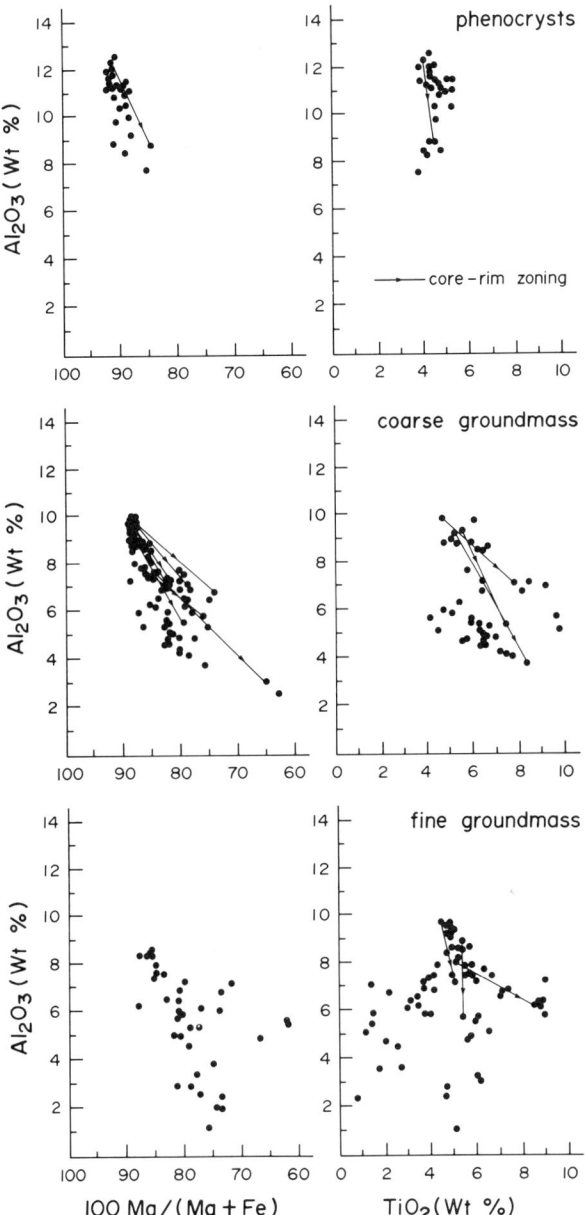

Figure 6.10. Compositional variation of phlogopites in olivine lamproites from the Ellendale field of the West Kimberley province (after Jaques *et al.* 1986).

Table 6.5. Representative Compositions of Phlogopites from Richterite Madupitic Lamproites from West Kimberley Province

	1	2	3	4	5	6	7	8	9	10
SiO_2	41.51	40.39	40.53	42.76	41.71	39.05	43.18	40.49	37.57	40.96
TiO_2	5.98	8.13	8.56	6.13	8.05	8.17	8.37	9.33	7.35	4.68
Al_2O_3	10.44	7.87	4.66	9.86	8.05	6.19	7.16	5.08	1.45	1.88
Cr_2O_3	0.17	n.d.	n.d.	0.35	0.02	0.01	n.d.	n.d.	n.d.	n.d.
FeO[b]	3.96	6.74	11.55	3.70	5.69	17.48	8.10	14.31	29.18	28.94
MnO	0.04	0.05	0.10	n.d.	n.d.	n.d.	0.03	0.07	0.25	0.25
MgO	22.55	20.76	18.43	23.38	21.51	14.45	17.63	17.04	9.71	9.91
CaO	n.d.	0.03	0.02	n.d.	n.d.	n.d.	0.02	0.05	0.05	0.20
Na_2O	0.06	0.27	0.68	0.08	0.21	0.58	0.29	0.36	0.45	0.24
K_2O	11.01	10.10	9.83	11.12	10.67	10.57	10.26	9.35	9.53	9.25
BaO	0.39	0.76	0.71	0.23	0.63	0.93	0.76	0.78	0.38	0.09
NiO	0.07	0.07	0.07	n.d.	n.d.	n.d.	0.14	0.16	0.00	0.11
	95.08	95.16	95.13	97.60	96.54	97.43	95.94	97.07	95.92	96.25
Structural formulas based on 22 oxygens:										
Si	5.956	5.899	6.080	5.963	5.964	5.918	6.252	6.008	6.142	6.516
Al	1.766	1.356	0.823	1.623	1.356	1.105	1.222	0.888	0.280	0.354
Ti	0.538	0.893	0.966	0.642	0.866	0.931	0.911	1.041	0.903	0.560
Cr	0.019	—	—	0.038	0.003	0.001	—	—	—	—
Fe	0.463	0.823	1.450	0.432	0.681	2.216	0.981	1.776	3.989	3.875
Mn	0.004	0.006	0.012	—	—	—	0.004	0.008	0.035	0.034
Mg	4.822	4.520	4.120	4.860	4.585	3.264	3.806	3.769	2.367	2.365
Ca	—	0.004	0.004	—	—	—	0.004	0.008	0.008	0.034
Na	0.016	0.076	0.196	0.021	0.057	0.170	0.081	0.103	0.143	0.075
K	2.015	1.882	1.881	1.978	1.946	2.044	1.896	1.769	1.987	1.889
Ba	0.022	0.044	0.042	0.012	0.035	0.055	0.043	0.045	0.024	0.005
Ni	0.007	0.008	0.007	—	—	—	0.016	0.019	—	0.015
CAT	15.628	15.511	15.581	15.569	15.493	15.704	15.216	15.440	15.878	15.725

[a] 1,2,3 and 4,5,6, cores, inner, and outer mantles, respectively, Mt. North; 7,8,9, center, intermediate and rim areas of zoned groundmass plate, Rice Hill; 10, groundmass tetraferriphlogopite, Rice Hill. All data from Mitchell (1981). CAT, cation sum.
[b] Total iron calculated as FeO; n.d., not determined.

6.1.3.3. Diopside Leucite Phlogopite Lamproite

Relatively fine-grained lamproites, characterized by phenocrysts of phlogopite, leucite, and diopside, comprise the bulk of the West Kimberley province occurrences. The micas have been described by Mitchell (1981) and Jaques et al. (1984, 1986). Large poikilitic plates of phlogopite are absent and groundmass micas occur rarely as small ragged strongly pleochroic flakes. Phlogopite lamproites contain abundant phenocrysts of phlogopite and scarce groundmass mica, while leucite diopside lamproites contain only scarce groundmass mica. Some examples contain zoned phenocrysts mantled by groundmass-type micas. Phlogopite lamproites from Mount North and Mount Gytha illustrate the range of composition displayed by micas as a whole from this group of rocks (Table 6.6).

Phenocryst cores are of relatively uniform composition (Table 6.6, Figures 6.14 and 6.15) with TiO_2 (5–6 wt %), Al_2O_3 (10–12 wt %), FeO_T (3.4–4.5 wt %), Na_2O (<0.2 wt %), BaO (0.2–0.45 wt %), and Cr_2O_3 (0.2–0.8 wt %). Strongly pleochroic mantles are enriched in TiO_2 (6–7.5 wt %), FeO_T (6.3–10.0 wt %), Na_2O (0.3–0.8 wt %), BaO (0.55–1.33 wt %) and depleted in Al_2O_3 (4–7 wt %) and Cr_2O_3 (<0.05 wt %) relative to the cores (Table 6.6,

MINERALOGY OF LAMPROITES

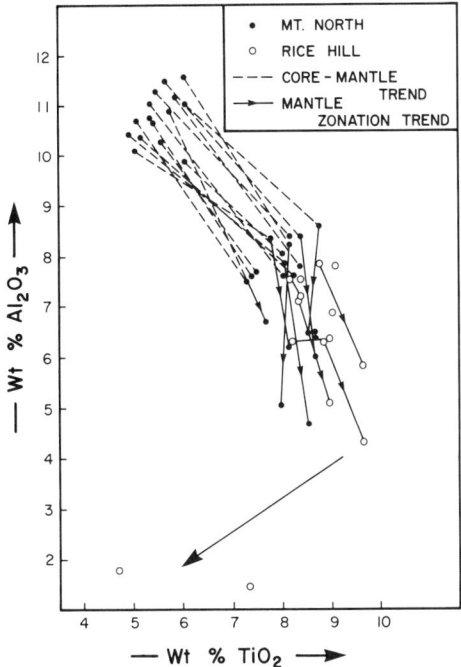

Figure 6.11. Al_2O_3 versus TiO_2 for mantled phlogopites and groundmass micas in the Mount North and Rice Hill madupitic lamproites, respectively (after Mitchell 1981).

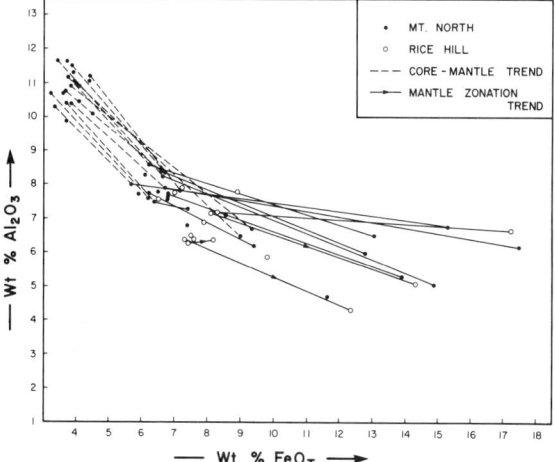

Figure 6.12. Al_2O_3 versus FeO_T for mantled phlogopites and groundmass micas in the Mount North and Rice Hill madupitic lamproites, respectively (after Mitchell 1981).

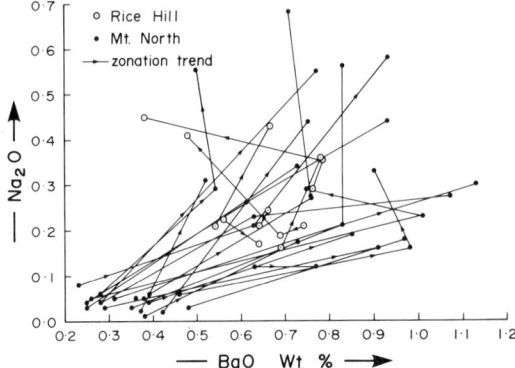

Figure 6.13. Na$_2$O versus BaO for mantled phlogopites and groundmass micas in the Mount North and Rice Hill madupitic lamproites, respectively (after Mitchell 1981).

Table 6.6. Representative Compositions of Micas from Diopside Leucite Phlogopite Lamproites, West Kimberley Province[a]

	1	2	3	4	5	6	7	8	9	10
SiO$_2$	40.78	42.97	41.94	44.15	44.29	43.37	40.58	42.21	41.48	44.35
TiO$_2$	5.31	6.18	5.84	6.52	5.83	6.37	6.45	7.56	7.07	6.06
Al$_2$O$_3$	10.83	4.69	10.70	3.67	3.05	1.84	9.53	5.04	2.48	1.28
Cr$_2$O$_3$	0.59	0.03	0.29	0.06	0.06	n.d.	n.d.	n.d.	n.d.	n.d.
FeO[b]	3.63	7.38	3.81	7.99	7.56	9.96	6.50	10.16	12.70	8.87
MnO	n.d.	0.05	n.d.	n.d.	0.05	n.d.	0.05	0.08	0.14	n.d.
MgO	22.91	21.96	23.10	21.73	22.43	20.89	21.51	19.46	18.82	19.22
CaO	0.03	0.04	n.d.	n.d.	0.06	0.03	0.01	n.d.	n.d.	0.10
Na$_2$O	0.04	0.62	0.06	0.81	0.81	1.00	0.12	0.89	1.16	1.76
K$_2$O	11.04	10.53	10.24	10.13	10.59	9.96	10.00	9.83	9.67	10.90
BaO	0.38	0.75	0.39	0.60	0.91	0.87	n.d.	n.d.	n.d.	0.54
NiO	0.18	0.06	n.d.	n.d.	0.08	n.d.	n.d.	n.d.	n.d.	n.d.
	95.72	95.25	96.37	95.66	95.70	94.29	94.76	95.25	93.51	93.09
Structural formulas based on 22 oxygens:										
Si	5.827	6.294	5.904	6.423	6.469	6.503	5.880	6.214	6.328	6.724
Al	1.844	0.809	1.776	0.629	0.525	0.325	1.624	0.873	0.443	0.229
Ti	0.527	0.681	0.618	0.714	0.640	0.718	0.701	0.834	0.811	0.692
Cr	0.067	0.003	0.032	0.006	0.006	—	—	—	—	—
Fe	0.433	0.905	0.449	0.973	0.923	1.249	0.787	1.249	1.340	1.124
Mn	—	0.006	—	—	0.006	—	0.004	0.008	0.013	—
Mg	4.879	4.794	4.848	4.712	4.884	4.668	4.643	4.429	4.277	4.342
Ca	0.004	0.006	—	—	0.010	0.005	—	—	—	0.016
Na	0.009	0.176	0.015	0.228	0.228	0.292	0.034	0.255	0.340	0.518
K	2.012	1.968	1.840	1.881	1.974	1.904	1.845	1.845	1.881	2.108
Ba	0.021	0.043	0.022	0.034	0.052	0.051	—	—	—	0.032
Ni	0.021	0.006	—	—	0.009	—	—	—	—	—
CAT	15.624	15.961	15.504	15.600	15.190	15.175	15.518	15.546	15.710	15.785

[a] 1–2, 3–4, cores and margins of mantled micas; 5–6, groundmass mica, Mt. North; 7,8,9, core mantle and groundmass mica, Mt. Gytha; 10, groundmass mica Mt. Cedric. All data from Mitchell (1981). CAT, cation sum; n.d., not determined.
[b] Total iron calculated as FeO.

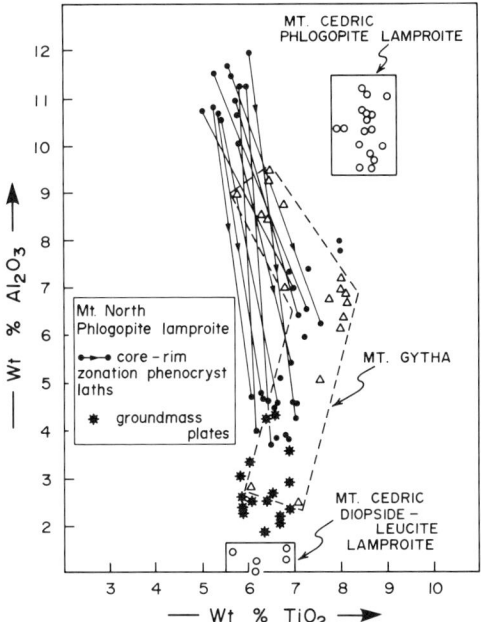

Figure 6.14. Al$_2$O$_3$ versus TiO$_2$ for phlogopites in lamproites from Mount North, Mount Gytha, and Mount Cedric, West Kimberley province (after Mitchell 1981).

Figures 6.14 and 6.15). The mantle zonation is one of Fe enrichment and Al depletion at essentially constant MgO (21–23 wt %). The compositions of the outermost mantles overlap those of the groundmass micas. These latter micas (Table 6.6) are characterized by high Na$_2$O (0.5–1.5 wt %) and BaO (0.7–1.2 wt %) and lower Al$_2$O$_3$ (2.0–4.5 wt %), relative to the mantles, although they possess similar Ti, Fe, and Mg contents.

The principal difference between the compositional trend of these micas and that exhibited by the coarse-grained madupitic lamproites is the extreme depletion in Al coupled with Fe enrichment at constant Mg content, i.e., a trend from titanian phlogopite to sodian titanian tetraferriphlogopite.

Figures 6.14 and 6.15 show that mica from the Mount Gytha phlogopite lamproite is zoned from cores containing 8–10 wt % Al$_2$O$_3$ and 5.5–6.5 wt % TiO$_2$ towards margins enriched in TiO$_2$ (6.5–8.0 wt %) and depleted in Al$_2$O$_3$ (6.0–7.5 wt %). Groundmass micas are poor in Al$_2$O$_3$ (2–3 wt %) and enriched in Na$_2$O (>1.0 wt %) relative to earlier micas (<0.4 wt % Na$_2$O) (Table 6.6). The evolutionary trend of compositions is similar to that shown by the Mount North phlogopite lamproite.

Other diopside leucite phlogopite lamproites contain micas with compositions corresponding to the initial or final portions of the titanian phlogopite to titanian tetraferriphlogopite trend described above. Thus, phenocrysts in the Mount Cedric phlogopite lamproite are relatively unevolved Al-rich titanian phlogopites (9.5–11.5 wt % Al$_2$O$_3$, 8–9 wt % TiO$_2$, 4.3–6.4 wt % FeO$_T$, mg = 0.85–0.89) whereas groundmass micas in the Mount Cedric diopside leucite lamproite are sodian titanian tetraferriphlogopites (1.2–1.6 wt % Al$_2$O$_3$, 5.5–7 wt % TiO$_2$, 7.8–9.4 wt % FeO$_T$, 1.2–1.8 wt % Na$_2$O, mg = 0.83–0.79) (Figures 6.14 and 6.15).

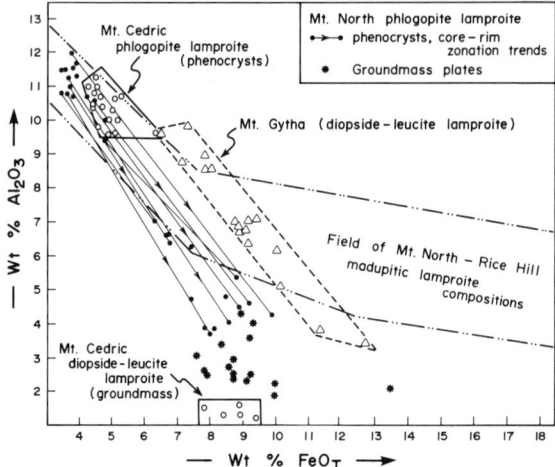

Figure 6.15. Al$_2$O$_3$ versus FeO$_T$ for phlogopites in lamproites from Mount North, Mount Gytha, and Mount Cedric, West Kimberley province (after Mitchell 1981).

Mitchell (1981) has presented data that suggest that each lamproite intrusion is characterized by phenocrystal phlogopite having a distinct Ti and Al content (Figure 6.16). This variation is in marked contrast to the uniform composition of similar phenocrysts in the Leucite Hills lamproites.

Jaques *et al.* (1984, 1986) have confirmed the mica compositional trends described above, and noted that the phlogopite to tetraferriphlogopite trend also occurs in the fine-grained richterite-bearing lamproites found at Dadja Hill and Old Leopold Hill.

Jaques *et al.* (1986) have shown that micas in the diopside leucite lamproites contain 0.3–4.35 wt % F. Phenocryst cores exhibit a compositional trend of increasing F coupled with increasing Ti and decreasing *mg* ratios. In contrast, mantles upon these cores show decreasing F with decreasing *mg* ratios and increasing Ti (Figure 6.17).

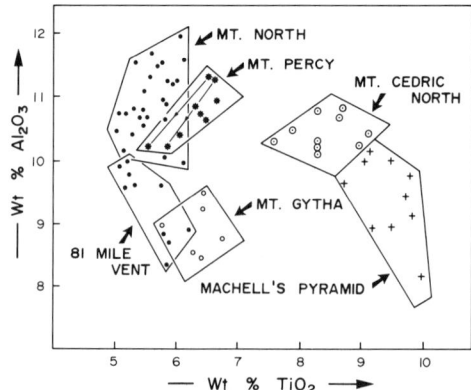

Figure 6.16. Al$_2$O$_3$ versus TiO$_2$ for phlogopite phenocrysts from diverse West Kimberley province lamproites (Mitchell 1981, this work).

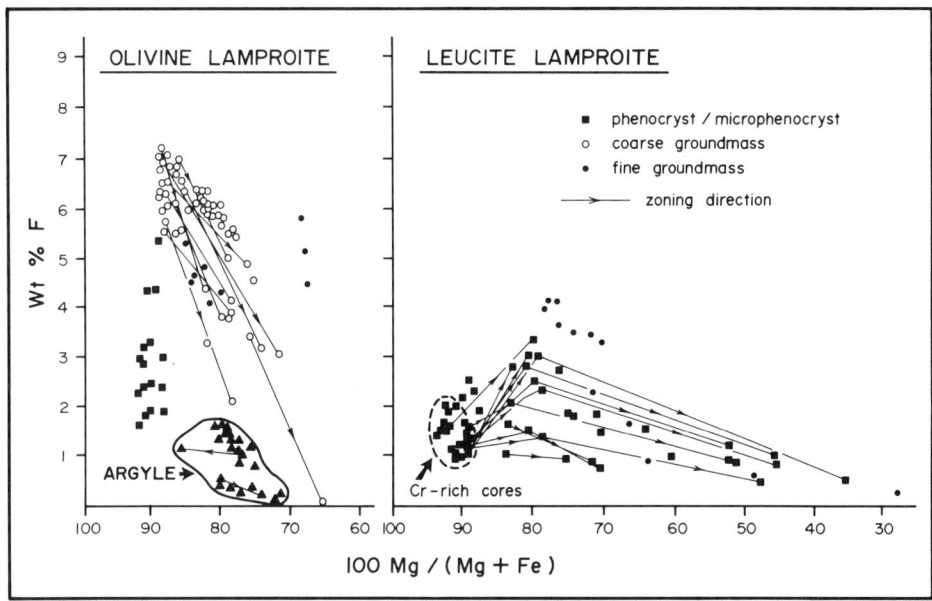

Figure 6.17. F versus 100 Mg/(Mg + Fe) for phlogopites in olivine and leucite lamproites (*sensu lato*) from the West Kimberley province and from the Argyle lamproite (after Jacques *et al.* 1986, 1989a).

6.1.3.4. Summary

Micas in this province follow two divergent trends of compositional evolution:

1. A trend characterized by strong total Fe enrichment, moderate Al and Mg depletion, and limited Fe^{3+} formation, i.e., from titanian phlogopite to Al-poor titanian phlogopite to titanian biotite. This trend, previously termed the wolgidite trend (Mitchell 1981), is found only in the coarse-grained madupitic lamproites.
2. A trend characterized by strong Al depletion coupled with moderate total Fe enrichment at essentially constant Mg content, i.e., from titanian phlogopite to titanian tetraferriphlogopite. This trend, previously termed the fitzroyite trend (Mitchell 1981), is found in fine-grained diopside leucite phlogopite lamproites. Micas in olivine lamproites follow this trend.

Mitchell (1981) considers that the differences between the trends are due to crystallization under different redox conditions. Trend 1 represents mica evolution under relatively reducing conditions with Fe being distributed between Fe^{2+} and Fe^{3+}. In trend 2 the Fe-enrichment is not as pronounced, but strong oxidation results in the bulk of the iron being present as Fe^{3+}.

At Mount North, micas belonging to trend 1 or 2 are developed upon phenocrysts of similar composition. This observation suggests that the madupitic lamproites and the phlogopite lamproites were derived from a common magma, which at the time of emplacement contained phenocrysts of Al-rich titanian phlogopite. The different trends thus reflect the postemplacement conditions of crystallization.

Phenocrystal Al-rich micas from each lamproite locality appear to have slightly different compositions with respect to their Ti and Al contents. These differences suggest that the lamproites were formed either from a single parental magma that was undergoing differentiation or that several batches of parental magma of slightly different composition existed. The West Kimberley province is thus unlike the Leucite Hills province in which all phenocrystal micas have similar compositions (see Section 6.1.2.1).

6.1.4. Argyle, West Australia

Micas in olivine lamproite dikes associated with the Argyle AK1 vent occur as microphenocrysts and as small groundmass plates (Jaques *et al.* 1986, 1989a). Microphenocrysts are zoned from pale yellow/orange cores to darker reddish brown rims that are similar to the groundmass micas. The micas have a compositional range of 2–8 wt % TiO_2, 3–9 wt % Al_2O_3, 9–14 wt % FeO_T, mg = 0.80–0.70, and have low Cr_2O_3 (<0.2 wt %) and BaO (<0.6 wt %) contents (Table 6.7). The zonation trend is principally one of strong Al depletion coupled with moderate total Fe (and Fe^{3+}) increase and slightly decreasing Mg, i.e., a tetraferriphlogopite trend (Figure 6.18). The Al–Ti relationships are complex and Ti

Table 6.7. Representative Compositions of Mica in Olivine Lamproite, Argyle, East Kimberley[a]

	1	2	3	4	5
SiO_2	40.50	40.64	38.38	39.78	39.63
TiO_2	5.80	6.67	6.51	6.34	7.71
Al_2O_3	5.34	5.32	5.95	5.68	4.40
Cr_2O_3	0.02	0.01	0.03	0.02	n.d.
FeO[b]	9.64	10.09	11.61	9.56	11.76
MnO	0.07	0.04	0.07	0.03	0.07
MgO	21.50	21.04	20.41	21.38	20.22
CaO	n.d.	0.03	0.10	0.04	0.05
Na_2O	0.05	n.d.	n.d.	0.03	0.05
K_2O	10.41	10.45	9.81	9.85	9.12
BaO	0.33	0.33	0.56	0.44	0.62
NiO	0.09	0.11	0.08	0.07	0.08
	93.75	94.73	93.51	93.22	93.71
Structural formulas based on 22 oxygens:					
Si	6.092	6.062	5.866	6.009	6.017
Al	0.947	0.935	1.072	1.011	0.787
Ti	0.656	0.748	0.748	0.720	0.880
Cr	0.002	0.001	0.004	0.002	—
Fe	1.213	1.259	1.484	1.208	1.493
Mn	0.009	0.005	0.009	0.004	0.009
Mg	4.821	4.678	4.650	4.814	4.576
Ca	—	0.005	0.016	0.006	0.008
Na	0.015	—	—	0.009	0.015
K	1.998	1.989	1.913	1.898	1.766
Ba	0.019	0.019	0.034	0.026	0.037
Ni	0.011	0.013	0.010	0.008	0.010
CAT	15.783	15.715	15.804	15.717	15.599

[a]n.d., not detected; CAT, cation sum. All data from Jaques *et al.* (1989a).
[b]Total Fe calculated as FeO.

may increase or decrease with decreasing Al. The F contents (<2 wt %) of the micas are lower than those found in the West Kimberley olivine lamproites (Figure 6.17).

The Argyle micas are richer in Fe than the majority of the West Kimberley lamproite micas and exhibit a relatively restricted range of compositional variation. In particular, high-Al, Cr-rich phenocrysts and groundmass sodian titanian tetraferriphlogopites are absent.

6.1.5. Prairie Creek, Arkansas

At Prairie Creek, early lamproite tuffs and breccias are intruded by hypabyssal pyroxene madupitic lamproite. The tuffs and breccias contain only phenocrystal phlogopite which is richer in Al_2O_3 (8.5–12 wt %) and Cr_2O_3 (up to 1 wt %), but poorer in FeO_T (4–7.5 wt %), BaO (<0.15 wt %), F (<1 wt %), and Na_2O (<0.25 wt %), than poikilitic micas in the hypabyssal madupitic phase (Table 6.8, Scott Smith and Skinner 1984b).

Individual crystals of poikilitic groundmass mica in the madupitic lamproites are typically homogeneous, although considerable intersample compositional variation is evident. The micas contain 5–8 wt % Al_2O_3, 2.5–7 wt % TiO_2, 0.3–1 wt % Na_2O, 0.4–1.3 wt % BaO, <0.1 wt % Cr_2O_3, and 4–6 wt % F (Table 6.8, Scott Smith and Skinner 1984b, Lewis 1977, Mitchell unpublished data). Crystals are rarely zoned towards margins enriched in Fe at the expense of Al.

The overall compositional trend found at Prairie Creek is one of strong Al depletion coupled with moderate total Fe enrichment, at essentially constant Mg and slightly decreasing Ti (Figures 6.19 and 6.20). The evolutionary trend is one from titanian phlogopite towards titanian fluorotetraferriphlogopite. Although tetraferriphlogopite has not yet been reported from the Prairie Creek intrusion, it has been recognized in the olivine lamproites of the Kimberlite Mine, an adjacent contemporaneous minor intrusion (Scott Smith and

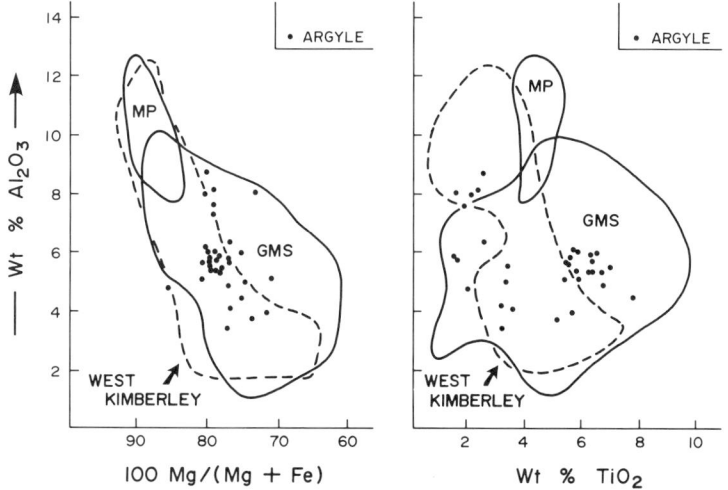

Figure 6.18. Compositional variation of Argyle groundmass mica compared with that of microphenocrystal (MP) and groundmass (GMS) mica in West Kimberley olivine lamproites (after Jaques *et al.* 1989a). The compositional range of phenocrystal and groundmass mica in West Kimberley leucite phlogopite lamproites after (Jaques *et al.* 1986) is enclosed by the dashed line.

Table 6.8. Representative Compositions of Phlogopites from Prairie Creek, Arkansas

	1	2	3	4	5	6	7	8	9	10
SiO_2	40.23	41.95	40.65	41.85	41.36	42.50	41.25	40.49	40.70	41.84
TiO_2	5.42	5.68	5.24	5.29	5.04	5.60	7.07	6.69	3.48	2.94
Al_2O_3	12.07	8.84	11.38	8.50	8.36	5.97	7.39	7.52	2.07	1.82
Cr_2O_3	1.05	0.18	0.96	0.20	0.03	0.03	0.10	0.10	n.d.	n.d.
FeO^b	3.76	5.09	4.42	5.97	6.28	7.70	6.66	7.47	15.52	14.00
MnO	0.02	0.03	0.03	0.03	0.04	0.06	0.02	0.11	0.12	0.09
MgO	22.75	22.96	22.67	23.23	22.76	21.87	20.95	21.67	22.12	22.03
CaO	n.d.	n.d.	0.01	0.05	0.03	0.02	n.d.	0.03	0.02	0.02
Na_2O	0.05	0.08	0.11	0.15	0.47	1.02	0.59	0.64	0.54	0.88
K_2O	10.58	10.59	10.28	9.33	10.09	9.70	10.04	9.81	9.66	10.06
BaO	0.32	0.25	0.29	0.18	0.77	1.86	0.91	2.01	1.04	0.58
NiO	0.21	0.11	0.18	0.16	0.07	0.11	0.07	0.08	0.05	0.02
F	0.94	0.74	0.58	0.60	5.32	n.d.	n.d.	n.d.	2.38	2.89
	97.41	96.50	96.80	95.54	100.62	96.44	95.05	96.62	97.72	97.18
Structural formulas based on 22 oxygens:										
Si	5.694	5.995	5.771	6.017	5.999	6.187	6.021	5.905	6.237	6.417
Al	2.013	1.489	1.903	1.441	1.430	1.024	1.272	1.293	0.371	0.330
Ti	0.576	0.609	0.561	0.572	0.550	0.613	0.776	0.734	0.400	0.337
Cr	0.117	0.002	0.106	0.022	0.004	0.003	0.012	0.012	—	—
Fe	0.444	0.609	0.524	0.719	0.763	0.937	0.813	0.911	1.987	1.797
Mn	0.004	0.004	0.004	0.004	0.004	0.007	0.003	0.014	0.015	0.011
Mg	4.800	4.891	4.796	4.976	4.921	4.746	4.562	4.710	5.053	5.034
Ca	—	—	—	0.007	0.004	0.003	—	0.005	0.004	0.004
Na	0.015	0.022	0.029	0.040	0.132	0.288	0.167	0.181	0.161	0.260
K	1.910	1.932	1.863	1.712	1.866	1.801	1.871	1.825	1.888	1.969
Ba	0.018	0.015	0.015	0.011	0.044	0.106	0.052	0.015	0.062	0.036
Ni	0.026	0.001	0.022	0.018	0.007	0.013	0.001	0.009	0.007	0.004
CAT	15.617	15.569	15.594	15.539	15.724	15.730	15.550	15.712	16.188	16.199

^{a}CAT, cation sum; n.d., not determined. 1–2, phenocrysts, tuffaceous lamproites; 3–4, phenocrysts breccia-form lamproites; 5–7, groundmass mica, madupitic lamproites; 8, reaction rim mica (Mitchell and Lewis, 1983); 9–10, core and margin groundmass mica, Kimberlite Mine olivine lamproite. Data sources 1–5, 9–10, Scott Smith and Skinner; 6–8, Mitchell (this work).
^{b}Total iron expressed as FeO.

Skinner 1984b,c). The mica compositional trends demonstrate that madupitic micas are more evolved than phenocrystal micas (Figures 6.19 and 6.20).

Micas forming reaction coronas around microinclusions of diopside richterite lamproite (Mitchell and Lewis 1983) are of identical composition to the poikilitic micas occurring in the host madupitic lamproite (Table 6.8, Figures 6.19 and 6.20).

6.1.6. Murcia-Almeria, Spain

Despite the importance of the Spanish lamproite province there have not been any systematic studies of mica compositional variation. The few data available (Fuster *et al*. 1967, Lopez Ruiz and Rodriguez Badiola, 1980, Mitchell 1985, Wagner and Velde 1986a, Venturelli *et al*. 1984a, 1988) suggest the existence of compositional trends similar to those found in the Leucite Hills and West Kimberley provinces.

Phlogopite lamproites contain phenocrysts with a compositional range of 10.7–14.3 wt

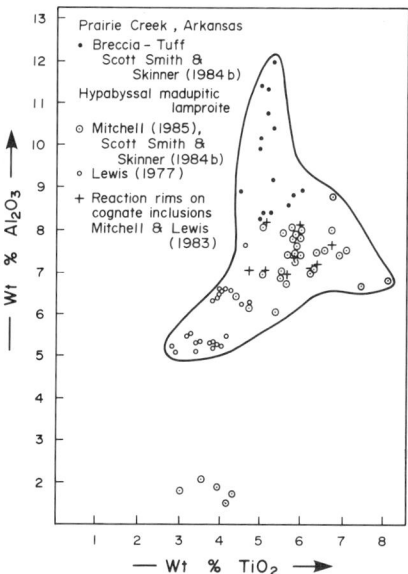

Figure 6.19. Al_2O_3 versus TiO_2 for phlogopites in the Prairie Creek lamproites.

% Al_2O_3, 2–5 wt % TiO_2, 3–7 wt % FeO_T and which have low Na_2O (<0.5 wt %) and BaO (<0.2 wt %) contents (Table 6.9). Insufficient data are available to determine whether or not particular volcanic centers are characterized by phenocrysts of distinct composition. Mitchell (1985) has presented data (Figure 6.21) that suggest that phenocrysts in hyalo-enstatite phlogopite lamproites are on average richer in Al, i.e., less evolved, than those in hyalo-olivine diopside phlogopite lamproites.

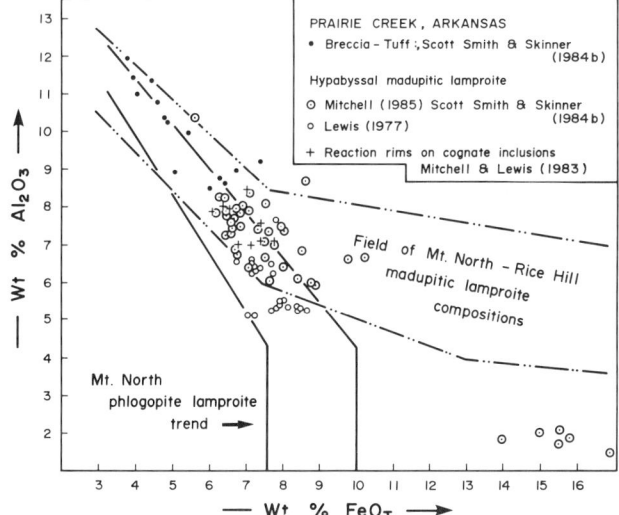

Figure 6.20. Al_2O_3 versus FeO_T for phlogopites in the Prairie Creek lamproites.

Table 6.9. Representative Compositions of Micas from the Murcia-Almeria[a]

	1	2	3	4	5	6	7	8	9	10
SiO_2	42.90	37.86	39.98	40.92	41.50	41.42	39.71	38.68	38.7	36.1
TiO_2	1.98	2.72	4.22	4.08	6.88	9.51	8.40	11.19	4.12	3.22
Al_2O_3	11.92	14.61	12.86	12.19	11.00	10.24	9.90	8.79	12.8	16.8
Cr_2O_3	0.76	1.08	0.96	0.30	0.24	0.09	0.00	0.03	n.d.	n.d.
FeO[b]	3.87	4.86	4.63	6.48	5.11	7.63	7.39	11.49	10.4	16.6
MnO	0.03	n.d.	0.07	n.d.	0.03	0.01	0.06	0.08	n.d.	0.18
MgO	24.60	22.20	21.45	20.56	20.79	18.19	19.45	16.24	16.9	13.4
CaO	n.d.	n.d.	0.02	n.d.	n.d.	n.d.	0.00	n.d.	0.17	0.09
Na_2O	0.45	0.23	0.11	n.d.	0.07	0.14	0.11	0.46	0.45	0.38
K_2O	9.87	10.06	10.42	10.23	9.42	9.33	9.01	8.30	8.90	9.02
BaO	0.08	0.39	0.60	0.57	0.84	0.61	2.05	2.11	n.d.	n.d.
NiO	n.d.	n.d.	0.03	n.d.	n.d.	n.d.	0.00	n.d.	n.d.	n.d.
	96.48	94.01	95.36	95.33	95.89	97.15	96.10	97.37	92.44	95.79
Structural formulas based on 22 oxygens:										
Si	5.994	5.521	5.742	5.896	5.898	5.872	5.759	5.766	5.815	5.404
Al	1.964	2.511	2.176	2.070	1.842	1.710	1.693	1.518	2.267	2.964
Ti	0.208	0.298	0.457	0.442	0.736	1.014	0.916	1.232	0.466	0.362
Cr	0.084	0.125	0.076	0.034	0.028	0.010	—	0.002	—	—
Fe	0.454	0.593	0.559	0.781	0.606	0.904	0.897	1.408	1.307	2.078
Mn	0.004	—	0.008	—	0.004	0.001	0.007	0.010	—	0.023
Mg	5.124	4.825	4.594	4.415	4.406	3.844	4.208	3.546	3.785	2.990
Ca	—	—	0.004	—	—	—	—	—	0.027	0.014
Na	0.124	0.065	0.031	—	0.020	0.038	0.031	0.130	0.131	0.110
K	1.758	1.871	1.910	1.880	1.708	1.686	1.668	1.552	1.706	1.722
Ba	0.006	0.022	0.035	0.032	0.046	0.034	0.117	0.122	—	—
Ni	—	—	0.004	—	—	—	—	—	—	—
CAT	15.720	15.831	15.629	15.550	15.294	15.114	15.296	15.186	15.504	15.668

[a]CAT, cation sum; n.d., not determined. 1–2, phenocrysts phlogopite lamproite, Fortuna; 3–4, phenocrysts hyalo-phlogopite lamproite, Vera; 5–6, richterite sanidine lamproites, Cancarix; 7–8, groundmass mica madupitic lamproites, Jumilla; 9–10, sieve-textured biotites (Venturelli et al. 1984). Data sources: 1, 5, 6, 8, Wagner and Velde, 1986; 3,4,7, Mitchell, 1985, this work.
[b]Total Fe iron calculated as FeO.

Richterite sanidine madupitic lamproites contain groundmass micas that exhibit significant intercrystal variations with respect to Ti and Fe (Figure 6.21), although individual crystals are of uniform composition. Groundmass micas are richer in TiO_2 (6–11.2 wt %), FeO_T (6–11.5 wt %), and BaO (>1 wt %) and poorer in Al_2O_3 (<11 wt %) than phenocrystal micas (Table 6.9). Zoned phenocrystal micas in transitional madupitic rocks have cores that are rich in Al and Mg and poor in Ti and Fe relative to their margins (Lopez Ruiz and Rodriguez Badiola 1980).

The overall compositional trend (Figure 6.21) in the province is one of decreasing Al and increasing Ti and Fe. The Al depletion is not extreme and tetraferriphlogopites do not occur. The trend is similar to trend 1 of the West Kimberley province. The Murcia-Almeria micas contain 1.2–4.4 wt % F (Venturelli et al. 1988).

Biotites, forming euhedral "xenocrysts" and groundmass microlites, have been described from the Vera, Barqueros, and Zaneta lamproites by Venturelli et al. (1984a, 1988). Commonly phenocrystal phlogopite and biotite coexist. The biotites may be mantled by phlogopite or vice versa. Many of the crystals have undergone breakdown to spinels (see Section 6.9.4) and glass. Their compositional range is 12.8–18.7 wt % Al_2O_3, 2.2–4.1 wt %

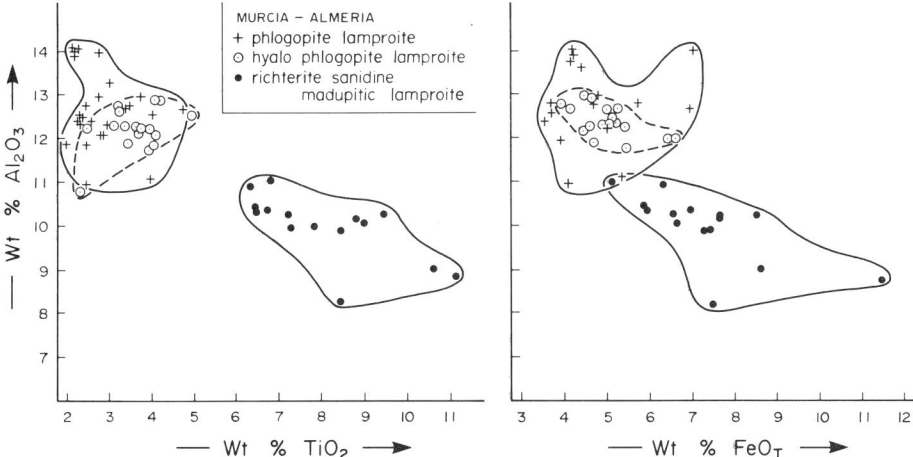

Figure 6.21. Compositional variation for phlogopites in the Murcia-Almeria lamproites. Data from Mitchell (1985), Venturelli *et al.* (1984a, 1988).

TiO$_2$, and 8.3–16.6 wt % FeO$_T$ (Table 6.9 analyses 9–10). The origin of these biotites is unknown, although there is high probability that they are crustally derived xenocrysts. Juxtaposition of these diverse micas is considered by Venturelli *et al.* (1984a) to result from magma mixing.

6.1.7. Smoky Butte, Montana

Only phenocrystal micas are present in the Smoky Butte lamproites. A detailed study of their compositional variation has been presented by Mitchell *et al.* (1987). Other data are given by Velde (1975) and Wagner and Velde (1986a). Representative compositions are given in Table 6.10.

The micas are continuously zoned and exhibit considerable variation with respect to TiO$_2$ (6–12.6 wt %) accompanied by lesser concomitant variation in Al$_2$O$_3$ (12.5–9.5 wt %) and FeO$_T$ (4–8.5 wt %) content. The zonation trend from core to the margin of the phenocrysts is one of increasing Ti, Fe, and BaO (1–3%) at the expense of Al and Cr (Figure 6.22).

Mica of a particular composition is not associated with a particular petrographic unit of the intrusion. Smoky Butte micas are exceptionally rich in Ti and BaO for a given Al content as compared with phenocrystal micas in other lamproites. Mitchell *et al.* (1987) have suggested that the parental magma was derived from a source relatively richer in Ti and Ba.

The micas exhibit a relatively restricted range of composition compared to those found in other lamproite provinces, and tetraferriphlogopite is absent even in the most evolved richterite-bearing rocks.

6.1.8. Kapamba, Zambia

Phlogopite occurs only as groundmass phase in the magmatic lamproites (Scott Smith *et al.* 1989). The overall composition range is 5–9 wt % TiO$_2$, 4–11.5 wt % Al$_2$O$_3$, and 7–15

Table 6.10. Representative Compositions of Smoky Butte Phlogopite[a]

	1	2	3	4	5	6	7	8
SiO_2	40.81	38.63	38.71	39.05	39.14	39.87	39.69	39.42
TiO_2	7.71	8.48	9.37	10.38	8.51	10.25	9.13	10.39
Al_2O_3	11.00	10.59	10.89	10.14	11.04	10.20	11.02	10.81
Cr_2O_3	0.96	0.20	0.13	0.05	0.39	0.21	0.44	0.13
FeO[b]	3.99	5.81	6.41	7.40	5.46	6.48	5.11	6.00
MnO	n.d.	0.02	0.09	0.13	0.09	0.08	0.04	0.11
MgO	20.55	18.40	17.93	17.13	19.18	18.37	19.82	18.63
CaO	n.d.	0.07	n.d.	n.d.	0.05	0.04	0.02	0.02
Na_2O	0.27	0.22	0.22	0.21	0.18	0.14	0.20	0.16
K_2O	10.60	9.71	8.62	8.53	9.23	9.34	9.17	8.81
BaO	0.90	3.07	2.61	2.91	1.68	2.51	1.98	2.39
NiO	0.17	0.10	0.20	0.15	0.17	0.15	0.13	0.07
	96.95	95.29	95.18	96.69	95.12	97.64	96.75	96.93
Structural formulas based on 22 oxygens:								
Si	5.782	5.702	5.678	5.711	5.695	5.713	5.673	5.653
Al	1.837	1.842	1.883	1.749	1.893	1.722	1.856	1.827
Ti	0.821	0.941	1.034	1.142	0.931	1.104	0.981	1.121
Cr	0.108	0.024	0.015	0.006	0.045	0.024	0.050	0.015
Fe	0.473	0.717	0.786	0.905	0.664	0.776	0.611	0.720
Mn	—	—	0.011	0.016	0.011	0.010	0.005	0.013
Mg	4.340	4.048	3.920	3.734	4.160	3.923	4.223	3.982
Ca	—	0.011	—	—	0.008	0.006	0.003	0.003
Na	0.074	0.063	0.063	0.060	0.051	0.039	0.055	0.044
K	1.916	1.828	1.613	1.591	1.713	1.707	1.672	1.612
Ba	0.050	0.178	0.150	0.167	0.096	0.141	0.111	0.134
Ni	0.019	0.012	0.024	0.018	0.020	0.017	0.015	0.008
CAT	15.419	15.369	15.177	15.096	15.286	15.183	15.256	15.133

[a]CAT, cation sum; n.d., not detected. 1-2, 3-4, 5-6, 7-8 cores and margin, respectively, of phlogopite phenocrysts (this work).
[b]Total iron expressed as FeO.

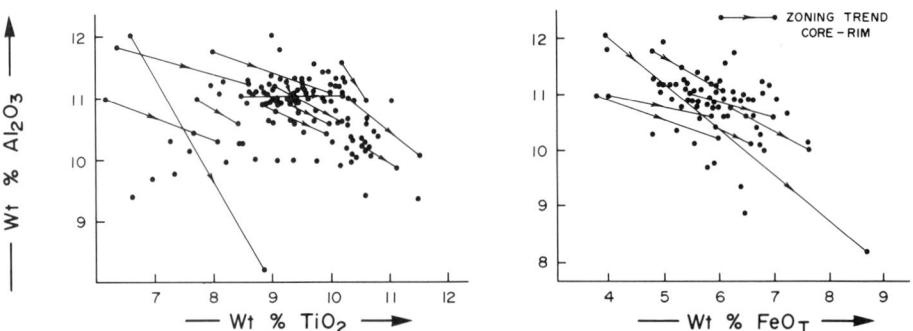

Figure 6.22. Compositional variation of phlogopites in the Smoky Butte lamproites (after Mitchell et al. 1987).

wt % FeO$_T$. In common with other madupitic micas the BaO (0.9–3.2 wt %), F (1–5 wt %), and Na$_2$O (0.3–1.3 wt %) contents are relatively high, with Cr being below detection limits (Table 6.11). The small size of the crystals precludes the determination of zonation trends. The few data available indicate that Ti and Al decrease with increasing Fe (Figure 6.23).

Each volcanic center appears to be characterized by phlogopite of a distinct composition, although intracenter compositional variation is extensive (Figure 6.23). These relationships are similar to those observed for the madupitic micas of the Leucite Hills lamproites (see Section 6.1.2.3).

Data are not available for phlogopites occurring as laths in lapilli in the lamproite tuffs. Consequently it is not yet possible to determine the overall trend of phlogopite evolution in the Kapamba lamproites.

6.1.9. Sisimiut, Greenland

Leucite phlogopite lamproite dikes from this area contain mica phenocrysts with a composition range of 10–12.5 wt % Al$_2$O$_3$, 6–8 wt % TiO$_2$, and 5–9 wt % FeO$_T$, and have low Na$_2$O (<0.2 wt %) and Cr$_2$O$_3$ (<0.5 wt %) contents (Table 6.12, Scott 1981, Thy *et al.*

Table 6.11. Representative Compositions of Phlogopites from Kapamba[a]

	1	2	3	4	5	6
SiO$_2$	38.57	38.15	41.47	42.11	36.51	38.47
TiO$_2$	8.35	7.83	7.44	7.01	6.46	9.09
Al$_2$O$_3$	9.59	9.09	7.39	6.35	4.88	10.71
FeO[b]	12.12	9.46	10.73	11.40	33.12	11.30
MnO	0.07	0.07	0.06	0.05	0.28	0.10
MgO	16.48	18.85	17.74	17.48	4.98	16.05
CaO	0.02	0.09	0.03	0.03	0.09	0.09
Na$_2$O	0.45	0.83	0.74	1.06	0.52	0.56
K$_2$O	9.26	8.73	9.34	9.52	9.18	8.88
BaO	1.04	2.80	1.59	1.20	0.39	1.25
NiO	0.03	0.04	0.06	0.04	0.02	0.07
F	1.78	3.89	2.51	2.73	0.38	1.25
	97.01	98.19	98.04	97.83	96.65	97.29
Structural formulas based on 22 oxygens:						
Si	5.702	5.598	6.078	6.239	6.123	5.550
Al	1.672	1.573	1.277	1.109	0.965	1.823
Ti	0.955	1.003	0.815	0.708	0.615	1.162
Fe	1.499	1.162	1.316	1.413	4.647	1.364
Mn	0.009	0.009	0.008	0.006	0.040	0.012
Mg	3.635	4.126	3.879	3.863	1.246	3.454
Ca	0.003	0.014	0.005	0.005	0.016	0.014
Na	0.129	0.236	0.211	0.305	0.169	0.157
K	1.748	1.635	1.748	1.801	1.965	1.636
Ba	0.060	0.161	0.080	0.070	0.026	0.071
Ni	0.002	0.003	0.004	0.003	0.001	0.005
CAT	15.414	15.520	15.421	15.522	15.813	15.248

[a]CAT, cation sum. 1, subhedral lath, D3; 2, poikilitic grain P6; 3–4, core and margin of poikilitic grain P6; 5, margin P10; 6, equant grain P12. All data from Scott Smith *et al.* (1989).
[b]Total iron expressed as FeO.

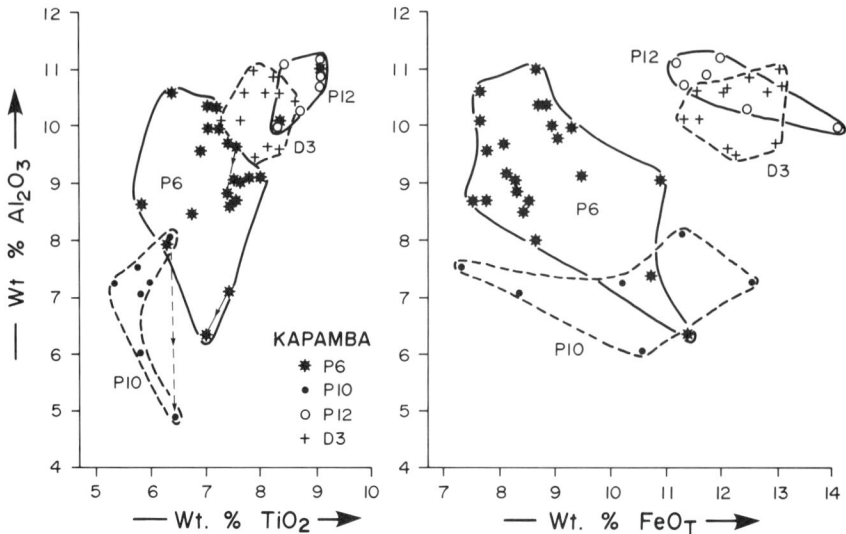

Figure 6.23. Compositional variation in the phlogopites in the Kapamba lamproites (after Scott Smith *et al*. 1989). Alphanumeric designations refer to individual intrusions.

1987). A single zonation trend, reported by Thy *et al*. (1987), is one of increasing Fe coupled with decreasing Al and Mg at approximately constant Ti (Table 6.12, Figures 6.24 and 6.25), i.e., a phlogopite to biotite trend. Tetraferriphlogopite is common as narrow mantles upon the phenocrysts, and is notably poor in Ti (Table 6.12) compared with tetraferriphlogopites found in other lamproites.

Amphibole-bearing madupitic lamproites contain TiO_2 (10–11 wt %) and Al_2O_3 (12–13 wt %)-rich micas, which are zoned to Fe-rich margins (9–15% FeO_T). The high Al content is anomalous relative to that of other lamproite groundmass micas and suggests that these rocks have no simple differentiation relationship to the associated phlogopite lamproites.

6.1.10. Hills Pond—Rose Dome, Kansas

Richterite diopside madupitic lamproites contain strongly zoned micas that exhibit extreme Al_2O_3 (8–1 wt %) depletion and moderate FeO_T (6–13.5 wt %) enrichment at essentially constant TiO_2 contents (Table 6.13, Figures 6.24 and 6.25). The BaO, Na_2O, and Cr_2O_3 contents are all low (<0.5 wt %) (Mitchell 1985).

Fresh micas are not preserved in the Rose Dome lamproites (Coopersmith and Mitchell 1989).

6.1.11. Bobi, Ivory Coast

The Bobi dike contains relatively fresh, corroded, and zonation-free phenocrysts. These micas contain 4.8–5.3 wt % TiO_2, 10.5–11.6 wt % Al_2O_3, 3.4–4.5 wt % FeO_T, 0.3–0.6 wt % Cr_2O_3, <0.1 wt % Na_2O, and 0.6 –1.8 wt % BaO (Table 6.13, Figures 6.24 and 6.25) (Mitchell 1985).

Table 6.12. Representative Compositions of Sisimiut Micas[a]

	1	2	3	4	5	6	7	8	9
SiO_2	39.61	39.28	38.79	41.57	38.19	37.56	45.26	37.87	37.15
TiO_2	7.12	7.51	6.68	3.33	8.13	8.36	0.65	10.21	10.34
Al_2O_3	11.67	10.49	10.51	4.24	12.13	9.69	0.13	12.67	12.38
Cr_2O_3	0.40	n.d.	0.13	0.05	0.27	n.d.	0.02	0.13	0.03
FeO[b]	5.66	8.25	7.63	13.65	8.55	19.82	17.82	9.57	15.19
MnO	0.02	0.07	0.06	0.06	0.02	0.02	0.06	0.04	0.12
MgO	21.09	19.38	19.61	20.00	17.57	11.55	19.39	15.27	11.30
CaO	n.d.	0.05	n.d.	n.d.	n.d.	n.d.	n.d.	n.d.	n.d.
Na_2O	0.09	0.13	0.15	0.04	0.10	0.11	0.09	0.18	0.21
K_2O	9.98	9.88	9.91	10.04	9.14	9.21	9.17	9.16	8.90
BaO	0.80	0.60	n.d.	n.d.	n.d.	n.d.	n.d.	n.d.	n.d.
NiO	0.15	0.04	n.d.	n.d.	n.d.	n.d.	n.d.	n.d.	n.d.
	96.59	95.68	93.47	92.98	94.10	96.32	92.59	95.10	95.62
Structural formulas based on 22 oxygens:									
Si	5.655	5.723	5.747	6.377	5.614	5.714	7.035	5.537	5.548
Al	1.963	1.801	1.835	0.767	2.102	1.737	0.024	2.183	2.179
Ti	0.764	0.823	0.744	0.384	0.899	0.956	0.076	1.123	1.161
Cr	0.045	—	0.015	0.006	0.031	—	0.002	0.015	0.004
Fe	0.676	1.005	0.945	1.751	1.051	2.522	2.316	1.170	1.897
Mn	0.002	0.009	0.008	0.008	0.002	0.003	0.008	0.005	0.015
Mg	4.488	4.209	4.331	4.573	3.850	2.619	4.492	3.328	2.516
Ca	—	0.008	—	—	—	—	—	—	—
Na	0.025	0.037	0.043	0.012	0.029	0.032	0.027	0.051	0.061
K	1.817	1.836	1.873	1.965	1.714	1.787	1.818	1.708	1.696
Ba	0.045	0.034	—	—	—	—	—	—	—
Ni	0.017	0.005	—	—	—	—	—	—	—
CAT	15.498	15.490	15.541	15.841	15.292	15.371	15.799	15.121	15.077

[a]1, phenocryst; 2, groundmass; 3–4, core and rim of phenocryst; 5–6, core and rim of phenocryst; 7, groundmass tetraferriphlogopite; 8–9, core and rim of groundmass phlogopite in amphibole lamproite. Data sources 1–4 Scott (1981); 5–9 Thy *et al.* (1987).
[b]Total iron expressed as FeO.

Leucite phlogopite lamproite from the Kohue River area contains zonation-free phenocrysts, similar in composition (Al_2O_3 = 10.7 –11.6 wt %, FeO_T = 3.7–5.5 wt %, Na_2O <0.2%, Cr_2O_3 = 0.2–0.8%, BaO = 0.3–1.8 wt %) to the Bobi micas except that they are richer in TiO_2 (6.9–7.9 wt %) (Mitchell 1985).

6.1.12. Francis, Utah

Micas are found only as phenocrysts at this locality. The few data available (Wagner and Velde 1986a, this work) indicate that these phlogopites are zoned with slightly decreasing Al_2O_3 (12.8 –11.3 wt %) at essentially constant TiO_2 (2.2–2.3 wt %) and increasing FeO_T (2.8–6.8 wt %) from core to margin (Table 6.14, Figures 6.26 and 6.27). The Cr_2O_3 (0.3–1.0 wt %), Na_2O (<0.4 wt %) and BaO (<0.2 wt %) contents are all low. The Francis micas are similar in composition to phenocrystal micas found in the Leucite Hills lamproites but are Ti-poor relative to phenocrysts in other lamproites.

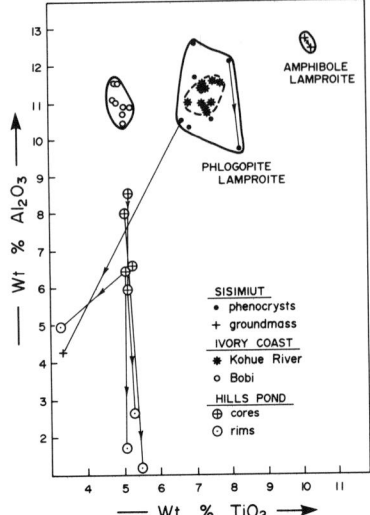

Figure 6.24. Al_2O_3 versus TiO_2 for phlogopites in lamproites from Sisimiut (Thy *et al.* 1987, Scott 1981), the Ivory Coast (Mitchell 1985), and Hills Pond (Mitchell 1985).

6.1.13. Yellow Water Butte, Montana

Phlogopite phenocrysts from Yellow Water Butte contain 7–13 wt % Al_2O_3, 0.6–2.0 wt % TiO_2, 3–8 wt % FeO_T and <1.5 wt % BaO and 1–2 wt % F (Table 6.14). Core to rim zoning trends (Figures 6.26 and 6.27) exhibit Al depletion and Ba, Ti, and Fe enrichment typical of lamproite phlogopite (this work).

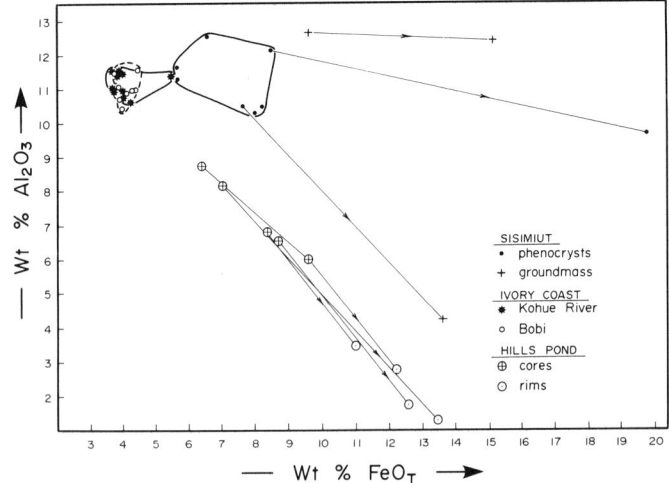

Figure 6.25. Al_2O_3 versus FeO_T for phlogopites in lamproites from Sisimiut, the Ivory Coast and Hills Pond. Data sources as in Figure 6.24.

Table 6.13. Representative Compositions of Micas from Hills Pond (Kansas) and the Ivory Coast[a]

	1	2	3	4	5	6
SiO_2	40.99	39.73	40.82	41.09	39.86	40.89
TiO_2	5.20	5.24	5.11	5.40	5.20	7.35
Al_2O_3	8.55	2.55	8.00	1.05	10.99	11.41
Cr_2O_3	0.18	0.16	0.31	0.09	0.60	0.56
FeO[b]	6.32	12.21	6.94	13.46	3.84	3.86
MnO	0.09	0.15	0.11	0.14	0.00	0.12
MgO	22.49	21.66	22.18	22.19	21.34	21.70
CaO	n.d.	n.d.	n.d.	n.d.	0.06	n.d.
Na_2O	n.d.	0.11	n.d.	n.d.	n.d.	0.06
K_2O	9.99	9.41	10.21	9.62	9.65	10.31
BaO	0.36	0.39	0.17	0.52	1.80	0.27
NiO	n.d.	n.d.	n.d.	n.d.	0.08	0.12
	94.17	91.61	93.85	93.56	93.48	96.64
Structural formulas based on 22 oxygens:						
Si	5.990	6.220	6.011	6.319	5.855	5.755
Al	1.470	0.468	1.387	0.189	1.901	1.892
Ti	0.570	0.588	0.562	0.623	0.573	0.777
Cr	0.017	0.118	0.034	0.009	0.009	0.061
Fe	0.771	1.599	0.851	1.729	0.470	0.451
Mn	0.008	0.008	0.013	0.018	—	0.012
Mg	4.892	5.053	4.866	5.087	4.669	4.533
Ca	—	—	—	—	0.009	—
Na	—	—	—	—	—	0.016
K	1.861	1.874	1.914	1.887	1.806	1.851
Ba	0.017	0.023	0.008	0.027	0.099	0.012
Ni	—	—	—	—	0.009	0.016
CAT	15.597	15.893	15.674	15.895	15.400	15.396

[a]CAT, cation sum; n.d., not detected. 1-2, 3-4, cores and rims Hills Pond; 5, phenocrysts Bobi dike; 6, phenocryst Kohue River lamproite. All data from Mitchell (1985).
[b]Total Fe expressed as FeO.

6.1.14. Sisco, Corsica

Micas in the Sisco phlogopite lamproite contain 12.3–10.6 wt % Al_2O_3, 3.0–5.1 wt % TiO_2, 2.8–6.1 wt % FeO_T, 0.2–1 wt % Cr_2O_3, <0.4 wt % Na_2O, and <0.3 wt % BaO (Table 6.14, Figures 6.26 and 6.27) (Wagner and Velde 1986, this work). Zonation trends are of decreasing Al and Ti coupled with increasing Fe. Wagner and Velde (1986) note the occurrence of one mica containing 11.8 wt % FeO_T.

6.1.15. Majhgawan, India

Phenocrystal micas contain 5.3–7.2 wt % TiO_2, 11.0–12.8 wt % Al_2O_3, and 4.4–6.7 wt % FeO_T (Table 6.14, Figure 6.26, 6.27, Scott Smith 1989, Middlemost and Paul 1984, Paul 1980, Gupta et al. 1986). Groundmass micas are too fine grained to analyze (Scott Smith 1989). Minor differences in the composition of macrocrystal and phenocrystal micas have been noted by Scott Smith (1989), with the former being relatively enriched in Cr, Ti, and Al.

Table 6.14. Representative Compositions of Micas from Francis (Utah), Sisco (Corsica), Yellow Water Butte (Montana), and Majhgawan (India)[a]

	1	2	3	4	5	6	7	8
SiO_2	40.46	38.64	40.58	40.43	39.62	38.78	38.85	39.29
TiO_2	2.27	2.32	3.35	3.35	1.18	0.09	7.24	5.33
Al_2O_3	12.00	11.63	12.15	12.31	10.73	0.29	12.81	10.97
Cr_2O_3	0.64	0.44	0.71	0.83	—	0.04	1.24	—
FeO[b]	2.80	6.79	3.09	3.23	6.09	15.75	4.53	6.74
MnO	0.01	0.04	0.06	0.04	n.d.	n.d.	0.03	0.06
MgO	24.74	23.05	23.52	22.23	25.38	26.22	21.22	23.90
CaO	—	—	—	—	0.05	0.34	0.02	0.05
Na_2O	0.14	0.25	0.14	0.15	0.35	0.13	0.08	0.08
K_2O	10.55	9.90	10.71	10.74	10.04	9.98	10.16	9.09
BaO	0.20	1.36	—	—	0.88	1.16	n.d.	n.d.
NiO	0.34	0.30	0.15	0.12	n.d.	n.d.	0.13	0.07
F	n.d.	n.d.	n.d.	n.d.	1.87	0.85	n.d.	n.d.
	94.13	94.70	94.47	94.44	96.19	93.63	96.10	95.58
Structural formulas based on 22 oxygens:								
Si	5.823	5.691	5.818	5.810	5.810	6.180	5.514	5.631
Al	2.031	2.020	2.053	2.083	1.856	0.055	2.144	1.854
Ti	0.245	0.255	0.357	0.358	0.130	0.011	0.773	0.575
Cr	0.072	0.047	0.080	0.092	—	0.005	0.139	—
Fe	0.333	0.834	0.370	0.388	0.747	2.100	0.538	0.808
Mn	—	—	0.004	0.004	—	—	0.004	0.007
Mg	5.308	5.060	5.024	4.975	5.552	6.233	4.493	5.110
Ca	—	—	—	—	0.008	0.058	0.003	0.008
Na	0.038	0.069	0.038	0.038	0.100	0.040	0.022	0.022
K	1.934	1.859	1.957	1.964	1.880	2.030	1.841	1.663
Ba	0.008	0.078	—	—	0.010	0.073	—	—
Ni	0.038	0.034	0.017	0.012	—	—	0.015	0.007
CAT	15.828	15.951	15.718	15.723	16.093	16.785	15.486	15.685

[a]CAT, cation total; n.d., not determined. 1–2, 3–4, 5–6, cores and rims of phenocrysts from Francis, Sisco, and Yellow Water Butte, respectively (this work) 7–8, groundmass micas Majhgawan (Scott Smith 1989).
[b]Total iron expressed as FeO.

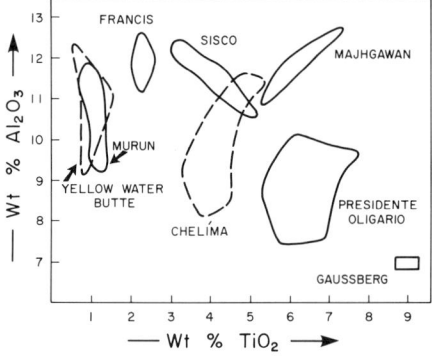

Figure 6.26. Al_2O_3 versus TiO_2 for phlogopites in the Francis (Wagner and Velde 1986a, this work), Sisco, Wagner and Velde 1986a, this work), Yellow Water Butte (this work), Murun (this work), Presidente Oligario (this work), Gaussberg (Sheraton and Cundari 1980), and Majhgawan (Scott Smith 1989) lamproites.

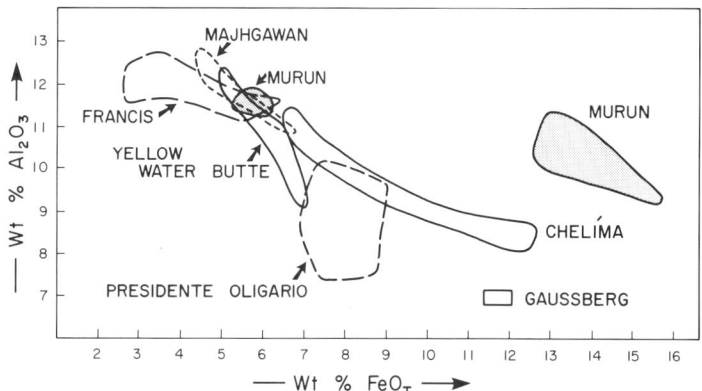

Figure 6.27. Al_2O_3 versus FeO_T for phlogopites in the Francis, Sisco, Yellow Water Butte, Presidente Oligario, Murun Gaussberg and Majhgawan lamproites. Data sources as in Figure 6.26.

6.1.16. Gaussberg, Antarctica

Mica in leucite-bearing potassic rocks from Gaussberg, Antarctica occurs as late crystallizing phases (Sheraton and Cundari 1980). They are TiO_2-rich (8.7–9.2 wt %), Al_2O_3-poor (6.8–7.1 wt %) phlogopites that contain 11.4–12.1 wt % FeO_T, 0.8–1.2 wt % BaO, 1.4–2.4 wt % F, and 0.3–0.5 wt % Na_2O. Their average composition is given in Table 6.15 and the range depicted in Figures 6.26 and 6.27. These micas are compositionally similar to moderately evolved madupitic phlogopites found in the Leucite Hills and West Kimberley provinces.

A relatively Al_2O_3-rich (15.9 wt %), TiO_2-rich (6.2 wt %) macrocryst containing 8.0 wt % FeO_T and 3.3 wt % BaO that is similar to some of the aluminous phlogopites found in the Leucite Hills lamproites is interpreted by Sheraton and Cundari (1980) as a xenocryst.

6.1.17. Probable Lamproites—Murun, Presidente Oligario, and Chelima

These occurrences of potassic rocks are considered to be lamproites or to be derived from closely related magmas.

Dike rocks that are apparently similar to phlogopite lamproites occur in the Murun potassic complex (Vladikin 1985). The rocks contain strongly pleochroic mantled phenocrysts and irregular groundmass plates. The cores and mantles of the phenocrysts define a compositional trend of decreasing Al_2O_3 (11.8–10.1 wt %) and TiO_2 (1.3–0.8 wt %) and increasing FeO_T (5.4–13.9 wt %). The trend culminates in relatively Al_2O_3-poor (9–10 wt %), FeO_T-rich (15–16 wt %) groundmass micas (Table 6.15, Figures 6.26 and 6.27) (this work).

Titanian tetraferriphlogopites are not found and the overall compositional trend is similar to that found in the coarse-grained transitional madupitic lamproites from the West Kimberley province (see Section 6.1.3.2).

Poikilitic groundmass mica occurs in rocks described as lamproites from the Presidente Oligario region, Brazil (Leonardos and Ulrich 1987, Meyer *et al.* 1988). These are phlogopites containing 7.6–10 wt % Al_2O_3, 5.4–7.7 wt % TiO_2, 7.1–9.6 wt % FeO_T, 0.2–

Table 6.15. Representative Compositions of Micas from Murun (USSR), Presidente Oligario (Brazil), Gaussberg (Antarctica), and Chelima (India)[a]

	1	2	3	4	5	6	7	8	9	10	11
SiO_2	41.29	39.83	41.21	39.56	39.95	41.21	41.83	40.66	34.7	39.58	40.01
TiO_2	1.02	1.06	0.86	1.00	1.02	6.46	6.82	9.0	6.15	4.85	3.95
Al_2O_3	11.68	10.81	11.59	10.61	10.19	8.00	8.52	6.9	15.9	11.33	8.36
Cr_2O_3	0.54	0.19	0.56	0.24	0.19	0.09	0.07	n.d.	n.d.	0.09	n.d.
FeO[b]	6.05	12.67	5.38	14.14	15.59	7.46	8.75	11.9	7.99	6.60	11.77
MnO	0.08	0.26	0.03	0.30	0.36	0.11	0.09	0.05	0.15	0.06	0.07
MgO	24.70	19.75	25.00	18.67	18.10	20.49	19.95	17.6	18.2	21.77	20.50
CaO	n.d.	n.d.	n.d.	n.d.	n.d.	0.09	0.15	n.d.	0.10	0.05	0.03
Na_2O	n.d.	0.02	0.17	0.04	n.d.	0.71	0.96	0.41	0.12	0.04	0.02
K_2O	10.55	10.00	10.27	9.83	10.22	10.43	10.07	9.3	9.0	10.11	9.87
BaO	n.d.	n.d.	n.d.	n.d.	n.d.	n.d.	n.d.	1.0	3.27	0.66	0.32
NiO	n.d.	0.04	0.07	0.04	0.09	0.04	n.d.	0.04	n.d.	n.d.	n.d.
	95.91	94.63	95.14	94.43	95.71	95.84	96.25	96.80	95.58	95.14	95.90
Structural formulas based on 22 oxygens:											
Si	5.897	6.950	5.909	5.962	6.056	5.958	6.052	5.958	5.160	5.741	5.950
Al	1.963	1.903	1.956	1.892	1.664	1.499	1.291	1.193	2.787	1.938	1.466
Ti	0.110	0.119	0.093	0.113	0.115	0.700	0.739	0.993	0.688	0.529	0.442
Cr	0.061	0.022	0.063	0.029	0.022	0.008	0.004	—	—	0.010	—
Fe	0.723	1.583	0.645	1.782	1.976	0.899	1.057	1.496	0.994	0.801	1.465
Mn	0.010	0.039	0.003	0.038	0.044	0.013	0.008	0.006	0.019	0.007	0.009
Mg	5.258	4.397	5.243	4.194	4.089	4.413	4.298	3.850	4.034	4.711	4.548
Ca	—	—	—	—	—	0.013	0.021	—	0.016	0.008	0.005
Na	—	0.060	0.047	0.012	—	0.119	0.267	0.117	0.035	0.011	0.006
K	1.922	1.906	1.876	1.810	1.976	1.922	1.860	1.741	1.707	1.872	1.874
	—	—	—	—	—	—	0.058	0.191	0.038	0.019	—
Ni	—	—	0.008	0.005	0.009	0.004	—	0.005	—	—	—
CAT	15.932	15.924	15.950	15.915	15.951	15.619	15.597	14.417	15.631	15.666	15.784

[a]CAT, cation total; n.d. not detected. 1–2, 3–4, core and rim phenocrysts; 5, groundmass, Murun (this work); 6–7, phlogopite groundmass Presidente Oligario (this work); 8, Gaussberg, Sheraton and Cundari (1980); 9–10, core and rim phenocryst; 11, groundmass, Chelima (this work).
[b]Total iron expressed at FeO.

1.0 wt % Na$_2$O, and <0.1 wt % Cr$_2$O$_3$ (Table 6.15, Figures 6.26 and 6.27, this work). Their habit and composition is identical to that of madupitic micas in the Leucite Hills and Prairie Creek lamproites.

Micas in the Chelima dikes have the composition (Al$_2$O$_3$ 11–8 wt %, TiO$_2$ 5.5–3.5 wt %) and zoning characteristics (Figures 6.26, 6.27) typical of those of lamproites (Bergman 1987).

6.1.18. Minor and Trace Elements

Lamproite phlogopites are enriched in F and may contain from 1–7 wt % F. The most detailed investigation of F abundances to date is that of Jaques *et al.* (1986) for the West Kimberley province. Microphenocrysts in the olivine lamproite increase in F content from approximately 1.6 wt %, in the most Mg-rich micas, to 5.5 wt % in the Mg-poor varieties. The most magnesian groundmass micas are richer in F (5.5–7.3 wt %) than the phenocrysts and are zoned towards decreasing F with decreasing *mg* ratio. Phenocrysts in the leucite phlogopite lamproites show an increase in F with decreasing *mg* ratio. Mantles upon such cores show a decrease in F with increasing Ti and decreasing *mg* ratios (Figure 6.17).

Micas in the Argyle olivine lamproite (Jaques *et al.* 1989) have lower F (<2 wt %) contents than those of the West Kimberley olivine lamproites and are compositionally similar to microphenocryst mantles in the West Kimberley leucite phlogopite lamproites (Figure 6.17).

Scott Smith and Skinner (1984b) have shown that phenocryst micas in the Prairie Creek tuffs and breccias have lower F (<1 wt %) contents than the more evolved madupitic micas (F = 4–6 wt %). In common with the West Australian lamproite micas, F decreases with decreasing *mg* ratio.

Few other data are available. Phlogopites from the Leucite Hills contain 1.1–4.9 wt % F (Cross 1897, Kuehner *et al.* 1981, Edgar personal communication). Scott Smith *et al.* (1989) give a range of 0.4–3.9 wt % F in the Kapamba lamproites and Edgar (personal communication) reports 0.9–3.4 wt % F in micas from Smoky Butte. Gaussberg (Foley *et al.* 1986b), Priestly Peak (Sheraton and England 1980) and Murcia-Almeria (Venturelli *et al.* 1984, 1988) micas contain 1.4–19.95, 3.9–4.1, and 1.0–4.4 wt % F, respectively. Micas from Sisimiut are relatively poor in F (0.2–0.9 wt % F) (Scott 1977, Scott 1981).

The majority of micas in lamproites contain from 0.1–1.5 wt % BaO. Typically phenocrysts are poorer in Ba than the groundmass phlogopites. The micas that are richest in BaO (1–3 wt %), however, occur as phenocrysts in the Smoky Butte lamproites (Mitchell *et al.* 1987). Ba and Na increase concomitantly towards the margins of zoned groundmass micas in the West Kimberley province (Mitchell 1981). The low Ba contents of lamproite micas sets them apart from Ba-rich micas found in some other potassic rocks (Section 6.1.21) and is surprising given their association with priderite and occurrence in a Ba-rich rock.

Sodium contents are low in phenocrystal micas (0.1–0.8 wt % Na$_2$O) compared to groundmass types (0.5–1.8 wt % Na$_2$O). Sodic titanian tetraferriphlogopites are known only from lamproites. This mineral is one of most sodic phases found in lamproites.

Chromium contents are typically less than 1.5 wt % Cr$_2$O$_3$ in phenocrysts and are not detectable by electron microprobe (<0.05 wt %) in groundmass varieties. Aluminous micas in a microinclusion from Hallock Butte, Leucite Hills, are unusual in containing 2–3 wt % Cr$_2$O$_3$. Kuehner *et al.* (1981) report 3354–5555 ppm Cr in Leucite Hills micas with the lowest Cr contents being found in the madupitic groundmass mica. This Cr depletion is in accord with the conclusion (Section 6.1.2.5) that these micas are more evolved than phenocrystal micas. Manganese contents of lamproite phlogopites are uniformly low, averaging 0.06 wt % and typically <0.10 wt % MnO. Nickel is the only other minor element present in sufficient quantities to be detected by electron microprobe analysis. The abundance ranges from <0.01–25 wt % NiO. The NiO contents are not significantly different from those found in other mantle-derived micas. Nickel contents are lower in groundmass micas than in phenocrystal types.

Very few studies of the trace element content of lamproite phlogopite have been undertaken. Micas (5 samples) from the Leucite Hills were found by Kuehner *et al.* (1981) to have the following concentrations (ppm) of $Zr = 122–164$; $Sr = 135–273$; $Rb = 250–287$; $Co = 54–57$; $Pb = 2–8$. Henage (1972), for the Francis lamproites (14 samples), reported $Rb = 272–1200$; $Cs = 0.04–12.5$; $Sc = 2.9–5.6$; $Co = 54–63$; $Ta = 0.6–1.2$; $Hf = 2.1–19.3$; $Th = 1.2–10.1$. Mooney (1984) reported for the West Kimberley lamproites (5 samples), $Rb = 358–672$; $Cs = 0.45–1.30$; $Co = 65–73$; $Sc = 2.5–2.9$.

The data are in broad agreement and do not point to any significant interprovincial differences in the trace element content of phlogopite. A particular hinderance to the study of the trace element geochemistry of lamproite mica is the common occurrence of microinclusions of spinel, apatite, wadeite, and diopside. Their presence may preclude the preparation of pure mineral separates and the data quoted above must be considered bearing this problem in mind.

6.1.19. Summary of Mica Compositional Variations

1. Individual lamproite provinces contain phenocrystal micas of distinct composition with respect to their Ti, Al, and Fe contents.

2. Within a lamproite province phenocrysts may exhibit relatively little compositional variation, e.g., Leucite Hills, or they may vary considerably in composition from vent to vent, e.g., West Kimberley, Kapamba.

3. Groundmass micas and mantles of phenocrysts are typically depleted in Cr and Al and enriched in Fe, Ti, Ba, and Na relative to phenocrysts, and thus may be considered to be more evolved than the latter.

4. Compositional evolutionary trends are typically of decreasing Al and increasing Fe at constant or increasing Ti. Slight to moderate Al depletion coupled with increasing Fe and decreasing Mg, i.e., reflecting Fe^{2+} increase, represents evolution towards titanian phlogopite–biotite solid solutions. Strong Al-depletion associated with increasing Fe at essentially constant Mg contents, i.e., reflecting increasing Fe^{3+}, represents evolution from titanian phlogopite toward titanian tetraferriphlogopite. The compositional trend exhibited by micas in any given lamproite may lie anywhere between these two extremes and reflects the local postemplacement crystallization environment. At present it is not possible to correlate the trends with either cooling rate and/or water content of the magma, although they clearly reflect crystallization under different redox conditions.

5. Compositional and zonation trends with respect to the Al, Fe, and Ti contents of micas may be used to assess the degree of evolution of the magma from which they crystallized. Thus micas with high Al_2O_3 (10–14 wt %) and high TiO_2 (4–10 wt %) typically occur as phenocrysts. These are considered to be relatively primitive unevolved micas which may have formed in the magma prior to eruption and possibly at high pressures (see

Section 6.1.19). Micas with low Al_2O_3 contents (<10 wt %) and relatively high FeO_T contents (>3 wt %) form mantles upon phenocrysts and occur as groundmass poikilitic plates. The micas formed at relatively low temperatures at low pressures after eruption or emplacement of the magma.

6. Complex mantling and reverse zoning patterns commonly interpreted to be indicative of magma mixing are not seen in lamproites. Their absence suggests that subvolcanic magma chambers, in which pooling of magma may occur, are typically absent or infrequently developed (see below).

7. Phlogopites that are Al-rich relative to phenocryst compositions are so far known only from the Leucite Hills. Their wide compositional variation with respect to their Fe, Cr, Ti, and Al contents precludes derivation from a single source. These micas were clearly present in their transporting magma prior to the formation of the phenocrysts, as evidenced by mantles of the latter upon them. Relatively Fe-poor varieties may represent crystallization at higher pressures and/or temperatures than the phenocrysts. Relatively Fe-rich varieties may represent rare examples of cumulates formed from evolved batches of magma. Inclusions of phlogopite pyroxenite found in the Leucite Hills lavas may be analogous to the olivine biotite pyroxenite suite found in the potassic lavas of the Western Rift Valley of Africa (Holmes 1950).

6.1.20. Solid Solutions in Lamproite Mica

The majority of the compositional data for mica have been obtained by electron microprobe analysis. Water and the halogens are usually not determined, consequently structural formulas are calculated on the basis of 22 oxygens/formula unit.

The nature of the solid solutions present in these titanian phlogopites is difficult to evaluate owing to the lack of knowledge of the Fe^{3+}/Fe^{2+} ratios, the site preferences of Ti^{4+}, Fe^{3+}, and Mg^{2+}, and the number of lattice site vacancies created by coupled substitutions involving Ti^{4+}. A rigorous assignment of cation distribution is in fact impossible without Mössbauer, visible, and infrared spectra and knowledge of the Fe_2O_3, F, Cl, and H_2O contents of individual micas.

Lamproite phlogopites as a group have insufficient Si and Al present to fill the tetrahedral sites. Consequently, Al_{vi} is absent from phenocrystal and groundmass micas. (N.B. Aluminous phlogopites of cryptogenic origin may contain Al_{vi}.) Within the overall compositional variation shown by lamproite mica, Si increases as Al decreases, and Si commonly exceeds the ideal value of six atoms of Si/formula unit (Figure 6.28). The Al deficiency is a direct reflection of the peralkalinity of the parental magma and increases with differentiation. The resulting tetrahedral site deficiency may be remedied by the entry of Ti^{4+}, Fe^{3+}, or Mg^{2+} to this site. It has not yet been established whether any Ti^{3+} is present in these micas.

Hartmann (1969) has proposed the existence of tetrahedrally coordinated Ti^{4+}, but experimental evidence presented by Robert (1976) and Forbes and Flower (1974) indicates that this possibility is remote. Consequently, it is considered that Ti exclusively occupies octahedral sites replacing divalent or trivalent cations. The following substitutional schemes have been proposed:

$$2Mg^{2+}_{vi} = Ti^{4+}_{vi} + \square \quad \text{Forbes and Flower (1974)} \quad (6.1)$$

$$Mg^{2+}_{vi} + 2Si^{4+}_{iv} = Ti^{4+}_{vi} + 2Al^{3+}_{iv} \quad \text{Robert (1976)} \quad (6.2)$$

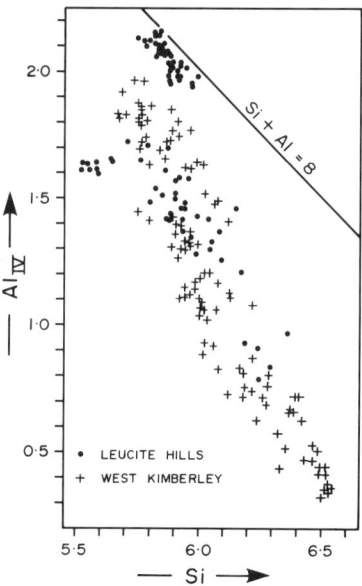

Figure 6.28. Si versus tetrahedrally coordinated Al in phlogopites from the Leucite Hills and West Kimberley lamproites.

$$Mg^{2+}_{vi} + Si^{4+}_{iv} = Ti^{4+}_{vi} + Mg^{2+}_{iv} \qquad \text{Wagner and Velde (1986)} \qquad (6.3)$$

$$Mg^{2+}_{vi} + 2OH^- = Ti^{4+}_{vi} + 2O^{2-} \qquad \text{Arima and Edgar (1981)} \qquad (6.4)$$

Figure 6.29 illustrates the positive correlation between octahedral site occupancy, a measure of the number of lattice site vacancies, and Ti content for some lamproite micas. The data suggest that substitution (6.1) plays an important role in accommodating Ti^{4+} in these micas. However, the displacement of the data from the ideal substitution line suggests that substitution (6.2) may be significant.

Substitution (6.3) has been endorsed by Wagner and Velde (1986a) on the basis of spectrographic evidence presented by Robert (1981) and by the synthesis of Mg_{iv}-bearing mica (Tateyama *et al.* 1974). Robert (1981) has deduced from the infrared spectrum of a Smoky Butte mica that this example contains Mg_{iv} but no Fe_{iv} and that all the Ti^{4+} is octahedrally coordinated. The existence of substitution (6.3) may be demonstrated by the existences of a negative correlation between Ti and (Si + Al + Cr). Such a relationship has been found for Smoky Butte mica (Wagner and Velde 1986a, Mitchell *et al.* 1987) and for micas from Murcia-Almeria and Sisco (Wagner and Velde 1986a). In contrast, data for the Leucite Hills and West Kimberley provinces (Figure 6.30) suggest that substitution (6.3) has no general applicability. Different correlations are found for different lamproites within and between provinces, and as noted by Wagner and Velde (1986a) are related to Al contents. The deviation from the proposed substitution results from the failure to include Fe_{iv} in the model.

While it is not possible to evaluate substitution (6.4) lacking H_2O contents, Arima and Edgar (1981) suggest that it is probably unimportant as they observed no correlation between Ti and H_2O content.

Wendlandt (1977), in contrast to Robert (1981), has proposed on the basis of Mössbauer

MINERALOGY OF LAMPROITES

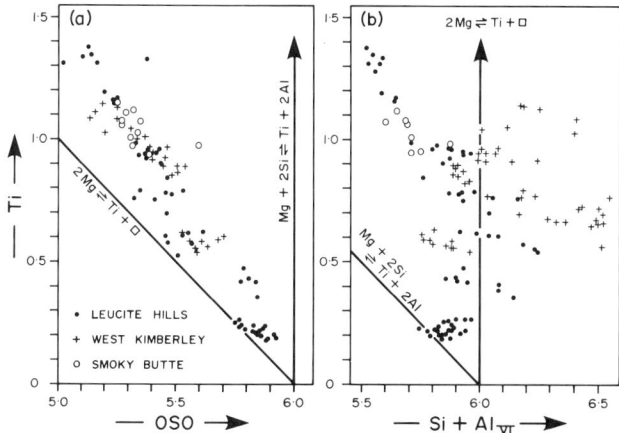

Figure 6.29. Ti versus (a) the octahedral site occupancy (OSO) and (b) Si + Al_{vi} for phlogopites in the Smoky Butte, Leucite Hills and West Kimberley lamproites.

spectra that the tetrahedral site is occupied by Fe^{3+}. Mössbauer spectra obtained by Arima and Edgar (1981) for a Leucite Hills phlogopite suggest that about 60 mol % of the Fe is present as tetrahedral Fe^{3+}. Mössbauer studies of synthetic ferriphlogopite (Annerstein *et al.* 1971) and other Al-poor phlogopites (Hogarth *et al.* 1970) demonstrate the presence of Fe_{iv}. Optical absorption spectra of mantle-derived phlogopites presented by Farmer and Boettcher (1981) suggest that the tetrahedral site preference is Si > Al > Ti > Fe^{3+}, and indicate that the development of reversed pleochroism in the micas may be related to the

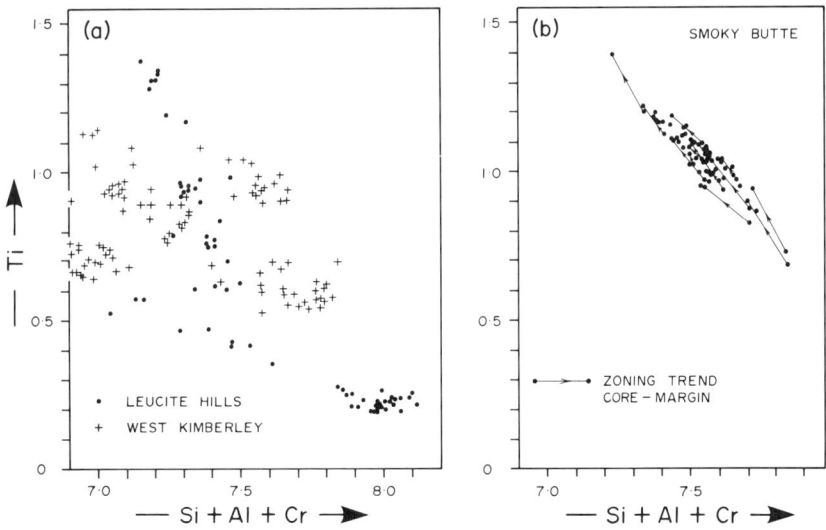

Figure 6.30. Ti versus Si + Al + Cr for phlogopites in (a) the Leucite Hills and West Kimberley and (b) Smoky Butte lamproites.

presence of Ti_{iv}. The above studies thus indicate that the dominant cation accommodating the tetrahedral site deficiency is probably Fe^{3+} and that some Ti_{iv} may occur.

Further complications arise owing to substitution at the 12-fold interlayer sites. Typically (Ba + K + Na) is close to 2 atoms/formula unit. However, samples with significant Ba contents commonly contain less than this ideal value. Substitution of Ba^{2+} for K^+ results in the creation of interlayer lattice site vacancies, e.g.,

$$2K^+ = Ba^{2+} + \square \qquad (6.5)$$

This simple substitution model does not fit the actual compositional variation and the following coupled substitution schemes have also been proposed:

$$3Mg^{2+}_{vi} = Ti^{4+}_{vi} + Ba^{2+}_{xii} + \square_{xii} \qquad \text{Velde (1979)} \qquad (6.6)$$

$$Ba^{2+}_{xii} + 2Ti^{4+}_{vi} + 3Al^{3+}_{iv} = 3(Mg,Fe)^{2+}_{vi} + 3Si^{4+}_{iv} \qquad \text{Mansker } et\ al.\ (1979) \qquad (6.7)$$

Mitchell (1981) has shown that micas from the West Kimberley lamproites approximate to substitution (6.7) (Figure 6.31) but that the scatter of data about the ideal correspondence results from the neglect of Fe_{iv}. A further plausible scheme modified from that of Mitchell (1981) is:

$$3K^+_{xii} + 2(Mg,Fe)^{2+}_{vi} + Al^{3+}_{iv} = Ti^{4+}_{vi} + \square_{vi} + \square_{xii} + Fe^{3+}_{iv} + (Ba^{2+},Na^+)_{xii} \qquad (6.8)$$

This model combines substitutions (6.1) and (6.5) with solid solution towards tetraferriphlogopite from phlogopite. Unfortunately this model cannot be assessed without complete analysis of the mica.

In summary, whereas the exact solid solution schemes cannot yet be determined, it appears that the compositional variation exhibited by lamproite mica is between the following end-member molecules:

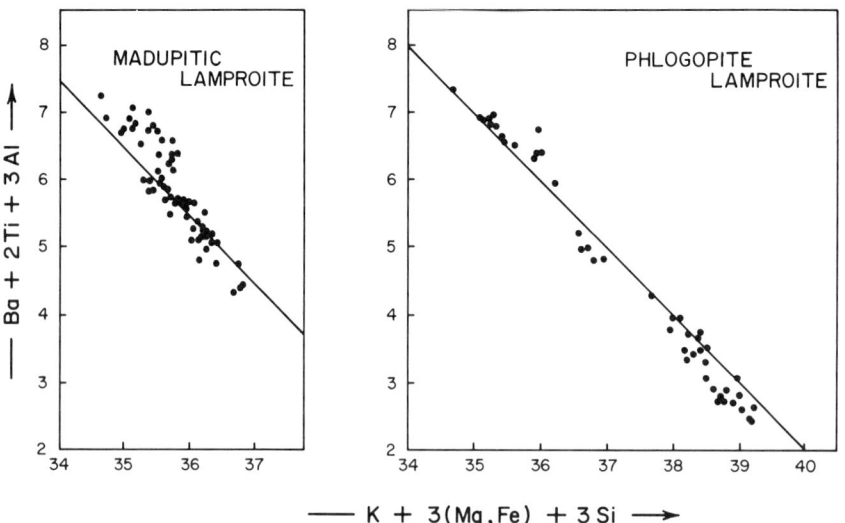

Figure 6.31. Distribution of cations in phlogopites from madupitic and phlogopite lamproites from West Kimberley.

1. $K_2(Mg,Fe)_6Si_6Al_2O_{20}(OH)_4$
   Phlogopite–biotite–annite
2. $K_2(Mg,Fe)_4TiSi_6Al_2O_{20}(OH)_4$
   Octahedral site deficient Ti-phlogopite
3. $K_2(Mg,Fe)_5TiSi_4Al_4O_{20}(OH)_4$
   Ti eastonite
4. $K_2(Mg,Fe)_6Si_6Fe_2O_{20}(OH)_4$
   Tetraferriphlogopite

Solid solution towards eastonitic micas containing Al_{vi} is not a characteristic of lamproite micas.

6.1.21. Comparisons with Mica in Other Potassic Rocks

6.1.21.1 Minettes

Micas in minettes invariably exhibit normal and reversed continuous zoning and light yellow-brown to red-brown pleochroism that reflects variations in their Ti and Fe content. Phenocrysts show wide variations in composition within and between minette occurrences. Bachinski and Simpson (1984) state that it is not possible to predict whether a given mica will have a particular composition based upon its paragenesis. Phenocrysts with discrete cores are common in many minettes. Typically, the cores are relatively Fe rich and exhibit a more intense pleochroism than their mantles. Cores in some cases may be compositionally and optically similar to groundmass micas. Such complex compositional variations are not typical of lamproites and are commonly interpreted as being indicative of mixing of crystal-bearing magmas of broadly similar composition, which have undergone differing amounts of fractional crystallization (Bachinski and Simpson 1984). Mantled micas of the type found in transitional madupitic lamproites are not equivalent to the reversely zoned mantled micas from minettes.

The compositional variation with respect to Ti, Al, and Fe content shown by some minette micas is given in Table 6.16 and illustrated in Figures 6.32 and 6.33. The characteristic compositional trend is one of increasing Fe with slightly increasing or constant Al. The Ti content may increase or decrease slightly with respect to Al. Many minette micas contain sufficient Al to accommodate all of the tetrahedral site deficiency and are thus characterized by the presence of Al_{vi}, indicating solid solution towards eastonitic end members. Bachinski and Simpson (1984) suggest that if a tetrahedral site deficiency exists, it is remedied by the entry of Ti^{4+} rather than Fe^{3+}. Titanian tetraferriphlogopite has not been reported from minettes and none of the micas have compositions equivalent to those of madupitic micas.

Figure 6.32 shows that only a few relatively unevolved minette mica phenocrysts (Devon, Bohemia) overlap the compositions of the least evolved lamproite micas in terms of their Al and Fe contents. This diagram thus provides an effective means of discriminating between the two groups of micas. Compositional trends diverge from a common unevolved phlogopite, with the minette mica always being Al rich relative to lamproite mica of equivalent Fe content. The minette mica compositional trend is similar to that of the Roman province micas (Section 6.1.21.3).

Figures 6.32 and 6.33 demonstrate that the compositions of phenocrysts in relatively

Table 6.16. Composition of Micas from Minettes[a]

	Al_2O_3	TiO_2	FeO_T	Dominant zoning trend
1. Gentengus, Celebes	15.6–16.2	4.7–7.4	9.10–19.3	Increasing FeO[b] with constant Al_2O_3; normal and reverse TiO_2 at constant Al_2O_3.
2. Holmead Farm, Devon	12.6–13.5	1.6–2.3	2.5–4.0	Increasing FeO[b] with increasing Al_2O_3 and TiO_2.
3. Dale Head, Yorkshire	13.4–15.2	1.1–4.1	3.6–13.5	Increasing FeO[b] and TiO_2 with increasing Al_2O_3.
4. Wattle Grill, Cumbria	14.0–15.8	2.7–4.7	4.6–11.6	Increasing FeO[b] and TiO_2 with decreasing Al_2O_3.
5. Navajo, Arizona	12.5–14.8	2.2–8.0	3.9–11.7	Increasing or decreasing FeO[b] and TiO_2 with increasing or decreasing Al_2O_3.
6. Devon, U.K.	10.8–15.8	1.9–6.0	2.6–23.4	As above
7. Linhaisai, Borneo	12.5–14.6	3.0–7.5	4.6–11.6	Increasing TiO_2 and FeO[b] with increasing or decreasing Al_2O_3.
8. Colima, Mexico	11.5–14.7	2.0–5.6	4.9–9.6	Increasing FeO[b] decreasing or increasing TiO_2 with increasing Al_2O_3.
9. Shaws Cove, New Brunswick	14.2–16.3	2.9–6.4	5.6–13.4	Increasing FeO[b] decreasing or increasing TiO_2 with increasing Al_2O_3.
10. Bohemia (alkaline)	11.4–13.5	1.4–4.0	2.5–23.7	Increasing FeO[b], decreasing or increasing TiO_2 with decreasing Al_2O_3.
11. Bohemia	12.1–16.5	2.1–5.8	3.6–17.6	Increasing FeO[b] increasing or decreasing TiO_2 with increasing or decreasing Al_2O_3.

[a]Data sources: 1–4, this work; 5, Jones and Smith (1983); 6, Jones and Smith (1985); 7, Bergman et al. (1988); 8, Allan and Carmichael (1984); 9, Bachinski and Simpson (1984); 10–11, Schulze et al. (1985).
[b]Total Fe calculated as FeO.

unevolved lamproites and minettes are similar, and that evolved micas from the two groups may be recognized on the basis of their Al content.

Minor elements, F (0.2–5.0 wt %), BaO (0–2 wt %), and Cr_2O_3 (typically 0–1 wt % and rarely up to 3 wt %) are found at similar levels to those of lamproite micas and cannot be used to discriminate between the two groups.

6.1.21.2. Kimberlites

Micaceous (Wagner 1914), or phlogopite kimberlites (Skinner and Clement 1979), exhibit some petrographic similarities to phlogopite lamproites. These kimberlites consist primarily of microphenocrystal micas of diverse habit, composition, and mantling styles. Detailed studies of their compositional variation (Smith et al. 1978, Mitchell and Meyer 1989a) have revealed the presence of two distinct varieties of microphenocryst. Relatively rare iron-rich types (10–22 wt % FeO_T, 3–6 wt % TiO_2, 11.5–16.5 wt % Al_2O_3) typically with no tetrahedral site deficiency are designated type I micas (Smith et al. 1978). The majority of the microphenocrysts and groundmass micas, designated type II, are poorer in FeO_T (3–10 wt %), TiO_2 (0.3–4.0 wt %), and Al_2O_3 (6–15 wt %) and may have tetrahedral site deficiencies. Typically type II micas form discrete mantles upon type I micas. Type II micas in turn are zoned to thin rims of tetraferriphlogopite. The extremely complex mantling and zonation patterns with respect to Ti, Fe, Ba, Cr, etc., are attributed to magma mixing (Mitchell and Meyer 1989a). The Fe-rich micas may represent high-pressure phenocrysts derived from differentiated batches of kimberlite. They do not occur as a groundmass phase.

MINERALOGY OF LAMPROITES

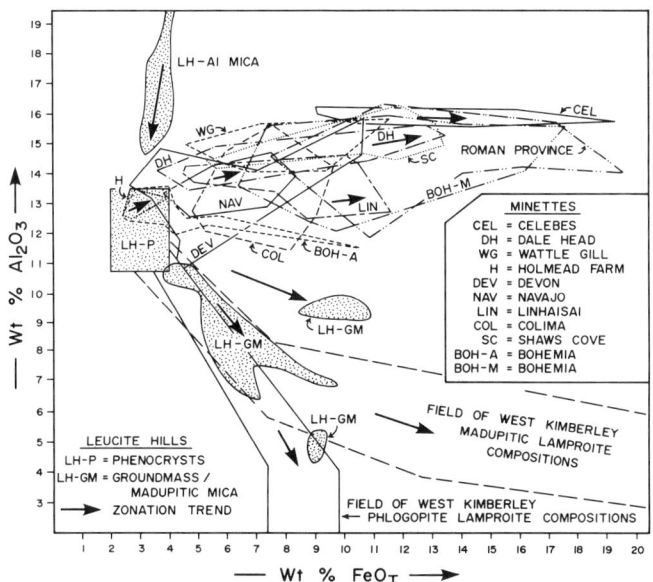

Figure 6.32. Variation of Al_2O_3 versus FeO_T for phlogopites in lamproites compared with that of micas from minettes and Roman province lavas. Data sources for minettes: Gentengus (Celebes), Dale Head (U.K.), Wattle Gill (U.K.), Holmead Farm (U.K., this work), Devon (Jones and Smith 1985), Navajo (Jones and Smith 1983), Linhaisai (Bergman et al. 1988), Colima (Luhr and Carmichael 1981, Allen and Carmichael 1984), Shaws Cover (Bachinski and Simpson 1984), and Bohemia (Schulze et al. 1985; BOH-A = alkaline minettes, BOH-B = other minettes). Data for Roman province lavas from Thomson (1977), Cundari (1975, 1979), Ghiara and Lirer (1977) and Holm (1982).

Type I micas have similar Fe contents to lamproite micas but may be distinguished from the latter on the basis of their lower Ti and higher Al contents. Type II micas have similar compositions to phenocrysts in some phlogopite lamproites, e.g., Leucite Hills. However, they may be distinguished from these on the basis of their evolutionary trend of compositions which is towards TiO_2-poor (<2 wt %) tetraferriphlogopite (Figure 6.34).

Serpentine calcite monticellite (Mitchell 1986, 1989a) kimberlites contain megacrystal and/or macrocrystal phlogopites which may be of xenocrystal or phenocrystal origin. Some varieties are modally enriched in phlogopite and superficially resemble group 2 phlogopite kimberlites. Colorless poikilitic groundmass laths and plates of eastonitic phlogopite are commonly present.

The megacrysts/macrocrysts exhibit a wide range of composition and each kimberlite intrusion is characterized by a distinctive assemblage of micas. The overall trend of compositional evolution (Figure 6.34) is from micas relatively poor in Ti, Fe, and Al to types that are relatively rich in these elements. The trend of Al-enrichment culminates with the formation of late stage groundmass eastonitic phlogopite, which is Ti-poor and Al-rich relative to the megacryst/macrocryst assemblage. Tetraferriphlogopite is not common, but when present is Ti-poor (Mitchell 1986).

The occurrence of phlogopites relatively rich in Al and poor in Ti serves to distinguish these kimberlites from most lamproites (Figure 6.34), although compositional overlap with some low Ti lamproite phlogopites does occur.

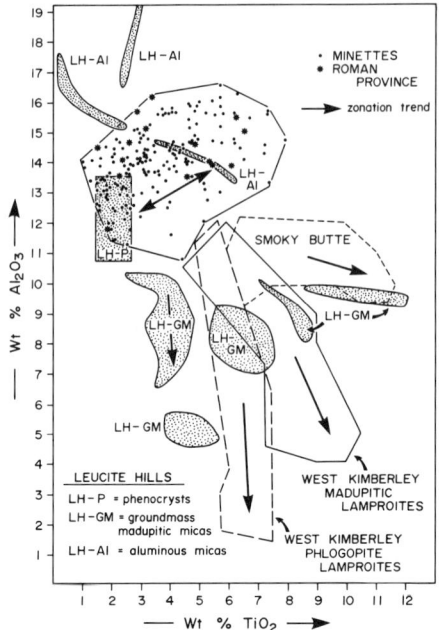

Figure 6.33. Variation of Al_2O_3 versus TiO_2 for phlogopites in lamproites compared with that of micas from minettes and Roman province lavas. Data sources as in Figure 6.32.

6.1.21.3. Roman Province-Type Lavas and Leucitites

Surprisingly few data are available for the composition of micas occurring in Roman province rocks (Thomson 1977, Cundari 1975, 1979, Ghiara and Lirer 1977, Holm 1982). In contrast to lamproite micas they are relatively rich in FeO_T (7.5–19.4 wt %) and Al_2O_3 (11.8–16.3 wt %). These micas typically contain sufficient Al to satisfy all of the tetrahedral site deficiency. The presence of Al_{vi} reflects solid solution towards eastonite. Tetraferriphlogopites are absent and the Fe-enrichment represents solid solution towards biotite. The TiO_2 contents range from 1.6 to 6.6 wt %. Extreme BaO enrichments of 7.3 wt % and 6.9 wt % have been noted by Thomson (1977) and Holm (1982), respectively. The high Al content of these micas serves to set them apart from lamproite phenocrystal and groundmass micas of similar Fe or Ti content (Figure 6.33). In terms of their Al and Fe contents, the Roman province micas are similar to those occurring in the phlogopite pyroxenite inclusions found in the Leucite Hills lavas.

Leucitites from New South Wales (Cundari 1973) contain Ti- and Ba-rich micas (TiO_2 = 8–11 wt %, BaO = 2.6–7.1 wt %, Mitchell 1985). The extreme Ba enrichment differentiates these micas from those of lamproites.

Phlogopites from leucite basanites of the Sierra Nevada contain 2.1–8.6 wt % TiO_2, 12.1–15.0 wt % Al_2O_3, 4–9 wt % FeO_T, and 0.2–2.5 wt % BaO (Van Kooten 1980). Groundmass phases have similar compositions to Smoky Butte phlogopites and phenocrysts are similar to Roman province phlogopites. Although these micas overlap in composition with selected phlogopites from the Leucite Hills and Smoky Butte lamproites they are in general more aluminous than lamproite phlogopites.

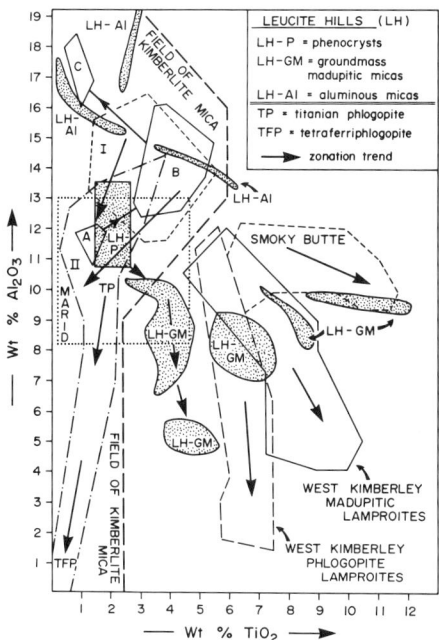

Figure 6.34. Variation of Al_2O_3 versus TiO_2 for phlogopites in lamproites compared with that of micas in kimberlites (Mitchell 1986), micaceous kimberlines (Mitchell and Meyer 1989a, Mitchell 1986, Smith *et al.* 1978) and MARID-suite xenoliths (Dawson and Smith 1977). Also shown are compostiion fields (I and II) of group I and II micas in micaceous kimberlites (Smith *et al.* 1978), and micas (A,B,C) in the Jos kimberlite (Mitchell and Meyer 1980).

Phlogopites from a Ugandan mafurite lava appear to be the only micas yet found that have compositions and evolutionary trends similar to those of lamproite mica. Edgar (1979) found that phenocrystal phlogopite contained 13–13.6 wt % Al_2O_3 and 7.9–8.2 wt % TiO_2, whereas groundmass mica contained 4.6–6.0 wt % Al_2O_3 and 3.6–4.1 wt % TiO_2. However, these micas are FeO_T-rich (14.3–15.1 wt %) relative to lamproite micas. Groundmass micas are richer in MgO (20.3–21.6 wt %) than phenocrysts (MgO = 13.9–14.2 wt %) and are tetraferriphlogopite—phlogopite solid solutions.

6.1.22. Conditions of Crystallization

The composition of lamproite mica implies crystallization from a $MgO–TiO_2$-rich, H_2O-F-bearing magma. Titanian phlogopites are, however, stable over a very wide range of pressures and temperatures. Hence, estimation of the conditions under which they formed is difficult.

The only experimental studies of lamproitic compositions are those of Arima and Edgar (1983a). Phlogopites (2.2–5.6 wt % TiO_2, 10.3–11.8 wt % Al_2O_3, 3.8–5.7 wt % FeO_T) that are identical to phenocrystal lamproite mica were crystallized from a madupitic lamproite (wolgidite) composition at 950–1200°C at 10–40 kbar. The TiO_2 content of the micas was found to decrease with increasing pressure from 10–20 kbar and to be independent of pressure from 20–40 kbar.

Forbes and Flower (1974) synthesized Ti-rich phlogopites between 7 and 30 kb and found them to be stable at 1350°C at 30 kb with excess water. Robert (1976) determined the solubility of Ti in phlogopite between 600 and 1000°C and showed that the solubility increased with increasing temperature but decreased with pressure reaching a maximum of 6.6 wt % TiO_2 at 1000°C at 1 kbar.

Edgar et al. (1976) and Ryabchikov and Green (1978) studied a biotite mafurite composition between 10 and 30 kb under different volatile conditions (H_2O, CO_2) and oxygen fugacities. The studies showed that the TiO_2 content of phlogopite increased with increasing temperature and oxygen fugacity but decreased with increasing pressure. The phlogopites synthesized contained 1–8 wt % TiO_2, 13.0–14.6 wt % Al_2O_3 and 3.6–8.3 wt % FeO_T. Arima and Edgar (1983a), using a katungite composition, found no correlation between the Ti content of phlogopite and temperature and that Ti contents decreased with increasing pressure. Phlogopites formed between 8 and 10 kbar at 1025–1200°C contained 0.7–2.4 wt % TiO_2, 12.5–14.1 wt % Al_2O_3, and 4.3–6.0 wt % FeO_T.

Trönnes et al. (1985) synthesized a variety of Ti-bearing micas between 825 and 1300°C at 10–30 kbar under vapor-absent conditions and concluded that there is a systematic increase in Ti solubility with increasing temperature and decreasing pressure for a given total TiO_2 content. The synthesized micas were solid solutions between phlogopite, eastonite, octahedral site deficient Ti-phlogopite, and Ti-eastonite and contained 0.3–7.1 wt % TiO_2 and 13.7–17.7 wt % Al_2O_3.

Esperanca and Holloway (1987), using a minette as starting composition, determined that between 10 and 20 kbar and 1070–1200°C the TiO_2 content of phlogopites increased with decreasing temperature and pressure and increasing oxygen fugacity. Phlogopites richest in TiO_2 (7.8 wt %) were formed at 1100°C at 10 kb at oxygen fugacities equivalent to those of the QFM-buffer. Phlogopites synthesized contained 3.3–7.8 wt % TiO_2, 14.7–16.3 wt % Al_2O_3, and 6.2–11.2 wt % FeO_T.

The above studies demonstrate that Ti-bearing phlogopites are stable under pressure and temperature conditions ranging from those of the crust to the upper mantle. The data of Arima and Edgar (1983b) suggest that Ti-phlogopites may be high-pressure phenocrysts in phlogopite lamproites. Phlogopites of differing Ti content may be formed from the same magma depending upon the temperature, pressure, and oxygen fugacity at the time of crystallization. Unfortunately the Ti content of phlogopite cannot as yet be used as a geothermobarometer because of the pronounced bulk composition effects noted by Trönnes et al. (1985). Consequently it is not yet possible to determine whether the differing compositions of phenocrysts within and between provinces are related to the bulk composition of the source or to crystallization of similar magmas under different conditions.

Robert's (1976) data are in accord with the occurrence of poikilitic groundmass Ti-rich micas as a late stage low-pressure (1 kbar) phase in madupitic lamproites. The commonly observed increase in the Ti content of mica during low-pressure crystallization with decreasing temperature is, with the exception of the data of Esperanca and Holloway (1987), not in accord with the experimental studies cited above. The observed compositional trends therefore suggest that increases in oxygen fugacity during groundmass crystallization play a greater role in determining the Ti content of phlogopite than temperature variations. This conclusion is supported by the evolutionary trend at this stage of crystallization towards tetraferriphlogopite and hence increasing Fe^{3+} contents.

Tetraferriphlogopites are clearly low pressure phases. Their formation may imply an increase in the oxygen fugacity of the magma as crystallization progresses as suggested above. Introduction of groundwaters to the magma might result in such an effect. An alternative is that increasing Fe^{3+} content may be due to changes in the $(K_2O + Na_2O)/Al_2O_3$ ration of the magma (Thornber *et al.* 1980). In such cases the $Fe^{3+}/(Fe^{2+} + Fe^{3+})$ ratio rises while oxygen fugacity falls. This effect is observed during the crystallization of pyroxenes and amphiboles in many alkaline rocks, where late stage phases enriched in Fe^{3+} are formed at low oxygen fugacities (Mitchell and Platt 1978). In addition, Mysen and Virgo (1978) and Mysen *et al.* (1985) have shown that tetrahedrally coordinated Fe^{3+} occurs in Na-bearing silicate melts with Na/Al > 1. Arima and Edgar (1981) propose that potassium has the same effect and leads to the formation of tetraferriphlogopites in Al-deficient liquids.

As a final point it should be noted that the phlogopites formed in most of the experimental studies are richer in Al and Fe than lamproite phenocryst micas. Bearing in mind the obvious bulk compositional differences, the data suggest that high-pressure lamproite micas may be relatively Al rich. Fractionation of such micas could lead to the Al deficiency, and hence peralkalinity, of the residual liquids. The presence of Al-rich early crystallizing micas mantled by relatively Al-poor types together with phlogopite (Al-rich) pyroxenites in the Leucite Hills province lends support to such a hypothesis. The inability of Arima and Edgar (1983a) to synthesize Al-rich micas at high pressure may be simply a bulk compositional effect related to the unusually low Al_2O_3 content (2.75 wt %) of their starting material.

Further discussion of experimental studies involving phlogopite can be found in Chapter 8.

6.2. AMPHIBOLE

An amphibole, with a pleochroism unlike that of any other amphibole, was initially reported by Cross (1897) and Osann (1906) from the Leucite Hills and Spanish lamproites, respectively. Subsequently, Prider (1939) noted that a similar amphibole was a common constituent of the West Kimberley lamproites. Analysis of large crystals obtained from the pegmatitic rocks of the Walgidee Hills demonstrated that this amphibole was Ti-rich and contained K in excess of Na. The amphibole was considered to be a new species and was named magnophorite by Prider (1939). This amphibole is, however, merely a potassium-rich variety of richterite, $Na(Na,Ca)Mg_5Si_8O_{22}(OH)_2$, and recent revisions to amphibole nomenclature (Leake 1978) have recommended that the name magnophorite be abandoned in favor of titanian potassium richterite.

Although recent studies of lamproites have revealed the existence of a much greater amphibole compositional variation than previously supposed, the presence of titanian potassium richterite may be considered as one of the characteristic features of lamproites. The amphibole may be readily identified by its unusual lemon yellow to pink or red pleochroism (Figure 2.15).

6.2.1. Classification

The amphibole terminology employed in this work follows that proposed by the International Mineralogical Association (Leake 1978). Essentially all compositional data for lamproite amphiboles have been obtained by electron microprobe methods and consequently their ferric iron contents are unknown. Attempts to estimate Fe_2O_3 contents by recalculation of the composition to 23 oxygens and 13 cations as recommended by Droop (1987) fail, as the cation total exclusive of Ca, Na, K, and Ba is typically less than 13. This is a consequence of Ti substitutions creating C-site lattice vacancies. Consequently for the majority of the compositions Fe_2O_3 cannot be assessed and total iron is expressed as FeO_T. Hence *mg* ratios used in the amphibole classification are based upon FeO_T. Amphibole compositions reported by Hernandez-Pacheco (1964) and Wade and Prider (1940) contain only 0.94% and 0.58% Fe_2O_3, respectively. These amphiboles are classified as potassium richterite regardless of how iron is distributed. Estimation of Fe_2O_3 is commonly successful only for magnesio-arfvedsonitic compositions.

The majority of lamproite amphiboles are titanian potassium richterites and titanian potassium magnesio-katophorites. An outstanding feature of their composition is their low Al_2O_3 and high TiO_2 contents. The Al_2O_3 content is typically between 0.2 and 1.0 wt % but may reach up to 1.5 wt %. Characteristically, there is insufficient Al to occupy all of the tetrahedral sites. This deficiency may be accommodated by Ti^{4+} or Fe^{3+}. Thy (1982) has argued that Ti^{4+} remedies the tetrahedral site deficiency as insufficient Fe^{3+} is commonly present. Following Thy (1982) and Thy *et al.* (1987), we suggest that Ti is distributed between tetrahedral (T) and octahedral (C) sites. Substitution in the latter is associated with the creation of lattice vacancies and typically there is a decrease in the sum of the T and C site cations as Ti increases. The substitution is, however, not simple as there is no regular relationship between C-site Ti and the octahedral site deficiency.

6.2.2. Paragenesis

Potassium richterite occurs principally as groundmass poikilitic plates and is one of the last phases to crystallize. Optically continuous plates (up to 0.5 mm in size) that enclose euhedral crystals of leucite, diopside, apatite, wadeite, and priderite are common in the coarser-grained varieties of lamproite, e.g., Rice Hill (West Kimberley), Jumilla (Murcia-Almeria), and Deer Butte (Leucite Hills). The coarsest grains (up to 2–15mm) occur in the pegmatitic phases of the Walgidee Hills (West Kimberley).

Groundmass plates may be zoned from pale yellow-pink cores through reddish-pink regions to dark red-brown margins (Figures 2.15, 2.26, 2.28). A variety of pleochroic schemes have been reported, e.g., X = colorless to pale yellow, Y = reddish, Z = yellow (Wade and Prider 1940); X = pale yellow to colorless, Y = salmon pink, Z = lemon-yellow; X = lemon-yellow, Y = pinkish-brown, Z = greenish-yellow (Jaques *et al.* 1986). The absorption is Y > Z > X.

A second paragenesis of potassium richterite is a small euhedral prisms lining vesicles in extrusive lamproites. This mode of occurrence is particularly common in the Leucite Hills lamproites. The amphiboles are similar in composition and pleochroism to the groundmass types. Some lamproites contain both varieties of amphibole, e.g., Hague Hill, Leucite Hills.

Titanian magnesio-arfvedsonites occur as mantles upon groundmass or vesicle types of potassium richterite. These amphiboles are minor phases and are not developed in all lamproites. In the groundmass they range from fibrous irregular mantles to aggregates of very small euhedral crystals. The textures suggest that they are for the most part secondary phases formed by reaction of previously formed amphibole with residual fluids. Their genesis is analogous to the development of acmitic rims upon preexisting pyroxenes.

Velde (1975) has noted mantles of purple to dark prussian blue, titanian potassium arfvedsonite (described by her as potassic riebeckite) upon potassium richterite at Smoky Butte, and ascribed their genesis to vapor-phase crystallization.

6.2.3. Composition

6.2.3.1. West Kimberley, Australia

Amphiboles in the West Kimberley province exhibit extensive compositional variation (Mitchell 1985, Tables 6.17 and 6.21). They occur principally as groundmass plates that range in composition from titanian potassium magnesio-katophorite to titanian potassium richterite. In the coarse-grained madupitic lamproites from Mount North and Rice Hill they are strongly zoned towards enrichment in Fe, Na, and Ti with concomitant decrease in K and Mg at their margins relative to their cores. Iron enrichment corresponds mainly to compositional evolution from titanian potassium magnesio-katophorite to titanian potassium richterite, although the reverse trend may be found (Figure 6.35). The least evolved amphiboles have Na<K but the trend of Na and Fe enrichment (Figure 6.36) results in Na>K in the Fe-rich varieties (Table 6.17). The majority of the amphiboles do not contain any Na in the A-site and it is only in the Fe-rich, Ca-poor varieties that this site becomes occupied (Table

Table 6.17. Representative Compositions of Sodic-Calcic Amphiboles from the West Kimberley Province[a]

		1	2	3	4	5	6	7
SiO_2		53.70	53.84	52.87	53.05	51.11	51.27	51.42
TiO_2		4.57	4.79	6.20	6.08	7.92	8.01	7.51
Al_2O_3		0.35	0.27	0.28	0.30	0.33	0.45	0.40
Cr_2O_3		n.d.	n.d.	0.06	0.05	n.d.	n.d.	n.d.
FeO[b]		2.32	3.00	4.31	6.09	7.13	8.70	10.97
MnO		0.03	0.06	0.04	0.14	0.11	0.10	0.14
MgO		21.37	20.85	18.12	16.92	16.75	15.04	14.02
CaO		6.57	6.43	5.89	5.27	5.04	4.59	4.27
Na_2O		3.43	3.32	3.63	4.24	4.59	4.70	5.14
K_2O		5.13	4.89	5.31	5.14	5.30	5.16	5.04
		97.47	97.45	96.71	97.28	98.29	98.02	98.97
Si	T	7.608	7.633	7.624	7.659	7.395	7.467	7.494
Al		0.058	0.045	0.048	0.051	0.056	0.077	0.069
Ti		0.334	0.322	0.328	0.290	0.549	0.456	0.437
Ti	C	0.153	0.189	0.344	0.371	0.313	0.421	0.386
Cr		—	—	0.007	0.006	—	—	—
Fe		0.275	0.356	0.520	0.735	0.863	1.060	1.337
Mn		0.004	0.007	0.005	0.017	0.013	0.012	0.017
Mg		4.513	4.406	3.895	3.641	3.612	3.265	3.046
Ca	B	0.997	0.977	0.910	0.815	0.781	0.716	0.667
Na		0.442	0.913	1.015	1.185	1.219	1.284	1.333
Na	A	—	—	—	0.002	0.069	0.043	0.119
K		0.927	0.884	0.977	0.947	0.978	0.959	0.937
mg[c]		0.943	0.925	0.882	0.832	0.807	0.755	0.695

[a] n.d., not detectable. 1–3, Mt. North; 4–7, Rice Hill (this work).
[b] Total Fe calculated as FeO.
[c] mg = Mg/(Fe + Mg).

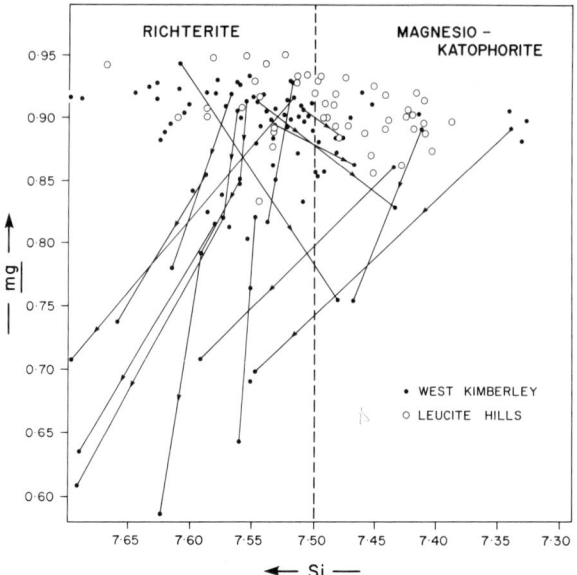

Figure 6.35. Compositional variation and zoning trends of *mg* versus Si of amphiboles in phlogopite and madupitic lamproites from West Kimberley and the Leucite Hills. Data from Mitchell (1985) and this work.

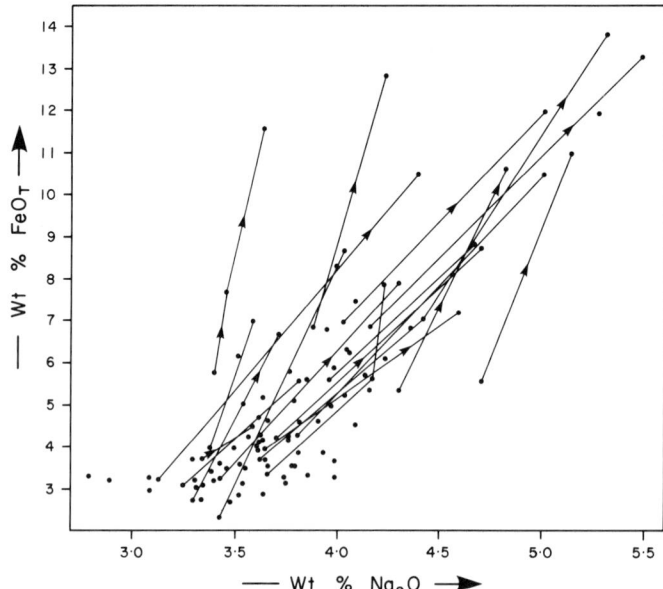

Figure 6.36. FeO_T and Na_2O for amphiboles from the West Kimberley phlogopite and madupitic lamproites. Data from Mitchell (1985) and this work. Arrows indicate core to margin zonation trends.

MINERALOGY OF LAMPROITES

6.17). The compositional trends culminate with the formation of titanian potassium magnesio-arfvedsonite (Table 6.21).

Amphiboles occurring in the olivine lamproites are indistinguishable from those in the phlogopite and madupitic lamproites of similar *mg* number (Figure 6.37, Jaques *et al.* 1986).

6.2.3.2. Leucite Hills

Titanian potassium richterite and titanium potassium magnesio-katophorite (Figure 6.37), occurring as euhedral prisms in vugs and as poikilitic groundmass plates, are of generally similar composition (Table 6.18). The few data available suggest that vug amphibole may be slightly poorer in FeO_T (1.9–3.7 wt %) than the groundmass varieties (FeO_T = 3.1–6.2 wt %). Individual crystals in vugs are homogeneous but show wide variations in Ti content (Table 6.18). Groundmass crystals also are of uniform composition or very weakly zoned towards Ti- and Fe-enriched margins.

All of the amphiboles are rich in Na relative to similar unevolved amphiboles from the West Kimberley province (Figure 6.38). Typically, Na is present in excess of K and commonly there are 0.1–0.3 atoms Na/formula unit in the A-site.

6.2.3.3. Murcia-Almeria

Amphiboles in these lamproites are all titanium potassium richterites that are Na-rich relative to amphiboles in the West Kimberley and Leucite Hills lamproites (Table 6.18, Figure 6.36). Typically, they have high Na contents, commonly 0.1–0.6 atoms of A-site Na. Few compositional data are available (Hernandez-Pacheco 1964, Carmichael 1967, Venturelli *et al.* 1984, 1988, Mitchell 1985, Wagner and Velde 1986a), and it is not possible to assess in detail the compositional variation between lamproite types. Data given by Wagner and Velde (1986a) suggest that olivine sanidine madupitic lamproite contains amphiboles that are Na-rich and K-poor relative to those in other Spanish lamproites (Table 6.18). Figure 6.37 shows the general trend of compositional evolution as being toward increasing Fe and Na and is analogous to that found for the West Kimberley lamproites. This trend culminates with the formation of titanian potassium arfvedsonite (Table 6.21). Titanium contents (2.5–

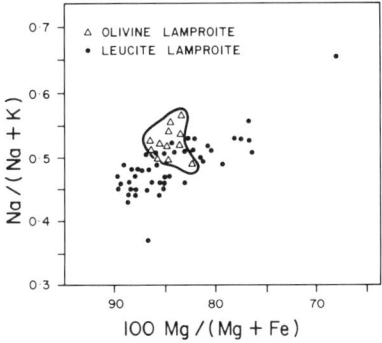

Figure 6.37. Compositional variation of amphiboles in olivine and leucite (*sensu lato*) lamproites from the West Kimberley province (after Jaques *et al.* 1986).

Table 6.18. Representative Compositions of Sodic-Calcic Amphiboles from the Leucite Hills and Murcia-Almeria[a]

	1	2	3	4	5	6	7	8	9	10	11	12
SiO_2	52.28	52.57	53.12	52.85	52.27	51.52	52.63	53.25	52.89	51.79	52.45	53.22
TiO_2	4.56	7.35	5.68	7.85	6.35	8.48	6.65	4.06	5.14	5.66	5.73	4.19
Al_2O_3	0.70	0.37	0.79	0.55	0.93	0.97	0.41	0.41	0.41	0.36	0.67	0.32
Cr_2O_3	0.06	0.03	0.09	0.07	0.01	0.09	0.03	n.d.	n.d.	n.d.	n.d.	n.d.
FeO[b]	1.91	3.66	2.71	4.57	3.14	4.38	6.21	3.06	4.51	6.62	7.54	9.55
MnO	0.05	0.09	0.10	0.13	0.03	0.08	0.10	0.06	n.d.	0.10	0.13	0.14
MgO	20.46	17.79	20.01	17.14	19.74	17.96	17.56	20.87	18.65	17.24	16.75	16.70
CaO	6.96	5.88	6.14	5.48	6.37	5.82	4.73	6.73	6.68	5.79	4.79	4.05
Na_2O	4.71	4.79	4.01	4.57	4.32	4.79	5.00	4.24	4.24	4.25	5.61	6.43
K_2O	4.50	4.64	4.61	4.73	4.29	4.16	4.86	3.93	3.64	3.69	3.36	2.94
BaO	0.20	0.12	n.d.	n.d.	n.d.	n.d.	n.d.	n.d.	n.d.	n.d.	n.d.	n.d.
	96.59	97.29	97.26	97.94	97.45	98.25	98.18	96.61	96.16	95.50	97.03	97.54
Si ⎤	7.523	7.534	7.548	7.533	7.441	7.330	7.543	7.613	7.632	7.577	7.597	7.710
Al ⎦ T	0.119	0.062	0.132	0.092	0.156	0.163	0.069	0.069	0.070	0.062	0.114	0.055
Ti	0.358	0.404	0.320	0.374	0.403	0.507	0.388	0.318	0.298	0.361	0.289	0.235
Ti ⎤	0.135	0.389	0.287	0.467	0.277	0.400	0.328	0.119	0.259	0.262	0.335	0.221
Cr	0.007	0.003	0.010	0.008	0.001	0.010	0.003	—	—	—	—	—
Fe ⎦ C	0.230	0.439	0.322	0.545	0.374	0.521	0.744	0.366	0.544	0.810	0.913	1.157
Mn	0.006	0.001	0.012	0.016	0.004	0.001	0.012	0.007	—	0.012	0.016	0.017
Mg ⎤	4.388	3.800	4.238	3.641	4.188	3.809	3.751	4.447	4.011	3.759	3.616	3.606
Ca ⎦ B	1.073	0.903	0.935	0.837	0.972	0.887	0.726	1.031	1.033	0.908	0.743	0.629
Na	0.927	1.097	1.065	1.163	1.028	1.113	1.274	0.969	0.967	1.092	1.257	1.371
Na ⎤	0.387	0.234	0.039	0.100	0.164	0.209	0.116	0.206	0.219	0.255	0.319	0.435
K ⎦ A	0.826	0.848	0.836	0.860	0.779	0.755	0.888	0.717	0.670	0.689	0.621	0.543
Ba	0.011	0.007	—	—	—	—	—	—	—	—	—	—
mg[c]	0.950	0.897	0.929	0.870	0.918	0.880	0.834	0.924	0.881	0.823	0.798	0.757

[a] n.d., not determined. 1–2, vug, North Table Mountain; 3–4, prismatic groundmass, Endlich Hill; groundmass plates, 5–6, Hague Hill; 7, Hallock Butte (1–7, this work); 8–9, Cancarix (Wagner and Velde 1986a); 10, Calasparra (Wagner and Velde 1986a); 11, Jumilla (this work); 12, Jumilla (Wagner and Velde 1986a).
[b] Total Fe expressed as FeO.
[c] $mg = Mg/(Fe + Mg)$.

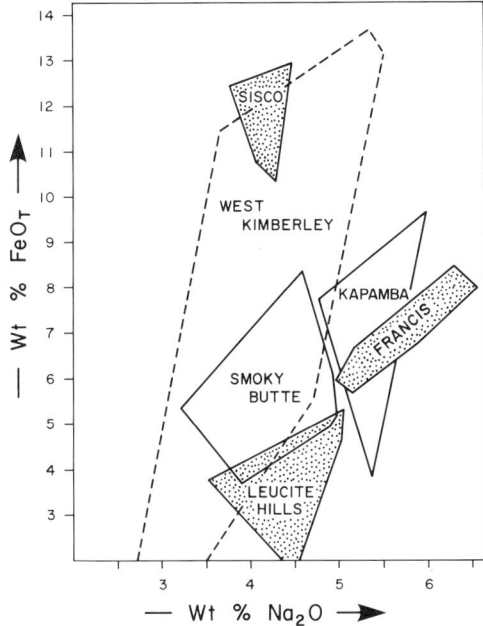

Figure 6.38. FeO$_T$ versus Na$_2$O for amphiboles from the Sisco (Wagner and Velder 1986a, this work), Kapamba (Scott Smith *et al.* 1989), Francis (Wagner and Velde 1986a, this work), Smoky Butte (Mitchell *et al.* 1987) and the Leucite Hills (this work) lamproites.

7.2 wt % TiO$_2$) vary widely. Wagner and Velde (1986a) indicate that amphiboles from Calasparra and Jumilla contain on average less Ti than those from Cancarix.

6.2.3.4. Smoky Butte

Amphiboles forming groundmass plates in these rocks are all titanian potassium richterites (Mitchell *et al.* 1987). Individual plates are not zoned, although considerable intergrain variation is present (Table 6.19). The least evolved amphiboles do not contain any A-site Na. This site becomes occupied as Fe and Na increase until Na exceeds K in the relatively evolved Fe-rich Amphiboles. The amphiboles are similar in terms of their FeO$_T$ (3.7–8.4 wt %) and Na$_2$O (3.7–5.0 wt %) contents to slightly evolved amphiboles from West Kimberley but are distinctly richer in Fe than most amphiboles from the Leucite Hills (Figure 6.37). Titanium contents vary from 3.4 to 6.9 wt % TiO$_2$, levels not significantly different from those found in other provinces and in marked contrast to the Ti-rich phlogopite found in this lamproite. Rapid quenching has prevented the development of late stage liquidus amphibole in all but the sanidine-bearing rocks and precluded the development of extreme Fe-enrichment.

Amphiboles occurring in vugs described by Velde (1975) are slightly poorer in FeO$_T$ (3.7–4.1 wt %) than the poikilitic groundmass plates. Velde (1975) also reported "potassium riebeckite" mantles upon these vug richterites. Recalculation of their composition does not however indicate any Fe^{3+} and as (Na + K) > 0.5 these amphiboles are best termed titanian potassium arfvedsonite (Table 6.21).

Table 6.19. Representative Compositions of Sodic-Calcic Amphiboles from Smoky Butte and Prairie Creek[a]

	1	2	3	4	5	6	7	8	9	10	11	12
SiO_2	53.38	54.06	54.07	54.69	54.41	55.03	51.86	54.45	51.83	53.15	53.76	53.76
TiO_2	5.39	5.58	6.59	5.90	6.89	1.43	4.42	3.98	5.07	5.30	4.37	5.17
Al_2O_3	0.31	0.80	n.d.	0.59	0.72	0.33	0.65	0.27	0.62	0.42	0.79	0.42
Cr_2O_3	n.d.	n.d.	n.d.	n.d.	n.d.	0.09	0.07	0.07	n.d.	n.d.	n.d.	n.d.
FeO[b]	4.10	5.17	6.15	7.10	8.38	2.36	3.67	5.14	3.57	4.44	4.10	6.42
MnO	n.a.	0.16	n.a.	n.a.	n.a.	0.04	0.09	0.11	0.06	0.12	0.04	0.11
MgO	18.75	17.61	16.96	15.55	14.59	21.86	21.29	19.18	21.23	20.54	20.20	17.99
CaO	6.21	5.39	4.61	4.67	3.36	6.43	6.08	4.47	5.91	5.04	6.70	5.96
Na_2O	4.04	3.93	4.78	4.56	4.60	4.18	4.25	5.17	4.07	4.90	3.77	4.20
K_2O	4.26	4.52	4.42	4.54	4.39	5.38	4.96	5.30	5.02	5.00	4.72	4.79
	96.44	97.22	97.58	97.60	97.34	97.13	97.34	98.14	97.38	98.92	98.45	98.82
Si ⎤	7.679	7.718	7.731	7.826	7.818	7.819	7.418	7.749	7.425	7.505	7.590	7.641
Al ⎦ T	0.052	0.135	—	0.100	0.122	0.055	0.110	0.045	0.105	0.070	0.131	0.070
Ti	0.268	0.147	0.269	0.075	0.060	0.125	0.472	0.205	0.470	0.425	0.278	0.289
Ti ⎤	0.312	0.452	0.439	0.560	0.684	0.028	0.005	0.221	0.076	0.138	0.186	0.264
Cr ⎦	—	—	—	—	—	0.010	0.008	0.008	—	—	—	—
Fe ⎤ C	0.491	0.617	0.735	0.850	1.007	0.280	0.440	0.612	0.428	0.525	0.484	0.763
Mn ⎦	—	0.019	—	—	—	0.005	0.011	0.013	0.007	0.016	0.005	0.013
Mg	4.005	3.747	3.614	3.317	3.125	4.630	4.554	4.069	4.533	4.325	4.251	3.811
Ca ⎤ B	0.954	0.824	0.706	0.716	0.517	0.979	0.935	0.682	0.907	0.763	1.014	0.908
Na ⎦	1.046	1.088	1.294	1.265	1.282	1.021	1.065	1.318	1.093	1.237	0.986	1.092
Na ⎤ A	0.076	—	0.031	—	—	0.131	0.117	0.108	0.038	0.105	0.046	0.065
K ⎦	0.779	0.859	0.806	0.829	0.805	0.975	0.908	0.962	0.917	0.901	0.850	0.868
mg[c]	0.891	0.859	0.831	0.796	0.756	0.943	0.912	0.869	0.913	0.892	0.898	0.833

[a] n.d., not detected; n.a., not analyzed. 1–5, Smoky Butte (Mitchell et al. 1987); 6–8, groundmass, Prairie Creek (this work); 9–10, pyroxene-richterite inclusion (Mitchell and Lewis 1983); 11–12, American Mine (Scott-Smith and Skinner 1984b).
[b] Total Fe expressed as FeO.
[c] $mg = Mg/(Fe + Mg)$.

6.2.3.5. Prairie Creek

Amphiboles occurring as groundmass plates in madupitic lamproites (Scott Smith and Skinner 1984b) or within richterite diopside microinclusions (Mitchell and Lewis 1983) are titanian potassium richterites and titanian potassium magnesio-katophorites of similar composition (Table 6.19). Typically K exceeds Na. Although individual plates are homogeneous, some intergrain compositional variation with respect to TiO_2 (2.7–5.1 wt %), FeO_T (2.4–5.1 wt %) and Na_2O (3.5–5.2 wt %) is present. Prairie Creek amphiboles are similar in composition to those found in the Leucite Hills lamproites (Figure 6.39).

Amphiboles occurring in the associated American Mine olivine lamproite are distinctly enriched in TiO_2 (4.4–6.1 wt %) and FeO_T (4.1–6.4 wt %) relative to those found in the Prairie Creek lamproite. (Figure 6.39, Scott Smith and Skinner 1984b).

6.2.3.6. Sisco, Francis, Kapamba

Amphiboles from these localities have been insufficiently studied and the few compositional data available allow only the broadest generalizations to be made. Representative compositions are given in Table 6.20.

Groundmass amphiboles from Sisco are low TiO_2 (2.5–3.4 wt %), relatively FeO_T-rich (10.5–12.9 wt %) titanian potassium richterites that evolve to TiO_2-rich (6.5 wt %) titanian

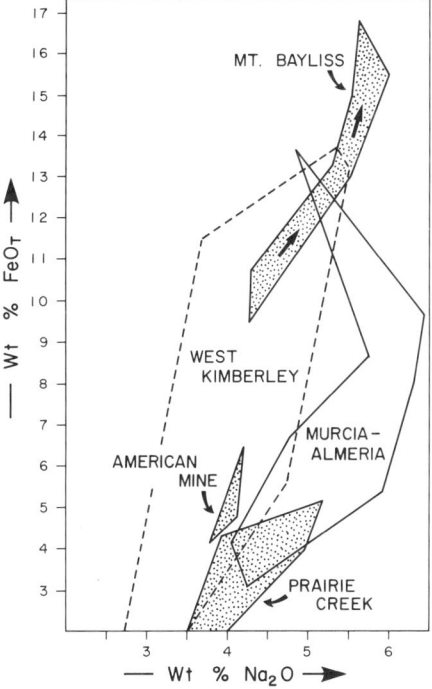

Figure 6.39. FeO_T versus Na_2O for amphiboles from the Mount Bayliss (this work), Murcia—Almeria (Venturelli et al. 1984a, 1988, Mitchell 1985, Wagner and Velde 1986a), Prairie Creek and American Mine (Scott Smith and Skinner 1984b, Mitchell 1985, Mitchell and Lewis 1983) lamproites.

Table 6.20. Representative Compositions of Sodic-Calcic Amphiboles from Sisco, Francis, and Kapamba[a]

		1	2	3	4	5	6	7	8
SiO_2		52.68	51.85	51.65	52.04	51.12	50.06	52.44	50.81
TiO_2		2.52	2.67	6.11	7.09	7.44	8.47	4.08	4.70
Al_2O_3		0.42	0.13	1.20	0.84	0.51	0.25	1.06	1.62
Cr_2O_3		n.d.	n.d.	0.04	0.03	n.d.	n.d.	n.d.	n.d.
FeO[b]		10.72	12.91	5.96	8.52	5.53	8.29	4.78	8.02
MnO		0.21	0.21	0.07	0.17	0.12	0.11	0.09	0.13
MgO		15.80	14.14	18.38	15.04	17.06	14.68	20.32	17.19
CaO		5.83	5.24	5.68	4.10	6.38	4.30	6.78	5.77
Na_2O		4.27	5.43	4.95	6.25	4.33	4.96	5.33	5.74
K_2O		4.28	3.92	3.63	3.74	4.22	3.94	2.91	3.02
		96.73	96.5	97.67	97.82	96.71	95.06	97.79	97.00
Si	T	7.783	7.825	7.411	7.538	7.432	7.470	7.466	7.412
Al		0.071	0.023	0.203	0.143	0.087	0.044	0.178	0.279
Ti		0.146	0.152	0.386	0.319	0.481	0.486	0.356	0.309
Ti	C	0.134	0.151	0.273	0.454	0.332	0.464	0.081	0.206
Cr		—	—	0.005	0.003	—	—	—	—
Fe		1.324	1.629	0.715	1.032	0.672	1.035	0.569	0.978
Mn		0.026	0.027	0.009	0.021	0.015	0.014	0.011	0.016
Mg		3.497	3.181	3.931	3.247	3.697	3.265	4.312	3.738
Ca	B	0.923	0.847	0.873	0.636	0.994	0.687	1.034	0.902
Na		1.077	1.153	1.127	1.364	1.006	1.313	0.966	1.098
Na	A	0.146	0.143	0.250	0.392	0.214	0.122	0.506	0.525
K		0.807	0.755	0.664	0.691	0.783	0.750	0.529	0.562
mg^c		0.724	0.661	0.846	0.759	0.846	0.759	0.883	0.793

[a] n.d., not detected. 1–2, Sisco (this work, Wagner and Velde 1986a); 3–6, Francis, (3–4, this work, 5–6, Wagner and Velde 1986a); 7–8, Kapamba (Scott Smith et al. 1988).
[b] Total Fe calculated as FeO.
[c] mg = Mg/(Fe + Mg).

potassium magnesio-arfvedsonites (Table 6.20, Wagner and Velde 1986a). These amphiboles are similar in composition to the more evolved West Kimberley amphiboles (Figure 6.37).

Amphiboles from Kapamba (TiO_2 = 2.5–6.3 wt %, FeO_T = 3.9–9.5 wt %, Na_2O = 4.8–6.0 wt %, Scott Smith et al. 1989) and Francis (TiO_2 = 5.4–8.5 wt %, FeO_T = 5.5–8.5 wt %, Na_2O = 4.3–6.6 wt %, this work) are of similar composition (Table 6.20) and are notably rich in Na compared to those from the Leucite Hills and West Kimberley (Figure 6.38) but are similar to amphiboles found in the Spanish lamproites (Figure 6.39).

6.2.3.7. Sisimiut

In the leucite lamproites, amphibole occurs as brown-green tabular groundmass crystals (Scott 1981, Thy et al. 1987). Unlike other lamproites, the assemblage is dominated by highly evolved varieties of amphibole. These range in composition from minor Fe-rich titanian potassium richterite to dominant Ti-poor and titanian magnesio-arfvedsonite (Table 6.21, Figure 6.40).

Amphibole lamproites (Thy 1982, Thy et al. 1987) contain brown-to-yellow green

MINERALOGY OF LAMPROITES

Table 6.21. Representative Compositions of Alkaline Amphiboles[a]

	1	2	3	4	5	6	7	8	9	10	11	12
SiO_2	51.22	51.12	50.66	55.68	52.31	54.4	50.59	54.44	54.14	51.03	49.45	53.09
TiO_2	5.07	4.98	4.25	1.32	5.42	2.81	4.50	1.07	0.88	4.69	6.92	2.72
Al_2O_3	0.38	0.85	0.21	0.14	1.01	n.d.	0.04	0.17	0.09	0.11	0.08	0.38
Cr_2O_3	n.d.	n.d.	n.d.	n.d.	n.d.	n.d.	n.d.	0.05	0.01	0.07	0.02	0.05
FeO[b]	13.12	16.67	20.89	10.99	7.99	18.0	30.56	11.86	14.35	17.18	29.19	15.24
MnO	0.16	0.28	0.26	0.12	0.13	0.30	n.d.	0.11	0.11	0.09	0.50	0.13
MgO	12.99	10.33	7.96	16.73	17.09	10.3	1.51	16.50	14.80	9.92	0.67	12.44
CaO	3.21	1.96	0.98	0.89	3.51	n.d.	0.06	0.46	2.02	3.21	0.16	1.88
Na_2O	5.49	6.22	6.71	6.88	6.33	7.74	6.47	6.98	5.89	4.87	6.81	6.37
K_2O	4.91	5.27	4.88	5.37	2.94	3.68	4.76	5.12	4.94	4.91	3.25	3.08
	96.67	97.68	96.80	98.12	96.73	97.23	98.44	96.76	97.23	96.08	97.15	95.38
Fe_2O_3	—	—	—	5.89	1.03	1.83	—	3.79	6.88	—	—	3.70
FeO	—	—	—	5.69	7.07	16.35	—	8.97	8.16	—	—	11.91
	—	—	—	—	—	97.41	—	97.62	97.92	—	—	95.75
Si ⎤	7.690	7.720	7.837	7.948	7.569	8.113	8.014	7.856	7.902	7.829	7.886	7.931
Al ⎥ T	0.067	0.151	0.038	0.024	0.172	—	0.007	0.029	0.015	0.020	0.015	0.067
Ti ⎦	0.243	0.129	0.105	0.028	0.259	—	—	0.115	0.083	0.151	0.099	0.002
Ti ⎤	0.329	0.437	0.391	0.114	0.331	0.315	0.536	0.001	0.014	0.390	0.731	0.304
Cr ⎥ C	—	—	—	—	—	—	—	0.006	0.001	0.008	0.003	0.006
Fe^{3+} ⎥	—	—	—	0.633	0.112	0.205	—	0.974	0.755	—	—	0.416
Fe^{2+} ⎦	1.662	2.105	2.709	0.679	0.855	2.039	4.048	0.458	0.996	2.204	3.893	1.488
Mn	0.020	0.036	0.034	0.015	0.016	0.038	—	0.013	0.014	0.012	0.068	0.016
Mg	2.907	2.325	1.840	3.560	3.686	2.289	0.357	3.549	3.220	2.268	0.159	2.770
Ca ⎤ B	0.516	0.317	0.163	0.136	0.544	—	0.010	0.071	0.316	0.528	0.027	0.301
Na ⎦	1.484	1.683	1.837	1.864	1.456	2.000	1.987	1.929	1.667	1.449	1.973	1.699
Na ⎤ A	0.114	0.138	0.180	0.040	0.320	0.238	—	0.013	—	—	0.133	0.146
K ⎦	0.940	1.015	0.965	0.978	0.533	0.700	0.962	0.942	0.920	0.961	0.682	0.587
mg[c]	0.636	0.525	0.404	0.840	0.812	0.529	0.081	0.886	0.764	0.507	0.039	0.651

[a] n.d., not detected. 1–3, Rice Hill (Mitchell 1985); 4–5, Jumilla (Wagner and Velde 1986a); 6, Cerro de Monagrillo (Venturelli *et al.* 1984); 7, Sisimiut (Scott 1981); 9–12, Sisimiut (Thy *et al.* 1987).
[b] Total Fe expressed as FeO.
[c] $mg = Mg/(Fe + Mg)$.

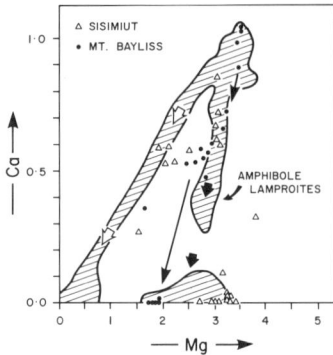

Figure 6.40. Composition variations (Ca versus Mg) of amphiboles in the Sisimiut (Thy *et al.* 1987) and Mount Bayliss (this work) lamproites. Open triangles and the shaded field are for leucite and amphibole lamproites, respectively, from Sisimiut.

richterites occurring in an interstitial or granular texture. They are strongly zoned toward greenish-blue titanian magnesio-arfvedsonite or nearly opaque titanian potassium arfvedsonite. Blue fibrous riebeckitelike minerals form mantles upon earlier varieties.

Thy *et al.* (1987) consider that two zoning trends are exhibited by these amphiboles. One is an arfvedsonite trend showing extensive Ca–Na exchange at the B-structural site accompanied by increasing Ti and decreasing Mg at approximately constant K. The other is a trend towards riebeckite involving Fe^{3+} exchange as Ca decreases and Na increases (Figure 6.40).

6.2.3.8. Mount Bayliss

Amphiboles from Mount Bayliss are similar in composition to evolved amphiboles from the West Kimberley and Sisimiut lamproites (Figures 6.39 and 6.40, Sheraton and England 1980, this work). They are characterized by extensive compositional zoning from brown cores through green intermediate regions to opaque rims. This marked change in pleochroism represents a compositional change from Fe-rich titanian potassium richterite via titanian potassium magnesio-arfvedsonite to titanian potassium magnesio-riebeckite and titanian potassium riebeckite (Table 6.22). This compositional trend lies between the two alkali amphibole trends noted by Thy *et al.* (1987) in the Sisimiut lamproites (Figure 6.40).

6.2.3.9. Summary

1. Individual lamproite provinces are characterized by the presence of titanian potassium richterite and titanian potassium magnesio-katophorite of distinct composition with respect to their Fe and Na contents and K/Na ratios. Amphiboles from Murcia-Almeria, Kapamba, and Francis are Na-rich relative to those from the Leucite Hills, West Kimberley, Smoky Butte, and Prairie Creek.

2. Compositional trends in all provinces are similar and characterized by increasing Fe and Na contents, commonly at constant K content. This trend represents evolution from

MINERALOGY OF LAMPROITES

Table 6.22. Representative Compositions of Amphiboles from Mount Bayliss[a]

	1	2	3	4	5	6	7	8	9
SiO_2	50.99	51.53	50.85	51.83	51.13	50.02	51.65	52.83	53.55
TiO_2	4.66	3.46	6.16	3.45	5.15	3.70	2.41	2.80	2.03
Al_2O_3	0.98	0.57	0.48	0.34	0.26	0.34	0.29	0.10	0.11
Cr_2O_3	0.08	0.07	0.0	0.16	0.05	0.13	0.17	0.05	0.02
FeO[b]	9.66	10.83	13.29	15.75	16.82	19.64	20.90	22.86	24.11
MnO	0.12	0.14	0.23	0.21	0.23	0.21	0.19	0.10	0.19
MgO	16.24	16.13	13.24	12.54	10.89	10.37	8.53	8.37	8.42
CaO	6.68	5.77	3.82	3.02	3.31	2.19	n.d.	n.d.	n.d.
Na_2O	4.30	4.56	5.35	5.96	5.66	5.69	7.75	7.98	8.00
K_2O	4.66	4.94	4.73	4.61	4.96	4.50	4.46	1.36	0.70
	98.37	98.00	98.15	97.87	98.46	97.19	96.35	96.45	97.13
Fe_2O_3	—	—	—	0.73	—	4.27	2.07	8.15	11.40
FeO	—	—	—	15.09	—	15.80	19.04	15.53	13.85
	—	—	—	97.94	—	97.62	96.56	97.27	98.27
Si ⎤	7.446	7.582	7.538	7.745	7.675	7.617	7.966	7.915	7.902
Al ⎬ T	0.169	0.099	0.084	0.060	0.046	0.061	0.053	0.018	0.019
Ti ⎦	0.385	0.320	0.378	0.195	0.279	0.322	—	0.067	0.079
Ti ⎤	0.127	0.063	0.309	0.193	0.303	0.102	0.281	0.248	0.147
Cr	0.009	0.008	—	0.019	0.006	0.016	0.021	0.006	0.002
Fe^{3+} ⎬ C	—	—	—	0.082	—	0.489	0.240	0.918	1.266
Fe^{2+}	1.180	1.333	1.648	1.886	2.112	2.012	2.456	1.946	1.709
Mn	0.015	0.017	0.029	0.027	0.029	0.027	0.025	0.013	0.024
Mg ⎦	3.535	3.537	2.925	2.793	2.437	2.354	1.961	1.869	1.852
Ca ⎤ B	1.045	0.910	0.607	0.484	0.532	0.357	—	—	—
Na ⎦	0.955	1.090	1.393	1.516	1.468	1.643	2.000	2.000	2.000
Na ⎤ A	0.263	0.210	0.144	0.210	0.180	0.037	0.317	0.318	0.289
K ⎦	0.868	0.927	0.894	0.879	0.950	0.952	0.877	0.260	0.132
mg[c]	0.750	0.726	0.640	0.597	0.536	0.539	0.444	0.490	0.520

[a] n.d., not detected. All data this work.
[b] Total Fe calculated as FeO_T.
[c] $mg = Mg/(Fe + Mg)$.

FeO_T-poor (2–3 wt %) titanian potassium richterite through FeO_T-rich (10–14 wt %) titanian potassium richterite to titanian potassian magnesio-arfvedsonite. Titanian riebeckite occurs only rarely.

3. Lamproite amphiboles are all Al_2O_3-poor (<1.5 wt %) and TiO_2-rich (2–9 wt %). The paucity of Al results in substantial tetrahedral site deficiencies and is a consequence of the formation of these amphiboles from a peralkaline magma.

4. Ferric iron-rich amphiboles are not characteristic of lamproites. Even rocks containing tetraferriphlogopite lack such amphiboles.

5. Calcic aluminous amphiboles such as kaersutite and pargasite are absent from lamproites.

6.2.4. Minor and Trace Elements

Little is known of the minor and trace element content of lamproite amphiboles. Barium contents are typically low (0–0.6 wt % BaO; Carmichael 1967, Jaques *et al.* 1986, this work) and commonly less than 0.2 wt % BaO. Carmichael (1967) and Mason (1977) have noted the presence of significant levels of strontium (0.15–0.74 wt % SrO).

Fluorine contents vary widely, e.g., Kapamba (0.9–2.3 wt % F, Scott Smith *et al.* 1988), West Kimberley (commonly 1–2 wt % F and rarely up to 4 wt % F, Jaques *et al.* 1986), Murcia-Almeria (0.1–2.7 wt % F, Wagner and Velde 1986a), Prairie Creek (0.3–3.25 wt % F, Edgar personal communication).

Henage (1972) has reported the following trace elements (ppm) in a richterite from Francis: Ba = 707, Th = 26.05, Sc = 28.1, Co = 82.4, Cr = 150, Ta = 5.03, Hf = 67.34, Zr = 1200, La = 184, Ce = 561, Nd = 176, Sm = 24.65, Eu = 4.90, Tb = 0.88.

Mason (1977) found the following trace elements (ppm) in a richterite from the Walgidee Hills lamproite pegmatite: V = 30, Cr = 20, Co = 20, Ni = 300, Rb = 33, Y = 1.5, Zr = 620, Nb = 4, Ba = 440, La = 1.9, Ce = 3.6, Pr = 0.46, Nd = 2.0, Sm = 0.57, Eu = 0.16, Gd = 0.59. The rare earth element distribution pattern shows light REE enrichment and no Eu anomaly. The vastly different rare earth element contents reported by Henage (1972) and Mason (1977) may reflect contamination of the Francis sample by inclusions of apatite.

6.2.5. Comparisons with Amphiboles in Other Potassic Rocks

Amphiboles similar in composition to those occurring in lamproites are rare. Potassium richterites are found in some metasomatized lherzolites and MARID-suite xenoliths in kimberlite (Dawson and Smith 1977). In terms of their major element composition they are very similar to lamproite potassium richterites but may be distinguished from these (Figure 6.41) on the basis of their lower Ti contents (typically <1 wt % TiO_2) and paragenesis (Table 6.23).

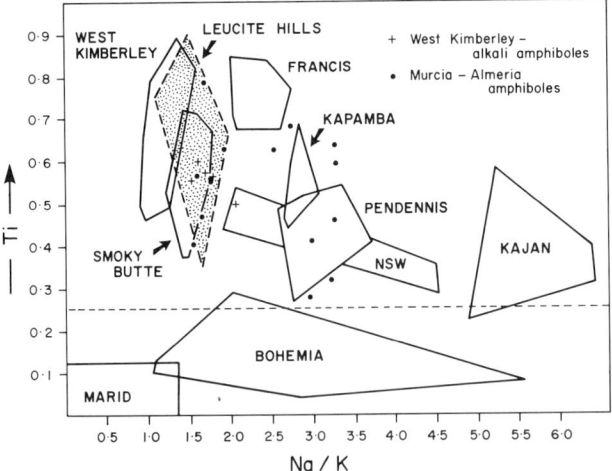

Figure 6.41. Compositional variation (Ti versus Na/K) of amphiboles in lamproites compared with that of amphiboles in minettes and other potassic rocks. Data sources: Pendennis minette (Hall 1982), Bohemian minettes (Němec 1988), New South Wales (NSW) leucitites (this work), "kajanite" (Wagner 1986), MARID-suite xenoliths (Dawson and Smith 1982).

Table 6.23. Representative Compositions of Amphiboles in MARID-Suite Xenoliths, Minettes, and Leucitites[a]

		1	2	3	4	5	6	7	8	9	10	11	12	13
SiO_2		55.0	53.4	53.49	51.60	53.96	55.90	53.70	52.48	54.16	50.96	53.39	52.6	50.73
TiO_2		0.45	1.03	3.36	3.17	0.37	1.67	0.94	2.52	0.74	4.93	2.86	4.1	3.29
Al_2O_3		0.59	1.71	0.23	0.30	0.57	0.25	0.60	0.21	0.32	2.75	1.58	2.0	1.95
Cr_2O_3		0.09	0.26	n.a.	n.a.	n.a.	n.a.	n.a.	n.a.	n.a.	n.a.	n.a.	n.a.	0.05
FeO[b]		4.18	3.95	12.83	18.29	23.48	9.00	11.18	17.27	21.08	7.83	9.05	6.2	12.07
MnO		0.05	0.03	0.35	0.55	0.10	0.19	0.36	0.43	0.61	0.12	0.13	0.07	0.18
MgO		21.5	21.4	13.58	10.12	8.60	17.43	17.58	12.79	10.32	16.61	16.65	19.0	14.47
CaO		6.22	7.18	1.22	0.92	0.18	4.10	6.48	1.52	1.18	5.11	6.52	5.3	4.16
Na_2O		3.91	2.83	7.03	7.44	7.19	4.64	2.91	5.55	7.23	6.75	6.57	4.7	7.02
K_2O		5.28	5.88	3.81	3.53	0.06	5.00	4.14	3.70	1.97	1.91	1.55	2.7	2.39
		97.25	97.70	95.90	95.92	95.01	98.18	97.90	96.46	97.65	96.97	98.3	97.67	96.34
Fe_2O_3		1.60	2.66	1.71	2.42	14.08	1.87	7.35	10.87	9.69	—	—	2.66	0.10
FeO		1.84	1.56	11.29	16.12	10.81	7.32	4.57	7.49	12.36	—	—	3.80	11.98
		97.53	97.93	96.07	96.16	95.92	98.37	98.63	97.56	98.58	—	—	97.94	96.32
Si	T	7.798	7.577	7.942	7.875	8.022	7.975	7.674	7.710	7.945	7.379	7.641	7.465	7.549
Al		0.099	0.286	0.040	0.054	0.100	0.042	0.101	0.036	0.055	0.469	0.266	0.335	0.342
Ti		0.048	0.111	0.018	0.071	—	—	0.103	0.254	—	0.152	0.093	0.201	0.109
Ti	C	—	—	0.357	0.293	0.100	0.180	—	0.024	0.082	0.385	0.215	0.237	0.260
Cr		0.010	0.029	—	—	—	—	—	—	—	—	—	—	—
Fe^{3+}		0.278	0.284	0.191	0.277	1.575	0.201	0.790	1.202	1.070	—	—	0.284	0.011
Fe^{2+}		0.218	0.185	1.402	1.057	1.344	0.873	0.546	0.920	1.516	0.948	1.083	0.452	1.491
Mn		0.006	0.004	0.044	0.071	0.013	0.030	0.044	0.054	0.076	0.015	0.016	0.008	0.023
Mg		4.544	4.526	3.005	2.302	1.906	3.707	3.744	2.801	2.256	3.585	3.552	4.019	0.210
Ca	B	0.945	1.092	0.194	0.150	0.029	0.627	0.992	0.239	0.185	0.793	1.000	0.806	0.663
Na		1.055	0.779	1.806	1.850	1.971	1.284	0.806	1.581	1.815	1.207	1.000	1.194	1.337
Na	A	0.020	—	0.218	0.352	0.101	—	—	—	0.242	0.688	0.823	0.099	0.689
K		0.955	1.064	0.722	0.687	0.011	0.910	0.755	0.693	0.369	0.353	0.283	0.670	0.454
mg[c]		0.954	0.961	0.682	0.523	0.586	0.809	0.833	0.753	0.598	0.791	0.766	0.899	0.683

[a] n.a., not analyzed. 1–2, MARID-xenoliths (Dawson and Smith 1977); 3–4, Pendennis (Hall 1982); 5–9, Bohemia (Němec 1988); 10–11, Kajan River (Wagner 1986); 12, Byrock (Cundari 1973); 13, Bygalorie (Mitchell 1985).
[b] Total Fe calculated as FeO.
[c] mg = Mg/(Fe + Mg).

Some minettes contain a suite of amphiboles whose compositions approach those of lamproites. In the Pendennis minette (Hall 1982) they range from titanian potassium magnesio-arfvedsonite to potassian magnesio-riebeckite and magnesio-riebeckite (Table 6.23). The least evolved amphiboles are similar to the most evolved amphiboles found in lamproites. In Bohemian minettes and peralkaline potassic dikes (Němec 1988) amphibole ranges in composition from potassium richterite through Ti-bearing and titanian potassium magnesio-arfvedsonite to potassian magnesio-arfvedsonite (Table 6.23). The assemblage is notably Ti-poor relative to lamproite amphiboles (Figure 6.41). Many minettes contain pargasite or hornblende (Rock 1984, Bergman et al. 1988).

Leucite-bearing rocks from New South Wales (Cundari 1973) contain amphiboles whose composition ranges from titanian potassium richterite and magnesio-katophorite to titanian potassium magnesio-arfvedsonite (Table 6.23). They differ from lamproite amphiboles in being more aluminous (approximately 2.0 wt % Al_2O_3) and poorer in K_2O (2–4 wt %).

The potassic rock termed kajanite (Lacroix 1926) consists of phenocrysts of olivine, phlogopite, and diopside set in a groundmass of nepheline, leucite, amphibole, and phlogopite. Kajanites are petrographically similar to some lamproites, although the presence of nepheline demonstrates clearly that they are not members of this clan (Wagner 1986). The amphiboles range in composition from titanian potassian magnesio-katophorite to titanian potassian richterite (Table 6.23). The principal compositional differences relative to lamproite amphiboles are their higher Na, lower K, and greater Al_2O_3 (1–3 wt %) contents (Figure 6.41).

In summary, it appears that Al-poor titanian potassium richterites containing more than 0.3 atoms Ti/formula unit (23 oxygens) and with Na/K < 3.0 (Figure 6.41) are characteristic only of lamproites. These amphiboles may evolve to titanian potassium magnesio-arfvedsonites; however, such amphiboles are not typically major phases and commonly appear to have formed during deuteric alteration of preexisting amphiboles. Amphiboles in minettes and leucitites are distinguished from those in lamproites by their lower Ti contents (<0.3 atoms Ti/formula unit) and higher Na/K ratios (>2.0). The overlap in amphibole compositions that occurs in some cases represents a convergence of evolutionary trends in amphiboles forming from magmas that are not genetically related. Such compositions are not diagnostic of any particular magma type and must be considered in conjunction with the nature and composition of associated phases, when they are used in classifying any given rock. Thus richterites in the New South Wales leucitites are not only poorer in K and richer in Al than those in lamproite but occur in association with nepheline and Al-rich pyroxenes.

6.2.6. Conditions of Crystallization

Amphiboles are a late crystallizing phase in lamproite magmas. Charles (1977) demonstrated that K-free richterite–ferrorichterite solid solutions, similar to those of lamproite amphiboles, are stable at temperatures below 1000°C over a wide range of pressure and oxygen fugacity conditions. The maximum thermal stability decreases with increasing Fe content but increases for any given Fe content with decreasing oxygen fugacity. Assuming crystallization at 1 kbar fluid pressure, amphiboles similar to those occurring at Rice Hill, West Kimberley may have formed over a temperature range of 900–1000°C at oxygen fugacities between 10^{-15} and 10^{-5} bars.

Few data are available on the stability of potassium richterite. Gilbert and Briggs (1974) have shown that the introduction of potassium leads to decreased thermal stability relative to that of the sodic endmember at low pressure but enhances the high-pressure stability. The effect of Ti is unknown.

Ernst (1962) has shown that arfvedsonitic amphiboles form only at low temperatures (approximately 500–725°C) at low oxygen fugacities (10^{-17}–10^{-22} bars). The stability of magnesio-arfvedsonite is unknown but can be expected to be higher than that of arfvedsonite given the relative stabilities of riebeckite and magnesio-riebeckite (Ernst 1960, 1962). The effect of Ti and K on their stabilities is unknown.

It is not yet possible to define the exact temperature–oxygen fugacity path followed during crystallization of lamproite amphiboles. However, the available data on relatively similar compositions are in agreement with the change from potassic to sodic richterite to magnesio-arfvedsonite as temperature and oxygen fugacity decrease.

6.3. CLINOPYROXENE

6.3.1. Paragenesis

Clinopyroxenes of differing composition are found in three parageneses:

1. Phenocrysts and groundmass crystals of diopside;
2. Green salites occurring as single crystals mantled by diopside and as a major constituent of clinopyroxenite xenoliths;
3. Colorless augites in olivine biotite pyroxenite inclusions in the Leucite Hills lavas.

Paragenesis (1) is by far the most important as primary phenocrystal (0.5–5 mm), microphenocrystal (0.5 mm–100 μm), and groundmass (<100 μm) diopside occurs in the majority of lamproites. Diopside crystallization appears to have been contemporaneous with leucite, after phenocrystal mica and prior to sanidine formation. Clinopyroxene is found enclosed in mica only in transitional and madupitic lamproites. The pyroxenes are typically euhedral and lack resorption features. They occur mainly as single crystals, occasionally as glomeroporphyritic clusters and rarely as acicular quench crystals. The latter are common in olivine and madupitic lamproites. Flow alignment is common.

The diopsides are colorless to very pale green and weakly pleochroic. Optical zonation is typically absent, but when present is from colorless cores to pale green margins. Reverse zoning is very rare. Twinning is common but not ubiquitous; when present it occurs as a simple twin with composition plane parallel to (010). Twinning is rare in the rapidly cooled Leucite Hills lamproites, in contrast to its common occurrence in the more slowly cooled West Kimberley lamproites.

In the Leucite Hill lamproites, thin, bright green, strongly pleochroic mantles of aegirine-augite are found upon diopside substrates. Such mantles are common on the pyramidal faces of euhedral diopside occurring in vesicles (Carmichael 1967, Barton and Van Bergen 1981). These sodic pyroxenes may have formed by vapor phase crystallization (Kuehner 1980, Mitchell 1985) and are analogous to the sodic amphiboles found in a similar paragenesis (Section 6.2.2).

Green salitic pyroxenes belonging to paragenesis (2) form single crystals 1–2.5 mm in length and are wholly or partially surrounded by thin rims of diopside, or occur as the cores to diopside phenocrysts. The green cores are of variable size ranging from a small proportion (<5%) to over 50% of the total crystal. The cores have embayed irregular margins suggestive of resorption or skeletal crystallization. The boundary between the core and mantle is sharp and commonly decorated with apatite, phlogopite, and Fe–Ti oxide inclusions. Many of the crystals exhibit strain lamellae and undulose extinction.

Clinopyroxenite xenoliths are composed of euhedral to subhedral green salitic pyroxenes together with apatite, phlogopite, and magnesian ilmenite. The pyroxene is identical to that forming the resorbed cores of diopside phenocrysts. Disaggregation of these xenoliths is undoubtedly the source of the mantled green pyroxenes.

Pyroxenes from paragenesis (2) are so far known only from the Leucite Hills (Hatcher Mesa) and have been described by Barton and van Bergen (1981) and Barton (1988). Their relationship to their host lamproite remains unclear (see Section 6.3.3).

Pyroxenes of paragenesis (3) are colorless anhedral augites intergrown with olivine and aluminous Fe-rich phlogopite. The assemblage is equivalent to the olivine biotite pyroxenite (OBP) suite of Holmes (1950). These xenoliths are known only from the Emmons Mesa cinder cone (this work) and possibly Hatcher Mesa (Barton and Van Bergen 1981).

6.3.2. Composition of Phenocrystal and Groundmass Pyroxenes

Representative compositions of phenocrystal pyroxenes are given in Tables 6.24 and 6.25. These tables and Figure 6.42 indicate that the most remarkable feature of lamproitic pyroxenes is their general lack of extensive compositional variation. This is in pronounced contrast to the pyroxenes of most alkaline rocks, which commonly show complex zonation trends that are particularly useful in establishing petrological relationships within and between differentiated rock series. The lack of zonation in lamproite pyroxenes is due either to rapid quenching of the magma soon after the onset of pyroxene crystallization and/or its early replacement as a liquidus phase by amphiboles.

Phenocrystal pyroxenes are diopsides that exhibit vary little solid solution towards enstatite or hedenbergite (Tables 6.24, 6.25, Figure 6.42). Inter- and intrasuite compositional variation is very limited. The majority of the pyroxenes are poor in TiO_2 (0.5–2.5 wt %), Al_2O_3 (0.05–0.5 wt %), Cr_2O_3 (<1 wt %), FeO_T (1.5–4.0 wt %), and Na_2O (0.05–0.6 wt %). Exceptions to the generally low Al contents are found in diopsides from Murcia-Almeria and Kapamba (see below). Zonation when present is towards decreasing Cr and increasing Fe and Na at constant, decreasing or increasing Ti and Al contents.

The exceptionally low Al content of the diopside commonly results in insufficient Al to fill the tetrahedral sites. Thus, (Al + Si) is characteristically less than 2.0 and commonly Na > (Al + Cr).

Figure 6.42 shows that pyroxenes from several lamproite suites are essentially of identical composition with respect to their major element composition. On average, pyroxenes from Kapamba are slightly richer in Fe, and those from Smoky Butte and Francis are poorer in Ca, than the majority of pyroxenes from other provinces.

Despite this overall uniformity, results from the few detailed studies that have been undertaken suggest that minor intraprovincial compositional variations exist. In the West Kimberley province, Mitchell (1985) has noted that pyroxenes from transitional madupitic lamproites are Ca-poor, relative to those occurring in phlogopite lamproites (Figure 6.43a). Jaques *et al.* (1986) have shown that groundmass diopside in olivine lamproites is of similar composition to phenocryst diopside found in phlogopite lamproites but exhibits less compositional variation. The phenocryst diopside shows greater solid solution towards enstatite at essentially constant Mg/Fe ratios (Figure 6.43a). Groundmass and microphenocrystal pyroxenes in the phlogopite lamproites overlap the composition of diopside phenocrysts

Table 6.24. Representative Compositions of Lamproite Pyroxene (West Kimberley, Leucite Hills, Smoky Butte, Francis, Prairie Creek)[a]

	1	2	3	4	5	6	7	8	9	10	11	12	13	14	15	16
SiO_2	54.41	53.43	53.13	53.74	51.68	54.25	54.25	53.80	52.96	52.30	54.28	54.01	53.36	52.34	53.14	51.99
TiO_2	0.74	1.20	2.03	1.04	2.76	0.73	0.95	1.25	1.41	1.84	1.69	2.18	0.40	0.82	1.36	2.08
Al_2O_3	0.04	0.27	0.04	0.06	0.15	0.32	0.60	0.31	0.26	0.28	0.22	0.25	0.26	0.41	0.23	0.41
Cr_2O_3	n.d.	0.45	0.14	0.31	0.22	0.06	0.05	0.39	0.05	n.d.	0.38	n.d.	0.18	0.05	0.16	0.19
FeO[b]	2.05	1.84	2.47	2.68	2.86	2.44	3.34	5.14	9.31	8.23	3.18	3.92	2.50	3.80	2.53	3.21
MnO	0.11	0.03	0.06	0.10	0.08	0.13	0.13	0.04	0.17	0.22	n.d.	n.d.	0.14	0.08	0.08	0.10
MgO	17.96	17.50	17.29	16.63	16.42	17.87	16.80	15.90	13.00	13.70	17.83	17.54	19.43	19.35	17.48	16.56
CaO	24.49	24.39	23.74	24.93	25.28	24.11	23.70	20.83	17.36	19.10	22.10	22.10	22.84	22.31	24.80	24.57
Na_2O	0.29	0.20	0.40	0.38	0.27	0.34	0.62	1.63	3.74	3.30	n.d.	n.d.	0.18	0.21	0.22	0.46
	100.09	99.31	99.30	99.87	99.72	100.25	100.44	99.29	98.26	98.97	99.68	99.99	99.29	99.20	100.00	99.57

Structural formula based on six oxygens:

	1	2	3	4	5	6	7	8	9	10	11	12	13	14	15	16
Si	1.977	1.959	1.953	1.970	1.910	1.971	1.973	1.987	2.010	1.973	1.975	1.964	1.955	1.915	1.945	1.922
Al	0.002	0.012	0.002	0.003	0.007	0.014	0.026	0.014	0.012	0.013	0.009	0.011	0.011	0.018	0.010	0.018
Ti	0.020	0.033	0.056	0.029	0.077	0.020	0.026	0.035	0.040	0.052	0.046	0.060	0.011	0.023	0.037	0.058
Cr	—	0.013	0.004	0.009	0.006	0.002	0.001	0.011	0.002	—	0.011	—	0.005	0.001	0.005	0.006
Fe	0.062	0.056	0.076	0.082	0.088	0.074	0.102	0.159	0.296	0.260	0.097	0.119	0.077	0.116	0.077	0.099
Mn	0.003	0.001	0.002	0.003	0.003	0.004	0.004	0.001	0.006	0.007	—	—	0.004	0.003	0.003	0.003
Mg	0.973	0.956	0.947	0.909	0.905	0.968	0.911	0.876	0.735	0.771	0.967	0.950	1.061	1.055	0.953	0.913
Ca	0.954	0.958	0.935	0.979	1.001	0.938	0.923	0.824	0.709	0.772	0.861	0.861	0.896	0.914	0.973	0.973
Na	0.020	0.014	0.028	0.027	0.019	0.024	0.044	0.117	0.275	0.241	—	—	0.013	0.015	0.016	0.033

[a] n.d., not detected. 1–3, West Kimberley lamproite (this work); 4–5, West Kimberley olivine lamproite (Jaques et al. 1986); 6–9, Leucite Hills, phenocrysts (this work); 10, acmitic overgrowth on diopside, Leucite Hills (Barton and van Bergen 1981); 11–12, Smoky Butte (Mitchell et al. 1987); 13–14, Francis (this work); 15–16, Prairie Creek (Scott Smith et al. 1984b).
[b] Total Fe calculated as FeO.

Table 6.25. Representative Composition of Lamproite (Murcia-Almeria, Sisco, Sisimiut, Kapamba) and Xenolithic Pyroxenes[a]

	1	2	3	4	5	6	7	8	9	10	11	12	13	14	15	16
SiO_2	54.48	54.6	53.28	54.0	52.15	53.09	52.65	52.51	52.45	51.85	52.16	51.88	53.2	51.0	50.6	53.06
TiO_2	0.39	0.94	0.72	0.51	0.66	0.99	1.43	1.10	20.68	1.17	1.48	1.81	0.23	0.41	0.30	0.21
Al_2O_3	0.51	0.33	0.30	0.38	0.10	0.48	0.64	0.85	1.12	1.27	0.59	0.78	1.51	3.65	2.92	2.88
Cr_2O_3	n.d.	n.d.	0.57	n.d.	0.08	0.76	0.50	0.71	0.93	0.92	n.d.	n.d.	n.d.	n.d.	—	0.41
FeO[b]	2.98	4.20	5.30	8.05	3.97	3.03	3.95	5.19	2.43	3.51	4.58	5.05	6.88	8.09	11.3	6.50
MnO	0.08	0.14	0.22	0.21	0.19	0.04	0.08	0.13	0.05	0.06	0.08	0.09	n.d.	n.d.	0.74	0.20
MgO	18.68	17.7	17.64	18.0	19.19	17.11	17.16	17.84	17.33	16.52	16.06	15.56	15.1	13.5	11.7	17.14
CaO	23.26	22.5	21.16	19.30	21.94	23.27	23.15	20.06	24.10	23.40	23.61	24.17	22.8	22.2	21.9	19.46
Na_2O	n.d.	0.18	0.21	0.11	n.d.	0.40	0.44	0.39	0.48	0.44	0.61	0.37	0.94	0.83	0.70	n.d.
	100.38	100.59	99.4	100.56	98.28	99.17	100.00	98.78	99.57	99.14	99.17	99.71	100.66	99.68	100.16	99.86

Structural formula based on six oxygens:

	1	2	3	4	5	6	7	8	9	10	11	12	13	14	15	16
Si	1.972	1.979	1.967	1.976	1.942	1.957	1.934	1.946	1.928	1.922	1.940	1.926	1.958	1.906	1.916	1.941
Al	0.022	0.014	0.013	0.016	0.004	0.021	0.028	0.037	0.049	0.056	0.026	0.034	0.066	0.161	0.130	0.124
Ti	0.011	0.026	0.020	0.014	0.019	0.027	0.040	0.031	0.019	0.033	0.041	0.051	0.006	0.012	0.009	0.006
Cr	—	—	0.017	—	0.002	0.022	0.015	0.021	0.027	0.027	—	—	—	—	—	0.012
Fe	0.090	0.127	0.164	0.246	0.124	0.093	0.121	0.161	0.075	0.109	0.143	0.157	0.212	0.253	0.358	0.199
Mn	0.003	0.004	0.007	0.007	0.006	0.001	0.003	0.004	0.002	0.002	0.003	0.003	—	—	0.024	0.006
Mg	1.008	0.957	0.970	0.982	1.065	0.940	0.939	0.985	0.950	0.913	0.890	0.861	0.828	0.752	0.660	0.935
Ca	0.902	0.874	0.836	0.757	0.875	0.919	0.911	0.796	0.949	0.929	0.941	0.961	0.899	0.889	0.888	0.763
Na	—	0.013	0.015	0.008	—	0.029	0.031	0.028	0.034	0.032	0.044	0.027	0.067	0.060	0.051	—

[a] n.d., not detected. 1–4, Murcia-Almeria, 1, Jumilla (Lopez Ruiz and Rodriguez-Badiola 1980); 2, Calasparra (Venturelli et al. 1984); 3–4, Barqueros (Wagner and Velde 1986, Venturelli et al. 1984); 5, Sisco (Wagner and Velde 1986); 6–7, leucite lamproite, Sisimiut (Scott 1981, Thy et al. 1987); 8, amphibole lamproite Sisimiut (Thy et al. 1987); 9–12, Kapamba (Scott et al. 1988); 13–14, green core pyroxenes; 15, green pyroxene in clinopyroxenite xenolith, Leucite Hills (Barton and van Bergen 1987); 16, colorless pyroxene in OBP xenolith, Leucite Hills (this work).
[b] Total Fe calculated as FeO.

MINERALOGY OF LAMPROITES

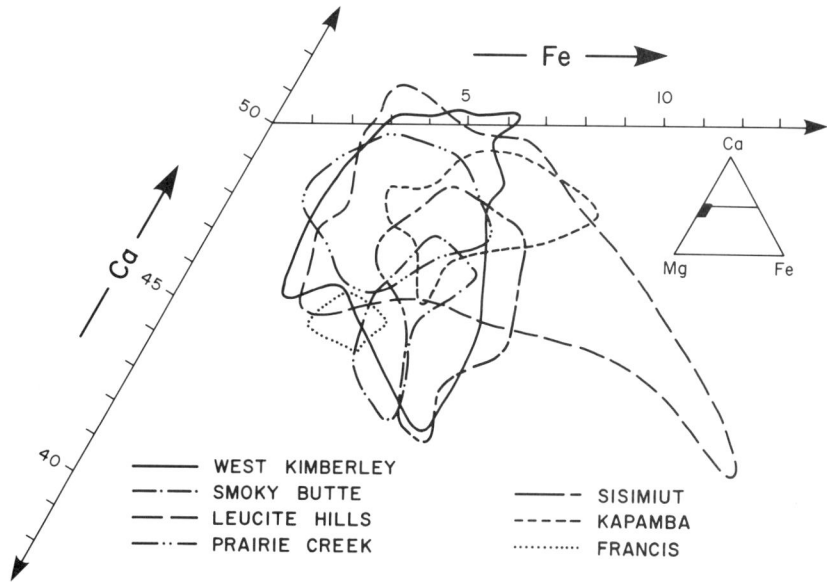

Figure 6.42. Compositional fields of pyroxenes in diverse lamproites (atomic proportions). Data sources given in Section 6.3.2.

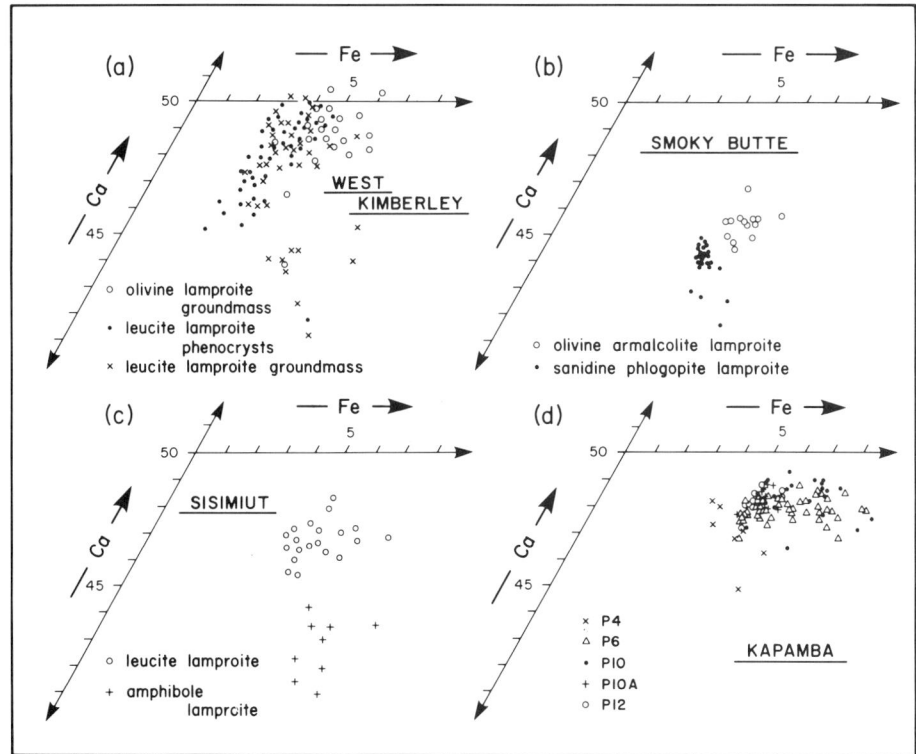

Figure 6.43. Compositions of pyroxenes in lamproites from West Kimberley, Smoky Butte, Sisimiut and Kapamba (atomic proportions). Data sources given in Section 6.3.2.

but extend towards slightly more Fe-rich salitic compositions (Figure 6.43a). In general groundmass diopsides are richer in Ti than phenocrysts.

Mitchell *et al.* (1987) have shown that pyroxenes from the glassy sanidine-free phlogopite lamproites at Smoky Butte are richer in Ca than pyroxenes in the sanidine-bearing varieties (Figure 6.43b).

Thy *et al.* (1987) have noted that the Sisimiut lamproite groundmass diopsides are slightly richer in MgO ($En_{48.4}$) than phenocrysts ($En_{47.3}$). Clinopyroxene megacrysts in the amphibole lamproites are endiopsides that are significantly richer in Mg and Fe than pyroxenes in the leucite phlogopite lamproites (Figure 6.43c).

Scott Smith *et al.* (1989) have shown that diopsides from each of the Kapamba intrusions are of slightly different composition (Figure 6.43d). Those from P4 show a range in Ca content compared to those in the other intrusions. Pyroxenes from P6 and P10 exhibit a relatively wide range of Fe content, that result from late state Fe enrichment in the rims and small groundmass crystals. Scott Smith *et al.* (1989) suggest that the Mg/(Mg + Fe) ratios can be used to indicate the degree of evolution of the magma, and that intrusion P4 is therefore less evolved than P6 and P10.

The diopsides from Kapamba are unusual in that they contain up to 2.2 wt % Al_2O_3 (Table 6.25). Although these levels are higher than those found in pyroxenes from lamproites (<0.5 wt %), they are not as high as those found in pyroxenes from kamafugites and leucitites (Section 6.3.4). Similar Al_2O_3 contents have been reported only in a diopside from the Aljorra lamproite (Murcia-Almeria) by Venturelli *et al.* (1984a).

Insufficient data are available for diopsides from the Murcia-Almeria lamproites (Venturelli *et al.* 1984a, 1988) to allow delineation of any compositional trends. Lopez Ruiz and Rodriguez Badiola (1980) suggest that pyroxenes in madupitic varieties are richer in Mg and Ca ($En_{51}Fs_4Wo_{45}$–$En_{47}Fs_6Wo_{47}$) than those in the phlogopite lamproites ($En_{51}Fs_6Wo_{43}$–$En_{46}Fs_{13}Wo_{41}$).

Pyroxenes in the Leucite Hills madupitic rocks are on average slightly richer in Ca (Wo_{47}–Wo_{49}) than those in the phlogopite lamproites (Wo_{45}–Wo_{49}) according to data obtained in this work and by Carmichael (1967) (Figure 6.44). In contrast, Kuehner (1980) and Kuehner *et al.* (1981) find a significant difference between the two groups, with pyroxenes in the madupitic lamproites being significantly richer in Ca (Wo_{49}–Wo_{51}) than those in the phlogopite lamproites (Wo_{45}–Wo_{49}; Figure 6.44). Given that all other lamproitic pyroxenes so far analyzed contain no more than 50 at. % Wo, consideration must be given to the possibility of a systematic 2% error in the Wo content in the data presented by Kuehner (1980) and Kuehner *et al.* (1981).

Pyroxenes from Hatcher Mesa show the greatest compositional variation of the Leucite Hills primary pyroxenes, and stubby euhedral groundmass pyroxene in this lamproite may contain up to 9.3 wt % FeO_T. Late stage aegirine–augite mantles on diopsidic pyroxenes are relatively rich in Na and Fe (Table 6.24).

Little is known of the trace element content of lamproite pyroxene. Kuehner *et al.* (1981) indicate the following concentration (ppm) for Leucite Hills pyroxenes; Ba = 1454–3884, Ni = 106–220, Nb = 6–40, Zr = 410–689, Y = 1–5, Sr = 2013–2875, Rb = 15–39, Pb = 2–24, Co = 22–40. Henage (1972), in a Francis pyroxene, found the following: Th = 4.84, Sc = 54.9, Co = 36.32, Hf = 9.74, La = 68.7, Ce = 170, Nd = 85.1, Sm = 1.88, Eu = 4.05, Tb = 0.8, Yb = 1.37, Lu = 0.27 (La/Yb = 50). Mason (1977), in a Walgidee Hills pyroxene, found V = 20, Co = 30, Ni = 400, Sr = 3000, Y = 5.5, Zr = 60, Nb = 1, Ba = 60, La = 6.8, Ce = 9.4, Pr = 1.1, Nd = 4.3, Sm = 1.2, Eu = 0.40, Gd = 1.4, Tb = 1.9, Dy = 1.1, Ho = 0.20, Er = 0.48, Yb = 0.33 (La/Yb = 21). Venturelli *et al.* (1984) note the

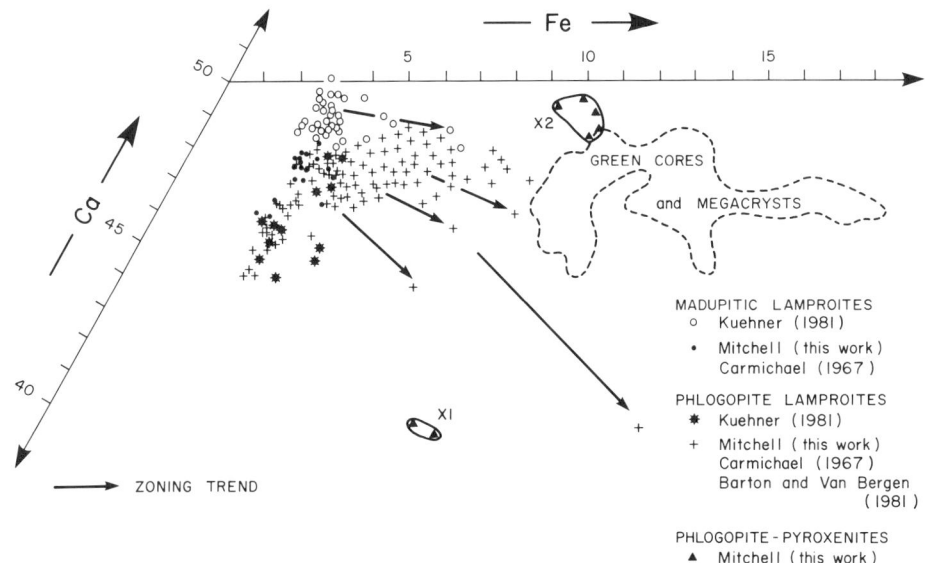

Figure 6.44. Compositions of microphenocrystal pyroxenes in madupitic and phlogopite lamproites, pyroxenes in phlogopite-pyroxenite inclusions (X1, X2) and pyroxene megacrysts from the Leucite Hills (atomic proportions).

presence of 800 ppm Ni in two Murcia-Almeria diopsides. Clearly, further studies are required to resolve the discrepancies in the abundances of most of these trace elements.

6.3.3. Compositions of Pyroxenes from Parageneses 2 and 3

Green pyroxenes belonging to paragenesis (2) have been shown by Barton and van Bergen (1981) to be salites (Table 6.25, Figure 6.44). They are compositionally unlike any of the phenocrystal pyroxenes, and in addition to being relatively Fe rich, contain significantly higher levels of Al_2O_3 (0.8–3.65 wt %) and Na_2O (0.5–1.1 wt %). Wide intergrain compositional variations occur and no systematic zonation pattern is evident. The trend of compositions is parallel to the diopside–hedenbergite join in Figure 6.4. with Fe-rich varieties not showing the trend of solid solution towards enstatite observed in the Fe-rich phenocrystal pyroxenes. This observation suggests that the salites were not derived by the fragmentation of pyroxene-rich rocks formed during the low-pressure differentiation of lamproite magma. Barton and Van Bergen (1981) suggest that the associated clinopyroxenites are fragments of metasomatized rocks that are genetically unrelated to their host lamproite. Barton and Van Bergen (1981) rule out a high-pressure cognate relationship on the grounds that it is improbable that salites can crystallize from such Mg-rich magmas as formed their current hosts. However, they do not discuss the possibility of high-pressure differentiation of lamproite magma to give Fe-rich cumulates which are disaggregated by subsequent batches of primitive magma. In support of such an origin is the experimental data of Barton and Hamilton (1979), which shows that the Al content of clinopyroxenes increases with increasing pressure.

Pyroxenes of broadly similar compositions have been described from a wide variety of alkaline rocks, including the Toro-Ankole kamafugites (Holmes and Harwood 1932) and Roman Province lavas (Ghiara and Lirer 1979, Thompson 1977, Brotzu *et al.* 1977). Their origins have been reviewed by Brooks and Printzlau (1978), Duda and Schminke (1985), and Lloyd (1987). Although no consensus has yet been reached, most petrologists favor a cognate relationship with the pyroxenes crystallizing from magmas genetically related to, but not identical to, their hosts.

Pyroxenes belonging to paragenesis (3) are augites (Table 6.25, Figure 6.44) that are relatively aluminous (2.0 wt % Al_2O_3), have low TiO_2 content, and no detectable Na_2O. They occur in association with aluminous Fe-rich phlogopite and olivine (Fo_{80}). Their relationship to their host lamproites is as yet unknown.

6.3.4. Comparison with Pyroxenes in Other Potassic Rocks

Pyroxenes that are compositionally identical to those of lamproites are found in the groundmass and as microphenocrysts in kimberlites (Mitchell 1986). The latter paragenesis is characteristic of group 2 kimberlites. These pyroxenes are essentially pure diopsides (Table 6.26) that exhibit very limited solid solution towards other endmember pyroxenes. Typically they contain less than 1 wt % TiO_2, Al_2O_3, and Na_2O and are notably Cr_2O_3 poor (<0.5 wt %). In common with lamproite pyroxenes insufficient Si and Al is present to fully occupy all of the tetrahedral sites. Clinopyroxene compositions cannot therefore be used to differentiate between lamproites and group 2 kimberlites.

Pyroxenes of very similar composition occur in MARID-suite xenoliths. They may be easily distinguished from lamproite pyroxene on the basis of their paragenesis and by their higher Na_2O contents (0.7–2.3 wt %, Dawson and Smith 1977, Jones et al. 1982).

Table 6.26. Representative Compositions of Pyroxenes in Other Potassic Rocks[a]

	1	2	3	4	5	6	7	8	9	10
SiO_2	54.39	53.75	51.9	46.0	48.6	42.8	54.6	53.0	53.6	49.8
TiO_2	0.22	0.79	0.54	1.43	2.23	5.55	0.43	0.37	0.37	1.16
Al_2O_3	0.01	0.43	3.60	9.0	5.20	8.76	0.69	0.93	1.20	4.40
Cr_2O_3	n.d.	0.15	0.07	n.d.	n.d.	n.d.	0.06	0.08	0.35	n.d.
FeO[b]	1.64	3.52	4.19	7.50	7.1	7.57	3.42	4.11	3.87	8.49
MnO	0.13	0.11	0.08	0.09	0.14	0.13	0.11	0.13	0.09	0.22
MgO	17.78	17.21	15.3	11.1	12.7	11.0	17.5	16.2	18.4	13.7
CaO	25.58	24.56	24.9	24.4	23.1	23.5	22.4	24.0	21.2	21.7
Na_2O	0.14	0.36	0.17	0.17	0.32	0.66	0.15	0.37	0.37	0.49
	99.89	100.88	100.75	99.69	99.39	99.97	99.46	99.19	99.45	99.96
Structural formulas based on six oxygens:										
Si	1.982	1.956	1.896	1.733	1.825	1.624	1.993	1.962	1.960	1.861
Al	0.004	0.018	0.155	0.400	0.230	0.392	0.030	0.041	0.052	0.194
Ti	0.006	0.022	0.015	0.041	0.063	0.158	0.012	0.010	0.010	0.003
Cr	—	0.004	0.002	—	—	—	0.002	0.002	0.010	—
Fe	0.050	0.107	0.128	0.236	0.223	0.240	0.104	0.127	0.118	0.265
Mn	0.004	0.003	0.003	0.003	0.005	0.004	0.003	0.004	0.003	0.007
Mg	0.966	0.933	0.833	0.623	0.711	0.622	0.952	0.894	1.003	0.763
Ca	0.999	0.957	0.974	0.985	0.929	0.956	0.876	0.952	0.831	0.869
Na	0.010	0.025	0.012	0.012	0.023	0.049	0.018	0.027	0.026	0.036

[a] n.d., not detectable. 1,2, group II kimberlites (Mitchell 1986); 3–4 Roman Province leucitites (Cundari and Ferguson 1982); 5–6, Ugandan leucitites (Cundari and Ferguson 1982); 7–8, Navajo minette (Jones and Smith 1983); 9–10, Linhaisai minette (Bergman et al. 1988).
[b] Total Fe expressed as FeO.

Pyroxenes in other leucite-bearing rocks are strikingly different from those found in lamproites with respect to their Al_2O_3 contents (Table 6.26). Barton (1979) was the first petrologist to recognize that this difference could be used to discriminate between rocks formed from different types of potassic magmas, and this feature has been emphasized by Mitchell (1985) and Bergman (1987). Some potassium-rich rocks such as the leucite basanites from the Sierra Nevada, California (Van Kooten 1980) contain low Al diopsides (<0.5 wt % Al_2O_3). Nevertheless, most diopsides from these rocks are more aluminous with $>1-3$ wt % Al_2O_3.

Figure 6.45 illustrates Ti and Al abundances in some lamproite pyroxenes and shows that they typically contain less than 0.03 atoms Al/6 oxygens. The only exceptions are pyroxenes from Kapamba (Figure 6.46), which are relatively Al rich (0.015–0.095 atoms Al/6 oxygens; Scott Smith *et al.* 1989). Figure 6.45 also indicates that some interprovincial differences exist with respect to the Ti–Al contents, the West Kimberley and Smoky Butte pyroxenes being Ti rich and Al poor relative to those from the Leucite Hills and Murcia-Almeria.

Figure 6.45 illustrates Ti and Al abundances in pyroxenes from some Roman Province lavas and Ugandan leucitites. Although Ti contents are similar to those of lamproite pyroxenes, these pyroxenes are characterized by notably higher Al contents and contain more than 0.05 atoms Al/6 oxygens (typically >0.1). Sufficient Al is present to satisfy any tetrahedral site deficiency and consequently both tetrahedrally and octahedrally coordinated Al is present. Figures 6.45 and 6.46 demonstrate that the Al content of clinopyroxene can be used to discriminate between lamproite and these leucite-bearing rocks, as the Al contents of the pyroxenes are a direct reflection of the alkalinity of their parent magmas.

Figure 6.46 also illustrates the compositions of pyroxenes found in some minettes. Those occurring in the Navajo minettes most closely approximate those in lamproites in terms of their major element and Ti and Al contents (Table 6.26, Figure 6.46). Pyroxenes in other minettes are enriched in Al and therefore similar to Roman province pyroxenes. Although few data are available, it appears that lamproite and minette pyroxenes may be distinguished on the basis of their Al contents.

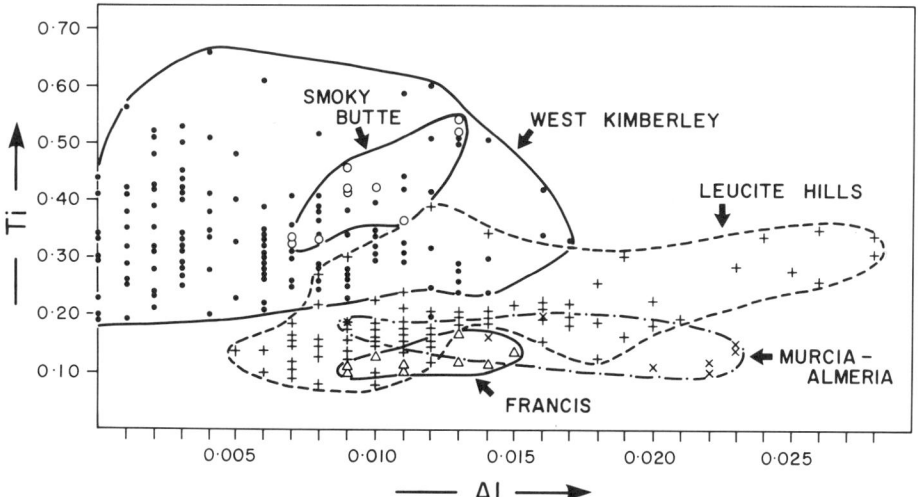

Figure 6.45. Ti versus Al in pyroxenes from diverse lamproites (atoms per six oxygens).

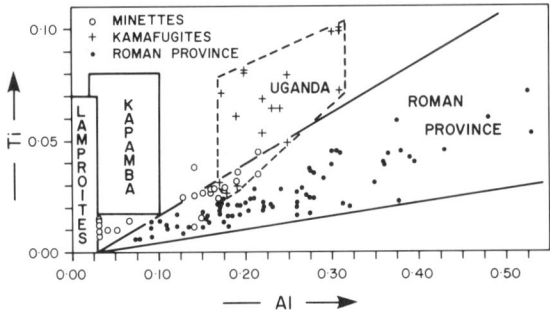

Figure 6.46. Composition variation (Ti versus Al) for pyroxenes in lamproites compared with that of pyroxenes in minettes, Roman province lavas and kamafugites (atoms per six oxygens).

6.4. ORTHOPYROXENE

The Spanish lamproites are the only examples that contain orthopyroxenes of apparently magmatic origin. They are commonest in the lamproites found at Fortuna, where they may comprise up to 10 modal % of the rocks. Xenoliths containing orthopyroxene are known from the Leucite Hills (Barton and van Bergen 1981) and the West Kimberley (Jaques *et al*. 1986) provinces.

6.4.1. Paragenesis

In the Murcia-Almeria lamproites, orthopyroxene occurs as:

1. Anhedral deformed crystals that may contain exsolved clinopyroxene, and have reaction mantles of phlogopite and clinopyroxene. Such orthopyroxenes are referred to as xenocrysts by Venturelli *et al*. (1984a).
2. Discrete subhedral-to-euhedral apparently phenocrystal pyroxenes (this work) and euhedral groundmass microlites (Venturelli *et al*. 1984a, 1988).
3. Microinclusions consisting of granular intergrowths of orthopyroxene and phlogopite (Mitchell and Platt, unpublished data).
4. Euhedral crystals lining miarolitic cavities (Pellicer 1973).
5. As a reaction phase mantling olivine (Venturelli *et al*. 1984a).

Commonly the orthopyroxenes are partially to completely altered to serpentine (Fuster *et al*. 1967, Lopez Ruiz and Rodriguez Badiola 1980).

6.4.2. Composition

Anhedral (? xenocrystal) pyroxenes (Table 6.27) are enstatite–bronzites (En_{91}–En_{87}) containing 0.7–0.8 wt % CaO, 0.3–0.4 wt % Cr_2O_3, and 0.5–4.0 wt % Al_2O_3 (Venturelli *et al*. 1984a, 1988). Allotriomorphic crystals described by Lopez Ruiz and Rodriguez Badiola (1980), which may correspond to this paragenesis, range from $En_{85}Fs_{12}Wo_3$ to $En_{72}Fs_{25}Wo_3$.

Pyroxenes in a phlogopite–orthopyroxene microinclusion from the Fortuna lamproite (Mitchell and Platt, unpublished data) have similar compositions (Table 6.27) to the

Table 6.27. Representative Compositions of Orthopyroxenes[a]

	1	2	3	4	5	6	7	8	9	10	11	12	13	14
SiO_2	55.75	55.7	55.1	57.91	57.6	58.63	57.58	56.82	55.00	55.00	54.1	54.00	56.10	54.3
TiO_2	n.d.	0.14	0.19	0.18	n.d.	0.10	0.20	0.17	0.46	0.21	0.54	0.33	0.64	0.86
Al_2O_3	4.05	1.67	1.56	0.81	0.36	0.16	0.17	0.41	0.04	n.d.	0.18	0.22	0.55	0.84
Cr_2O_3	0.31	0.33	0.38	0.24	n.d.	0.14	0.04	0.14	0.07	0.10	n.d.	n.d.	0.07	0.36
FeO[b]	5.54	6.54	8.03	8.29	3.96	6.34	8.71	11.44	8.34	13.68	15.3	13.17	6.17	7.44
MnO	0.13	0.12	0.13	0.21	0.13	0.09	0.17	0.20	0.20	0.36	0.21	0.37	0.00	0.13
MgO	33.09	34.3	33.2	31.71	36.39	34.13	31.83	29.55	32.33	27.99	27.3	26.01	34.80	32.7
CaO	0.68	0.74	0.78	0.71	0.41	0.55	1.02	1.11	1.55	1.27	1.80	2.80	0.79	0.79
Na_2O	0.07	n.d.	n.d.	n.d.	n.d.	0.08	n.d.	0.11	0.08	0.14	n.d.	0.16	0.05	n.d.
	99.62	99.54	99.37	100.06	98.85	100.22	99.72	99.95	98.07	98.75	99.43	97.06	99.17	97.47
Structural formula based on six oxygens:														
Si	1.924	1.938	1.935	2.009	1.987	2.013	2.011	2.007	1.965	1.995	1.968	1.998	1.955	1.942
Al	0.165	0.069	0.065	0.033	0.015	0.007	0.007	0.017	0.002	—	0.008	0.010	0.023	0.035
Ti	—	0.004	0.005	0.005	—	0.003	0.005	0.005	0.012	0.006	0.015	0.009	0.017	0.023
Cr	0.009	0.009	0.011	0.007	—	0.004	0.001	0.004	0.002	0.003	—	—	0.002	0.010
Fe	0.160	0.190	0.236	0.241	0.114	0.182	0.254	0.338	0.249	0.415	0.466	0.408	0.179	0.223
Mn	0.004	0.004	0.004	0.006	0.004	0.003	0.005	0.006	0.006	0.011	0.007	0.012	—	0.004
Mg	1.702	1.779	1.738	1.639	1.871	1.746	1.657	1.556	1.722	1.513	1.480	1.435	1.808	1.744
Ca	0.025	0.028	0.029	0.026	0.152	0.020	0.038	0.042	0.059	0.049	0.070	0.111	0.030	0.030
Na	0.005	—	—	—	—	0.005	—	0.008	0.006	0.010	—	0.012	0.003	—

[a] n.d., not detected. 1–3, xenocrysts (Venturelli et al. 1984, 1988); 4, phlogopite-orthopyroxenite (Mitchell and Platt, unpublished); 5—8, phenocrysts (Mitchell and Platt unpub.); 9–11, microlitic pyroxene (Venturelli et al. 1984, 1988); 12, miarolitic pyroxene (Pellicer 1973); 13–14, pyroxene synthesized at 25 kbar, 1100°C and 40 kbar, 1150°C, respectively (Arima and Edgar 1983a).
[b] Total Fe expressed as FeO.

anhedral pyroxenes. The orthopyroxene in this inclusion coexists with unevolved Fe-poor phlogopite of similar composition to the phenocrystal phlogopite in the host rock.

Phenocrystal orthopyroxenes from Fortuna (Table 6.27) are enstatite–bronzites ranging in composition from $En_{93}Fs_6Wo_1$ to $En_{80}Fs_{18}Wo_2$ (Mitchell and Platt, unpublished data). They are typically poor in Cr_2O_3 and Al_2O_3 (<0.5 wt %) and contain 0.5–1.5 wt % CaO. The crystals are not typically zoned, although one example of reverse zoning was noted. These phenocrysts are also similar in composition to the anhedral (? xenocrystal) orthopyroxenes.

Microlitic pyroxenes are bronzites ranging in composition from $En_{85}Fs_{12}Wo_3$ to $En_{73}Fs_{23}Wo_4$ (Table 6.27), that are relatively rich in CaO (1.3–1.9 wt %), poor in Cr_2O_3 (<0.2 wt %), and of variable Al_2O_3 content (0–1.4 wt %; Venturelli et al. 1984a, 1988). A pyroxene in a miarolitic cavity was determined by Pellicer (1973) to be a high-Ca bronzite (Table 6.27).

All of the Murcia-Almeria orthopyroxenes are Fe rich relative to orthopyroxenes occurring in mantle-derived lherzolites and harzburgites (Nixon 1987) and derivation from such a source seems unlikely.

The compositional data suggest that the "xenocrystal" pyroxenes may have been derived by the fragmentation of phlogopite-orthopyroxenite inclusions. Petrographic evidence suggests that orthopyroxene was a primary liquidus phase. These pyroxenes evolved from enstatite to bronzite as crystallization progressed.

The relatively Fe-rich nature of the orthopyroxenes would initially seem to preclude a cognate origin for the phlogopite–orthopyroxenite inclusions. However, Arima and Edgar (1983a) have shown that orthopyroxenes alone or coexisting with phlogopite are primary liquidus phases in melts of lamproitic composition at pressures above 20 kbar. These orthopyroxenes have similar compositions (Table 6.27) to those occurring in the phlogopite–orthopyroxenite inclusions. Clearly, the experimental data support the hypothesis that orthopyroxenes in the Spanish lamproites could be phenocrysts.

6.5. OLIVINE

Although olivine is a common constituent of lamproites, it is neither ubiquitous nor a characteristic phase. The olivine content is highly variable, reaching the maximum amounts (30–40 modal %) in olivine lamproites of the Ellendale field (Jaques et al. 1986) and at Prairie Creek (Scott Smith and Skinner 1984b). Olivine in other lamproites ranges from nil or trace quantities up to 11 modal% (Scott 1981, Carmichael 1967, Wade and Prider 1940).

6.5.1. Paragenesis

Two texturally distinct varieties of olivine are found in most olivine-bearing lamproites:

1. Anhedral macrocrysts (Scott Smith and Skinner 1984b) or xenocrysts (Jaques et al. 1986);
2. Subhedral to euhedral phenocrysts and microphenocrysts.

Anhedral macrocrysts are from 1 to 10 mm in diameter. They typically exhibit undulose extinction and/or kink banding. Commonly they are embayed and partially resorbed at their margins. Some grains are subhedral and have complex serrate margins, consisting of parallel growth aggregates of euhedral olivine (Figure 2.19). Such features may represent overgrowths of primary liquidus olivine that has nucleated upon an anhedral substrate (this work) or be an imposed morphology produced during resorption of macrocrysts (Scott Smith and Skinner 1984b, Scott Smith *et al.* 1989).

The deformation features, anhedral habit, and association with microxenoliths of peridotite, dunite, and harzburgite (Jaques *et al.* 1986) suggest that this fraction of the olivine population is of xenocrystal origin.

Phenocrysts (>1 mm) and microphenocrysts or groundmass (<1 mm) olivines are typically euhedral single crystals of simple habit. Solid state deformation features are absent. Skeletal or hopper-type olivines are found in some glassy lamproites from Ellendale (Jaques *et al.* 1986), Oscars Plug, West Kimberley (this work), and Smoky Butte (Mitchell *et al.* 1987). In the Ellendale lamproites, the olivines typically contain euhedral inclusions of chrome spinel.

An olivine population consisting of two generations (or a continuum) of primary phenocrysts together with cryptogenic macrocrysts is analogous to that found in kimberlites (Mitchell 1986). The two populations cannot be confused as they are petrographically different. Macrocrysts in kimberlite are typically rounded and never exhibit the complex resorption and/or parallel growth aggregates that appear to be characteristic of lamproites.

A third olivine paragenesis is known primarily from the Leucite Hills (Cross 1897, Carmichael 1967). Here olivine sanidine phlogopite lamproites contain discrete anhedral olivines mantled by phlogopite laths (Figure 6.2). Such olivines are not known from all of the Leucite Hills localities and appear to be confined to the geographically closely related occurrences at North and South Table Mountains and Endlich Hill. Diopside leucite phlogopite lamproites rarely contain olivine. The olivines have irregular embayed margins and are alteration-free except for the occurrence of red-orange iddingsitic material along fractures (Kuehner 1980). Mantled olivines have also been noted to occur at Francis (Wagner and Velde 1986).

The mantled olivines coexist with microphenocrystal euhedral olivines. The latter are commonly altered at their margins to red-orange iddingsite. Microphenocrystal olivines also form reaction rims around phlogopite. Kuehner (1980) has described chains of such olivine crystals penetrating phlogopites along cleavage planes.

Olivine in many lamproites is partially or totally altered to serpentine, talc, nontronite (Prider and Cole 1942), celadonite (Jaques *et al.* 1986), carbonate minerals (Scott 1981), and chlorite (Jaques *et al.* 1986). Commonly, alteration is more extensive among the smaller microphenocrystal olivines than in the macrocrystal types.

6.5.2. Composition

Olivines in lamproites are all Mg-rich and are forsterites and chrysolites poor in MnO and CaO. Few detailed studies of their compositional variation have been undertaken. Representative compositions are given in Table 6.28.

6.5.2.1. West Kimberley

Xenocrystal (mg = 0.93–0.88) and phenocrystal (mg = 0.93–0.82) olivines are essentially of identical composition with respect to their major element content, with the majority of both varieties having compositions in the range mg = 0.93–0.90. Xenocrystal

Table 6.28. Representative Compositions of Lamproite Primary Olivine[a]

	1	2	3	4	5	6	7	8	9	10	11	12	13	14	15
SiO_2	40.80	40.09	40.82	40.67	40.26	40.29	39.97	41.83	41.10	42.03	40.46	40.75	39.23	40.20	39.82
FeO[b]	7.14	11.17	6.95	8.70	9.00	10.36	11.31	7.37	7.55	7.36	14.09	8.79	20.05	9.58	14.02
MnO	0.09	0.17	0.08	0.14	0.15	0.11	0.19	0.09	0.16	n.d.	n.d.	n.d.	n.d.	0.09	0.24
NiO	0.64	0.29	0.40	0.38	0.34	0.37	0.26	0.36	0.18	0.47	0.37	0.43	n.d.	0.43	0.28
MgO	51.02	47.51	51.36	50.39	50.32	48.94	47.81	50.46	50.88	49.58	46.67	50.54	41.56	49.86	46.00
CaO	0.12	0.23	0.10	0.16	0.31	0.15	0.36	n.d.	0.18	n.d.	n.d.	n.d.	n.d.	0.16	0.20
	99.81	99.46	99.71	100.44	100.38	100.22	99.9	100.11	100.05	99.44	101.59	100.51	100.84	100.32	100.56
Structural formula based on four oxygens:															
Si	0.993	0.996	0.992	0.990	0.983	0.990	0.990	1.011	0.997	1.022	0.955	0.991	0.998	0.984	0.992
Fe	0.145	0.232	0.141	0.177	0.184	0.213	0.234	0.149	0.153	0.150	0.290	0.179	0.427	0.196	0.292
Mn	0.002	0.004	0.002	0.003	0.003	0.002	0.004	0.002	0.003	—	—	—	—	0.002	0.005
Ni	0.010	0.005	0.006	0.006	0.005	0.006	0.004	0.005	0.003	0.007	0.006	0.007	—	0.007	0.004
Mg	1.851	1.760	1.862	1.828	1.832	1.793	1.766	1.819	1.841	1.797	1.712	1.832	1.577	1.820	1.708
Ca	0.003	0.006	0.003	0.004	0.008	0.004	0.010	—	0.005	—	—	—	—	0.004	0.005
mg[c]	92.7	88.4	92.9	91.2	90.9	89.4	88.3	92.4	92.3	92.3	85.5	91.1	78.7	90.3	85.4

[a] 1–2, core and rim, microphenocryst, Oscar, West Kimberley (Jaques et al. 1986); 3, microphenocryst, Ellendale; 9, West Kimberley (Jaques et al. 1986); 4–5, 6–7, core and rims microphenocrysts P4 and P6, respectively, Kapamba (Scott Smith et al. 1989); 8–9, phlogopite-mantled olivine and euhedral phenocryst, South Table Mountain, Leucite Hills (this work); 10–11, 12–13, phenocrysts and microphenocrysts, respectively, Murcia Almeria (Lopez Ruiz and Rodriguez Badiola, 1980); 14–15, core and rim phenocryst, Smoky Butte (Mitchell et al. 1987).
[b] Total Fe calculated as FeO.
[c] mg, Mg/(Mg + Fe).

olivines are characterized by uniform high NiO (0.3–0.5 wt %) and low TiO_2, CaO, and Al_2O_3 (<0.05 wt %) contents. Phenocrystal olivines typically contain higher CaO (up to 0.5 wt %) and NiO (up to 0.6 wt %). All olivines have low Cr_2O_3 (0.02–0.07 wt %) contents which correlate positively with increasing *mg* ratio (Jaques *et al.* 1986).

The majority of the olivines exhibit reverse and normal zoning. Xenocrystal olivines commonly have narrow (< 100 μm) reversely zoned mantles upon uniform cores, which in turn are rimmed by normally zoned material. Figure 6.47 illustrates Ni–*mg* ratio variations associated with the reverse zonation. In contrast, phenocrystal olivines are typically normally zoned (Figure 6.47) with Ni decreasing with decreasing *mg* ratio (Jaques *et al.* 1986).

6.5.2.2. Kapamba

The cores of microphenocrystal olivines range in composition from *mg* 0.86 to 0.93. Core compositions are different in each of the intrusions studied, e.g., P4 (*mg* = 0.90–0.93), P6 (*mg* = 0.89–0.91), P10 (*mg* = 0.88–0.89). The data are interpreted by Scott Smith *et al.* (1989) to indicate that P4 is the least evolved and P10 is the most evolved of the occurrences. The cores of the xenocrysts fall primarily in the range *mg* = 0.90–0.925, and thus overlap the compositions of the phenocryst cores.

Phenocrysts and xenocrysts are normally zoned to thin rims of relatively restricted composition. These rim compositions are different for each intrusion, e.g., P4 (*mg* = 0.90–0.915), P6 (*mg* = 0.86–0.89), P10 (*mg* = 0.88–0.90). The zonation pattern with respect to minor elements is one of decreasing NiO (0.56–0.16 wt %) with increasing CaO (0.09–0.46 wt %) from core to rim (Scott Smith *et al.* 1989).

6.5.2.3. Leucite Hills

Olivines mantled by phlogopite are of restricted composition (*mg* = 0.93–0.92). Individual crystals are not zoned and are characterized by very low CaO (0–0.04 wt %) contents (Kuehner *et al.* 1981, this work). NiO contents reported by Kuehner *et al.* (1981) range from 0.03 to 0.11 wt %, in contrast to 0.3–0.4 wt % found in this work. Olivine occurring in an olivine-phlogopite microxenolith is of identical composition (this work).

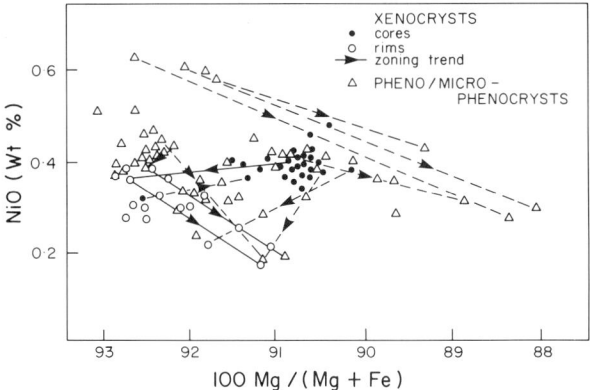

Figure 6.47. Compositional variation exhibited by xenocrystal and phenocrystal olivine in West Kimberley lamproites (after Jaques *et al.* 1986).

Microphenocrystal olivines range in composition from $mg = 0.93–0.87$, are enriched in CaO (0.11–0.62 wt %) and depleted in NiO (0.25–0.17 wt %) relative to the mantled variety (Kuehner *et al*. 1981, this work).

The mantled olivines are demonstrably not in equilibrium with their host magma and may represent upper-mantle-derived xenocrysts. The phlogopites forming the mantles are of identical composition to associated phenocrysts (Section 6.1.2.2). This similarity suggests that the contamination of the magma by the xenocrysts simply induced further crystallization of an existing liquidus phase, rather than reacting to generate a new phase.

Microphenocrystal olivines are clearly primary phases that crystallized after phlogopite, i.e., the reverse of the sequence found in most lamproites.

6.5.2.4. Murcia-Almeria

Xenocrysts containing brown chrome spinel inclusions range in composition from $mg = 0.94$ to 0.90 and contain 0.47–0.7 wt % NiO (Venturelli *et al*. 1984a). Phenocrysts ($mg = 0.91–0.86$) are more magnesian than microphenocrysts ($mg = 0.88–0.83$; Venturelli *et al*. 1988, Lopez Ruiz and Rodriguez Badiola 1980). Wagner and Velde (1986a) report the compositions of diverse olivines to range from mg 0.92 to 0.77. Nickel contents vary widely (0.15–0.6 wt % NiO) and systematic differences between xenocrysts and phenocrysts are evident. The CaO content of all olivines is less than 0.2 wt %. Minor normal zoning at the margins of phenocrysts has been reported by Venturelli *et al*. (1984a).

Venturelli *et al*. (1988) have determined U contents by the fission track method of olivine xenocrysts and phenocrysts from Calasperra are 58–115 ppb and 11–22 ppb, respectively.

6.5.2.5. Smoky Butte

Olivines occur as embayed and corroded subhedral crystals. Multiple growth aggregates and xenocrysts are absent. These phenocrysts range in composition from mg 0.92 to 0.85, with the majority of compositions clustering at mg 0.92–0.90. The olivines are weakly continuously zoned towards margins enriched in Fe, Mn, and Ca and depleted in Ni relative to their cores. Calcium contents do not exceed 0.2 wt % CaO, whereas NiO ranges from 0.3 to 0.6 wt %. Olivines of a particular composition are not associated with a particular petrographic unit of the intrusion (Mitchell *et al*. 1987).

6.5.2.6. Prairie Creek

Anhedral, subhedral, and euhedral crystals, together with multiple parallel growth aggregates and polycrystalline microxenolithic olivines, range in composition from mg 0.92 to 0.89, with the majority falling in the range mg 0.92 to 0.91 (Scott Smith and Skinner 1984b). No apparent correlation exists between composition, size, or paragenesis. Many of the crystals are zonation-free, although minor reversed and normal zoning can be found. Nickel contents range from 0.23 to 0.44 wt % NiO. Narrow margins (<50 μm) depleted in Ni can be found around some crystals. Calcium contents are all low (<0.3 wt % CaO).

6.5.2.7. Sisimiut

Xenocrystal olivines fall within the compositional range mg 0.93–0.88 and contain 0.32–0.8 wt % NiO (Scott 1981, Thy *et al*. 1987). Zoning is not usually apparent but where present is normal. Euhedral phenocrysts are rare and typically completely altered. Scott (1981) reported one crystal of $mg = 0.87$. Thy et al. (1987) note that skeletal olivines found in the chilled margins of dikes are Mg rich ($mg = 0.93-0.91$).

6.5.3. Summary

Despite the paucity of data the following conclusions regarding lamproite olivine may be drawn:

1. Xenocrystal and primary olivines may be present. These are not distinguishable on the basis of their major element composition. However, phenocrysts are slightly richer in CaO (>0.1 wt %) than xenocrysts.
2. Primary olivines evolve towards Fe-rich compositions. Zonation is, however, very limited, possibly because olivine is not a low-pressure liquidus phase in these magmas.
3. Olivine compositions are essentially identical in different lamproites within and between provinces. The sole exception is the Kapamba province in which each lamproite locality appears to contain olivine of slightly different composition. Olivines in general are therefore of little use in determining the petrogenetic relationships between lamproite types.

Lamproite phenocrystal olivines are very similar in composition and habit to primary olivines occurring in kimberlites (Mitchell 1986). The olivine population of the latter is the result of mixing of batches of magma containing phenocrysts of slightly different composition. Normal and reverse zoning is characteristic and reflects attempts by the phenocrysts to equilibrate with the final hybrid magma. Such complex mantling is not found in the lamproitic phenocyrstal olivine population but is seen in the xenocrystal assemblage. Although the major element composition cannot be used as a discriminant, the former are slightly richer in NiO (0.3–0.6 wt %) than the latter (0.1–0.5 wt %; Mitchell 1986).

6.6. LEUCITE

Leucite is a characteristic component of many phlogopite and madupitic lamproites but is rarely present in olivine lamproites. In the latter rocks pseudomorphs after leucite are found only as very small (<50 μm) rounded crystals in the most evolved portions of the groundmass (Scott Smith and Skinner 1984b, Jaques et al. 1986).

6.6.1. Paragenesis

Leucite occurs as subhedral phenocrysts (0.5–5 mm) exhibiting an icositetrahedral habit and as anhedral commonly rounded microphenocrystal and groundmass crystals. The latter may be up to 0.5 mm in diameter but commonly are smaller than 0.1 mm. These small crystals commonly form "frog-spawn-like" aggregations of leucites which have apparently fused together. The rounded habit in some instances appears to result from the preferential resorption of icositetrahedral coigns.

Leucites from relatively rapidly quenched lamproites, e.g., Leucite Hills, are typically small (<0.1 mm), isotropic, and twin-free. Leucites that have crystallized in relatively slowly cooled lamproites, e.g., West Kimberley, commonly show a wider range in size and habit and are weakly anisotropic and complexly twinned. The development of twinning is associated with the inversion of leucite from cubic to tetragonal symmetry at approximately 690°C (Peacor 1968). The factors that determine the development and coarsening of the twins are not understood. Apparently twin-free leucites are probably twinned on a scale only resolvable by electron microscopy.

Leucite is an early phenocrystal phase in most lamproites. Crystallization occurs after phenocrystal phlogopite, and leucites are poikilitically enclosed only by groundmass micas. Commonly, phlogopites in transitional madupitic lamproites contain in their outer mantle microphenocrysts of rounded leucite together with euhedral apatite, wadeite, and priderite.

Leucite typically crystallizes before sanidine in most lamproites and is commonly poikilitically enclosed in groundmass plates of sanidine.

In the Leucite Hills lamproites, leucite may form either before or after sanidine (Cross 1897, Carmichael 1967). The proportions vary widely and either leucite or sanidine may dominate on a centimeter scale even within the same lava flow. The factors determining the order of crystallization are not well understood. The bulk composition of sanidine-bearing and sanidine-free rocks are similar, and Carmichael (1967) suggests that the presence or absence of volatiles plays a major role in determining the sequence of crystallization of felsic minerals.

Kuehner (1980) has noted that leucites are set in a glass matrix in diopside leucite phlogopite lamproites. Such glass is absent or rare in other lamproites, where leucite is enclosed by sanidine. Leucites in these rocks are irregular in habit and much smaller than leucites not enclosed by sanidine. Kuehner *et al*. (1981) and Kuehner (1980) suggest that this matrix sanidine is produced by reaction of leucite with the residual magma. Kuehner (1980) also notes that microphenocrystal olivine occurs in all examples of Leucite Hills lamproite in which sanidine forms before leucite and that such olivine is absent where sanidine forms after leucite.

The leucites contain varied amounts of spherical to elliptical, yellow-brown or pale green glass inclusions. These are typically arranged in a "clock face" or quasiradial pattern. Leucites from the Leucite Hills are inclusion-rich relative to those from the West Kimberley province. The inclusions represent quenched samples of trapped magma (Section 7.1.3.5). Jaques and Foley (1985) have noted the common occurrence of aluminous spinels in West Kimberley leucite (Section 6.9.4) and attributed their presence to exsolution from a Mg, Fe, Al-rich nonstoichiometric leucite precursor.

Fresh leucite is rare in most lamproites. Typically it is completely pseudomorphed by sanidine, analcite, quartz (chalcedony), carbonate, or zeolite (commonly harmotome, Prider and Cole 1942).

Sanidine is one of the more common pseudomorphing phases, and altered leucites may consist of uniform near-isotropic single crystals or polycrystalline aggregates of this mineral. Commonly the pseudomorphs have a brownish turbid appearance due to the release of Fe as hematite during the alteration process. Pseudomorphing commonly occurs preferentially along chains of glass inclusions. Occasionally such pseudomorphs are themselves resorbed by the fluids that formed the groundmass, leaving only cruciform residua of sanidine. Pseudomorphic feldspars may in turn be replaced by sericite.

Thy *et al*. (1987) have described a very fine-grained finger-print-like intergrowth of two unidentified phases in the Sisimiut lamproites. The pseudomorph has the bulk composition of potassium feldspar.

True pseudoleucites, i.e., aggregates of potassium feldspar and nepheline (Shand

1943), are absent from lamproites. Their absence clearly sets lamproites apart from other potassic rocks, such as the New South Wales leucitites (Cundari 1973), which characteristically contain pseudomorphs consisting of alkali feldspar and nepheline.

Gittins (1979) suggests that the formation of pseudoleucite is due to subsolidus ion exchange reactions with Na-rich fluids. The lack of pseudoleucite is not simply related to the low Na_2O contents of lamproitic relative to leucititic magmas, as Na-enriched amphiboles and pyroxenes form in lamproites during the final stages of crystallization or by vapor phase crystallization. The fluids responsible might reasonably be expected to promote pseudoleucite formation yet have failed to do so. No explanation for this observation has yet been advanced.

Also absent from lamproites are the kalsilite–potassium feldspar intergrowths that may result from the subsolidus breakdown of leucite (Gittins 1979, Scarfe et al. 1966).

6.6.2. Composition

Representative compositions of leucite are given in Table 6.29. These deviate significantly from ideal leucite stoichiometry being enriched in Si and K and depleted in Al. The Al-deficiency, which is accommodated by the entry of Fe^{3+} into the tetrahedral structural sites, is a reflection of crystallization from a peralkaline magma.

The leucites are notably poor in Na_2O (<0.2 wt %), CaO (<0.05 wt %), and BaO (<0.02 wt %) and enriched in Fe_2O_3 (0.6–2.7 wt %) (Kuehner et al. 1981, Mitchell 1985, Jaques et al. 1986, this work). The iron enrichment represents up to 10 mol % solid solution towards iron leucite. This extent of solid solution is in reasonable agreement with experimental studies quoted by Gupta and Yagi (1980), which indicate that up to 7.7 wt % $KFeSi_2O_6$ may be incorporated into leucite at 2 kbar water pressure. Most lamproite leucites contain minor amounts of MgO which do not seem to be artifacts of the analysis or related to

Table 6.29. Representative Compositions of Leucite[a]

	1	2	3	4	5	6	7	8
SiO_2	56.12	55.33	55.89	55.58	56.96	55.77	55.70	56.87
Al_2O_3	21.85	21.03	21.21	21.35	20.00	19.00	20.96	20.31
Fe_2O_3[b]	0.62	1.08	0.71	1.14	1.55	2.30	1.38	1.73
MgO	0.38	0.46	0.25	0.30	0.49	0.78	0.35	0.33
Na_2O	0.08	0.09	0.10	0.08	0.04	0.11	n.d.	n.d.
K_2O	21.40	21.25	20.92	21.44	21.21	21.01	21.08	21.16
	100.45	99.24	99.08	99.69	100.26	98.97	99.47	100.4
Structural formulas based on six oxygens:								
Si	2.030	2.032	2.046	2.026	2.067	2.061	2.038	2.061
Al	0.932	0.910	0.915	0.921	0.856	0.827	0.904	0.867
Fe	0.017	0.030	0.020	0.031	0.042	0.064	0.038	0.047
Mg	0.021	0.025	0.014	0.016	0.027	0.043	0.019	0.018
Na	0.006	0.006	0.007	0.006	0.003	0.008	—	—
K	0.988	0.996	0.977	1.007	0.982	0.990	0.984	0.978

[a] 1–4, West Kimberley; 5—8, Leucite Hills; 1–2, core and rim, Oscar (Jaques et al. 1986); 3, Mt. Cedric (this work); 4, Mt. Gytha (this work); 5, Spring Butte; 6, Deer Butte; 7, Middle Table Mountain; 8, Hatcher Mesa (5–8, this work).
[b] Total Fe expressed as Fe_2O_3.

glass inclusions. Experimental data presented by Jaques and Foley (1985) suggest that this MgO is present in solid solution.

Few data are available regarding inter- or intraprovince compositional variation. Figure 6.48 shows that leucites from the Leucite Hills and West Kimberley have different compositions with respect to their molar SiO_2 content and molar K_2O/Al_2O_3 ration. This division is also reflected in their Fe_2O_3 contents. Leucites from the Leucite Hills are enriched in Fe_2O_3 (1.0–3.0 wt %) relative to those from West Kimberley (0.6–1.2 wt % Fe_2O_3). Limited data (Table 6.29; this work) suggest that the more evolved rocks in the Leucite Hills (Middle Table Mountain, Hatcher Mesa) contain leucites that have lower molar K_2O/Al_2O_3 ratios (1.08–1.12) than those in the rapidly cooled lavas (1.11–1.24).

Leucites in the Gaussberg lavas contain from 0.85 to 1.3 wt % Fe_2O_3 and have high molar K_2O/Al_2O_3 ratios (average 1.1, Sheraton and Cundari, 1980). Data have not been published for leucite in Spanish lamproites and only one leucite composition has been reported from the Sisimiut lamproites (Thy et al. 1987). In other provinces, leucites are entirely pseudomorphed, e.g., by analcite at Smoky Butte (Mitchell et al. 1987) or by potassium feldspar at Kapamba (Scott Smith et al. 1989).

Trace element data for three Leucite Hills leucites (Kuehner et al. 1981), show the following concentrations (ppm): Cr = 17–31, Ba = 1549–2990, Ni = 4–7, Nb = 23–25, Zr = 306–439, Y = 5–11, Sr = 309–418, Rb = 436–555, Pb = 11–19, Co = 10–12. It is not clear what contribution the presence of glass inclusions has made to these data. This glass is rich in Ti and Ba (Section 7.1.3.5) and may account for the anomalously high Ba/Rb ratios of these leucites, which are the reverse of the relationship commonly found (Henderson 1965). The glass is also expected to be enriched in other incompatible elements.

Lamproite leucites are different in composition from leucites from other potassic volcanic rocks. Figure 6.48 shows that leucite in the Roman province lavas (Cundari 1975, Baldridge et al. 1981, Holm 1982) and New South Wales leucitites (Cundari 1973) typically contains excess Al and has molar K_2O/Al_2O_3 ratios ranging from 0.92–1.04. In addition these leucites are sodium-rich (0.04–0.79 wt % Na_2O) and iron-poor (0.2–1.0 wt % Fe_2O_3) relative to lamproite leucite. Barton (1979) was the first to note the relatively high Al content of leucitite leucite and to suggest that the Al content of leucite reflects the Al_2O_3 activity of the parental magma.

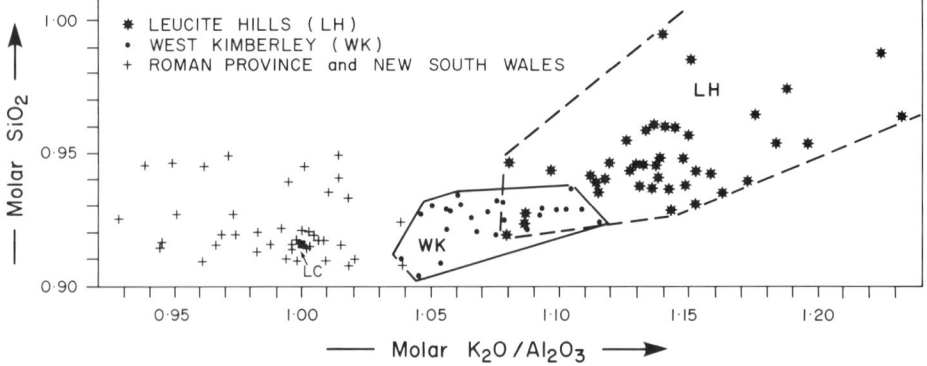

Figure 6.48. Compositions of leucites from lamproites and other potassic rocks. LC, stoichiometric leucite composition.

6.7. ANALCITE

Analcite is common in the Smoky Butte and Francis lamproites. At Smoky Butte it is found primarily in hyaloarmalcolite phlogopite lamproites and is a minor constituent of sanidine-rich varieties (Mitchell et al. 1987, Velde 1975). At Francis, analcites are also found in glassy lamproites (Henage 1972). Analcite also occurs in the Kapamba intrusions (Scott et al. 1988) and has been identified as a minor constituent of the Leucite Hills (Henage 1972, Kuehner 1980) and Murcia-Almeria (Lopez Ruiz and Rodriguez Badiola 1980) lamproites.

The analcite typically occurs as rounded to anhedral isotropic microphenocrysts commonly containing radial glass inclusions. This petrographic similarity to leucite has led to misidentification as leucite, e.g., Matson (1960), and perhaps to the apparent rarity of analcite in most occurrences of lamproite. Conclusive identification is provided only by electron microprobe analysis or X-ray diffraction.

Little is known of the composition of lamproitic analcite (Table 6.30). Data for Smoky Butte material (Mitchell et al. 1987) show the mineral to be uniform in composition and Si rich and K poor. These analcites contain significant levels of Fe (Table 6.30). Kapamba analcites have a turbid appearance due to hematite formation. Scott Smith et al. (1989) report variable composition arising perhaps in part from decomposition of the analcite during electron microprobe analysis. It should be noted that analcite is one of the most sodic minerals found in lamproites.

The petrographic similarity to leucite suggests that analcite is not a primary mineral and has pseudomorphed leucite microphenocrysts. This conclusion is supported by the experimental studies of Liou (1971) and Gupta and Fyfe (1975), who have demonstrated the ease with which Na-bearing water will transform leucite into analcite at low temperature (25°C).

The source of the Na-bearing fluids is problematical. The trend towards Na enrichment

Table 6.30. Representative Compositions of Analcite[a]

	1	2	3	4	5	6
SiO_2	58.63	59.05	54.86	52.04	53.21	53.81
Al_2O_3	21.33	21.28	23.56	23.58	24.46	23.05
Fe_2O_3[b]	0.93	1.00	0.63	0.10	0.07	0.06
CaO	0.05	0.05	0.51	0.48	1.65	0.03
Na_2O	9.17	9.54	9.44	11.89	8.82	12.63
K_2O	0.34	0.29	1.00	0.27	0.56	0.40
	90.45	91.21	90.00	88.36	88.77	89.98
Structural formulas based on seven oxygens:						
Si	2.485	2.485	2.364	2.304	2.322	2.340
Al	1.065	1.055	1.197	1.231	1.258	1.181
Fe	0.030	0.032	0.020	0.003	0.002	0.002
Ca	0.002	0.002	0.024	0.023	0.077	0.001
Na	0.753	0.778	0.788	1.021	0.746	1.065
K	0.018	0.016	0.055	0.015	0.031	0.022

[a] 1–2, Smoky Butte (Mitchell 1985); 3–5, Kapamba (Scott Smith, unpublished data); 6, Leucite Hills (Kuehner et al. 1981).
[b] Total Fe expressed as Fe_2O_3.

observed in the later stages of crystallization could result in the formation of Na-rich fluids which could deuterically alter preexisting leucite. Such an origin would seem adequate for the Kapamba analcites but would seem unlikely for Smoky Butte where analcite occurs in glassy rocks. In this occurrence leaching of K_2O and replacement by Na_2O of both glass and the leucite by ground waters at low temperatures has been proposed by Mitchell *et al.* (1987). Keuhner (1980) has suggested that fluids derived from the Na-rich Green River Formation were responsible for the replacement of leucite by analcite in the Badger's Teeth intrusion (Leucite Hills). Henage (1972) has noted that the tops and bottoms of lamproite "flows" at Francis are analcite-bearing while the centers are analcite-free. The formation of analcite is believed to result from the transfer of fluids to the margins of the "flows" during cooling.

6.8. SANIDINE

6.8.1. Paragenesis

Sanidine is the only feldspar found in lamproites. The abundance ranges from trace quantities to up to 50 modal %. The feldspar occurs as euhedral-to-subhedral prismatic microphenocrysts, as groundmass poikilitic plates, and as pseudomorphs after leucite.

Microphenocrysts exhibit a wide range in habit and form quench stellate rosettes, slender prisms, or blocky equant crystals. Plumose textures are common. The feldspar habit is related to the degree of cooling or quenching of the magma and the various quench textures are confined to the outer margins of flows or to glassy lamproites. Large optically continuous plates of sanidine are found only in the relatively more slowly cooled holocrystalline lamproites. Microphenocrystal sanidine may be unaltered or partially altered to sericite and hematite. Groundmass sanidine is commonly colorless, clear, and alteration-free, with the exception of sanidine in the groundmass of some West Kimberley madupitic lamproites. In these rocks the sanidine is difficult to recognize owing to extensive deuteric alteration to a turbid brownish amorphous material (Mitchell 1985, Jaques *et al.* 1986).

Sanidine may contain inclusions of phlogopite, olivine, diopside, leucite, and apatite, indicating that it is one of the last primary phases to crystallize. Amphiboles crystallize contemporaneously with and/or after sanidine. In some plumose-textured lamproites amphiboles occupy the interstices between the sanidine prisms as ophitic plates. In coarse-grained lamproites, sanidine and amphibole coexist as poikilitic allotriomorphic plates.

Twinning is typically absent. Rare Carlsbad twins have been noted in the sanidines in some Spanish lamproites (Borley 1967), and Keuhner (1980) notes that Carlsbad, Baveno, and penetration twinning is common in some of the Zirkel Mesa lamproites.

Carmichael (1967) and Henage (1972) consider that the feldspars have the optics of high sanidine [OAP $\parallel$ (010) 2V = 55–65°]. However, Fuster *et al.* (1956) have noted the presence of low sanidine [OAP $\perp$ (010), 2V = 46–51°] in the Cancarix lamproites. The characteristic absence of twinning and the presumably disordered structure attest to the rapid quenching and crystallization of lamproite magmas. High-temperature feldspars are preserved even in those rocks which, on textural evidence, appear to have cooled slowly, e.g., sanidine madupitic rocks. X-ray studies of Si, Al, Fe ordering in these feldspars have not yet been undertaken. Refined unit cell parameters of an analyzed sanidine from Francis (Best *et al.* 1968) differ widely from those of other K-feldspars and plot outside of the boundary of the *b–c* diagram (Stewart and Wright 1974) commonly used to estimate the degree of ordering. The *b* dimension in particular is excessive and when combined with the large *c* dimension, *a* seems too small. Bachinski (personal communication) suggests that these anomalous dimensions might be due in part to substitution of Fe^{3+} for Al^{3+} in the tetrahedral sites.

Sanidine appears to have been a primary liquidus phase, and there is no conclusive textual evidence for its formation by reaction between leucite and liquid. Many leucites are pseudomorphed by sanidine but it cannot be determined whether this is a true reaction product or a secondary replacement. Leucites with reaction coronas of potassium feldspar do not occur. Many leucite phenocrysts are corroded and it may be assumed that reaction with liquid has occurred to a limited extent. If the magma was already saturated with sanidine, as seems probable, the contribution to the total sanidine content from this source is unlikely to be volumetrically significant. Sanidines enclosing irregular leucite crystals in some Leucite Hills lavas have been interpreted by Kuehner *et al.* (1981) to have formed by leucite–liquid reaction (Section 6.6.1); however, textural evidence for this conclusion is ambiguous.

6.8.2. Composition

Representative compositions of sanidine are given in Table 6.31. All are highly potassic and contain only minor amounts of Na_2O (<2.5 wt %). The maximum albite content is found in the Murcia-Almeria and Kapamba lamproites (c. 23 mol % Ab), while the majority contain less than 5 mol % Ab. Significant amounts of BaO (0.1–1.7 wt %) and Fe_2O_3 (0–4.7 wt %) are present. Sanidines in lamproites contain up to 17 mol % $KFeSi_3O_8$ in solid solution and are the most iron-rich feldspars found in igneous rocks. Typically CaO contents are negligible (<0.1 wt %), although up to 0.8 wt % CaO is found in sanidines from the Murcia-Almeria lamproites. This CaO content corresponds to a maximum anorthite content of 4 mol %. Table 6.31 also shows that the glass in some hyalo-lamproites approximates potassium feldspar in composition and would have crystallized as sanidine given the correct cooling conditions. Some lamproite sanidines contain significant TiO_2 (0.3–0.6 wt %) and MgO (0.1–1.1 wt %) contents, which are not artifacts of analysis techniques. Wagner and Velde (1986a) have noted that lamproite feldspars appear to be characteristically enriched in Ti relative to feldspars from granitoid parageneses. It is possible that the high Ti and Mg contents may be related to submicroscopic inclusions of glass or Fe–Ti oxides (Mitchell *et al.* 1987) rather than being present in solid solution.

Figure 6.29 shows that individual lamproite provinces contain sanidines that differ with respect to their Fe and Na contents. Sanidines from the Leucite Hills, Francis, and Jumilla madupitic lamproite show a limited range in composition and are enriched in Fe relative to other localities, although they differ in their Na contents. In contrast, sanidines from the majority of the Murcia-Almeria localities and Kapamba exhibit a very wide range in Na content and have much lower Fe contents. Sanidines from West Kimberley are notably deficient in both Na and Fe. Only the Murcia-Almeria lamproites are known to exhibit intraprovincial sanidine compositional variation.

The differences in sanidine compositions are related to the bulk compositions of the parental magmas and not to the time of crystallization as previously suggested by Mitchell (1985), as Fe-rich feldspars from the Leucite Hills and Jumilla are microphenocrystal and poikilitic varieties, respectively.

Sanidines that pseudomorph leucite are similar in composition to primary sanidines (Table 6.31). Mitchell *et al.* (1987) have shown that pseudomorphic sanidines at Smoky Butte are poorer in Na_2O (<0.3 wt %) and BaO (<1 wt %) than primary plumose sanidines. At Kapamba and Sisimiut pseudomorphic sanidine similarly contains negligible Na and Ba relative to primary sanidine (Scott Smith *et al.* 1989, Thy *et al.* 1987).

Table 6.31. Representative Compositions of Sanidine[a]

	1	2	3	4	5	6	7	8	9	10	11	12	13	14	15	16	17	18
SiO_2	65.88	64.1	64.46	64.23	64.34	65.11	64.63	63.96	64.23	64.24	64.52	64.04	64.91	64.30	62.04	62.61	65.44	64.03
Al_2O_3	17.01	14.2	17.02	15.16	16.08	18.35	18.34	18.04	17.42	15.26	17.59	17.29	18.11	16.98	16.21	18.83	18.89	17.87
Fe_2O_3[b]	1.83	4.7	1.41	2.48	3.53	0.14	0.07	0.07	1.03	3.15	0.63	1.00	0.32	0.83	1.61	0.79	1.06	0.46
CaO	n.d.	0.08	n.d.	n.d.	0.04	0.77	0.60	n.d.	n.d.	n.d.	0.08	n.d.	0.03	0.13	0.73	0.02	n.d.	0.01
Na_2O	0.48	0.14	0.61	0.66	1.05	1.61	2.33	0.06	2.53	0.96	0.44	0.21	0.06	0.15	1.08	0.31	0.02	0.10
K_2O	15.64	15.19	15.11	14.88	13.69	13.70	13.05	16.61	12.74	15.16	15.95	16.50	17.03	16.74	12.75	15.43	14.97	16.37
BaO	0.48	0.66	1.09	1.02	n.a.	0.29	0.14	0.11	0.84	n.a.	n.a.	n.a.	n.a.	n.a.	0.23	0.41	n.a.	0.08
	101.32	99.07	99.70	98.43	98.73	99.97	99.16	98.85	98.79	98.77	99.21	99.04	100.46	99.13	94.65	98.40	100.38	98.92
Structural formula based on eight oxygens:																		
Si	3.014	3.033	3.006	3.042	3.007	2.987	2.983	2.994	2.994	3.025	3.004	2.999	2.994	3.027	3.009	2.946	2.986	2.995
Al	0.918	0.792	0.936	0.847	0.886	0.993	0.998	0.996	0.958	0.847	0.966	0.955	0.985	0.915	0.927	1.045	1.016	0.986
Fe	0.063	0.167	0.050	0.088	0.124	0.005	0.002	0.003	0.036	0.112	0.022	0.035	0.011	0.029	0.059	0.028	0.036	0.016
Ca	—	0.004	—	—	0.002	0.038	0.030	—	—	—	0.004	—	0.002	0.007	0.038	0.001	—	0.001
Na	0.043	0.013	0.055	0.061	0.095	0.143	0.209	0.006	0.229	0.088	0.040	0.019	0.005	0.014	0.102	0.028	0.002	0.009
K	0.914	0.918	0.900	0.900	0.817	0.803	0.769	0.993	0.758	0.911	0.948	0.987	1.003	1.006	0.790	0.927	0.872	0.977
Ba	0.009	0.012	0.020	0.019	—	0.005	0.003	0.002	0.015	—	—	—	—	—	0.004	0.008	—	0.002
Mol % end member composition																		
Cs	0.89	1.29	2.04	1.93	—	0.53	0.25	0.2	1.53	—	—	—	—	—	0.47	0.78	—	0.15
An	—	0.43	—	—	0.22	3.83	2.94	—	—	—	0.4	—	0.15	0.64	4.07	0.10	—	0.05
Ab	6.54	1.36	5.66	6.19	10.42	14.49	20.66	0.54	22.83	8.78	4.01	1.90	0.53	1.33	10.89	2.94	0.20	0.92
FeOr	4.42	17.69	5.08	9.03	13.59	0.49	0.24	0.25	3.61	11.18	2.23	3.51	1.10	2.87	6.30	2.90	4.17	1.64
Or	88.16	79.23	87.21	82.84	75.77	80.66	76.28	99.01	72.03	80.04	93.36	94.6	98.22	95.16	78.28	93.27	95.63	97.24

[a] Cs, $BaAl_2Si_2O_8$; An, $CaAl_2Si_2O_8$; Ab, $NaAlSi_3O_8$; FeOr, $KFeSi_3O_8$; Or, $KAlSi_3O_8$; n.d., not determined; n.a., not analyzed. 1–2, Leucite Hills (Kuehner 1981, Carmichael 1967); 3–4, Smoky Butte (Mitchell et al. 1987); 5, Jumilla, Murcia-Almeria (this work); 6, Fortuna, Murcia-Almeria (Mitchell and Platt unpub. data); 8–9, Kapamba (Scot Smith unpublished data); 10 Francis (this work); 11–12, Sisco (this work); 13–14 West Kimberley (Jaques et al. 1986); 15, glass, Fortuna (Mitchell 1985); 16–18, pseudomorphic sanidines after leucite; 16 Smoky Butte (Mitchell et al. 1987); 17, Sisimiut (Thy et al. 1987), 18, Kapamba (Scott Smith unpublished data).
[b] Total Fe calculated as Fe_2O_3.

The trace element content of sanidine is inadequately known. Henage (1972) has determined the following concentrations (ppm) in a Francis sanidine: Rb = 354, Cs = 1.03, Ba = 5391, Th = 9.02, Sc = 0.64, Co = 2.37, Cr = 7.1, Ta = 1.15, Hf = 10.48, Zr = 301, La = 25.1, Ce = 60.0, Nd = 18.9, Sm = 2.76, Eu = 0.57. Preferential incorporation of Eu into the feldspar does not occur. The contribution of microinclusions to these trace element abundances may be significant. Kuehner (1980), for example, notes that such inclusions are commonplace and preclude determining the trace element content of Leucite Hills sanidine.

Feldspar in other leucite-bearing rocks, such as the Roman province, Uganda and New South Wales tephrites and leucitites, are ternary feldspars, typically containing greater Na_2O (>2 wt %) and CaO (>0.5 wt %) and lesser Fe_2O_3 (<1 wt %) contents than lamproite sanidines (Barton 1979, Cundari 1975, Ferguson and Cundari 1982, Baldridge et al. 1981, Holm 1982). Some Sierra Nevada leucite basanites contain sanidines with 1.5–2.2 wt % Fe_2O_3, however, these sanidines also contain 2.2–3.3 wt % Na_2O (Van Kooten 1980) thus distinguishing them from lamproites sanidines. The Murcia-Almeria and Kapamba sanidines overlap the compositions of some of these feldspars and consequently in these provinces the alkali feldspar composition cannot be used to recognize their lamproitic character.

Members of the iron sanidine–sanidine series containing more than 1 wt % Fe_2O_3 (i.e., > 4 mol.% $KFeSi_3O_8$), less than 1% Na_2O and negligible CaO appear to occur exclusively in lamproites. Consequently the presence of such feldspars in a rock may be indicative of a lamproitic paragenesis. However, given the wide variation in feldspar composition shown in Figure 6.49 it is suggested that the composition of sanidine is not one of the critical components in the recognition of lamproites.

Sanidines from Mount Bayliss (Fe_2O_3 = 0.40–0.77 wt %, Na_2O = 0–0.2 wt %, <0.02 wt % CaO, Sheraton and England 1980, this work), Priestly Peak (Sheraton and England 1980) and the Murun phlogopite-rich dikes (Fe_2O_3 = 0.9–2.5 wt %, Na_2O = 0.04–0.13 wt %, <0.02 wt % CaO, this work) have a composition suggestive of a lamproitic paragenesis.

6.9. SPINEL

Primary spinels are common only in olivine lamproites from the Ellendale field, West Kimberley (Jaques et al. 1986) and in olivine-bearing madupitic lamproites from the Leucite Hills (Mitchell 1985), Prairie Creek (Scott Smith and Skinner 1984b, Mitchell 1985) and Murcia-Almeria (Mitchell 1985, Venturelli et al. 1988).

Spinels are absent in phlogopite lamproites and in these rocks the primary oxide phase is priderite, armalcolite, and/or ilmenite. Spinels and priderite do not crystallize contemporaneously. Jaques et al. (1986) note that chrome spinels are generally not present in rocks containing less than 8–10 modal% olivine.

Spinels produced by the decomposition and/or reaction of other minerals (Section 6.9.4.), together with xenocrystal spinels (Section 9.3.1.) are common in some lamproites.

6.9.1. Paragenesis—Primary Spinels

Primary spinels in the Ellendale olivine lamproites occur as small (<100 μm, mostly <40 μm) red-brown euhedral groundmass crystals. Commonly, similar spinels are found as inclusions in olivine and phlogopite (Jaques et al. 1986).

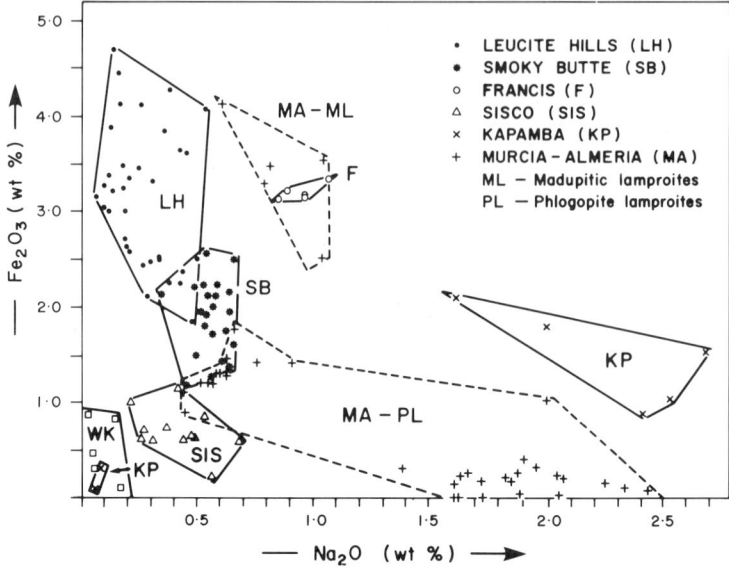

Figure 6.49. Compositional variation (Fe$_2$O$_3$ versus Na$_2$O) of sanidines from lamproites.

In olivine-bearing madupitic rocks, spinels occur as small (<50 μm) single-phase euhedral crystals or as two-phase crystals consisting of a discrete core and mantle. Both types are poikilitically enclosed by groundmass mica. Wagner and Velde (1986a) have noted the common occurrence of euhedral spinel inclusions in olivines in olivine-poor lamproites from Murcia-Almeria and Sisco. Spinels in all occurrences are devoid of exsolution features and skeletal and atoll-textures are absent.

6.9.2. Composition—Primary Spinels

Representative compositions of spinels are given in Tables 6.32 and 6.33. These primary spinels fall into four broad compositional groups (Mitchell 1985):

Group 1: Ti-poor (<1 wt % TiO$_2$) aluminous (>10 wt % Al$_2$O$_3$) magnesiochromites, characterized by moderate Cr/(Cr + Al) ratios (<0.75) and Fe^{2+}/(Fe^{2+} + Mg) ratios (<0.50).

Group 2: Titanian (1–5 wt % TiO$_2$) aluminous (1–10 wt % Al$_2$O$_3$) magnesiochromites, characterized by high Cr/(Cr + Al) ratios (0.75–0.90) and moderate Fe^{2+}/(Fe^{2+} + Mg) ratios (0.40–0.75).

Group 3: Alumina-poor (<1 wt % Al$_2$O$_3$) titanian (>5 wt % TiO$_2$) magnesian (<5 wt % MgO) chromites, characterized by very high Cr/(Cr + Al) ratios (>0.90) and high Fe^{2+}/(Fe^{2+} + Mg) ratios (>0.75).

Group 4: Magnesian (<5 wt % MgO) titaniferous (>10 wt % TiO$_2$) magnetite.

Spinels have not been extensively studied; however, sufficient data are available for the Ellendale olivine lamproites (Jaques *et al.* 1986, Mitchell 1985) and Prairie Creek madupitic lamproites (Mitchell 1985, Lewis 1977) to allow delineation of the evolutionary compositional trend of lamproite spinels.

Table 6.32. Representative Compositions of Spinels in the Ellendale and Prairie Creek Lamproites[a]

	1	2	3	4	5	6	7	8	9	10	11	12	13	14	15
TiO_2	0.48	0.77	1.00	2.15	3.91	4.35	12.89	15.75	0.30	0.58	1.68	4.55	9.95	11.88	13.11
Al_2O_3	16.67	14.37	9.71	11.37	2.78	1.51	n.d.	n.d.	18.16	11.65	7.35	0.58	0.13	0.10	0.12
Cr_2O_3	40.40	44.55	56.61	53.36	57.83	57.15	21.91	0.45	36.14	43.92	53.43	44.84	22.75	19.41	9.02
FeO[b]	28.32	26.69	19.56	17.88	22.59	27.25	57.58	74.57	29.22	28.72	22.48	42.21	60.10	59.37	69.48
MnO	0.28	0.35	0.43	0.27	0.27	0.31	1.06	1.23	0.34	0.38	0.49	1.21	1.27	1.21	1.45
MgO	13.08	12.59	11.20	14.08	11.86	8.95	3.58	2.81	13.93	12.76	12.95	5.57	4.00	4.79	3.94
	99.23	99.34	98.51	99.11	99.24	99.52	97.01	94.81	98.09	98.01	98.38	98.96	98.60	96.76	97.12
Recalculated analyses:															
Fe_2O_3	14.97	12.53	3.38	13.78	5.69	5.64	23.33	38.74	17.76	16.17	9.33	17.47	30.00	28.54	37.08
FeO	14.85	15.42	16.52	4.55	17.47	22.17	36.59	39.71	13.24	14.17	14.68	26.49	33.11	33.69	36.12
	100.73	100.57	98.85	99.57	99.81	100.09	99.35	98.69	99.87	99.63	99.31	100.71	101.21	99.62	100.83
Mol % end member spinel molecules:															
$MgAl_2O_4$	28.3	24.9	18.4	20.4	5.1	2.8	—	—	30.3	20.2	13.3	1.0	0.2	0.2	0.2
Mg_2TiO_4	1.6	2.6	3.6	7.4	13.7	15.5	11.0	7.9	1.0	1.9	5.8	15.5	13.1	14.3	11.1
Mn_2TiO_4	—	—	—	—	—	—	1.8	2.0	—	—	—	—	2.2	2.1	2.3
Fe_2TiO_4	—	—	—	—	—	—	27.2	34.8	—	—	—	—	14.9	19.7	24.2
$MnCr_2O_4$	0.7	0.9	1.2	0.7	0.7	0.8	—	—	0.8	1.0	1.3	3.1	—	—	—
$MgCr_2O_4$	25.7	26.9	30.4	33.7	31.6	18.7	—	—	27.2	33.2	38.1	3.4	—	—	—
$FeCr_2O_4$	19.5	24.0	40.3	29.9	38.8	52.0	23.8	0.5	12.4	16.9	25.3	47.1	24.2	20.6	9.1
Fe_3O_4	24.3	20.8	6.1	7.8	10.0	10.1	36.2	54.9	28.4	26.8	16.2	29.8	45.5	43.3	53.2

[a] n.d., not detected. 1–8, Ellendale, West Kimberley; 1–3, Mitchell (1985); 4–8, Jaques et al. (1986); 9–15, Prairie Creek; 9–11, Lewis (1977); 12–15, Mitchell (this work).
[b] Total Fe calculated as FeO.

Table 6.33. Representative Composition of Lamproite Spinels[a]

	1	2	3	4	5	6	7	8	9	10	11	12	13	14
TiO_2	1.23	4.52	1.03	6.83	12.28	10.41	12.09	2.45	2.75	3.49	10.29	15.29	24.00	23.96
Al_2O_3	3.04	0.23	2.00	0.13	0.22	n.d.	0.13	2.86	8.11	5.04	1.35	0.84	0.60	0.85
Cr_2O_3	65.34	51.41	55.78	0.08	0.02	0.07	0.71	57.96	46.70	43.31	27.10	15.15	0.11	0.11
FeO[b]	20.33	37.05	29.34	84.24	78.78	78.18	75.52	25.27	31.72	38.21	52.88	58.51	64.33	66.68
MnO	0.69	0.84	0.79	0.79	1.11	2.02	2.13	1.19	0.37	0.55	0.65	0.67	0.71	1.40
MgO	10.41	4.42	9.80	1.50	1.95	4.05	4.33	8.80	8.06	6.83	5.70	6.02	4.47	2.26
	101.04	98.47	98.74	93.57	94.11	94.73	94.91	98.57	97.71	97.43	97.97	96.48	95.26	94.22
Recalculated analyses:														
Fe_2O_3	3.22	9.77	13.32	55.79	45.29	50.89	46.84	6.18	10.17	15.29	23.29	25.72	21.76	21.03
FeO	17.43	28.26	17.35	34.04	38.03	32.39	33.37	19.71	22.57	24.45	31.93	35.43	44.75	44.76
	101.36	99.45	100.07	99.16	98.90	99.83	99.60	99.15	98.73	98.96	100.32	99.12	96.4	97.37
Mol % end member spinel molecules:														
$MgAl_2O_4$	5.8	0.4	3.7	0.2	0.3	—	0.2	5.5	14.9	9.1	2.2	1.3	0.9	1.3
Mg_2TiO_4	4.5	15.6	3.6	4.1	5.3	11.2	11.8	9.0	9.7	12.6	15.9	16.6	11.9	5.5
Mn_2TiO_4	—	0.8	—	1.3	1.8	3.2	3.3	—	—	—	1.1	1.1	1.1	2.3
Fe_2TiO_4	—	—	—	14.2	30.0	14.7	18.6	—	—	—	15.0	27.4	55.1	60.8
$MnCr_2O_4$	1.9	1.2	2.1	—	—	—	—	3.3	1.0	1.4	—	—	—	—
$MgCr_2O_4$	38.5	—	37.2	—	—	—	—	25.1	9.7	6.0	—	—	—	—
$FeCr_2O_4$	43.4	64.2	29.8	0.1	—	0.1	0.7	45.9	46.9	45.0	29.5	15.7	0.1	0.1
Fe_3O_4	5.9	17.8	23.6	80.1	64.6	70.9	65.4	11.3	17.9	26.4	36.3	37.9	30.9	30.1

[a] 1–5, Murcia-Almeria; 1–3, euhedral inclusions in olivines (Wagner and Velde, 1981); 4–5, Jumilla (Mitchell 1985); 6–7, Pilot Butte, Leucite Hills (Mitchell 1985); 8, Hills Pond, Kansas (Mitchell 1985); 9–14, Kapamba (Scott Smith unpublished data).
[b] Total Fe calculated as FeO.

MINERALOGY OF LAMPROITES

Spinels in the Ellendale olivine lamproite (Table 6.32) belong primarily to groups 1 and 2, with group 2 spinels predominating. In the more coarsely crystalline lamproites, group 2 spinels are rarely found to be mantled by group 3 spinels. The compositional trend is one of decreasing Al and Mg with increasing Cr and Fe^{2+} followed by increasing Ti, Fe^{2+}, and Fe^{3+} at very high Cr/(Cr + Al) ratios, i.e., from group 1 to group 3 via group 2. Rare tiny euhedral groundmass titaniferous magnetites (group 4) represent the culmination of this trend (Jaques et al. 1986).

In the groundmass of the Prairie Creek olivine-bearing madupitic rocks, spinels (Table 6.33) belonging to group 3 occur as discrete crystals and as mantles upon group 2 spinels (Mitchell 1985). Discrete crystals of group 3 spinels are zoned towards decreasing Cr and increasing Fe^{3+} from core to margin. The compositional trend is one of increasing Ti, Fe^{2+} and Fe^{3+} with decreasing Mg and Cr. The spinel assemblage as a whole is more evolved than that of the Ellendale lamproites. Spinels in heavy mineral concentrates analyzed by Lewis (1977) are predominantly group 1 spinels.

The compositional trends of the Ellendale and Prairie Creek spinels are shown in Figures 6.50 and 6.51, which are conventional representations of the "oxidized" and "reduced" spinel prisms respectively (Haggerty 1976, Mitchell 1986).

Initially, spinels evolve diagonally across the base of the prisms toward increasing Cr/(Cr + Al) and $Fe^{2+}/(Fe^{2+} + Mg)$ ratios at low and essentially constant Ti and Fe^{3+} contents. This trend of Cr enrichment and Al depletion results in the formation of evolved group 2 and 3 spinels with very high Cr/(Cr + Al) ratios. Further evolution is a trend of

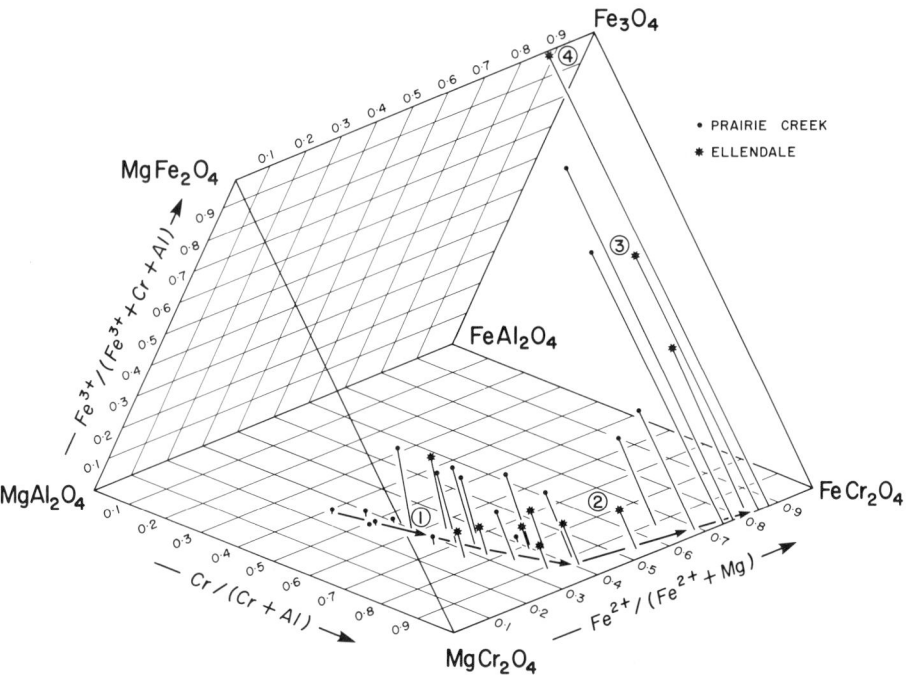

Figure 6.50. Compositions of spinels from the Prairie Creek and Ellendale lamproites plotted in the "oxidized" spinel prism. Circled numbers correspond to spinel groups described in the text.

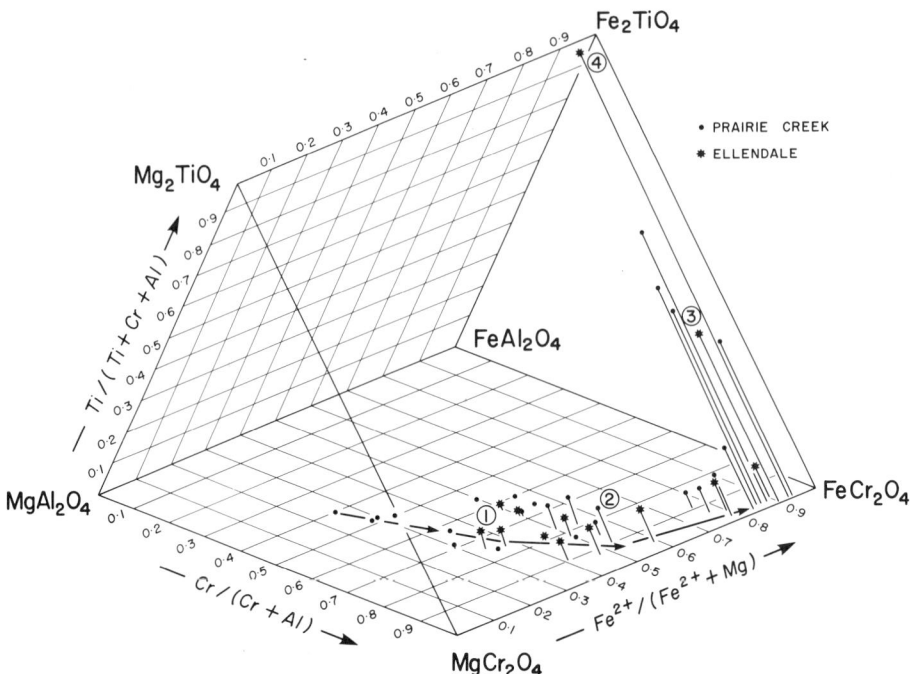

Figure 6.51. Compositions of spinels from the Prairie Creek and Ellendale lamproites plotted in the "reduced" spinel prism. Circled numbers correspond to spinel groups discussed in the text.

increasing Fe^{2+}, Fe^{3+}, and Ti with decreasing Cr leading to the formation of group 4 spinels. These evolved spinels plot essentially on the front faces of the prism.

The above mantling and zonation trends thus demonstrate that the lamproite spinels evolve from aluminous magnesiochromites, to titanian aluminous magnesiochromite, to titanian aluminous magnesiochromite, to titanian magnetite (ulvöspinel–magnetite series.)

In the Kapamba madupitic lamproites most of the groundmass spinels are group 4 ulvöspinel–magnetites. Some grains have group 3 spinel cores and the evolutionary trend is considered by Scott Smith *et al.* (1989) to be one of increasing Ti, Fe^{2+}, Fe^{3+}, and Mn with decreasing Cr and Mg. Group 2 spinels are found in heavy mineral concentrates. Spinels from intrusion P10 are anomalously enriched in MnO (up to 10 wt %) relative to other lamproite spinels.

Spinels enclosed in phlogopite in the Pilot Butte (Leucite Hills) and Jumilla madupitic lamproites belong entirely to group 4 (Table 6.33, Figures 6.52 and 6.53, Mitchell 1985). Spinels occurring as inclusions in olivines from Jumilla (Wagner and Velde 1986a) are group 2 spinels.

Spinels in the Kapamba, Leucite Hills and Jumilla lamproites are more evolved than those in the Prairie Creek and Ellendale lamproites (Figures 6.52, 6.53). The least evolved spinels are found in the Ellendale olivine lamproites. In the Leucite Hills and Murcia-Almeria occurrences the most evolved spinels are found in lamproites which, on the basis of phlogopite composition, are considered to be the most evolved members of the province.

Two main groups of spinels are present in the Rose Dome lamproite (Coopersmith and

MINERALOGY OF LAMPROITES

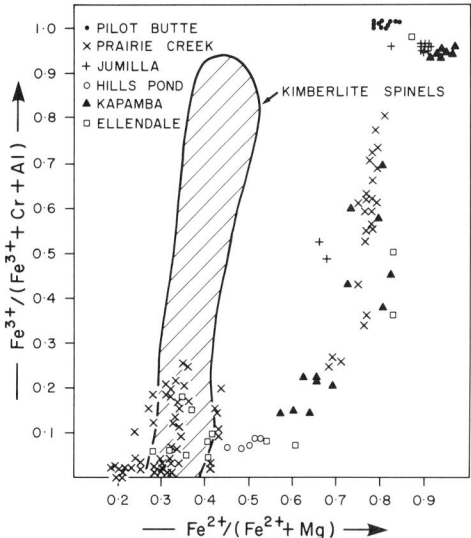

Figure 6.52. Composition of lamproite spinels projected onto the front face of the "oxidized" spinel prism. The shaded region is the compositional field of trend 1 kimberlite spinels.

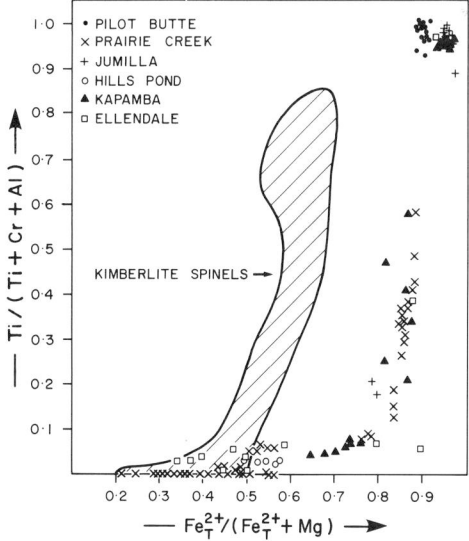

Figure 6.53. Composition of lamproite spinels projected onto the front face of the "reduced" spinel prism. The shaded region is the compositional field of trend 1 kimberlite spinels.

Mitchell 1989). Most abundant are low-Ti, low-Fe_2O_3 spinels, which range in composition from magnesian aluminous chromites to aluminous magnesian chromites (i.e., group 1). The second group of spinels are Fe_2O_3 rich and TiO_2 poor and represent solid solutions between $MgAl_2O_4$–$MgCr_2O_4$–Fe_3O_4–$FeCr_2O_4$, which may also contain a substantial amount of the $MgFe_2O_4$ molecule. All of the data for Rose Dome spinels were obtained on heavy mineral concentrates and there is high probability that the Al-rich and Fe-rich spinels are xenocrysts. These latter spinels are similar to spinels found as macrocrysts in alnöitic diatremes (Nixon et al. 1980, Mitchell 1979).

Spinels in the groundmass of the Hills Pond lamproite (Mitchell 1985) belong entirely to group 2 (Table 6.33, Figures 6.52, 6.53).

Group 1 spinels are similar in their composition to spinels occurring in a wide variety of environments, e.g., mantle-derived spinel lherzolites, ultrabasic, and basic intrusions (Haggerty 1976) and this group may include both xenocrystal and phenocrystal spinels. The composition of these spinels therefore cannot be used to distinguish between lamproites and other rock types.

Groups 2–4 are considered to be primary liquidus spinels whose composition may be used to assess the degree of evolution of their host lamproite. Their extremely high Cr/(Cr + Al) ratios (commonly >0.95) are characteristic of lamproites and recognition of such spinels may signify that their host rock belongs to the lamproite clan.

6.9.3. Comparison with Primary Spinels from Kimberlite and Lamprophyre

Figure 6.54 summarizes the two compositional trends found in the groundmass spinel assemblage of kimberlites. These are termed the "magnesian ulvöspinel" or magmatic trend 1 and the "Ti-magnetite" or magmatic trend 2 of Mitchell (1986).

Magmatic trend 1 is characteristic of phlogopite-poor monticellite–serpentine–calcite kimberlites. The compositional trend is across the spinel prism from the base near the $MgCr_2O_4$–$FeCr_2O_4$ join [Cr/(Cr + Al) = 0.80–0.95, Fe^{2+}/(Fe^{2+} + Mg) = 0.4–0.6] toward the rear rectangular face and upwards toward the Mg_2TiO_4–Fe_2TiO_4 or $MgFe_2O_4$–Fe_3O_4 apex. Spinels evolve from titanian magnesian aluminous chromite or titanian magnesian chromite containing 1–12 wt % TiO_2 towards members of the magnesian ulvöspinel–magnetite series (>15 wt % TiO_2). The trend is one of increasing Ti, Fe^{3+}/Fe^{2+}, total Fe and decreasing Cr at approximately constant Fe^{2+}/(Fe^{2+} + Mg) ratios. This trend culminates with the formation of Ti- and Mg-free magnetite. The presence of Ti-rich spinels containing a substantial proportion of the Mg_2TiO_4 molecule is the hallmark of this trend.

Magmatic trend 2 appears to be found only in micaceous kimberlites. Trend 2 spinels range in composition from aluminous magnesian chromites to titanian magnesian chromites to ulvöspinel–magnetites. The compositional trend is initially along the axis of the prism toward increasing Fe^{2+}/(Fe^{2+} + Mg) ratios at relatively low TiO_2 contents and high Cr/(Cr + Al) ratios (>0.80), followed by a rapid increase in Ti at high Fe^{2+}/(Fe^{2+} + Mg) ratio (>0.80) towards the Fe_2TiO_4 or Fe_3O_4 apex. The trend is characterized by Ti and Fe enrichment and spinels rich in the Mg_2TiO_4 component are not formed. Spinels belonging to trend 2 are Al-poor relative to those of trend 1. This feature has been ascribed to the depletion of the magma in Al by extensive phlogopite crystallization (Mitchell 1986).

Magmatic trend 1 appears to be unique to kimberlites and affords a simple means of distinguishing between them and petrographically similar mica peridotites, alnöites, lamprophyres, and lamproites (Mitchell 1986, 1988).

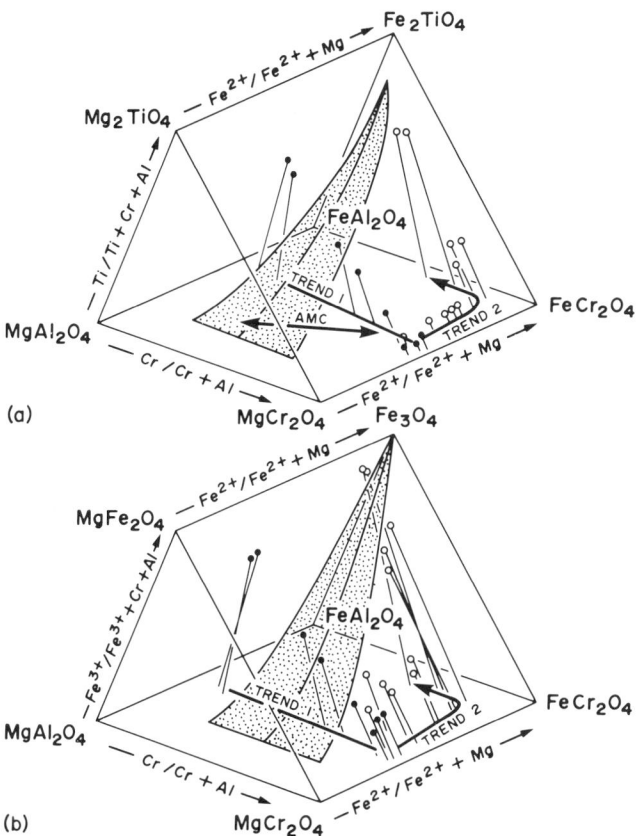

Figure 6.54. Compositional trends of spinels in kimberlites plotted in the "oxidized" (a) and "reduced" (b) spinel prisms (after Mitchell 1986).

Magmatic trend 2 is very similar to the lamproite spinel trend and in projection (Figure 6.55), the two trends are identical. The trends differ in that very high Cr/(Cr + Al) ratios (>0.95) are characteristic only of lamproite spinels.

The kimberlite trend 2 and the lamproite spinel trend appear to be Cr-rich variants of a common spinel trend found in a wide variety of rocks that include basalt, alnöites, and diverse lamprophyres. The trends are similar in that their evolution is approximately along the axis of the spinel prisms. The trends differ with respect to the Cr/(Cr + Al) ratios of the least evolved members of the series (Mitchell 1986). Consequently, spinels with high Cr/(Cr + Al) ratios may be distinguished from those occurring in lamprophyric rocks. Spinels may not readily be used to distinguish between micaceous kimberlite and lamproite unless a wide range in composition is found.

Note that all of the kimberlite and lamproite spinel compositions overlap close to the base of the spinel prisms (Figure 6.55). Spinels plotting in this region are not characteristic of a particular paragenesis and are known from a wide variety of igneous rocks (Haggerty 1976).

The low Al content of evolved lamproite spinels probably reflects the effects of extensive crystallization of phlogopite and leucite prior to their appearance as a liquidus

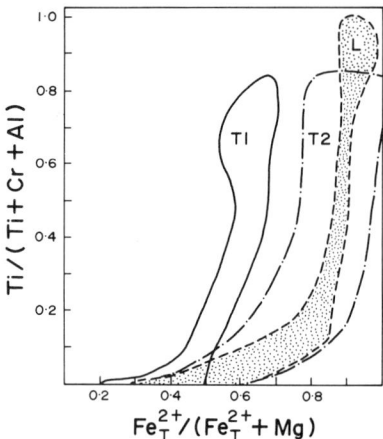

Figure 6.55. Compositional fields of trend 1 (T1) and trend 2 (T2) kimberlite and lamproite (L) spinels projected onto the front face of the "reduced" spinel prism (after Mitchell 1986).

phase. Lamproite spinels are poorer in Al than spinels of similar Fe/(Fe + Mg) ratio in kimberlites as a consequence of their crystallization from a peralkaline magma.

Systematic studies of spinels in other leucite-bearing rocks and minettes have not been undertaken. Spinels in Roman Province lavas are notably MgO poor (<1 wt %) and Al_2O_3 rich (2–5 wt %) relative to lamproite spinels of similar total Fe and TiO_2 content (Cundari 1975, 1979, Baldridge *et al.* 1981).

Spinels in the Agathla Peak (Arizona) and Devon minettes are very similar to evolved lamproite spinels, being titaniferous magnetites and magnetites (Jones and Smith 1983, 1985). In the Linhaisai minette, titaniferous magnetite is also the principal spinel. These magnetites are relatively homogeneous with 7–9 wt % TiO_2, 1–3 wt % MgO, 1–2 wt % Al_2O_3, <0.5 wt % Cr_2O_3, 1 wt % MnO and 77–81 wt % FeO_T (Bergman *et al.* 1988). The spinel evolutionary trend of minettes is unknown, as relatively unevolved Cr-bearing spinels have not been described from these occurrences.

6.9.4. Secondary Aluminous Spinels

Aluminous spinels (pleonaste–hercynite) occur as inclusions in leucite (Jaques and Foley 1985) and phlogopite (Wagner and Velde 1987) in the West Kimberley and Murcia-Almeria provinces respectively. The occurrence of aluminous spinel is unexpected in rocks that are characterized by the presence of Al-deficient phases. These aluminous spinels are believed to be secondary and formed by either exsolution or reaction processes.

In the West Kimberley rocks, aluminous spinels occur as small (20–40 μm) inclusions in aggregates of leucite phenocrysts in fine-grained to glassy leucite–phyric lamproites. The spinels occur as irregular clusters of discrete green euhedral crystals, or as elongate trains and clusters and schlierenlike aggregates of greenish-brown euhedral to subhedral grains of pleonaste to ferrian pleonaste composition. The largest discrete crystals are the most magnesian and aluminous and contain only minor amounts of magnetite in solid solution. Increasing development of the brownish color reflects increasing TiO_2 and magnetite contents (Table 6.34).

Table 6.34. Representative Compositions of Secondary Aluminous Spinels[a]

	1	2	3	4	5	6
TiO_2	0.21	0.62	0.96	0.38	0.20	0.24
Al_2O_3	64.87	59.62	50.91	61.02	61.23	60.85
Cr_2O_3	0.23	n.d.	n.d.	0.04	1.05	0.07
FeO[b]	13.23	20.74	28.37	24.04	17.12	12.66
MnO	0.05	0.12	0.19	0.18	0.10	0.16
MgO	21.77	18.88	18.18	14.43	18.40	22.84
	100.36	99.98	98.61	100.09	99.10	96.82
Recalculated analyses:						
Fe_2O_3	9.48	14.63	19.88	17.43	12.12	9.16
FeO	4.70	7.58	10.48	8.36	6.21	4.42
	101.31	101.45	100.60	101.84	100.31	97.74

[a]n.d., not detected. 1–3, inclusions in leucite, West Kimberley (Jaques *et al.* 1986); 4–6, inclusions in biotite (Wagner and Velde 1987); 4, Vera; 5, Fortuna; 6, Leucite Hills.
[b]Total Fe calculated as FeO.

All of the spinels are notably Cr_2O_3 poor (<0.2 wt %) in marked contrast to the Cr-rich primary groundmass spinels.

Jaques and Foley (1985) propose that the spinels originated by exsolution from a nonstoichiometric leucite containing Mg, Fe, and Ti in solid solution. In support of this hypothesis Jaques and Foley (1985) present experimental evidence that demonstrates that leucite is able to incorporate Mg in solid solution. Gupta and Yagi (1980) have also shown that leucites may be synthesized with up to 7.7 wt % $KFe^{3+}Si_2O_6$ in solid solution at 2 kbar water pressure. Naturally occurring leucites in lamproites are known to contain significant amounts of Fe^{3+} and Mg (see Section 6.6.2).

The formation of the Mg-rich, nonstoichiometric leucite is believed by Jaques and Foley (1985) to be related to rapid crystallization from a magma supersaturated with leucite or to crystallization of spinel from a melt included within leucite. In the latter case local supersaturation with respect to Al is required. This may be induced by dissolution of leucite or the contamination of the magma with pelitic country rock.

Aluminous spinels are widespread as pale blue-to-mauve, euhedral-to-subhedral, 20–150-μm crystals included in biotite in the Murcia-Almeria lamproites (Venturelli *et al.* 1984a, 1988, Wagner and Velde 1986a). Similar spinels are rarely present in the Leucite Hills lamproites (Wagner and Velde 1986a). The spinels vary in composition within the pleonaste–ferrian pleonaste series. Individual specimens are of uniform composition, but wide interspecimen variation is evident (Table 6.34). The Ti and Cr contents are notably low and the spinels are similar in their composition to those described by Jaques and Foley (1985). The micas that host the spinels are biotites and thus Fe and Al rich relative to the discrete spinel-free phlogopite phenocrysts present in the same lamproites. Wagner and Velde (1987) propose that the spinels are formed by the decomposition of xenocrystal biotite consequent upon its incorporation into the lamproite magma.

Aluminous spinels of unknown origin, previously described as pseudobrookite (Vila *et al.* 1974), occur in the groundmass of the Azzaba "hyalolamproite" (Wagner and Velde 1986a).

6.10. PRIDERITE

Priderite was initially identified as rutile by Cross (1897) and Wade and Prider (1940), but was subsequently shown to be a new mineral by Norrish (1951). Priderites are members of the hollandite group of minerals (Post *et al.* 1982). These minerals have the general formula $(A)_{1-2}(B)_{1-2}(C)_{7-6}O_{16}$, where A = K, Ba, Cs; B = Fe, Cr, Al, V, Ce; and C = Ti, Nb. They consist of paired chains of edge-sharing $(B,C)O_6$ octahedra extending along the crystallographic c-axis. The chains are linked along their margins to form a framework with continuous tunnels aligned along the c-axis. The tunnels are filled by A-site cations to varying degrees, hence hollandites, including lamproite-derived priderite, typically display nonintegral stoichiometries and superlattice ordering (Pring and Jefferson 1983, Kesson and White 1986).

Priderite plays a key role in lamproite recognition as its occurrence is essentially restricted to this paragenesis. The recent discoveries of other K–Ba–Ti-rich hollandites in carbonatites and kimberlites do not negate the value of priderite as a diagnostic mineral. Interestingly priderites have not yet been recognized from the Murcia-Almeria or Kapamba lamproites.

6.10.1. Paragenesis

Priderite occurs as elongate acicular crystals that typically range from 0.2 to 0.5 mm in length, but which may reach up to 4 mm in the Walgidee Hills pegmatitic lamproites. The abundance ranges from trace amounts in olivine lamproites to a maximum of 10 modal % in some phlogopite leucite lamproites (Jaques *et al.* 1986).

Priderite typically forms euhedral tetragonal prisms capped by tetragonal bipyramids. The crystals may show all degrees of resorption ranging from coign modification to deep embayments. Priderites have a high relief and birefringence. They may be virtually opaque or exhibit pleochroism in shades of yellow-orange, orange-red, or deep red. The color is commonly unevenly developed, and alternating bands of different hue give the crystals a barred or striped appearance (Figure 2.16). In reflected light a well-developed cleavage is evident. Characteristic triangular etch pits (Figure 6.56) are commonly developed along this cleavage.

Priderites form after phenocrystal phlogopite and prior to groundmass phlogopite. Inclusions of other phases in priderite have not been reported. Priderites mantled by ilmenite laths and jeppeite prisms are found in the Rice Hill and Walgidee Hills lamproites, respectively. Keuhner (1980) has noted the common occurrence of sagenitic grids of priderite on the (001) planes of resorbed phlogopite phenocrysts in sanidine-bearing lamproites from the Leucite Hills.

Jaques *et al.* (1989) have noted priderites of unusual composition (see below) occurring as small euhedral groundmass crystals in olivine phlogopite lamproite dikes from the Argyle intrusion. Here priderite occurs (1) in ovoid calcite globules with Mn-ilmenite and rutile, (2) as small needles with talc at the margins of irregular cavitylike structures, (3) in ovoid to spherical calcites, and (4) as macrocrysts.

6.10.2. Composition

Representative compositions of priderite are given in Table 6.35. Total iron is expressed as Fe_2O_3, as Pring and Jefferey (1983) have shown that ferric iron is dominant in

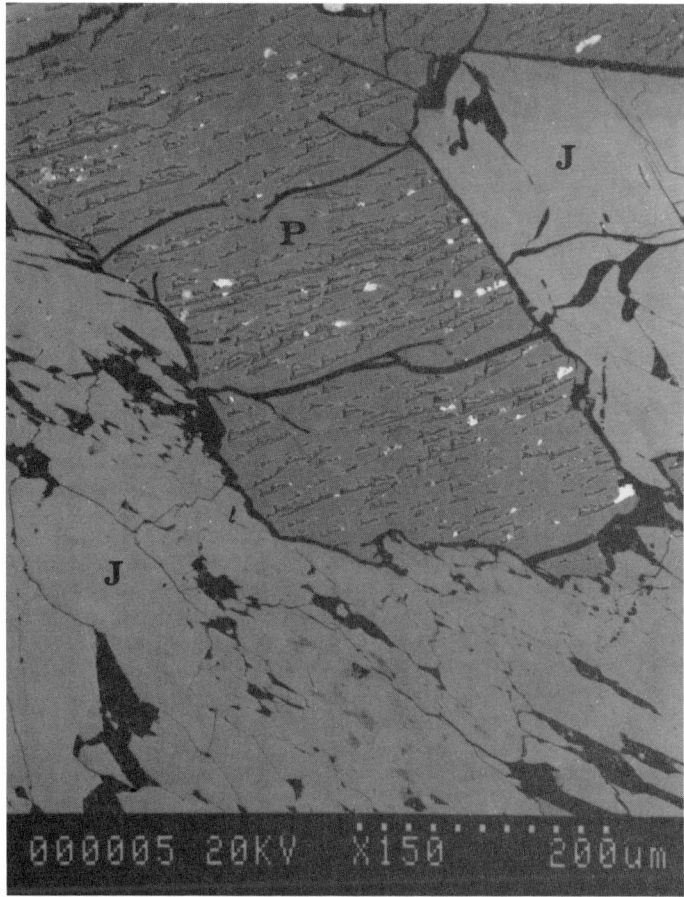

Figure 6.56. Priderite (P) exhibiting triangular etch pits mantled by jeppeite (J). Backscattered electron image.

natural priderites. However, it should be noted that Fe^{2+} is undoubtedly also present as Myrha et al. (1988) have demonstrated the presence of Ti^{3+} in priderite. The Fe^{3+}/Fe^{2+} ratios and the Ti^{4+}/Ti^{3+} ratios cannot be estimated by calculation as most hollandites are nonstoichiometric (Kesson and White 1986).

Priderite is an intermediate member of a complex range of solid solutions exhibited by TiO_2-based hollandites. In the natural environment the most important of these are between the $K_2(Mg,Fe^{2+})Ti_7O_{16}$–$Ba(Mg,Fe^{2+})Ti_7O_{16}$ series and the $K_2(M^{3+})_2Ti_6O_{16}$–$Ba(M^{3+})_2Ti_6O_{16}$ series where M^{3+} = Fe, Cr, V, Ce, Al (Mitchell and Haggerty 1986). Most priderites do not represent endmember compositions and belong to the $(K, Ba)_{1-2}(Fe^{3+})_2Ti_6O_{16}$–$(K,Ba)_{1-2}(Mg,Fe^{2+})Ti_7O_{16}$ series. The presence of significant amounts of Cr_2O_3 (typically <1 wt %, but ranging up to 9 wt %) in some priderites (Argyle, Endlich Hill, Smoky Butte) indicates solid solution towards the redledgeite–potassian redledgeite series ($BaCr_2Ti_6O_{16}$–$K_2Cr_2Ti_6O_{16}$). Vanadium (<1 wt % V_2O_3) and cerium (<0.5 wt % Ce_2O_3) contents are typically low in most priderites, although Jaques et al. (1989) report priderites with relatively high V_2O_3 (1.3–1.7 wt %) and Ce_2O_3 (0.2–1.1 wt %) in strongly zoned priderites from Argyle.

Table 6.35. Representative Compositions of Priderite[a]

	1	2	3	4	5	6	7	8	9	10	11	12	13
TiO_2	73.26	69.50	65.83	74.26	72.14	66.60	70.83	78.01	74.08	79.36	78.71	77.56	81.56
Cr_2O_3	0.24	3.18	5.80	n.d.	0.30	5.16	0.26	n.d.	n.d.	n.d.	0.99	1.55	0.42
V_2O_3	0.57	0.98	1.05	0.94	1.12	n.a.	n.a.	0.28	0.90	0.58	0.45	0.82	n.a.
Fe_2O_3[b]	11.69	9.54	9.84	8.72	9.31	7.66	9.59	11.37	11.06	9.21	8.97	7.22	10.56
MgO	0.50	0.19	1.05	0.84	n.d.	0.76	0.03	0.37	0.88	0.94	0.81	0.83	n.a.
K_2O	4.39	4.19	3.73	2.92	3.24	2.67	1.37	8.33	7.60	6.54	6.85	5.57	8.65
BaO	8.46	12.26	11.93	11.75	13.97	16.93	17.08	1.50	5.38	3.24	2.96	6.00	0.26
	99.11	99.84	99.23	99.33	100.08	99.78	99.16	99.86	99.90	99.87	99.74	99.55	101.45

[a] n.d., not detected; n.a., not analyzed. 1, Zirkel Mesa; 2, North Table Butte; 3, Endlich Hill; 4–6, Smoky Butte; 7, Sisco; 8–9, Prairie Creek; 10, Rice Hill; 11, Mt. North; 12, Seltrust 2; 13, Sisimiut. Analyses 1–5, 8–12 (this work); 6–8 (Wagner and Velde 1986); 13 (Scott 1981).
[b] Total Fe calculated as Fe_2O_3.

The principal compositional variation in lamproite priderite is with respect to their Ba, K, and Fe contents. As expected for the solid solutions noted above the Ba and K contents are inversely correlated. None of the major constituents show any correlation with the Cr, V, and Ce contents. Most priderites are not zoned.

Figure 6.57 illustrates the BaO–Fe_2O_3 compositional variation of priderite and demonstrates that different lamproite localities are characterized by priderites of markedly different Ba content. Thus the Leucite Hills and Smoky Butte priderites are significantly enriched in Ba relative to those from West Kimberley and Prairie Creek. Such interprovincial differences are probably a reflection of the differences in the composition of the parental magmas, which in turn reflect the Ba contents of their source regions. Figure 6.57 also shows that within the Leucite Hills and West Kimberley provinces there are significant intraprovincial compositional variations in priderites derived from different localities. These slight differences, especially with respect to the Fe content, may be related to differences in the postemplacement crystallization history of individual lamproites.

Mason (1977) has reported the following concentrations (ppm) of trace elements in Walgidee Hills priderite: $V = 300$, $Cr = 700$, $Mn = 300$, $Co = 50$, $Ni = 200$, $Sr = 70$, $Y = 1.7$, $Zr = 180$, $Nb = 92$, $Ho = 0.1$, $Er = 0.19$, $Yb = 0.45$.

The stability relationships of priderites of lamproitic composition are not known. Dubeau and Edgar (1985) have investigated solid solution between $K_2MgTi_7O_{16}$ and $BaMgTi_7O_{16}$ and shown that "priderite" corresponding to compositions along this join may be readily synthesized at 1200°C and 30 kbar.

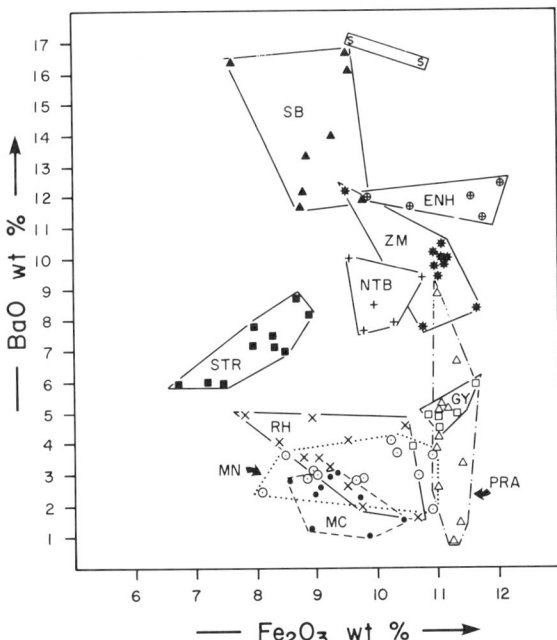

Figure 6.57. Composition variation of priderite in lamproites from the Leucite Hills (ENH, Endlich Hill; ZM, Zirkel Mesa; NTB, North Table Butte); West Kimberley (GY, Mount Gytha; RH, Rice Hill; MN, Mount North; MC, Mount Cedric; STR, Seltrust Pipe No. 2); Smoky Butte (SB), and Prairie Creek (PRA), all data this work. Sisco (S) data from Wagner and Velde (1986a).

6.10.3. Ba-Titanates Related to Priderite

Jaques et al. (1989a) have described a Ba–Cr-titanate which occurs in association with subhedral dolomite crystals and talc, replacing olivine (Table 6.36). The composition represents significant solid solution toward the mannardite–redledgeite series.

Priderites within microinclusions in the Prairie Creek lamproite are in some instances mantled by a Ba–Fe-titanate (Table 6.36) that is K poor and similar to the hollandites described from the Kovdor and Murun complexes (see below).

6.10.4. Comparison with Hollandites in Other Potassic Rocks

TiO_2-based hollandites related to priderite are rare minerals that are known from only four parageneses. Representative compositions are given in Table 6.36, which shows that these minerals are significantly different to those occurring in lamproites.

Mitchell and Haggerty (1986) and Mitchell and Meyer (1989b) have shown that hollandites in micaceous kimberlites (Star, New Elands) are V_2O_3 (2–9 wt %), Ce_2O_3 (0.5–2.0 wt %) and Nb_2O_5 (0.5–6.5 wt %) rich. The minerals exhibit a wide range in K/(K + Ba) and $Fe^{3+}/(Fe^{3+} + V)$ ratios (Table 6.36, Figure 6.58), that represents solid solution toward mannardite–potassian mannardite $[(Ba,K)_{1-2}V_2Ti_6O_{16}]$ and the $(Ba,K)_{1-2}Fe_2Ti_6O_{16}$ series. Such hollandites differ from those in lamproites in having $Fe^{3+}/(Fe^{3+} + V)$ less than 0.8 (Figure 6.58).

A hollandite mineral (Table 6.36) in the Benfontein calcite kimberlite (Scatena-Watchel and Jones 1984) is similar to the high-Cr priderite described by Jaques et al. (1989a) from the Argyle lamproite, but differs in being relatively poor in BaO. The composition $K_{1.45}Ba_{0.05}Cr_{0.80}Fe_{0.44}Ti_{6.53}O_{16}$ represents solid solution towards potassian redledgeite $(K_2Cr_2Ti_6O_{16})$.

Hollandite occurring in the Kovdor carbonatite (Zhuravleva et al. 1978) is Ba rich (Table 6.36) and associated with geikielite, zirkelite, and clinohumite. The mineral probably formed by the reaction of early crystallizing Ti-minerals with carbonatite magma (Mitchell 1985). The composition $K_{0.06}Ba_{0.74}Fe_{1.08}Ti_{6.81}O_{16}$ is unlike that of any lamproite hollandite, with the exception of the Prairie Creek material (see Section 6.10.3).

Table 6.36. Representative Compositions of Titanates[a]

	1	2	3	4	5	6	7	8
TiO_2	60.77	70.13	71.30	66.16	51–54	72.00	66.91	83.86
Nb_2O_5	n.d.	0.18	4.57	0.53	1.6	n.a.	n.a.	n.a.
Cr_2O_3	6.02	n.d.	0.71	0.34	n.a.	n.a.	n.a.	n.a.
V_2O_3	4.78	0.28	2.79	9.45	n.a.	n.a.	n.a.	n.a.
Fe_2O_3[b]	6.43	8.01	6.62	4.12	15.6–16.1	11.39	12.20	4.84
Ce_2O_3	3.97	n.d.	n.a.	n.a.	0.5	n.a.	n.a.	n.a.
MgO	0.34	n.d.	0.21	0.10	2.1–3.4	n.d.	n.a.	n.a.
K_2O	n.d.	n.d.	4.81	1.32	n.d.	0.40	1.70	8.55
BaO	20.86	20.45	8.36	16.36	16.1–16.7	15.00	16.51	0.68
	103.17	99.05	99.37	98.38	—	98.79	97.32	97.93

[a] n.d., not detected; n.a., not analyzed. 1, Argyle (Jaques et al. 1989a); 2, Prairie Creek (this work); 3–4, Star kimberlite (Mitchell and Meyer 1989b); 5, Benfontein kimberlite (Scatena-Wachel and Jones 1984); 6, Kovdor carbonatite (Zhuravleva et al. 1978; 7–8 Murun potassic complex (Lazebnik et al. 1985).
[b] Total Fe calculated as Fe_2O_3.

MINERALOGY OF LAMPROITES

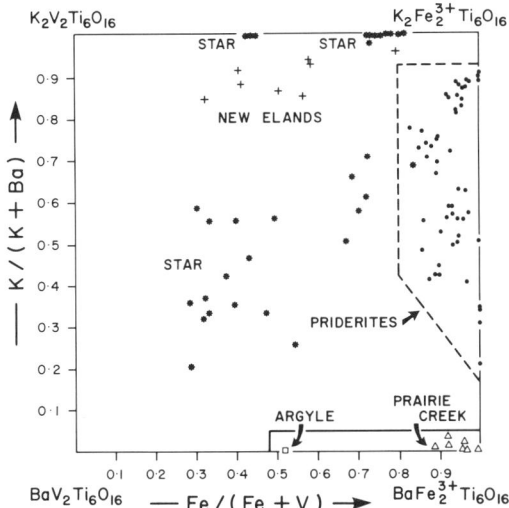

Figure 6.58. Composition of hollandite-group minerals in kimberlites and lamproites. Data sources: Star and New Elands micaceous kimberlites (Mitchell and Meyer 1989b, Mitchell and Haggerty 1986), Argyle lamproite (Jaques *et al.* 1989), Prairie Creek Ba–Fe titanate and lamproite priderites (this work).

Lazebnik *et al.* (1985) have described hollandites from the Murun potassic complex that exhibit a wide range in composition (Table 6.36). Some examples are Ba and Fe rich and similar to the Kovdor hollandites (Zhuravleva *et al.* 1978); others are Ba and Fe deficient and Ti rich, while further examples are of intermediate composition. All are unlike lamproite priderite in composition. These hollandites coexist with biotite, diopside, olivine, apatite, sphene, barite, zircon, and ilmenite. Priderites have not been reported in the rocks described as lamproites (Vladikin 1985) from the Murun complex.

In summary, although hollandites are known from other potassic rocks they may be distinguished from those found in lamproites on the basis of their higher Ba, V, Ce, and Nb contents and/or their different parageneses. These compositional differences allow a real distinction to be made between hollandite occurring in lamproite and in petrographically similar micaceous kimberlite.

6.11. JEPPEITE

A potassium barium titanate, isostructural with synthetic $K_2Ti_6O_{13}$, was reported by Bagshaw *et al.* (1977) to occur in lamproite pegmatites at the Walgidee Hills. Subsequently, this phase has been shown to be a new mineral which has been named jeppeite by Pryce *et al.* (1984). In the Walgidee Hills occurrences, jeppeite is found as prismatic to acicular crystals mantling priderite (Figure 6.56) and in association with titanian potassium richterite, diopside, wadeite, shcherbakovite, and perovskite. A second occurrence in priderite-bearing diopside richterite lamproite microinclusions found within Prairie Creek madupitic lamproite (Mitchell and Lewis 1983) also occurs as overgrowths on priderite (Mitchell 1985).

Jeppeite may be distinguished from priderite by its pleochroism: X = slate grey, Y = dark green-brown, Z = olive buff, $Y > Z > X$ (Jaques *et al.* 1986, Pryce *et al.* 1984).

Jeppeite is a potassium barium titanate containing minor quantities of iron and sodium. Although individual crystals are homogeneous, minor intergrain compositional variation is evident. Representative compositions are given in Table 6.37, which shows that the Prairie Creek jeppeite is much richer in Ba than the Walgidee Hills variety.

Figure 6.59 indicates that jeppeites are intermediate members of the solid solution series between $K_2Ti_6O_{13}$ and $Ba_3Ti_5O_{13}$. Hervieu *et al.* (1979) note that Fe-free synthetic phases can be considered to be nonstoichiometric titanates with the approximate general formula $K_{2-2x}Ba_{3x}Ti_{6-x}O_{13}$, where $0 < x < 0.225$. Considerable nonstoichiometry is evident in jeppeites due to vacancies at the (K,Na) and Ba sites as $x < 0.1$ (Mitchell 1985). Figure 6.59 shows that jeppeites are Fe and Ti poor and Ba rich relative to priderites.

Preliminary experiments by Arima (personal communication) have shown that jeppeite can be synthesized at 800°C at atmospheric pressure and that at higher temperatures it breaks down into priderite and unidentified titanates (? rutile). The results of these experiments are in accord with the paragenesis of jeppeite. The rarity of this low-temperature phase corresponds to the scarcity of slowly cooled lamproite intrusions.

6.12. IRON TITANIUM OXIDES

6.12.1. Armalcolite

Armalcolite is known only from lamproites occurring at Smoky Butte, Francis, Jumilla, Cancarix, and Barqueros (Wagner and Velde 1986a, Mitchell *et al.* 1987, Venturelli *et al.* 1988).

Armalcolite is particularly abundant in the Smoky Butte lamproites, where it forms euhedral prisms that exhibit a dark red to purple-red weak pleochroism (Mitchell *et al.* 1987). Green armalcolite that does not differ in composition from the red varieties is also present (Wagner and Velde 1986a). At Francis, armalcolite exhibits a golden brown pleochroism and is optically similar to rutile (Wagner and Velde 1986a). Armalcolite occurs in the groundmass of the Murcia-Almeria lamproites as subhedral deep red crystals (Venturelli *et al.* 1988). At Smoky Butte armalcolite is especially abundant in glassy rocks and appears to have crystallized after olivine and phlogopite phenocrysts and prior to diopside. Many of the crystals are resorbed.

Table 6.37. Representative Compositions of Jeppeite[a]

	1	2	3	4	5
SiO_2	0.19	0.06	n.a.	0.03	n.d.
TiO_2	70.55	69.42	69.29	69.30	67.89
Al_2O_3	0.18	0.26	n.a.	n.d.	n.d.
Fe_2O_3[b]	4.82	4.56	4.73	5.03	4.04
MgO	0.44	0.45	n.d.	0.33	1.36
K_2O	10.54	10.08	8.47	8.60	7.27
Na_2O	0.54	0.53	n.a.	0.41	n.d.
BaO	14.00	16.11	17.35	16.76	20.14
	101.26	101.47	99.84	100.43	100.70

[a]n.d., not detected; n.a., not analyzed. 1–4, Walgidee Hills; 5, Prairie Creek; 1–2, 5 (Mitchell 1985); 3 Pryce *et al.* (1984); 4 Jaques *et al.* (1986).
[b]Total Fe calculated as Fe_2O_3.

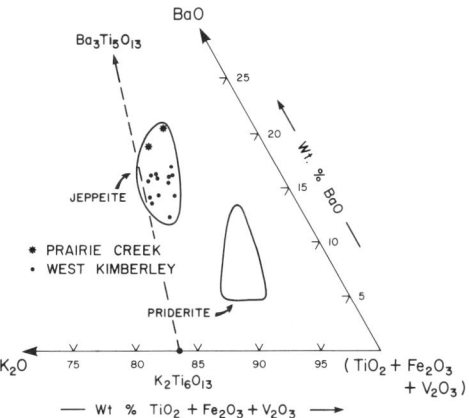

Figure 6.59. Composition of jeppeite from West Kimberley and Prairie Creek lamproites (after Mitchell 1985).

Representative compositions are given in Table 6.38. The crystals are relatively uniform both within and between grains. Minor elements present in significant amounts include V_2O_3 (0–0.5 wt %), Al_2O_3 (0.2–0.5 wt %), and Cr_2O_3 (0.7–1.4 wt %). Recalculation of the total iron content into FeO and Fe_2O_3 on a stoichiometric basis (Table 6.38) indicates that the minerals are poor in Fe_2O_3 and are essentially members of the ferropseudobrookite ($FeTi_2O_5$)–karroite ($MgTi_2O_5$) series with less than 10 mol % solid solution towards pseudobrookite (Fe_2TiO_5). According to the terminology of Bowles (1988) they

Table 6.38. Representative Compositions of Armalcolite and Pseudobrookite[a]

	1	2	3	4	5	6	7	8
TiO_2	68.79	69.45	68.70	65.99	69.04	66.32	60.33	69.02
Al_2O_3	0.31	0.22	0.12	0.88	0.30	0.06	0.02	0.47
V_2O_3	n.d.	0.25	0.44	n.a.	n.a.	n.a.	n.a.	n.a.
Cr_2O_3	1.28	0.98	0.89	0.67	0.99	0.05	0.82	0.43
MnO	0.05	0.11	0.19	0.02	0.14	0.20	0.33	0.15
MgO	9.67	9.32	9.67	10.53	8.92	7.61	8.65	9.39
FeO[b]	17.54	17.36	17.53	19.65	19.86	23.46	28.97	19.66
	97.64	97.69	97.54	97.74	99.25	97.70	99.12	99.12
Recalculated analyses:								
Fe_2O_3	6.38	4.77	6.55	13.68	7.70	11.02	26.36	8.56
FeO	11.80	13.07	11.63	7.34	12.93	13.54	5.25	11.96
	98.28	98.17	98.20	99.11	100.02	98.80	101.76	99.98
Mol % end member molecules:								
Fe_2TiO_5	6.18	4.60	6.38	13.58	7.42	10.87	27.68	8.22
$MgTi_2O_5$	55.68	53.40	55.89	62.12	51.05	44.59	53.96	53.54
$FeTi_2O_5$	38.13	42.00	37.73	24.30	41.56	44.57	18.36	38.25

[a] n.d., not detected; n.a., not analyzed. 1–5, Smoky Butte; 6, Moon Canyon; 7, Cancarix; 8, Barqueros. Data sources: 1–3, Mitchell et al. (1987); 4, Velde (1975); 5–8, Wagner and Velde 1986a).
[b] Total Fe calculated as FeO.

are armalcolites, hence reference to these oxides as pseudobrookites, e.g., Wagner and Velde (1986a), is inappropriate. Figure 6.60 shows that their composition is close to that of lunar armalcolite, and unlike that of most terrestrial pseudobrookites, which are Fe_2O_3-rich ternary "kennedyites" or members of the pseudobrookite–karroite series (Von Knorring and Cox 1961, El Goresy 1976, Van Kooten 1980). Only the sample from Barqueros contains relatively high Fe_2O_3 contents and Wagner and Velde (1986a) interpret this to be a xenocryst.

Smoky Butte armalcolites, while showing minor resorption, exhibit no evidence of reaction with the magma to form magnesian ilmenite and rutile as observed in lunar rocks, and quenching from high temperatures must have been rapid. Although the stability limit of Fe^{3+}-bearing armalcolite is unknown it is probably not significantly different from that of ideal armalcolite which is not stable below 1010°C (Lindsley et al. 1974). Thus the presence of armalcolite implies a high temperature of formation at low oxygen fugacities.

6.12.2. Ilmenite

Ilmenite is not a characteristic mineral of most lamproites and it is relatively common as subhedral prisms only in the Jumilla (Murcia-Almeria) and Oscar (West Kimberley) intrusions. In most West Kimberley lamproites, ilmenite occurs rarely as small (<20 μm) groundmass grains (Jaques et al. 1986, Bergman 1987). In the Sisco lamproite it forms euhedral crystals 5–50 μm in length (Wagner and Velde 1986a).

Ilmenite crystallizes after olivine, diopside, and phenocrystal phlogopite. Mantles of small ilmenite prisms are found upon priderites in the Rice Hill (West Kimberley) intrusion (this work).

Jaques et al. (1986, 1989a) have noted the occurrence of ilmenite of irregular habit in the Ellendale and Argyle olivine lamproites. In the latter, the ilmenite is closely associated with carbonate minerals. Jaques et al. (1989) interpret such ilmenites to be late-forming phases of either magmatic or deuteric origin.

Representative ilmenite compositions are given in Table 6.39, which demonstrates that groundmass ilmenites have significant MgO (1–7.5 wt %) and MnO (1–2 wt %) contents.

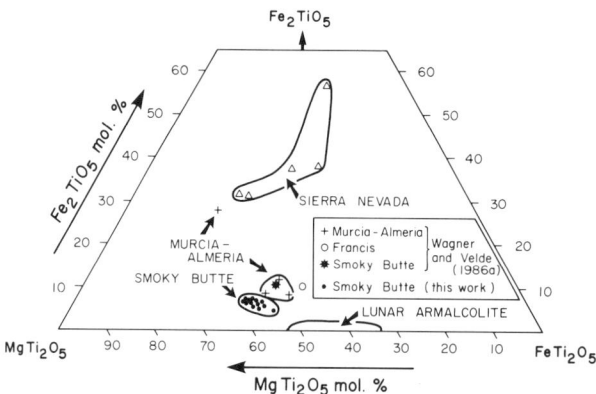

Figure 6.60. Composition of armalcolites from the Smoky Butte (Mitchell et al. 1987), Murcia-Almeria (Wagner and Velde 1986a, Venturelli et al. 1988), Francis (Wagner and Velde 1986a) lamproites. Data for lunar armalcolites Sierra Nevada leucitites from El Goresy (1976) and Van Kooten (1980), respectively.

MINERALOGY OF LAMPROITES

Table 6.39. Representative Compositions of Ilmenite[a]

	1	2	3	4	5	6	7	8	9	10	11	12
TiO_2	48.84	48.80	51.64	50.18	53.65	55.78	53.75	53.16	54.30	55.03	51.44	51.78
Al_2O_3	0.25	0.21	0.07	n.d.	n.d.	n.d.	n.d.	n.d.	n.d.	n.d.	n.d.	n.d.
Cr_2O_3	0.07	0.02	0.50	0.58	0.09	0.31	n.d.	0.28	0.16	n.d.	0.09	n.d.
FeO[b]	45.37	45.12	44.69	40.50	41.15	33.51	44.00	41.58	40.41	35.77	43.61	41.16
MnO	1.08	1.92	0.72	0.56	1.25	0.78	1.59	2.88	1.55	0.56	2.95	6.27
MgO	3.19	2.55	1.92	6.41	3.64	9.58	0.68	1.99	3.55	7.40	0.46	0.60
	98.80	98.62	99.54	98.18	99.78	99.96	100.02	99.89	99.77	98.76	98.55	99.81
Recalculated analyses:												
Fe_2O_3	9.15	8.59	2.67	8.19	0.73	1.35	—	0.27	1.51	0.04	1.29	2.24
FeO	37.14	37.39	42.29	33.13	40.49	32.30	44.00	41.34	39.06	35.73	42.45	39.14
	99.72	99.48	99.81	99.05	99.85	100.09	100.02	99.92	99.93	98.76	98.68	100.03
Mol % end member molecules												
$FeTiO_3$	77.3	78.3	88.6	67.7	83.3	63.5	94.0	86.2	83.2	72.2	90.6	82.3
$MgTiO_3$	11.8	9.5	7.2	23.3	13.4	33.6	2.6	7.4	12.3	26.6	1.8	2.2
$MnTiO_3$	2.3	4.1	1.5	1.2	2.6	1.6	3.4	6.1	3.2	1.2	6.4	13.4
Fe_2O_3	8.6	8.1	2.5	7.5	0.7	1.2	—	0.3	1.3	—	1.2	2.1

[a] n.d., not detected. 1–2, Jumilla; 3–4, Cancarix; 5–6, Francis; 7, Sisco; 8, Mt. North; 9, Ellendale; 10, Brooking Creek; 11–12, Argyle. Data sources: 1–2, Mitchell (1985); 3–8, (Wagner and Velde 1986a); 9–10, Jaques et al. 1986; 11–12, Jaques et al. (1989a).
[b] Total Fe calculated as FeO.

Recalculated compositions demonstrate that they are poor in Fe_2O_3 and typically contain less than 15 mol % hematite. Figure 6.61 shows that their composition is similar to that of ilmenites occurring in a wide variety of rocks, including leucitites, and hence is not diagnostic of their lamproitic paragenesis. The ilmenites are unlike those found in kimberlites (Mitchell 1986) in being on average Mg and Cr poor.

Jaques *et al.* (1989) have shown that late-stage carbonate-associated ilmenites at Argyle (Table 6.39) are MnO rich (3–8 wt %) and MgO poor (<1 wt %) and may have significant Nb_2O_5 contents (up to 1.2 wt %). Mn-rich ilmenite is also reported to occur in association with carbonate in the Walgidee Hills intrusion (Jaques *et al.* 1986). The association of Mn-ilmenites with late-stage carbonate minerals has also been noted in kimberlites and carbonatites (Tompkins and Haggerty 1985, Gaspar and Wyllie 1984). Haggerty *et al.* (1979) have suggested that the formation of Mn-rich ilmenite may be associated with the development of late-stage carbonate immiscibility.

6.12.3. Titanium Dioxides

Minerals composed essentially of TiO_2 are common in some lamproites. They are identified optically either as rutile (opaque-to-red-brown, uniaxial positive) or anatase (bluish-brown, uniaxial negative). X-ray identification of the polymorphs has not been undertaken and brookite has not been reported.

Few data are available regarding either paragenesis or composition. In many lamproites rutile/anatase crystals occur apparently at random, scattered throughout the groundmass. In the highly altered lamproites forming the Prince dike, Bobi, anatase occurs as anhedral platelike masses or aggregates of small irregular crystals in association with chlorite and calcite (Mitchell 1985). Thy *et al.* (1987) describe rutile as a groundmass phase intergrown with ilmenite in amphibole lamproites, or as the only opaque groundmass phase in leucite

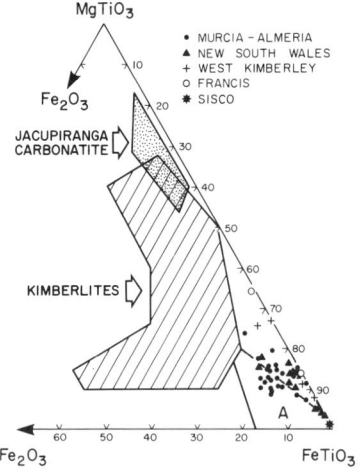

Figure 6.61. Composition of ilmenites from lamproites compared with those of ilmenites from New South Wales leucitites and other igneous rocks (Mitchell 1985). Field A is for ilmenites from carbonatites, granites and basaltic rocks. Lamproite data from Venturelli *et al.* (1988), Wagner and Velde (1986a), Jaques *et al.* (1986), and Mitchell (1985).

lamproites from Sisimiut. Large irregular masses of anatase occur adjacent to titanian alkali amphiboles in the Mount Bayliss rocks (Sheraton and England 1980).

Anatase in the Bobi and Mount Bayliss rocks contains more than 98 wt % TiO_2 and has low Fe_2O_3 (<1 wt %) and Cr_2O_3 (<0.5 wt %) contents (this work). Sisimiut rutile contains more than 99 wt % TiO_2 with 0.4–0.8 wt % and 0.1–2.1 wt % Fe_2O_3 in leucite and amphibole lamproites, respectively (Thy *et al.* 1987). Rutile in the latter contains up to 2.2 wt % Cr_2O_3. Scott (1981) notes that Sisimiut rutile may also contain up to 10 wt % FeO.

It has not yet been determined whether these TiO_2 polymorphs are primary phases or entirely secondary phases formed by the alteration of preexisting titanates (i.e., perovskite, priderite, ilmenite). The latter possibility would seem likely for the anatase found in the altered Bobi and Mount Bayliss rocks.

6.13. POTASSIUM ZIRCONIUM SILICATES

6.13.1. Wadeite

Wadeite, $K_2ZrSi_3O_9$, was discovered in the West Kimberley lamproites by Prider (1939) and subsequently recognized in the Leucite Hills and Francis lamproites by Carmichael (1967) and Henage (1972), respectively. Wadeite has not yet been reported from other lamproite suites, perhaps in part owing to its similarity to small crystals of apatite in thin section. Large crystals are, however, morphologically (Figure 6.62) and optically distinct from apatite in birefringence and optic sign. Wadeite is readily identified by its bright blue fluorescence under electron beams (Carmichael 1967) or by backscattered electron imagery. Wadeite is a cyclosilicate consisting of a framework of $(Si_3O_9)^{6-}$ rings with K and Zr occupying the interstitial sites (Henshaw 1955).

Wadeite typically occurs as hexagonal euhedral prisms, which may be up to 1.5 mm in length in the pegmatitic lamproites of the Walgidee Hills (Jaques *et al.* 1986). The crystals commonly exhibit a characteristic rectangular cross section (Figures 2.17 and 6.62). In the Middle Table Mountain (Leucite Hills) and some Ellendale lamproites, wadeite also forms poikilitic groundmass crystals (Figure 6.63). Wadeite is not found as inclusions in early crystallizing phases and forms before groundmass phlogopite and titanian potassium richterite. Poikilitic wadeites enclose perovskite, apatite, and diopside. Wadeite is thus a late-crystallizing groundmass phase. Commonly wadeite is partially replaced by calcite and rarely by barite.

Representative compositions are given in Table 6.40, which shows that wadeites are essentially pure potassium zirconium silicates. Other elements, with the exception of TiO_2, are not present in significant amounts. Although insufficient data are available to assess intra/interprovincial compositional variations it appears that West Kimberley wadeites may be richer in TiO_2 (0.1–2.8 wt %) than the Leucite Hills variety (0.8 wt % TiO_2). Carmichael (1967) has noted that there are K-site deficiencies in natural wadeites. Recent analyses of wadeite (Mitchell 1985, Jaques *et al.* 1986) confirm Carmichael's (1967) conclusion that wadeite does not contain calcium, and that Prider's (1939) sample was contaminated by apatite.

Mason (1977) has determined the following concentrations (ppm) of trace elements in the Walgidee Hills wadeite: Na = 300, Ca = 200, V = 7, Cr = 50, Mn = 10, Fe = 300, Co < 1, Ni = 20, Sr = 900, Nb = 10.

Figure 6.62. Characteristic morphology of euhedral wadeite (W), Mount North (West Kimberley). Also illustrated are priderite (P) and apatite (A). Backscattered electron image.

Arima and Edgar (1980) have shown that, by itself, wadeite is stable over a very wide range of temperatures and pressures (from 1 bar to 25 kbar; 800–1250°C). The presence or absence of this mineral is therefore of little use in constraining the conditions under which it forms in lamproites. The data, however, are in agreement with the observed late-stage, presumably low-temperature, paragenesis.

In contrast to the common occurrence of wadeite in lamproites, it has been reported only from two other parageneses, i.e., the Khibina alkaline complex (Tikhonenkev *et al.* 1960) and the Kovdor carbonatite (Kapustin 1963). At these localities wadeite is a very rare mineral and occurs in late-stage carbonate veins in association with labuntsovite, phlogopite, orthoclase, and thorite. Unlike lamproite wadeite it exhibits a bright violet color (Kapustin 1980). The composition of the Kovdor material is similar to that of lamproite wadeite with the exception of having a higher Na_2O content (Table 6.40). The composition of the Khibina material is so unlike any other wadeite that contamination of the sample analyzed must be suspected.

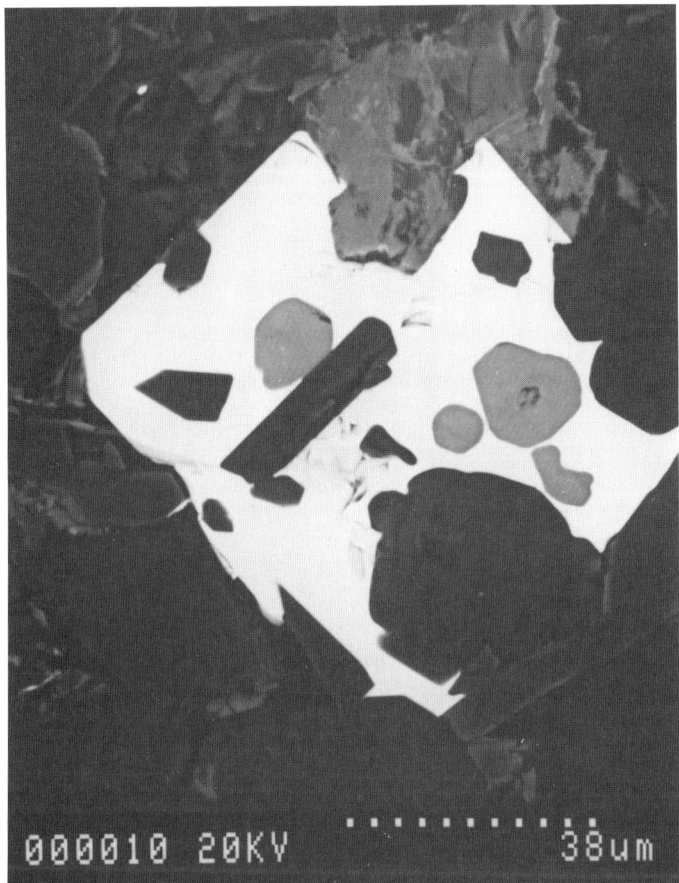

Figure 6.63. Poikilitic groundmass wadeite (light grey), Middle Table Mountain (Leucite Hills). Backscattered electron image.

The recognition of wadeite in these carbonatites, where it occurs in a completely different paragenesis and is a very rare mineral, does not diminish the importance of wadeite as a typomorphic phase for lamproite recognition.

6.13.2. Dalyite

Dalyite, $K_2ZrSi_6O_{15}$, has recently been recognized in the Sierra de Cabras (Murcia-Almeria) lamproite (Linthout *et al.* 1988). Here, euhedral dalyite crystals (ca. 150 μm) border the margins of miarolitic cavities in association with potassium richterite and aegirine. The paragenesis indicates a very late, perhaps vapor phase crystallization origin. The dalyite shows little compositional variation (Table 6.40) and is similar to dalyites described from peralkaline granitoids. The formation of dalyite is related to the development of late-stage SiO_2-rich fluids (Linthout 1984).

Linthout (1984) has noted that peralkaline rocks do not typically contain zircon

Table 6.40. Representative Compositions of Wadeite and Dalyite[a]

	1	2	3	4	5	6
SiO_2	48.8	48.85	46.08	47.4	46.36	62.72
TiO_2	2.8	0.13	1.60	0.8	1.03	1.03
ZrO_2	27.9	28.53	31.27	28.5	29.62	19.39
Nb_2O_5	n.a.	n.a.	n.d.	n.a.	n.a.	n.a.
Al_2O_3	0.2	0.02	n.d.	0.8	n.d.	0.08
Fe_2O_3[b]	0.4	0.44	n.d.	0.6	n.d.	0.19
MgO	n.d.	0.02	0.01	n.d.	n.a.	0.02
CaO	0.1	0.49	0.08	0.1	0.47	n.d.
BaO	0.1	0.29	0.08	0.1	n.d.	0.32
SrO	n.d.	n.a.	n.d.	n.d.	0.06	n.a.
Na_2O	0.1	n.d.	n.d.	0.1	1.28	0.14
K_2O	19.7	23.72	21.46	21.5	20.99	16.04
	100.1	100.09	100.50	99.9	99.81	99.93

[a]n.d., not detected; n.a., not analyzed. 1–3, Wadeite, West Kimberley; 1, Carmichael (1967); 2, Mitchell (1985); 3, Jaques *et al.* 1986; 4, Wadeite, Leucite Hills (Carmichael 1967); 5, Wadeite, Kovdor carbonatite (Kapustin 1963); 6, Dalyite, Sierre de Cabras, mean of six analyses (Linthout 1988).
[b]Total Fe calculated as Fe_2O_3.

($ZrSiO_4$) and that the presence of wadeite or dalyite is related to their silica content. Thus relatively SiO_2-poor lamproites contain wadeite, in contrast to the presence of dalyite in granitoids. This difference may be related to the silica activity of the magma via the relation:

$$K_2ZrSi_3O_9 + 3SiO_2 = K_2ZrSi_6O_{15}$$
$$\text{wadeite} \qquad\qquad\qquad \text{dalyite}$$

The absence of zircon in lamproites is believed to be related to the higher degree of polymerization of lamproite melts. Zircon has been recognized only as a very rare phase in the Smoky Butte lamproite (this work) and has not been reported to occur in any other lamproite.

6.14. APATITE

6.14.1. Paragenesis

Apatite is a ubiquitous phase, occurring as phenocrysts and microphenocrysts. The abundance ranges from trace quantities in olivine lamproites to several modal percent in some coarse grained evolved varieties (Jaques *et al.* 1986). The crystals exhibit a very wide range in size but are typically 50–500 μm in length. Jaques *et al.* (1986) note that crystals up to 2 mm in length occur in the Walgidee Hills pegmatites.

The apatite forms euhedral hexagonal prisms, that commonly have a hollow core (Figure 6.64). This habit may signify rapid growth from a supersaturated melt. Many of the apatite crystals are strongly resorbed and exhibit rounded and embayed habits. In some examples preferential enlargement of the hollow cores has occurred (Figure 6.65).

Apatite forms after phenocrystal phlogopite and is poikilitically enclosed by groundmass phlogopite, potassium richterite and sanidine. Crystallization appears to be contemporaneous with priderite, wadeite, and perovskite.

Figure 6.64. Hollow-cored apatite, Zirkel Mesa (Leucite Hills). Scanning electron micrograph.

6.14.2. Composition

Representative compositions are given in Table 6.41 which demonstrates that lamproite apatites are fluor-apatites rich in Sr and Ba. The following ranges in F content have been reported: 2.2–7.1% wt %, Prairie Creek (Scott Smith and Skinner 1984b); 1.9 wt %, Sisimiut (Thy *et al.* 1987); 2.8–3.0 wt %, Murcia-Almeria (Venturelli *et al.* 1988); 2.6–6.4 wt %, Smoky Butte (Edgar, personal communication); 0–5.9 wt %, Leucite Hills (Edgar, personal communication); <0.10–5.4 wt %, West Kimberley (Edgar 1989).

A notable feature of lamproite apatites is their significant Ba contents. Edgar (1989) has noted intra- and interprovincial variations in Ba abundances. Thus in the West Kimberley province apatites from 81 mile Vent, Mount North, and the Walgidee Hills contain 0.27–0.49 wt %, 0.49–0.96 wt %, and 2.1–12.2 wt % BaO, respectively. Many of the Ba-rich apatites occur as sheaflike masses whose habit is reminiscent of quench textures. The apatites exhibit compositional zoning with decreasing Ba and increasing Sr from core to margin. Leucite Hills apatites typically have very low BaO (0.16–0.4 wt %) contents

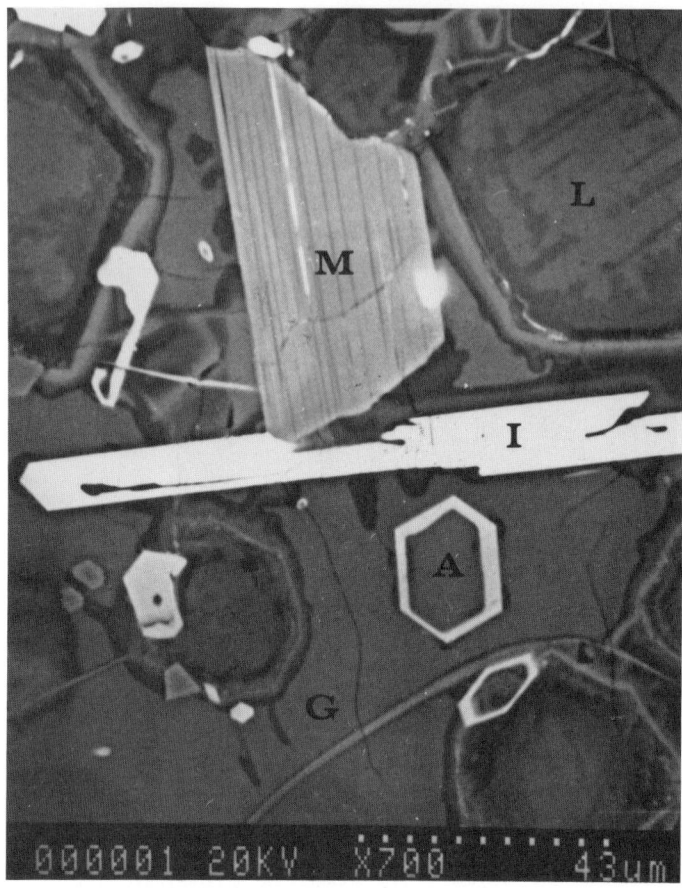

Figure 6.65. Resorbed crystals of hollow-cored apatite (A) and ilmenite (I), Oscar dike (West Kimberley). Also present are leucite (L), mica (M), and glass (G). Backscattered electron image.

(Carmichael 1967). However Ba- and Sr-rich apatites (Table 6.41) that mantle low-Ba varieties are common in the Middle Table Mountain lamproite. Apatites with high Ba contents appear to be unique to lamproites.

Apatite from Smoky Butte is apparently anomalous in having negligible BaO (<0.1 wt %) contents (Edgar, personal communication), despite occurring in host rocks containing 8860 ppm Ba and Ba-rich priderites. Apatites from Prairie Creek contain 0.2–2.8 wt % BaO (Scott Smith and Skinner 1984b).

The Cl content of lamproite apatite is inadequately known. Less than 0.3 wt % Cl has been reported in the Sisimiut and Smoky Butte apatites by Thy *et al.* (1987) and Edgar (personal communication), respectively. Apatites in the West Kimberley lamproites contain negligible Cl, the maximum amount found by Edgar (1989) being 0.04 wt %. Venturelli *et al.* (1988), however, found 0.18–1.09 wt % Cl in those from Murcia-Almeria.

Carmichael (1967) and Kuehner *et al.* (1981) have reported significant levels of the rare

Table 6.41. Representative Compositions of Apatite[a]

	1	2	3	4	5	6	7	8	9	10	11	12	13	14	15
SiO_2	n.a.	n.a.	n.d.	0.38	0.71	0.30	2.20	0.01	0.26	n.d.	0.86	n.d.	0.26	n.d.	0.02
TiO_2	n.a.	n.a.	n.d.	n.d.	n.d.	n.d.	0.05	0.01	0.13	0.14	0.56	0.77	0.10	0.12	0.09
FeO[b]	0.11	0.13	n.d.	0.37	0.12	0.60	0.27	0.08	0.29	0.27	0.19	0.27	0.17	0.26	0.28
MgO	0.40	0.23	n.a.	0.51	0.27	0.24	1.27	0.13	0.63	0.40	n.d.	n.d.	0.06	0.05	0.03
CaO	53.02	52.68	35.46	53.53	52.63	49.58	46.70	51.13	53.69	50.71	51.12	48.31	50.76	52.96	54.68
BaO	0.09	0.19	18.17	n.a.	n.a.	n.a.	2.76	0.22	n.a.	n.a.	6.25	12.54	0.64	0.27	0.57
SrO	n.a.	n.a.	12.35	n.a.	n.a.	2.08	5.15	2.22	n.d.	n.d.	1.14	1.74	4.33	2.94	2.33
Na_2O	n.a.	n.a.	n.a.	n.d.	n.d.	n.d.	0.74	n.d.	n.d.	n.d.	n.d.	n.d.	n.d.	n.d.	n.d.
K_2O	0.07	0.06	n.a.	n.d.	n.d.	n.d.	0.22	0.05	0.08	0.11	0.21	0.07	0.08	0.11	0.24
P_2O_5	39.79	36.65	34.03	40.94	40.72	39.93	35.75	34.49	40.54	40.62	39.21	36.61	38.60	41.19	40.12
La_2O_3	0.12	0.27	n.d.	n.a.	n.a.	n.a.	n.a.	n.a.	n.a.	n.d.	n.d.	n.a.	n.a.	n.a.	n.a.
Ce_2O_3	0.41	0.65	n.a.	n.a.	n.a.	n.a.	n.a.	n.a.	n.a.	n.a.	n.a.	n.a.	n.a.	n.a.	n.a.
Pr_2O_3	n.d.	0.09	n.a.	n.a.	n.a.	n.a.	n.a.	n.a.	n.a.	n.a.	n.a.	n.a.	n.a.	n.a.	n.a.
Nd_2O_3	0.24	0.41	n.a.	n.a.	n.a.	n.a.	n.a.	n.a.	n.a.	n.a.	n.a.	n.a.	n.a.	n.a.	n.a.
F	n.a.	n.a.	n.a.	2.85	3.04	1.95	2.57	7.09	2.94	6.40	n.d.	n.d.	4.44	1.59	2.72
Cl	n.a.	n.a.	n.a.	1.09	0.18	0.08	n.a.	n.a.	0.02	n.d.	n.a.	n.a.	n.a.	n.a.	n.a.
	94.25	91.36	100.01	99.47	97.48	94.76	94.76	95.45	98.58	98.65	99.54	100.31	99.44	99.49	101.08

[a] n.d., not detected; n.a., not analyzed. 1–2, Leucite Hills (Kuehner et al. 1981); 3, Leucite Hills (this work); 4, Murcia-Almeria (Venturelli et al. 1988); 6, Sisimiut (Thy et al. 1987); 7–8, Prairie Creek (Scott Smith and Skinner 1984b); 9–15, Edgar (personal communication); 9–10, Smoky Butte; 11–12, Walgidee Hills; 13, 81 Mile Vent; 14, Mt. North; 15, Machells Pyramid.
[b] Total Fe calculated as FeO.

earth elements in Leucite Hills apatite (Table 6.41). Mason (1977) found the following trace elements (ppm) in apatite from the Walgidee Hills pegmatite: V = 40, Cr = 5, Mn = 30, Co < 1, Ni = 5. Rb < 1, Y = 150, Zr = 10, Nb = 1, La = 350, Ce = 560, Pr = 61, Nd = 210, Sm = 37, Ho = 5.0, Er = 8.5, Yb = 3.9. The rare earth element contents of apatite on the basis of the existing data do not appear to be present in sufficient amounts to control the very high rare earth element contents found in whole rock compositions.

6.15. PEROVSKITE

6.15.1. Paragenesis

Perovskite is found in low-SiO_2 lamproites of madupitic character. In the West Kimberley province, it is abundant as small (<50 μm) euhedral crystals in the Ellendale olivine lamproites and in the olivine-rich lamproites of the Noonkanbah field, e.g., Mount Ibis, Mount Cedric (Jaques et al. 1986). The Walgidee Hills pegmatite contains large (1–2 mm) euhedral, greenish-yellow, strongly zoned crystals exhibiting anisotropism and complex polysynthetic twinning (Wade and Prider 1940).

In the Leucite Hills province, perovskite is known to occur only at Pilot Butte (Carmichael 1967), the Badger's Teeth (Keuhner et al. 1981), and Middle Table Mountain (this work). In the latter occurrence it forms poikilitic groundmass plates (Figure 6.66). Perovskite has also been reported from olivine madupitic lamproite at Prairie Creek (Scott Smith and Skinner 1984b) and Kapamba (Scott Smith et al. 1989), but is apparently absent from the madupitic rocks found at Jumilla (Murcia-Almeria).

6.15.2. Composition

Few compositional data are available. Table 6.42 shows that Leucite Hills perovskite is enriched in Fe relative to other provinces. All of the perovskites contain significant levels of Sr, with particularly high contents being found in groundmass perovskites from Middle Table Mountain (Table 6.42). Scott Smith et al. (1989) note that each intrusion in the Kapamba province contains perovskites of differing Na_2O and SrO content. Significant Nb contents (up to 1 wt % Nb_2O_5) are found in the Walgidee Hills perovskite (Jaques et al. 1986). Lamproite perovskites are also very rich in the light rare earth elements (Figure 6.67) (Carmichael 1967, Kuehner et al. 1981, Mason 1977, Jaques et al. 1986, Mitchell and Reed 1988) and consequently have very high La/Yb ratios (733: Mason 1977). Groundmass perovskites from Middle Table Mountain are highly enriched in rare earth elements (Table 6.42, Figure 6.67). Europium anomalies are not seen in the two rare earth element distribution patterns available (Mason 1977, Mitchell and Reed 1988). Perovskite is clearly a major host for the rare earths in madupitic rocks but cannot play that role in other varieties of lamproite.

Lamproite perovskites differ from those found in kimberlites, alnöites, and perovskite pyroxenites (Figure 6.67) in containing higher levels of Sr and Ce. Iron contents as high as those found in the Leucite Hills perovskites (Table 6.42) have never been reported from kimberlitic varieties. The rare earth contents and distribution patterns are, however, similar (Mitchell and Reed 1988).

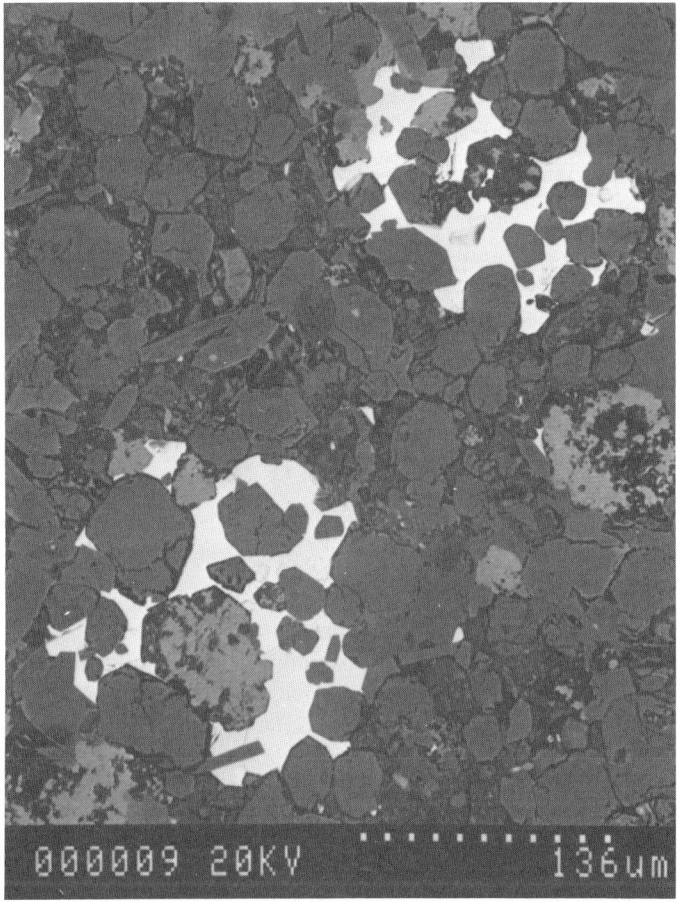

Figure 6.66. Poikilitic groundmass perovskite (light grey) enclosing leucite and pyroxene (dark grey), Middle Table Mountain (Leucite Hills). Backscattered electron image.

6.16. TITANOSILICATES

6.16.1 Shcherbakovite

A potassium barium titanosilicate from the Walgidee Hills was originally named noonkanbahite by Prider (1965). This name was subsequently discredited by the IMA as the mineral is merely a variety of the shcherbakovite (K)–batisite (Ba) series (BaK)(K,Na)Na (Ti,Nb,Zr,Fe)$_2$Si$_4$O$_{14}$ (Schmahl and Tillmanns 1987).

Shcherbakovite occurs in lamproites in two parageneses. In the pegmatites of the Walgidee Hills it is found as large (1 mm) prisms exhibiting a strong colorless to golden yellow pleochroism. It occurs in association with priderite, jeppeite, wadeite, perovskite, phlogopite, and potassium richterite, and is commonly partially replaced by barite (Prider 1965).

In the Leucite Hills shcherbakovite occurs as small (<15 × 2 μm) euhedral blue green

Table 6.42. Representative Compositions of Perovskite[a]

	1	2	3	4	5	6	7	8	9
TiO_2	55.50	48.2	51.88	52.75	54.60	57.10	56.97	58.05	55.52
FeO[b]	0.98	11.4	0.62	2.71	1.81	0.42	0.74	0.18	0.19
MgO	n.a.	n.d.	n.a.	0.06	0.08	n.d.	0.02	0.13	n.d.
CaO	31.37	25.1	16.07	38.38	37.24	32.67	38.28	37.63	37.83
SrO	2.53	3.8	6.12	0.22	0.13	3.88	0.97	1.42	2.81
Na_2O	n.a.	1.0	n.a.	0.25	0.45	2.01	0.77	0.62	n.d.
K_2O	n.a.	0.3	n.a.	0.03	0.03	0.05	0.09	0.11	n.d.
La_2O_3	2.91	2.0	6.33	n.a.	n.a.	n.a.	n.a.	0.52	1.19
Ce_2O_3	3.41	5.1	12.13	n.a.	n.a.	n.a.	n.a.	0.81	1.68
Pr_2O_3	n.d.	1.5	0.83	n.a.	n.a.	n.a.	n.a.	0.05	0.14
Nd_2O_3	1.21	1.9	3.67	n.a.	n.a.	n.a.	n.a.	0.15	0.39
Eu_2O_3	n.a.	0.1	n.a.	n.a.	n.a.	n.a.	n.a.	—	0.02
Gd_2O_3	n.a.	n.d.	n.a.	n.a.	n.a.	n.a.	n.a.	0.02	0.02
	97.91	100.40	97.65	94.40	94.34	96.13	97.84	99.69	99.79

[a] n.d., not detected; n.a., not analyzed. 1–2, Pilot Butte, Leucite Hills (this work, Carmichael 1967); 3, Middle Table Mountain, Leucite Hills (also contains 1.05 wt % Nb_2O_5, this work); 4–5, Prairie Creek (Scott-Smith and Skinner 1984b); 6–7, Kapamba (Scott-Smith et al. 1988); 8, Walgidee Hills (Mason 1977); 9, Walgidee Hills (Mitchell and Reed 1988, this work).
[b] Total Fe calculated as FeO.

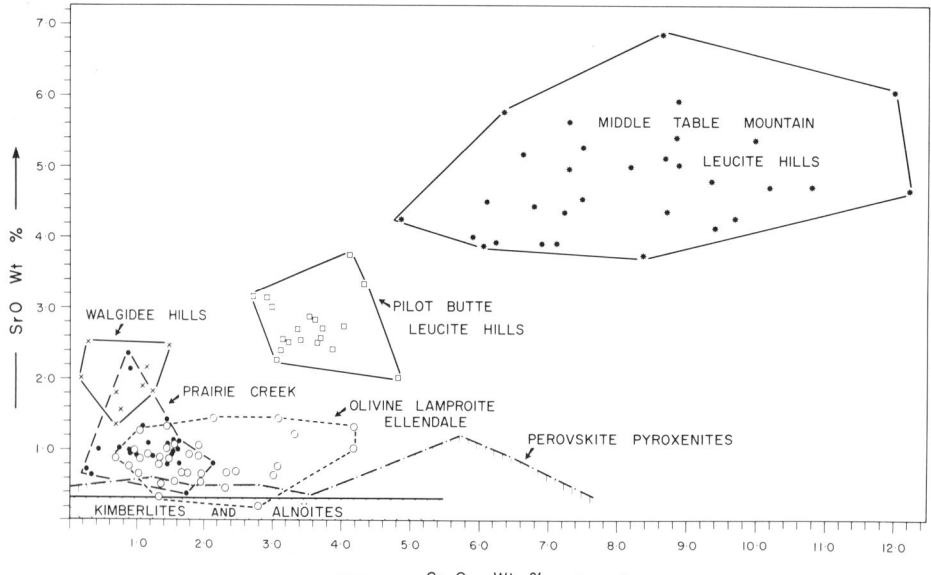

Figure 6.67. Composition variation (SrO versus Ce_2O_3) of perovskites from lamproites, kimberlites, alnöites, and perovskite pyroxenites. All data this work.

MINERALOGY OF LAMPROITES

pleochroic prisms lining the walls of vesicles (Figure 6.68) in association with potassium richterite (this work).

Shcherbakovites from the two localities are of broadly similar composition (Table 6.43), with the Leucite Hills examples being richer in Ba and poorer in K than the Walgidee Hills variety. Minor intergrain compositional variations with respect to the Na, Fe, and Ca contents are found.

6.16.2. Davanite

Davanite $K_2TiSi_6O_{15}$, the Ti analogue of dalyite, has only been recognized from the Smoky Butte lamproite (Wagner and Velde 1986b), where it is associated with Ca–Sr carbonates (Section 6.17.2).

Figure 6.68. Shcherbakovite crystals lining vesicles in lamproite from Emmons Mesa (Leucite Hills). Scanning electron micrograph.

Table 6.43. Representative Compositions of Titanosilicates[a]

	1	2	3	4	5	6	7
SiO_2	43.25	40.72	41.05	39.88	66.70	67.47	31.62
TiO_2	21.15	24.36	24.06	23.94	14.48	15.07	37.59
ZrO_2	0.85	0.03	n.a.	n.a.	0.09	n.d.	n.a.
Al_2O_3	0.07	n.d.	0.27	0.07	n.a.	n.a.	0.84
Fe_2O_3	1.70	—	—	—	—	—	—
FeO	0.61	1.79[b]	1.48[b]	1.31[b]	0.15[b]	n.d.	1.68[b]
MgO	n.d.	0.56	0.51	0.30	n.a.	n.d.	0.64
CaO	1.77	3.01	3.04	0.97	n.a.	n.a.	0.43
BaO	14.47	14.14	13.57	16.84	n.a.	n.a.	16.60
Na_2O	4.50	3.77	4.32	5.99	0.02	n.d.	0.11
K_2O	10.83	11.19	10.99	8.21	17.67	17.40	5.95
Nb_2O_5	n.a.	0.30	n.a.	0.51	n.a.	n.a.	n.a.
H_2O	0.94	n.a.	n.a.	n.a.	n.a.	n.a.	n.a.
	100.14	99.87	99.29	98.02	99.11	99.59	95.44

[a] n.d., not detected; n.a., not analyzed. 1–4, Shcherbakovite; 1–3, Walgidee Hills, Prider (1965), Jaques et al (1986), Mitchell (1985), respectively; 4, Emmons Mesa (this work); 5, Davanite, Smoky Butte (Wagner and Velde 1986); 6, Davan Springs, Murun (Lazebnik et al. 1984); 7, unnamed titanosilicate, Rice Hill, W. Kimberley (this work).
[b] Total Fe calculated as FeO.

The davanite forms small (130 × 40 μm) euhedral crystals of hexagonal habit. Wagner and Velde (1986b) consider that the davanite is of magmatic origin and propose, on the basis of the composition of the associated carbonates, a crystallization temperature of about 800°C. However, the paragenesis is analogous to that of shcherbakovite and dalyite, and a late-stage hydrothermal or vapor phase crystallization origin from SiO_2-rich fluids is also probable.

The composition of the Smoky Butte davanite (Table 6.43) is identical to that of davanite from the type locality in the contact rocks of the Murun complex (Lazebnik et al. 1984). The paragenesis of the latter is, however, different as the davanite is associated with aegirine, pectolite, and sphene in a fenitized granite.

6.16.3. Unnamed K-Ba-Titanosilicate

An unnamed titanosilicate is common as a late-stage groundmass phase in some of the more evolved rocks of the West Kimberley (Rice Hill) and Leucite Hills (Middle Table Mountain) provinces (this work). The mineral occurs as small brownish to opaque laths. These commonly are found as mantles upon priderite in the Rice Hill example. A representative composition is given in Table 6.43. The mineral differs from shcherbakovite in being poorer in Si, richer in Ti, and lacking Na. Low totals are believed to reflect the presence of 4–5 wt % H_2O.

6.17. MINOR ACCESSORY AND SECONDARY MINERALS

A wide variety of minor accessory and secondary minerals have been reported. Few of these have been the subject of detailed investigations with regard to either their paragenesis or composition.

6.17.1. Zeolites

Zeolites are common, filling vugs or as veins traversing the groundmass. Ogden (1979) has described a Ba-bearing zeolite $(K,Na,Ca_{0.5}Ba_{0.5})_{12}(Fe,Al)_{15}Si_{32}O_{94}\cdot nH_2O$ (Table 6.44) occurring as amygdaloidal fillings and within shale xenoliths in the madupitic rocks of Pilot Butte (Leucite Hills). The zeolite has an X-ray diffraction pattern identical to the synthetic zeolite K-M (Barrar and Baynham 1956). Zeolites richer in BaO (Table 6.44) occur as veins in the groundmass of the Pilot Butte lamproites (this work).

Zeolites filling amygdales (Figure 6.69) at the Badger's Teeth (Leucite Hills) are also Ba rich (Table 6.44; this work). Scott Smith and Skinner (1984b) report a barian zeolite (Table 6.44) occurring in poollike segregations in the groundmass of the Prairie Creek madupitic rocks. Jaques et al. (1986) refer to the presence of harmotome, $BaAl_2Si_6O_{16}\cdot 6H_2O$, as a secondary phase in the West Kimberley lamproites.

6.17.2. Carbonate Minerals

Carbonates are not characteristic accessory phases of lamproites but do occur in the groundmass and as vug fillings. Calcite and dolomite have been noted in the groundmass of the Prairie Creek rocks (Scott Smith and Skinner 1984b). Vug fillings in the Leucite Hills lamproites include calcite and less commonly witherite (Black Butte, Badger's Teeth, this work). Wagner and Velde (1986b) have noted the coexistence of orthorhombic Ca-bearing strontianite (SrO = 42.1–56.2 wt %, BaO = 1.7–3.0 wt %, CaO = 9.2–11.9 wt %) and rhombohedral Sr-bearing calcite (SrO = 7.1–23.5 wt %, BaO = 0–0.7 wt %, CaO = 37.1–52.1 wt %) and davanite lining vugs in the Smoky Butte lamproite. Wagner and Velde (1986b) using experimentally determined phase relationships in the system $SrCO_3$–$CaCO_3$ (Chang 1965) suggest that the assemblage crystallized at about 800°C.

6.17.3. Roedderitelike Phases

Wagner and Velde (1986a) have noted that the Francis and Cancarix lamproites contain small prismatic crystals of a blue pleochroic mineral. This phase has a composition (SiO_2 =

Table 6.44. Representative Compositions of Ba-Bearing Zeolites[a]

	1	2	3	4	5	6
SiO_2	54.2	48.74	47.68	48.22	50.40	47.20
Al_2O_3	18.9	23.17	21.12	22.92	22.54	19.43
Fe_2O_3[b]	0.8	0.53	1.03	0.83	0.68	0.79
MgO	n.a.	0.05	0.04	0.04	0.04	1.69
CaO	0.3	0.93	0.41	3.42	3.05	1.16
Na_2O	1.4	0.57	0.21	0.85	0.84	0.27
K_2O	13.6	12.88	13.77	6.92	7.03	1.01
BaO	1.0	5.72	8.82	5.53	4.85	20.52
	90.2	92.59	93.08	88.76	89.43	92.07

[a] n.a., not analyzed. 1, vesicle filling, Pilot Butte (Ogden 1979); 2–3, groundmass, Pilot Butte (this work); 4–5, vesicle filling, Badger's Teeth (this work); 6, groundmass, Prairie Creek (Scott Smith and Skinner 1984b).
[b] Total Fe expressed as Fe_2O_3.

Figure 6.69. Ba-rich zeolite in vesicles in the Badger's Teeth dike (Leucite Hills). Scanning electron micrograph.

69.6–70.3 wt %, FeO_T = 9.8–12.6 wt %, MgO = 11.4–13.9 wt %, K_2O = 4.5–5.1 wt %), and is similar to minerals of the roedderite–eifelite series (Abraham *et al.* 1983). The lamproite variety is different in being notably poor in Na_2O (<0.3 wt %) and rich in Fe. Wagner and Velde (1986) suggest that the minerals represent solid solutions between roedderite $(Na,K)_2Mg_5Si_{12}O_{30}$ and the endmember molecule $KFe^{3+}Fe^{2+}Mg_3Si_{12}O_{30}$. The blue pleochroism is considered to be related to the presence of Fe^{3+} as the intensity of absorption increases with increasing Fe content. These minerals may be formed by the alteration of richterite as Gilbert *et al.* (1982) have shown that roedderite is one of its low pressure breakdown products.

6.17.4. Cerium-Bearing Minerals

Lucasite-(Ce), $CeTi_2(O,OH)_6$, has been found in heavy mineral concentrates derived from the Argyle olivine lamproite tuff. It occurs as brown subhedral grains associated with calcite, talc, sphene, dolomite, amphibole, Mn-ilmenite, and bastnasite-(Ce) and was recognized by Nickel *et al.* (1987).

MINERALOGY OF LAMPROITES

The groundmass of the Smoky Butte lamproites contains small (<10 μm) euhedral to subhedral prisms of rhabdophane-(Ce), $(Ce,Ca)_5(PO_4)_3(OH,F)$ [3.70 wt % CaO, 28.58 wt % P_2O_5, 13.0 wt % La_2O_3, 29.6 wt % Ce_2O_3, 8.6 wt % Nd_2O_3] and rare anhedral crystals of CeO_2 (this work).

6.17.5. Silicon Dioxide

Silicon dioxide is a common constituent of vug fillings in association with carbonate and zeolite. The bulk of the SiO_2 is probably alpha quartz, although X-ray investigations have not been undertaken to verify this assumption.

At Machell's Pyramid (West Kimberley), bluish euhedral crystals lining vesicles have a morphology (Figure 6.70) that suggests the presence of metastable beta-cristobalite (this work).

Figure 6.70. Crystals of beta-cristobalite (?) lining a vesicle in diopside leucite lamproite from Machells Pyramid (West Kimberley). Scanning electron micrograph.

6.17.6. Barite

Bluish-green barite is very common as late stage (?) hydrothermal veins and vesicle fillings. Carmichael (1967) considers that the barite is very rich in Sr, although compositions were not determined. Barite commonly replaces preexisting barium-bearing minerals such as priderite and shcherbakovite.

6.17.7. Other Minerals

Barton (1988) has reported a Na–Fe-titanate (TiO_2 = 73.1 wt %, FeO_T = 16.1 wt %, MgO = 1.9 wt %, Na_2O = 8.45 wt %) associated with Mg-ilmenite in pyroxenite xenoliths from Hatcher Mesa (Leucite Hills). The mineral is considered by Barton (1979) to be a sodian priderite.

Sylvite, KCl, occurs as very rare euhedral crystals in the Smoky Butte and Francis Canyon lamproites (this work).

Rounded small (<5 μm) crystals of an unidentified Ca–Zr silicate occur in the groundmass of the Smoky Butte lamproites.

Potash nitre (K_2O = 44.91 wt %, N_2O_5 = 51.49 wt %, CaO = 1.09 wt %, H_2O = 0.62 wt %) was reported by Cross (1897) from the Boar's Tusk (Leucite Hills). Corundum, moissanite (SiC), native metals (Cu, Zn, Fe, Co, Ni), cohenite (Fe_3C), and khamrabayevite (TiC) have been reported by Novgorodova *et al.* (1987) and Eremeev *et al.* (1989) in "lamproitelike" (*sic*) rocks from localities in central Asia. We suspect an anthropogenic source for this unusual suite of "minerals" namely, grinding wheels used to polish drill steel. Novgorodova *et al.* (1987) suggest that the minerals have been formed by reactions involving fluids and the parent rock.

6.17.8. Secondary Phases

Secondary phases replacing earlier formed minerals include:

- Serpentine, talc, nickel and copper sulphides, and nontronite after olivines and pyroxenes.
- Celadonite, talc, clay minerals, and illite after micas.
- Nepheline (Scott Smith *et al.* 1989), sanidine, montmorillonite, and leucite after feldspars.

One should always be a little improbable.
Oscar Wilde

The Geochemistry of Lamproites

7.1. MAJOR ELEMENT GEOCHEMISTRY

7.1.1. General Characteristics

Relative to other mafic alkaline rocks lamproites are distinguished by their high K_2O, TiO_2, P_2O_5, and MgO contents, K_2O/Na_2O ratios and low Al_2O_3 and CaO contents. The general characteristics of the lamproite clan have been outlined in Section 2.6. In this section the compositional averages and ranges of lamproites are compared with those of other potassic rocks to demonstrate the major differences in their compositions. Representative compositions of petrographically diverse lamproites from particular provinces can be found in Section 7.1.3.

The average compositions of lamproite suites from several provinces are given in Tables 7.1 and 7.2. To facilitate interprovincial comparisons the original data have been recalculated to 100% on a CO_2-free basis with total iron expressed as FeO. The CO_2 content (typically less than 0.5 wt % in a fresh rock) is excluded, as calcite, the major contributor to the analysis, is not a typical primary phase of lamproites and its presence is indicative of alteration. Compositions not included in this presentation are those that have been judged on the basis of petrographic or compositional evidence to be silicified, carbonatized, or contaminated by xenolithic material (i.e., tuffs, breccias). Also excluded are rocks rich in Na_2O and poor in K_2O such as those found at Vera (Murcia-Almeria) and Smoky Butte. These rocks are hyalo-lamproites that have been devitrified or analcitized by secondary weathering processes.

Separate averages are given for olivine and madupitic lamproites and phlogopite–leucite–diopside–sanidine–richterite lamproites (referred to in this section simply as phlogopite lamproites) as a simple average for geochemically and mineralogically bimodal provinces, such as the Leucite Hills or West Kimberley, is petrologically meaningless.

Although CIPW norms are listed in Tables 7.1 and 7.2 the reader must bear in mind the limitations of this norm, given the unusual composition and mineralogy of these rocks. Lewis (1987) has attempted to eliminate this problem by devising a lamproite norm. This scheme calculates normative minerals similar in composition to the modal mineralogy. However, an inability to calculate normative phlogopite or amphibole limits its usefulness.

Table 7.1 demonstrates that phlogopite lamproites may be regarded as intermediate rocks on the basis of their SiO_2 contents. This observation is important as it demonstrates

Table 7.1. Average Compositions (wt %) of Phlogopite Lamproites[a]

	1	2	3	4	5
SiO_2	53.64 ± 3.40	54.51 ± 1.71	57.36 ± 3.24	57.29 ± 0.50	52.08 ± 1.74
TiO_2	6.29 ± 0.88	2.48 ± 0.24	1.59 ± 0.35	2.34 ± 0.27	6.10 ± 0.49
Al_2O_3	8.13 ± 1.09	10.22 ± 0.64	10.49 ± 1.33	10.08 ± 0.41	10.54 ± 1.35
FeO	6.78 ± 0.87	4.21 ± 0.45	4.89 ± 0.98	4.93 ± 0.30	4.94 ± 0.41
MgO	7.82 ± 2.52	7.61 ± 1.55	9.85 ± 3.40	7.10 ± 0.33	5.30 ± 2.17
CaO	3.23 ± 1.15	4.33 ± 1.07	3.43 ± 0.93	4.63 ± 0.63	5.83 ± 2.20
Na_2O	0.49 ± 0.27	1.34 ± 0.28	1.46 ± 0.46	1.08 ± 0.18	0.62 ± 0.29
K_2O	9.60 ± 1.51	11.59 ± 0.81	8.02 ± 0.99	10.14 ± 0.71	10.28 ± 1.11
P_2O_5	1.23 ± 0.59	1.64 ± 0.54	0.90 ± 0.25	1.23 ± 0.11	2.49 ± 0.30
H_2O+	2.64 ± 0.99	1.96 ± 0.89	1.94 ± 0.13	1.13 ± 0.21	1.73 ± 1.03
(n)	(228)	(42)	(55)	(6)	(9)
Alk	19.6	8.6	5.5	9.4	16.6
P.I.	1.4	1.4	1.5	1.3	1.2
P.C.	1.3	1.2	1.3	1.1	1.1
CIPW norm					
Ap	2.7	3.6	2.0	2.7	5.5
Il	12.0	4.7	3.0	4.5	10.4
Or	44.4	55.9	47.5	55.1	57.4
Ab			9.3		
Di	6.3	0.6	8.9	12.0	8.2
Hy	18.6	18.6	26.4	17.2	9.4
Ol			0.2		
Q	8.8	0.4		3.9	3.7
Ns	1.0	2.7	0.7	2.1	1.2
Ks	3.4	3.5		1.4	0.9

[a](n), number of samples; Alk, K_2O/Na_2O (wt %); P.I., peralkalinity index; P.C., K_2O/Al_2O_3 (molar). N.B. All compositions have been recalculated on a CO_2-free basis with total Fe expressed as FeO. 1, West Kimberley (Jaques et al. 1986); 2, Leucite Hills (Cross 1897, Carmichael 1967, Ogden 1979, Kuehner (1980); 3, Murcia-Almeria (Fuster et al. 1967, Venturelli et al. 1984a); 4, Francis (this work); 5, Smoky Butte (Mitchell et al. 1987, this work).

that the majority of phlogopite lamproites found as lavas or hypabyssal rocks are not silica-undersaturated, despite containing abundant modal leucite. The relatively high silica contents result in some cases in normative quartz, and typically in normative hypersthene (Bergman 1987, Carmichael 1967). Individual provinces are characterized by lamproites exhibiting a relatively restricted range in SiO_2 content. Lamproites from Murcia-Almeria and Francis are significantly richer in silica than other provinces. This enrichment may be a reflection of crustal contamination or be related to the volatile composition of the source (see Sections 7.8, 7.9.1, 8.3).

Lamproites have high K_2O/Na_2O (wt) ratios (>5) and are simultaneously peralkaline and perpotassic. These features set them apart from all other varieties of potassic rock. Low Al_2O_3 coupled with high K_2O contents leads to the appearance of alkali metasilicate in the CIPW norm.

Figure 7.1 is a plot of discriminant function scores (Le Maitre 1982) which illustrates the intra- and interprovincial compositional variation. The plot is based upon two discriminant functions which account for 95% of the variance in the compositional data. The figure shows that individual provinces exhibit relatively restricted ranges in composition and that *each province is characterized by phlogopite lamproites of a particular composition.* The

Table 7.2. Average Composition (wt %) of Olivine and Madupitic Lamproites[a]

	1	2	3
SiO_2	42.31 ± 2.21	45.47 ± 1.16	48.86 ± 2.04
TiO_2	3.75 ± 0.82	2.34 ± 0.32	1.46 ± 0.13
Al_2O_3	3.92 ± 0.87	8.89 ± 0.67	8.16 ± 0.89
FeO	8.27 ± 0.54	5.99 ± 0.27	6.19 ± 0.50
MgO	24.42 ± 3.56	11.15 ± 0.94	16.45 ± 1.34
CaO	5.00 ± 0.95	11.84 ± 1.79	7.01 ± 1.65
Na_2O	0.50 ± 0.25	0.83 ± 0.15	1.62 ± 0.30
K_2O	4.01 ± 1.09	7.75 ± 1.49	5.28 ± 1.15
P_2O_5	1.59 ± 0.48	2.08 ± 0.66	1.68 ± 0.33
H_2O+	6.07 ± 1.88	3.49 ± 1.19	3.16 ± 1.17
(n)	(105)	(6)	(13)
Alk	8.0	9.3	3.3
P.I.	1.3	1.1	1.0
P.C.	1.1	0.9	0.7
CPIW norm			
Ap	3.5	4.6	3.7
Il	7.1	4.5	2.8
Or	21.4	0.2	31.2
Ab			7.9
Di	11.4	36.1	19.0
Hy	9.6		
Ol	39.1	12.8	29.3
Ne		1.4	2.5
Lc		35.8	
Ns	0.9	1.0	0.3
Ks	0.6		

[a] (n), number of samples; Alk, K_2O/Na_2O (wt %); P.I., peralkalinity index; P.C., K_2O/Al_2O_3 (molar). N.B. All compositions have been calculated on a CO_2-free basis with total Fe expressed as FeO. 1, olivine lamproite, West Kimberley (Jaques et al. 1986); 2, madupitic lamproite, Leucite Hills (Carmichael et al. 1967); 3, olivine madupitic lamproite, Murcia-Almeria (Carmichael 1967, Fuster et al. 1967, Venturelli et al. 1984a).

West Kimberley and Smoky Butte phlogopite lamproites are separated from the Leucite Hills, Francis, and Murcia-Almeria phlogopite lamproites primarily on the basis of the former's higher TiO_2 contents. The West Kimberley province rocks show the greatest modal variation and consequently the greatest compositional variation.

Table 7.2 shows that olivine and madupitic lamproites are ultrabasic to basic in composition. They have high K_2O/Na_2O (wt) ratios (3–9), are peralkaline but not typically perpotassic. Only the West Kimberley olivine lamproites are perpotassic. In contrast to phlogopite lamproites (Table 7.1) these rocks exhibit much greater interprovincial compositional variation (Table 7.2). These differences are primarily a reflection of their modes. Hence West Kimberley rocks are magnesia-rich owing to their high modal olivine contents. As many of these olivines are xenocrysts (see Section 6.5.1) it is important to note that the compositions of all of this group of lamproites are unlikely to be representative of those of their parental magmas.

Figure 7.2 is a plot of discriminant function scores for olivine and madupitic lamproites

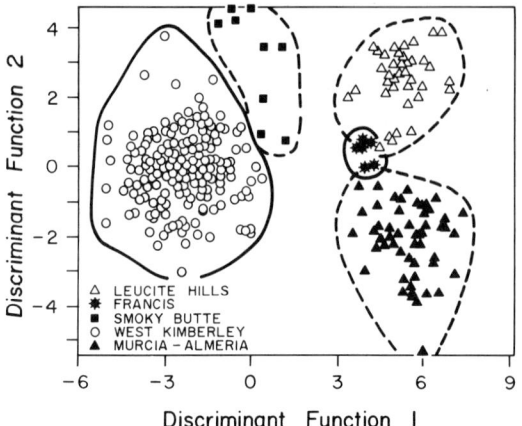

Figure 7.1. Plot of discriminant function scores for phlogopite lamproites. Function 1 = $-0.15\,SiO_2 - 0.84\,TiO_2 + 0.44\,Al_2O_3 - 0.18\,FeO + 0.42\,MgO - 0.14\,CaO + 0.45\,Na_2O + 0.18\,K_2O + 0.48\,P_2O_5 - 0.13\,H_2O$ (83% total variance); Function 2 = $-1.21\,SiO_2 - 0.16\,TiO_2 - 0.60\,Al_2O_3 - 0.74\,FeO - 1.13\,MgO + 0.18\,CaO - 0.06\,Na_2O + 0.57\,K_2O + 0.23\,P_2O_5 = 0.07\,H_2O$ (12% total variance). Data sources given in Table 7.1.

based upon two discriminant functions accounting for 100% of the variance in the compositional data. The figure demonstrates that each lamproite suite is compositionally distinct. Olivine lamproites from West Kimberley are TiO_2-rich relative to other provinces (Table 7.2).

7.1.2. Compositional Relationships to Other Potassic Lavas

The average compositions of three representative high potassium volcanic suites from the Roman province (RPT-lavas) are given in Table 7.3. None of these suites is peralkaline or

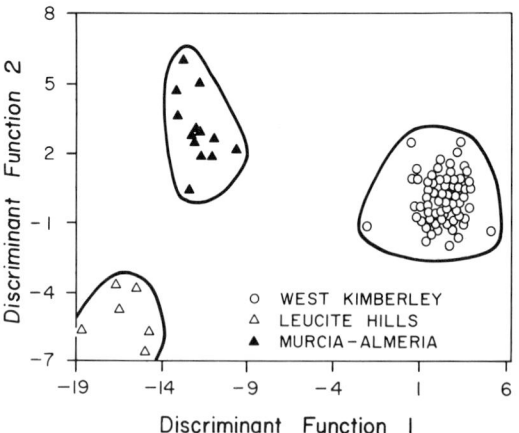

Figure 7.2. Plot of discriminant function scores for olivine and madupitic lamproites. Function 1 = $0.45\,SiO_2 + 1.16\,TiO_2 - 1.19\,Al_2O_3 + 1.03\,FeO + 0.47\,MgO - 0.50\,CaO - 0.02\,Na_2O - 0.21\,K_2O - 0.16\,P_2O_5 - 0.07\,H_2O$ (94% total variance); Function 2 = $-0.51\,SiO_2 - 0.59\,TiO_2 - 0.24\,Al_2O_3 - 0.03\,FeO - 0.83\,MgO - 0.93\,CaO + Na_2O - 1.00\,K_2O + 0.04\,P_2O_5 - 0.37\,H_2O$ (6% total variance). Data sources given in Table 7.2.

THE GEOCHEMISTRY OF LAMPROITES

Table 7.3. Average Compositions (wt %) of Roman Province and Kamafugitic Lavas[a]

	1	2	3	4	5
SiO_2	52.01 ± 2.86	51.75 ± 3.58	53.79 ± 1.98	41.98 ± 1.59	38.53 ± 2.71
TiO_2	0.66 ± 0.15	0.72 ± 0.12	0.68 ± 0.12	0.99 ± 1.59	4.62 ± 0.96
Al_2O_3	19.73 ± 1.38	17.86 ± 2.15	18.11 ± 1.43	10.70 ± 1.81	7.92 ± 1.07
FeO	5.86 ± 1.61	6.14 ± 1.34	5.72 ± 1.04	6.94 ± 1.18	11.17 ± 0.99
MgO	2.34 ± 1.38	3.58 ± 2.23	3.19 ± 1.31	11.66 ± 1.84	13.91 ± 3.55
CaO	6.45 ± 2.26	8.42 ± 3.58	6.08 ± 1.73	15.77 ± 1.16	14.42 ± 2.92
Na_2O	2.76 ± 0.71	2.11 ± 0.77	2.13 ± 0.48	0.97 ± 0.42	1.36 ± 0.42
K_2O	8.75 ± 1.42	7.69 ± 1.35	8.31 ± 1.23	8.48 ± 0.71	4.73 ± 1.41
P_2O_5	0.33 ± 0.14	0.35 ± 0.11	0.48 ± 0.12	0.76 ± 0.46	0.85 ± 0.29
H_2O	0.96 ± 0.41	1.24 ± 0.51	1.34 ± 0.31	1.63 ± 1.05	2.21 ± 0.75
(n)	(88)	(94)	(34)	(10)	(22)
Alk	3.2	3.6	3.9	8.7	3.5
P.I.	0.7	0.7	0.7	1.0	0.9
P.C.	0.5	0.5	0.5	0.9	0.7

[a](n), number of samples; Alk, K_2O/Na_2O (wt %); P.I., peralkalinity index; P.C., K_2O/Al_2O_3 (molar). N.B. All compositions have been recalculated on a CO_2-free basis with total Fe expressed as FeO. 1–3, Roman Province lavas; 1, Roccamonfina (Appleton 1970); 2, Vulsini (Holm *et al.* 1982); 3, Vico (Cundari and Mattias 1974); 4–5, Kamafugitic lavas; 4, San Venanzo and Cupaello (Peccerillo *et al.* 1988), Western Rift Uganda (El-Hinnawi 1965).

perpotassic. Although of similar SiO_2 content they differ markedly from lamproites in being richer in Al_2O_3, CaO and Na_2O and poorer in TiO_2, K_2O, MgO, P_2O_5, and H_2O. Alumina contents are higher by a factor of 2, reflecting the common presence of plagioclase in these rocks.

Figure 7.3 plots discriminant function scores of phlogopite lamproites and RPT-lavas derived from two discriminant functions accounting for 97% of the variance in the data base. The figure shows clearly their differing character with respect to the major elements and

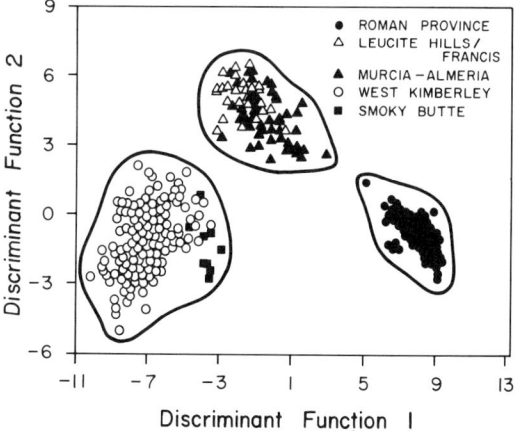

Figure 7.3. Plot of discriminant function scores for phlogopite lamproites, and Roman province type lavas. Function 1 = $0.89SiO_2 - 0.30TiO_2 + 1.68Al_2O_3 + 0.22FeO + 0.94MgO + 1.39CaO + 0.21Na_2O + 0.08K_2O + 0.18P_2O_5 - 0.01H_2O$ (88% total variance); Function 2 = $-0.74SiO_2 - 1.04TiO_2 - 1.21Al_2O_3 - 0.28FeO - 0.24MgO - 0.83CaO + 0.09N_2O + 0.34K_2O + 0.32P_2O_5 - 0.13H_2O$ (9% total variance). Data sources given in Tables 7.1 and 7.3.

demonstrates that there is no continuum of compositions between the two groups. These observations amplify our contention, based upon the petrographic character of each suite (see Section 2.5.1), that these rocks originate from genetically different magma types and, therefore, probably distinct sources.

Using the data given in Tables 7.1 and 7.3 we suggest that fresh rocks, which are neither peralkaline nor perpotassic, containing >2 wt % Na_2O, >12 wt % Al_2O_3, <1 wt % TiO_2, and <1 wt % P_2O_5 are unlikely to be members of the lamproite clan. Olivine and madupitic lamproites cannot be confused with RPT-lavas on the basis of the much lower SiO_2 and higher MgO contents of these lamproites.

The average composition of kamafugitic lavas from Uganda and Italy are given in Table 7.3. They may be easily distinguished from phlogopite lamproites (and RPT-lavas) on the basis of their much lower SiO_2 and higher CaO contents. Kamafugitic lavas are characteristically silica-undersaturated and not perpotassic. The undersaturation is reflected by the presence of melilite and kalsilite in the mode.

Olivine and madupitic lamproites have similar SiO_2 contents to kamafugites and compositional overlap between the two groups is evident in Niggli k versus mg or SiO_2 versus peralkalinity plots (see Figures 2.1 and 2.7). Figure 7.4 is a plot of discriminant function scores for olivine/madupitic lamproites and kamafugites derived from two discriminant functions accounting for 95% of the variance in the data base. This figure shows clearly that kamafugites from Uganda and Italy, and SiO_2-poor lamproites are all compositionally distinct. The Italian lavas bear some compositional similarities to madupitic lamproites from Pilot Butte (Leucite Hills), but differ in being richer in Al_2O_3 and CaO and poorer in TiO_2 and P_2O_5. However, it should be borne in mind that comparison of hypabyssal SiO_2-poor lamproites, which have undergone crystal accumulation, with lavas is undesirable. Thus, we do not agree with the conclusion of Peccerillo *et al.* (1988) that the lavas of San Venanzo and Cupaello should be called madupites. In conclusion, we suggest that lamproites and kamafugites are derived from genetically different magma types.

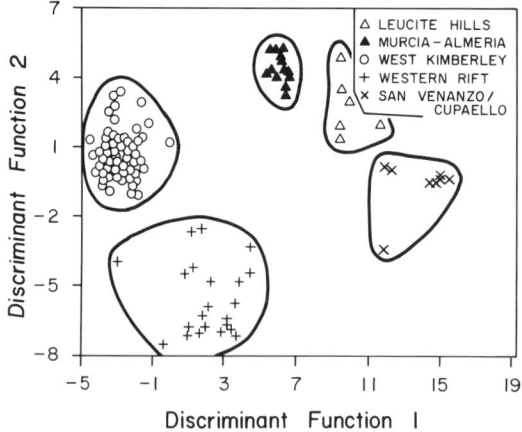

Figure 7.4. Plot of discriminant function scores for olivine and madupitic lamproites and kamafugitic lavas. Function 1 = $0.93SiO_2 - 0.64TiO_2 + 0.91Al_2O_3 - 0.24FeO - 0.98MgO - 1.48CaO + 0.18Na_2O + 0.95K_2O + 0.26P_2O_5 + 0.97H_2O$ (76% total variance); Function 2 = $0.21SiO_2 - 0.45TiO_2 - 0.09Al_2O_3 - 0.61FeO - 0.38MgO - 0.40CaO + 0.07Na_2O - 0.22K_2O + 0.48P_2O_5 + 0.9H_2O$ (20% total variance). Data sources given in Tables 7.2 and 7.3.

Although group 2 (micaceous) kimberlites bear a superficial resemblance to some phlogopite lamproites, they are compositionally distinct. Table 7.4 shows that in terms of their average composition they are poorer in SiO_2, Al_2O_3, Na_2O, and K_2O relative to most lamproites. Characteristic features are the presence of significant amounts of CO_2, reflecting the occurrence of primary calcite, and the extraordinary low Na_2O contents. The rocks have very high K_2O/Na_2O (wt) ratios and are peralkaline and perpotassic. This apparent affinity with lamproites is, however, misleading in that the major element geochemistry of group 2 kimberlites is controlled by the high modal content of microphenocrystal phlogopite. Mitchell and Meyer (1989a) have shown that these micas are a transported assemblage and that their concentration results from differentiation processes. Consequently whole rock compositions do not even remotely resemble those of their parent magmas. Mitchell and Meyer (1989a) thus caution that direct comparison of their composition with that of lamproites (cf. Dawson 1987) may be inappropriate.

Minettes are the only potassic rocks similar in composition (Table 7.4) to phlogopite lamproites of intermediate silica content. The principal differences in average compositions are: minettes have higher Na_2O and Al_2O_3 coupled with lower K_2O contents; the majority of minettes are neither peralkaline nor perpotassic; however, some K_2O-rich alkali minettes (Table 7.4) cannot be distinguished from phlogopite lamproites on the basis of their major element geochemistry alone. Additional mineralogical data are required if such rocks are to be classified correctly.

7.1.3. Intraprovincial Characteristics

7.1.3.1. West Kimberley

More compositional data (Wade and Prider 1940, Prider 1960, Nixon *et al.*, 1984, Appendix 2 of Jaques *et al.*, 1986, Lewis 1987, Tainton 1987) are available for this province than any other. When using these data it is important to bear in mind the caution of Jaques *et al.* (1986) that the majority of samples analyzed have been affected by minor alteration, such as serpentinization of olivine, replacement of leucite, and groundmass replacement by secondary phases. Other samples have experienced silification, carbonization, oxidation, or leaching of sodium. (N.B. This proviso is also relevant to lamproite compositions in other provinces.) The presence of secondary barite and calcite is common in many of the Ellendale olivine lamproites and the rocks of the Walgidee Hills intrusion. Data used in this work are based upon samples showing minimal alteration and having low CO_2 (<1 wt %) contents. Data for the unusual Walgidee Hills intrusion have not been included in any of the averages.

The main features of the geochemistry have been summarized by Jaques *et al.* (1984, 1986), who note the very high K_2O (3–12 wt %) and low Na_2O (typically <0.5 wt %) contents resulting in these lamproites having very high K_2O/Na_2O (>10, average 30) ratios. Some of the highest K_2O/Na_2O ratios may result from the subaerial leaching of Na_2O (Jaques *et al.* 1986). As molar K_2O exceeds Al_2O_3 the rocks are typically peralkaline and perpotassic. Their TiO_2 contents are notably higher than those of other lamproite provinces (Table 7.1 and 7.2).

The province as a whole exhibits a wide range in composition. Representative examples of this range are given in Tables 7.5 and 7.6. In general SiO_2, Al_2O_3, K_2O, TiO_2, and H_2O increase as MgO and CaO decrease. Total Fe appears to decrease with MgO. Jaques *et al.* (1986, p. 223) suggest "that the West Kimberley lamproite suite forms a chemical continuum from ultrabasic peridotitic compositions represented by the olivine lamproites, to the

Table 7.4. Compositions (wt %) of Micaceous Kimberlites and Minettes

	1	2	3	4	5	6
SiO_2	36.45	49.10	51.1	57.5	50.59	52.6
TiO_2	1.16	2.05	1.2	1.5	1.58	1.9
Al_2O_3	3.71	10.10	10.8	10.3	9.71	12.6
FeO	7.39	7.11	6.2	4.6	6.12	7.6
MgO	25.29	10.77	10.5	7.4	6.07	8.0
CaO	8.01	8.66	6.8	3.7	5.28	7.9
Na_2O	0.18	0.80	1.0	1.2	0.47	2.0
K_2O	3.76	5.35	7.2	8.6	9.22	6.0
P_2O_5	1.24	0.94	—	—	1.73	1.2
H_2O+	4.97	2.83	1.9	1.3	0.82	2.2
CO_2	4.73	0.11	1.7	—	6.91	1.8
Alk	20.9	6.7	7.2	7.2	19.6	3.0
P.I.	1.2	0.7	0.9	1.1	1.1	0.8
P.C.	1.1	0.6	0.7	0.9	1.0	0.5

Alk, K_2O/Na_2O (wt %); P.I., peralkalinity index, P.C., K_2O/Al_2O_3 (molar). Total Fe expressed as FeO. 1, Average of 16 micaceous kimberlites (Smith et al. 1985); 2–6 Minettes: 2, Arizona (Rogers et al. 1982); 3, pyroxene alkali minette, Bohemia (Němec 1975); 4, amphibole alkali minette, Bohemia (Němec 1975); 5, Pendennis peralkaline minette (Hall 1982); 6, average minette (Bergman 1987).

Table 7.5. Representative Compositions (wt %) of Olivine Lamproites, West Kimberley[a]

	1	2	3	4	5	6
SiO_2	34.80	37.85	40.09	41.90	44.10	41.50
TiO_2	3.26	3.97	3.54	2.72	3.20	3.62
Al_2O_3	3.10	3.91	4.62	3.70	3.50	3.64
Fe_2O_3	6.80	5.77	4.61	3.70	4.50	—
FeO	1.96	2.84	4.38	4.82	3.09	8.10
MnO	0.12	0.13	0.15	0.13	0.12	0.13
MgO	24.70	23.20	22.50	25.00	21.30	25.00
CaO	4.55	4.51	5.75	4.91	3.74	4.99
SrO	0.15	0.17	0.16	0.13	0.11	0.06
BaO	0.68	1.42	1.95	0.47	1.05	1.15
Na_2O	0.17	0.28	0.60	0.45	0.93	0.46
K_2O	2.61	4.23	4.58	4.02	6.51	4.12
P_2O_5	1.77	1.68	1.02	1.22	1.13	1.68
H_2O+	8.65	7.20	4.52	3.95	3.09	6.36
H_2O-	4.00	2.02	1.17	1.16	2.54	—
CO_2	0.29	0.15	0.15	0.14	0.28	0.45
	97.61	99.33	99.79	98.42	99.19	

[a] 1–2, Ellendale 4; 3 and 5, Ellendale 9; 4, Ellendale 11; 6, average of 89 olivine lamproites from Ellendale 4 and 9. (All data from Jaques et al. 1986.)

Table 7.6. Representative Compositions (wt %) of Phlogopite Leucite Diopside Lamproites from the Ellendale and Noonkanbah Fields, West Kimberley[a]

	1	2	3	4	5	6	7	8	9	10	11	12	13	14	15
SiO_2	41.33	45.28	45.65	45.80	48.10	50.10	51.75	52.58	53.12	55.42	56.05	57.55	59.80	51.81	53.20
TiO_2	3.95	4.96	6.64	7.02	4.69	7.02	6.08	6.10	5.89	4.62	5.19	5.43	3.99	6.29	5.84
Al_2O_3	5.02	6.53	7.11	5.50	6.70	8.90	7.47	8.23	9.38	9.02	7.92	6.91	6.80	9.72	8.34
Fe_2O_3	4.78	6.98	7.21	5.60	4.40	6.60	5.11	4.67	5.44	3.94	6.21	5.21	5.60	6.96	—
FeO	2.02	1.55	0.86	1.68	2.31	0.71	0.84	1.88	1.38	1.95	0.93	1.00	1.11	0.88	6.60
MnO	0.10	0.10	0.11	0.07	0.10	0.06	0.07	0.09	0.11	0.06	0.07	0.08	0.08	0.06	0.08
MgO	21.28	13.54	10.50	11.70	13.80	6.62	7.34	6.84	5.33	6.45	3.56	5.91	4.76	5.11	7.80
CaO	2.49	4.66	3.57	3.90	4.25	1.88	2.39	3.46	2.34	1.93	3.47	3.64	4.61	0.52	3.22
SrO	0.12	0.14	0.19	0.20	0.12	0.16	0.21	0.15	0.13	0.16	0.06	0.12	0.10	0.57	—
BaO	0.66	1.30	1.97	1.45	1.55	0.71	1.45	1.09	0.91	1.17	0.19	1.03	0.44	1.61	—
Na_2O	0.15	0.30	0.49	0.73	0.84	0.22	0.52	0.68	0.68	0.68	0.15	0.58	0.44	0.09	0.57
K_2O	3.16	8.13	9.64	7.58	6.72	11.45	10.12	10.45	11.52	11.15	7.88	8.67	7.16	11.15	9.89
P_2O_5	0.23	1.07	1.84	1.41	0.69	1.45	1.18	0.49	0.71	0.71	2.81	0.78	1.09	0.43	0.98
H_2O+	6.48	3.30	2.20	2.14	3.41	2.33	2.25	1.70	1.79	1.41	2.18	1.29	2.01	2.52	2.98
H_2O-	5.83	0.45	0.60	3.27	1.45	1.10	0.66	1.03	0.53	0.24	1.70	0.54	1.56	1.70	—
CO_2	0.84	0.28	0.25	0.14	0.14	0.09	0.32	0.21	0.50	0.28	0.28	0.45	0.16	0.28	0.50
	98.44	98.57	98.83	98.19	99.27	99.31	97.76	99.65	99.76	99.19	98.65	99.19	99.71	99.70	100.00

[a] 1, Di-Ph-TKr-Lc-Ol lamproite, Walgidee Hills; 2, Ol-Di-Lc lamproite, Walgidee Hills; 3, Ph-Lc lamproite, Mount Percy; 4, Ol-Di-TKr-Ph-Lc transitional maduphitic lamproite, Mount North; 5, Ol-Di-Lc lamproite, Mount Cedric; 6, Di-Ol-Lc lamproite, Fishery Hill; 7, Ph-Lc lamproite, 81 Mile Vent; 8, Di-Lc lamproite, P Hill; 9, Di-TKr-Lc lamproite, Mamilu Hill; 10, Ph-Lc lamproite, Djada Hill; 11, Ol-Lc lamproite, Mount Gytha; 12, Ph-Di-TKr-Lc lamproite, Old Leopold Hill; 13, Ol-Di-Lc lamproite, Machells Pyramid; 14, hyalo-Ol-Lc-Ph lamproite, Oscar; 15, Mean of 100 leucite lamproites from the Noonkanbah field; Di, diopside; Ph, phlogopite; TKr, titanium potassic richterite; Lc, leucite; Ol, olivine. All data from Jaques et al. (1986).

more basic compositions represented by the leucite lamproites." This continuum is apparent in standard variation diagrams (Figure 7.5) where data are plotted without regard to the petrographic character of the rocks. However, when examined in more detail it is evident that the compositional range results from the presence of two petrographically distinct suites of rocks whose compositional fields partially overlap. The two suites are the olivine lamproites of the Ellendale field and the phlogopite–leucite–diopside–richterite lamproites of the Ellendale and Noonkanbah fields. Further, there is a clear spatial and temporal relationship with respect to composition, with the southern, younger (18–20 Ma) Noonkanbah field being on average richer in SiO_2 (Tables 7.5 and 7.6) than the northern, older (20–22 Ma) Ellendale field (Jaques *et al.* 1986). Figure 7.6, based upon the data used to construct Figure 7.5, shows clearly the bimodality of compositions within the province. The low SiO_2, high MgO peaks in the histograms represent principally the compositions of the olivine lamproites of the Ellendale field. The compositions of these rocks are controlled primarily by the presence of phenocrystal and xenocrystal olivines. The high-SiO_2, low-MgO peaks in the histograms represent the hypabyssal crystal-phyric lamproites of the Noonkanbah field together with transitional madupitic and phlogopite-rich lamproites from the Ellendale field (e.g., Mt. North, Mt. Percy, 81 Mile Vent). Jaques *et al.* (1986) attribute the compositional variation within these rocks to shallow level fractional crystallization that results in variation in the proportions of the principal phenocrystal and groundmass phases. Samples whose

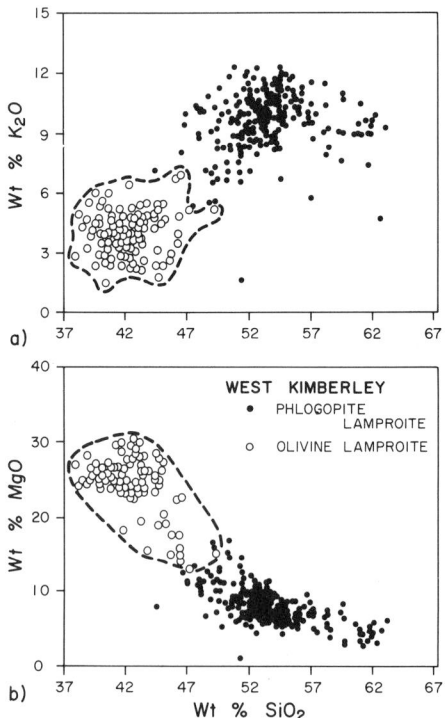

Figure 7.5. Compositional variation of West Kimberley olivine and phlogopite lamproites (a) K_2O versus SiO_2 and (b) MgO versus SiO_2. Data from Jaques *et al.* (1986).

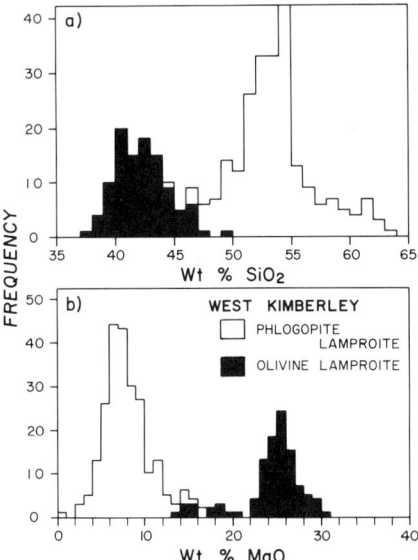

Figure 7.6. Distribution of (a) SiO$_2$ and (b) MgO in West Kimberley olivine and phlogopite lamproites. Data from Jaques *et al.* 1986).

compositions bridge the gap between these major groups represent lamproites containing major amounts of altered macrocrystal olivine, e.g., Mt. Ibis, Mt. Cedric. Jaques *et al.* (1986) refer to such rocks as being transitional between the two groups.

Individual lamproite occurrences within the province do not exhibit a simple differentiation trend, as illustrated by the following examples. At Mt. North there is no clear relationship between the composition of the late hypabyssal transitional madupitic lamproites filling the vent core, and those of the earlier phlogopite lamproite sills. The former contain 45–47 wt % SiO$_2$ (volatile-free basis), 5–7 wt % Al$_2$O$_3$, 5–8 wt %, K$_2$O, and 10–12 wt % MgO. The Mg/Mg + Fe ratio increases from 67 in the lower portions of this unit to 76 at the top, and shows no correlation with K$_2$O, Ni, Cr, or Zr abundances. The sills, in contrast, are SiO$_2$ (49–55 wt %) and K$_2$O (8–11 wt %) rich and MgO (6.5–8 wt %) poor. The rocks do not define smooth curves on variation diagrams (not plotted). At Mt. Cedric a bimodal suite may be distinguished whose compositions cannot be related by simple crystal fractionation. Neither group defines a trend on variation diagrams (not plotted). The SiO$_2$-poor rocks form a group that is transitional to olivine lamproite. The SiO$_2$-rich group is identical in composition to other intrusions in the Noonkanbah field, e.g., Machells Pyramid.

Jaques *et al.* (1986) have shown that some of the petrographically complex intrusions, e.g., P Hill, Mamilu Hill, Noonkanbah Hill, are of similar composition and they concluded that the petrographic variation is due to different cooling rates rather than differing bulk compositions. Within the Noonkanbah field there is some minor interintrusional compositional variation. However, the bulk of this variation is similar to the intraintrusional variation and reflects modal variations resulting from the very limited differentiation and crystallization of a single parental magma. The wide compositional range that might be expected if the rocks represented a strongly fractionated sequence is not present.

Important points to bear in mind when considering compositional variation within this province are as follows:

(1) At each intrusive center there is no apparent sequence of consanguineous rocks whose compositions define a standard differentiation trend or liquid line of descent originating from a parental basic magma.

(2) Olivine lamproites are hybrid rocks whose compositions may be viewed as being those of mixtures. They may represent the compositions of phlogopite lamproites, or their differentiates, to which olivine has been added. The compositions are not representative of their parental magmas and when plotted on variation diagrams cannot be used to indicate liquid lines of descent from some primitive MgO-rich magma.

(3) Holocrystalline lamproites have compositions that may approximate those of their parental magmas, but the absence of aphyric rocks or glassy *lavas* (Jaques *et al*. 1986, p. 72) precludes the exact identification of this magma.

(4) The parental magma for this province has not been identified. Variation diagrams such as Figure 7.5 (and those presented by Jaques *et al*. 1986, e.g., Figure 151), create the impression that there is continuous compositional variation within this province. Jaques *et al*. (1986) on the basis of such diagrams, consider the trends to have genetic significance and state that the rocks follow an orenditic differentiation trend of SiO_2-enrichment. Such an interpretation suggests that the parental magma must be of basic or ultrabasic composition. However, given points (1–3) above it is our contention that these diagrams do not represent major differentiation trends and are in fact trends of mixing, coupled with minor trends of fractional crystallization involving a relatively SiO_2-rich parental magma (see Section 7.1.3.6). This conclusion is supported by isotopic and trace element data (Jaques *et al*. 1986, McCulloch *et al*. 1983) which shows clearly that the rocks cannot be related by any *simple* fractional crystallization scheme (see Section 7.9).

7.1.3.2. Leucite Hills

Compositional data have been presented by Cross (1897), Schulz and Cross (1912), Smithson (1959), Johnson (1959), Yagi and Matsumoto (1966), Carmichael (1967), Barton and Hamilton (1978), Ogden (1979), Kuehner (1980), and Fraser (1987). The preponderance of these data have been provided by Ogden (1979), but unfortunately many of these analyses are incomplete or have poor analytical totals. Representative compositions are given in Table 7.7. All of the rocks have high K_2O/Na_2O ratios (>8) and are peralkaline. Lamproites from Steamboat Mountain represent the most potassic volcanic rocks ever reported (Carmichael 1967, Ogden 1979). The province does not exhibit as great a compositional range as the West Kimberley province and is notably poorer in TiO_2 (Tables 7.1 and 7.2).

Variation diagrams (Figure 7.7) are bimodal with respect to silica, and reflect the petrographic division of the suite into madupitic and phlogopite lamproites. Madupitic lamproites are low SiO_2, commonly peraluminous rocks occurring as hypabyssal plugs and dikes at only three localities (Pilot Butte, Middle Table Mountain, Badger's Teeth). The province consists primarily of relatively SiO_2-rich perpotassic vesicular hyalo-phlogopite lamproite lavas.

Individual phlogopite lamproite vents exhibit a limited range of composition (Figure 7.7). Ogden (1979) did not find any significant compositional variation within and between

THE GEOCHEMISTRY OF LAMPROITES

Table 7.7. Representative Compositions (wt %) of Leucite Hills Lamproites[a]

	1	2	3	4	5	6	7	8	9	10
SiO_2	43.56	47.6	43.16	50.23	52.98	55.12	54.17	53.6	55.5	53.07
TiO_2	2.31	2.4	2.35	2.30	2.55	2.58	2.67	2.3	2.5	2.41
Al_2O_3	7.85	8.1	8.37	10.15	10.49	10.35	10.16	10.4	11.1	8.96
Fe_2O_3	5.57	—	4.85	3.65	2.64	3.27	3.34	—	—	3.86
FeO	0.85	5.0	0.84	1.21	1.94	0.62	0.65	4.3	4.0	0.91
MnO	0.15	0.1	0.14	0.09	0.07	0.06	0.06	0.1	—	0.08
MgO	11.03	8.8	8.21	7.48	7.23	6.41	6.62	6.0	6.2	11.17
CaO	11.89	8.3	12.84	6.12	4.31	3.45	4.19	4.1	2.8	3.56
SrO	0.40	0.3	0.57	0.32	0.22	0.26	0.18	0.2	0.2	0.27
BaO	0.66	0.9	1.38	0.61	0.60	0.52	0.59	0.7	0.6	0.34
Na_2O	0.74	—	2.23	1.29	1.29	1.29	1.21	1.3	1.3	1.15
K_2O	7.19	8.8	4.31	10.48	11.15	11.77	11.91	12.1	12.7	10.72
P_2O_5	1.50	3.0	2.55	1.81	1.37	1.40	1.59	3.0	1.8	1.24
H_2O^+	2.89	4.3	4.65	2.34	—	1.23	1.01	1.4	—	1.16
H_2O^-	2.09	—	—	0.93	—	0.61	0.52	—	—	0.40
CO_2	—	—	2.76	—	—	0.20	0.49	—	—	—
LOI	—	—	—	—	2.86	—	—	—	—	—
	98.68	97.7	99.21	99.01	99.70	99.06	99.36	99.5	98.7	99.30

[a] 1–3, Madupitic lamproites; 1, Pilot Butte (Carmichael 1967); 2, Middle Table Mountain (Ogden 1979); 3, Badger's Teeth (Kuehner 1980); 4–6, leucite phlogopite lamproites; 4, Boar's Tusk (Carmichael 1967); 5, Emmons Mesa (Kuehner 1980); 6, Steamboat Mountain (Carmichael 1967); 7–9, sanidine leucite lamproites; 7, North Table Mountain (Carmichael 1967); 8, Zirkel Mesa (Ogden 1979); 9, Steamboat Mountain (Ogden 1979); 10, olivine sanidine phlogopite lamproite, South Table Mountain (Carmichael 1967).

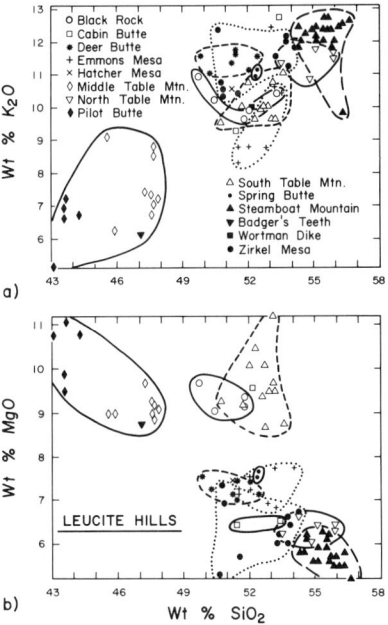

Figure 7.7. Compositional variation of lamproites from the Leucite Hills, (a) K_2O versus SiO_2 and (b) MgO versus SiO_2. Data sources given in Section 5.1.3.2.

flows. The observed ranges reflect primarily the proportions of crystals to glass. Sanidine phlogopite lamproites and leucite phlogopite lamproites are identical in composition (Table 7.7), clearly demonstrating that they are heteromorphs, and in accord with observations that the two varieties may be found within the same flow (Carmichael 1967, Ogden 1979). Although there is considerable overlap, it is evident from Figure 7.7 that individual vents differ slightly in their overall composition. These differences probably reflect the effects of minor differentiation and/or small changes in the source composition or degree of melting. Phlogopite lamproites from South Table Mountain are notably MgO rich by virtue of the presence of phenocrystal and xenocrystal olivine.

Madupitic lamproites from Pilot Butte and Middle Table Mountain differ in their compositions, but both are SiO_2 poor and MgO rich relative to other Leucite Hills rocks, by virtue of the presence of phlogopite and oxides (not olivine). The Middle Table Mountain lamproites are SiO_2 rich relative to those from Pilot Butte, as the former contain more leucite, diopside, and wadeite in the groundmass.

Important points stemming from mineralogical studies (Chapter 6) to be emphasized when considering compositional variation in this province are the following:

1. Phlogopite lamproites do not define differentiation trends at each volcanic center i.e. there are no shallow evolving magma chambers underlying each vent.

2. There is no well-defined differentiation trend for phlogopite lamproites within the province as a whole. Crystal fractionation processes as documented in the Noonkanbah lamproites do not appear to have occurred. Thus hyalo-phlogopite lamproites, containing identical phlogopite phenocrysts (6.1.2.1), have erupted as relatively uniform lavas over a wide geographic area. Consequently the average composition of these lavas (Table 7.1) probably provides a reasonable estimate of the composition of the parental magmas of this province.

3. Compositional trends between madupitic and phlogopite lamproites have been interpreted by Ogden (1979) and Kuehner (1980) as differentiation trends. We suggest that such an interpretation is not valid as:

   a. Compositional trends established for the Leucite Hills micas (Section 6.1.2) demonstrate that the madupitic rocks are more evolved than the phlogopite lamproites.
   b. Madupitic lamproites contain abundant wadeite, Sr- and Ba-rich apatite and shcherbakovite in their groundmass. This enrichment in incompatible element-rich phases suggests that the rocks are highly evolved relative to the phlogopite lamproites.
   c. Madupitic and phlogopite lamproites are isotopically different (Section 7.9). This difference implies that processes other than crystal fractionation must have played some role in their genesis.
   d. Kuehner *et al.* (1981) note that madupitic and phlogopite lamproites cannot be related by any fractionation process involving the observed primary mineral phases as such fractionation requires the removal of excessive amounts of some minerals.

Ogden (1979) has noted that the composition of the madupitic lavas of Pilot Butte have been modified slightly by the assimilation of carbonate-rich xenoliths derived from the Green River Formation. Metasomatism of the xenoliths has resulted in the formation of sanidine, leucite, phlogopite, and Ba-rich zeolites. Contamination of phlogopite lamproite has been demonstrated at Black Butte. Here the earliest lamproites are Na-rich (5.2 wt % Na_2O) relative to later rocks (1–2 wt % Na_2O). Ogden (1979) suggests that the magmas have reacted with Na-rich groundwaters.

7.1.3.3. Murcia-Almeria

Compositional data have been provided by Osann (1906), Jeremine and Fallot (1928), Parga-Pondal (1935), San Miguel and de Pedro (1945), San Miguel et al. (1951), Fuster et al. (1954, 1967), Fuster and Gastesi (1964), Borley (1967), Carmichael (1967), Fernandez and Hernanadez-Pacheco (1972), Pellicer (1973), Lopez Ruiz and Rodriguez Badiola (1980), Venturelli et al. (1984), and Nixon et al. (1984a). Despite the plethora of investigators there have been no systematic studies of compositional variation and most studies report the composition of 1–5 random samples collected at single volcanic centers. Many of the rocks have been silicified and carbonatized (Lopez Ruiz and Rodriguez Badiola 1980). Glassy rocks, in particular the hyalo-olivine phlogopite lamproites, have been extensively altered. Their high Na_2O contents (1–4 wt %, Borley 1967), suggest natrolitization similar to that reported from Smoky Butte altered hyalo-lamproites (Mitchell et al. 1987). Some lamproites, e.g., Zaneta, may have been contaminated by crustal granitic rock microxenoliths (Venturelli et al. 1984a). Representative compositions are given in Table 7.8 and variation diagrams for contamination- and alteration-free rocks presented in Figure 7.8. Particularly notable is the wide range in SiO_2 content. Spanish lamproites on average are richer in SiO_2, and poorer in K_2O, TiO_2, and P_2O_5 than those from the Leucite Hills and West Kimberley provinces (Tables 7.1, 7.2).

Figure 7.8 shows that madupitic rocks are poorer in SiO_2 than phlogopite lamproites. The compositions of phlogopite (Section 6.1.6) suggest that the former are more evolved than the latter. This relationship is identical to that observed in the Leucite Hills and cannot be fortuitous. Individual phlogopite lamproites, with the exception of Zaneta, do not appear to differ significantly from one another in composition. Differentiation trends cannot be

Table 7.8. Representative Compositions of Murcia-Almeria Lamproites[a]

	1	2	3	4	5	6	7	8	9	10
SiO_2	45.53	47.07	53.4	52.80	53.39	55.1	56.24	57.2	61.1	68.5
TiO_2	1.57	1.32	1.44	1.51	1.76	1.51	1.32	1.78	0.96	1.01
Al_2O_3	8.5	7.2	9.88	9.18	10.93	9.18	10.17	8.98	12.80	12.3
Fe_2O_3	2.93	3.03	6.01	1.27	2.11	1.27	0.84	5.89	4.56	2.13
FeO	4.31	3.19	—	3.79	2.77	3.79	1.96	—	—	—
MnO	0.11	0.10	0.07	0.01	0.04	0.01	0.04	0.06	0.05	0.03
MgO	14.86	16.88	12.8	12.98	12.79	12.98	5.31	7.99	5.05	1.73
CaO	9.06	7.90	4.86	4.14	5.00	4.14	6.47	3.66	2.77	2.32
SrO	0.22	0.19	0.08	—	—	—	—	0.09	0.07	0.06
BaO	0.41	0.48	0.27	—	—	—	0.19	0.22	0.21	0.21
Na_2O	1.5	1.40	1.90	1.00	1.49	1.00	1.86	1.22	2.22	2.01
K_2O	3.6	5.0	6.54	8.30	8.53	8.30	7.00	8.72	5.78	7.28
P_2O_5	1.82	1.78	0.98	1.05	0.63	1.05	1.55	0.83	0.88	0.91
H_2O+	4.1	2.90	—	—	0.68	—	4.85	—	—	—
H_2O-	1.13	0.52	—	—	0.45	—	1.50	—	—	—
CO_2	0.6	1.20	—	—	—	—	0.85	—	—	—
LOI	—	—	2.05	2.97	—	2.97	—	3.45	3.73	1.63
	100.25	100.16	100.28	99.00	100.57	101.30	100.15	100.09	100.18	100.12

[a] 1–3, Olivine diopside richterite madupitic lamproites, Jumilla (1 and 2, Borley 1967; 3, Venturelli et al. 1984); 4–10, phlogopite lamproites, Calasparra (Fuster et al. 1967); 5, Cancarix (Parga Pondal 1935); 6, Calasparra (Venturelli et al. 1984); 7, Vera (Fuster and de Pedro 1953); 8, Cancarix; 9 and 10, Zeneta (Venturelli et al. 1984a).

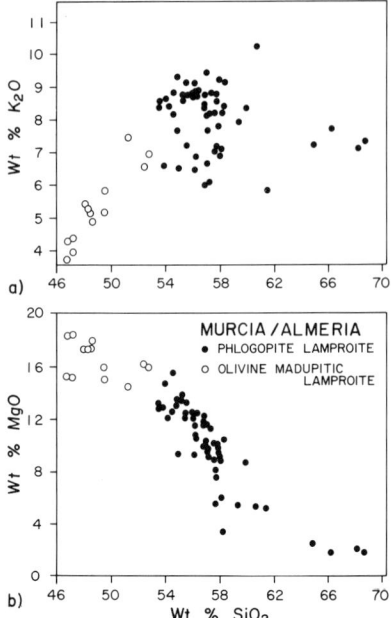

Figure 7.8. Compositional variation of lamproites from Murcia-Almeria (a) K_2O versus SiO_2 and (b) MgO versus SiO_2. Data sources given in Section 5.1.3.3.

established for individual centers or for the province as a whole. In common with the Leucite Hills lavas, it appears that the phlogopite lamproite lavas provide a reasonable estimate of the primitive magma composition.

7.1.3.4. Other Provinces

Representative compositions of lamproites from other provinces are given in Table 7.9. Lamproites from these occurrences have been insufficiently studied and/or are highly altered. Consequently, intra- and interprovincial differentiation trends have not been established.

Lamproite lavas whose compositions may be indicative of those of the primitive magmas occur at Francis, Utah (Henage 1972, Best *et al.* 1968, this work) and Gaussberg, Antarctica (Sheraton and Cundari 1980). These rocks are relatively SiO_2-rich (Table 7.9). Those from Francis differ in their composition to phlogopite lamproites from the near-by Leucite Hills province (Table 7.1).

Hypabyssal hyalo-lamproite dikes found at Smoky Butte Montana (Velde 1975, Mitchell *et al.* 1987), are commonly natrolitized and deficient in K_2O. Relatively unaltered rocks indicate that the parental magma for this suite was SiO_2 rich (Tables 7.1, 7.9). Hypabyssal lamproites from Sisco, Corsica (Velde 1967), are richer in silica than most other lamproites (Table 7.9; Peccerillo *et al.* 1988) suggesting geochemical affinities with the Murcia-Almeria lamproites.

Hypabyssal lamproites (Table 7.9) from Prairie Creek, Arkansas (Scott Smith and

Table 7.9. Representative Compositions of Diverse Lamproites[a]

	1	2	3	4	5	6
SiO_2	56.1	50.9	52.94	57.97	39.46	41.76
TiO_2	2.20	3.42	5.14	2.29	2.61	3.12
Al_2O_3	9.51	9.86	8.55	10.71	3.53	7.17
Fe_2O_3	5.12	2.46	5.65	1.73	8.78	1.85
FeO	0.70	3.78	—	2.69	—	5.85
MnO	—	0.09	0.07	0.07	0.13	0.10
MgO	6.57	7.95	8.38	6.75	26.67	11.84
CaO	4.07	4.63	4.51	3.34	5.14	7.85
SrO	0.13	0.21	0.38	0.09	0.14	—
BaO	0.21	0.62	0.86	0.14	0.23	—
Na_2O	0.96	1.65	0.96	1.06	0.29	1.42
K_2O	10.6	11.61	8.84	10.48	2.56	7.05
P_2O_5	1.14	1.48	2.21	0.73	0.29	1.14
H_2O+	0.8	1.17	—	—	7.70	2.49
H_2O-	0.6	0.05	—	—	1.75	—
CO_2	0.22	0.05	—	—	0.21	5.72
LOI	—	—	0.83	2.31	—	—
	98.93	99.93	99.32	100.36	99.49	97.36

[a] 1, phlogopite lamproite, Francis (this work); 2, average of 11 lavas from Gaussberg volcano (Sheraton and Cundari 1980); 3, hyalo armalcolite phlogopite lamproite, Smoky Butte (Mitchell et al. 1987); 4, average of five Sisco lamproites (Peccerillo et al. 1988); 5, olivine madupitic lamproite, Prairie Creek (Scott Smith and Skinner 1984); 6, average of 52 olivine lamproites, Sisimiut (Scott 1979).

Skinner 1984b) are unlikely to represent liquid compositions as their whole rock geochemistry is dominated by the presence of macrocrystal and xenocrystal olivines. Similar conclusions can be drawn for the Sisimiut dikes. Scott (1979) has shown that their compositions are related by an olivine fractionation (or mixing) line. The Sisimiut lamproites are unusually rich in CO_2 (Table 7.9).

7.1.3.5. Lamproite Glasses

Many lamproites have a glass matrix. The compositions of these glasses should indicate stages along the liquid line of descent resulting from the fractional crystallization of the parent magmas. Systematic investigations of this compositional variation have not been undertaken although such studies may be hampered by the extreme susceptibility of the glass to alteration. Although color variation is common, glasses typically show no obvious signs of devitrification and recrystallization. However, many glasses, assessed as fresh on a petrographic basis, are actually highly altered and may display extreme compositional variation on a millimeter scale. Alteration commonly results in loss of K_2O and Al_2O_3 and addition of Na_2O and H_2O (Mitchell et al. 1987, Edgar personal communication). Alteration of glass is the origin of the high Na_2O contents reported in some whole rock compositions of lamproites, e.g. Borley (1967). Jaques et al. (1986) claim that the low K_2O contents of some glasses are due to volatilization of potassium during electron microprobe analysis. However, Mitchell et al. (1987) and Edgar (personal communication) have analyzed both high and low K_2O glasses using the same analytical conditions. These data indicate that alkali volatilization cannot be the primary explanation for the low K_2O contents of the lamproite glasses. Mitchell et al. (1987) have suggested that ground water plays a major role in the alteration process.

Few glass compositions have been published. Table 7.10 demonstrates their very wide range in composition. Smoky Butte glasses (Table 7.10, analyses 1–2) are typically altered and consequently have high Na_2O and low K_2O contents (Mitchell et al. 1987), although K_2O-rich glasses have been reported by Velde (1975). All are SiO_2-rich and low in FeO, MgO and CaO. Glass in enstatite phlogopite lamproite (Table 7.10, analysis 3) from Fortuna (Murcia-Almeria) has a composition approximating potassium feldspar and, given slower cooling, would have crystallized as sanidine. Mitchell et al. (1987) have suggested that the Smoky Butte glasses were of similar composition prior to alteration.

Glass in the Leucite Hills lamproites (Table 7.10, analyses 4–6) exhibits a very wide range of composition within and between samples (Edgar, personal communication). Leucite phlogopite lamproites contain high SiO_2 glasses which may have low or high K_2O contents, some of the latter approaching potassium feldspar compositions. Glasses are typically richer in SiO_2 than their host whole rock compositions (Carmichael 1967, Edgar, personal communication). Low SiO_2 glasses are also present, these may have high Al_2O_3 and Na_2O contents which are indicative of natrolite formation, or low Na_2O contents coupled with high Al_2O_3 and K_2O.

Glass from the Oscar dike, West Kimberley is extremely rich in TiO_2 in addition to its high SiO_2 content (Table 7.10, analysis 7). Both low and high K_2O glasses have been reported by Prider (1982) and Jaques et al. (1986), respectively. Glass in other West Kimberley lamproites shows a very wide range of composition and in some cases is rich in TiO_2, FeO_T and MgO (Table 7.10, Edgar, personal communication). Glass from Gaussberg (Table 7.10, analysis 10) has high SiO_2, TiO_2, BaO and low Al_2O_3 contents and is slightly more siliceous than the whole rock composition (Table 7.9) reported by Sheraton and Cundari (1980).

Sobolev et al. (1985, 1989) have described primary glass inclusions in microphenocrystal olivines from West Kimberley olivine lamproites. These glasses coexist with chromite, orthopyroxene, kalsilite, magnesian ilmenite (12.8 wt % MgO), apatite, perovskite, phlogopite (5.81 wt % TiO_2, 5.50 wt % Al_2O_3), titanian potassium richterite and (?)fluorite. The glass is low in SiO_2, Al_2O_3, and CaO and rich in K_2O, TiO_2, F, Ba, Sr, and Zr (Table 7.10, analysis 11). Fluid portions of the inclusions consist predominantly of CO_2, (90–95 mol %) with up to 3.5 and 2.5 mol % N_2 and CO, respectively. The H_2O content is estimated to be <5 mol %. The glass begins to melt to a low-viscosity fluid at 600°C and homogenization temperatures range from 950 to 1050°C. Sobolev et al. (1985) have noted that the most common inclusions in microphenocrystal olivines have irregular shapes, typically form a continuous framework within the olivine grain, generally coalesce with the groundmass, and that many apparently primary inclusions are open. The compositions of the glasses thus may have been modified by reaction with groundmass-forming magma and also by resorption of preexisting olivines. The mineral assemblage found within the inclusions is in part typical of lamproite groundmass but the origin of kalsilite and orthopyroxene remains enigmatic. The latter may have formed by reaction of olivine with relatively Si-rich magma.

Inclusions in leucites in lamproites from the Leucite Hills and the Oscar dike (West Kimberley) consist of two discrete glasses that probably represent quenched immisicible liquids (Mitchell, in press). Representative compositions (Table 7.10) show that both glasses are rich in TiO_2 and BaO, but differ with respect to their K_2O contents.

Table 7.10. Representative Compositions of Glass[a]

	1	2	3	4	5	6	7	8	9	10	11	12	13
SiO_2	61.52	55.93	63.08	55.73	58.43	40.14	61.97	39.52	40.57	54.27	44.7	52.7	53.0
TiO_2	3.47	2.21	1.34	2.78	1.95	3.56	10.32	7.68	2.11	4.45	6.20	7.6	7.9
Al_2O_3	10.24	8.87	15.06	12.04	11.48	11.20	2.69	3.34	3.19	9.50	10.24	2.6	3.2
FeO	3.47	5.53	1.92	4.07	3.13	6.76	7.95	13.39	19.98	6.36	3.47	7.4	5.9
MnO	0.09	0.00	—	0.10	0.09	0.09	—	0.08	0.16	0.12	—	—	—
MgO	1.55	5.25	1.92	5.60	2.69	15.57	3.65	18.19	14.34	1.37	8.00	7.6	7.1
CaO	0.61	0.78	0.73	1.79	0.78	0.14	—	—	0.23	0.88	4.21	3.1	1.5
Na_2O	2.64	1.10	0.92	1.63	1.05	1.27	—	0.36	0.17	3.18	2.41	2.2	1.0
K_2O	1.55	8.74	11.96	13.02	13.56	15.24	2.23	9.32	8.46	14.46	10.5	10.4	3.8
BaO	1.64	—	0.25	0.41	—	1.24	1.96	0.55	0.61	0.85	3.25	3.5	2.0
F	—	—	—	0.75	—	3.21	—	0.88	0.56	0.12	1.43	—	—
	86.78	88.41	97.18	97.92	93.16	98.42	90.77	93.30	90.38	95.57	94.41	97.2	85.3

[a] 1–2, Altered and "fresh" glass, Smoky Butte, Mitchell et al. (1987) and Velde (1975) respectively; 3, enstatite phlogopite lamproite, Fortuna (Mitchell et al. 1987); 4–5, leucite phlogopite lamproites, Zirkel Mesa and Steamboat Mountain (Edgar, personal communication); 6, madupitic lamproite, Pilot Butte (Edgar, personal communication); 7, hyalo-olivine leucite lamproite, Oscar (Prider 1982); 8–9, West Kimberley; 10, Gaussberg (Edgar, personal communication); 11, glass inclusion in olivine, Ellendale 11. Also contains SrO = 0.43%, P_2O_5 = 2.13%, ZrO_2 = 0.51% (Sobolev et al. 1985); 12–13, high and low K immiscible glasses in leucite, Oscar (this work).

7.1.3.6. Summary

Lamproites occurring as lavas (Leucite Hills, Gaussberg) or glassy hypabyssal intrusions (Oscar, Smoky Butte) probably have compositions that are closest to those of the parental lamproite magma. These compositions (Tables 7.6–7.9) are relatively SiO_2 rich (50–55 wt %) with similar to the average composition of phlogopite lamproites given in Table 7.1. Lamproites richer in SiO_2 (Murcia-Almeria, Francis) may have resulted from crystal fractionation and/or contamination.

Hypabyssal SiO_2-poor, commonly olivine-rich, lamproites are not representative of liquid compositions and these rocks were not derived from the magmas which were parental to the SiO_2-rich rocks in the same province. Processes other than fractional crystallization have played a major role in determining their composition.

The liquid line of descent of lamproite magmas has not yet been established. Glass compositions and the limited fractional crystallization exhibited by the Noonkanbah field lamproites suggests that differentiation results in slight SiO_2 enrichment. The relatively limited compositional variation exhibited within a given province suggests that extensive differentiation does not typically occur. This conclusion is supported by the lack of evidence for continuously evolving magma chambers, complex volcanic sequences and the absence of plutonic complexes.

In summary, we believe that lamproite magmas that are parental to a given province are of small volume, are rapidly emplaced, and contain 52–55 wt % SiO_2 (whole rock composition recalculated to 100 wt % on a CO_2-free basis with total Fe expressed as FeO; e.g., analyses 1,2, or 5 in Table 7.1). The origin of the SiO_2-poor lamproites must be sought in processes involving differentiation of these magmas followed by hybridization and contamination of these differentiates with mantle and crustal material (see Chapter 10).

7.2. COMPATIBLE TRACE ELEMENTS

7.2.1. First Period Transition Elements

Scandium is hosted principally by phlogopite and pyroxene. In olivine lamproites it may also be incorporated into chromite and potassium richterite (Lewis 1987). The Sc contents of all lamproites are low (Tables 7.11 and 7.12) and there is no significant difference in the Sc contents of phlogopite and olivine/madupitic lamproites. Data given by Jaques *et al*. (1986) indicate that intraprovincial differences in the West Kimberley suite are negligible. Significant correlations with other trace elements are not found. Lamproites cannot be distinguished from RPT-lavas, group 2 kimberlites or minettes on the basis of their Sc contents (Tables 7.11–7.13).

Vanadium is hosted by phlogopite, chromite, and priderite. Table 7.11 shows that most phlogopite lamproites have similar V contents, although those from West Kimberley are exceptional with respect to their wide range in, and relatively higher, V contents. Olivine lamproites exhibit a wide range in V content, the highest levels being found in the Sisimiut and Barakar suites (Table 7.12). Within the West Kimberley province Jaques *et al*. (1986) and Lewis (1987) have noted that the V content of olivine lamproites is typically less than that of phlogopite/leucite lamproites (Tables 7.11 and 7.12). However, wide intraintrusion variation in V content is found, e.g., 81 Mile Vent (120–678 ppm), Mt. Percy (72–445 ppm),

Table 7.11. Average and Ranges of Compatible Trace Element Abundances (ppm) in Phlogopite Lamproites[a]

	W. Kimberley	Smoky Butte	Leucite Hills	Francis	Murcia-Almeria	Sisco	Gaussberg
Sc	17; 8–45	15; 12–19	13; 12–14	13; 12–13	15; 11–23	12; 10–13	106; 100–112
V	213; 72–678		82; 41–111	128; 110–150	95; 78–112	90; 85–93	306; 272–338
Cr	373; 125–867	499; 310–583	391; 257–565	495; 490–500	582; 350–816	417; 400–425	
Co	34; 19–52	29; 18–42	38; 22–61	28; 25–30	26; 10–34	23; 19–27	233
Ni	343; 30–858	309; 140–440	275; 135–429	362; 310–340	392; 108–719	264; 239–283	
Cu	66; 20–130	32; 24–41		48; 43–56	45; 24–82		29; 26–32
Zn	80; 52–132	95; 73–140		105; 94–110	65; 28–174		79; 72–93
(n)	154–229	24	5–27	2–5	30–42	3–6	11

[a]West Kimberley, Jaques et al. (1986); Smoky Butte, Mitchell et al. (1987), this work; Leucite Hills, Fraser (1987); Kuehner (1980); Francis, Henage (1972), this work; Murcia-Almeria, Venturelli et al. (1984a), Nixon et al. (1984); Sisco, Peccerillo et al. (1988); Gaussberg, Sheraton and Cundari (198). (n), number of samples.

Table 7.12. Averages and Ranges of Compatible Trace Element Abundances (ppm) in Olivine/Madupitic Lamproites[a]

	West Kimberley	Leucite Hills	Murcia-Almeria	Sisimiut	Prairie Creek	Kapamba	Barakar
Sc	21; 9–39	19; 18–20	18; 13–24		15; 14–16	14; 12–16	21; 15–24
V	82; 20–267	66; 10–113	108; 96–105	316; 219–407	46; 27–68	133; 84–151	185; 81–240
Cr	1014; 379–1703	430; 305–607	927; 827–1086	468; 164–155	1447; 1391–1500	490; 350–672	575; 300–977
Co	69; 31–92	41; 33–57	40; 37–43		96; 95–97	60; 49–72	
Ni	968; 401–1500	152; 120–171	715; 669–737	340; 54–782	1356; 1285–1443	490; 366–739	422; 213–930
Cu	55; 39–93		78; 37–257	37; 26–53	52; 47–57	84; 75–111	44; 26–67
Zn	73; 58–107		74; 53–91	87; 73–111	73; 71–74	102; 74–119	
(n)	102–109	2–5	4–8	12	2–5	6	6

[a]West Kimberley, Jaques et al. (1986); Leucite Hills, Fraser (1987); Kuehner (1980); Mucia-Almeria, Venturelli et al. (1984), Nixon et al. (1984); Sisimiut, Scott (1979); Prairie Creek, Fraser (1987), Scott Smith and Skinner (1984a); Kapamba, Scott Smith et al. (1989); Barakar, Middlemost et al. (1988). (n), number of samples.

Table 7.13. Averages and Ranges of Compatible Trace Element Abundances (ppm) in Some Potassic Rocks[a]

	Navajo	Colima	France	Northern England	Linhaisai	Minette	Vulsini	Sabatini	Group 2 Kimberlites
Sc	16; 9–24	21; 13–28	20; 14–23		20; 14–30	21	15; 1–46		20; 15–39
V	156; 96–240	187; 109–265		165; 126–219	209; 160–260	200	210; 37–346		85; 40–225
Cr	408; 95–1000	484; 30–805	376; 200–480	442; 246–843	353; 280–520	340	101; 1–888	139; 47–368	1800; 1033–2251
Co	42; 26–76	22; 18–28	26; 14–36	40; 25–50	47; 19–80	37	26; 5–62	23; 6–34	85; 62–112
Ni	248; 123–350	264; 30–506	257; 48–300	259; 107–416	297; 120–470	155	37; 4–222	34; 7–61	1400; 543–2022
Cu	64; 22–96	63; 41–100			135; 80–170	56	38; 4–116	59; 5–124	30; 11–72
Zn	79; 59–92	95; 76–162			150; 130–190	110	81; 14–120	63; 53–78	60; 46–88
(n)	5–21	3–10	7	16	4–10	43–125	63–123	11–15	16

[a]Navajo, Roden (1981), Rogers *et al.* (1982), Alibert *et al.* (1986); Colima, Luhr, and Carmichael (1981), Allan and Carmichael (1984); France, Turpin *et al.* (1988); Northern England, MacDonald *et al.* (1985); Linhaisai and average minette from Bergman *et al.* (1988); Vulsini, Holm *et al.* (1982); Sabatini, Cundari (1989); Group 2 kimberlites, Smith *et al.*(1985). (*n*), number of samples.

Machells Pyramid (76–578 ppm), Ellendale 4 (20–205 ppm, with isolated high values of 410 and 805 ppm), Ellendale 9 (32–221 ppm). Madupitic lamproites from the Leucite Hills contain less V than associated phlogopite lamproites. Vanadium does not exhibit any significant correlations with other trace elements. Table 7.13 shows that the V contents of minettes, kamafugites, and RPT-lavas are typically greater than those of phlogopite lamproites and group 2 kimberlites cannot be distinguished from lamproites on the basis of their V content.

Chromium is hosted primarily by chromite and to a lesser extent by phenocrystal phlogopite and pyroxene. Nickel is sequestered principally by olivine. Despite these different hosts both elements exhibit similar distributions in lamproites. Table 7.11 shows that different phlogopite lamproite suites have similar means but wide ranges in their Cr and Ni contents. Olivine lamproites also exhibit wide ranges in Cr and Ni content but averages for each province are different (Table 7.12). As expected, chromite-bearing olivine lamproites have much higher Cr and Ni contents than associated phlogopite lamproites in the West Kimberley and Murcia-Almeria provinces. Madupitic lamproites from the Leucite Hills which lack olivine have lower Ni contents than phlogopite lamproites in this province. In the West Kimberley province Cr and Ni exhibit significant intraintrusion variation, e.g., Mt. Cedric (190–627 ppm Ni, 198–718 ppm Cr), Machells Pyramid (121–512 ppm Ni, 192–516 ppm Cr), Ellendale 4 (889–1464 ppm Ni, 742–1257 ppm Cr). Few West Kimberley lamproites contain less than 200 ppm Ni and only the most Mg-poor leucite lamproites are depleted in Ni, e.g., Old Leopold Hill (77–143 ppm Ni). Cr and Ni exhibit a positive correlation with MgO and Mg/(Mg+Fe) ratios (Jaques *et al*. 1986, Lewis 1987).

Within each province Cr and Ni are not strongly correlated, e.g., West Kimberley phlogopite lamproites ($r = 0.38$), West Kimberley olivine lamproites ($r = 0.57$), Smoky Butte ($r = 0.09$). Lamproites from Sisimiut are exceptional in showing a strong ($r = 0.92$) Cr–Ni correlation. The Cr/Ni ratios on average are close to unity but range from approximately 0.3 to 5.

Figure 7.9 shows that phlogopite lamproites cannot be distinguished from minettes on the basis of their Cr and Ni contents as the compositional fields for the two groups completely overlap. Group 2 kimberlites have higher Cr and Ni contents than phlogopite lamproites, but overlap with the highest values found in olivine lamproites. RPT-lavas, as represented by Vulsini (Holm *et al*. 1982) and Sabatini (Cundari 1979), have significantly lower Cr and Ni contents than lamproites (Figure 7.9, Table 7.13).

Cobalt may be expected to reside primarily in olivines and chromites with lesser amounts in other ferromagnesian silicates. Table 7.11 shows that the Co contents of phlogopite lamproites are low (typically <50 ppm) and that there are no significant interprovincial differences. Olivine lamproites are relatively enriched in Co (Table 7.12). Cobalt does not show any significant correlations with other trace elements. Minettes and RPT-lavas have similar Co contents to phlogopite lamproites, whereas group 2 kimberlites are slightly richer in Co than olivine lamproites (Table 7.13).

The geochemistry of Cu and Zn in igneous rocks is normally controlled by their dominantly chalcophilic character. However, sulfides are notably absent from most lamproites and it must be assumed that these elements substitute for Mg and Fe in ferromagnesian silicates and oxides, with phlogopite and chromite being the principal hosts. But Cu and Zn contents are typically low (<100 ppm). Table 7.11 indicates that there are no significant differences in the Cu and Zn contents of phlogopite lamproites from different provinces. Olivine lamproites exhibit a greater range in Cu and Zn abundances (Table 7.12),

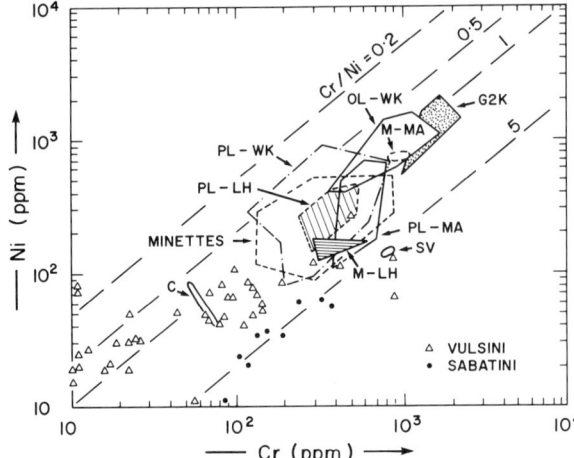

Figure 7.9. Ni versus Cr content of lamproites, group 2 kimberlites, minettes, Roman province-type lavas and kamafugites. Data sources given in in Tables 5.11–5.13. C, Cupaello; G2K, group 2 kimberlites; M-LH, madupitic lamproite Leucite Hills; M-MA, madupitic lamproite Murcia-Almeria; OL-WK, olivine lamproite West Kimberley; PH-LH, phlogopite lamproite Leucite Hills; PL-MA, phlogopite lamproite Murcia-Almeria; PL-WK, phlogopite lamproite West Kimberley; SV, San Vananzo.

with exceptionally high Cu contents being found in the Murcia-Almeria lamproites (Venturelli *et al.* 1984). Those from West Kimberley contain less Cu and Zn than associated phlogopite lamproites; however, the reverse relationship is found for the Spanish suite. Cu and Zn show no significant correlations with each other or other trace elements. The Cu and Zn contents of minettes, RPT-lavas, kamafugites, group 2 kimberlites and lamproites are similar (Tables 7.11–7.13).

7.2.2. Platinum Group Elements

Paul *et al.* (1979) report the following concentrations (ppb) of platinum group elements (PGE) in the Majhgawan lamproite: Pd = 4.1, 4.3; Ir = 5.9, 1.5; Au = 24, 2.3.

Data for the PGE in the West Kimberley lamproites have been provided by Lewis (1987). Concentrations are very low and typically below the detection limits of the NiS-AAS analytical method employed for three phlogopite leucite lamproites and one olivine lamproite. A single olivine lamproite (Ellendale 11) gave the following concentrations (ppb) of PGE: Ir = 7, Ru = 3, Rh = 1, Pt = 4, Pd = 56, Au 54. These limited data suggest that olivine lamproites are enriched in the PGE relative to phlogopite lamproites. Given the absence of sulfides in these rocks the PGE are probably hosted by chromite. The chondrite normalized PGE pattern for the Ellendale 11 sample is an asymmetric u-shape with the minimum at Pt (Lewis 1987). This differs from the positively sloping pattern proposed for kimberlites (Barnes *et al.* 1985). As the PGE contents of both kimberlites (Mitchell 1986) and lamproites are inadequately known, speculation as to the significance of these patterns (cf. Lewis 1987) is premature.

7.3. INCOMPATIBLE TRACE ELEMENTS—1: Ba–Sr, Zr–Hf, Nb–Ta, Th–U

7.3.1. Barium and Strontium

Barium contents, typically greater than 2000 ppm and commonly in excess of 5000 ppm, are a characteristic geochemical feature of most lamproites. Barium is hosted primarily by priderite and phlogopite and to a lesser extent by sanidine, Ba-bearing zeolites, Ba-rich apatite, and K–Ba-titanosilicates. Different averages and ranges in Ba content are found in lamproites from different provinces (Tables 7.14 and 7.15). Phlogopite lamproites from West Kimberley, Smoky Butte, and the Leucite Hills are relatively rich in Ba compared to those from Francis and Murcia-Almeria (Table 7.14). Jaques *et al.* (1986) have noted that significant intraintrusion variations exist in the West Kimberley province, e.g., Mt. North = 3607–22076 ppm Ba, Machells Pyramid = 5245–13252 ppm Ba, and that there is no systematic variation within the province as a whole. These variations in Ba content are a reflection of modal variations in the primary Ba-bearing phases.

Surprisingly, given the high modal abundance of Ba-free phases, olivine lamproites from the West Kimberley province have average Ba contents similar to those of West Kimberley phlogopite lamproites, although they exhibit a greater range in Ba content (Table 7.15). These data imply that the groundmass of these rocks is far richer in Ba than are phlogopite lamproites. The extreme Ba content of some samples might be related to the common presence of secondary barite in these rocks. However, both groups of West Kimberley lamproites contain Ba-rich rocks that are low in sulfur, and within the province there is no significant Ba–S correlation (Figure 7.10). The Ba content of olivine/madupitic lamproites varies widely and, in the case of those from the Leucite Hills and Jumilla, is greater than that of associated phlogopite lamproites. Corresponding to the wide range in Ba contents of all lamproites there is a wide range in K/Ba ratios (2–23) and no significant K–Ba correlations.

Lamproites are characterized by high Sr contents, that are typically greater than 1000 ppm. Strontium is hosted primarily by apatite and perovskite and to a lesser degree by wadeite, priderite, diopside, and richterite. Table 7.14 shows that Smoky Butte and Leucite Hills phlogopite lamproites are much richer in Sr than those from West Kimberley and Francis, and that the Spanish lamproites are depleted in Sr. Wider ranges in Sr contents are found within each province. Significant intra- and interintrusion variation is found in the West Kimberley province (Jaques *et al.* 1986). Phlogopite lamproites in the Ellendale field, e.g., 81 Mile Vent (757–2781 ppm Sr, Jaques *et al.* 1986; 1240–2441 ppm Sr, Tainton 1987), Mt. Percy (698–1917 ppm Sr) are much richer in Sr than lamproites in the Noonkanbah field, e.g., Hills Cone (725–1365 ppm Sr) P Hill (999–1291 ppm Sr). Carbonated lamproites within this province typically contain more than 2000 ppm Sr (Jaques *et al.* 1986). High Sr contents of some of the Smoky Butte lamproites may be related to the presence of strontianite.

Olivine/madupitic lamproites from different provinces, with the exception of the Leucite Hills madupitic rocks, have similar ranges and average Sr contents (Table 7.15). Madupitic lamproites from the Leucite Hills have exceptionally high Sr contents due to the presence of Sr-rich apatite and perovskite (see Sections 6.14 and 6.15). Olivine lamproites from the Ellendale field have similar average compositions but a lesser range in Sr content than those of associated phlogopite lamproites. In contrast, olivine/madupitic lamproites

320   CHAPTER 7

Table 7.14. Averages and Ranges of Incompatible Trace Element Abundances (ppm) in Phlogopite Lamproites[a]

	West Kimberley	Smoky Butte	Leucite Hills	Francis	Murcia-Almeria	Sisco	Gaussberg
Ba	10607; 3607–22102	9541; 6600–16534	6614; 3065–24800	2648; 1678–3407	1895; 1200–3055	1232; 1022–1329	5550; 5320–5970
Sr	1296; 484–2881	3094; 1900–3600	2213; 1652–2878	1232; 1060–1460	581; 384–898	814; 738–847	1808; 1710–1940
Zr	1401; 627–3193	2002; 1799–2359	1695; 1250–8139	1254; 1110–1320	711; 295–1045	1313; 1257–1384	1004; 890–1360
Hf	38; 24–64	53; 38–71	42; 38–47	39; 29–48		31; 25–35	
Nb	147; 81–324	113; 70–120	46; 33–64	58	34; 10–53	64; 61–66	90; 87–97
Ta	7.9; 5.5–15	5.7; 2.9–9.3	2.4; 2.1–2.7	2.8; 2–3.2		4.1; 3–5	
Th	30; 10–102	6.3; 4.7–8.0	17; 13–23	18; 11–27	110; 78–159	45; 41–53	30; 28–35
U	3.8; 1–32	3.3; 2.0–4.9	4.4; 2.8–7.4	2.3; 1.0–3.6	20; 15–29		2.5; 1.5–3.5
(n)	12–229	9–24	9–41	1–21	31–42	3–6	11

[a]West Kimberley, Jaques et al. (1986); Smoky Butte, Mitchell et al. (1987), this work; Leucite Hills, Fraser (1987) Kuehner (1980), Powell and Bell (1970); Francis, Henage (1972), this work; Murcia-Almeria, Venturelli et al. (1984a), Nixon et al. (1984), Nelson et al. (1986); Sisco, Peccerillo et al. (1988); Gaussberg, Sheraton and Cundari (1980), (n), number of samples.

Table 7.15. Averages and Ranges of Incompatible Trace Element Abundances (ppm) in Olivine/Madupitic Lamproites[a]

	West Kimberley	Leucite Hills	Murcia-Almeria	Sisimiut	Prairie Creek	Kapamba	Barakar
Ba	10584; 3007–41378	9831; 4319–14400	3262; 2450–4120	3760; 1730–7002	1971; 1624–2540	3059; 2173–4897	4160; 3165–5797
Sr	1325; 725–1889	3860; 2581–4904	1354; 709–2039	1798; 857–2549	1094; 972–1284	1377; 873–1577	2899; 2058–3729
Zr	1167; 564–3740	1302; 1152–1557	659; 575–705	741; 495–1648	718; 662–745	403; 339–448	1246; 829–2300
Hf	38; 32–49	42; 38–43			17; 16–18		
Nb	186; 104–309	99; 79–137	43; 30–51		101; 87–112	108; 93–116	58–179
Ta	9.9; 6.3–14	5.8; 4.9–7.1			6.1; 5.3–6.6		
Th	57; 18–98	37; 33–45	143; 105–166		12; 11–14		31; 11–53
U	2.4; 1–0	8.5; 8.2–8.7	15	1.7; 1.3–2.3	2.6; 2.5–2.7		4.3; 0–6
(n)	9–109	3–10	4–10	12	3–5	6	6

[a]West Kimberley, Jaques et al. (1986); Leucite Hills, Kuehner (1980), Powell and Bell (1970); Murcia-Almeria, Ventureli et al. (1984), Nelson et al. (1986); Sisimiut, Scott (1979), Nelson (1989); Prairie Creek, Fraser (1987), Scott Smith and Skinner (1984b); Kapamba, Scott Smith et al. (1989); Barakar, Middlemost et al. (1988). (n), number of samples.

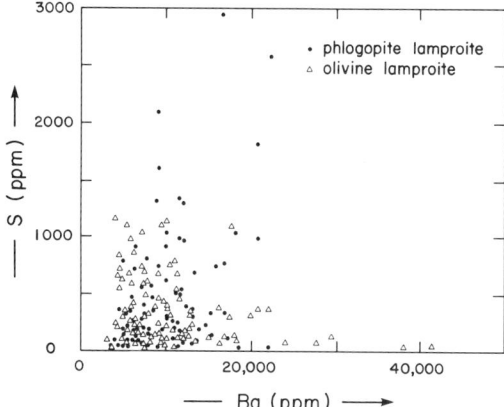

Figure 7.10. Ba versus S content of olivine and phlogopite lamproites from West Kimberley. Data from Jaques *et al.* (1986).

from the Leucite Hills and Jumilla are richer in Sr than phlogopite lamproites in these provinces.

Figure 7.11 shows that there is no significant correlation ($r = 0.3$) between Ba and Sr in the West Kimberley phlogopite lamproites, and in the olivine lamproites the Ba content is essentially independent of the Sr content ($r = 0.03$). The Ba/Sr ratios are all high (>8). Although few data are available, a similar lack of Ba–Sr correlation appears to exist for other provinces with the exception of the Murcia-Almeria province (Venturelli *et al.* 1984a) where a weak correlation ($r = 0.65$; Ba/Sr = 3–4) is found. The lack of significant Ba–Sr correlations shows that these elements are not behaving entirely as incompatible elements in lamproites.

Roman Province type (RPT-) lavas can be easily distinguished from most lamproites (Figure 7.12) on the basis of the former's lower Ba (av. 1496, range = 0–3659 ppm) and Sr

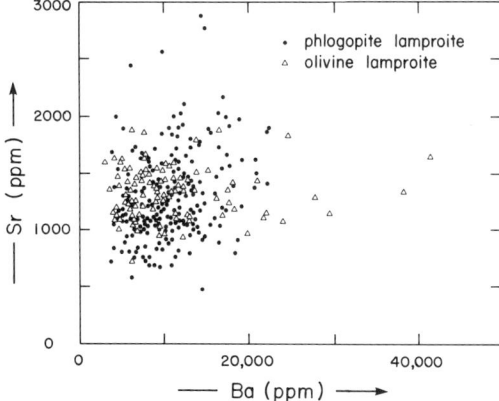

Figure 7.11. Ba versus Sr content of olivine and phlogopite lamproites from West Kimberley. Data from Jaques *et al.* (1986).

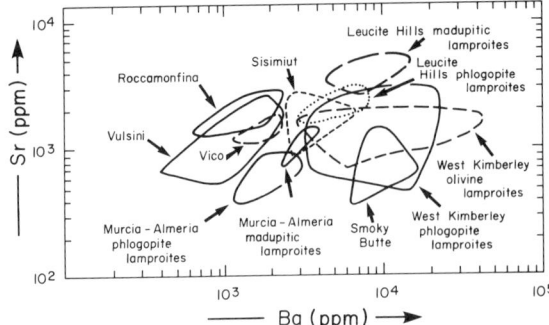

Figure 7.12. Ba versus Sr content of lamproites and Roman-province-type lavas. Data sources given in Tables 7.14–7.16.

(av. 1570, range 438–2740) contents (Table 7.16). Unlike lamproites Ba and Sr exhibit relatively well-defined correlations in RPT-lavas, e.g., Roccamonfina $r = 0.83$ (Appleton 1970), Vulsini $r = 0.75$ (Holm et al. 1982), Vico $r = 0.83$ (Cundari and Mattias 1974), suggesting that they are behaving as incompatible elements in these suites. The Ba/Sr ratios (0.5–2.0) of RPT-lavas are lower than those of most lamproites. Group 2 kimberlites have similar Ba (av. 3000, range 1322–4415 ppm) and Sr (av. 1140, range 601–1694 ppm) contents (Smith et al. 1985) to many lamproites. The elevated Ba/Sr ratios of lamproites are exceptional compared with those of kimberlites, alkali basalts, lamprophyres (1–1.4) and the primitive mantle (0.3, Bergman 1987).

The Ba and Sr contents of minettes vary widely (Figure 7.13, Table 7.17). In general, average concentrations are lower in minettes than those found in lamproites with the exception of those from Murcia-Almeria and Sisimiut. However, substantial overlap occurs with the main field of lamproite compositions shown in Figure 7.13, suggesting that minettes may not be easily distinguished from lamproites on the basis of their Ba–Sr relationships. Peralkaline minettes in particular are notably enriched (>5000 ppm) in Ba (Hall 1982, Němec 1987).

Table 7.16. Averages and Ranges of Incompatible Trace Element Abundances (ppm) in Other Potassic Rocks[a]

	Vulsini		Vico		Roccamonfina		Group 2 Kimberlite		Uganda	
Ba	1413;	0–2417	1963;	1165–3659	1430;	313–2301	3000;	1322–5024	3647;	1800–1700
Sr	1420;	481–2740	1374;	438–2590	1845;	772–2595	1140;	601–1694	1942;	862–4083
Zr	395;	187–1300	470;	311–813	270;	187–587	290;	123–615	524;	268–854
Hf	8.3;	6–17	13;	9–24			5.9;	2.1–13	6.5;	4.0–10
Nb	29;	9–64					120;	43–216	165;	92–214
Ta	1;	1–2	2.2;	0.9–4.6			6.1;	1.4–11	12;	8.1–19
Th	73;	19–162	119;	47–278	45;	14–197	30;	8–66	18;	7.7–29
U	16;	3–55					5;	2.6–11		
(n)		26–131		19–34		87		16–27		6–12

[a]Vulsini, Holm et al. (1982); Vico, Cundari and Mattias (1974), Barbieri et al. (1988); Roccamonfina, Appleton (1970); Group 2 kimberlites, Smith et al. (1985), Fesq et al. (1975), Mitchell and Brunfelt (1975); Uganda, kamafugites, Powell and Bell (1969), Mitchell and Bell (1976), Higazy (1954). (n), number of samples.

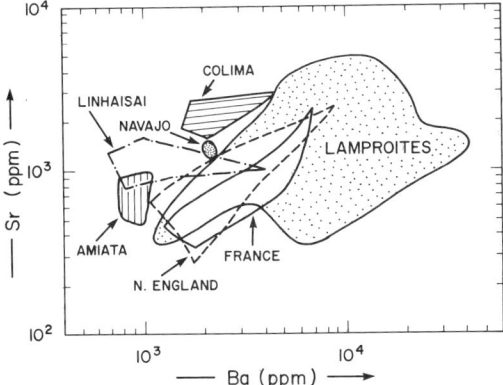

Figure 7.13. Ba versus Sr content of lamproites and minettes. Data sources for minettes given in Table 7.17. Field of lamproite compositions from Figure 7.12.

7.3.2. Zirconium and Hafnium

The Zr content of lamproites typically exceeds 1000 ppm. The principal host is wadeite, although significant amounts probably replace Ti in priderite, perovskite, and titanosilicates. Wide intraprovincial variations in the Zr content of the West Kimberley, Leucite Hills, and Murcia-Almeria provinces exist, with the latter province being exceptionally poor in Zr (Table 7.14). Jaques *et al.* (1986) note significant intra- and interintrusion variation in the West Kimberley province with phlogopite lamproites from the Ellendale field being rich in Zr, e.g., Mt. North (1595–2844 ppm), 81 Mile Vent (1485–1971 ppm) relative to those in the Noonkanbah field, e.g., Machells Pyramid (627–1412 ppm), Mt. Cedric (764–1428 ppm). The highest Zr contents are found in TiO_2-rich rocks, e.g., Ellendale 12 (3200 ppm Zr, 7.53 wt % TiO_2) and in wadeite-rich pegmatitic rocks from the Walgidee Hills (6000 ppm Zr, 12 wt % TiO_2).

Olivine lamproites from West Kimberley contain less Zr than phlogopite lamproites in this province. In contrast there are no significant differences in the Zr contents of phlogopite and olivine/madupitic lamproites in the Leucite Hills and Murcia-Almeria provinces. Zr contents of the Prairie Creek, Sisimiut, and Murcia-Almeria olivine/madupitic lamproites are a factor of 2 lower than those from West Kimberley. Lamproites from Kapamba have exceptionally low Zr contents (Table 7.15). The occurrence of Zr-rich olivine lamproites (West Kimberley) suggests that the low Zr contents of other olivine lamproites (Prairie Creek) do not result simply from dilution with olivine.

Few Hf data are available (Table 7.14). At Smoky Butte Zr and Hf behave relatively coherently ($r = 0.75$) with an average Zr/Hf ratio of 38. The average Zr/Hf ratios for phlogopite and olivine lamproites in the West Kimberley province are 37 and 31, respectively (Fraser 1987). In the Leucite Hills the Zr/Hf ratios of the phlogopite and madupitic lamproites are 40 and 31, respectively (Fraser 1987). Francis lamproites have a Zr/Hf ratio of 32 (this work). The Zr/Hf ratios of lamproites are similar to those of alkali basalts (38), kimberlites (36), alkaline and calc-alkaline lamprophyres (39), and the primitive mantle (35; Bergman 1987).

Table 7.17. Averages and Ranges in Incompatible Trace Element Abundances (ppm) in Some Representative Minettes[a]

	Navajo	Colima	France	Amiata	Northern England	Linhaisai	Minette
Ba	2214; 1843–2524	2428; 1570–4230	3696; 1300–6900	902; 775–1110	2892; 1070–8707	1500; 650–4000	1345
Sr	1373; 1209–1697	2386; 1670–3079	1007; 365–2500	688; 487–995	981; 281–2623	1100; 820–1660	1010
Zr	360; 350–370	384; 274–554		224; 180–267	294; 192–177	650; 230–970	350
Hf		12; 8–18	15; 8–22			26; 16–33	9
Nb	43; 15–74	16; 9–23		14; 7–22		15; 10–40	83
Ta		0.8; 0.6–1.0	1.8; 0.8–2.8		19; 12–30	1.0; 0.8–1.5	2
Th	51; 22–90	6.3; 3.2–8.6	37; 18–77		35; 22–50	19; 12–25	24
U	13; 10–16	1.7; 0.7–2.4	10; 3–28			5; 3–7	5
(n)	2–28	3–10	7	37	16	4–10	43–125

[a]Navajo, Roden (1981), Rogers *et al.* (1982), Alibert *et al.* (1986); Colima, Luhr and Carmichael (1981), Allan and Carmichael (1984); France, Turpin *et al.* (1988); Amiata, Van Bergen *et al.* (1983); Northern England, MacDonald *et al.* (1985); Linhaisai and average minette, Bergman *et al.* (1988). (*n*), number of samples.

7.3.3. Niobium and Tantalum

Niobium is hosted principally by perovskite and to a lesser degree by priderite and titanosilicates. Table 7.14 shows that phlogopite lamproites from West Kimberley have higher levels and wider ranges in Nb content than lamproites from other provinces. Jaques *et al.* (1986) note significant intraprovincial variations in the West Kimberley province, e.g., Machells Pyramid (102–167 ppm), Mamilu Hill (106–141 ppm), with the maximum Nb contents being found in the Ellendale field, e.g., Mt. North (120–308 ppm), 81 Mile Vent (117–324 ppm). Olivine lamproites from Ellendale contain on average higher Nb contents, e.g., Ellendale 4 (99–222 ppm), Ellendale 9 (140–252 ppm) than Noonkanbah phlogopite lamproites. Olivine/madupitic lamproites in general have higher Nb contents than associated phlogopite lamproites (Tables 7.14 and 7.15). This enrichment is a reflection of the common occurrence of perovskite in these rocks.

Few Ta data are available. These data (Tables 7.14 and 7.15) suggest that the Nb/Ta ratios of the phlogopite and olivine lamproites of the West Kimberley province are 18.6 and 18.8, respectively, (Fraser 1987). In the Leucite Hills the Nb/Ta ratios of the phlogopite and madupitic lamproites are 19.2 and 17.1, respectively (Fraser 1987). The Nb/Ta ratio for Francis is 20.7 (this work).

Zr and Nb behave moderately coherently ($r = 0.73$) in West Kimberley phlogopite lamproites, but significant correlations are not found in the olivine lamproites ($r = 0.37$). The difference may be due to Nb behaving as a compatible element in the latter suite. Zr/Nb ratios range from 4 to 10 in the olivine lamproites and from 6 to 20 in the phlogopite lamproites (Figure 7.14). Jaques *et al.* (1986) have suggested that this increase results from the fractionation of Nb-bearing perovskite. Given the nonparental status of the olivine lamproites (Section 7.1.3.6) we consider that such fractionation is not responsible for the relatively low Nb contents of phlogopite lamproites. Similar relationships are evident in the Leucite Hills rocks where madupitic lamproites have lower Zr/Nb ratios (av. 13) than phlogopite lamproites (av. 36) (Figure 7.14). Zr and Nb in the Spanish lamproites show a

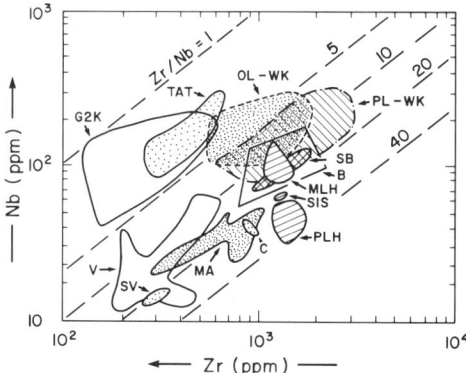

Figure 7.14. Nb versus Zr contents of lamproites, group 2 kimberlites, Roman-province-type lavas and kamafugites. Data sources given in Tables 7.14–7.16. B, Barakar; SB, Smoky Butte; SIS, Sisco; TAT, Toro-Ankole; V, Vulsini. Other abbreviations as in Figure 7.9.

moderate correlation ($r = 0.67$) with Zr/Nb ratios ranging from 10–30 (Figure 7.14). No significant difference in Zr/Nb ratios exists between madupitic and phlogopite lamproites.

Most lamproites can be easily distinguished from RPT-lavas on the basis of the latter's lower Nb, Zr, Ta, and Hf contents (Table 7.16). Only the Spanish lamproites have geochemical similarities to RPT-lavas. Group 2 kimberlites differ from lamproites in having lower Zr contents although Nb contents are similar. Kamafugites from Toro-Ankole and San Venanzo are relatively poor in Zr (Table 7.16) and only those from Cupaello have Nb and Zr contents similar to those of lamproites (Figure 7.14). The Nb/Zr ratios of lamproites are significantly higher than those of kimberlites, alkali basalts, and lamprophyres (0.4–4) but similar to that of the primitive mantle (13; Bergman 1987).

Table 7.17 and Figure 7.15 show that minettes may be distinguished from lamproites on the basis of their lower Zr, Nb, Hf, and Ta contents. A plot of Nb versus Zr (Figure 7.15), or Hf versus Ta effectively discriminates lamproites from minettes. Compositional overlap occurs only with respect to the most Zr-poor phlogopite lamproites from Murcia-Almeria. The Nb/Ta (15–20), Zr/Hf (25–32), and Zr/Nb (15–43) ratios of minettes are similar to those of lamproites.

7.3.4. Thorium and Uranium

Thorium is hosted primarily by perovskite and its geochemical behavior in lamproites is thus similar to that of Nb (see below). Thorium may also be present in priderite and apatite. Table 7.14 shows that phlogopite lamproites exhibit a wide range in their average Th contents, and that significant intraprovincial variations exist in the West Kimberley and Murcia-Almeria provinces. The Spanish lamproites are notably enriched in Th relative to other lamproite provinces. Olivine lamproites in the Ellendale field, e.g., Ellendale 4 (39–93 ppm), Ellendale 9 (30–97 ppm), are on average (Table 7.15) richer in Th than phlogopite lamproites, e.g., Machells Pyramid (14–35 ppm), Mamilu Hill (17–29 ppm) in the Noonkanbah field, although Ellendale phlogopite lamproites have the highest Th contents of the province, e.g., Mt. North (35–102 ppm), 81 Mile Vent (26–74 ppm) (Jaques *et al.* 1986). Similar relationships are evident in the Spanish and Leucite Hills provinces where madupitic lamproites have higher Th contents than phlogopite lamproites (Tables 7.14 and 7.15).

Significant Nb–Th correlations (Figure 7.16) exist for phlogopite lamproites from West Kimberley ($r = 0.85$) and Murcia-Almeria ($r = 0.82$) but are less well defined for Ellendale

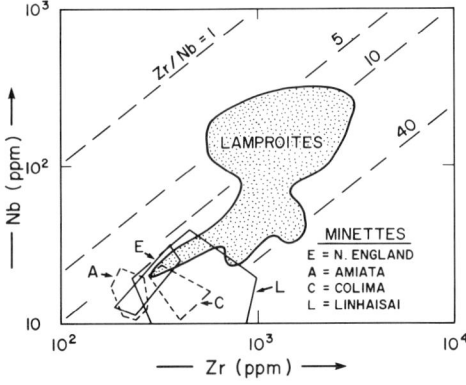

Figure 7.15. Nb versus Zr content of lamproites and minettes. Data sources for minettes given in Table 7.17. Field of lamproite compositions from Figure 7.14.

THE GEOCHEMISTRY OF LAMPROITES

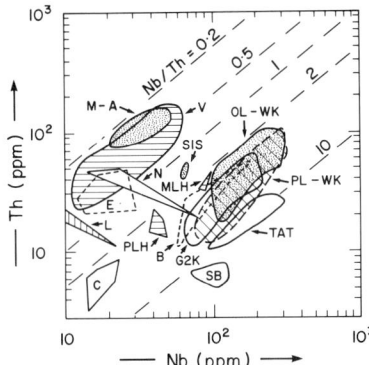

Figure 7.16. Th versus Nb content of lamproites, group 2 kimberlites, minettes, Roman province-type lavas and kamafugites. Data sources and abbreviations as in Figures 7.9, 7.14, and 7.15.

olivine lamproites ($r = 0.69$). Nb/Th ratios in all provinces with the exception of Murcia-Almeria lie between 1 and 10 (Figure 7.16). Spanish lamproites have low Nb/Th ratios (0.2–0.5) similar to those of RPT-lavas (0.2–0.8) from Vulsini (Holm *et al.* 1982) and kamafugites from San Venanzo (0.35–0.43) and Cupaello (0.34–0.35) (Peccerillo *et al.* 1988). Nb/Th ratios in Toro-Ankole kamafugites (Mitchell and Bell 1976) are higher (7–14) than in most lamproites, while those of group 2 kimberlites (3–6.1, Smith *et al.* 1985) are similar to those of West Kimberley phlogopite lamproites (Figure 7.16).

Uranium is hosted principally by apatite and perovskite. Tables 7.14 and 7.15 show that all provinces have similar U contents. Significant intraprovincial variation exists in the West Kimberley province and phlogopite lamproites have, on average, lower Th contents than olivine lamproites (Jaques *et al.* 1986). Th/U ratios on average are higher in olivine lamproites (24) than phlogopite lamproites (7.5), but significant Th–U correlations ($r = 0.2$) are absent in both suites. In contrast Th and U are highly correlated ($r = 0.93$) in the low Th/U ratio (1.9) lamproites from Smoky Butte.

RPT-lavas have higher U (16 ppm) contents and lower Th/U ratios (3.9) than lamproites. Group 2 kimberlites have similar U contents (5 ppm) but lower Th/U ratios (6). The Th/U ratios of most lamproites are higher than those found in most rocks (3–4) and that proposed for the primitive mantle (4; Bergman 1987).

The Th and U contents of minettes are similar to those of many lamproites (Table 7.17). The two suites may be distinguished on a Th–Nb diagram (Figure 7.16) primarily on the basis of the relatively low Nb content of minettes. Minettes have low (<5.0, av. 3.5) Nb/Th ratios.

7.4. INCOMPATIBLE TRACE ELEMENTS—2: RARE EARTH ELEMENTS AND YTTRIUM

7.4.1. Rare Earth Elements

The rare earth elements (REE) are hosted principally by perovskite (Section 6.15.2), apatite (Section 6.14.2), and cerium-rich phases (Section 6.17.4). Minor amounts may also be found replacing Ca in a variety of silicates.

Lamproites are typically enriched in the REE relative to nearly all compositionally allied rocks. Averages and ranges in REE content are given in Tables 7.18 and 7.19 and representative chondrite normalized distribution patterns are illustrated in Figures 7.17 to 7.20. Wide interprovincial differences in REE abundances exist. The West Kimberley province and the Smoky Butte intrusion have the highest, and the Murcia-Almeria province the least, REE contents. The distribution patterns emphasize the characteristic light REE enrichment which is reflected in the very high La/Yb ratios (100–300) of many lamproites.

Within the West Kimberley province olivine lamproites are on average (mean La = 412, range = 146–830 ppm; mean Ce = 711, range 260–1629 ppm, 109 samples, Jaques *et al.* 1986) richer in REE than phlogopite lamproites (mean La = 348, range 161–739 ppm; mean Ce = 549, range = 225–1734 ppm, 229 samples, Jaques *et al.* 1986). However, significant intraprovincial variations exist and phlogopite lamproites from the Ellendale field, e.g., Mt. North (La = 311–716, Ce = 525–1734 ppm), Mt. Percy (La = 327–660, Ce = 502–1047 ppm), 81 Mile Vent (La = 360–600, Ce = 594–1060 ppm) have similar REE contents to the Ellendale olivine lamproites, e.g., Ellendale 4 (La = 202–519, Ce = 460–1046 ppm), Ellendale 9 (La = 194–443, Ce = 362–816 ppm). In contrast phlogopite lamproites from the Noonkanbah field, e.g., Mt. Cedric (La = 197–434, Ce = 274–616 ppm), Machells Pyramid (La = 161–311, Ce = 255–464 ppm), are poor in REE relative to those from Ellendale. All West Kimberley lamproites are enriched in the light REE (500–2000 times chondrites) relative to the heavy REE (3–6 times chondrites). The REE distribution patterns (Figure 7.17) of olivine lamproites (mean La/Yb = 227, range = 128–316) are essentially identical to those of phlogopite lamproites (mean La/Yb = 160, 92–238). Low La/Yb ratios are reflections of analytical error in the determination of the heavy REE. Europium anomalies are absent (Lewis 1987, Fraser 1987, Jaques *et al.* 1986, Fraser *et al.* 1985).

Olivine lamproites from Argyle (Jaques *et al.* 1989c) are not as enriched in the light REE as those from Ellendale. Consequently, their La/Yb ratios are relatively low (Table 7.18, Figure 7.17a).

All lamproites from Smoky Butte have similar REE distribution patterns and abundances (Table 7.19, Figure 7.18a). The rocks are strongly enriched in the light REE with abundances being 1000–1400 times those of chondrites. Abundances of heavy REE are low (6–20 times chondrites), consequently La/Yb ratios are high, ranging from 184–281 (mean 226). The distribution patterns are linear and Eu anomalies are absent. Mitchell *et al.* (1987) found no correlation between petrographic type of lamproite and REE abundance and noted that REE distributions are not changed during alteration.

Figures 7.18b and 7.21 and Table 7.19 demonstrate that madupitic lamproites from the Leucite Hills are enriched in REE and have higher La/Yb ratios than associated phlogopite lamproites (Fraser 1987, Kay and Gast 1973). Higher REE abundances coupled with lower Ni contents confirm that madupitic lamproites cannot be parental to the phlogopite lamproites (Section 7.1.3.2.). Sanidine-bearing and sanidine-free phlogopite lamproites do not differ in their REE contents or La/Yb ratios.

Similar relationships are evident in the Murcia-Almeria province where olivine-bearing madupitic lamproites from Jumilla are enriched in REE relative to phlogopite lamproites (Table 7.18, Figure 7.19). The Spanish lamproites differ from the other major lamproite provinces in having relatively low La/Yb ratios (33–63). Figure 7.21 illustrates clearly the anomalously low La/Yb ratios and high Sm contents of the Spanish lamproites relative to other provinces. Distribution patterns are characterized by the presence of a small

Table 7.18. Averages and Ranges in REE Abundances (ppm) in Lamproites from West and East Kimberley, Francis, Murcia-Almeria and Gaussberg[a]

	1	2	3	4	5	6	7
La	348; 157–596	242; 110–425	129; 98–148	133; 105–175	87; 64–109	155; 137–171	215; 204–254
Ce	629; 237–1035	414; 181–825	269; 218–295	275; 223–354	242; 180–301	401; 360–450	398; 368–478
Pr	49; 33–77	33; 15–64		33; 28–41	31; 24–38	46; 42–50	45; 42–49
Nd	212; 92–371	146; 93–276	104; 89–112	128; 111–160	149; 116–184	224; 208–239	134; 127–160
Sm	25; 12–41	18; 8.8–29	15; 14–17	17; 14–20	27; 22–34	40; 37–42	16; 15–21
Eu	5.8; 3.1–9.9	4.3; 2.6–7.0	3.4; 2.9–3.8	3.7; 3.3–4.3	4.5; 3.5–5.5	6.9; 6.3–7.5	4.1; 3.3–4.9
Gd	9.5; 6.9–12	8.4; 6.6–15	8.7; 7.7–9.7	9.6; 8.6–11	14; 11–17	21; 19–23	11; 9–13
Tb	1.8; 1.1–2.8	1.4; 0.9–1.9	1.1; 1.0–1.3	0.9; 0.8–1.0			
Dy	4.8; 3.6–5.9	4.2; 4.1–6.9		4.8; 4.6–5.1	6.4; 5.9–7.6	9.6; 9.0–10	6.9; 6.6–7.2
Ho	0.9; 0.8–1.2	1.0; 0.8–1.6	0.9; 0.8–1.0		1.1; 1.0–1.3	1.6; 1.5–1.7	0.6; 0.5–0.7
Er	1.8; 1.1–2.5	1.9; 1.5–2.4			2.6; 2.4–3.1	3.7; 3.0–4.0	1.5; 1.1–1.8
Yb	1.6; 0.9–2.6	1.5; 1.1–2.1	1.4; 1.2–1.5	1.3; 1.2–1.5	1.9; 1.7–2.5	2.6; 2.4–2.7	0.8; 0.4–1.3
Lu	0.19; 0.08–0.33	0.17; 0.07–0.27	0.17; 0.15–0.18	0.16; 0.14–0.17	0.27; 0.23–0.34	0.36; 0.33–0.38	
La/Yb	227; 128–316	160; 92–238	93; 65–123	103; 70–146	47; 33–59	60; 57–63	269
(n)	6–17	6–16	4	5	11	4	2–13

[a] 1–2, Olivine and phlogopite lamproites, respectively, West Kimberley (Lewis 1987, Fraser 1987, Nixon et al. (1984); 3, olivine lamproites, Argyle, East Kimberley (Jaques et al. 1989c); 4, phlogopite lamproites, Francis (this work); 5–6, phlogopite and madupitic lamproites, respectively, Murcia-Almeria (Nixon et al. 1984); 7, Gaussberg, Collerson and McCulloch (1983), Foley et al. (1987). (n, number of samples.

Table 7.19. Averages and Ranges in REE Abundances (ppm) in Lamproites from Smoky Butte, Leucite Hills, Sisco, Hills Pond, Prairie Creek, Priestly Peak and Mt. Bayliss[a]

	1	2	3	4	5	6	7	8
La	368; 331–430	148; 112—182	297; 260–357	183; 172–193	116; 139–199	144	159; 151–167	151
Ce	805; 686–980	273; 232–366	597; 516–722	344; 302–400	312; 260–396	277	319; 273–358	309
Nd		117; 91–133	228; 204–251	138; 111–165		107	149; 133—164	117
Sm	33; 28–39	15; 12–17	27; 24–33	20; 19–21	14; 11–19	13	27; 16–36	18
Eu	8.1; 6.9–10	3.6; 3.0–4.1	6.2; 5.5–7.5	3.1; 2.5–3.6	3.1; 2.7–3.4	3.1	7.0; 3.9–9.0	5.1
Tb	2.1; 1.6–2.7	1.1; 0.9–1.2	1.6; 1.5–1.9	1.1; 0.8–1.3	0.6	0.9		
Yb	1.6; 1.3–2.3	1.3; 1.0–1.4	1.4; 1.3–1.5	1.2; 1.1–1.4	0.8; 0.7–0.8	0.9	0.9; 0.5–1.2	1.4
Lu		0.14; 0.12–0.15	0.18; 0.17–0.18		0.16; 0.14–0.18	0.12		
La/Yb	226; 184–281	118; 81–160	209; 188–240	153; 126–166	221; 190–250	153	177	108
(n)	15	9–10	2–3	2–5	1–5	3	4	2

[a]1, phlogopite lamproites, Smoky Butte (Mitchell et al. 1987); 2–3, phlogopite and maduptic lamproites, respectively, Leucite Hills (Fraser 1987); 4, phlogopite lamproites, Sisco (Peccerillo et al. 1988); 5, olivine maduptic lamproite, Hills Pond, (Cullers et al. 1985); 6, average of three olivine lamproites. Prairie Creek (Fraser 1987); 7, Priestly Peak (Foley et al. 1987); 8, Mt. Bayliss (Foley et al. 1987). (n), number of samples.

THE GEOCHEMISTRY OF LAMPROITES

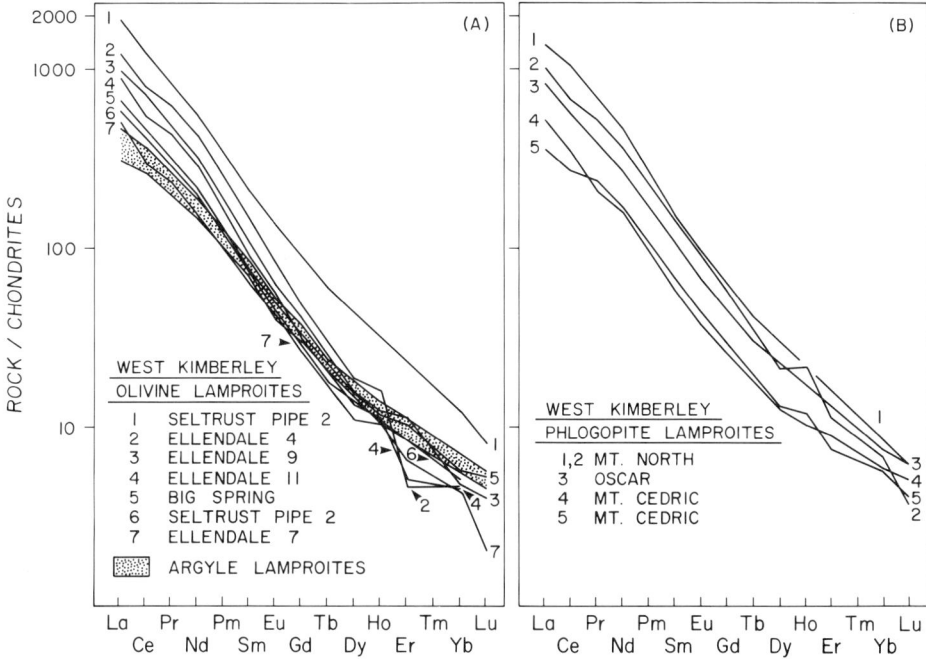

Figure 7.17. Rare earth element distribution patterns for (a) olivine lamproites from West Kimberley and Argyle and (b) phlogopite lamproites from West Kimberley. Data sources given in Table 7.18.

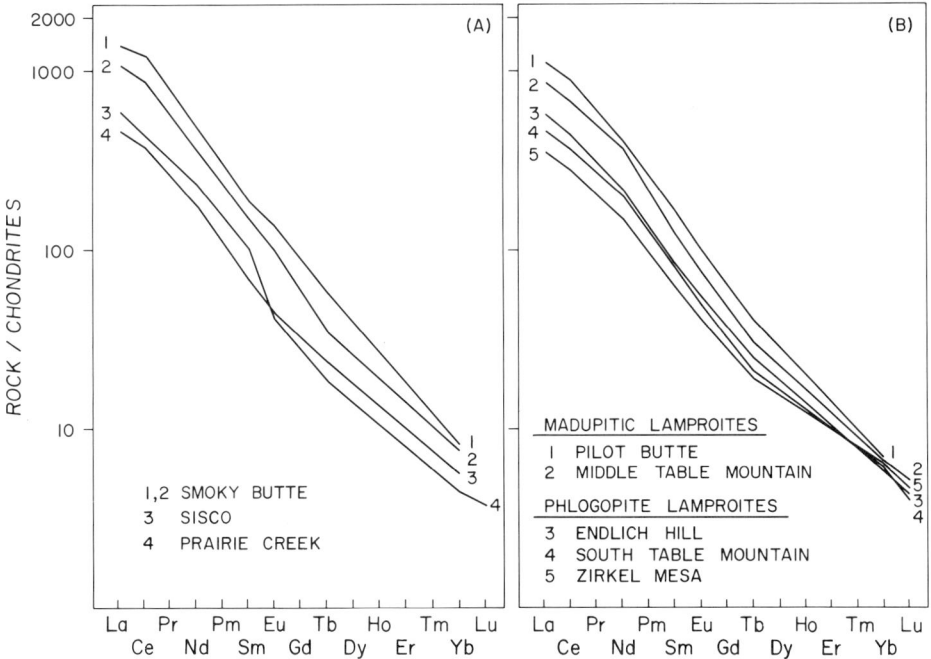

Figure 7.18. Rare earth element distribution patterns for lamproites from (a) Smoky Butte, Sisco and Prairie Creek and (b) the Leucite Hills. Data sources given in Table 7.19.

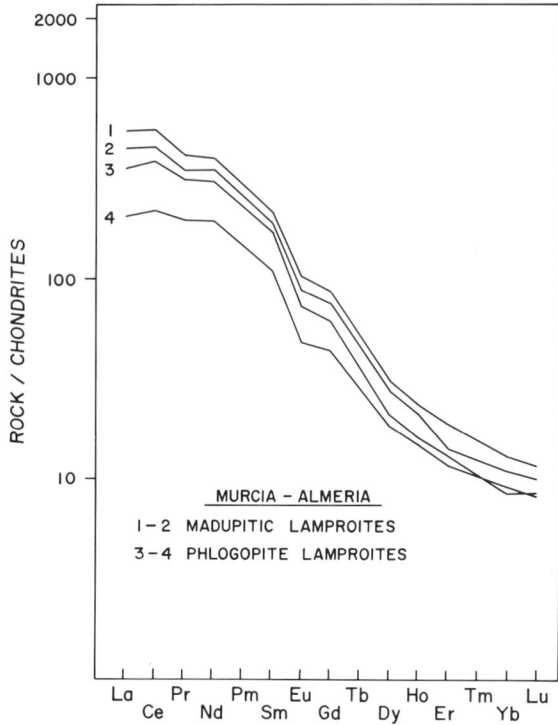

Figure 7.19. Rare earth element distribution patterns for lamproites from Murcia-Almeria (Nixon *et al.* 1984).

negative Eu anomaly (Figure 7.19). The origin of this anomaly has not yet been satisfactorily explained. Although feldspar fractionation is a plausible means of producing the Eu deficiency it should be noted that the anomaly is present in the most primitive hyalophlogopite lamproites. This suggests the anomaly may have been inherited from the source of the magma. Experimental error is not the cause of the anomaly as other lamproites analyzed by the same ICP-MS technique do not show any Eu deficiency (Nixon *et al.* 1984). The low light REE contents of the Murcia-Almeria lamproites probably are related to the derivation of these magmas from a REE-poor source. The relatively high, heavy REE abundances have been suggested by Mitchell *et al.* (1987) to reflect crustal contamination. The only other lamproites to show a negative Eu anomaly are from Sisco (Figure 7.18a, Peccerillo *et al.* 1988). These lamproites differ however in having high La/Yb ratios (Figure 7.18a, Table 7.19). Anomalies seen in the Francis lamproite distribution patterns (Figure 7.20a) are believed to result from overestimation of Tb (and Yb, Lu) abundances by Henage's (1972) INAA method as they are not present in ICP-MS derived data (this work).

REE abundances (Table 7.19) and distribution patterns (Figure 7.20b) for the Hills Pond and Rose Dome lamproites (Cullers *et al.* 1985) are similar to those of other olivine lamproites, e.g., Prairie Creek (Figure 7.18a), excepting that the heavy REE abundances are elevated, apparently due to crustal contamination.

Many micaceous or group 2 kimberlites have similar REE abundances and La/Yb ratios to lamproites, e.g., New Elands (La = 225–334 ppm, Ce = 405–647 ppm, La/Yb =

THE GEOCHEMISTRY OF LAMPROITES

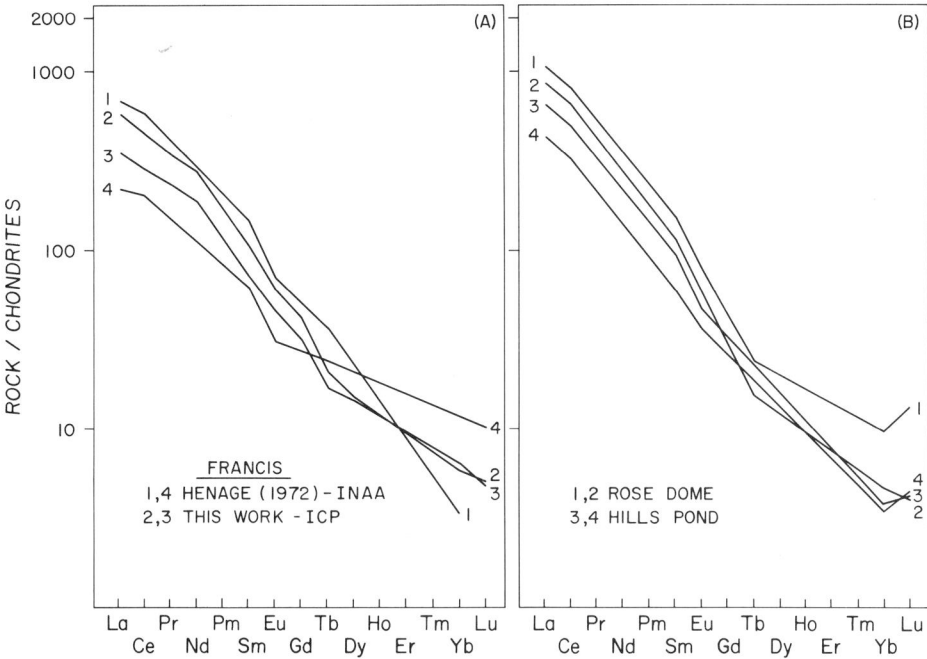

Figure 7.20. Rare earth element distribution patterns for lamproites from (a) Francis and (b) Hills Pond and Rose Dome (Cullers *et al.* 1985).

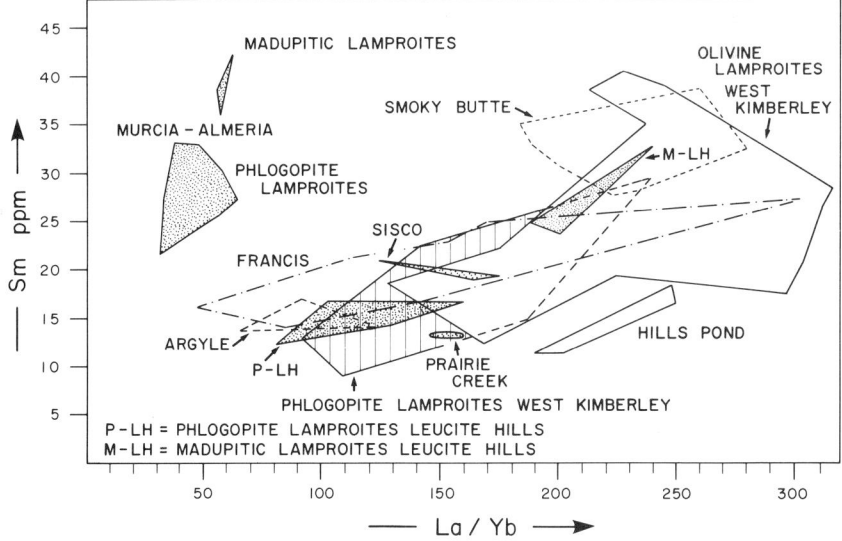

Figure 7.21. Sm content versus La/Yb ratio for diverse lamproites. Data sources given in Tables 7.18 and 7.19.

144–188, this work), Swartruggens (La = 230–259 ppm, Ce = 378–588 ppm, La/Yb = 96–192). Some micaceous kimberlites are extraordinarily rich in REE (e.g., the Main fissure of the Bellsbank group) and have REE abundances (La = 870–1120 ppm, Ce = 1910–2080 ppm, La/Yb = 235–254, Fesq *et al.* 1975) that far exceed those of the most REE-rich lamproites. Others, e.g., Finsch, are relatively poor in REE (La = 41–105 ppm, Ce = 90–227 ppm, La/Yb = 54–114, Fraser 1987). REE distribution patterns are similar to those of lamproites (Figure 7.22), with the exception that weak (Swartruggens) to strong (Bellsbank) negative Eu anomalies are found in some micaceous kimberlites. Figures 7.22 and 7.23 demonstrate that micaceous kimberlites and lamproites cannot be distinguished from each other on the basis of their REE geochemistry.

Minettes, including peralkaline varieties, generally have lower REE abundances, e.g., Navajo (La = 108–228 ppm, Ce = 248–455 ppm, La/Yb = 70–125, Rogers *et al.* 1982, Alibert *et al.* 1986); Colima (La = 40–89 ppm, Ce = 91–188 ppm, La/Yb = 28–70, Luhr and Carmichael 1982, Allan and Carmichael 1984); Linhaisai (La = 47–167 ppm, Ce = 94–340 ppm, La/Yb = 40–75, Bergman *et al.*, 1988) than most lamproites. The REE are less fractionated and distribution patterns have a shallower slope than those of lamproites. Some distribution patterns are characterized by the presence of a weak negative Eu anomaly (Figure 7.24). Figure 7.23 shows that minettes can be distinguished from most lamproites on the basis of their lower La/Yb ratios.

RPT-lavas generally have lower REE contents than lamproites, e.g., Vico (La = 103–

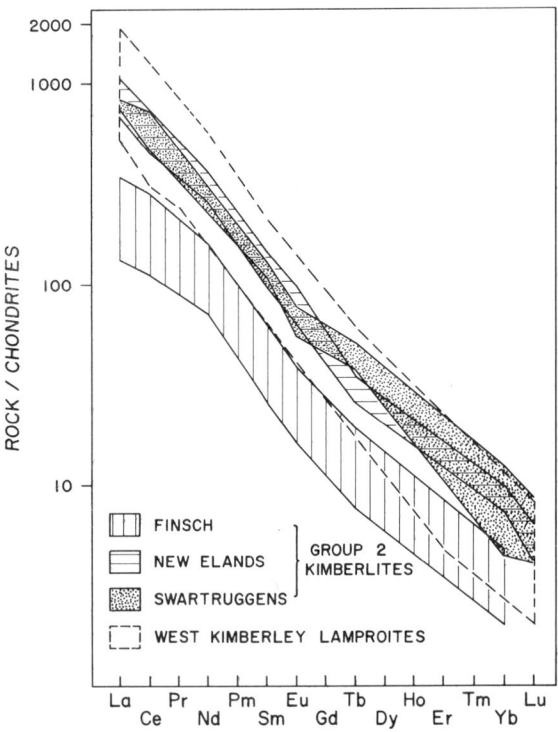

Figure 7.22. Rare earth element distribution patterns of West Kimberley lamproites compared with those of group 2 kimberlites from Finsch (Fraser *et al.* 1985), New Elands (this work) and Swartruggens (Mitchell and Brunfelt 1975).

THE GEOCHEMISTRY OF LAMPROITES

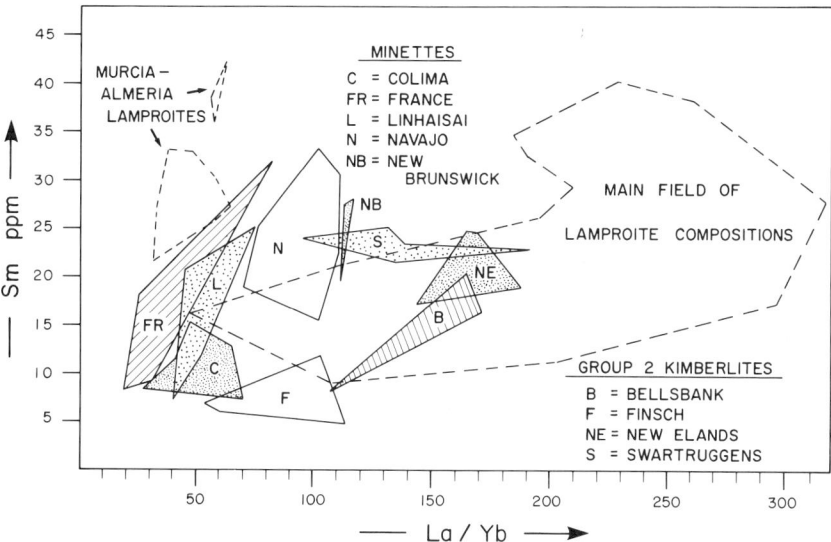

Figure 7.23. Sm content versus La/Yb ratio of lamproites, group 2 kimberlites and minettes. Data sources given in Section 7.4.1.

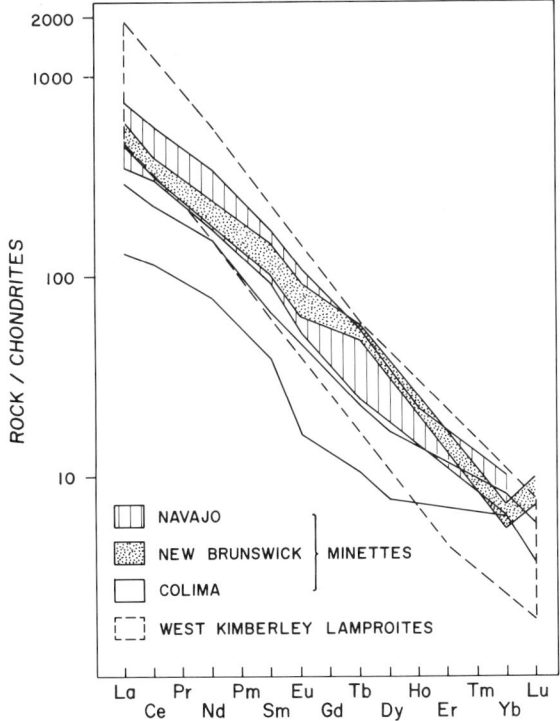

Figure 7.24. Rare earth element distribution patterns of West Kimberley lamproites compared with those of minettes. Data sources given in Section 7.4.1.

184 ppm, Ce = 223–378 ppm, La/Yb = 45–65, Barbieri *et al.* 1988); Ernici (La = 80–98 ppm, Ce = 170–212 ppm, La/Yb 31–40, Civetta *et al.* 1981). REE are not strongly fractionated and RPT-lavas typically have low La/Yb ratios (<70). Distribution patterns (Figure 7.25) thus have shallower slopes than those of lamproites and in addition are commonly characterized by a moderate negative Eu anomaly. Some extreme differentiates having high La/Yb ratios (>100) have strong Eu anomalies, reflecting extensive feldspar fractionation. Figures 7.25 and 7.26 show that RPT-lavas can be distinguished easily from lamproites on the basis of their REE geochemistry.

Kamafugitic lavas from Toro-Ankole have REE abundances that are generally lower, i.e., (La 49–236 ppm, Ce = 102–486 ppm, Mitchell and Bell 1976) than those found in lamproites. Katungite, ugandite, and mafurite form a group characterized by very high La/Yb ratios (125–319). REE distribution patterns (Figure 7.25) are linear and similar to those found in kimberlites and lamproites. Figures 7.25 and 7.26 suggest that these rocks cannot be distinguished from lamproites on the basis of their REE geochemistry.

Kamafugitic lavas from San Venanzo and Cupaello (Peccerillo *et al.* 1988) have very different REE characteristics from the African kamafugites and lamproites. Figure 7.25 shows that they have REE distribution patterns which have shallow slopes relative to lamproites and significant negative Eu anomalies. Figure 7.26 illustrates their relatively low La/Yb ratios. The REE geochemistry of these rocks suggests geochemical affinities with RPT-lavas rather than with lamproites.

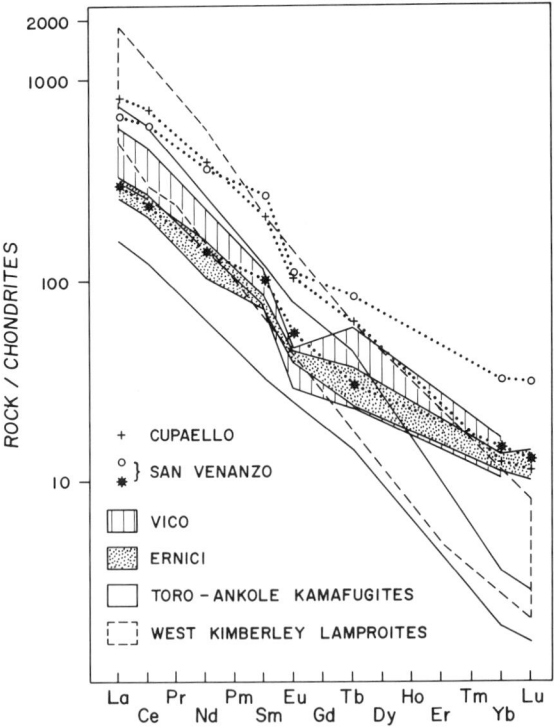

Figure 7.25. Rare earth element distribution patterns of West Kimberley lamproites compared with those of Roman province type lavas (Vico, Ernici) and kamafugitic lavas from Italy (Cupaello, San Venanzo) and Uganda (Toro-Ankole). Data sources given in Section 7.4.1.

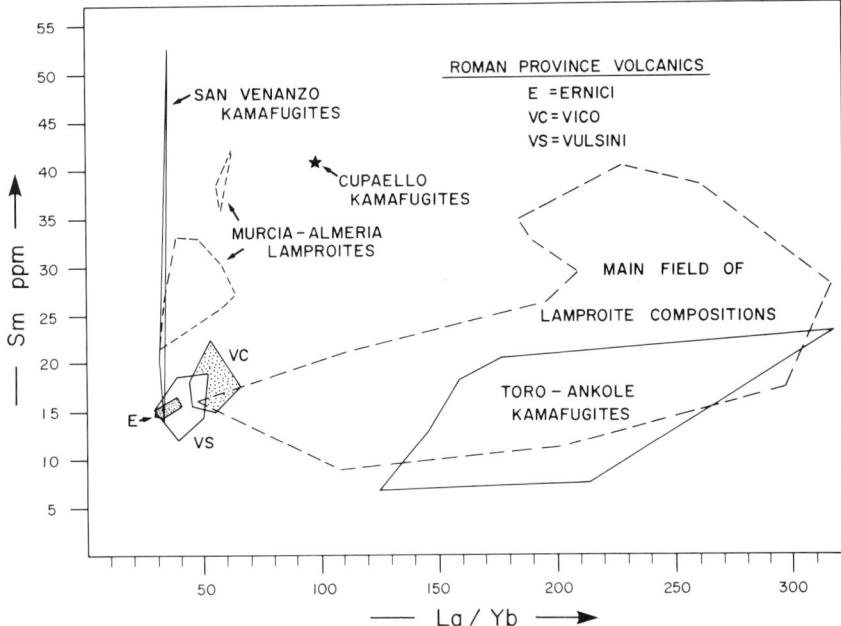

Figure 7.26. Sm contents versus La/Yb ratios of lamproites. Roman province type lavas and kamafugites.

Further discussion of REE distribution patterns and La/Yb ratios with reference to partial melting models of the upper mantle can be found in Chapter 10.

7.4.2. Yttrium

The yttrium contents of lamproites are low and no significant differences have been demonstrated within, e.g., phlogopite and olivine lamproites from West Kimberley contain 2–91 ppm Y (mean 21) and 6–38 ppm Y (mean 16), respectively, or between provinces, e.g., Leucite Hills (range 14–27, mean 17 ppm), Smoky Butte range (22–61, mean 29 ppm), Murcia-Almeria (range 24–46, mean 30 ppm). Minettes, e.g., Colima (14–29, mean 22 ppm, Luhr and Carmichael 1981, Allan and Carmichael 1984), Linhaisai (10–60, mean 29 ppm, Bergman *et al.* 1988), Northern England (22–40, mean 32 ppm, Macdonald *et al.* 1985), and group 2 kimberlites (6–32, mean 16 ppm, Smith *et al.* 1985) have similar Y contents to lamproites. RPT-lavas have slightly higher Y contents, e.g., Vulsini (10–58, mean 34 ppm, Holm *et al.* 1982), Roccamonfina (14–40, mean 51 ppm, Appleton 1970) than lamproites.

7.5. INCOMPATIBLE TRACE ELEMENTS—3: ALKALI ELEMENTS

7.5.1. Lithium

Lithium abundances are inadequately known but apparently vary between provinces. Thus, phlogopite lamproites from Murcia-Almeria contain more Li (mean 42, range 21–73 ppm, Venturelli *et al.* 1984) than phlogopite lamproites from West Kimberley (mean 9.5, range 4–37 ppm, Jaques *et al.* 1986), Smoky Butte (mean 20, range 12–33 ppm, this work) and Francis (mean 15, range 10–20 ppm, this work). Madupitic lamproites from Jumilla (mean 51, range 33–77 ppm, Venturelli *et al.* 1984a) and olivine lamproites from West Kimberley (mean 11, range 3–61 ppm, Jaques *et al.* 1986) have similar Li contents to associated phlogopite lamproites.

7.5.2. Rubidium

Rubidium is hosted primarily by phlogopite, sanidine, and leucite. Rb contents vary widely within and between provinces. The highest levels are found in the West Kimberley province (Jaques *et al.* 1986, Lewis 1987) where phlogopite lamproites contain 135–8626 (mean 457) ppm Rb. Extreme Rb contents (>1000 ppm) are found only in lamproites from 81 Mile Vent (315–8626 ppm), Mt. Percy (248–2042) and the Oscar intrusion (242–3694 ppm). The origin of the high Rb contents is unknown as there are no obvious mineralogical differences relative to low Rb lamproites. The Rb contents of other intrusions also varies widely, e.g., Mt. Cedric (143–429 ppm), Machells Pyramid (135–472 ppm). Phlogopite lamproites in general have higher Rb contents, e.g., Murcia-Almeria (mean 487, range 259–658 ppm, Venturelli *et al.* 1984a, Nixon *et al.* 1984), Leucite Hills (mean 275, range 143–460 ppm, Kay and Gast 1973, Fraser 1987), Francis (mean 347, range 172–808 ppm, Henage 1972, this work), Sisco (mean 356, range 356–380 ppm, Peccerillo *et al.* 1988) than olivine/madupitic lamproites, e.g., Murcia-Almeria (mean 145, range 83–238 ppm), Leucite Hills (mean 191, range 156–218), Sisimiut (mean 162, range 125–216 ppm, Scott 1979), Prairie Creek (mean 189, range 166–2111 ppm, Scott Smith and Skinner 1984b, Fraser 1987), Barakar (mean 87, range 38–170 ppm, Middlemost *et al.* 1988). Lamproites from Smoky Butte have low Rb contents (mean 97, range 18–180, this work). The lowest Rb contents are found in analcitized samples which are also low in K. Similar low Rb contents in some of the Murcia-Almeria and Barakar lamproites undoubtedly also result from secondary alteration. Olivine lamproites from West Kimberley are exceptional in having Rb contents (mean 471, range 208–764 ppm) similar to associated phlogopite lamproites.

In the West Kimberley province, K/Rb ratios range from as low as 5 (81 Mile vent high Rb sample) to approximately 400 in phlogopite lamproites, and from 22 to 117 in olivine lamproites (Lewis 1987). Figure 7.27 shows that there is a trend of increasing K/Rb ratio with decreasing MgO content of phlogopite lamproites. Lewis (1987) suggests that the K/Rb of 400 for the most leucite-rich lamproites is close to the K/Rb ratio of leucite, while a K/Rb ratio of 75 is close to that of phlogopite in the olivine lamproites. Lewis further suggests that the Rb content of phlogopite required to generate this K/Rb ratio should be greater than 1000 ppm. However, Mooney (1984) finds only 358 ppm in West Kimberley micas (see Section 6.1.15). K/Rb ratios of lamproite phlogopites range from 113 to 179 at Francis (Henage 1972) and are about 307 in the Leucite Hills (Kuehner 1980).

Figure 7.28 shows that the K/Rb ratios of the Leucite Hills and Francis phlogopite lamproites and the Sisimiut olivine lamproites are similar and do not correlate with MgO content. The Murcia-Almeria and Prairie Creek lamproites have low K/Rb ratios. Figure 7.28 also shows that the K/Rb ratios of Leucite Hills madupitic lamproites are less than those of associated phlogopite lamproites. Such a trend of decreasing K/Rb is to be expected if the madupitic rocks are the more evolved lamproites. This conclusion also explains the low K/Rb ratios of the West Kimberley olivine lamproites. The trend in K/Rb ratios versus MgO, observed in Figure 7.27, is thus interpreted to be one of differentiation *and* mixing of olivine

THE GEOCHEMISTRY OF LAMPROITES

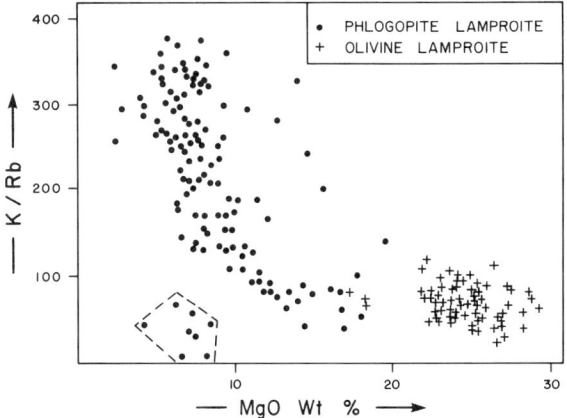

Figure 7.27. K/Rb ratio versus MgO content of olivine and phlogopite lamproites from West Kimberley (after Lewis 1987). Field enclosed by dashed line is for lamproites with unusually high Rb contents (e.g., 81 Mile Vent).

with the differentiates, and not one of simple fractionation from high to low MgO contents, as proposed by Lewis (1987). In contrast, the limited data for the Spanish province exhibit the reverse relationship. The K/Rb ratios of unaltered high K Smoky Butte lamproites are exceptionally high (480–932, Fraser 1987, this work). In all provinces there is no correlation between Li and Rb abundances.

Group 2 kimberlites have, in general, lower Rb abundances and K/Rb ratios than many lamproites, e.g., Finsch (42–182 ppm Rb, K/Rb = 156–230, Fraser 1987), Bellsbank (4–120 ppm Rb, K/Rb = 98–124, Fesq *et al.* 1975), diverse South African examples (40–273 ppm, K/Rb = 108–262, Smith *et al.* 1985). The overlap in Rb contents and K/Rb ratios precludes the general use of these parameters in distinguishing between lamproites and

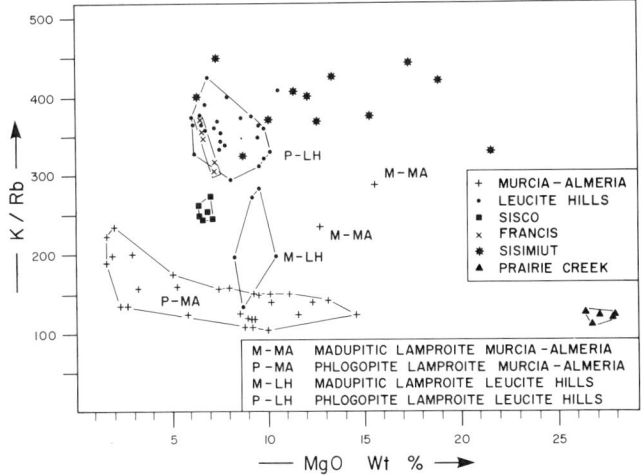

Figure 7.28. K/Rb ratio versus MgO content of diverse lamproites. Data sources given in Section 7.5.2.

kimberlites, although high values may be considered as being suggestive of a lamproite paragenesis.

Minettes exhibit a wide range in Rb contents and K/Rb ratios, e.g., Linhaisai (Rb = 30–250 ppm, K/Rb = 185–594, Bergman *et al*. 1988), Navajo (Rb = 96–237 ppm, K/Rb = 231–352, Roden 1981, Alibert *et al*. 1986), Colima (Rb = 31–60 ppm, K/Rb = 332–1039, Luhr and Carmichael, Allan and Carmichael 1984), various French localities (Rb = 117–313 ppm, K/Rb = 159–427, Turpin *et al*. 1988), Northern England (Rb = 64–239 ppm, K/Rb = 161–504, Macdonald *et al*. 1985). Insufficient data are available for minettes to permit their Rb geochemistry to be used with confidence as a discriminating parameter, although on average they appear to have lower Rb contents than those of lamproites.

RPT-lavas have high Rb contents, e.g., Vico (mean 572, range 416–785 ppm, Cundari and Mattias 1974), Vulsini (mean 439, range 246–872, Holm *et al*. 1982), Roccamonfina (mean 584, range 329–1270, Appleton 1970) that overlap with those of lamproites. Their K/Rb ratios are, however, typically low, e.g., Vico (72–183, Cundari and Mattias 1974, Barbieri *et al*. 1988), Ernici (131–196, Civetta *et al*. 1981). Similar Rb contents and K/Rb ratios are found in the nearby kamafugitic lavas of San Venanzo (Rb = 277–483 ppm, K/Rb = 145–211) and Cupaello (Rb = 445–596 ppm, K/Rb = 114–137) (Peccerillo *et al*. 1988; Holm *et al*. 1982). The Ugandan kamafugites have typically low Rb (78–200 ppm) relative to lamproites but similar K/Rb ratios (177–335, Bell and Doyle 1971). None of these potassic lavas thus have any significant differences in their Rb geochemistry from that of lamproites.

7.5.3. Cesium

Cesium abundances are inadequately known. Available data (<10 samples/province) indicate that the Cs content of phlogopite lamproites typically ranges from about 0.5 to 4.0 ppm, e.g., Leucite Hills (mean 2.06, range 1.80–2.34 ppm, Fraser 1987), West Kimberley (mean 11.3, range 0.48–2.09 ppm, Fraser 1987), Smoky Butte (mean 1.03, range 0.34–2.3, Fraser 1987, this work), Francis (mean 1.75, range 0.67–3.98 ppm, Henage 1972, this work). Wide ranges in Cs content occur in olivine/madupitic lamproites from West Kimberley (0.7–44 ppm) and the Leucite Hills (3–7 ppm). These data suggest that the Cs content of olivine/madupitic rocks is greater than that of phlogopite lamproites. However, the Cs contents of Prairie Creek madupitic lamproites is low (0.8–1.12 ppm, Fraser 1987). High Cs contents (up to 11 ppm) are found in analcitized lamproites at Smoky Butte (this work). No correlations exist between Cs and Rb contents.

K/Cs ratios are erratic and typically very high, e.g., Leucite Hills (phlogopite lamproites = 38500–53100, madupitic rocks = 8500–24400), Smoky Butte (30300–214600), West Kimberley (phlogopite lamproites = 9800–212400, olivine lamproites = 1900–102500), Prairie Creek (14900–27200). The data suggest that olivine/madupitic lamproites have low K/Cs ratios relative to phlogopite lamproites. High whole rock K/Cs ratios reflect high ratios (33200–102500) in lamproitic phlogopite (Henage 1972).

Group 2 kimberlites have similar Cs contents to lamproites, but their lower potassium contents result in much lower K/Cs ratios, e.g., Finsch (Cs = 1.75–3.0, mean 2.43 ppm, K/Cs = 8800–15400, Fraser 1987), Bellsbank group (Cs = 0.4–4.4 ppm, K/Cs = 975–3790, Fesq *et al*. 1975). The Cs content of minettes is inadequately known. Available data suggest that they have Cs contents ranging from 0.6 to 51 ppm Cs and with K/Cs ratios of approximately 900–65000 (Roden 1981, Luhr and Carmichael 1981, Allan and Carmichael 1984, Turpin *et al*. 1988, Bergman *et al*. 1988). The generally lower K/Cs ratios of group 2 kimberlites and minettes suggest that this ratio may be of use in discriminating between these rocks and phlogopite lamproites. RPT-lavas are enriched in Cs, e.g., Vico (83–499 ppm, Barbieri et al. 1988), Vulsini (10–66 ppm, Holm et al. 1982). K/Cs ratios are typically less than 4000 and thus very different from those of lamproites. Kamafugites from Uganda are the only other potassic rocks with K/Cs ratios (51000–960000) similar to those of lamproites. These high ratios result primarily from their low Cs contents (0.4–0.9 ppm, this work). A kamafugite from San Venanzo has a low K/Cs ratio (1912, Holm et al. 1982).

7.6. VOLATILE TRACE ELEMENTS: FLUORINE, SULFUR, AND CHLORINE

The high F content of lamproite micas (Section 6.1.15) results in whole rocks having high F contents (ppm), e.g., West Kimberley (phlogopite lamproite = 1100–8380, mean 3100, olivine lamproites = 1700–8610, mean 4806, Jaques et al. 1986), Smoky Butte (2500–4800, mean 4011, this work), Steamboat Mountain, Leucite Hills (7620, Aoki et al. 1981). These F contents are in general higher than those found in continental and oceanic basalts (<2000 ppm, Aoki et al. 1981, Sigvaldson and Oskarson 1986), although the absence of phlogopite in the latter makes direct comparison of little value. However, the presence of fluorphlogopite must signify a real enrichment of flourine in lamproite magmas.

Sulfur occurs principally as sulfate, and sulfides are very rare. Sulfur contents are generally low, e.g., West Kimberley (phlogopite lamproites = 40–2960, mean 501 ppm, olivine lamproites = 30–3510, mean 386 ppm, Jaques et al. 1986), Smoky Butte (100–2000, mean 644 ppm, this work). High sulfur contents may in some instances reflect the presence of barite, but Ba and S in general do not show any significant correlations (Figure 7.10).

Chlorine may be hosted by mica and apatite. The abundances are low, e.g., West Kimberley (phlogopite lamproites = 10–601, mean 127 ppm, olivine lamproites = 20–256, mean 110 ppm, Jaques et al. 1986), Smoky Butte (50–350, mean 144 ppm, this work).

7.7. OTHER TRACE ELEMENTS

In the West Kimberley province the Ga contents of olivine lamproites (1–11, mean 4 ppm) are less than those of phlogopite lamproites (7–35, mean 17 ppm). Ga × 10/Al ratios of the phlogopite lamproites (mean 2.2) are twice those of olivine lamproites (mean 2.2, Lewis 1987). Lamproites from Francis contain 18.3–20.4 (mean 19.6) ppm Ga (this work).

The Pb contents of lamproites vary between provinces, e.g., West Kimberley (phlogopite lamproites = 20–120, mean 60 ppm, olivine lamproites = 17–124, mean 51 ppm, Jaques et al. 1988), Leucite Hills (phlogopite lamproite = 24–32, mean 37 ppm, madupitic lamproites = 33–57, mean 41 ppm, Kuehner 1980, Fraser 1987), Smoky Butte (0.2–44, mean 18 ppm, Fraser 1987, this work), Murcia-Almeria (64–117, mean 86 ppm, Nelson et al. 1986).

The concentrations of elements not discussed above are inadequately known. The following concentrations (ppm) have been noted: Be 8–10 (Francis); 3–33 (West Kimberley, Lewis 1987); B 20–30 (Francis); 1–31 (West Kimberley, Lewis 1987); As 0.6–11 (Francis); 1–7 (Smoky Butte); 1–3 (West Kimberley, Jaques et al. 1986); Se 0.6–4.1 (Smoky Butte); Sn

3–5 (Francis); 1–27 (West Kimberley, Lewis 1987); Sb 0.2–10 (Francis); 0.1–0.7 (Smoky Butte); W 3–8 (Smoky Butte); Tl 0.2–0.6 (Francis). All data for Francis and Smoky Butte are from this work. Reliable data for Ge, Br, Mo, Ag, Cd, In, Te, I, Hg, and Bi are not available as their abundances in lamproites are low and at the detection limits of routine analytical methods.

7.8. INTERELEMENT RELATIONSHIPS

Average abundances of hygromagmatophile elements in phlogopite lamproites normalized to estimate primitive mantle abundances are compared in Figure 7.29. The diagram illustrates the similarities and differences between provinces and in particular emphasizes the anomalous geochemical character of the Murcia-Almeria province. The distribution patterns are in general similar being characterized by significant positive Ba, K, La, and Ce anomalies and by negative Ta, Nb, Sr, P, and Ti anomalies. Suites having the least depletion in Ta and Nb relative to K, have the highest La, Sr, and P contents. Increasing degree of Ta and Nb depletion is accompanied by decreasing La and Sr contents. The Murcia-Almeria province is different from all other provinces in exhibiting a pronounced negative Ba

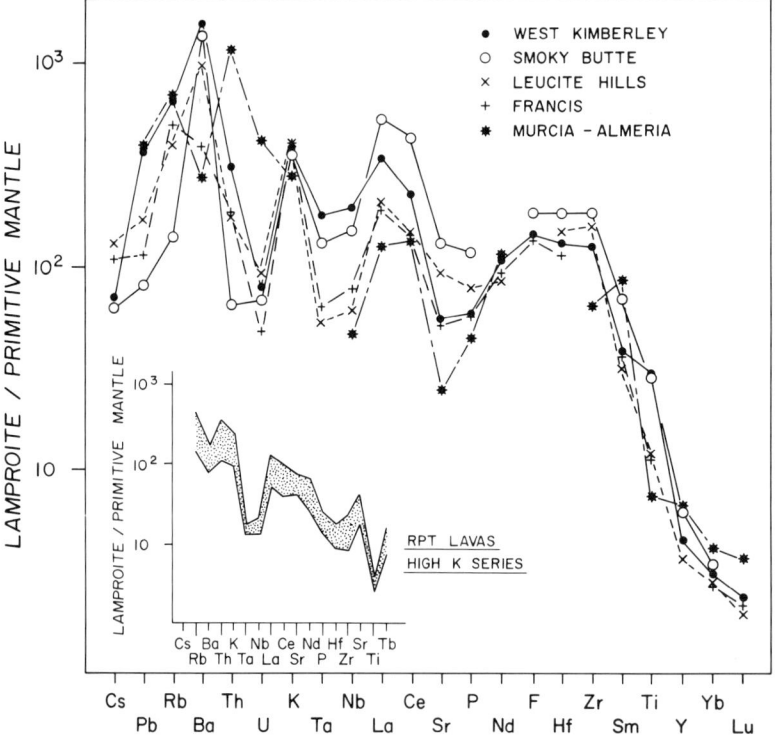

Figure 7.29. Hygromagmatophile element distribution patterns (Wood 1979) for phlogopite lamproites. Representative pattern (inset) for Roman province high-potassium-series lavas from Monti Ernici and Vulsini after Peccerillo *et al.* (1988).

anomaly and positive Th and U anomalies relative to K in addition to the greatest depletions in Ta, Nb, Sr, P, and Ti.

The anomalies in the hygromagmatophile distribution patterns may reflect either the geochemical characteristics of the sources of the magmas, or processes that have operated to different degrees upon magmas that were initially of similar geochemical character. The depletion of Ta, Nb, and Ti observed in lamproites is similar to that observed in island arc lavas (Green 1980, Arculus and Johnson 1981) and minettes such as those of the Colorado Plateau (Alibert *et al.* 1986) and Borneo (Bergman *et al.* 1988). It has been inferred by Nelson *et al.* (1986, 1989) that the presence of this geochemical feature in lamproites indicates that subduction processes have played some role in the generation of lamproitic magmas. Significant depletion in Nb and Ti in the Roman province potassic lavas (Peccerillo *et al.* 1988, Peccerillo and Manetti 1985) which are demonstrably associated with plate convergence magmatism reinforces this interpretation. However, there is no *a priori* way to separate the geochemical features that are characteristic of active subduction processes from those of ancient subduction events which have modified the mantle composition. Only the Murcia-Almeria province is located in an area where subduction has been recently active and it is in this province that the Nb and Ti depletions are greatest. The presence of similar anomalies in other provinces requires, according to this interpretation, the presence of ancient subduction zones (see Section 4.3.4).

An alternative explanation of the Ta, Nb, and Ti anomalies requires that a titanium-rich phase, not formed by subduction processes, remains in the mantle source regions of the magmas. K–Ba–REE-rich titanates, which may be metasomatically introduced components of the source rocks, could upon breakdown during partial melting give rutile as a residual phase. Retention of Ta and Nb in this phase would result in depletion of these elements and Ti in the derivative magmas. This hypothesis is the more attractive for provinces which cannot easily be related to either ancient or modern subduction zones.

Variation diagrams presented by Jaques *et al.* (1988), Scott (1979), and Venturelli *et al.* (1984) demonstrate that only Ni and Cr show significant positive correlations with major elements (Figures 7.30, 7.31). The highest Ni and Cr contents occur in chromite-bearing olivine lamproites and the positive correlation with MgO suggests that the trends represent mixing lines. Incompatible elements do not exhibit any significant correlations with major elements (Figures 7.30, 7.31), or with Ni and Cr. The absence of correlations reinforces the conclusion (Section 7.1.3.6) that the geochemistry of lamproites cannot be interpreted in terms of a simple fractional crystallization sequence.

7.9. ISOTOPIC COMPOSITION

7.9.1. Strontium and Neodymium

Table 7.20 shows that each lamproite province is characterized by particular ranges in Sr isotopic composition and that the West Kimberley and Murcia-Almeria provinces are much richer in radiogenic Sr than the North American and Greenland lamproites. Lamproites from West Kimberley exhibit the greatest range in initial ratios with phlogopite lamproites being, in general, enriched in radiogenic Sr relative to olivine lamproites. Significant differences in isotopic composition occur between and within intrusions, e.g., Mamilu Hills ($^{87}Sr/^{86}Sr$ = 0.7190–0.7215, Powell and Bell 1970), Mt. North ($^{87}Sr/^{86}Sr$ =

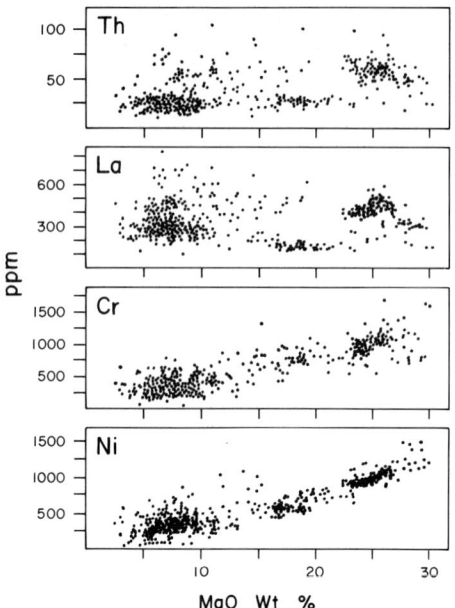

Figure 7.30. Variation diagrams for some trace elements versus MgO for lamproites from West Kimberley (after Jaques *et al.* 1986).

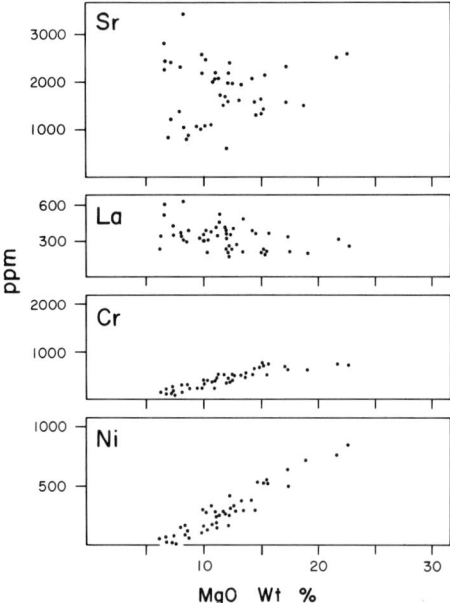

Figure 7.31. Variation diagrams for some trace elements versus MgO for lamproites from Sisimiut (afater Scott 1979).

Table 7.20. Sr and Nd Isotopic Composition of Lamproites[a]

$^{87}Sr/^{86}Sr$	$^{143}Nd/^{144}Nd$	M	ϵ_{Nd}	Ref.	(n)
West Kimberley, phlogopite lamproites:					
0.7125–0.7215				1	(8)
0.71385–0.71865	0.51103–0.51115	A	−13.1 to −15.4	2	(5)
0.71453–0.72065	0.51180–0.51203	B	−11.3 to −15.8	3	(8)
	0.51197–0.51214	B	−9.4 to −12.7	4	(3)
West Kimberley, olivine lamproites:					
0.71037–0.71482	0.51111–0.51144	A	−7.4 to −13.9	2	(8)
0.70228–0.71855	0.51163–0.51210	B	−10.0 to −19.1	3	(9)
	0.51198–0.51225	B	−7.3 to −12.6	4	(3)
Leucite Hills, phlogopite lamproites:					
0.7056–0.7070				1	(15)
0.70530–0.70779	0.51177–0.51194	B	−13.7 to −17.0	5	(37)
0.70564–0.70587	0.51172–0.51188	B	−14.6 to −17.9	6	(9)
Leucite Hills, madupitic lamproites:					
0.7057–0.7066				1	(2)
0.70539–0.70564	0.51201–0.51210	B	−10.5 to −12.3	5	(16)
0.70556–0.70560	0.51203–0.51207	B	−11.1 to −11.9	6	(2)
Smoky Butte:					
0.70583–0.70631	0.51127–0.51149	B	−21.6 to −25.9	3	(6)
Murcia-Almeria, phlogopite lamproites:					
0.71733–0.72073	0.51119–0.51126	A	−11.2 to −12.6	7	(7)
Murcia-Almeria, madupitic lamproites:					
0.7136–0.7158				1	(6)
0.71680	0.51122	A	−12.1	7	(1)
Prairie Creek:					
0.7064–0.7132				8	(34)
0.70667–0.70703	0.51193–0.51196	B	−10.6 to −11.2	6	(3)
0.70772	0.51197	B	−11	9	(1)
Gaussberg:					
0.70923–0.70978	0.51111–0.51117	A	−13.0 to −14.3	10	(7)
Sisimiut:					
0.7045–0.7061	0.51016–0.51037	A	−9.9 to −13.0	11	(4)

[a]M, Nd isotopic composition relative to BCR-1; A, 0.51183; B, 0.51262; (n), number of samples. Ref, Reference. 1, Powell and Bell (1970); 2, McCulloch et al. (1983); 3, Fraser et al. (1985); 4, Nixon et al. (1984); 5, Vollmer et al. (1984); 6, Fraser (1987); 7, Nelson et al. (1986); 8, Bolivar (1977); 9, Alibert and Albarade (1988); 10, Collerson and McCulloch (1983); 11, Nelson (1989).

0.71385–0.71453, McCulloch et al. 1983, Fraser 1987). Large differences between olivine lamproite ($^{87}Sr/^{86}Sr = 0.71037$) and leucite lamproite ($^{87}Sr/^{86}Sr = 0.71677$) are found in Ellendale 7 (McCulloch et al. 1983). Similar relationships exist in the Murcia-Almeria province where olivine/madupitic lamproites from Jumilla are of variable isotopic composition and not as enriched in radiogenic Sr as phlogopite lamproites (Table 7.20, Powell and Bell 1970, Nelson et al. 1986). In contrast, phlogopite lamproites from the Leucite Hills exhibit a remarkably restricted range in Sr isotopic composition. With one exception from the Boars Tusk agglomerate ($^{87}Sr/^{86}Sr = 0.70779$), the range in $^{87}Sr/^{86}Sr$ for sanidine-bearing and sanidine-free rocks is only 0.70530–0.70609. Madupitic lamproites have Sr isotopic compositions that are indistinguishable from phlogopite lamproites (Vollmer et al. 1984).

Table 7.20 lists the range in Nd isotopic compositions observed in lamproites. Measured and calculated initial ^{143}Nd/^{144}Nd ratios obtained in different laboratories are not directly comparable owing to the use of different standards and analytical techniques (Hawkesworth and van Calsteren 1984). To overcome this problem initial Nd isotopic compositions are commonly expressed as the deviation from the bulk earth chondritic value of ^{143}Nd/^{144}Nd at the time of formation (T) of the sample, by the relation

$$\epsilon_{Nd} = \left[\frac{^{143}\text{Nd}/^{144}\text{Nd sample initial ratio (T)}}{^{143}\text{Nd}/^{144}\text{Nd CHUR (T)}} - 1 \right] \times 10^4$$

where CHUR is the isotopic composition at time (T) of a chondritic uniform reservoir that is used to represent the Sm/Nd and isotopic composition of the bulk earth (DePaolo and Wasserburg 1976, O'Nions et al. 1979). Values of ϵ_{Nd} of zero or near zero in mantle-derived rocks indicate undifferentiated primitive mantle sources in terms of their Sm/Nd ratios. Positive or negative values require that at least one episode of fractionation has increased or decreased the source Sm/Nd ratio relative to the chondritic ratio.

Table 7.20 shows that ϵ_{Nd} values range from -7.4 to -25.9 for all lamproites, indicating that these magmas are derived from, or contain a contribution from, an old source enriched in the light REE (low Sm/Nd ratios) relative to the bulk earth. The least radiogenic Nd compositions occur in the Smoky Butte lamproites. Within the West Kimberley province phlogopite lamproites have in general less radiogenic Nd than olivine lamproites (Table 7.20). In common with Sr, there are distinct variations in the isotopic composition of Nd between and within intrusions, e.g., Ellendale 7 (olivine lamproite $\epsilon_{Nd} = -7.8$, phlogopite lamproite $\epsilon_{Nd} = -15.4$, McCulloch et al. 1983), Mt. North $\epsilon_{Nd} = -12.7$ to -14.4, McCulloch et al. 1983, Fraser 1987). Insufficient data are available for the Murcia-Almeria province to determine whether or not similar relationships exist. In the Leucite Hills, madupitic lamproites are richer in radiogenic Nd ($\epsilon_{Nd} = -10.5$ to -12.3) than phlogopite lamproites ($\epsilon_{Nd} = -13.7$ to -17.9, Vollmer et al. 1984, Fraser 1987). Slight variations in initial Nd ratios are found at each locality, e.g., South Table Mountain ($\epsilon_{Nd} = -16.2$ to -17.0), Steamboat Mountain ($\epsilon_{Nd} = -15.4$ to -15.6), Zirkel Mesa ($\epsilon_{Nd} = -15.4$ to -17.9).

Current opinion (McCulloch et al. 1983, Vollmer et al. 1984, Fraser et al. 1985, Nelson et al. 1986, Bergman 1987) holds that due to the high REE and Sr contents of lamproites they are unlikely to have been subjected to significant crustal contamination, and the isotopic variations observed reflect those of their mantle sources. Thus, Figure 7.32 suggests that the West Kimberley and Murcia-Almeria lamproites originated from old sources enriched in Rb with low Sm/Nd ratios. In contrast, the Smoky Butte, Leucite Hills, and Sisimiut lamproites were derived from sources with relatively low Rb/Sr and Sm/Nd ratios. Gaussberg and Prairie Creek isotopic compositions fall between these limiting trends. The source regions of the Smoky Butte lamproites were apparently even more enriched in light REE than those of any other province. Intraprovincial isotopic variations reflect either derivation of the magmas from isotopically different sources or that they are the products of mixing of two (or more) components of radically different isotopic characteristics. Thus, McCulloch et al. (1983) interpret the spread in isotopic compositions observed in the West Kimberley province to be due to mixing between enriched and depleted mantle. The former corresponds to the composition of midoceanic ridge basalts (MORB, $\epsilon_{Nd} = +10$) and the latter to the most radiogenic Sr and least radiogenic Nd compositions in the West Kimberley rocks. In Figure 7.32 madupitic and phlogopite lamproites from the Leucite Hills occupy different fields. Vollmer et al. (1984) suggested that the observed compositions result from mixing of

THE GEOCHEMISTRY OF LAMPROITES

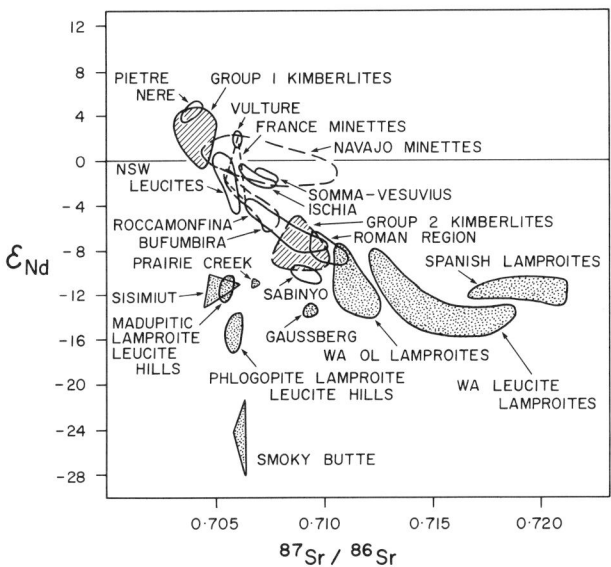

Figure 7.32. Isotopic composition of Sr and Nd in lamproites, kimberlites, Roman province type lavas, leucitites and minettes. Data sources given in Table 7.20 and Section 7.9.1.

a component with $\epsilon_{Nd} = 0$ and one with a highly negative ϵ_{Nd} value. Mitchell *et al.* (1987) have suggested that the Smoky Butte source meets the latter requirements. The above mixing models differ only in the isotopic compositions of the depleted and enriched end-member components. The sources of these components may be sought in the asthenosphere and the lithosphere, respectively. Importantly, the Leucite Hills data suggest, in contradiction to the model of McCulloch *et al.* (1983), that lamproites with Sr and Nd isotopic compositions corresponding to those of the enriched source end member may not be found in a given province.

An alternative explanation is that the large variation in Sr and Nd isotopic composition in a given lamproite province is due to derivation from a heterogeneous source (Bergman 1987). In such a source, discrete domains of diverse Rb/Sr and Sm/Nd ratio would have existed for long periods of time (>1–2 Ga.). Partial melting without mixing of the derivative magmas is required to explain the observed isotopic compositional variation.

Group 2 kimberlites have Sr and Nd isotopic compositions (Smith *et al.* 1983, Fraser *et al.* 1985) that also indicate derivation from an old light REE-enriched source. In contrast, group 1 kimberlites are enriched in radiogenic Nd suggesting derivation from an old light REE-depleted asthenospheric source. Other potassic volcanic rocks define a trend of compositions (Figure 7.32) between bulk earth and those of West Kimberley olivine lamproites. These data suggest that these rocks are derived from sources similar in character to those of lamproites but not as enriched in light REE. The few data for minettes (Turpin *et al.* 1988, Alibert *et al.* 1986) plot close to the bulk earth composition (Figure 7.32) demonstrating either derivation from different sources than those of lamproites or a greater contribution of depleted mantle in their genesis. Figure 7.32 indicates that Sr and Nd isotopic compositions cannot be used with confidence to discriminate between lamproites and other potassic rocks, although lamproites are characterized in general by less radiogenic Nd (lower ϵ_{Nd}) compositions.

7.9.2. Lead

Table 7.21 and Figure 7.33 demonstrate that each lamproite province is characterized by Pb of a particular isotopic composition, implying that the $^{238}U/^{204}Pb$ ratios (μ values) of their source regions were very different. Insufficient data are available to ascertain intraprovincial isotopic characteristics, although olivine and phlogopite lamproites in the West Kimberley province appear to have similar isotopic compositions. All of the data, with the exception of Pb from Murcia-Almeria, are characterized by low $^{206}Pb/^{204}Pb$ and high and variable $^{207}Pb/^{204}Pb$ ratios. They plot to the left of the geochron (Figure 7.33), implying that they cannot have had a single-stage evolution and that their sources must have experienced ancient fractionation events which reduced their U/Pb relative to those of midoceanic ridge and oceanic island basalts. Isotopic variation within each province may be explained by mixing of a low $^{206}Pb/^{204}Pb$, high $^{207}Pb/^{204}Pb$ component with lead derived from a region with higher U/Pb ratios, e.g., a MORB-type reservoir. The former component may reside in the lithosphere and represent regions that have had a multistage history. Nelson et al. (1986) and Fraser et al. (1985) suggest that this component was produced in three stages. These

Table 7.21. Isotopic Composition of Pb in Lamproites[a]

$^{206}Pb/^{204}Pb$	$^{207}Pb/^{204}Pb$	$^{208}Pb/^{204}Pb$	Ref.	(n)
West Kimberley phlogopite lamproites:				
17.24–17.54	15.72–15.77	37.80–38.41	1	(6)
17.23–17.58	15.71–15.76	37.87–38.31	2	(5)
West Kimberley olivine lamproites:				
17.41–17.88	15.69–15.80	38.05–38.59	1	(7)
17.27–17.57	15.69–15.75	38.05–38.42	2	(4)
Leucite Hills phlogopite lamproites:				
17.15–17.28	15.43–15.47	37.16–37.34	3	(7)
17.23–17.30	15.46–15.49	37.15–17.31	6	(?)
Leucite Hills madupitic lamproite:				
17.54	15.50	37.55	3	(1)
17.56–17.59	15.46–15.49	37.53–37.60	6	(?)
Murcia-Almeria phlogopite lamproites:				
18.66–18.82	15.69–15.74	39.03–39.20	1	(7)
Murcia-Almeria madupitic lamproite:				
18.80	15.72	39.15	1	(1)
Smoky Butte:				
16.02–16.47	15.19–15.28	36.20–36.58	2	(6)
Prairie Creek:				
16.70–16.77	15.31–15.35	36.66–36.80	5	(3)
16.64	15.40	35.77	3	(1)
Gaussberg:				
17.59–17.60	15.65–15.67	38.40–38.48	1	(2)
Sisimiut:				
14.23–15.01	15.01–15.12	35.84–36.46	4	(3)

[a](n), number of samples. Ref, Reference. 1, Nelson et al. (1986); 2, Fraser et al. (1985); 3, Fraser (1987); 4, Nelson (1989); 5, Alibert and Albarede (1988); 6, Salters and Barton (1985).

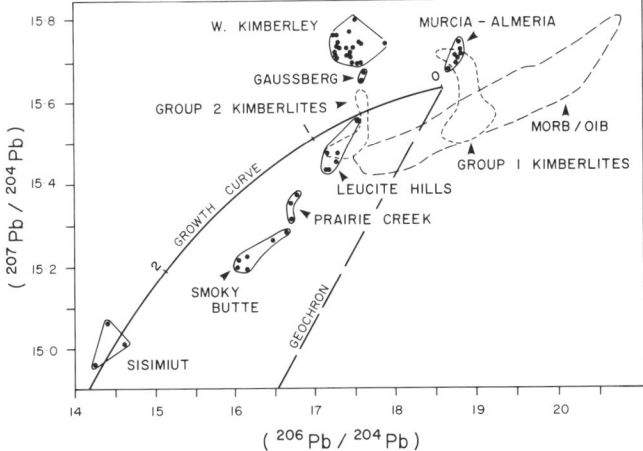

Figure 7.33. Isotopic composition of Pb in lamproites and kimberlites. Data sources for lamproites given in Table 7.21. Kimberlite data from Smith (1983) and Fraser (1987). Growth curve is from Stacey and Kramers (1975). Field of compositions of basalts from midoceanic ridges (MORB) and oceanic islands (OIB) after Fraser (1987).

models are required to account for the West Kimberley (together with the Gaussberg, Priestly Peak, and Mt. Bayliss) lamproites plotting above the average mantle Pb growth curve (Figure 7.33). Fraser *et al*. (1985) suggest that μ values in the three stages were 8.3, 9.2, and 5.6, respectively, and that fractionation occurred at 3.5 Ga and 1.8 Ga. Nelson *et al*. (1986) suggest that U/Pb ratios changed such that $\mu_0 = 8 < \mu_1 > \mu_2 > 8$, and that the initial fractionation events must have occurred about 2.1 Ga ago. Lamproites that plot below the mantle Pb growth curve have isotopic compositions that may be interpreted in terms of two-stage models. Fraser *et al*. (1985) attribute the Smoky Butte isotopic composition to U and Th depletion of the mantle taking place 2.5 Ga ago and resulting in a second-stage μ value of approximately 4.1. Similar, though less extreme, depletions can account for the Leucite Hills, Prairie Creek, and Sisimiut Pb isotopic compositions. The times of these fractionation events are compatible with the time scales required to generate the observed Sr and Nd isotopic compositions. However, simultaneous depletion of U and Th and addition of light REE would seem improbable given the similar geochemical character of all of these incompatible elements. Hence, it is considered that the U and Th depletion event is unlikely to have coincided with an event leading to light REE enrichment. The latter must have occurred subsequent to the U and Th depletion and not involved further addition of these elements.

The Spanish lamproites are anomalous relative to other lamproites in that they exhibit a large range in ^{207}Pb/^{204}Pb accompanied by a very limited range in ^{206}Pb/^{204}Pb ratios, and plot to the right of the geochron in Figure 7.34. These relationships are considered by Nelson *et al*. (1989) to be due to mixing of highly radiogenic Pb (continental crust or sediments) with Pb having a low ^{207}Pb/^{204}Pb ratio such as the depleted mantle source of MORB (see below).

Heaman (1989) has shown that perovskite isolated from Prairie Creek lamproite contains Pb that is more radiogenic (initial ^{206}Pb/^{204}Pb = 17.15, ^{207}Pb/^{204}Pb = 15.41) than Pb found in whole rock samples from this intrusion (Table 7.20). These data show that there is more than one isotopic component in the total rock Pb. Data for Sr and Nd in the same perovskite are within the ranges reported previously for whole rock samples.

Group 2 kimberlites have Pb isotopic compositions similar to those of the Leucite Hills lamproites and share with lamproites in general the characteristic of being derived from ancient sources of low U/Pb ratio (Figure 7.33). In contrast, group 1 kimberlites are apparently derived from asthenospheric sources (Fraser *et al.* 1985, Smith 1983).

Figure 7.34 shows that lamproites, with the exception of Murcia-Almeria, have Pb isotopic compositions unlike those of RPT-lavas and minettes, these latter rocks plotting to right of the geochron. Their isotopic compositions can be explained in terms of mixing of a depleted mantle component with a recycled crustal component (Vollmer and Hawkesworth 1980, Hawkesworth and Vollmer 1979, Alibert *et al.* 1986, Turpin *et al.* 1988). The Pb isotopic compositions of the Murcia-Almeria lavas are identical to those of RPT-lavas. They are, however, richer in radiogenic Sr and lie on an extension of the Sr–Pb isotopic correlations observed in the province (Nelson *et al.* 1986). These data suggest that both provinces contain the same Pb component of continental crustal origin (Nelson *et al.* 1986). The Spanish lamproites may, however, contain at least three isotopic components originating in the asthenosphere, lithosphere, and the crust respectively, and unambiguous interpretation of the isotopic data is difficult.

7.9.3. Oxygen

Systematic oxygen isotopic studies of lamproites have not yet been undertaken. Kuehner (1980) determined the whole rock $\delta^{18}O$ of five Leucite Hills lamproites to range

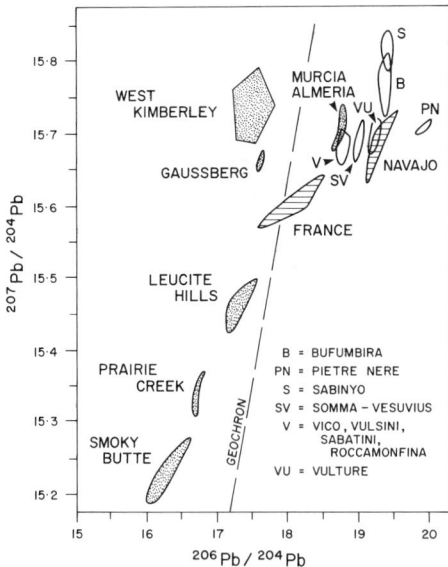

Figure 7.34. Isotopic composition of Pb in lamproites compared with that of Roman province type lavas, minettes, and potassic rocks from Bufumbira and Sabinyo. Lamproite data sources given in Table 7.21. Other data from Hawkesworth and Vollmer (1979), Vollmer (1976, 1977), Vollmer and Hawkesworth (1980), Vollmer and Norry (1983), Alibert *et al.* (1986), and Turpin *et al.* (1988).

from +8.8 to +12‰, and the $\delta^{18}O$ of phenocrystal phlogopite to be +8.8‰. These rocks are RPT-lavas ($\delta^{18}O = +6$ to +16‰, Taylor and Turi 1976, Taylor et al. 1979) and are enriched in ^{18}O relative to typical mantle-derived mafic rocks ($\delta^{18}O = +5.5$ to +7.4‰, Taylor 968). The enrichment in ^{18}O in the RPT-lavas is ascribed to assimilation of crustal material (Taylor and Turi 1976). Such a process is unlikely to be involved in the formation of the Leucite Hills magmas. The high $\delta^{18}O$ of the phlogopite phenocrysts suggests that the high $\delta^{18}O$ of the rocks is a primary feature (Kuehner 1980). Isotopic variation in the province may also result from rock–water interactions. The Gaussberg lamproites display primary mantle $\delta^{18}O$ signatures (+6.5‰) when corrected for near surface alteration effects (Taylor et al. 1984). More data are required before it can be shown whether or not the source regions of lamproites are anomalous in their $\delta^{18}O$ composition compared with those of other mantle-derived rocks.

7.10. SUMMARY

Relative to most alkalic mafic rocks, lamproites are enriched in the light REE, Ba, Sr, Zr, F, Cr, and Ni. Excluding the Murcia-Almeria suite they may be distinguished from RPT-lavas on the basis of their higher Ba, Sr, Nb, Zr, Ta, Hf, Cr, Ni, La, Ce contents and K/Cs ratios, and lower V, U, and Th contents. Lamproites typically have higher Ba, Zr, Nb, Hf, and Ta contents and lower V contents than minettes. Group 2 kimberlites have higher Cr, Ni, and Co contents and lower K/Cs ratios than lamproites.

Minettes and RPT-lavas have lower La/Yb ratios than lamproites and their REE distribution patterns typically exhibit negative Eu anomalies. Only the Murcia-Almeria suite exhibits such anomalies. Lamproites, group 2 kimberlites, and kamafugites have very similar REE distribution patterns and high La/Yb ratios.

Madupitic lamproites from the Leucite Hills and Murcia-Almeria are enriched in Ba, Sr, Th, REE, depleted in Ni and have lower K/Rb ratios than associated phlogopite lamproites. This observation suggests that madupitic lamproites are derivatives of phlogopite lamproites. The relationship of the West Kimberley olivine lamproites to the phlogopite lamproites is not clearly defined by geochemical data, although the former appear to be enriched in REE, Ba, and Rb and have lower K/Rb ratios than the latter, suggesting that they are derived from the more evolved fractions of lamproite magma.

Lamproites, with the exception of the Murcia-Almeria suite, have Sr, Nd, and Pb isotopic compositions which indicate derivation from old light REE-enriched sources with low U/Pb ratios. Interprovincial differences in isotopic composition indicate that the degree of light REE enrichment and the Rb/Sr ratio of these sources differed within and between cratons. Mixing of Sr, Nd, and Pb from 2 or 3 distinct sources is required to explain the isotopic variation found within each province.

Lamproites from Murcia-Almeria are anomalous with respect to their depleted Ba, Nb, Ta, and Ti contents, enrichment in Th and U and radiogenic Pb and Sr isotopic signatures. They have many geochemical similarities with RPT-lavas and minettes and it may be concluded that the Murcia-Almeria suite has been contaminated by crustal material.

Visita interiora terrae rectificando invenies occultum lapidem
　　　　　　　　Basil Valentine-Azoth

Experimental Studies Relevant to the Formation and Crystallization of Lamproites

Experimental studies of lamproites are useful in determining low-pressure crystallization sequences and inferring the mineralogy of the source regions of these magmas. In such studies the composition of the rock used as a starting material must be as close as possible to that of an initial partial melt whose composition has not been modified by differentiation, assimilation, or hybridization. The compositions of phlogopite lamproites may come close to meeting this criterion, but it is unlikely that xenocryst-bearing olivine/madupitic lamproites represent unmodified primary partial melts (see Chapter 7).

8.1. LOW-PRESSURE STUDIES OF LAMPROITES

8.1.1. Anhydrous Melting Relationships

The first experimental study of lamproites, undertaken by W.S. Fyfe and reported by Carmichael (1967), demonstrated the wide melting interval and high liquidus temperatures of lavas from the Leucite Hills. Leucite phlogopite lamproites and sanidine phlogopite lamproites were determined to have solidus and liquidus temperatures from <1000 to 1010°C and 1165 to 1275°C, respectively, at atmospheric pressure. The highest liquidus temperatures were found for olivine-phyric sanidine-bearing lamproite, while a madupitic lamproite gave 1245°C and 1040°C for the liquidus and solidus temperature. The initial liquidus phases were leucite in the leucite phlogopite and madupitic lamproites, and olivine or olivine plus leucite in the sanidine leucite phlogopite lamproite. Phlogopite was not found as a low-pressure initial liquidus phase, due to its instability at high temperature in these open system experiments. Carmichael (1967) interpreted the data to indicate that the Leucite Hills rocks could not be the products of crustal fusion.

In a series of similar melting experiments Sobolev *et al*. (1975) reported significantly higher liquidus temperatures (1332°C) for a Leucite Hills phlogopite lamproite. Diopside was the initial liquidus phase, followed by leucite at 1233°C, with complete cystallization occurring at about 1000°C.

Studies of melt inclusions in minerals by Bazarova and Krasnov (1975) and Sobolev *et al*. (1975) also demonstrate the wide melting interval of leucite phlogopite lamproites from the Leucite Hills. Phenocrystal phlogopite and diopside contain 10–50 μm primary melt inclusions, consisting of gas, plus glass with and without crystals, which homogenize from 1240 to 1270°C. Inclusions in the cores of host crystals gave the highest homogenization temperatures. Euhedral crystals within the inclusions, optically identified as leucite, begin to melt at 1100°C. Primary leucites contain 5–10 μm, one-phase glass inclusions. On heating these develop a small (2–3% volume) gas bubble which rehomogenizes at 1150–1250°C. Pseudosecondary two-phase inclusions, occurring in fractures in phlogopite phenocrysts, homogenize at 1040–1100°C. The data are interpreted by Bazarova and Krasnov (1975) to indicate initial high-temperature crystallization of phlogopite, followed by diopside then leucite. Although the postulated melting interval of 1270–1040°C is in broad agreement with Carmichael's (1967) study, the temperatures inferred for phlogopite crystallization are not in accord with the upper stability limit of 1150°C for phlogopite in these lamproites as reported by Barton and Hamilton (1978). Consequently, the temperatures reported by Bazarova and Krasnov (1975) must be regarded as overestimations.

8.1.2. Water-Saturated Melting Relationships

Barton and Hamilton (1978) determined the phase relationships of three Leucite Hills lamproites under water-saturated conditions up to 5 kbar total pressure (Figure 8.1). A wide melting interval and high liquidus temperatures were found for all compositions. Madupitic lamproites exhibited the highest liquidus and solidus temperatures. Water-saturation reduces the liquidus temperatures at 1 kbar by 100–140°C relative to the 1 bar anhydrous liquidus.

Figure 8.1 shows that leucite is a primary liquidus phase in lamproites only at low pressures. The maximum stability of leucite is dependent upon silica content and varies from 0.5 kbar in sanidine phlogopite lamproite (55 wt % SiO_2) to 2 kbar in madupitic lamproite (44 wt % SiO_2). Leucite may be the sole initial liquidus phase or it may coexist with olivine and/or clinopyroxene. In phlogopite lamproites with decreasing temperature leucite reacts with liquid to form sanidine. Low-temperature, low-pressure near-solidus assemblages consist of sanidine, amphibole, clinopyroxene, and phlogopite with or without leucite. Differences in the low-temperature phase relationships shown in Figure 8.1 stem from differences in bulk composition and possibly from the presence of quench or metastable phases. Sanidine does not appear at all in the phase diagram for the madupitic rock as the bulk composition is inappropriate for its formation.

For phlogopite lamproite compositions, olivine is a primary phase only at high temperatures and low pressures. With increasing pressure olivine becomes unstable and the onset of phlogopite crystallization is marked by a decrease in the abundance of olivine. The primary liquidus phase assemblage above 1 kbar is dominated by phlogopite and clinopyroxene. With decreasing temperature, previously formed olivine reacts with the liquid to form phlogopite and clinopyroxene. Phase relations of a madupitic lamproite composition show that olivine and clinopyroxene persist as liquidus phases to 5 kbar and that phlogopite is never an initial liquidus phase. Clinopyroxenes are stable over a wide range of pressure and temperature. Amphiboles are confined to low temperatures but remain stable up to 5 kbar pressure.

Of the above experiments, those on phlogopite lamproites are of most relevance to the low-pressure crystallization history of lamproitic magmas, as these rocks are closest in composition to parental magmas. The experimentally determined order of crystallization is

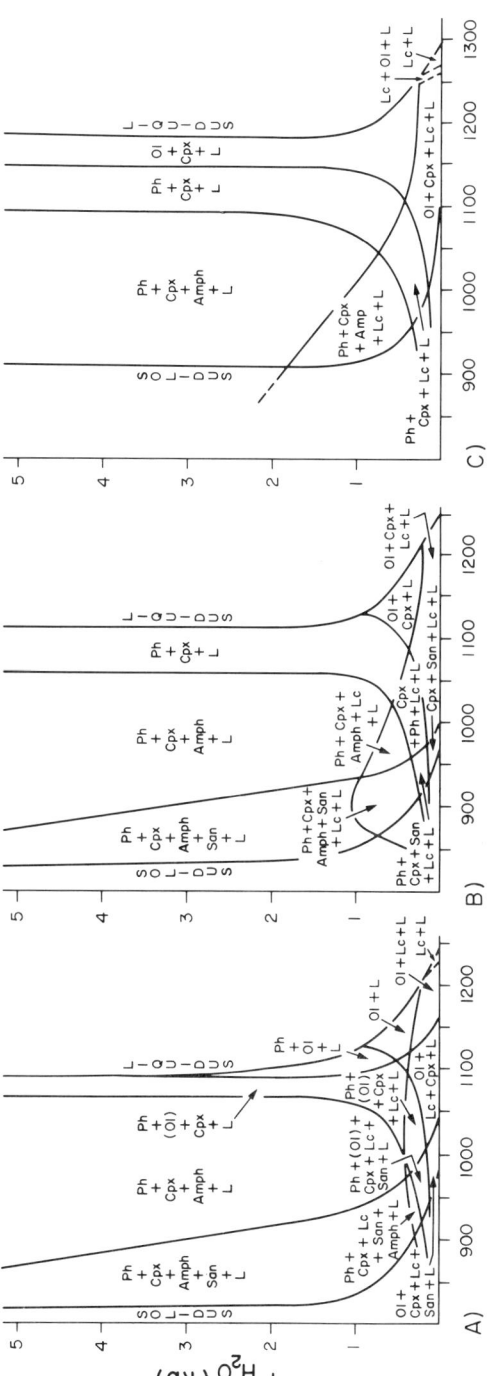

Figure 8.1. Water-saturated phase relationships of (a) sanidine phlogopite lamproite, (b) leucite phlogopite lamproite, (c) madupitic lamproite (after Barton and Hamilton 1978). Ph, phlogopite; Cpx, clinopyroxene; Amph, amphibole; San, sanidine; Ol, olivine; Lc, leucite; L, liquid.

in reasonable agreement with that observed in lamproite lavas, as phlogopite is predicted to be an early formed phenocrystal phase followed by diopside. Leucite is expected to form only at low pressure immediately before or subsequent to eruption. Depending upon the local cooling history and bulk composition, lavas are predicted to have a wide range in the proportions of leucite and sanidine present. Rapid quenching will result in the assemblage phlogopite, diopside, leucite, plus glass. Slower cooling may result in sanidine-bearing assemblages which may or may not contain leucite. The phase relationships explain the common heteromorphy observed in silica-rich phlogopite lamproites from the Leucite Hills. The occurrence of amphibole at low temperatures explains the restriction of this phase to the groundmass where it coexists with sanidine in many lamproites, e.g., Smoky Butte, Jumilla, Francis. It is cautioned, however, that the above conclusions are not necessarily indicative of the actual crystallization paths of lamproites, as there is no evidence that their evolution has taken place under such high water pressures.

The data also explain the absence of olivine in many lamproites as it is shown to crystallize only over a relatively restricted range of temperature and pressure. Euhedral primary phenocrystal olivine is thus demonstrated to be a high-temperature low-pressure phase. Reaction of olivine with liquid is predicted and observed in the skeletal phenocrystal olivines in Smoky Butte hyalo-olivine phlogopite lamproites, and in many olivine madupitic lamproites from Prairie Creek and Ellendale. Anhedral olivines are undoubtedly xenocrystal phases with the common occurrence of phlogopite coronas about these crystals resulting from their assimilation, promoting the precipitation of liquidus phlogopite (see Sections 6.1.2.2 and 8.3). Formation of priderite, a common microphenocrystal phase in Leucite Hills lavas, was not observed in the experimental study.

The compositions of madupitic lamproites from the Leucite Hills are not representative of their parental magmas. Their low silica contents result from the modal predominance of phlogopite. Moreover, the geochemistry and mineralogy of these rocks suggest that they crystallized from evolved liquids. Thus the phase relationships established by Barton and Hamilton (1978) above 1 kbar have no relevance to the crystallization of lamproites in general, as the bulk composition is not that of primary lamproitic magma. The low-pressure experimental data are in accord with the observed primary silicate mineralogy of the rocks. However, perovskite, apatite, and spinel, important constituents of these rocks, were not observed to form in the experimental study. Barton and Hamilton (1978) ascribe the absence of olivine in these rocks to resorption of this phase when phlogopite crystallizes. As an alternative we suggest that olivine, assuming it was ever a liquidus phase, may have been removed by fractional crystallization prior to eruption. A resorption hypothesis is thus unnecessary.

Barton and Hamilton (1978) note that the very high liquidus temperatures of lamproite liquids, even at 5 kbar pressure, are much higher than the solidus temperatures of crustal material and these magmas cannot be partial melts of such material. The data also show that cooling of lamproitic magma at pressures above 1 kbar would result in the formation of phlogopite pyroxenites. This observation can account for the occurrence of inclusions of this type in some of the Leucite Hills lavas (Emmons Cone, Hatcher Mesa).

8.2. HIGH-PRESSURE PHASE RELATIONSHIPS OF NATURAL LAMPROITES

Near-liquidus assemblages of minerals in a lamproite melt at high pressure represent phases that may have been in equilibrium with the melt and the mantle. Thus determination

FORMATION AND CRYSTALLIZATION OF LAMPROITES

of this phase assemblage at a specific temperature, pressure, and volatile composition is a means of inferring the mineralogy of the source regions of that magma.

8.2.1. Phlogopite Lamproites

Barton and Hamilton (1982) have determined the water-undersaturated melting relationships of a sanidine leucite phlogopite lamproite from the Leucite Hills at pressures up to 5 kbar. The experiments were not buffered with respect to oxygen fugacity and carried out using Pt capsules. Barton and Hamilton (1982) estimate that 50%–80% of the Fe in the starting material is lost to the container during the experimental runs. Loss of Fe is not considered by Barton and Hamilton (1982) to alter the relative order of appearance of phases but it is important to note that the phase diagram shown in Figure 8.2 is for the Fe-depleted material. Consequently minerals crystallized in the experiments are anomalously Mg-rich and their compositions are not directly comparable to those of minerals in the natural rock.

Figure 8.2 shows that the near liquidus phase assemblage up to 27 kbar is dominated by olivine, orthopyroxene, and clinopyroxene. Olivine is the sole primary liquidus phase up to 12 kbar. At this pressure it is joined by orthopyroxene. Barton and Hamilton (1982) note that there is a reaction relationship between olivine and liquid to produce orthopyroxene. Olivine is considered to continue to crystallize at higher pressures, although Barton and Hamilton

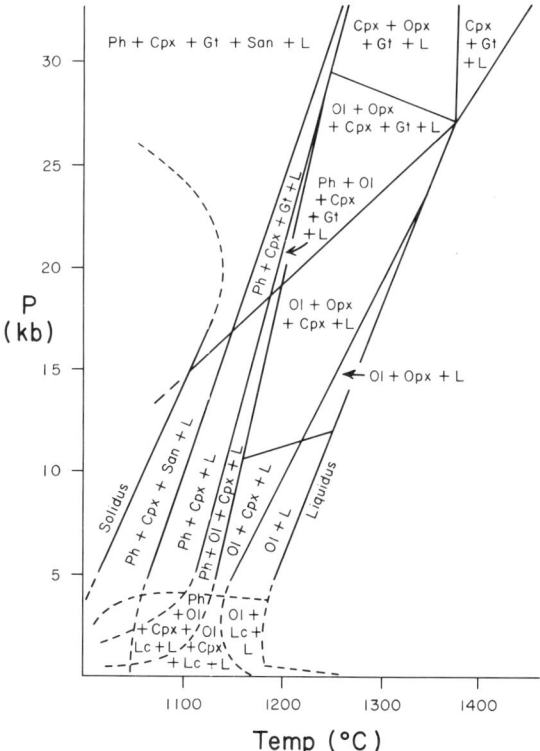

Figure 8.2. Water-undersaturated phase relationships of sanidine phlogopite lamproite (after Barton and Hamilton 1982). Gt, garnet; Opx, orthopyroxene; other abbreviations as in Figure 8.1.

(1982) stress that identification of the very small crystals makes determination of its presence or absence difficult. Clinopyroxene appears on the liquidus at 25 kbar and Barton and Hamilton (1982) contend that olivine, two pyroxenes and garnet occur together at the liquidus 27 kbar over a limited temperature range. Above 27 kbar pressure only clinopyroxene and minor garnet are found as liquidus phases. Phlogopite is not found at any pressure as a near-liquidus phase.

Phlogopite occurs primarily as a near-solidus phase where is may be produced by a variety of reactions involving olivine and pyroxenes. Figure 8.2 shows that possible cumulates from a fractionating magma at 5–20 kbar pressure are phlogopite clinopyroxenites, wehrlites and lherzolites.

Barton and Hamilton (1982) conclude from their experiments that phlogopite lamproite may be produced by the partial melting, under H_2O-rich conditions, of a garnet lherzolite source enriched in K and incompatible elements. The source mineralogy is based upon the inferred presence of olivine as a near-liquidus phase at 27 kbar pressure. However, given the difficulty of determining the presence or absence of olivine this conclusion may be incorrect. Further, Barton and Hamilton (1982) did not report the composition of the garnet formed in their experiments and it is thus unknown whether or not it was a typical mantle-type chrome pyrope. As the near-liquidus assemblage above 27 kbar is dominated by clinopyroxene, a phlogopite pyroxenite source would seem preferable. Phlogopite is required in the source to account for the high K content of the derivative magmas.

Arima and Edgar (1983a) determined the liquidus and subliquidus phase relations (Figure 8.3) of a diopside leucite richterite transitional madupitic lamproite (wolgidite) from Mt. North in the Ellendale field of the West Kimberley province. Experiments were carried out in Ag–Pd capsules to limit Fe loss to less than 20% of the total Fe. One set of experiments was run without adding H_2O, the volatiles representing those of the rock itself (3.22 wt % H_2O, 0.93 wt % CO_2). In the other set H_2O was added such that the charge contained 13 wt % H_2O and 0.84 wt % CO_2. In both sets of experiments the liquids formed were vapor undersaturated.

Phase relations for the H_2O-added runs are given in Figure 8.3a. Olivine is the liquidus phase up to 24 kbar at which pressure it is replaced by orthopyroxene. Phlogopite crystallizes about 50°C below the liquidus at 10 kbar and its crystallization temperature and proximity to the liquidus increase with increasing temperature. Rutile coexists with phlogopite and orthopyroxene at pressures above about 16 kbar. Below this pressure it is replaced by armalcolite. Clinopyroxenes form only below 1025°C and at low pressures are associated with priderite.

Phase relations for the runs without added H_2O are given in Figure 8.3b, which shows that orthopyroxene and olivine are the high (16–30 kbar) and low (<16 kbar) pressure liquidus phases, respectively. Phlogopite crystallizes together with orthopyroxene at temperatures of about 30°C below the liquidus at 30 kbar.

Arima and Edgar (1983a) suggest that the near-liquidus phase relationships indicate that this lamproite could have been derived by the partial melting of a highly metasomatized mantle containing rutile, phlogopite, orthopyroxene, and olivine, i.e., phlogopite harzburgite. The experiments demonstrate that orthopyroxene and phlogopite could crystallize as phenocrysts during ascent of the magma and olivine will form only as a relatively low pressure phase. The data agree relatively well with the observed crystallization sequence of phenocrysts in this lamproite with the exception that armalcolite is not present in the natural assemblage. Lamproites from Smoky Butte contain phenocrysts that have crystallized in the

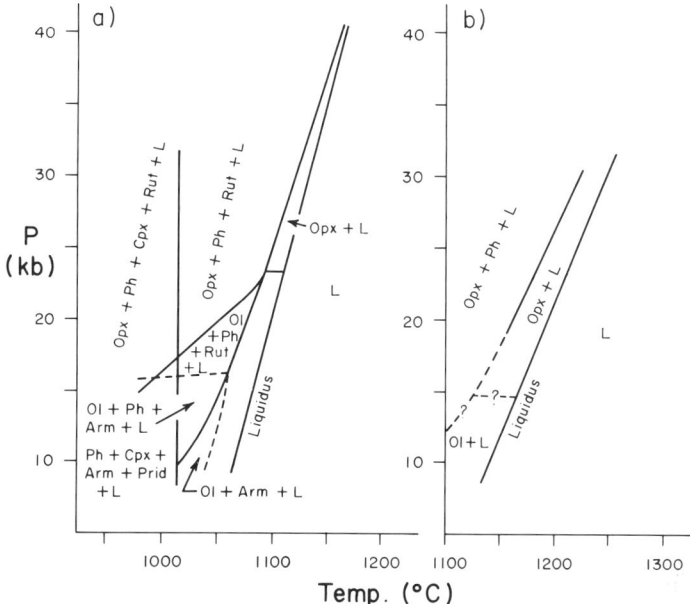

Figure 8.3. Phase relationships of a transitional madupitic lamproite as determined by Arima and Edgar (1983a) with (a) H$_2$O added and (b) no H$_2$O added. Opx, orthopyroxene; Ol, olivine; Ph, phlogopite; Cpx, clinopyroxene; Rut, rutile; Arm, armalcolite; prid, priderite.

sequence olivine, phlogopite, armalcolite, and diopside. Priderite forms after these phases and prior to groundmass richterite and sanidine. The phenocrystal assemblage is similar to that forming at temperatures of 1025–1075°C and 10–15 kbar pressure in the experimental study. Although the Smoky Butte rocks have a different bulk composition to that used by Arima and Edgar (1983a), the experimental study demonstrates the correct relative stability of priderite and armalcolite, and the possibility of crystallizing at near-liquidus temperatures, mineral assemblages equivalent to those observed in the Smoky Butte lamproites. The study demonstrates that orthopyroxene may be high-pressure phenocryst in lamproites, accounting for the common occurrence of this phase in hyalo-enstatite phlogopite lamproites from the Murcia-Almeria province.

Arima and Edgar's (1983a) study cannot be compared directly with that of Barton and Hamilton (1982) as the bulk compositions of the lamproites used differ significantly. The sample chosen by Arima and Edgar (1983a) is relatively low in SiO$_2$ (44.7 wt %) and Al$_2$O$_3$ (2.8 wt %) in addition to being rich in MgO (14.4 wt %). This atypical phlogopite lamproite composition may not be representative of the parental liquid as it reflects the high modal content of phlogopite.

Phlogopite leucite lamproites from the Gaussberg volcano, are likely to represent relatively unevolved lamproite liquids, and their liquidus phase relationships (Figure 8.4) have been determined by Foley (1986, 1989). The experiments were carried out using graphite capsules enclosed in Ag–Pd capsules. Oxidation conditions were kept just below those of the carbon–water buffer by means of equilibration with an iron–wustite mixture. Under these conditions the vapor composition is dominated by water although minor

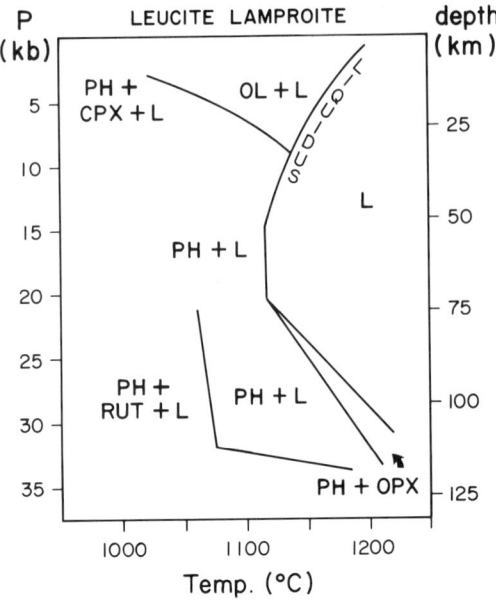

Figure 8.4. Phase relationships for a synthetic lamproite composition based upon that of a leucite lamproite from Gaussberg (after Foley 1989). PH, phlogopite; CPX, clinopyroxene; OPX, orthopyroxene; RUT, rutile; OL, olivine; L, liquid.

amounts of methane are present. Figure 8.4 shows that orthopyroxene is the liquidus phase above 25 kbar. Mica is a near-liquidus phase at high and low pressures but only coexists with orthopyroxene at 25 kbar. With decreasing temperature phlogopite and rutile become the near-liquidus phases. Olivine and clinopyroxene are not present as liquidus phases at the pressures and temperatures of the experiments. Foley (1986) contends that the absence of olivine on the high pressure liquidus is due to the lamproite not representing a primary magma, i.e., it has undergone olivine fractionation, or the experiments contain more water than is present in the natural melting conditions. Foley (1986, 1989) concludes that in natural systems, at pressures greater than 20 kbar, the liquid is saturated in phlogopite, olivine, and orthopyroxene and, therefore, the lamproite may be derived by the partial melting of a phlogopite harzburgite under H_2O-rich reduced conditions. The high-pressure near-liquidus phase relations of the Gaussberg lamproite are remarkably similar to those of the Mt. North lamproite (Arima and Edgar 1983a) despite the difference in bulk composition.

Phase relationships for a leucite diopside lamproite from Mt. Gytha in the Noonkanbah field of the West Kimberley province have been determined by A.D. Edgar (personal communication). Clinopyroxene and phlogopite are the high-pressure (30 kbar) liquidus phases, suggesting derivation of the magma from a phlogopite pyroxenite source. Olivine appears together with phlogopite and clinopyroxene as a liquidus phase at pressures below 20 kbar.

8.2.2. Olivine and Madupitic Lamproites

Experimentally determined phase relationships for olivine and madupitic lamproites are briefly described below. However, it should be clearly understood that we do not believe

FORMATION AND CRYSTALLIZATION OF LAMPROITES

these rocks represent the composition of the initial melt produced by the partial melting of the source (see Sections 6.1 and 7.1.3.6). Consequently, the high-pressure near-liquidus phase assemblage cannot be used to infer the nature of the source. The data may be useful in determining how evolved or hybrid melts may crystallize at a variety of temperatures and pressures.

Barton and Hamilton (1979) have determined the water-undersaturated melting relationships of a madupitic lamproite from Pilot Butte, Leucite Hills up to 30 kbar pressure. At low pressures (<5 kbar) leucite is the dominant liquidus phase (Figure 8.5). Leucite is replaced by clinopyroxene and olivine at <5–7 kbar, and by clinopyroxene at pressures above 7 kbar. At all pressures there is a reaction relationship with falling temperature between melt, olivine, and clinopyroxene to produce phlogopite. Apatite is a near-solidus phase at all pressures. Garnet occurs in very small amounts above 25 kbar pressure. As expected for this low SiO_2 bulk composition, sanidine and orthopyroxene do not crystallize under the experimental conditions. Barton and Hamilton (1979) concluded that this lamproite was derived from a mica clinopyroxenite source. The data indicate that crystallization of this madupitic lamproite over a wide range of pressures could give rise to apatite phlogopite clinopyroxenite cumulates (see Section 8.1.2).

The high proportions of xenocrystal olivine in West Kimberley olivine lamproites

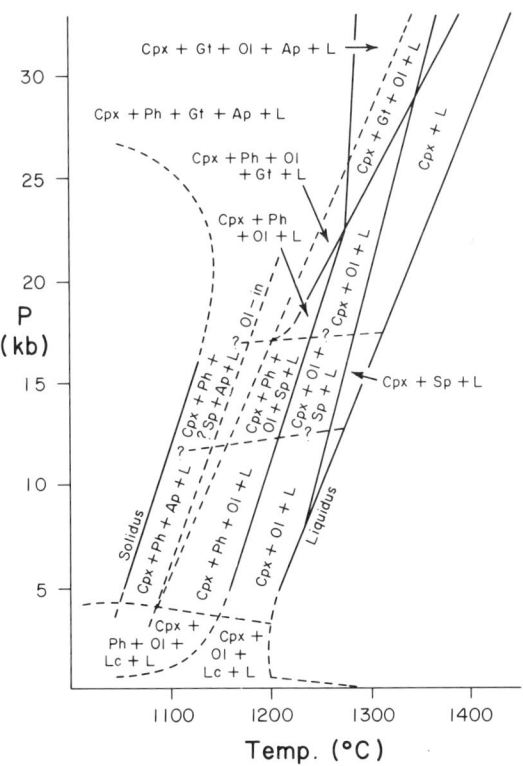

Figure 8.5. Water-undersaturated phase relationships to 30 kb for a madupitic lamproite from the Leucite Hills (after Barton and Hamilton 1979). Cpx, clinopyroxene; Gt, garnet; Ol, olivine; Ap, apatite; Ph, phlogopite; sp, spinel; Lc, leucite; L, liquid.

makes such rocks particularly inappropriate choices for experimental studies. Nevertheless, Foley (1986, 1989) has determined the near-liquidus phase relationships of such an olivine lamproite under H_2O-rich reduced conditions (Figure 8.6). Not surprisingly olivine is the sole liquidus phase over a wide pressure range (1–40 kbar) for this MgO-rich, SiO_2-poor composition. With decreasing temperature, olivine and phlogopite are subliquidus phases up to 35 kbar. Above this pressure orthopyroxene and phlogopite coexist. Clinopyroxene crystallizes only as a low temperature (<1025°C) and low pressure (<20 kbar) phase. Extrapolating the phase relationships to 55 kbar, Foley (1986, 1989) claims that phlogopite, orthopyroxene, and olivine will coexist on the liquidus at that pressure, thus suggesting derivation from a phlogopite harzburgite source.

Walker and Edgar (1989), in a study of a madupitic lamproite containing 6.4 wt % H_2O from Prairie Creek, Arkansas, have determined the phase relationships at 10–40 kbar and 1000–1500°C, with an oxygen fugacity estimated between Ni–NiO and QFM. Olivine is a near liquidus phase over the $P-T$ range examined and is joined by clinopyroxene, phlogopite, chromite, and perovskite with decreasing temperature. The crystallization sequence agrees with that observed in the natural rock, except that leucite and richterite were not produced in the experiments. Addition of water (15 wt %) to the rock results in a reversal of the crystallization sequence of phlogopite and clinopyroxene relative to that of the natural rock. Walker and Edgar (1989), on the basis of these data, claim that the lamproite was hydrated during eruption, this process increasing the original volatile content and oxygen fugacity. Walker and Edgar (1989) in this preliminary study were unable to determine the source rock assemblage for this magma.

8.3. SYNTHETIC SYSTEMS

Ruddock and Hamilton (1978) have investigated the system leucite–diopside–silica–H_2O to 4 kbar pressure. Diopside and phlogopite are the first two liquidus phases followed

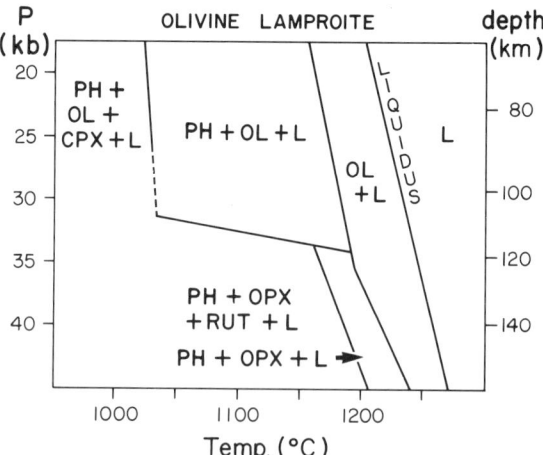

Figure 8.6. Phase relationship to 40 kb for an olivine lamproite composition based upon that of a West Kimberley olivine lamproite (after Foley 1989). PHL, phlogopite; CPX, clinopyroxene, OPX, orthopyroxene; OL, olivine; RUT, rutile; L, liquid.

FORMATION AND CRYSTALLIZATION OF LAMPROITES

by sanidine and quartz. This work explains the observed crystallization sequence of many lamproites and is in agreement with Luth's (1967) observations that sanidine and silica are late-forming phases.

Studies of the system $KAlSiO_4$–Mg_2SiO_4–SiO_2 (Ks–Fo–Qtz) with H_2O, CO_2 or F are of use in modeling the low-pressure crystallization of potassic rocks and the high-pressure melting relationships of a potassic mantle. *The major limitation of this system with regard to natural lamproites is that it does not include any Ca-bearing phases.* Phase relationships in the anhydrous system (Wendlandt and Eggler 1980a) are not directly applicable to lamproites because of the absence of phlogopite. Kuehner *et al.* (1981) have, however, used the anhydrous and CO_2-bearing systems as models for lamproite petrogenesis (see Chapter 10).

Luth (1967) studied the system Ks–Fo–Qtz with H_2O at pressures up to 3 kbar. Phase relationships are extraordinarily complex as they involve 35 ternary and quaternary univariant reaction curves, and 5 quaternary invariant points. Figure 8.7 depicts the *inferred* phase relationships on the H_2O-saturated liquidus surface projected onto the anhydrous ternary base of the quaternary system. Liquids on this surface are saturated with H_2O and coexist with crystals and vapor. The most important conclusion of the work is that phlogopite is not a low-pressure liquidus phase below 500 bars. Above this pressure phlogopite appears on the liquidus and its stability field expands with increasing pressure at the expense of the leucite field. Luth (1967) recognized three important reactions in the system that probably also occur in natural rocks: olivine plus liquid reacting to form phlogopite; phlogopite plus liquid reacting to give leucite and pyroxene; and the formation of sanidine by reaction of leucite and liquid.

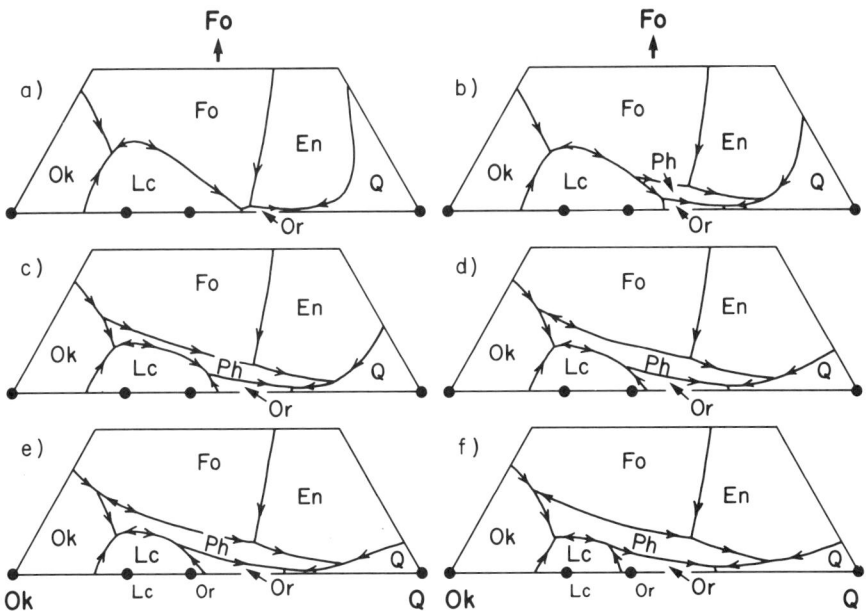

Figure 8.7. Inferred phase relationships on the saturation surface of the system $KAlSiO_4$-Mg_2SiO_4-SiO_2-H_2O represented as projections of the saturation surface onto the anhydrous base of the tetrahedron from the H_2O apex. (a)–(f) are isobaric-polythermal sections from 0.5 kb to 3 kb at 0.5 kb intervals (after Luth 1967). Fo, forsterite; En, enstatite; Q, quartz; Ph, phlogopite; Or, potash feldspar; Lc, leucite; Ok, orthorhombic $KAlSiO_4$.

The boundary curve on the saturation surface involving phlogopite, olivine, liquid, and vapor is thus a reaction curve. Liquids that initially precipitate olivine will reach this curve and olivine will react to produce phlogopite. Depending upon the bulk composition and crystallization conditions a large number of crystallization paths may be followed. For example, olivine may be completely reacted out and the liquid may leave the boundary curve, subsequently traversing the phlogopite field to one of the boundary curves involving phlogopite and sanidine or leucite. Reaction of leucite with liquid may eliminate this phase and the final assemblage may consist of potash feldspar and phlogopite. Other paths may result in the precipitation of pyroxene prior to phlogopite and potash feldspar formation. Fractional crystallization may result in the failure of early olivine to react with the liquid, and in some instances olivine rimmed by phlogopite may be preserved. Addition of olivine xenocrysts to a magma precipitating phlogopite would also result in the formation of phlogopite-rimmed olivines. The restriction of phlogopite to relatively high pressures implies that phlogopite phenocrysts in a magma will, upon eruption, undergo resorption to form leucite and pyroxene. During fractional crystallization early formed leucite may also be preserved. Luth's (1967) study mirrors, in a general manner, the low-pressure crystallization history of phlogopite lamproites, the formation of leucite prior to potash feldspar, the restriction of potash feldspar to the later stages of crystallization, and the common occurrence of resorbed phlogopite phenocrysts. The work also suggests that madupitic micas must form at relatively high H_2O pressures. *A major difference with respect to natural lamproites is the absence of diopside and amphibole in the experimental studies.*

Phase relations in the system Ks–Fo–Qtz–H_2O at 20 kbar (Sekine and Wyllie 1982) and 28 kbar (Gupta and Green in preparation from Foley *et al.* 1986a) are shown in Figure 8.8. The topology of the H_2O-saturated liquidus, and thus crystallization paths, are similar to those found at low pressures, except for the absence of leucite. Liquids that coexist with phlogopite, enstatite, and olivine at the peritectic point A, range from being very potassic, SiO_2-rich and MgO-poor at 20 kbar to silica-saturated and MgO-rich at 28 kbar. Foley *et al.*

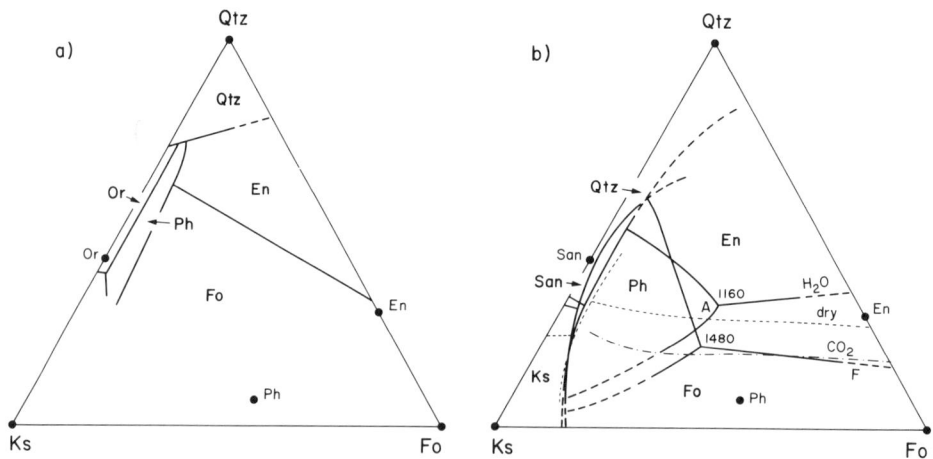

Figure 8.8. Phase relationships in the systems (a) Ks-Fo-Qtz-H_2O at 20kb (after Sekine and Wyllie 1982) and (b)Ks-Fo-Qtz, Ks-Fo-Qtz-CO_2, and Ks-Fo-Qtz-F at 28 kb (after Foley *et al.* 1986a). Qtz, quartz; En, enstatite; Phl, phlogopite; Fo, forsterite; Or, orthoclase; San, sanidine; Ks, kalsilite.

(1986a) note that the peritectic point is a simple analogue for a phlogopite harzburgite. Partial melting of such material at 1160°C and 28 kbar pressure under H_2O-rich conditions will generate a relatively silica-rich MgO-rich magma, i.e., phlogopite lamproite. Phase relationships at higher pressures are not yet known but it may be expected that the peritectic will retreat further towards Fo and that liquids will be correspondingly richer in MgO. [N.B. Foley et al. (1986b) consider that the position of the peritectic given by Sekine and Wyllie (1982) is erroneous owing to excessive extrapolation of Luth's (1967) low pressure data.]

Figure 8.8 also depicts phase relationships for Ks–Fo–Qtz–F at 28 kbar (Foley et al. 1986a). In these experiments 4 at. % of the oxygen in the starting compositions was replaced by fluorine. This amounts to about 4 wt % F. The system differs from the H_2O-saturated system in that the enstatite field is enlarged. The peritectic point A lies on the Ks side of the sanidine–forsterite join and liquids coexisting with phlogopite, orthopyroxene, and olivine are now silica-undersaturated. The increased stability of fluorphlogopite relative to hydroxyphlogopite also results in an increase in the temperature of the peritectic point to 1480°C.

The position of the enstatite–forsterite field boundary in the fluorine-bearing system is similar to that in the system Ks–Fo–Qtz–CO_2 (Figure 8.8, Foley et al. 1986a), although this latter system does not contain a phlogopite field (Wendlandt and Eggler 1980a). Phase relationships in the system Ks–MgO–Qtz–H_2O–CO_2 (Wendlandt and Eggler 1980b) indicate that phlogopite is stable over a wide range of pressures. The position of the peritectic point as a function of CO_2/H_2O ratios has not been determined. Melting relations are more complicated in CO_2-bearing systems as a consequence of carbonation reactions resulting in the appearance of magnesite and dolomite. Studies of the melting of carbonated systems at their vapor buffered solidii suggest that melts may vary from carbonatite to melilitite to kimberlite with increasing pressure (Wendlandt and Eggler 1980b, Wyllie 1980, Brey 1978). These studies are not described here, as CO_2-bearing systems are not directly relevant to the genesis of lamproite magmas in which CO_2 is an insignificant component. This observation is based upon the very low CO_2 contents of fresh phlogopite and olivine/madupitic lamproites and the typical absence of calcite as a primary phase. The lack of primary carbonates is in marked contrast to their common occurrence in kimberlites, melilitites, and kamafugitic rocks. Lamproite genesis is believed to occur in a F-rich regime where $H_2O/H_2O + CO_2$) ratios are very high (Foley et al. 1986a,b). Primary silica-rich magmas may only be formed under these conditions as the polymerization effects of high CO_2 contents are expected to lead to the formation of low silica melts (Bergman 1987). Foley et al. (1986b) suggest that the presence of HF and CH_4 (see Section 8.4) will assist in the formation of silica-rich liquids, as in H_2O-rich conditions their presence promotes depolymerization. Foley et al. (1986b) further suggest that lamproites of differing SiO_2 content result from changes in the position of the peritectic. The same source may thus give rise to SiO_2-rich or SiO_2-poor lamproites at different pressures without any difference in the volatile composition. A greater depth of origin for olivine lamproite is postulated than for phlogopite lamproite. (See Chapter 10 for a discussion of this hypothesis.)

Isotopic studies of lamproites (Section 7.9) have indicated that they are unlikely to be derived from simple phlogopite garnet lherzolite sources. This conclusion is supported by the high-pressure near-liquidus phase relations described above. Consequently studies of melting and phase relations of the system phlogopite–lherzolite–H_2O–CO_2 (Bravo and O'Hara 1975, Modreski and Boettcher 1973, Wendlandt and Eggler 1980b) are not of direct relevance to lamproite genesis. These systems demonstrate the persistence of phlogopite to high pressure (>50 kbar) in the upper mantle; however, partial melts of such sources are typically K_2O poor relative to lamproites, although they may be quartz and/or hypersthene normative (Bravo and O'Hara 1975).

Modreski and Boettcher (1972, 1973) studied the stability of phlogopite in the presence of enstatite and/or diopside between 2 and 35 kbar under vapor-present and vapor-absent conditions (Figure 8.9). These experiments are relevant to lamproites, if they are derived from phlogopite harzburgite or phlogopite clinopyroxenite sources. The studies show that phlogopite is stable to greater depths under vapor-absent conditions than when vapor is present. Introduction of vapor to the assemblage phlogopite–pyroxene will result in melting, because phlogopite, enstatite (or diopside), and vapor react to form forsterite plus liquid. Figure 8.9 shows that phlogopite is not consumed at the solidus. Modreski and Boettcher (1972) extrapolated the reaction curves to higher pressures and suggested that, in regions of low geothermal gradient, phlogopite may be stable to depths greater than 175 km. Liquids produced during melting under vapor-present conditions vary widely in their composition, depending upon the proportions of phases and the pressure and temperature of melting. None of the liquid compositions correspond to those of lamproites in that they are typically richer in SiO_2, Al_2O_3, and K_2O and poorer in MgO and CaO. The studies are important as they demonstrate that silica-rich potassic liquids may be formed in the upper mantle by the incongruent melting of enstatite and phlogopite. The work shows that simple

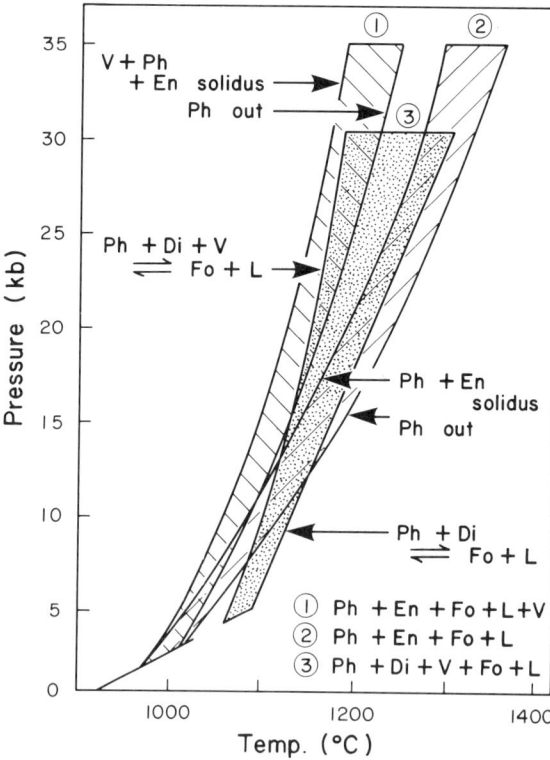

Figure 8.9. Stability of phlogopite (phl) in association with diopside (Di) or enstatite (En) with and without vapor (V) and Modreski and Boettcher (1972, 1973). Fo, forsterite.

phlogopite pyroxenites or harzburgites cannot be the source of lamproites. The presence of other phases in these sources, as suggested by isotopic and geochemical studies, is required to account for the major and trace element geochemistry of lamproite magmas.

In an attempt to explain the petrogenesis of lamproites, Zyryanov (1986) and Zyryanov and Zharikov (1985) have reacted mixtures of garnet harzburgite and alkali basalt with KOH, K_2CO_3, KCl, and $KFHF \cdot 2H_2O$ at 1100°C and 5 kbar pressure, followed by decompression to 2.5 kbar and then quenching. Run products included olivine, phlogopite, kalsilite, ortho- and clinopyroxene, and glass. Phase relationships were not determined and the low pressures of the experiments are clearly unrealistic with regard to lamproite genesis. On the basis of these experiments it is claimed that lamproite melts are formed as a result of "fluid-magmatic" interactions between the mantle and high K_2O fluids.

Detailed discussion of experimental studies of other potassic lavas is beyond the scope of this work. A useful summary is provided by Edgar (1987). Experiments on kamafugitic rocks in the presence of H_2O and CO_2 up to 40 kbar pressure by Edgar *et al.* (1976, 1980), Arima and Edgar (1983b), and Ryabchikov and Green (1978) found that these magmas could also be derived from orthopyroxene-free phlogopite wehrlite sources. The low silica contents of kamafugites result from partial melting under CO_2-rich H_2O-poor conditions (Wendlandt and Eggler 1980b). Foley *et al.* (1986b) suggest that the sources are oxidized relative to those of lamproites (see Section 8.4) and the presence of fluorine is required to maintain a phlogopite phase field in order that derivative melts have appreciable MgO and K_2O contents.

8.4. THE OXIDATION STATE OF LAMPROITE MAGMAS

Foley (1985) has studied the crystallization of a synthetic glass, equivalent in composition to the Gaussberg leucite lamproite, as a function of oxygen fugacity at atmospheric pressure. Figure 8.10A illustrates the liquidus phase relations for a melt with 0.045 wt % Cr_2O_3. Olivine is the initial liquidus phase at all oxygen fugacities studied and varies in composition with Fo_{98}, at the hematite–magnetite (HM) buffer, to Fo_{89} at the wustite–magnetite (WM) buffer. Leucite is stabilized by increasing oxygen fugacity, and has Fe_2O_3 contents ranging from 0.9 wt % at WM to 3.6 wt % at HM. Spinel crystallization was found to be a function of oxygen fugacity but the crystals formed were too small for analysis. Spinels are not found in these low Cr melts at the liquidus, suggesting that some fractionation of spinel has reduced the Cr content of the erupted lava. Accordingly, Foley (1985) undertook a second series of experiments using a melt containing 0.2 wt % Cr_2O_3. In these runs spinel occurs as the sole liquidus phase at all oxygen fugacities studied. Subliquidus assemblages consist of spinel and olivine with and without leucite (Figure 8.10B). Leucite crystals contain exsolved Cr-free aluminous spinel inclusions similar to those found in West Kimberley lamproites (Jaques and Foley 1985; see Section 6.9.4). Liquidus spinel compositions are a function of oxygen fugacity. Their Cr content decreases and Fe^{3+} content increases as oxygen fugacity increases. Using these experimental data and the compositions of natural spinel, leucite, and olivine in the Gaussberg lamproites, Foley (1985) concludes that the magma crystallized at oxygen fugacities less than 0.5 log units below the nickel–nickel oxide (NNO) buffer. The oxygen fugacity ($\log f_{O_2} = -6.7$ at 1280°C), calculated using empirically derived equations relating the molar ratio Fe_2O_3/FeO of the liquid to oxygen fugacity and melt composition (Kilinc *et al.* 1983), is in excellent agreement with the

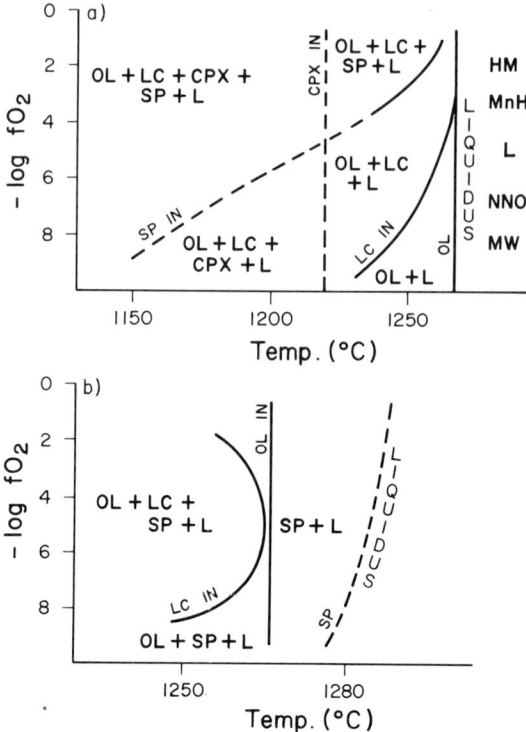

Figure 8.10. Near-liquidus phase relationships with respect to temperature and oxygen fugacity for a synthetic lamproite composition based on that of a leucite lamproite from Gaussberg (after Foley 1985). Oxygen fugacity buffers are hematite–magnetite (HM), manganosite–haussmanite (MnH), nickel–nickel oxide (NNO), and magnetite–wustite (MW). OL, olivine; LC, leucite; CPX, clinopyroxene; SP, spinel; L, liquid.

above estimate. The composition of spinels included in olivines in other lamproites suggests oxygen fugacities ranging from well above NNO (Leucite Hills) to the WM buffer (West Kimberley, Murcia-Almeria). Relatively high oxygen fugacities for the Leucite Hills magmas are also suggested by the high Fe_2O_3 contents of leucites in these lamproites. It should be noted that Foley's (1985) data are applicable only to early crystallizing spinels and not to low-temperature groundmass spinels in olivine/madupitic lamproites.

Oxygen fugacities of the Murcia-Almeria lamproites have been calculated by Venturelli et al. (1988) using the methods of Kilinc et al. (1983). Calculated redox conditions cover a very wide range of oxygen fugacities, e.g., 5 log units at 1000°C (WM to above NNO) and 3 log units at 1300°C (QFM/IW to above NNO). This wide range may reflect the poor calibration of the empirical equations relating melt composition and f_{O_2} for alkaline rocks, the effects of near-surface oxidation, and/or the use of contaminated and hybrid magma compositions in the calculations. The data therefore do not provide useful constraints on lamproite oxygen fugacity.

Coexisting ilmenite and magnetite in olivine madupitic lamproites from Jumilla (Murcia-Almeria) give equilibration temperatures and oxygen fugacities (log) of 740–820°C and -14.8 and -12.9. Redox conditions are thus slightly above those of the NNO buffer and are in agreement with Foley's (1985) conclusion that the Gaussberg lamproite crystallized at the NNO buffer.

There is no direct evidence regarding the oxygen fugacity of lamproite magmas in their source regions. Inferences may be made based upon intrinsic f_{O_2} measurements of mantle-derived xenoliths (Arculus et al. 1984), calculation from heterogeneous phase equilibria (Eggler and Baker 1982, Taylor and Green 1986), Fe–Ti oxide equilibria (Haggerty and Tompkins 1983) or the compositions of fluid inclusions in early formed phenocrysts or cognate xenocrysts (Bergman and Dubessy 1985). Estimates of mantle redox conditions range from mildly oxidizing, near the quartz–fayalite–magnetite (QFM) buffer, to highly reducing at the iron–wustite (IW) buffer. It is probable that these may represent extremes of a continuum of oxygen fugacities, and that a single oxygen fugacity cannot be expected for the whole of the upper mantle. Haggerty (1988) has suggested that vertical and lateral variations in redox conditions have evolved with time, recycling of crustal components and tectonic setting. Experimental studies of model mantle compositions with varying CO_2/H_2O ratios indicate that primary silica-rich magmas must be formed under H_2O-rich conditions, and that high CO_2 contents lead to the formation of silica-poor melts (Mysen and Boettcher 1975a,b, Wendlandt and Eggler 1980b, Wyllie 1980). Low CO_2 contents in magmas are indicative of low oxygen fugacities. Foley et al. (1986b) thus suggest that redox conditions of lamproite magmas at their upper mantle sources are best represented by the carbon–water (CW) buffer. This is the locus of points on the carbon (graphite or diamond) saturation surface where the H_2O content is maximized. Oxygen fugacities for CW lie between the IW and WM buffers at temperatures and pressures in the mantle where diamond is stable (Figure 8.11). At these redox conditions, the fluid phase will be mixtures of CH_4, H_2O, H_2, and C_2H_6, and crystalline carbonates will not be stable (Taylor and Green 1986). The H_2O/CH_4 ratio will vary greatly with f_{O_2}, temperature and pressure, but H_2O/CO_2 ratios will always be

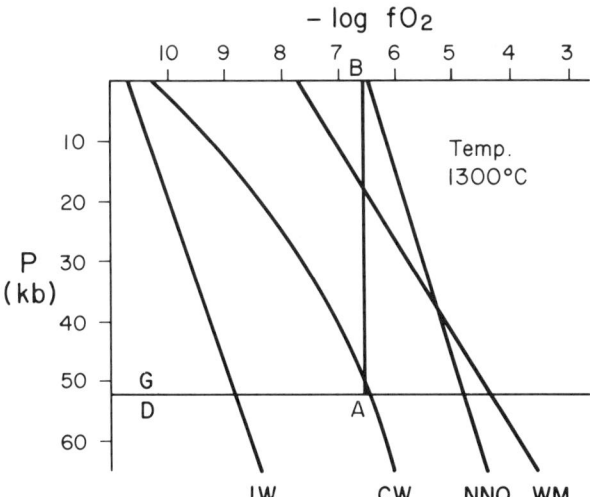

Figure 8.11. A possible emplacement path (A–B) compared to diverse oxygen buffers for a lamproite magma of constant oxygen fugacity ascending from depths equivalent to the pressure of the diamond/graphite (D/G) transition to the surface (after Foley et al. 1986b). The constant oxygen fugacity is maintained by H_2O dissociation. Oxygen buffers are iron-wustite (IW), graphite-water (CW), nickel-nickel oxide (NNO) and wustite-magnetite (WM).

high. Figure 8.11 illustrates a possible isothermal open system ascent path for a lamproite magma assuming that the near-surface f_{O_2} corresponds to that of the NNO-buffer (Foley 1985). Foley *et al*. (1986b) calculate that dissociation of less than 0.1 wt % H_2O driven by H_2 loss is sufficient to account for oxidation from CW to NNO during magma ascent. Foley *et al*. (1986b) calculate that half of the water dissociation occurs above 15 kbar pressure, implying that oxygen fugacities may remain low, even at relatively high levels in the mantle. Other f_{O_2}–T ascent paths are of course possible, but the constraint that CO_2 contents be kept low requires lamproite magmas at their sources to have low oxygen fugacities, i.e., between those of the IW and NNO buffers.

8.5. SUMMARY

Experimental studies of lamproites at low and high pressure indicate that:

1. Primary olivine is a low-pressure, high-temperature phenocryst. Reaction of olivine with the liquid to form phlogopite may occur upon cooling.
2. Phlogopite is stable at high or low pressures and may crystallize as a phenocryst during ascent of the magma. Under some conditions, early formed phlogopite may react with the liquid at low pressures to form diopside and leucite.
3. Orthopyroxene may be a high-pressure phenocryst.
4. Leucite, sanidine, and amphibole are all low-pressure phases.
5. The near-liquidus phase assemblage of lamproite melts at high pressures suggests that they may be derived from phlogopite harzburgite or phlogopite wehrlite sources under F- and H_2O-rich conditions.
6. The sources of silica-rich lamproites are reduced and volatiles are dominated by H_2O and CH_4, whereas the sources of silica-poor kamafugites, melilitites, and alkali basalts in general are oxidized and dominated by CO_2.

Considerable further work on the phase relationships of lamproites at high and low pressures is required. In particular, water undersaturated experimental studies at low pressure, with controlled oxygen fugacities and no Fe-loss, are required to establish the crystallization conditions of richterite and groundmass micas. High-pressure studies of natural lamproites in H_2O- and F-bearing systems are required to determine whether or not the genetic hypotheses based upon studies of synthetic systems are valid.

One rock is as good as another to be wrecked on.
George Bernard Shaw

Diamonds, Xenoliths, and Exploration Techniques

9.1. DIAMONDS AND XENOLITHS: ALIEN, YET BENEFICIAL, COMPANIONS OF LAMPROITES

Diamonds and ultramafic xenoliths found in lamproites represent samples of the mantle occurring near, and overlying, the source rocks of these magmas. Despite their rarity, these samples contribute significantly to the economic and scientific importance of lamproites. There are seven known diamondiferous lamproite provinces, fields, or occurrences on five continents: West Kimberley, Argyle, Prairie Creek, Bobi, Majhgawan, Kapamba, and Coromandel. These diamondiferous lamproites range in age from Proterozoic (1.2 Ga) to Miocene (20–22 Ma). The lamproites from the Ellendale field, West Kimberley province, represent the youngest-recognized, near-economic sources of primary diamond in the world.

Mantle-derived xenoliths or discrete xenocrysts are found in most lamproites, although less frequently than in kimberlites or alkali basalts. In general, distinct lithologic groups of mantle xenoliths occur in each of these three magma types.

9.2. DIAMONDS

Diamonds in lamproites are considered to be xenocrysts and derived from parts of the lithospheric mantle that lie above the regions of lamproite genesis. Although the scientific analysis of lamproite diamonds is still in its infancy, much has been learned in the last 20 years by the study of diamonds from Prairie Creek (Giardini and Melton 1975a,b,c, Giardini *et al.* 1974, Melton *et al.* 1972, Melton and Giardini 1975, 1976, 1980, Pantaleo *et al.* 1979), Argyle and Ellendale (Hall and Smith 1984, Harris and Collins 1985, Jaques 1989, Jaques *et al.* 1986, 1989b, Tombs and Sechos 1985, O'Neill *et al.* 1986, Richardson 1986, Honda *et al.* 1987, Griffin *et al.* 1988).

Studies of the Prairie Creek diamonds have concentrated upon their volatile content, and it is suggested that they contain various combinations of gases in the system C–H–O–N. These data are not discussed here as Roedder (1984) suggests that the measurements must be repeated by an independent group before they can be accepted. Honda *et al.* (1987) were unable to confirm the argon isotopic compositions reported by Melton and Giardini (1980). The bulk of our knowledge of lamproite diamonds stems from studies of those from Ellendale and Argyle.

9.2.1. Type, Morphology, Color, and Size

Diamonds are classified into two major groups, type I and type II, respectively, on the basis of the presence or absence of infrared-detectable nitrogen (Robertson *et al.* 1934). Type I diamonds are by far the most common in kimberlites, forming over 95% of the total population. Studies by Harris and Collins (1985) showed that Argyle diamonds are mostly type I stones. In a sample of over a hundred diamonds, 75%–85% contain 90–260 ppm N. The diamonds are unusual in having most (typically 80%) of their nitrogen in B-aggregates, indicating extensive high-temperature annealing. Ellendale diamonds, in contrast, have most of their nitrogen in A-aggregates (Jaques 1989).

The crystal morphology and color of lamproite diamonds have been reported by Hall and Smith (1985), Jaques (1989), and Jaques *et al.* (1986, 1989b). Large (>1 mm) Ellendale diamonds are dominantly yellow dodecahedra, with shiny smooth surfaces, which have been resorbed and rounded to varying degrees. Small (<1 mm) Ellendale stones are mostly colorless to pale brown, frosted, unresorbed, step-layered octahedra. Argyle diamonds are dominantly frosted and etched, irregularly shaped stones. Macles or aggregates are subordinate. Dodecahedral or resorbed octahedral–dodecahedral forms are common in all Argyle stones. The dominant color of all samples is brown with less than 20% yellow and colorless stones. Green and pink stones are rarely present. A characteristic feature of the Argyle diamonds is the presence of flat-bottomed hexagonal etch pits, commonly with striations or trigons on their floors. Small brown diamonds (<0.4 mm) are present in peridotite xenoliths occurring in lapilli tuffs at Argyle. In contrast to the lamproite diamonds that are unresorbed strongly frosted octahedra.

Lamproite diamonds from Argyle and Ellendale possess typical kimberlitic diamond log-normal size distributions (Hall and Smith 1985). The mean size of Argyle diamonds is <0.1 carat (for stones not passing a 0.8-mm sieve), while those from Ellendale are 0.1–0.2 carat. The largest stones recovered from Ellendale and Argyle are 6 and 16 carats, respectively.

Argyle and Ellendale diamonds are relatively small in size compared to those from kimberlites, perhaps as a result of strong resorption of the lamproite diamonds in their transporting magma. They exhibit the same morphologies and colors as observed in kimberlite-derived diamonds (Hall and Smith 1985).

9.2.2. Mineral Inclusion Suite

Mineral inclusions in diamonds are varied and have been classified into two major groups: peridotitic or ultramafic (various combinations of purple Cr-pyrope, Cr-diopside, forsterite, and enstatite) and eclogitic (various combinations of orange pyrope-almandine, omphacite, kyanite, coesite, sanidine, and rutile) (Harris and Gurney 1979, Meyer 1987). Mineral inclusions within lamproite diamonds have been studied by Pantaleo *et al.* (1979), Hall and Smith (1984), Harris and Collins (1985), Jaques (1989), Jaques *et al.* (1986, 1989a), Richardson (1986), and Griffin *et al.* (1988).

Eclogitic and peridotitic suites of inclusions are present in roughly equal proportions in Ellendale diamonds. Primary inclusions in Argyle diamonds are dominantly (>76%) of the eclogitic paragenesis. The inclusion suite in the Prairie Creek diamonds is as yet inadequately characterized. The summary of the available data by Pantaleo *et al.* (1979) indicates that peridotitic and eclogitic inclusions are present.

Syngenetic inclusions reported in lamproite diamonds include olivine, Cr-pyrope, Cr-diopside, enstatite, Cr-spinel, almandine-pyrope, omphacitic clinopyroxene, coesite, rutile, pyrrhotite, kyanite, ilmenite, and moissanite. Probable epigenetic inclusions include orthoclase, phlogopite, talc, calcite, quartz, hematite, magnetite, chlorite, anatase, rutile, kaolinite, and rare priderite.

The major, minor, and trace element contents of eclogitic (E-type) and peridotitic (P-type) suite inclusions from Argyle and Ellendale stones were reported by Griffin *et al.* (1988), and Jaques *et al.* (1989b). Argyle E-type clinopyroxenes show a wide range in *mg* (ca. 0.65–0.80). They are jadeite-rich (<9.5 wt % Na_2O, <20 wt % Al_2O_3), enriched in K_2O (0.3–1.3 wt %), Rb (5–16 ppm), Sr (300–800 ppm), Zr (3–70 ppm), with high K/Rb (245–1410) ratios and TiO_2 (0.04–0.5 wt %) contents. These pyroxenes are the most Al and K rich ever reported from diamonds. Ellendale E-type clinopyroxenes have lower jadeite (<4.9 wt % Na_2O, <14.2 wt % Al_2O_3) and K_2O (<0.6 wt %) contents and are similar to most omphacites reported from kimberlite-derived diamonds.

Argyle E-type garnets show a wide compositional range, from Ca-poor almandine-pyrope to Ca- and Fe-rich pyrope-almandine. They have high Na_2O (<0.7 wt %), TiO_2 (<1.45 wt %), and P_2O_5 (<0.39 wt %) contents, which exhibit negative correlations with increasing *mg* (0.35–0.75). The garnets are poor in Ni (<60 ppm), with the Na- and Fe-rich types having high Y (30–70 ppm and Zr (60–180 ppm) contents. Ellendale E-type garnets are mostly pyrope almandines with lower Na_2O (<0.2 wt %) and TiO_2 (<0.4 wt %) contents than the Argyle garnets. Griffin *et al.* (1988) calculated equilibration temperatures of 1085 to 1575°C for E-type garnet-clinopyroxene pairs from individual Argyle diamonds. E-type rutiles are similar to previously reported compositions, although some are enriched in Nb_2O_5 (<1.77 wt %) and ZrO_2 (0.9 wt %).

P-type olivine inclusions from Argyle are similar in composition to those found in kimberlite-derived diamonds, and are forsterites (*mg* = 0.92–94) with high NiO (0.31–0.47 wt %) and Cr_2O_3 contents. Ellendale P-type olivines have a wider range in composition (*mg* = 0.88–0.97) than those from Argyle. Most Ellendale olivines are richer in Fe (*mg* < 0.92) than those in kimberlite-derived diamonds. P-type enstatites are slightly enriched in Fe (*mg* = 0.91–0.94) relative to Argyle (*mg* = 0.93) and most kimberlite-derived (*mg* = 0.91– 0.95) inclusions. Cr-diopsides are relatively common in Ellendale diamonds compared to other diamonds. Their major element composition is not unusual [(Ca/(Ca+Mg) = 0.43–0.45], although they exhibit a wide range in Al_2O_3 (0.5–1.5 wt %) and Na_2O (0.3–1.1 wt %) contents.

Chrome pyropes from Argyle and Ellendale have a wide range in Cr_2O_3 (1.8–14.6 wt %). In contrast to Cr-pyropes from southern African diamonds, none of the lamproite pyropes have low CaO (<5.0 wt %) contents.

Jaques *et al.* (1989a) recognized a correlation between diamond crystal morphology and mineral inclusion type in Argyle stones. They found that sharp-edged octahedral stones with etched and frosted surfaces contain only peridotitic inclusions and resemble diamonds recovered from peridotitic xenoliths at Argyle. In contrast, rounded, resorbed, dodecahedral stones contain eclogitic inclusions.

9.2.3. Isotopic Composition

Carbon isotope measurements of lamproite diamonds display a wide range of $^{12}C/^{13}C$ ratios. Hall and Smith (1985) reported $\delta^{13}C$ (PDB) values for three Argyle stones ($-7.5‰$, $-7.8‰$, $-11.1‰$) and three Ellendale stones ($-2.8‰$, $-3.5‰$, $-5.7‰$). These ratios overlap those observed in diamonds from kimberlites, which range from $-34‰$ to $+5‰$, but are mostly from $-6‰$ to $-5‰$ (Harris 1987). A more extensive ($n > 70$ stones) study by Jaques et al. (1989b) found that stones with eclogitic inclusions from Argyle are markedly depleted in $\delta^{13}C$ ($-5‰$ to $-16‰$) relative to most Ellendale stones ($-4‰$ to $-6‰$) and those from Argyle which contain peridotitic inclusions (mostly from $-5‰$ to $-6‰$, but ranging from $-4‰$ to $-9‰$). There is no significant correlation of isotopic composition of the Argyle E-type diamonds with color, shape, inclusion composition, equilibration temperature, or provenance within the pipe.

Argyle and Prairie Creek diamonds exhibit wide variations in $^{3}He/^{4}He$ ratios (ca. 10^4). Some have solarlike He and Ne isotopic compositions and lack radiogenic components. Others are enriched in radiogenic and fissionogenic components (Honda et al. 1987).

9.2.4. Comparison with Kimberlite Diamonds

Lamproite diamonds exhibit both similarities with and differences from kimberlite diamonds. Peridotitic- and eclogitic-suite inclusions occur in both groups; however, lamproite diamonds apparently have more eclogitic inclusions than the average kimberlite-derived diamond. The compositions of silicate inclusions in lamproite diamonds overlap those of kimberlitic diamonds, although Argyle inclusions are notably richer in K, Al, and Fe. Inclusion compositions and the abundance of clinopyroxene in the Argyle and Ellendale inclusion suites suggest that their source regions were less refractory than the roots of many Archean cratons (Jaques 1989). Detailed studies of inclusions in diamonds from the Prairie Creek, Bobi, and Majhgawan lamproites are necessary in order to determine whether or not all lamproite diamonds are derived from such sources. Lamproite diamonds are of poorer quality than typical kimberlitic diamonds, as the majority of lamproite stones are grey, yellow, or brown and of industrial quality. Lamproite diamonds do not have any unique morphological characteristics or colors than enable them to be unambiguously distinguished from kimberlite diamonds.

9.2.5. Discussion

On the basis of the stone morphology, inclusion suite and carbon isotopic composition Jaques et al. (1989b) suggested that two types of diamonds, with distinct paragenesis, comprise the Argyle suite. Sharp, frosted octahedral stones with peridotitic inclusions are thought to represent refractory Archean lower lithospheric mantle beneath the Kimberley craton. In contrast, rounded, resorbed dodecahedral diamonds with eclogitic inclusions might have been derived from recycled crustal material added to the lower lithosphere during Halls Creek Mobile zone orogenic activity. This conclusion is supported by the Sm–Nd model age (1580 ± 80 Ma) of Argyle eclogitic inclusions in diamonds (Richardson 1986). This age is older than that of the intrusion (1177 ± 47 Ma, Pidgeon et al. 1989) but younger than the inferred age of cratonization (1.7–1.8 Ga, Plumb et al. 1981). The recycled crustal material could be the source of the >2 Ga enriched component in the Argyle intrusion identified by Jaques et al. (1989c).

9.3. XENOLITHS AND XENOCRYSTS

Xenoliths are generally grouped into two main classes, accidental and cognate. Accidental xenoliths and their disaggregated components can be incorporated at any point during the ascent of mafic magmas from their mantle source regions to the near-surface crustal environments. Cognate xenoliths bear some magmatic relationship to the host rock and, in the case of mafic melts, are generally thought to represent crystallized samples of magmas from the same mantle source region as their hosts. It is possible that cognate xenoliths and xenocrysts could have crystallized from the host melt. However, the mechanics of achieving the timing necessary for the incorporation of crystallized products into the same ascending melt are rather exceptional, as extremely high crystal growth rates are required.

This section considers only upper mantle-derived xenoliths or xenocrysts incorporated in lamproite magmas. Lower and near-surface crustal assemblages provide important constraints on the nature of crustal basement but will not be considered in this work. These latter xenoliths are especially abundant in the lamproites at Hatcher and Zirkel Mesas, Leucite Hills (Johnston 1959, Ogden 1979).

Kimberlites and alkali basalts are characterized by a varied and abundant xenolith population (Nixon 1987). In contrast, lamproites contain only a rare and limited suite of xenoliths. These differences are undoubtedly due to contrasts in the mechanics of magma ascent, including: crack propagation rate, yield strength, viscosity, and ascent rate of the various magma types. It is also possible that the structural competence and nature of the lithospheric mantle associated with lamproite melts is distinct from that of the asthenospheric mantle associated with kimberlite and alkali basalt magmas.

The types of xenoliths and xenocrysts found in olivine lamproites are distinct from those present in phlogopite- and leucite-rich lamproites. The latter are dominated by a cognate suite of phlogopite clinopyroxenites and diopsides, whereas olivine lamproites are characterized principally by dunite xenoliths and "dogtooth" olivine xenocrysts. Lherzolite and harzburgite xenoliths are rarely present in both lamproite types, but are more abundant in the mafic varieties. Some olivine lamproites contain both varieties of xenolith.

9.3.1. Xenoliths in Olivine Lamproites

Ultramafic xenoliths in olivine lamproites (primary from Argyle and Ellendale) have been described by Atkinson *et al.* (1984), Jaques *et al.* (1984, 1986, 1989a), Nixon *et al.* (1987), and O'Neill *et al.* (1986). Lithologies are most commonly coarse-grained dunites composed of large (2–8 mm), strained olivine porphyroclasts with small (<0.03–0.3 mm) polygonal olivine neoblasts. Larger olivine grains are characteristically strained with undulatory extinction and kink bands. Harzburgitic and lherzolitic assemblages are rarely found. All olivine-rich xenoliths are commonly altered to talc, serpentine, and other hydrous phases. The presence of pseudomorphs or relict primary minerals indicates the suite consists of garnet-bearing dunites, garnet harzburgites, diopside-bearing garnet harzburgites, chromite-bearing garnet harzburgites, and chromite dunites.

The few available mineral compositional data of these xenoliths demonstrate that they are similar in composition to minerals in kimberlite-derived xenoliths. All phases are magnesium-rich and titanium-poor. Jaques *et al.* (1986) suggest equilibration temperatures of 880–1100°C.

O'Neill *et al.* (1986) have described a suite of diamondiferous (7.5 carats/tonne) garnet and chromite-garnet lherzolites and harzburgites from Argyle. In all examples the original garnet has been replaced by Al-spinel and Al-pyroxene symplectite intergrowths and calcite. The primary phases are forsterite (mg = 0.92–0.94), low Al enstatite (0.5–0.7 wt % Al_2O_3, 0.8–0.11 wt % CaO) and Cr-diopside (0.7–1.5 wt % Cr_2O_3, 1.2–1.5 wt % Al_2O_3). Primary spinels are magnesiochromites [Cr/(Cr+Al) = 0.7–0.8]. Equilibration temperatures and pressures for the primary assemblage are in the range 47–57 kbar and 1100–1300°C. Secondary phases are believed to represent a decompression assemblage, as they have equilibration pressures of about 15–20 kbar, and temperatures similar to the primary phases. The highly refractory nature of the primary mineral assemblage is characteristic of depleted mantle. The unusual complete elimination of primary garnet during transport of the xenoliths could reflect higher magmatic temperatures in the lamproite host than are typically found in kimberlites.

Olivine lamproites contain a rich assortment of xenocrysts dominated by olivine and including diverse Fe–Mg–Al–Cr oxides, a wide variety of garnet types, clinopyroxene, orthopyroxene, mica, rutile, and moissanite.

Olivines occur as anhedral single and composite grains, easily recognized by their characteristic strained extinction and habit. Many represent microxenoliths composed of two to three crystals. Such olivines are petrographically similar to macrocrystal olivines found in kimberlites. Commonly, olivine xenocrysts are mantled by multiple parallel overgrowths of euhedral olivine of "dog tooth" habit (see Section 6.5.1). Olivine compositions (mg = 0.88–0.93) are identical to those in mantle-derived lherzolites.

Garnet xenocrysts range in composition from pyrope to almandine-pyrope and correspond to groups 1,3 5,9 of the Dawson and Stevens (1975) classification. Most pyropes are poor in TiO_2 (<0.4 wt %) and all are Ca-saturated, although they contain 2–10 wt % Cr_2O_3. *The low Ca- Cr-rich pyropes (G10 garnets) typical of southern African kimberlites are absent from the Prairie Creek, and Argyle diamond-bearing lamproites.* N.B. Such garnets are extremely rare in the Ellendale 7 and 9 lamproites (Lucas *et al.* 1989) and only one has been found as an inclusion in an Ellendale diamond (Hall and Smith 1985).

Red-brown translucent and opaque spinels are very common in heavy mineral concentrates from olivine lamproites. They range in composition from magnesian aluminous chromite to magnesiochromite, and are characterized by low TiO_2 (<1 wt %) contents. Consequently, their compositions [Mg/(Mg+Fe^{2+}) = 0.40–0.7, Cr/(Cr+Al) = 0.02–0.93] plot on the base of the reduced and oxidized spinel prisms. The spinel compositions resemble those from spinel lherzolites and harzburgites, and a wide variety of basic and ultrabasic rocks, but differ from the Ti-rich primary spinels occurring in lamproites and kimberlites (see Section 6.9).

Cr-diopside and enstatite xenocrysts are rare and have compositions identical to those of lherzolites.

Significantly absent from olivine (and phlogopite) lamproites are megacrysts characteristic of kimberlite, i.e., Ti-pyrope, magnesian ilmenite, subcalcic diopside, bronzite, and zircon.

Cognate xenoliths have also been recognized in olivine lamproites. Mitchell and Lewis (1983) described priderite-bearing diopside–richterite xenoliths from the Prairie Creek olivine lamproite. They interpreted these xenoliths as fragments of high-pressure cumulates derived from ultrapotassic (lamproite) magmas that are consanguineous with the host lamproite.

9.3.2. Xenoliths in Leucite and Phlogopite Lamproites

Xenoliths in leucite and phlogopite lamproites consist of olivine-rich, mostly accidental xenoliths, and clinopyroxene- and phlogopite-rich cognate xenoliths. The xenoliths have been described by Kuehner (1980), Barton and van Bergen (1981), and Jaques et al. (1986). Olivine-rich lithologies are primarily spinel harzburgites, whereas the cognate population includes phlogopite clinopyroxenites and phlogopitites (glimmerites). All of these xenolith types display a variety of magma–xenolith interaction textures ranging from wholesale infiltration of lamproite magma along cracks, to corrosion and reaction of marginal phases (Barton and van Bergen 1981). Note that Barton and van Bergen (1981) suggested that the green clinopyroxenes (present as discrete crystals and in phlogopite pyroxenite xenoliths) from the Leucite Hills (see Section 6.3), were derived from accidental fragments of wall rocks, metasomatized by previous igneous events in the upper mantle.

Xenocrysts of clinopyroxene, orthopyroxene, olivine, spinel, and biotite and aluminous phlogopite dominate the xenocryst populations of phlogopite and leucite lamproites, although the volumetric proportion of these xenocrysts is very small. Many phlogopite or leucite lamproites contain no recognizable xenocrysts. Mantle-derived garnet xenocrysts are virtually nonexistent in phlogopite or leucite lamproites, except in the more tuffaceous lithologies at vents such as Prairie Creek, and other intrusives which also contain olivine lamproites.

9.3.3. Discussion

The use of xenoliths to deduce the nature of the mantle source regions of lamproites and their petrogenetic evolution has two severe drawbacks: rare in lamproites, they are found in profusion in kimberlites and alkali basalts; as the systematic electron microprobe analysis of xenolith phases has not yet been undertaken, there are significantly fewer analytical data (a factor 10^3) available for xenoliths derived from lamproites than those from kimberlites and alkali basalts. Given the meagre data base, it is possible to conclude only that lamproite-derived xenoliths are apparently more restricted in their mineralogical assemblages than those found in kimberlites and alkali basalts.

Xenoliths and xenocrysts found in olivine lamproites are mainly dunites and harzburgites (and their disaggregated components) derived from deformed portions of the mantle lithosphere. Leucite and phlogopite lamproites contain a sparse population of harzburgite xenoliths. Xenoliths occurring in lamproites are generally much smaller than those found in kimberlites and alkali basalts. The xenolith and xenocryst populations are generally quite distinct from those of the kimberlites, being more refractory and depleted in clinopyroxene, supporting the view that the mantle source regions of the two magma types are distinct. It is probable that the mechanics of xenolith incorporation and transportation also differ.

The presence of phlogopite clinopyroxenite xenoliths in leucite phlogopite lamproites indicates that crystallization of some lamproite magmas occurs at high pressures, possibly in the mantle. Detailed studies of these xenoliths that would permit some insights regarding this process have not yet been undertaken.

9.4. EXPLORATION TECHNIQUES FOR DIAMONDIFEROUS LAMPROITE

Since the discovery in 1979 of the world's largest and richest primary igneous diamond deposit, at Argyle, West Australia, exploration geologists around the world have been seeking potentially diamondiferous lamproite bodies. The Argyle lamproite was discovered three years after the recognition of subeconomic diamondiferous lamproites in the Ellendale region of the well-known West Kimberley lamproite province. Note that these discoveries occurred prior to the reclassification of several important diamondiferous "anomalous kimberlites" at Bobi, Prairie Creek, and Majhgawan as lamproites. Many of the lamproites containing diamond, or associated closely with crater facies diamondiferous lamproites, possess some geochemical and mineralogical similarities with kimberlites. Principal among these are the presence of high modal amounts of olivine, Cr-rich spinels and similar enrichments in Cr, Ni, and incompatible elements (see Chapters 6 and 7). Thus, despite the existence of several other key distinguishing features, one might expect that similar geophysical and geochemical exploration methods could be used to locate both rock groups. *However, the significant tectonic, mineralogical, compositional, and morphological differences between lamproites and kimberlites result in different exploration philosophies and techniques being required for their location.* The West Kimberley diamondiferous lamproites not found by Stockdale, Selection Trust, BRGM, or other companies during their investigations of the region, using conventional kimberlite exploration techniques, illustrates the importance of these differences.

Among the general exploration techniques for kimberlite are: indicator/heavy mineral surveys of soils and stream sediments; soil and rock geochemistry; and geophysical surveys. The same methods are applicable to lamproite exploration with some modifications, bearing in mind the aforementioned comments.

The following sections of this chapter give a brief overview of diamond exploration techniques, with emphasis on those methods most relevant to exploration for lamproite. This area of mineral exploration has been extensively investigated by many commercial concerns (e.g., DeBeers, C.R.A., Ashton, etc.) and the coverage of the subject is by no means representative of their in-house expertise.

Perhaps the most fundamental aspect of diamond exploration is the selection of a region in which exploration efforts should be concentrated. The geological and tectonic framework of domains most likely to contain lamproites have been summarized in Chapter 4. Kimberlites and lamproites are restricted to portions of the earth in which continental crust and relatively thick lithospheric mantle form the basement.

Diamondiferous kimberlites are generally restricted to portions of crust underlain by ancient cratons which have been stable for at least the last 1.5 Ga (Clifford 1966). These cratons are generally characterized by deep lithospheric roots and low geothermal gradients. Whereas diamondiferous kimberlites are found in the central regions of these cratons, diamondiferous lamproites occur on their margins in crustal domains which experienced Proterozoic to Archean accretionary and/or other orogenic events. Kimberlites occurring in these mobile belts are typically subeconomic or barren of diamonds.

In addition, areas that have been only slightly uplifted or depressed and containing deep-seated fracture networks have a greater potential for containing kimberlites and lamproites than those that have been excessively uplifted. Bearing these features in mind, the areas most favorable for lamproite exploration can be placed in the following order of priority:

1. Regions with known diamondiferous lamproites;
2. Areas containing commercial alluvial diamond deposits with no known source;
3. Mobile belts with apparently diamond-free lamproites;
4. Areas with known lamproite indicator minerals in stream sediments or soils;
5. Areas containing noncommercial alluvial diamond occurrences;
6. Areas having a tectonic history and basement framework characteristics suitable for the generation and emplacement of lamproites.

A plethora of techniques are available with which to restrict exploration efforts and delineate kimberlite or lamproite intrusive bodies. These techniques are relevant to a variety of scales, from regional exploration involving remote sensing surveys (1:1,000,000 to 1:100,000 scales), to intrusion delineation using ground mapping and other surveys (1:100 to 1:1,000 scales). Recent general reviews on the subject have been contributed by Nixon (1980), Glover and Phillips (1980), Gregory (1984), and Atkinson (1989), and the interested reader is referred to these works for a more detailed treatment of this extensive subject.

9.4.1. Remote Sensing Techniques

Remote sensing refers to the acquisition and interpretation of data obtained from great distances. We restrict the present usage to electromagnetic radiation used to infer the nature of the Earth's surface, including: Multispectral Scanner (MSS) and Thematic Mapper (TM) satellite data; Side-looking Radar (SLAR) imaging; and conventional visible or infrared aerial photography. Once a prospective region or area has been identified on the basis of the presence of known lamproites, remote sensing techniques are primarily most useful because of the scales involved. Remotely sensed electromagnetic radiation data analysis can be grouped into several categories of relevance to diamondiferous lamproite (and kimberlite) exploration: general lineament/structural trend/geomorphological analysis, lithologic mapping, and geobotanical anomaly mapping. Structural and topographic lineaments and circular features representing intrusive/extrusive lamproite bodies are commonly resolvable on high-resolution aerial photography or modern satellite (e.g., TM, SPOT) imagery. As lamproites readily weather in most surface environments and produce a rather exotic suite of secondary phyllosilicate and alteration minerals (e.g., nontronite, smectite, barite), recently developed multispectral systems, in which many spectral regions (bands) can be transformed into indicators of specific mineralogical features, possess great potential for mapping. For example, Kingston (1989) has applied this technique to characterize the reflectance (absorption) spectra signatures of kimberlites and carbonatites. Remote sensing imagery can be used to extend the known distribution of mapped lamproite intrusive/extrusive rocks into adjacent unmapped areas, on the basis of regional structural attributes and geobotanical anomalies. Lamproites possess anomalous contents of elements relative to the local country rocks. Consequently, growth of different flora, or the presence of stressed vegetation, can lead to geobotanical anomalies identifiable on a variety of remotely sensed images. For example, infrared photography shows the Big Spring vent (West Kimberley) to be characterized by anomalous faint pink tones, which reflect the growth pattern of grasses on the vent (Jaques *et al.* 1986). Conventional aerial photography has the resolution to delineate some of the larger low-relief lamproites, e.g., 81 Mile vent, Ellendale 4. In all cases, remote sensing must be used in combination with basic surface geologic mapping and the other techniques described below.

9.4.2. Geophysical Techniques

Geophysical techniques useful in the delineation and exploration of lamproite bodies include gravity, magnetics, resistivity, electrical methods, seismic, and radiometric (gamma-ray) techniques. Each of these methods is based on the tenet that lamproite bodies contrast in some physical property (density, magnetic susceptibility, mineral and chemical composition, sonic velocity, etc.) with their country rocks. Some techniques have historically been used more for regional exploration, whereas others are more applicable to the detailed delineation of a given intrusive or extrusive mass.

9.4.2.1. Magnetic Methods

The magnetic susceptibility of lamproites is controlled primarily by the presence of Ti-magnetite, other spinels, and Fe- and Ti-oxides. Lamproites may possess a wide range in magnetic susceptibility ($180-1800 \times 10^6$ emu) depending upon the abundance of these phases (Atkinson 1989). The existence of an aeromagnetic or ground magnetic anomaly associated with a lamproite body is controlled by the relative susceptibility difference between the lamproite and country rocks. Magnetic surveys were first applied to diamondiferous lamproites over half a century ago by Stearn (1932), and the Crater of Diamonds vent at Prairie Creek was recognized to possess an anomaly of over 2000 nT. Magnetic surveys have since been widely used to delineate and locate lamproites in the Prairie Creek field (Waldman et al. 1987, Bolivar and Brookins 1979), Kansas (Coopersmith and Mitchell 1989), and West Kimberley province (Jaques et al. 1986, Atkinson 1989).

Aeromagnetic surveys may provide the best means of delineating most of the lamproites within a given province, as demonstrated by the success of the method in the Ellendale field (Jenke 1983, Jaques et al. 1986). In this case dipolar anomalies are associated with most of the olivine lamproites. Importantly, some of the large lamproites, visible in aerial photographs (Mt. North, 81 Mile vent), do not produce a magnetic anomaly. Such regional surveys obviously require the magnetic signature of the country rock to be relatively simple. Steady magnetic backgrounds are found for thick sequences of shale, sandstone, and limestone. An aeromagnetic survey showed that the background of the region around the Argyle vent was complex and that the vent did not produce a significant or easily identifiable magmatic anomaly (Drew 1986). Flight line spacings and elevations of regional magnetic surveys are typically 250 m and 50–100 m, respectively.

Ground magnetic surveys provide an important follow-up procedure once potential lamproites have been located by aeromagnetic methods. The purpose of such surveys is to delineate accurately the dimensions of the vent or intrusion. The magnetic signatures of lamproite bodies are complex and not easily related to the surface geology; however, the anomaly as a whole corresponds fairly closely to the outcrop (or subcrop) of the body (Sharma and Nandi 1964, Bolivar and Brookins 1979, Jenke 1983).

9.4.2.2. Electrical Methods

Various electrical methods (EM) are excellent delineation tools for lamproites, given the correct weathering environment and contrasts with country rocks. Aerial EM are relatively expensive compared to other airborne techniques, whereas ground-based EM methods are fast and relatively inexpensive compared with other geophysical techniques.

The electrical resistivity of weathered lamproites is generally thought to be less than most country rocks, owing for the most part to the conductive nature of smectitic clay relative to illite, kaolinite, or other clay minerals (Gerryts 1970, McNae 1979, Jenke 1983). Fresh lamproite may show higher resistivity than many country rocks. Lamproite vents, such as Argyle, possess moderate to strong resistivity anomalies of 40–100 Ω m, these being less than those (200 Ω m) of the country rocks (Drew 1986).

Lamproites may be located and delineated using regional aerial (INPUT) and ground-based (EM) surveys respectively. Atkinson (1989) has noted that there is a higher success rate in finding lamproites by aeromagnetic (100%) than INPUT EM (60%) surveys. An INPUT survey of the Argyle pipe (Drew 1986) located a small anomaly over the northern widest end of the vent where the topographic relief was least. Ground-based EM 34 surveys at Argyle, however, showed that the lithic tuffs were the more conductive rocks of the vent and that intrusion-wall rock contacts could be easily delineated (Drew 1986). Coopersmith and Mitchell (1989) showed that the Rose and Hills Pond lamproites could be delineated using ground-based EM surveys due to the highly weathered nature of the lamproite outcrop. Despite the success of the method in these cases, it is concluded that, in general, EM surveys do not appear to offer any distinct advantages over magnetic methods.

9.4.2.3. Seismic and Gravity Techniques

The application of seismic techniques to regional lamproite exploration is precluded by the elevated costs involved and the small size of lamproite pipes. Seismic sections would clearly be capable of resolving pipe morphology and internal stratigraphy, given pipe margins and internal beds with sufficient velocity contrast. Housel et al. (1979) described the use of seismic refraction surveys in delineating the internal stratigraphy of State Line kimberlites (Colorado–Wyoming). No data have been published describing the application of this technique to lamproite vents.

Gravity surveys are of little use in exploration for, or delineation of, lamproites owing to the small size of the bodies. The low amplitude of the response (<1 milligal) cannot be distinguished from small variations in the regional field, and may be less than the local terrain corrections. Bolivar and Brookins (1979) found that the total gravity anomaly over the Prairie Creek vent was only 0.5 milligal. A small anomaly of 1.8 milligals was found over the hypabyssal lamproite. The tuffs and breccias did not exhibit significant anomalies.

9.4.2.4. Radiometric Techniques

Lamproites possess on average 5 ppm U, 46 ppm Th, and 7.2 wt % K_2O (Bergman 1987), and therefore are expected to have higher gamma-ray intensities than typical continental crustal rocks, with the exception of relatively uraniferous or potassic granitic rocks. However, radiometric techniques have received only limited use in lamproite (Jaques et al. 1986) or kimberlite (Paterson et al. 1977) exploration.

Atkinson (1989) notes that the usefulness of the method depends to a great degree on the amount of exposure or residual soil cover and the size and radioactivity of the lamproite target. In the Ellendale field only 6 out of 26 vents tested by this method produced a significant radiometric response. All of the anomalous vents were exposed at the surface and were greater than 7 hectares in area. Lamproites in this field showing no anomalies are covered by a 10-m-thick aeolian sand. Radiometric anomalies are not found over the Argyle vent (Drew 1986).

9.4.3. Geochemical Surveys

Lamproites, in common with many other mantle-derived alkaline mafic to ultramafic rocks, possess elevated contents of compatible—i.e., Ni, Cr, Co, V, Sc—and incompatible—i.e., Rb, Sr, Ba, Zr, Hr, Ti, P, Nb, REE, Y, Th, U—elements relative to average crustal rocks. This distinctive geochemical signature allows lamproites to be distinguished easily from most common ultramafic and mafic rocks which are not enriched in incompatible elements.

Stream sediment geochemistry does not, on the basis of the limited published data, appear to be very useful in locating near-surface (buried) or weathered lamproite bodies. Gregory and Tooms (1969) investigated dispersion from the Prairie Creek lamproites and found that Mg, Ni, and Nb anomalies extended for only 1 km from the vent. Similar limited stream sediment transport of lamproitic material was noted by Haebig and Jackson (1986).

Soil geochemical anomalies (Ni, Co, Cr, Nb) are developed in sand and soil overburden immediately above lamproite vents in the West Kimberley province (Haebig and Jackson 1986). Olivine lamproites are commonly marked by a topographic depression, consequently there is little dispersion of the geochemical anomaly and the vent contacts are well defined. On sloping ground anomalies do not extend for more than the diameter of the vent. Anomalies may be found where the overburden is up to 7 m in thickness. Haebig and Jackson (1986) concluded that soil geochemistry is not as effective a method of locating or delineating lamproite intrusions as heavy mineral or magnetic methods.

Coopersmith and Mitchell (1989) note that the ratio of the content of two diagnostic signature elements is sufficient for the recognition of lamproite soil: at the Rose lamproite these elements are Ni and Nb. Geochemical anomalies (Ni, Cr, Ti, Ba, Zr, La, Nb) over this body delineate the intrusion and can be used as a mapping tool.

Geobotanical or biogeochemical prospecting techniques have not been applied to lamproite exploration, although they have had limited use in kimberlite exploration (Alexander 1983, 1986, Alexander *et al*. 1984).

Many exploration programs routinely utilize lithogeochemistry as a means of rock identification. This technique is most useful in cases where rocks have been extensively altered and/or weathered (Coopersmith and Mitchell 1989). The method is also useful in examining drill cuttings (Atkinson 1989). Gregory (1984) has used lithogeochemistry to map complex lamproite intrusions and states that olivine lamproites may be distinguished from leucite lamproites where the two occur together using ratios of Mg, Ni, Cr, and Co.

The greatest utility of lithogeochemistry is as an adjunct to petrographic or mineralogical studies which have identified a rock as being a lamproite. Geochemical criteria for the identification of lamproites are given in Chapters 2 and 7.

Bergman (1987) has suggested that diamondiferous lamproites (i.e., olivine lamproites) are generally enriched in compatible elements relative to barren varieties, mostly due to the abundance of xenocrystal olivine in the former. Barren lamproites (i.e., phlogopite lamproites) contain elevated alkali and LILE contents (K, Na, Y, U, Th, Zr) relative to diamondiferous varieties. Despite wide ranges, diamondiferous lamproites possess about twice the Mg, Ni, Cr, Co, and Nb and half the K, Na, and Al compared to nondiamondiferous lamproites. *However, it should be clearly realized that lithogeochemis-*

try is not an accurate guide as to the diamond potential of a given lamproite. In the West Kimberley province, the rocks richest in diamonds are commonly silica-rich lithic tuffs whose composition bears little relation to that of diamond-poor magmatic olivine lamproite. It is the identification of olivine lamproites that is important in exploration, as all known diamondiferous lamproites are associated with such rocks. *Recognition of olivine lamproite is only a guide to the possible presence of potentially diamondiferous rocks.*

9.4.4. Indicator/Heavy Mineral Sampling

Diamond exploration can be performed by exploring exclusively for diamonds (both microdiamonds and larger stones) and many groups have effectively utilized this approach. The rarity of diamond in even economic deposits makes this technique costly, time consuming, and cumbersome at best. It is therefore desirable to use lamproite minerals that occur in much larger concentrations than diamonds and may be concentrated by alluvial/fluvial processes downstream, downwind, or downcurrent of their primary igneous source rocks.

A key aspect pertaining to heavy mineral surveys in diamond exploration is the recognition that kimberlites and lamproites possess distinct, yet somewhat similar, heavy mineral suites. The principal similarities lie in the presence of xenocrysts derived from the disaggregation of mantle xenoliths [e.g., Cr-pyrope (G9), Cr-rich spinels, diamond], and early crystallizing primary Cr-rich spinels. Garnets (G3, G6) derived from mantle eclogitic sources may also be present in both suites. The principal difference between the suites is that megacrystal/macrocrystal Ti–Cr-pyrope (G1, G2) and Mg–Cr-rich ilmenite, the hallmark indicators of kimberlite sources, are rare- to- nonexistent in lamproites.

Lamproite indicator minerals include Ti-rich phlogopite, K–Ti-richterite, low-Al diopside, chrome spinel, forsterite, perovskite, priderite, wadeite, and shcherbakovite. The latter three minerals are very rare, yet highly diagnostic of lamproites. It is shown in Chapter 6 that lamproite-derived phlogopites, richterites, diopsides, and chrome spinels are distinctive, especially in core-rim zoning trends, when compared with the compositions of similar minerals in nearly all other rocks. Cr-pyropes (G9) are very rare, but have been reported from Prairie Creek (Lewis 1977), Rose Dome (Coopersmith and Mitchell 1989), Argyle, Ellendale (Jaques *et al.* 1986, Lucas *et al.* 1989) and Majhgawan (Mathur and Singh 1971). Magnesian ilmenites are extremely rare in lamproites (see Section 6.12.2). Lamproite ilmenites are typically poor in Mg relative to most kimberlite-derived ilmenite.

The compositions tabulated in Chapter 6 provide a summary of the expected ranges in composition for phlogopite, richterite, diopside, chrome spinel, and other phases. The main differences between kimberlite and lamproite indicator minerals can be established using these and intragrain zoning trends.

Heavy mineral dispersion trains or anomalies may be identified by the field or laboratory processing of the samples by standard mineral separation techniques. Typical procedures are outlined by Jaques *et al.* (1986). Commonly, specific indicators are hand-picked from the concentrates produced by these methods.

Heavy minerals may be obtained from stream sediments and gravels or by loam sampling. Stream sampling has been effective in locating many of the West Kimberley lamproites (Jaques *et al.* 1986, Atkinson 1989). Recognizable lamproite-derived pyrope and spinel are found at distances of up to 14 km from their sources. Chromite has proven to be the most useful heavy mineral indicator for olivine lamproites in the Ellendale field because of its high abundance and distinctive composition. The Argyle vent was located by tracing diamonds and chromite 20 km upstream from the Smoke Creek alluvial diamond deposit (Jaques *et al.* 1986).

Aeromagnetic anomalies are typically investigated by loam or soil sampling. In the Ellendale field lamproite-derived heavy mineral suites are typically present in the soil overlying the vents (Jaques *et al.* 1986). Similar results for the Rose lamproite were documented by Coopersmith and Mitchell (1989).

9.4.5. Summary

On the basis of the recent (within the last two decades) discovery of diamondiferous lamproites and reclassification of some kimberlites as lamproites, it is probable that new lamproites will eventually be identified.

Lamproites are easily distinguished from kimberlites and other alkaline ultramafic rocks on the basis of their bulk composition and mineral chemistry. Conventional kimberlite exploration techniques are only partially applicable to lamproites. The best exploration program should involve an integrated approach, utilizing remote sensing, geophysical surveys, alluvial heavy mineral, and rock geochemical techniques. Whereas any one technique may be capable of locating prospective bodies, their combination is much more effective and efficient. Coopersmith and Mitchell (1989) note that no single technique proved to be totally adequate in the exploration of the Kansas lamproites. A combination of aeromagnetic and heavy mineral methods has been demonstrated as useful in the West Kimberley province. Unfortunately, case histories for other provinces are not available.

The inclusion of lamproites in the lamprophyre clan (Rock 1989) is of no value in exploration for these rocks. Diamonds apparently occur in source rocks that are neither kimberlites nor lamproites (Haggerty and Nagieb 1989, Nixon and Bergman 1987, Kaminskii 1984). Other unconventional sources appear to be lamprophyres (Rock 1986, 1989). The only common factor linking these diverse groups is the presence of diamond, and exploration techniques devised for monchiquites or minettes will not be valid for lamproites.

Our understanding of the physical properties of lamproites is incomplete. Detailed studies on the weathering and alteration mineralogy, as well as changes in density, magnetic susceptibility, and resistivity are required to improve exploration techniques in general. Much work is also needed on the quantitative mineralogy and hydraulics of alluvial systems draining lamproite sources. The theoretical predictions of Slingerland (1977) require confirmation by empirical studies. Documentation of the effectiveness of the exploration methods described above is required for other known lamproite provinces and especially for areas characterized by extensive weathering.

For Nature is a perpetuall circulatory worker, generating fluids out of solids, and solids out of fluids, fixed things out of volatile, and volatile out of fixed, subtile out of gross and gross out of subtile. Some things to ascend and make the upper terrestriall juices.

Isaac Newton (1675)

Petrogenesis of Lamproites

10.1 INTRODUCTION

The development of realistic hypotheses for the petrogenesis of potassic rocks has been hindered, until recently, by the tendency of petrologists to consider all leucite-bearing K-rich rocks as being related. Sufficient geological, geochemical, and mineralogical evidence (see Chapter 2) has now accumulated to show that this assumption is not correct, and that at least three distinct clans of potassic rocks exist. These diverse magmas cannot be derived from the same source or be related to each other by processes such as fractional crystallization or assimilation. Consequently, many of the older petrogenetic hypotheses, summarized and discussed by Turner and Verhoogen (1960) and Gupta and Yagi (1980), have no relevance to the genesis of the lamproite clan as defined in this work. In particular the hypotheses of Holmes (1950) and Marinelli and Mittempherger (1966), involving reaction of granitic rocks with carbonatite or limestone, respectively, must be regarded as irrelevant to lamproite genesis.

Discussion of the petrogenesis of lamproites revolves around three problems: the nature of the source; the extent of melting of that source; and the interrelationship between the various members of the clan. Each of these aspects of lamproite genesis have been addressed to varying degrees by petrologists, but commonly specific authors have concentrated upon one aspect to the exclusion of the others.

The first section of this chapter summarizes and discusses the merits and detractions of the various genetic models that have been proposed for lamproites. Subsequent sections outline the preferred model for the genesis of the clan in the context of current knowledge and the relationships of lamproites to other potassic rocks.

10.2 PREVIOUS PETROGENETIC MODELS

10.2.1. Fractionation of Peridotite and Kimberlite

Wade and Prider (1940) devised a complicated scheme for the genesis of the West Kimberley lamproites. It was proposed that an initial parental magma of peridotitic composition was modified by fractional crystallization of aluminous phases (garnet, kyanite) to produce a mica peridotite magma. Crystallization of aluminous phases was required to provide enrichment in alkalies and depletion in alumina. The mica peridotite magma was considered to be equivalent to kimberlite. Subtraction of olivine from this magma was believed to result in the formation of a magma rich in Ti, Zr, P, and Ba, with K > Al and relatively high silica contents. This second derivative was termed an orenditic magma. All of the rocks exposed in the West Kimberley province were then derived from orenditic magma by its subsequent fractional crystallization. Early crystallization of phlogopite and diopside led to the formation of residual magmas poor in Fe, Mg, and Ca. Partial resorption of previously crystallized phlogopite was considered to lead to an increase in the K content of residual relatively silica-rich liquid. The sequence leucite diopside lamproite (cedricite), phlogopite leucite lamproite (fitzroyite), richterite leucite lamproite (mamilite) was considered to reflect the liquid line of descent of orendite magma.

Applying this scheme to the Leucite Hills, Wade and Prider (1940) suggested that the madupitic rocks—i.e., those richest in Mg and poorest in Si—represent crystal accumulations of phlogopite and diopside from an orenditic magma. The residual liquid produced by this process is parental to the leucite (wyomingite) and sanidine (orendite) phlogopite lamproites. The latter were believed to be produced by the subtraction of phlogopite and leucite from leucite lamproite to give a residual liquid richer in Si and capable of crystallizing sanidine.

The above hypothesis is clearly inappropriate in the light of modern petrological knowledge, as it is now known that the source regions of kimberlites and lamproites are vastly different (see Chapter 7). Major drawbacks are the need to introduce a peridotite magma of undetermined origin and to call upon a relationship with kimberlites. The latter magma was introduced principally because Wade and Prider (1940), following the accepted practices of the time, used Harker diagrams to explain compositional and genetic relationships between diverse rock types. They concluded that the average composition of kimberlite lay on the extrapolated liquid line of descent for the West Kimberley rocks and hence was a potential parental magma. Two factors: the enrichment of incompatible elements and the high K/Al ratios of kimberlite were used to justify this conclusion.

The schemes relating the diverse members of the lamproite clan are now otiose, given the discovery of olivine lamproites in West Kimberley and the realization that madupitic rocks in the Leucite Hills cannot be parental to rocks richer in silica (Carmichael 1967, this work).

Prider (1960) modified the original Wade and Prider hypothesis because of lack of evidence for the crystallization of aluminous phases, i.e., alumina-rich cognate cumulates, from the peridotitic magma. The enrichment in K was instead ascribed to gas transfer processes. These were considered to lead to successively more alkaline and silica-saturated melts at higher levels in the volcanic vents. The original, peridotite, kimberlite, and orendite "parental magmas" were retained in this scheme. The different rock types in the West Kimberley province were derived by fractional crystallization from a mica peridotite (= kimberlite) magma generated in small vents developed above a peridotitic mass. Prider's (1960) hypothesis was accepted by Fuster et al. (1967) in their explanation of the compositional variation found in the Murcia-Almeria lamproites.

The revised hypothesis obviously retains the flaws inherent in the original hypothesis. An additional problem is that there is no real evidence for the operation of gas transfer processes in the genesis of the West Kimberley intrusions, even assuming they are capable of causing K enrichment. N.B. Much of the silicification of country rocks bordering the vents may be related to gas or fluid transfer from lamproite magmas subsequent to their emplacement. Prider (1960) appears to have introduced the process as a result of uncritical acceptance of Saether's (1950) paper on the topic and the misconception that the West Kimberley vents are "gas-drilled pipes" (Prider 1960, p. 109). Carmichael (1967) has noted that the formation of identical potassic magmas in different provinces is unlikely to be the result of such random processes as gas transfer.

From a historical viewpoint, Wade and Prider's ideas are interesting, in that they inspired the original exploration for diamonds in the West Kimberley region. Given that the hypothesis is now proven to be petrologically unsound, it is not surprising that exploration techniques designed to locate kimberlites did not find diamondiferous lamproites (see Chapter 9).

Prider's long-term influence upon hypotheses of lamproites genesis is indicated by the reference of Jaques *et al.* (1986) to the existence in the West Kimberley province of an "orenditic differentiation trend", i.e., formation of silica-rich derivatives from a silica-poor parental magma, by low-pressure crystal fractionation. Jaques *et al.* (1986) thus consider that all of the various lamproites in this province are derived from an olivine lamproite parental magma. Chapter 5 and Section 10.3.3 outline the reasons why we believe this hypothesis is incorrect and that the parental magma is actually silica rich.

10.2.2. Contamination of Kimberlite

Lopez Ruiz and Rodriguez Badiola (1980) have assumed that the Murcia-Almeria lamproites were formed by the mixing of kimberlitic liquids with shoshonitic magmas. The latter were believed to be derived by the partial melting of subducted oceanic crust at depths of 150 km. Initial partial melts were further modified by mixing with incompatible element-enriched fluids of unspecified origin and assimilated continental crust. The kimberlites were also believed to originate from subducted material at 300 km depth. Mixing with a variety of shoshonitic magmas (toscanites, banakites) during their ascent resulted in lamproite formation. Lopez Ruiz and Rodriguez Badiola (1980) calculated that the amount of kimberlitic liquid involved in mixing varied from about 60% in the most silica-poor lamproites, to 14% in the silica-rich types.

Apart from tectonic problems related to melt formation and the disposition of the subducted slab, the hypothesis is untenable in that kimberlites are now known to be derived from asthenospheric sources unrelated to subduction. Further, invoking mixing with a magma, whose origins are also controversial, is an unsatisfactory solution to the problem of lamproite formation. Of greater significance is that there is no mineralogical evidence for mixing with kimberlite even in the most silica-poor Spanish lamproites. The hypothesis cannot be extended to provinces lacking recent subduction or association with shoshonitic rocks. Perhaps in recognition of these problems, the hypothesis has been slightly modified by Hartogen *et al.* (1985). The kimberlite component was eliminated and it was considered that the lamproites arose by partial melting of normal mantle, contaminated with shoshonitic melt released from subducted continental sediments.

10.2.3. Partial Melting of Garnet Lherzolite

During the 1960s it was realized that garnet lherzolite xenoliths found in kimberlites represent samples of the continental lithospheric mantle. In consequence, petrologists at that time based most petrogenetic hypotheses upon such mantle sources.

O'Hara and Yoder (1967) and O'Hara (1968) suggested that liquids with high K/Na ratios could be produced by the fractional crystallization of omphacite and garnet, i.e., eclogite, from a liquid formed by the partial fusion of garnet lherzolite. The hypothesis was advanced to explain the origin of incompatible element-enriched magmas such as kimberlites, but applies equally well to lamproites or any K-rich magma. Carmichael (1967), echoing Wade and Prider (1940), has suggested that only the madupitic lamproites from the Leucite Hills might be formed from liquids derived by the high-pressure fractionation of a partial melt of garnet peridotite. Carmichael (1967) noted that leucite and sanidine-bearing rocks could not be derived from madupitic rocks by fractional crystallization, as phlogopite is not an early crystallizing phase in the latter. Moreover, derivation of silica-rich magmas from silica-poor partial melts is improbable, as crystallization of silicates would perpetuate the silica impoverishment. Carmichael (1967) considered that fractional crystallization was involved in the evolution of the suite, as cumulates of apatite, mica, diopside, and amphibole are common in the rocks. However, the compositions of leucite- and sanidine-bearing rocks cannot be derived by simply subtracting these minerals from madupitic liquids (Ogden 1979).

Applied to lamproites, the hypothesis is untenable, in that there is no evidence for the postulated high-pressure fractionation event, as eclogite xenoliths or Al-rich xenocrysts do not occur in lamproites. Least-squares fractional crystallization models show that the major element composition of lamproites, especially their K and Al contents, cannot be produced by any simple fractionation scheme involving eclogite and picritic liquids formed by the partial melting of four-phase garnet lherzolite mantle. Carmichael (1967) recognized such problems, and implied, though he did not explicitly state, that the source must contain additional K-bearing phases, if high levels of K and incompatible elements are to be present in derivative magmas. In addition, serious problems result from the paucity of REE, Ba, Ti, etc., in the starting material. For example, Mitchell and Brunfelt (1975) and Frey *et al.* (1977) have shown that eclogite fractionation from REE-poor melts cannot significantly increase the La and Yb contents of derivative melts to the levels found in kimberlites or lamproites. Cullers *et al.* (1977) have also pointed out that eclogite fraction will result in unacceptable depletions in Co, Sc, and Cr. The hypothesis of O'Hara and Yoder (1967), which was developed prior to the concept of mantle metasomatism playing a significant role in the genesis of alkaline rocks, must now be regarded as improbable.

Another approach assumes that lamproites represent unmodified partial melts of a garnet lherzolite source, and the degree of melting of the source rocks may be inferred from their REE distribution patterns. Thus, Kay and Gast (1973) proposed that 0.35–0.5 wt % melting of a mantle composed (wt %) of olivine (55%), orthopyroxene (25%), garnet (17%) and clinopyroxene (3%) could generate the REE distribution patterns and La/Yb ratios of the Leucite Hills lamproites. In this model, major element abundances and possible REE-bearing phases in the mantle are totally ignored. It is doubtful that such small amounts of melting can explain the observed major element variation. In addition, there may be difficulties in extracting such low-volume melts from their source rocks.

Cullers *et al.* (1985) noted that the concentrations of many trace elements predicted by the Kay and Gast (1973) model are low compared to those actually found. To overcome this problem they appealed to cryptic metasomatism (see Section 10.3.5) as means of increasing the trace element content of the source peridotite. It was also found necessary to add phlogopite to account for the K and Rb contents of the partial melts. Cullers *et al.* (1985)

determined that the best match between the predicted and observed REE, Ba, Rb, and Sr contents of the Hills Pond lamproites required about 2 wt % melting of an enriched phlogopite garnet lherzolite source. The problem of the major element composition of the melt was not addressed. It was assumed, following experimental studies of the system peridotite–H_2O–CO_2 (see Chapter 8), that under high H_2O/CO_2 ratios, the melt would be a silica-saturated, olivine hypersthene normative magma. The Cullers *et al.* (1985) model can be applied to other lamproites, but in every case very low degrees of partial melting of variably metasomatized source rocks are required.

Other petrologists, discussing the genesis of lamproites in general terms (Borley 1967, Sheraton and Cundari 1980), have also favored low degrees of partial melting of a phlogopite garnet lherzolite mantle source. Sheraton and Cundari (1980) note that the composition of Gaussberg lavas is unrelated to invariant points in the system CaO–K_2O–MgO–Al_2O_3–SiO_2–H_2O (Bravo and O'Hara 1975) at 15 and 30 kbar. The liquid composition at the 30 kbar invariant point, coexisting with residual phlogopite garnet lherzolite, contains only 3.6 wt % K_2O and is not peralkaline. Lamproites cannot be generated from such liquids by any reasonable crystal fractionation schemes (Sheraton and Cundari 1980).

Although the above hypotheses fail to explain the genesis of lamproites from simple garnet lherzolite, they are important in that this failure led to models invoking lamproite derivation from a phlogopite-bearing metasomatized mantle.

10.2.4. Incongruent Melting of Phlogopite

Henage (1972) suggested that the high K content of the Francis lamproites was due to the incongruent melting of phlogopite. To support this claim a mixing model was presented in which the compositions of the phenocrystal phlogopite, diopside, and olivine, were assumed to approach the composition of minerals in the source mantle. A melt composition, matching that observed for the Francis lamproites, was then calculated using this hypothetical mantle assemblage as a source rock. Agreement could only be achieved by the addition of SiO_2 to the source composition. Formation of leucite sanidine lamproite from these components requires the bulk of the melt to be made up of phlogopite and a large amount of olivine to be subtracted from the system. The latter can be accounted for by the fractionation of olivine formed by the incongruent melting of phlogopite.

Henage's (1972) hypothesis is untenable in that the postulated source consisting of phlogopite, forsterite, diopside, and silica is unrealistic. The assumption that the composition of the phenocrysts is the same as that of minerals in the source is undoubtedly incorrect. Moreover, the model ignores the common occurrence of phenocrystal leucite. Its presence or fractionation will have significant effects on the K and Al budget not addressed by Henage (1972). Finally, it is not surprising that Henage was able to obtain agreement between observed and calculated liquid compositions as the hypothesis rests upon circular reasoning, i.e., the requirements of the result are imposed upon the starting material.

Ogden (1979) has devised a model for the Leucite Hills lamproites that assumes the suite represents a series of primary partial melts. Following ideas advanced by Flower (1971) and Beswick (1976), it is suggested that the compositional variation within the province reflects different degrees of melting of a phlogopite-bearing, garnet-free peridotitic source. The basis for this assertion is that whole rock compositions define linear trends on Pearce (1968) mole ratio plots and Harker diagrams. Madupitic rocks are considered to represent the earliest melt fraction as they are poorest in K and richest in Mg, Ca, and incompatible elements. Using Luth's (1967) study of the system Ks–Qtz–Fo (see Section 8.3) as a guide, Ogden predicts that there will be a eutectic point in the quaternary system Ks–Qtz–Fo–Di. At this eutectic, phlogopite melts incongruently, together with diopside to form forsterite plus enstatite and a liquid with the composition of average madupitic lamproite (43.9 wt % SiO_2, 6.3 wt %, K_2O, 10.3 wt % MgO). Melting at this eutectic continues until all the diopside is consumed. Leucite- and sanidine-bearing lamproites are formed by further melting of the residual phlogopite harzburgite. This occurs at a peritectic point, where phlogopite melts incongruently to forsterite and liquid (Yoder and Kushiro 1969), and continues until all the phlogopite is consumed. Ogden suggests that the melting process is fractional rather than equilibrium in character, as the Leucite Hills suite is essentially bimodal with respect to composition.

Ogden's (1979) hypothesis is interesting in that he was the first to suggest that phlogopite harzburgite rather than garnet lherzolite might be a source for some lamproites. The origin of the phlogopite is not discussed. The model is unlikely to be valid as:

1. There is no experimental evidence for the existence of the quaternary eutectic at high pressures or that liquids produced at the invariant points discussed correspond to lamproite compositions.
2. Mineralogical evidence suggests that the madupitic rocks are more evolved than the other lamproites in this province.
3. Madupitic and leucite phlogopite lamproites have distinct isotopic compositions (Vollmer et al. 1984), implying that processes other than fractional melting must have played a role in their genesis.

10.2.5. Fractional Fusion and Diapiric Uprise

Kuehner et al. (1981) have modeled the genesis of the Leucite Hills lamproites on the basis of phase relations in the systems Ks–Fo–Qtz–CO_2 and Ks–MgO–Qtz–CO_2–H_2O (see Section 8.3). Madupitic compositions recalculated into the Ks–Fo–Qtz subsystem at 28–34 kbar plot (Figure 10.1a) near an invariant point (P) involving enstatite, forsterite, and sanidine. Leucite and sanidine phlogopite lamproites similarly plot near the enstatite–forsterite–sanidine eutectic at 18.5–19.5 kbar pressure. Kuehner et al. (1981) thus suggest that partial melting of a source material, whose compositions can be represented within this subsystem may produce liquids comparable to madupitic or leucite–sanidine lamproites at different pressures. Liquids produced at higher pressures are undersaturated relative to those formed at lower pressures due to the expansion of the primary phase fields of sanidine and enstatite.

Figure 10.1b represents the P–T section of Figure 10.1a and illustrates how diapiric uprise might generate lamproitic melts of differing composition from the same source mantle. Rising solid material will begin to melt at some pressure and temperature above that of invariant point I_{16} (about 34 kbar, 1350°C). With further ascent and decreasing pressure, liquid of madupitic composition will continue to be produced until the diapir reaches a depth corresponding to a pressure of slightly less than I_{16}. At this point the solidus is reached and melting ceases. The diapir P–T path then traverses the sanidine plus forsterite field until the solidus is encountered again between the singular points S_5 and S_6 at 18.5–19.5 kbar. At this stage a liquid, corresponding in composition to leucite and sanidine lamproites, will be

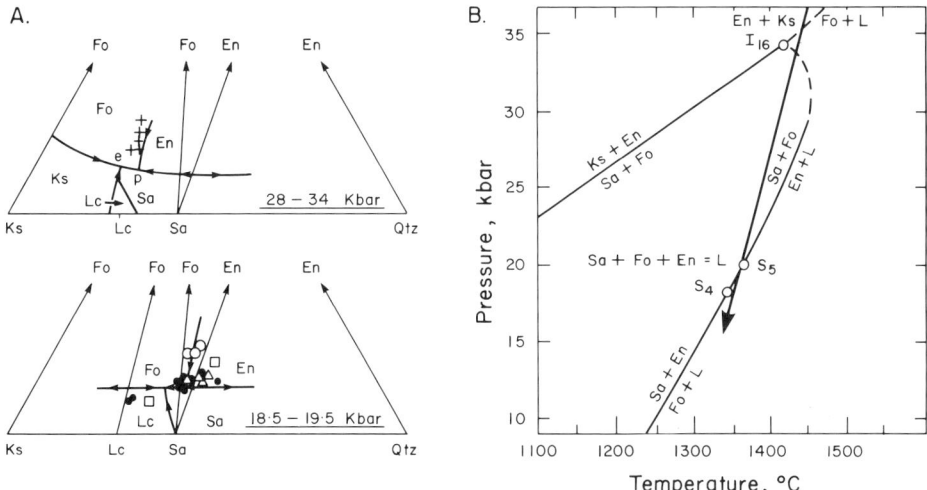

Figure 10.1. (a) Recalculated normative compositions of Leucite Hill lamproites plotted in the system $KAlSiO_4$–Mg_2SiO_4–SiO_2 (Wendlandt and Eggler 1980a); crosses, madupitic rocks; open circles, olivine-bearing sanidine phlogopite lamproites; open triangles, sanidine phlogopite lamproites; solid circles, leucite phlogopite lamproites; open squares, transitional madupitic lamproites (after Kuehner et al. 1981). (b) P–T diagram showing the path of diapiric uprise (arrowed line) of the source material of the Leucite Hills lamproites. Madupitic liquids form around 34 kb (I_{16}) and phlogopite lamproites between 18.5–19.5 kb (S_4-S_5) (after Kuehner et al. 1981).

produced. The generation of the two compositionally distinct liquids is caused by the negative slope of the solidus curve at pressures slightly below I_{16}. Similar phase relationships are present in the CO_2- and H_2O-bearing systems (Wendlandt and Eggler 1980b). Under conditions of CO_2 saturation with H_2O buffered at low activities, madupitic lamproites might form at 24 kbar and leucite sanidine lamproite at 14–17 kbar (Kuehner et al. 1981).

Kuehner et al.'s (1981) hypothesis is useful in that it provides an explanation of the compositional bimodality of the Leucite Hills province. The model assumes that all of the liquids are primary partial melts. This may be correct in the case of the leucite and sanidine phlogopite lamproites, but is unlikely to be true for the madupitic rocks. Kuehner et al. (1981) recognize that the low Ni contents (160 ppm) of the latter indicate they cannot represent an unmodified low-melting fraction of the mantle, but disregard this problem in their genetic model. Further problems arise in that isotopic data suggest both lamproite types cannot be derived from the same source (Vollmer et al. 1984), and mineralogical data indicate that madupitic rocks do not represent liquid compositions (Chapters 6 and 7). The mineralogy of the source mantle is not discussed beyond speculation that it is a K-enriched peridotite which may contain phlogopite, apatite, and wadeite. Whether the bulk composition of such a mantle would fall in the phase fields required by the model is unknown. Moreover, there is no guarantee that a mineralogically realistic source would melt in the manner suggested by the experimental work. An important argument against the model is that the pressures and temperatures of melting are very low and below those of any reasonable continental shield geotherm. Derivation of some lamproites from within the diamond stability field mitigates against their genesis by the Kuehner et al. (1981) model.

10.2.6. Partial Melting of Enriched Mantle

The realization that anhydrous K-poor mantle could not be the source material for a variety of alkaline basaltic rocks led to the conclusion that many portions of the mantle are enriched in incompatible elements, relative to estimates of primitive mantle compositions. This conclusion is supported by isotopic studies demonstrating that many alkaline rocks can only be derived from mantle sources enriched in Rb and depleted in Sm relative to the bulk earth composition. These advances, based upon geochemical arguments, were taken by petrologists as further evidence for the existence of a "metasomatized" mantle. This concept had been growing in favor from about 1975 to 1980 as a result of the description of a variety of upper mantle-derived xenoliths containing additional minerals to those "normally" found in garnet or spinel lherzolites. The reader is referred to Menzies and Hawkesworth (1987) for a detailed discussion of these ideas.

Application of isotopic studies to lamproites has resulted in a radical revision of ideas concerning the nature of their source rocks and of interrelationships within a given province. Consequently many hypotheses of lamproite genesis, proposed since 1980, call upon an enriched mantle source. This source is commonly referred to as being metasomatized. In many cases it is not explicitly stated whether the "metasomatism" is the result of the migration of incompatible element-enriched, H_2O-rich fluids, or veining by silicate melts. Menzies and Hawkesworth (1987) correctly point out that the latter case is not actually metasomatism, although the process may lead to enrichment of large portions of the mantle in incompatible elements. The terms "modal" (Harte 1983) and "patent" (Dawson 1984) are commonly used to describe the metasomatic process which causes mineralogical, e.g., formation of phlogopite, amphibole, apatite, rutile, K–Ba-titanates etc., as well as chemical modification of the mantle. Enrichment processes not resulting in the development of new mineral phases are termed cryptic metasomatism. Patent metasomatic processes are more important than cryptic ones with regard to lamproite genesis, given the requirement that phlogopite or potassium richterite be present in their source mantle.

Models explaining the isotopic composition of lamproites are discussed in Sections 7.9 and 10.3. These models deal primarily with the long-term history of the source. Its mineralogical character and the melting process are commonly ignored, e.g., McCulloch *et al.* (1983), Collerson and McCulloch (1983), Vollmer *et al.* (1984).

Hawkesworth *et al.* (1985) and Fraser *et al.* (1985) consider that the sources of the West Kimberley and Smoky Butte lamproites were of the melt-enriched variety, and that they contained sufficient H_2O to stabilize phlogopite. Mitchell *et al.* (1987) suggested that the Smoky Butte source could be either patently metasomatized or veined and contained phosphates and K–Ba-titanates in addition to phlogopite. Other authors (Wagner and Velde 1986a, Venturelli *et al.* 1984a, Jaques *et al.* 1984, 1986, Nelson *et al.* 1986) are less specific and simply refer to "metasomatized" phlogopite-bearing lherzolite, peridotite or harzburgite mantle sources.

Most of the models noted above do not discuss the amount of partial melting in any detail. This lack of specificity stems from the procrustean nature of the model, as the source composition can be preset to any desired incompatible element abundance. Mitchell (1986) and Mitchell *et al.* (1987) have discussed this problem with regard to kimberlites and lamproites and noted that the high REE contents and La/Yb ratios of melts derived from metasomatized sources are not due to REE fractionation during melting, but are reflections of the La/Yb ratios and REE contents of such phases as apatite. La/Yb ratios are relatively insensitive to the degree of melting and hence large volumes of melt of high La/Yb ratio can be produced. For example, Mitchell (1986) concluded that the REE characteristics of kimberlites could be reproduced by 1%–8% partial melting of an apatite, potassian titanian richterite-bearing lherzolite source, a conclusion that is also valid for lamproites. Although the melt extraction problems incurred due to the small degree of melting required of cryptically enriched sources (Section 10.2.3) are avoided in metasomatic models, the degree of melting cannot be unambiguously determined.

The major benefit of metasomatic models is that they provide a means of explaining some of the geochemical features of lamproites. A detraction is that the models are not specific to lamproites, and almost identical models, based upon incompatible trace element and isotopic arguments, can be advanced to explain the geochemistry of other potassic rocks, kimberlites, carbonatites, melilitites, and minettes. Thus, without considering the nature of the controls on the major element geochemistry, metasomatic models based upon isotopic or trace element data alone are inadequate.

10.2.7. Partial Melting of Harzburgitic Sources

A key objective of lamproite petrogenetic models is to explain why many members of the clan are relatively siliceous, K_2O and MgO rich, and CaO poor. The compositional data presented in Chapter 7 indicate that lamproites are depleted in Na, Al, and Ca compared with alkaline basalts and other primitive mantle-derived melts, suggesting that they represent partial melts of refractory material, i.e., harzburgitic sources. This conclusion is supported by the experimental work that forms the basis of the petrogenetic hypotheses described below. Metasomatized harzburgitic sources are now favored over lherzolitic in most of the recent hypotheses of lamproite genesis, e.g., Jaques *et al.* (1984, 1986), Foley *et al.* (1986b, 1987), Mitchell *et al.* (1987), Bergman (1987), Venturelli *et al.* (1988), Foley (1989). These hypotheses rely for the most part on concepts introduced by Arima and Edgar (1983a) and Foley *et al.* (1986b).

Arima and Edgar (1983a) suggested, on the basis of their studies of the melting relationships of a transitional madupitic lamproite (see Section 8.2.1), that lamproitic magmas, at their source, could have been in equilibrium with a phlogopite rutile harzburgite. Such a mantle could produce magmas with the requisite characteristics of high K, Mg, and Ti by the melting of phlogopite and titanates. The congruent melting of enstatite would result in relatively silica-rich magmas. The style of mantle enrichment and the amount of partial melting are not specified. The Ca contents of lamproites are not readily explained by this model, although apatite and/or minor amounts of clinopyroxene may provide suitable sources.

Foley *et al.* (1986b) and Foley (1989) have addressed the origins of silica-rich and silica-poor lamproites on the basis of experimental studies of the system Ks–Fo–Qtz–H_2O–CO_2–F (Section 8.3) and melting relationships of synthetic lamproite compositions (Sections 8.2.1, 8.2.2). The synthetic system contains a peritectic point, involving phlogopite, enstatite, olivine, and liquid, whose position is a function of volatile composition. The point lies on the silica saturated side of the forsterite–sanidine join under H_2O-, HF-, or CH_4-rich conditions and migrates to the undersaturated side under CO_2- and F-rich conditions (Figure 8.8). The position of the point is also dependant upon the stability of phlogopite, and the composition of the first melt to form from a phlogopite harzburgite is expected to vary greatly in both silica content and Ks/Fo ratio with volatile composition. The addition of fluorine expands the phase volume of phlogopite and increases the Fo/Ks ratio at which the peritectic is found, leading to more magnesian melts regardless of silica content. The presence of abundant F in the melt will also lead to the formation of F-rich phlogopite with K > Al. Melting of such micas will give rise to perpotassic Al-poor melts. Experimental studies of the water-saturated system Ne–Fo–Qtz, demonstrate that the Fo–En phase boundary moves to more silica-undersaturated compositions with increasing pressure. Foley et al. (1986b) consider that an analogous phase boundary shift occurs in the Ks–Fo–Qtz system and the peritectic moves towards forsterite, but remains on the silica-rich side of the sanidine-forsterite join (Foley 1989). Thus, it is possible to derive silica-rich and silica-poor, Mg-rich melts from the same source at different pressures under the same HF-bearing reduced volatile conditions. This hypothesis is considered to be in accord with a greater depth or origin for diamond-bearing olivine lamproites than diamond-free leucite phlogopite lamproites.

Foley (1989) considers that partial melting occurs in response to the introduction of reduced volatiles to a geochemically depleted upper mantle. The depletion results from the long-term operation of redox melting processes (Taylor and Green 1989). The volatiles are believed to be juvenile and derived from incompletely degassed deep mantle, although some modification of their composition by fluids derived from deep subduction may have occurred. A consequence of the hypothesis is that all events leading to the formation of lamproites, e.g., ancient mantle enrichment, recent enrichment, partial melting etc., must take place under reduced conditions.

Foley (1989) also suggests that relatively calcic lamproites, such as the Leucite Hills madupitic rocks, represent partial melts formed under relatively more oxidized conditions. This conclusion is based on experiments that suggest CO_2 stabilizes clinopyroxene in the source mantle.

Foley et al. (1986b) do not discuss the mechanism that generates lamproites at different depths, or why a spectrum of primary melts is not found in lamproite provinces. The latter may be expected if melting occurs in response to volatile introduction over the wide pressure range (28–50 kbar) required to generate the range in silica contents.

Problems exist with the interpretation of the experimental studies of natural lamproite compositions cited by Foley in support of the hypothesis. In particular, the phase relationships of the olivine lamproite composition are extrapolated to high pressure. Foley (1989) himself notes that there is no evidence for an origin by melting of a phlogopite harzburgite at the maximum pressure (40 kbar) of the investigation. It is only suggested that mica and orthopyroxene will occur at the liquidus at higher pressure. Further, Foley (1989) assumes that the olivine lamproites represent primitive liquid compositions. However, if the rocks are hybrids, as suggested by mineralogical and isotopic evidence, then the experimental data have no bearing on the nature of the source rocks. Similarly, for the leucite lamproite composition there is no pressure (5–35 kbar) at which there is multiple saturation of the liquidus with phlogopite, orthopyroxene, and olivine. The data can only be reconciled with a harzburgitic source by inferring that the composition studied is not actually primitive and has undergone some olivine fractionation.

In summary, geochemical and experimental evidence suggests that partial melting of harzburgitic sources provides an attractive mechanism for producing melts of the required major element composition. However, knowledge of the high-pressure (>50 kbar) phase relationships of primitive lamproites, e.g., leucite phlogopite lamproites, under reduced volatile conditions is required to confirm or refute the hypothesis.

10.2.8. Subduction-Related Models

The occurrence of potassic rocks in recent island arcs, together with the postulated correlation between depth to the Benioff zone and K_2O content of arc-related magmas (Hatherton and Dickinson 1969), has led to suggestions that lamproites may be related in some manner to deeply subducted material. Petrogenetic theories based on this assumption are colored by failure to appreciate that lamproites form a distinct clan of rocks unrelated to shoshonites or Roman province-type lavas. For example, Thompson and Fowler (1986, p. 516) incorrectly state that "the shoshonitic and ultrapotassic rocks are arbitrary groups which grade into each other and also into calcalkaline and alkaline magmatic suites."

The association with subduction is commonly loosely defined, and even Thompson and Fowler (1986) find it difficult to envision any direct connection with subduction for *bona fide* lamproites. Nevertheless, they state that "there is obviously a potential transition between this group and subduction-related magmatism around the edges and fringes in space and time of subduction zones" (Thompson and Fowler 1986, p. 516). Petrogenetic models are based on the general concept of the addition of subducted components to asthenospheric (Thompson and Fowler 1986) or lithospheric (Varne 1985) mantle. The details of the melting process with reference to lamproites are typically not considered.

Rowell and Edgar (1983) suggested that potassic magmatism in the western U.S.A. was related to deep subduction and that the age of volcanism decreases from northeast to southwest. Unfortunately, the relatively young lamproites from the Leucite Hills do not fit this simple model. The discovery of the Smoky Butte lamproites makes the model even less reasonable. Curiously, Edgar (1983) relates western U.S. potassic volcanism to the proposed Yellowstone plume but does not integrate this work with Rowell and Edgar (1983).

Schreyer *et al.* (1987) have predicted that common rocks of the upper continental crust, such as granites and acid gneisses will, upon deep subduction, develop K- and Mg-rich fluids at the expense of biotite and K-feldspar. This hypothesis is based upon experimental studies of the system $K_2O-MgO-Al_2O_3-SiO_2-H_2O$ which show that K-feldspar and mica become unstable in the presence of excess H_2O at 15–20 kbar and 400–700°C. The fluids liberated by these reactions are believed to interact with ultramafic mantle rock to produce phlogopite and potassium richterite at the expense of olivine and clinopyroxene. Subsequent partial melting of these patently metasomatized rocks is believed to result in the formation of lamproites. Schreyer *et al.* (1987) suggest that lamproite melts are formed considerably later than the metasomatic event. They suggest that there should be a close relationship between lamproites and former collision zones, and cite the Murcia-Almeria, Sesio-Lanzo, and Sisco lamproites as evidence of such a relationship.

This hypothesis differs from most petrogenetic schemes involving subduction by invoking very deeply subducted granitoid as a source of metasomatizing fluids. The scheme has the merit of suggesting a source for such fluids and bears some similarities to ideas advanced by Sekine and Wyllie (1982) for the formation of phlogopite pyroxenites above subducted oceanic crust (see Section 10.3.1).

The major weakness of the hypothesis, as stated, is that the lamproites discussed in support of the scheme are atypical, and their geochemistry is best considered to result from the interaction of a previously generated lamproite magma with crustal material (Section 10.3.5). However, as noted in Section 4.3.4. we believe there is an association of lamproites with paleo-subduction zones, i.e., mobile belts, and therefore do not preclude that fluid interactions of the type described by Schreyer *et al.* (1987) play some role in the development of the source regions of lamproites.

In summary, as detailed in Chapter 4, we find no evidence for any relationship of lamproite magmatism to contemporaneous subduction and thus conclude that simple subduction-related hypotheses of the type advanced for island arc potassic rocks are improbable. This conclusion does of course not rule out the possibility that material subducted down to the 670-km discontinuity can be recycled (Ringwood 1989) and provide a source of incompatible elements.

10.3. GENESIS OF THE LAMPROITE CLAN

10.3.1. Character of the Source

Hypotheses regarding the development and nature of the sources of lamproites must account for the following observations:

1. Lamproites are derived from sources with time integrated Rb/Sr ratios and Sm/Nd ratios that are higher and lower, respectively, than those of bulk earth.
2. Lamproites of similar major element composition are derived from sources with different Rb/Sr and Sm/Nd ratios.
3. Multistage and/or multicomponent models are required to explain the Sr–Nd–Pb isotopic compositions.
4. Lamproites are enriched in incompatible elements (REE, Ba, Zr) and have high La/Yb ratios.
5. The geochemical signature of lamproites is distinctive in having low Nb, Ta, and Sr and high Ba contents relative to primitive mantle melts.
6. The depletion of Na, Al, Ca, and enrichment in K, Mg, and Ti, of primitive lamproites, suggests that the source is most likely of a metasomatized harzburgitic composition.
7. Lamproite magmatism is associated with fossil Benioff zones and mobile belts surrounding ancient cratons.
8. Most lamproites are derived from subcontinental lithospheric sources, with in some instances, contributions from the asthenosphere or the crust.

The first stage in the development of the source is the formation of depleted lithosphere. This process is related to the long-term development of continental cratons, the details of which are beyond the scope of this work. Currently, depletion of primitive or fertile mantle is believed to occur by the extraction of basaltic or other Ca-, Na-, Al-enriched magmas. Depleted harzburgitic mantle may be generated by decompression melting, consequent to diapiric uprise of fertile asthenospheric material, or by long-term redox melting of fertile lithosphere (Taylor and Green 1989, Green *et al.* 1987). This refractory harzburgitic lithosphere is believed to be the principal control on the major element composition of lamproitic magmas. Harzburgitic lithosphere may form at any time, but isotopic data require that involved in lamproite genesis to be older than 1–2 Ga. Continental lithospheric roots consisting principally of depleted mantle may extend to depths of 150 to 400 km (Anderson 1987, Jordan 1975).

The second stage of source preparation involves addition of incompatible element-rich components to a harzburgitic substrate. How this is accomplished is unknown, making this stage of the source evolution one of the more speculative aspects of lamproite genesis.

One potential source of incompatible elements lies in supra-subduction zone metasomatism or melt infiltration. Subducted oceanic lithosphere, MORB, and continental-derived sediments may be to expected to underplate the continental lithosphere. Dehydration and melting processes occurring in the subducted material will affect the overlying mantle in a variety of ways. Fyfe and McBirney (1975) suggest that aqueous fluids rising from subducted oceanic crust will result in the formation of phlogopite in the mantle. Sekine and Wyllie (1982) have noted that aqueous fluids may also be expelled during hybridization and solidification of siliceous melts formed by partial melting of subducted material. At depths on the order 150 km such melts may rise into the overlying mantle wedge and develop around them an aureole, consisting of a variety of phlogopite-bearing rocks, including phlogopite orthopyroxenites. Clearly, such reactions and those proposed by Schreyer *et al.* 1987 (see Section 10.3.8) can generate a K-enriched environment suitable for generation of lamproite from patently metasomatized mantle.

Alternatively, the depleted harzburgites forming the roots of cratons may be enriched by metasomatism or veining with fluids and magmas of juvenile origin. The latter may represent limited partial melts of pockets of fertile lithospheric mantle or be derived from deeper asthenospheric material. It may be expected that lamproites formed from lithospheric mantle will have different geochemical signatures depending upon the amount subduction-related or asthenospheric components in the source. This may be the origin of the varying degrees of Nb depletion seen in different lamproite suites. The association of many lamproites with mobile belts and their distinctive geochemical signatures suggests, but does not prove, that enriched mantle above paleo-subduction zones has played a role in their genesis (see Section 4.3.4).

Patent metasomatism or vein formation will result in the formation of minerals that will act as repositories for incompatible elements. The most important with respect to K, Rb, Ba, and volatiles are phlogopite and richterite. Both are stable in the continental mantle at depths of up to 300 km (Tronnes *et al.* 1989). Minerals that have been described from metasomatized xenoliths, such as Nb-rutile, Cr–Mg-ilmenite, lindsleyite-mathiasite, yimengite-hawthorneite (Haggerty *et al.* 1986, 1989) may provide sites for K, Ba, Sr, Nb, U, REE, etc. Phosphates that are stable at high pressure (Murayama *et al.* 1986) may also provide suitable sites for the sequestration of REE and Sr. Zr may be held in wadeite (Arima and Edgar 1980).

It is to be expected that the mineralogy of the enriched mantle will not be everywhere identical. Modally different portions of the mantle will develop different isotopic compositions over long time periods. Such modal variation explains why the West Kimberley lamproites appear to have been derived from sources with relatively higher Rb/Sr and Sm/Nd ratios than those of the Leucite Hills or Smoky Butte lamproites. This could reflect different mica/amphibole ratios in these sources, with the West Kimberley source perhaps containing more phlogopite than that of the Leucite Hills. Differences in the Sm/Nd ratios undoubtedly reflect modal variations of other phases. However, it is also important to realize that the observed isotopic composition of lamproites does not necessarily reflect that of their source. Observed isotopic compositions might result from the mixing of Sr, Pb, Nd from a variety of sources (McCulloch *et al.* 1983, Vollmer *et al.* 1984, Nelson *et al.* 1986, Mitchell *et al.* 1987) and are developed during partial melting (see below). It is possible that other geochemical signatures, such as the interprovincial differences in Ti and Ba contents, are developed during this ancient stage of enrichment.

Mantle enrichment might not be a single-stage process, although all stages must occur long before the generation of the partial melt. For example, Nelson *et al.* (1986) have interpreted the Pb isotopic composition of the West Kimberley lamproites as resulting from a three-stage process (see Section 7.9.2). Mitchell *et al.* (1987) have noted that events leading to U, Th, and Pb depletion cannot coincide with addition of REE, given the similar geochemical character of these elements, and therefore multiple metasomatic events are commonplace.

As a final point it must be understood that the mantle enrichment events described above do not control the major element composition of partial melts derived from such mantle. This is clearly demonstrated by the occurrence of lamproites of similar major element chemistry, but different trace element and isotopic signatures.

10.3.2. Melting of the Source

Partial melting of the source may be initiated by any of the following processes:

1. Simple thermally induced melting. No new components are introduced and ancient metasomatized lithosphere and/or old Benioff zones melt directly at the fringes of thermal plumes to form lamproite magmas (Section 4.2.4).
2. Decompressional melting, resulting from uplift, with no new components being introduced.
3. Introduction of volatiles alone, with melting resulting from reduction of the mantle solidus temperature. Volatile influx may result from degassing of the earth, e.g., the ascent of reduced fluids as advocated by Foley (1989) and Taylor and Green (1989) or by the thermally induced decomposition of subducted hydrous phases at great depth in the mantle (Ringwood 1989).
4. Introduction of volatiles and other components. In this case the volatile flux contains alkali (albeit Na poor) and other incompatible elements. These may be scavenged from the mantle during transit or be present in H_2O-rich limited partial melts derived from asthenospheric sources, i.e., they are similar to the melts that caused the original enrichment of harzburgitic mantle. This may be considered as a further enrichment event.

Initiation of partial melting may result from a combination of these processes. The relative contribution of each will determine the geochemical character of the ultimate lamproite, and thus, in part may be responsible for their observed diversity. Note that processes 2 and 3 may be interrelated, as Bailey (1983) considered that the introduction of volatiles will lead to the formation of low-density phases and ultimately to uplift. Process 4 has important consequences in that there is the possibility of addition of material of different isotopic composition to that of the ancient source (see Section 10.3.4), as well as other trace elements.

Details of the actual melting event are poorly understood owing to a lack of experimental data on plausible sources. Unfortunately, nothing is known about the partial melting of phlogopite-, richterite-, titanate-, phosphate-bearing harzburgitic sources under any volatile conditions or oxygen fugacities at 40–60 kbar pressure. The degree of partial melting cannot be unambiguously estimated from geochemical parameters, such as the REE content and La/Yb ratio, and may range from 1 to 10 wt % depending upon the mode. Tectonic constraints suggest that diapiric uprise of lamproite sources is unlikely to occur, and therefore that fractional or polybaric fusion processes, as advocated by Kuehner *et al.* (1981) and Foley *et al.* (1986b), respectively, are improbable.

Melting probably occurs at a single invariant point, involving phlogopite and enstatite, and results in a primitive lamproite magma. The major element composition of the melt is clearly constrained to a particular limited compositional range, as demonstrated by the occurrence of phlogopite lamproites of similar composition (Table 7.1) on a worldwide basis. We thus consider that primitive lamproite magmas are relatively rich in silica (52–55 wt % SiO_2) and similar to those forming the Gaussberg, Leucite Hills, and Smoky Butte lavas. The randomly introduced metasomatic components control, in part, the trace element and isotopic character of such magmas. Derivation of such relatively siliceous magmas from a phlogopite harzburgite source under H_2O-, CH_4-, HF-rich conditions is supported by Foley's (1989) experimental study of a leucite phlogopite lamproite composition (Section 8.2.1) based on that of Gaussberg lavas.

All other members of the clan are derived by fractionation and hybridization of this unevolved primitive melt. It should be noted that the bulk of the lamproites in the major provinces consist of relatively silica-rich rocks. It is difficult to understand how these could have originated by fractionation of less voluminous silica-poor magmas. The origins of the latter are discussed below.

Extensive differentiation is not a characteristic of phlogopite lamproite magmas (Chapter 7). Limited amounts of low-pressure fractional crystallization adequately explain the compositional variation found in the Leucite Hills and West Kimberley leucite diopside sanidine phlogopite lamproites (Kuehner *et al.* 1981, Jaques *et al.* 1986). Petrographic differences between lamproites in a given province in many instances result from heteromorphism (Jaques *et al.* 1986, Mitchell *et al.* 1987).

10.3.3. Relationships between Olivine and Phlogopite Lamproites and the Origins of the Isotopic Signatures of the West Kimberley Lamproites

One conclusion of this work is that olivine lamproites such as those found in the West Kimberley province are not primitive magmas that are parental to the more silica-rich phlogopite lamproites. To account for the genesis of the olivine lamproites from a siliceous parent requires that they are hybrid rocks, formed by the mixing of the differentiates of primitive lamproites with mantle material. The two compositions are not related by the simple addition or subtraction of olivine. Least-squares mixing calculations suggest that the average West Kimberley olivine lamproite may be derived from phlogopite lamproite if there is an initial period of crystallization of phlogopite and orthopyroxene, followed by mixing of the derivative magma with olivine and minor clinopyroxene. Trace elements do not fit this type of model, although an excellent fit can be obtained for the major elements. This point may be appreciated by noting that the average Ba content of West Kimberley olivine lamproites is similar to that of the phlogopite lamproites, despite the former containing 30–40 wt % Ba-free olivine. Thus, a complex fractionation–hybridization origin, involving mixing of evolved and primitive melts, that is not easily modeled is indicated.

The wide range in Sr and Nd isotopic composition may reflect melting of different portions of the mantle of discrete Rb/Sr and Sm/Nd ratio (Bergman 1987). According to this model the isotopic compositions of individual intrusions are direct reflections of those of their source mantle. A major difficulty with the model lies in explaining why partial melts formed from a large volume of mantle do not mix and homogenize. However, it may be possible to preserve the heterogeneities if the partial melts are of limited volume and if they migrate directly and rapidly to upper crustal levels without magma mixing taking place.

In contrast, McCulloch *et al.* (1983) have interpreted the data in terms of a mixing model, in which a small amount of enriched component derived from a high-Rb/Sr, low-Sm/Nd source, is added to a depleted MORB-type component. This model implies that the major element composition of the rocks is controlled by the depleted component and the trace element content by the enriched mantle. As an alternative, we suggest the reverse process, in which partial melts of ancient enriched phlogopite harzburgite are contaminated by trace-element-rich limited partial melts derived from asthenospheric sources, with Sr and Nd isotopic compositions similar to those of oceanic island basalts. It is possible that introduction of the asthenospheric component initiates partial melting. Either model requires that the olivine lamproites not be related to phlogopite lamproites by any simple fractional crystallization process.

Interpretation of the Pb isotopic data for the West Kimberley lamproites requires at least a three-stage evolutionary model (Nelson *et al.* 1986). The estimated times of depletion of U and Th cannot easily be related to the metasomatic event leading to the Rb and light REE-enrichment of the mantle. Nelson *et al.* (1986) suggest that the Pb isotopic compositions result from the mixing of an ancient (>2.1 Ga) high-^{207}Pb/^{204}Pb, low-^{206}Pb/^{204}Pb component with typical MORB component. The model agrees with the interpretation of the Sr and Nd isotopic data, although there are no significant differences between the Pb isotopic compositions of olivine and phlogopite lamproites.

In summary, we suggest that olivine lamproites are fractionated magmas that have been hybridized with depleted asthenospheric material and contaminated with olivine derived from the lithospheric mantle. Olivine lamproites may represent the first partial melts derived from a pregnant source, their contamination being related to clearing of an ascent channel through the upper mantle. This may account for their association with diamonds, as later batches of melt probably ascend more rapidly through the conduits swept clean of diamonds and xenolithic material. Diamond preservation does not imply rapid ascent of these magmas if oxygen fugacities are such that the volatile phase remains reduced and H_2O and CH_4 are the dominant components.

10.3.4. Relationship of Madupitic Lamproites to Phlogopite Lamproites and the Origins of Isotopic Signatures of the Leucite Hills Lamproites

Carmichael (1967), Ogden (1979), and Kuehner *et al.* (1981) have shown that madupitic lamproites cannot be related to phlogopite lamproites by fractional crystallization. Consequently, madupitic whole rock compositions are regarded as being representative of their parent melt compositions. Thus, Ogden (1979) and Kuehner *et al.* (1986) consider the madupitic rocks to be derived from Si-poor, Mg-rich primitive liquids generated from mantle sources (Sections 10.3.4, 10.3.5). In contrast, we regard madupitic rocks as not representing melt compositions and their apparent ultrabasic character to be simply an artifact, resulting from the high modal abundance of phlogopite and oxides. Mineralogical (Section 6.1.2) and geochemical (Sections 7.1.3.2, 7.4) data clearly demonstrate that they are evolved rocks.

The different Sr and Nd isotopic compositions of the Leucite Hills shows that there can be no simple relationship between the madupitic and phlogopite lamproites. It seems improbable that they are derived from discrete sources because of their mineralogical and chemical affinities, and mixing models have been proposed to account for their isotopic compositions (Vollmer *et al.* 1984, Mitchell *et al.* 1987). In one case it may be assumed that the isotopic composition of the phlogopite lamproites represents that of the source. Madupitic rocks represent mixing of Nd derived from this source with Nd derived from a depleted source, possibly asthenospheric mantle. Alternatively, it is possible that none of the isotopic compositions are representative of the source. In this case they represent mixing between Nd derived from a reservoir having a strongly negative ϵ_{Nd} signature with asthenosphere Nd. The isotopic composition of Smoky Butte lamproites (Fraser *et al.* 1985) demonstrates that sources of extremely low Sm/Nd ratio exist in the Wyoming craton. Significant differences exist in the Pb isotopic composition of the two types of lamproites, with madupitic rocks having lower $^{206}Pb/^{204}Pb$ and $^{208}Pb/^{204}Pb$ ratios than those of phlogopite lamproites (Table 7.21). These data suggest mixing of Pb from asthenospheric with U- and Th-depleted lithospheric sources.

In summary, the madupitic rocks appear to be evolved hybrids. Unfortunately, there is no evidence preserved in their petrology of how they arrived at this condition. Olivine and other possible mantle-derived xenocrysts are absent, as are cognate cumulates of high-pressure liquidus phases. Such materials must have been completely removed by fractionation or resorbed leaving no trace of their former presence. The petrogenesis of the madupitic rocks must be considered as unsolved, although their evolution must have taken place in the upper mantle, and involves the fractionation and contamination of primitive phlogopite lamproite magma.

10.3.5. The Anomalous Murcia-Almeria Lamproites

Geochemical data presented in Chapter 7 demonstrate that the Murcia-Almeria lamproites are anomalous relative to other lamproite provinces. Nelson *et al.* (1986) suggest that they contain a component with the isotopic characteristics of continental crust or subducted sediments. The lamproites contain anomalously low contents of most incompatible elements and their constituent minerals are characteristically more sodic and aluminous than found elsewhere. Hartogen *et al.* (1985) note that a sedimentary signature is reflected in the high Th/Ta, Th/La ratios and the negative Eu anomaly that is unrelated to plagioclase fractionation. All of the above observations are in accord with extensive contamination of a lamproite magma during ascent through the crust. Contamination may occur when magma ascend from the upper mantle and impinge on the subduction zone (Foden and Varne 1980). Alternatively, contamination may result from emplacement in a region of recently, complexly folded and thrusted rocks (Section 4.4.3). The extensive faulting probably limits and controls their ascent to a degree not found in more homogenous ancient mobile belts, and provides a regime that favors slow tortuous ascent and extensive magma–crust interaction. The maximum and minimum amounts of contamination are found in the extrusive rocks of the Barqueros and Zaneta volcanoes and the hypabyssal rocks at Jumilla and Fortuna, respectively. Interestingly, some of the first rocks to be described as lamproites are now known to be atypical of the clan with respect to their mineralogy and geochemistry.

10.4. RELATIONSHIPS TO KIMBERLITES

The incorrect designation of diamond-bearing rocks from West Australia (see Section 1.6) as kimberlites prompted a number of studies into the relationships between kimberlites and lamproites (Mitchell 1981, Scott Smith and Skinner 1984a,b, Dawson 1987, Mitchell, 1986, 1989a).

Comparison of the mineralogy and geochemistry of group 1 kimberlites (Mitchell 1986, 1989a, Skinner 1989) with that of rocks of the lamproite clan shows clearly they have no mineralogical affinities. In particular, lamproites lack the characteristic megacryst suite (Mg-ilmenite, Ti-pyrope, subcalcic diopside, etc.), monticellite, serpophitic primary polygonal serpentine, and Mg–Ti-rich spinels. Compositional trends of spinels are different in the two groups of rocks (Section 6.9.3). Olivine lamproites bear a superficial petrographic resemblance to some macrocrystal hypabyssal group 1 kimberlites, but may be easily distinguished from the latter on the basis of their detailed mineralogy and petrography (Scott Smith and Skinner 1984a,b, Mitchell 1986). Sanidine, titanian potassium richterite, and leucite are absent from group 1 kimberlites.

In contrast, group 2 kimberlites have some mineralogical affinities with lamproites, as they contain:

1. Abundant mica of broadly similar composition to that of lamproites, although differing with respect to Ti-content and compositional trends (Section 6.1.21.2);
2. Spinels of similar composition and evolutionary trend (Section 6.9.3);
3. Microphenocrystal primary diopside identical in composition to that of lamproites (Section 6.3.4);
4. Primary groundmass Ti-Ba-V-rich hollandites related in composition to priderite (Section 6.10.4).

Group 2 kimberlites differ from lamproites in that they contain abundant calcite and typically lack sanidine, titanian potassium richterite, and leucite. The dikes occurring at Swartruggens (Skinner and Scott 1979) are an exception in containing sanidine and altered leucite (?), and may represent rocks that are mineralogically transitional to the lamproite group or an extreme variant of a distinct magma type (see below). Other occurrences, such as Pniel, which contain richterite are regarded as possible lamproites. Compared to group 1, the group 2 kimberlites are poor in megacrysts and typically lack Mg-ilmenite.

Group 2 kimberlites are geochemically similar to lamproites in being rich in K, Ba, Zr. They have similar REE distribution patterns, contents, and La/Yb ratios. Direct comparison of lamproite compositions with those of group 2 kimberlites (Dawson 1987) is not useful as the latter are not representative of their parental magmas (Mitchell 1989a).

Isotopic studies (Smith 1983, Fraser et al. 1985) have revealed that group 1 and 2 kimberlites are derived from asthenospheric and lithospheric sources, respectively (Section 7.9.1). In common with lamproites, group 2 kimberlites are apparently derived from ancient enriched sources. The derivation of group 1 and 2 kimberlites from different sources implies that they are petrogenetically unrelated. These different origins are reflected in their mineralogy. Both groups are termed kimberlites principally for historical reasons and because they both contain diamond. In the light of the modern geochemical and mineralogical studies there seems little rationale for perpetuating this terminology. Certainly if group 2 kimberlites were discovered today it is doubtful that they would be called kimberlites. In order to emphasize the different character of these rocks, Mitchell (1989a) has suggested that group 2 kimberlites constitute a third variety of mantle-derived diamond-bearing rocks that are neither kimberlites nor lamproites. This heresy led Mitchell and Meyer (1989a) to introduce the term "orangeite" to describe this suite of rocks. While introduction of new rock names is generally undesirable we believe that the genetic and mineralogical characteristics of group 2 kimberlites are so different from those of group 1 kimberlites that a distinct name is necessary. The rocks are not simply modal variants of group 1 kimberlites. Mica-rich varieties of the latter with some superficial petrographic resemblance to group 2 rocks and lamproites do occur. However, detailed mineralogical study and the nature of the spinel compositional trend will reveal their true character.

It should be particularly noted that mica-rich kimberlites, analogous to the South African group 2 kimberlites, are not found anywhere else in the world. Their genesis is clearly related to the long-term evolution of the Kapvaal craton. Conditions that are prerequisites for the genesis of the group have not developed within other cratons. Significantly, abundant lamproitic magmatism is absent in Southern Africa, yet common in other cratons. This antipathy suggests that the magmatic expression of K-rich mantle metasomatism, or enrichment, results in orangeites in Southern Africa and lamproites elsewhere. As noted in Section 10.3.6 metasomatism is likely to be a random process, and we suggest that it is this randomness, superimposed on the unique character of the Kapvaal craton that has led to the apparently unique orangeite magmatism.

Although the composition of the magmas that were parental to the orangeites is not known, it is evident they formed from ancient lithospheric sources prior to group 1 kimberlite genesis. The garnet and subcalcic diopside megacrysts present in some orangeites may be primary, or more likely of xenocrystal origin. In the latter case, they would be derived from earlier episodes of kimberlite magmatism. The Cretaceous (80–90 Ma) group 1 kimberlites of South Africa must have passed through the lithospheric sources of the older (120–150 Ma) orangeites, as they ascended from their deeper asthenospheric sources. This may explain the occurrence of MARID-suite xenoliths in Group 1 kimberlites, if the former represent high-pressure cumulates, heteromorphs, or metasomatic veins derived from orangeite magmas.

Experimental studies (Chapter 8) demonstrate that the conditions under which group 1 kimberlites and lamproites form are very different. Data presented by Brey (1978), Wendlandt and Eggler (1980b), and Wyllie (1980) suggest that partial melting of a magnesite phlogopite garnet lherzolite source with high CO_2/H_2O ratios will generate kimberlites at 40–50 kbar at 1000–1300°C (Mitchell 1986). In contrast, lamproites are apparently derived from phlogopite harzburgite sources with extremely low CO_2/H_2O ratios, the volatiles being dominated by H_2O, CH_4, and HF. The significant CO_2 contents of orangeites suggest they may be derived from sources with CO_2/H_2O ratios between these extremes. The origins of the anomalously CO_2-rich Sisimiut lamproites may lie in such mixed volatile sources.

10.5. RELATIONSHIPS TO MARID-SUITE XENOLITHS

A distinctive group of phlogopite-rich xenoliths found in some South African kimberlites were given the acronym "MARID" by Dawson and Smith (1977), to describe their characteristic mineralogy, i.e., *M*ica-*A*mphibole-*R*utile-*I*lmenite-*D*iopside. Details of their petrology may be found in Dawson and Smith (1987), Jones *et al*. (1982), Dawson (1987), and Waters (1987). Wagner and Velde (1986a) initially suggested a relationship of these xenoliths to lamproites on the basis that the bulk composition of the latter could be derived by fusion of a MARID-assemblage. Formation of lamproites requires the melting of about 50–70 wt % phlogopite and 30–50 wt % richterite, followed by olivine and clinopyroxene fractionation. Although such least-squares calculations are mathematically exact, they totally ignore the real melting behavior of the minerals involved. The scheme also is unsatisfactory in ascribing an origin for lamproites to a source of unknown derivation.

Waters (1987) has suggested on the basis of high-pressure experimental studies of potassic rocks, and the composition and textural appearance of the MARID-suite, that they are the high-pressure magmatic compositional equivalents of MgO-rich lamproites. The experimental studies cited by Waters (1987), summarized by Edgar (1987), suggest that potassic magmas crystallize at 25–30 kbar pressure as phlogopite clinopyroxenites. These data are used by Waters (1987) to infer that lamproitic magmas will behave similarly, and crystallize at high pressure to assemblages dominated by phlogopite, richterite, and diopside. Waters (1987) ascribes the compositional difference between these phases as found in MARID-rocks and lamproites, as being due to differences in the pressure of their formation. MARID-rocks are not as enriched in incompatible elements as lamproites. To explain this difference, Waters (1987) assumes that these elements are lost to the surrounding mantle rocks as late-stage vapor-rich melts. This material is thought to be responsible for the formation of veined phlogopite–potassium richterite peridotites (Erlank *et al*. 1987). Waters (1987) considers that the paucity of high-level lamproite volcanism in South Africa is due to failure of the magmas to penetrate the thick cratonic lithosphere.

Waters' (1987) hypothesis is highly speculative and not constrained by any experimental data on MARID-suite xenoliths or lamproites. The comparison of major and trace element compositions of rocks that may be cumulates with olivine lamproites which are hybrids, is petrogenetically unsound, and is the major flaw in the hypothesis. Any simple geochemical comparison between high-pressure cumulates and low-pressure rocks, postulated as being derived from the same putative parental magma, is doomed to be misleading. The hypothesis has the merit of proposing a relationship of the MARID-suite to a potassic magma.

MARID-xenoliths appear to be found only in South African kimberlites. Mica-rich rocks described from Siberian kimberlites (Frantsesson 1970) have not yet been verified as belonging to the MARID-suite. As isotopic group 2 kimberlites or orangeites (see Section 10.4) are also only found in South Africa, a case may be made for suggesting that the magmas that formed these rocks were also parental to the MARID-suite. Such an origin accounts for the temporal and spatial relationships of MARID-rocks (see Section 10.4). In this case, MARID-rocks are not the high-pressure equivalents of either lamproites or kimberlites. Their composition is undoubtedly far removed from that of their parents, as a consequence of high-pressure differentiation events similar to those proposed by Waters (1987). In summary, we consider that the problem of the origin of the MARID-suite will be resolved only by high-pressure experimental studies of lamproites and the parent magmas or orangeites.

10.6. RELATIONSHIPS TO OTHER POTASSIC ROCKS AND LAMPROPHYRES

The relationship of lamproites to other potassic magmas has been clarified during the past decade by the realization that kamafugites and Roman province type-lavas (RPT-lavas) are distinct in their mineralogy and geochemistry relative to each other, and to lamproites (see Chapters 2, 6, 7). Experimental studies (Edgar 1987) suggest that kamafugites may be derived by the partial melting of carbonated phlogopite clinopyroxenite or lherzolitic mantle, in an environment of high CO_2/H_2O ratios. The type examples from Uganda occur in an extensional tectonic environment associated with rifting. The extreme differences in the sources and compositions of kamafugites and lamproites precludes any genetic relationship. The only common aspect to their petrogenesis is that both clans are derived from metasomatized mantle enriched in phlogopite.

The origins of RPT-lavas remain obscure despite decades of study. Their source has been proposed to be located in the mantle wedge overlying (Holm et al. 1982, Civetta et al. 1981), or the asthenosphere, underlying (Foden and Varne 1980, Wheller et al. 1987) the Benioff zone. The only agreement between the various protagonists appears to be that all models require the presence of a subduction-related component. Much of the controversy perhaps stems from the significant petrological and tectonic difference existing between the Sunda arc and the Roman province. Potassic rocks of the Roman province in particular remain the subject of controversy (Peccerillo and Manetti 1985, Vollmer 1989). Current models suggest that mixing of three or four components is required to explain the spectrum of compositions (Varekamp and Kalamarides 1989, Ellam et al. 1989).

The mineralogical and compositional distinctions between RPT-lavas and lamproites precludes any genetic relationship. Interestingly, the anomalous geochemical signature, which suggests the presence of crustal or subduction-related material in the Murcia-Almeria lamproites, echoes that of the Roman province suite. Vollmer (1989) has suggested that the paleo-subduction zone underlying the Roman province leaves an isotopic and chemical imprint on later magmatism, although the igneous activity is not controlled by subduction. The K-rich component of the potassic rocks is considered to be derived from the fusion of metasomatic veins in the asthenospheric mantle underlying the subduction zone, with contamination of these melts occurring during ascent. This model is similar to our model explaining the contamination of the Murcia-Almeria lamproites (Section 10.3.5), and in agreement with hypotheses proposed for the potassic rocks of the Sunda arc (Wheller et al. 1987). It seems probable that all of the potassic magmatism in the northern Mediterranean region (Keller 1983) reflects this style of contamination, although the nature of the mantle-derived melts parental to the Roman and Afyon province lavas remains enigmatic.

Shoshonites occur in association with, and in the same tectonic setting as, RPT-lavas. They are mineralogically and chemically distinct from lamproites and hence genetically unrelated. Their origins must be sought in the context of island arc magmatism.

A central tenet of this work is that lamproites are not varieties of lamprophyres, as they form a distinct petrological clan of rocks (Section 2.3.5). We thus regard the statement of Rock (1989) that lamproites grade globally on the one hand into minettes and on the other into kimberlites, via olivine-lamproites, as totally misleading and petrologically unfounded. Rock's (1989) assertion is based upon simple petrographic arguments that totally ignore the assemblage and composition of phases present, the geochemistry of the rocks, their source and tectonic setting. For example, we fail to see how an olivine lamproite derived from a lithospheric source can possibly grade into a group 1 macrocrystal kimberlite derived from an asthenospheric source. It follows that we do not consider that diamonds are found in lamprophyres (*sensu* Rock 1989). Propagation of such petrogenetically incorrect assertions may lead the unwary exploration geologist astray. N.B. We agree that diamonds do occur in lamprophyres such as the Wandagee picritic monchiquites (Jaques et al. 1989d); however, these rocks belong to the basaltoid clan.

In a petrographic sense, minettes are the only lamprophyres that bear a superficial resemblance to lamproites. Notwithstanding the edicts of the IUGS subcommission concerning the nomenclature of lamprophyres (Streckeisen 1978), we believe that rocks consisting of mica (*sensu lato*) and K-feldspar cannot be unambiguously classified on a simple petrographic basis. Thus, lamproites consisting of this assemblage may be indiscriminately termed minette. However, detailed studies of the compositional trends of micas and other phases, together with geochemical studies, will demonstrate the true affinities of such rocks.

In conclusion, we consider that attempting to relate lamproites (and kimberlites) to lamprophyres is reactionary and will only compound the confusion surrounding this group of ill-defined rocks. In contrast to Rock (1989, p. 53), we believe we do not have to add to the contents of the lamprophyric "petrological garbage can," and further, that we now are able to dispose of "the can" entirely. Most rocks described as lamprophyres are now known to belong to groups of rocks that are distinct in terms of their genesis—i.e., kimberlites, lamproites—or can be related to other magma types—i.e., sannaites and camptonites to the basalt clan, alnöites to the melilitite clan, etc., thus eliminating the need for the term "lamprophyre," except as a field term for rocks of uncertain petrological affinity.

Finally, we wish to draw attention to the similarities between hypotheses advanced to explain the origin of boninites and current models of lamproite genesis. Boninites are high-Mg, low-Ca, relatively silica-rich (>53 wt % SiO_2) rocks found in the western Pacific island arcs (Meijer 1980). Green *et al.* (1987) have suggested that boninites are primitive melts derived by the volatile-induced (H_2O-rich) partial melting of a very refractory residual harzburgite source. The wide variation in the trace element content of individual suites has been interpreted to indicate addition of a minor enriched component to the source prior to melting. Moreover, the low-Nd isotopic signature of this material suggests derivation from an old light REE-enriched mantle (Hickey and Frey 1982).

Boninites are enriched in Al and Na relative to lamproites and we do not wish to suggest any petrogenetic relationship between these rocks. The importance of the comparison lies in the style of petrogenesis, as it appears that identical processes have occurred in both oceanic and continental lithosphere. Consequently, boninites might be regarded as the oceanic sodic equivalent of the lamproite suite.

10.7. SUMMARY

The lamproite clan is believed to result from the partial melting, under reduced (carbon–water buffer) H_2O-rich volatile conditions, of ancient enriched harzburgitic sources. The primitive magma from which all rocks of the lamproite clan are derived by fractional crystallization or contamination is relatively silica- (52–55 wt %) and Mg- (7–10 wt % MgO) rich. Olivine lamproites are hybrid rocks whose compositions do not represent those of their parental magmas. Lamproites are not related to kimberlites, other potassic lavas, or lamprophyres. Lamproites are melts derived from subcontinental mantle lithospheric sources. The presence of paleo-Benioff zones and multiple ancient "mantle metasomatic" events are prerequisites for the long-term development of this source. These conditions restrict the occurrence of lamproites to mobile belts surrounding continental cratons. Kimberlites, in contrast, appear to be relatively oxidized (wustite–magnetite buffer) CO_2-dominated asthenospheric melts. Their deeper source allows kimberlites to be emplaced in mobile belts or cratons. Geographical overlap of kimberlite and lamproite provinces is thus possible.

Postscript

Since 1980 significant advances have been made in our understanding of the mineralogy, geochemistry, and volcanology of the lamproite clan. However, many aspects of their petrology and evolution are poorly understood, primarily because of a lack of systematic field investigation, petrological examination, and laboratory-based experimental studies. Further information regarding the origin and evolution of the lamproite clan may be gained by pursuing the following areas of research:

1. *A reexamination of the volcanology and petrology of Gaussberg, the youngest example of lamproite volcanism.* Although the eruption is subglacial and atypical of lamproite volcanism, such a study would provide important information about the physical properties of lamproite magmas.

2. *Study of the physical volcanology and facies analysis of the Leucite Hills province.* This volcanic field, consisting of many slightly eroded cinder cones and lavas, would provide the best information for understanding subaerial lamproite volcanism.

3. *A detailed mineralogical and geochemical investigation of the relationship of olivine lamproites to phlogopite lamproites in the West Kimberley province and the Prairie Creek field.* This would establish the nature of the contamination of the former group by xenolithic components and determine whether or not they are low- or high-pressure differentiates of phlogopite lamproites.

4. *Further studies of radiogenic isotopic compositions (Sr, Nd, Hf, Pb, Os), concentrating upon mineral separates rather than whole rock samples.* This would provide data relevant to understanding the nature and evolution of the source lithospheric mantle and the extent of contamination of lamproite magmas during ascent and differentiation.

5. *Stable isotope studies (H, O, N, C) of rocks, glasses, and minerals.* This would allow estimation of the role of meteoric water in the formation of lamproite vents and recognition of recycled crustal components in the mantle source regions of lamproites melts.

6. *Experimental studies of the crystallization of realistic primitive lamproitic melt compositions (i.e., high silica phlogopite lamproites) over a wide range of pressures and temperatures at diverse oxygen fugacites and fluorine/water ratios.* The conclusions stated in Chapters 5, 6, 7, 8, and 10 indicate that olivine lamproites are unlikely to represent primitive lamproite magmas which are parental to other members of the clan. Unfortunately, much of the previous high-pressure experimental work (Arima and Edgar 1983a, Foley 1989) has been undertaken on such unrealistic melt compositions.

7. *Experimental studies of the partial melting of potential mantle source rocks at high pressures and varying volatile compositions.* The conclusions stated in Chapter 8 suggest that phlogopite-bearing harzburgite and similar assemblages containing K-Ba-titanates should be used as starting materials.

8. *Detailed studies of the physical properties of lamproite melts (density, viscosity) and rocks (magnetic susceptibility, electrical resistivity, thermal conductivity, density).* The former would improve our understanding of the volcanology of lamproite vents and lava flows and the physics of magma ascent. The latter would facilitate the exploration of lamproite fields.

9. *Experimental studies of the effects of pressure, temperature, and composition on the solubility of volatiles (H_2O, CO_2, CH_4, HF, CO, etc.) in lamproite melts.* This work is required for a better understanding of lamproite vent formation and magma ascent.

10. *Detailed comprehensive studies of the better known lamproite provinces (Leucite Hills, Prairie Creek, West Kimberley, Murcia-Almeria).* Integration of remote sensing, geophysical, geological, and petrological data would lead to a better understanding of the emplacement and evolution of lamproite fields.

11. *Systematic investigation of poorly characterized provinces*, e.g., Sisimiut, Francis, Yellow Water Butte, Kapamba, Bobi, Coromandel, Majhgawan, Chelima, Barakar, and Murun.

12. *Detailed investigations of provinces in which lamprophyres (minettes) or kimberlites appear to be associated with possible lamproites*, e.g., Montana alkaline province, Swartruggens, Barakar, Coromandel, to assess the petrological relationships (if any) between these diverse magmas.

Research on the topics listed above will ensure that the next decade of lamproite studies will be as fruitful as the previous in unraveling the mysteries of lamproite genesis.

References

Åberg, G. 1988. Middle Proterozoic anorogenic magmatism in Sweden and worldwide. *Lithos* **21**, 279–289.
Abraham, K., Gebert, W., Medenbach, O., Schreyer, W., Hentschel, G. 1983. Eifelite, $KNa_3Mg_4Si_{12}O_{20}$, a new mineral of the osmulite group with octahedral sodium. *Contrib. Mineral. Petrol.* **82**, 252–258.
Ackermann, E. 1962. Das Sockelstockwerk der Orogene in Ostafrika. *Geol. Rundsch.* **52**, 675–720.
Ahmad, F. 1956. On the source of the Panna diamonds and the nature of the Majhgawan plug. Proc. 43rd Indian Sci. Congr. Pt. 3, pp. 192–193.
Ajarzo, G. B. M. 1985. The Argyle diamond mine. *Syrian J. Geol.* **11–12**, 17–22.
Alexander, P. O. 1983. Looking for diamonds? Try geobotany. *De Beers Ind. Diamonds Q. London* **3**, 33–38.
Alexander, P. O. 1986. Preliminary study of soil bacterial populations over and adjacent to three kimberlite diatremes. 4th Int. Kimberlite Conf. Perth, Western Australia, Extended Abstr. 440–442.
Alexander, P. O., Shrivastava, V. K. 1984. Geobotanical expression of a blind kimberlite pipe, Central India. In Kornprobst (1984), q.v., vol. 1, pp. 33–40.
Alibert, C., Albarede, F. 1988. Relationships between mineralogical, chemical, and isotopic properties of some North American kimberlites. *J. Geophys. Res.* **93**, 7643–7671.
Alibert, C., Michard, A., Albarede, F. 1986. Isotope and trace element geochemistry of Colorado Plateau volcanics. *Geochim. Cosmochim. Acta* **50**, 2735–2750.
Allan, J. F., Carmichael, I. S. E. 1984. Lamprophyric lavas in the Colima graben, southwest Mexico. *Contrib. Mineral. Petrol.* **88**, 203–216.
Allsopp, H. L., Barrett, D. R. 1975. Rb-Sr determinations of South African kimberlite pipes. *Phys. Chem. Earth* **9**, 605–617.
Allsopp, H. L., Kramers, J. D. 1977. Rb-Sr and U-Pb age determinations of southern Africa kimberlites. 2nd. Internat. Kimberlite Conf. Santa Fe, New Mexico, Extended Abstracts (no pagination).
Almeida, F. F., Derze, G. R., Vinha, C. A. G. 1971. Mapa geologico de Brasil (1:5,000,000), Seĉao de Cartographia Geologica, Brazil.
Ancochea, E. 1982. Evolution espacial y temporal del vulcanismo reiente de España Central. Doc. Thesis, Fac. Sci. Univ. Madrid.
Ancochea, E., Nixon, P. H. 1987. Xenoliths in the Iberian Peninsula. In Nixon (1987) q.v., pp. 119–124.
Anderson, D. L. 1987. The depths of mantle reservoirs. In Mysen (1987) q.v., pp. 3–12.
Andrews, J. R., Emeleus, C., 1971. Preliminary account of kimberlite intrusions from the Frederikshåb district, southwest Greenland. Geol. Surv. Greenland, Report No. 31, 26 pp.
Aoki, K., Ishikawa, K., Kanisawa, S. 1981. Fluorine geochemistry of basaltic rocks from continental and oceanic regions and petrogenetic implications. *Contrib. Mineral. Petrol.* **76**, 53–59.
Apparadhanulu, K., 1966. Minette and riebeckite-kalisyenite dykes in some Upper Cuddapah rocks, Kurnool district, Andhra Pradesh. *Rec. Geol. Surv. India* **94**(2), 303–304.
Appleton, J. D. 1970. The petrology of the potassium-rich lavas of the Roccamonfina volcano, Italy. Ph.D. thesis, Univ. Edinburgh, U.K.
Appleton, J. D. 1972. Petrogenesis of potassium-rich lavas from the Roccamonfina volcano, Roman Region, Italy. *J. Petrol.* **13**, 425–456.

Annerstein, H., Devanarayanan, S., Haggstrom, S., Wappling, R. 1971. Mössbauer study of synthetic ferriphlogopite $KMg_3Si_3O_{10}(OH)_2$. *Phys. Status Solidi B*, K137.

Arculus, R. J., Dawson, J. B., Mitchell, R. H., Gust, D. A., Holmes, R. D. 1984. Oxidation state of the upper mantle recorded by megacrystal ilmenite in kimberlite and type A and B spinel lherzolites. *Contrib. Mineral. Petrol.* **85**, 85–94.

Arculus, R. J., Johnson, R. W. 1981. Island arc magma sources. *Geochem. J.* **15**, 109–133.

Arima, M., Edgar, A. D. 1980. Stability of wadeite ($Zr_4K_2Si_6O_{18}$) under upper mantle conditions. *Contrib. Mineral. Petrol.* **72**, 191–195.

Arima, M., Edgar, A. D. 1981. Substitution mechanisms and solubility of titanium in phlogopites from rocks of probable mantle origin. *Contrib. Mineral. Petrol.* **77**, 288–295.

Arima, M., Edgar, A. D. 1983a. A high pressure experimental study on a magnesian-rich leucite-lamproite from the West Kimberley area, Australia: Petrogenetic implications. *Contrib. Mineral. Petrol.* **84**, 228–234.

Arima, M., Edgar, A. D. 1983b. High pressure experimental studies of a katungite and their bearing on the genesis of some potassium-rich magmas of the West Branch of the African Rift. *J. Petrol.* **24**, 166–187.

Arkhangelskaya, V. V. 1974. *The Rare Metal Alkaline Rock Complexes on the Southern Margin of the Siberian Platform*. Nedra, Moscow (Russian).

Atkinson, W. J. 1989. Diamond exploration philosophy, practice and promises: A review. In Ross *et al.* (1989) q.v., Vol. 2, pp. 1075–1107.

Atkinson, W. J., Hughes, F. E., Smith, C. B. 1982. A review of the kimberlitic rocks of Western Australia. *Terra Cognita* **2**, 204 (abstract).

Atkinson, W. J., Hughes, F. E., Smith, C. B. 1983. A review of the kimberlitic rocks of Western Australia. Proc. 6th Aust. Geol. Conf. pp. 284–285 (abstract).

Atkinson, W. J., Hughes, F. E., Smith, C. B. 1984a. A review of the kimberlitic rocks of Western Australia. In Kornprobst (1984) q.v., Vol. 1, pp. 195–224.

Atkinson, W. J., Smith, C. B., Boxer, G. L., 1984b. The discovery and evaluation of the Ellendale and Argyle lamproite diamond deposits, Kimberley region, Western Australia. Soc. Min. Eng. AIME Preprint No. 84-384, 13 pp.

Aubouin, J. 1980. The main structural complexes of Europe. In *The Geology of Europe*. 26th. Int. Geol. Congr. Bordas, pp. xiii–xxii.

Bachinski, S. W., Simpson, E. L. 1984. Ti-phlogopites of the Shaw's Cove minette: A comparison with micas of other lamprophyres, potassic rocks, kimberlites and mantle xenoliths. *Am. Mineral.* **69**, 41–56.

Bagshaw, A. N., Doran, B. H., White, A. H., Willis, A. C. 1977. Crystal structure of a natural potassium barium hexatitanate isostructural with $K_2Ti_6O_{13}$. *Aust. J. Chem.* **30**, 1195–2000.

Bailey, D. K. 1974. Continental rifting and alkaline magmatism. In Sörensen (1974) q.v., pp. 148–159.

Bailey, D. K. 1977. Lithosphere control of continental rift magmatism. *J. Geol. Soc. London* **133**, 103–106.

Bailey, D. K. 1983. The chemical and thermal evolution of rifts. *Tectonophys.* **94**, 585–597.

Baker, P. E., Gass, I. G., Harris, P. G., LeMaitre, R. W. 1964. The volcanological report of the Royal Society expedition to Tristan da Cunha 1962. *Phil. Trans. R. Soc. London* **256**, 439–578.

Balasubrahmanyan, M. N., Murty, M. K., Paul, D. K., Sarkar, A. 1978. Potassium-argon ages of Indian kimberlites. *J. Geol. Soc. India* **19**, 584–585.

Baldridge, W. S., Carmichael, I. S. E., Albee, A. L. 1981. Crystallization paths of leucite-bearing lavas: Examples from Italy. *Contrib. Mineral. Petrol.* **76**, 321–335.

Banerjee, S. 1953. Petrology of the lamprophyres and associated rocks of Raniganj coal field. *Indian Mineral. J.* **1**, 9–29.

Banerjee, S. P. K., Agarawal, S. S. K. 1980. History of mining with special reference to Panna, M. P. (India). Seminar on Diamonds, Panna, Geol. Surv. India.

Barberi, F., Innocenti, F. 1967. Le rocce selagitiche di arciatico e Montecatini in Val di Cecina. *Atti Soc. Tosc. Sci. Nat. Mem. Ser. A* **74**, 139–180.

Barberi, F., Peccerillo, A., Poli, G., Tolomeo, L. 1988. Major, trace element and Sr isotopic composition of lavas from Vico volcano (Central Italy) and their evolution in an open system. *Contrib. Mineral. Petrol.* **99**, 485–497.

Barberi, F., Santacroce, R., Varet, J. 1982. Chemical aspects of rift magmatism. In Palmason, G. (ed.) *Continental and Oceanic rifts*. Am. Geophys. Union, Washington, D.C., pp. 223–258.

Barbosa, O., Braun, O. P. G., Dayer, R. C., Cunha, C. A. B. R. 1970. Geology of the Triangulo Mineiro region. Dept. Nacion da Produčao Mineral. Bull. 136 (Portuguese).

Bardet, M. G. 1974. *Geologie du Diamant*. II. Editions BRGM No. 83. Paris.

Bardet, M. G. 1977. *Geologie du Diamant*. III. Editions BRGM No. 83. Paris.

REFERENCES

Bardet, M. G., Vachette, M. 1966. Determination of the ages of kimberlites of West Africa. 3rd Symposium African Geol. Commonwealth. Geol. Lias. Office Report No. G.G.L.O. 88, pp. 15–17.
Barker, D. S. 1983. *Igneous Rocks*. Prentice Hall, Englewood Cliffs, New Jersey.
Barker, D. S., Mitchell, R. H., McKay, D. 1987. Late Cretaceous nephelinite to phonolite magmatism in the Balcones province, Texas. In Morris and Pasteris (1987) q.v., pp. 293–304.
Barnes, S. J., Naldrett, A. J., Gorton, M. P. 1985. The origin of the fractionation of platinum-group elements in terrestrial magmas. *Chem Geol.* **53**, 303–323.
Barrer, R. M., Baynham, J. W. 1956. The hydrothermal chemistry of the silicates Part VII. Synthetic potassium aluminosilicates. *J. Chem. Soc. London*, **1956**, 2882–2891.
Barton, E. S. 1983. Reconnaissance isotopic investigations in the Namaqua mobile belt and implications for Proterozoic crustal evolution—Namaqualand geotraverse. *Geol. Soc. South Africa Spec. Publ.* **10**, 45–66.
Barton, M. 1975. The origin of potassium-rich lavas. Ph.D. thesis, Univ. Manchester, U.K.
Barton, M. 1976. Melting relations of some ultrapotassic volcanic rocks. *Progr. Exp. Petrol. N.E.R.C. Pub. Ser. D.* **6**, 91–94.
Barton, M. 1979. A comparative study of some minerals occurring in the potassium-rich alkaline rocks of the Leucite Hills, Wyoming, the Vico Volcano, western Italy, and the Toro-Ankole region, Uganda. *N.J. Miner. Abh.* **137**, 113–134.
Barton, M. 1988. The occurrence and significance of xenocrysts of apatite, ilmenite and Na-Fe-Ti oxide in ultrapotassic lavas from the Leucite Hills, Wyoming. *Mineral. Mag.* **51**, 265–270.
Barton, M., Hamilton, D. L. 1978. Water-saturated melting relations to 5 kilobars of three Leucite Hills lavas. *Contrib. Mineral. Petrol.* **66**, 41–49.
Barton, M., Hamilton, D. L. 1979. The melting relationships of a madupite from the Leucite Hills, Wyoming, to 30 kb. *Contrib. Mineral. Petrol.* **69**, 133–142.
Barton, M., Hamilton, D. L. 1982. Water-saturated melting experiments bearing upon the origin of potassium-rich magmas. *Mineral. Mag.* **45**, 267–278.
Barton, M., van Bergen, M. J. 1981. Green clinopyroxenes and associated phases in a potassium-rich lava from the Leucite Hills, Wyoming. *Contrib. Mineral. Petrol.* **77**, 101–114.
Bazarova, T. Y., Krasnov, A. A. 1975. Temperatures and sequence of crystallization of some leucite-bearing basaltoids. *Dokl. Akad. Nauk SSSR* **222**, 177–180.
Bell, K., Doyle, R. J. 1971. K–Rb relationships in some continental alkalic rocks associated with the East African Rift valley system. *Geochim. Cosmochim. Acta* **35**, 903–915.
Bellon, H. 1976 Séries magmatiques néogènes et Quaternaires du Pourtour de la Méditerranée occidentale, comparees dans leur cadre geochronometrique-implications geodynamiques. Ph.D. thesis, Univ. Paris-Sud, Orsai, Series A No, 1750.
Bellon, H. 1981. Chronologique radiometrique (K–Ar) des manifestations magmatiques autour de la Méditerranée occidentale entre 33 M.A. et 1 M.A. Proc. Int. Conf. Urbino Univ. Italy, pp. 341–360.
Bellon, H., Brousse, R. 1977. Le magmatisme periméditerranéen occidental. Essai de synthèses. *Bull. Soc. Geol. France* **19**, 469–480.
Bellon, H., Bizon, G., Pedro-Calvo, J., Elizager, E., Gaudant, J., Lopez-Martinez, N. 1981. Le volcan du Cerro del Monagrillo (Province de Murcie): Age absolu et correlations avec les sediments néogènes du basin de Hellin (Espagne). *C. R. Acad. Sci. Paris D* **292**, 1035–1038.
Bellon, H., Bordet, P., Moutenat, G. 1983. Chronique du magmatisme néogène des cordilleres bétiques (Espagne meridionale). *Bull. Soc. Geol. France* **25**, 205–217.
Bellon, H., Letousey, J. 1977. Volcanism related to plate tectonics in the Western and Eastern Mediterranean.: In Bijou-Duval B. and Montadert, L. (eds), *International Symposium on Mediterranean Basins*. Editions Technique, Paris, pp. 165–184.
Berbert, C. O., Svisero, D. P., Sial, A. N., Meyer, H. O. A. 1981. Upper mantle material in the Brazilian shield. *Earth Sci. Rev.* **17**, 109–133.
Berendsen, P., Blair, K. P. 1987. Subsurface structural maps over the central North American rift system (CNARS), central Kansas with discussion. Kansas Geol. Surv. Subsurf. Geol Ser. 8.
Berendsen, P., Cullers, R. L., Mansker, W. L., Cole, G. P. 1985. Late Cretaceous kimberlite and lamproite occurrences in E. Kansas, U.S.A. *Geol. Soc. Am. Ann. Mtg. Abstr. with Progr.* **17**, 151.
Bergman, S. C. 1987. Lamproites and other potassium-rich igneous rocks: A review of their occurrence, mineralogy and geochemistry. In Fitton and Upton (1987) q.v., pp. 103–190.
Bergman, S. C., Baker, N. R. 1984. A new look at the Proterozoic dikes from Chelima, Andhra Pradesh India: Diamondiferous lamproites? *Geol. Soc. Am. Ann. Mtg. Abstr. with Progr.* **16**, 444.

Bergman, S. C., Dubessy, J. 1984. CO_2–CO fluid inclusions in a composite peridotite xenolith: Implications for upper mantle oxygen fugacity. *Contrib. Mineral. Petrol.* **85**, 1–13.

Bergman, S. C., Dunn, D. P., Krol, L. G. 1988. Petrology and geochemistry of the Linhaisai Minette, central Kalimantan and the origin of Borneo diamonds. *Can. Mineral.* **26**, 23–43.

Best, M. G., Henage, L. F., Adams, J. A. S. 1968. Mica peridotite, wyomingite, and associated potassic igneous rocks in northeastern Utah. *Am. Mineral.* **53**, 1041–1048.

Beswick, A. E. 1976. K and Rb relations in basalts and other mantle-derived materials: Is phlogopite the key? *Geochim. Cosmochim. Acta* **32**, 1167–1183.

Bhaskara Rao, B. 1976. A note on the micaceous kimberlite dyke in the Cumbum formations near Zangamrajupalle, Cuddapah district, Andra Pradesh. *Ind. Minerals* **30**, 55–58.

Bhattacharji, S., 1987. Asthenospheric upwelling, lineament reactivations, magmatic episodes and ore mineral localization in Proterozoic basin evolution on the Archean Indian Shield. *Internat. Basement Tectonics Assoc. Pub.* **5**, 187–200.

Bickford, M. E., Harrower, K. L., Nusbaum, R. L., Thomas, J. J., Nelson, G. E. 1979. Preliminary geologic map of the Precambrian basement rocks of Kansas. *Kansas Geol. Surv. Map* M-9. Scale 1:500,000.

Bickford, M. E., Mose, D. E., Wetherill, G. W., Franks, P. C. 1971. Metamorphism of Precambrian granitic xenoliths in a mica peridotite at Rose Dome, Woodson County, Kansas. Part 1, Rb-Sr Isotopic Studies. *Geol. Soc. Am. Bull.* **82**, 2863–2868.

Bickford, M. E., van Schmus, W. R., Zietz, I. 1986. Proterozoic history of the midcontinent region of North America. *Geology* **14**, 492–496.

Bijou-Duval, B., Dercourt, J., Le Pichon, X. 1977. From the Tethys ocean to the Mediterranean seas: A plate tectonic model of the evolution of the western Alpine system. In, Bijou-Duval, B., and Montadert, L. (eds.), *International Symposium on the Structural History of the Mediterranean Basins*. Editions Technique, Paris, pp. 199–214.

Black, L. P., Gulson, B. L. 1978. The age of the Mud Tank carbonatite, Strangways range, Northern Territory. *Aust. Bur. Miner. Res. J. Geol. Geophys.* **3**, 227–232.

Black, L. P., James, P. R. 1983. Geological history of the Napier complex of Enderby Land. In, Oliver, R. L., James, P. R., and Jago, J. B. (eds.). *Antarctic Earth Science*. Aust. Acad. Sci., Canberra, pp. 11–15.

Blackstone, D. L. 1972. Tectonic analysis of southwestern Wyoming from ERTS-1 imagery. Univ. Wyoming Remote Sensing Lab. Spec. Rep. NAS-5-21799.

Blatchford, T. 1927. The geology of portions of the Kimberley Division, with special reference to the Fitzroy Basin and the possibilities of the occurrence of mineral oil. *West. Aust. Geol. Surv. Bull.* **93**.

Blaxland, A. B., van Breemen, O., Emeleus, C. H., Anderson, J. G. 1978. Age and origin of the major syenite centers in the Gardar province of South Greenland: Rb–Sr studies. *Geol. Soc. Amer. Bull.* **89**, 231–244.

Bogatikov, V. A., Makhotkin, I. L., Kononova, V. A. 1985. Lamproites and their place in the systematics of high magnesium potassic rocks. *Izv. Akad. Nauk SSSR* **1985**(12), 3–10 (Russian).

Bogatikov, V. A., Eremeyev, N. V., Makhotkin, I. L., Kononova, V. A., Novgorodova, M. I., Laputina, I. L. 1986. Lamproites from the Aldan and Central Asia. *Dokl. Akad. Nauk. SSSR* **290**, 936–940 (Russian).

Bogatikov, V. A., Kononova, V. A., Makhotkin, I. L., Eremeyev, N. V., Savosin, S. I., Kerzin, A. L., Malov, Y., Tsepin, A. I. 1987. Rare earth and rare elements as indicators of the origin of lamproites of the central Aldan (USSR). *Vulkanol. Seismol.* **1**, 15–29 (Russian).

Bolivar, S. L. 1977. Geochemistry of the Prairie Creek, Arkansas and Elliott County, Kentucky intrusions. Ph.D. thesis, Univ. New Mexico, Albuquerque.

Bolivar, S. L. 1982. The Prairie Creek kimberlite, Arkansas. In, McFarland, J. D. (ed.), *Contributions to the Geology of Arkansas*. State of Arkansas Geol. Comm. Misc. Pub. 18, pp. 1–21.

Bolivar, S. L. 1984. An overview of the Prairie Creek intrusion, Arkansas. Soc. Min. Eng. AIME Fall Meeting, October 1984, Preprint No. 84-346, 12 pp.

Bolivar, S. L., Brookins, D. G. 1979. Geophysical and Rb–Sr study of the Prairie Creek, AR kimberlite. In, Boyd and Meyer (1979) q.v., Vol. 1, pp. 289–299.

Bolivar, S. L., Brookins, D. G. Lewis, R. D., Meyer, H. O. A. 1976. Geophysical studies of the Prairie Creek kimberlite, Murfreesboro, Arkansas, *EOS* **57**, 762 (abstract).

Bordet, P. 1985. Le volcanism miocene des Sierras de Gata et de Carboneras (Espagne du Sud-est). *Doc. Trav. Inst. Geol. Albert de Lapparent,* **8**, 1–70.

Bordet, P., De Larouzière, F. D. 1983. Particularities geochemiques des volcanites miocenes des Sierras de Gata et de Carboneras (Almeria, S. E. Espagne). *C.R. Acad. Sci. Paris* **296**, 449–452.

Borley, G. D. 1967. Potassium-rich volcanic rocks from southern Spain. *Miner. Mag.* **36**, 364–379.

REFERENCES

Borsi, S., Ferrara, G., Tongiorgi, E. 1967. Determinazioni con il metoda K/Ar delle rocce magmatiche della Toscana. *Boll. Soc. Geol. Italy* **86**, 403–410.
Boutwell, J. M. 1912. Geology and ore deposits of the Park City District, Utah. *U. S. Geol. Surv. Prof. Pap.* **77**, 231.
Bowen, C. F. 1915. Possibilities of oil in the Porcupine Dome, Rosebud County, Montana. *U. S. Geol. Surv. Bull.* **621**, 61–70.
Bowles, J. F. W. 1988. Definition and range of naturally occurring minerals with the pseudobrookite structure. *Am. Mineral.* **73**, 1377–1383.
Boxer, G. L., Lorenz, V., Smith, C. B. 1989. The geology and volcanology of the Argyle (AK1) lamproite diatreme, Western Australia. In Ross *et al.* (1989) q.v., Vol. 1, pp. 140–152.
Boyd, F. R., Meyer, H. O. A. (Eds.) 1979. *Proceedings of the Second International Kimberlite Conference.* Vol. 1. *Kimberlites, Diatremes and Diamond: Their Geology, Petrology and Geochemistry.* Vol. 2. *The Mantle Sample: Inclusions in Kimberlites and Other Volcanics.* Am. Geophys. Union, Washington D.C.
Bradley, W. H. 1964. Geology of Green River Formation and associated Eocene rocks in southwest Wyoming and adjacent parts of Colorado and Utah. *U. S. Geol. Surv. Prof. Pap.* **496-A**, A1-A86.
Braile, L. W., Hinze, W. J., Keller, G. R., Lidiak, E. G., Sexton, J. L. 1986. Tectonic development of the New Madrid Rift complex, Mississippi embayment, North America. *Tectonophysics* **131**, 1–21.
Branner, J. C., Brackett, R. N. 1889. The peridotite of Pike County, Arkansas. *Am. J. Sci.* **38**, 50–59.
Bravo, M. S., O'Hara, M. J. 1975. Partial melting of phlogopite-bearing synthetic spinel and garnet lherzolites. *Phys. Chem. Earth* **9**, 845–854.
Brey, G. 1978. Origin of olivine melilitites- chemical and experimental constraints. *J. Volcanol. Geotherm. Res.* **3**, 61–88.
Briqueu, L., Bougault, H., Joron, J. L. 1984. Quantification of Nb, Ta, Ti, and V anomalies in magmas associated with subduction zones: Petrogenetic implications. *Earth Planet. Sci. Lett.* **68**, 297–308.
Bristow, J. W., Smith, C. B., Allsopp, H. L., Shee, S. R., Skinner, E. M. W. 1986. Setting geochronology and geochemical characteristic of 1600 m.y. kimberlites and related rocks from the Kuruman Province, South Africa. 4th. Internat. Kimberlite Conf. Perth, Western Australia, Extended Abstracts, pp. 112–114.
Brookins, D. G. 1970. The kimberlites of Riley County, Kansas. *Kansas. Geol. Surv. Bull.* **200**, 1–32.
Brookins, D. G., Della Valle, R. S., Bolivar, S. L. 1979. Significance of uranium abundance in United States kimberlites. In Boyd and Meyer (1979) q.v., Vol. 1, pp. 280–288.
Brookins, D. G., Treves, S. B., Bolivar, S. L. 1975. Elk Creek, Nebraska, carbonatite. Strontium geochemistry. *Earth Planet. Sci. Lett.* **28**, 79–82.
Brooks, C. K., Printzlau, I. 1978. Magma mixing in mafic alkaline volcanic rocks: The evidence from relict phenocrystal phases and other inclusions. *J. Volcanol. Geotherm. Res.* **4**, 315–331.
Brooks, C. K., Noe-Nygaard, A., Rex, D. C., Roensbo, J. C. 1978. An occurrence of ultrapotassic dikes in the neighborhood of Holsteinsborg, West Greenland. *Bull. Geol. Soc. Denmark* **27**, 1–8.
Brotzu, P., Morbidelli, L., Traveisa, G. 1976. Petrological significance of the compositional variation characterizing phenocrysts of clinopyroxene of the alkalic-potassic lava of the Monterado area, eastern Vulsini. *Mem. Inst. Geol. Miner. Univ. Padova* **31**, 1–18.
Brouwer, H. A. 1909. Glimmerleucitbazalt van Oost-Borneo. *Versl. Kon. Akad. Wetensch. Amsterdam Wis. Nat. Afd.* **18**, 85–91.
Bryant, B., Nichols, D. J. 1988. Late Mesozoic and early Tertiary reactivation of an ancient crustal boundary along the Uinta trend and its interaction with the Sevier Orogenic Belt. In Perry, W. J., and Schmidt, C. J. (eds.), *Interaction of the Rocky Mountains Foreland and Cordilleran Thrust Belt.* Geol. Soc. Am. Mem. 171, pp. 411–430.
Bunker, B. J., Witzke, B. J., Watney, W. L., Ludvigson, G. A. 1988. Phanerozoic history of the central Midcontinent, United States. In Sloss, L. L. (ed.), *Sedimentary Cover, North American Craton*, Geol. Soc. Amer. Decade North Amer. Geol. D-2, pp. 243–260.
Burchett, R. R., Luza, K. V., van Eck, O. J., Wilson, F. W., 1983. Seismicity and tectonic relationships of the Nemaha uplift and the midcontinent geophysical anomaly. *U. S. Nuclear Reg. Comm. Report No. NUREG/CR-3117.*
Burke, K. C. 1985. Rift basins: Origin, history and distribution. Offshore Tech. Conf. paper 4844, Vol. 1, pp. 33–40.
Burke, K. C. 1977. Aulacogens and continental breakup. *Ann. Rev. Earth Planet Sci.* **5**, 371–396.
Burke, K. C., Wilson, J. T. 1972. Is the African plate stationary? *Nature (London)* **239**, 387–390.
Burke, K. C., Dewey, J. F. 1973. Plume generated triple junctions: Key indicators in applying plate tectonics to old rocks. *J. Geol.* **81**, 406–433.
Burke, K. C., Wilson, J. T. 1976. Hot spots on the earth's surface. *Sci. Am.* **235**, 46–57.

Burke, K. C., Delano, L., Dewey, J. F., Edelstein, A., Kidd, W. S. F., Nelson, K. D., Sengor, A. M. C., Stroup, J. 1978. Rifts and sutures of the world. Report No. NA55-24094 Goddard Space Flight Center, pp. 1–238.

Burri, C., Parga-Pondal, I. 1937. Die eruptivgesteine der Insel Alboran (Provinz Almeria, Spanien). *Schweiz. Mineral. Petrogr. Mitteil.* **17**, 230–268.

Cahen, L., Snelling, N. J., Delhal, J., Vail, J. R. 1984. *The Geochronology and Evolution of Africa.* Clarendon Press, Oxford.

Calvo, J. P., Elizaga, E., Lopez Martinez, N., Robles, F., Usera, J. 1978. El Mioceno superior continental del Prebetico externo: evolucion del Estrecho Nordbetico. *Boll. Geol. Min.* **89**, 407–426.

Caraballo, J. M. 1975. Geoquimica de las rocas lamproiticas españolas. Ph.D. thesis, Univ. Madrid.

Carey, B. D. 1955. A review of the geology of the Leucite Hills. 10th. Ann. Field Conf. Green River Basin Guidebook, Wyoming Geol. Assoc. pp. 112–113.

Carmichael, I. S. E. 1967. The mineralogy and petrology of the volcanic rocks from the Leucite Hills, Wyoming. *Contrib. Mineral. Petrol.* **15**, 24–66.

Carmichael, I. S. E., Turner, F. S., Verhoogen, J. 1974. *Igneous Petrology.* McGraw-Hill, New York.

Carr, S. G., Olliver, J. G. 1980. Olivine basalt dyke near Terowie. *Q. Geol. Notes, Geol Surv. S. Aust.* **73**, 2–7.

Carraro, F., Ferrara, G. 1968. Alpine "Tonalite" at Miagliano, Biella (zona Diorite-Kinzigitica). *Schweiz. Mineral. Petrogr. Mitteil.* **48**, 75–78.

Cas, R. A. F., Wright, J. V. 1987. *Volcanic Successions: Modern and Ancient.* Allen & Unwin, London.

Casey, J. N. 1958. Derby-4 mile geological series, sheet E51-7, Bur. Min. Resour. Australia Explan. Notes 8.

Chang, L. L. Y. 1965. Subsolidus phase relations in the systems $BaCO_3$–$SrCO_3$, $SrCO_3$–$CaCO_3$ and $BaCO_3$–$CaCO_3$. *J. Geol.* **73**, 346–368.

Chapman, D. S., Pollack, H. N. 1977. Heat flow and heat production in Zambia: Evidence for lithospheric thinning in central Africa. *Tectonophys.* **41**, 79–100.

Charles, R. W. 1977. The phase equilibria of intermediate compositions on the pseudobinary $Na_2CaMg_5Si_8O_{22}(OH)_2$–$Na_2CaFe_5Si_8O_{22}(OH)_2$. *Am. J. Sci.* **277**, 594–625.

Chatterjee, N. N. 1937. On a basic mica-trap dyke from the Raniganj coalfield. *Q. J. Mining. Metall. Soc. India* **9**, 51–59.

Chatterjee, S. C. 1974. *Petrography of the Igneous and Metamorphic Rocks of India.* Macmillan Co., Bombay.

Chattopadhyay, P. B., Venkataraman, K. 1977. Petrography and petrochemistry of the kimberlite and associated volcanic rocks of the Jungel Valley district, Mirzapur, U.P. India. *J. Geol. Soc. India* **18**, 653–661.

Civetta, L., Innocenti, F., Manetti, P., Peccerillo, A., Poli, G., 1981. Geochemical characteristics of potassic volcanics from Mts. Ernici (southern Latium, Italy). *Contrib. Mineral. Petrol.* **78**, 37–47.

Civetta, L., Orsi, G., Scandone, G. 1978. Eastwards migration of the Tuscan anatectic magmatism due to anticlockwise rotation of the Appenines. *Nature* **276**, 604–606.

Clarkson, P. D. 1981. The geology of the Shackleton Range: IV. The dolerite dykes. *British Antarct. Surv. Bull.* **53**, 201–212.

Clement, C. R., Reid, A. M. 1989. The origin of kimberlite pipes: An interpretation based on a synthesis of geological features displayed by southern African occurrences. In Ross *et al.* (1989) q.v., Vol. 1, pp. 632–646.

Clement, C. R., Skinner, E. M. W. 1979. A textural genetic classification of kimberlite rocks. Kimberlite Symposium II, Cambridge (Extended Abstract).

Clifford, T. N. 1966. Tectono-metallogenic units and metallogenic provinces of Africa. *Earth. Planet. Sci. Lett.* **1**, 421–434.

Coello, J., Castañon, A. 1965. Las sucesiones volcanicas de la zona de Carboneras (Almeria). *Estudios Geol.* **21**, 145–166.

Colchester, D. M. 1972. A preliminary note on kimberlite occurrences in South Australia. *J. Geol. Soc. Aust.* **19**, 383–386.

Colchester, D. M. 1982. Geology and petrology of some kimberlites near Terowie, S. Australia. M.Sc. thesis, N.S.W. Inst. Tech. Sydney.

Colchester, D. M. 1983. The geology and petrology of some kimberlites near Terowie, S. Australia, Proc. 6th Aust. Geol. Conv. pp. 282–283.

Collerson, K. D., McCulloch, M. T. 1983. Nd and Sr isotope geochemistry of leucite-bearing lavas from Gaussberg, East Antarctica. Proc. 4th. Symp. Antarctic Earth Sci., pp. 767–680.

Collerson, K. D., Williams, R. W., Gill, J. B. 1988. Leucitites with large initial ^{230}Th enrichment: Gaussberg volcano, Antarctica. *Chem. Geol.* **70**, 125 (abstract).

Condie, K. C. 1989. *Plate Tectonics and Crustal Evolution.* Pergamon Press, New York.

Coopersmith, H. C., Mitchell, R. H. 1989. Geology and exploration of the Rose lamproite, southeast Kansas, U.S.A. In Ross *et al.* (1989) q.v., Vol. 2, pp. 1179–1191.

Cordani, U. G., Teixeira, W., Tassinari, C. C. G., Kawashita, K., Sato, K. 1988. The growth of the Brazilian shield. *Episodes* **11**, 163–167.

Cosgrove, M. E., 1972. Geochemistry of the potassium-rich Permian volcanic rocks of Devonshire, England. *Contrib. Mineral. Petrol.* **36**, 155–170.

Crawford, A. R. 1969. India, Ceylon and Pakistan: New age data and comparisons with Australia. *Nature* **233**, 380–384.

Crawford, A. R., Compston, W. 1970. Age of the Vindhyan systems of Peninsular India. *Q. J. Geol. Soc. London* **125**, 351–371.

Crawford, A. R., Compston, W. 1973. An age of the Cuddapah and Kurnool systems, southern India. *J. Geol. Soc. Aust.* **19**, 4.

Crittenden, M. D., Kistler, R. W. 1966. Isotopic dating of intrusive rocks in the Cottonwood area, Utah. *Geol. Soc. Am. Spec. Pap.* **101**, 298–299.

Cross, C. W. 1897. The igneous rocks of the Leucite Hills and Pilot Butte, Wyoming. *Am. J. Sci.* **4**, 115–141.

Cross, C. W., Iddings, J. P., Pirsson, L. V., Washington, H.S. 1903. *Quantitative Classification of Igneous Rocks*. Univ. Chicago Press.

Crough, S. T. 1979. Hot spot epeirogeny. *Tectonophys.* **61**, 321–333.

Crough, S. T. 1981. Mesozoic hot spot epeirogeny in eastern North America. *Geology* **9**, 2–6.

Crough, S. T., Morgan, W. J., Hargraves, R. B. 1980. Kimberlites: Their relation to mantle hotspots. *Earth Planet. Sci. Lett.* **50**, 260–274.

Crowe, R. W. A., Towner, R. R. 1981. Noonkanbah, Western Australia. West Aust. Geol. Surv. 1:250,000 Geol. Series, explan. notes.

Cullers, R. L., Mullenax, J., Dimarco, M. J., Nordeng, S. 1982. The trace element content and petrogenesis of kimberlites in Riley County, Kansas, U.S.A. *Am. Mineral.* **67**, 223–233.

Cullers, R. L., Ramakrishnan, S., Berendsen, P., Griffin, T. 1985. Geochemistry and petrogenesis of lamproites, late Cretaceous age, Woodson County, Kansas, U.S.A. *Geochim. Cosmochim. Acta* **49**, 1388–1402.

Culver, S. J., Williams, H. R. 1979. Late Precambrian and Phanerozoic geology of Sierra Leone. *J. Geol. Soc. London* **136**, 605–618.

Cundari, A. 1973. Petrology of the leucite-bearing lavas in New South Wales. *J. Geol. Soc. Aust.* **20**, 465–491.

Cundari, A. 1975. Mineral chemistry and petrogenetic aspects of the Vico lavas, Roman volcanic region, Italy. *Contrib. Mineral. Petrol.* **53**, 129–144.

Cundari, A. 1979. Petrogenesis of the leucite-bearing lavas in the Roman region, Italy: The Sabatini lavas. *Contrib. Mineral. Petrol.* **70**, 9–21.

Cundari, A., Ferguson, A. K. 1982. Significance of the pyroxene chemistry from leucite-bearing and related assemblages. *Tschermaks Mineral. Petrol. Mitteil.* **30**, 189–204.

Cundari, A., Mattias, P. P. 1974. Evolution of the Vico lavas, Roman volcanic region, Italy. *Bull. Volcanol.* **38**, 98–114.

Currie, K. L., Curtis, L. W., Gittins, J. 1975. Petrology of the Red Wine alkaline complexes Central Labrador and a comparison with the Ilimaussaq complex, SW Greenland. *Geol. Surv. Canada Pap.* **75**(1A), 271–280.

Dal Piaz, G. V., Ernst, G. W. 1978. Areal geology and petrology of eclogites and associated metabasites of the Piemonte ophiolite nappe, Breuil-St. Jacques area, Italian Western Alps. *Tectonophysics* **51**, 99–126.

Dal Piaz, G. V., Hunziker, J. C., Martinotti, G. 1973. Excursion to the Sesia zone. *Schweiz. Mineral. Petrogr. Mitteil.* **53**, 447–490.

Dal Piaz, G. V., Venturelli, G., Scolari, A. 1979. Calc-alkaline to ultrapotassic postcollisional volcanic activity in the internal northwestern Alps. *Mem. Sci. Geol. Italy* **32**, 4–16.

Daly, M. C. 1986. Crustal shear zones and thrust belts: their geometry and continuity in Central Africa, *Phil. Trans. R. Soc. London* **A317**, 111–128.

Das, K. N., Lakshmanan, S. 1971. Repositories of the Panna diamond deposits and age of the Majhgawan volcanic pipe. *Geol. Surv. India Misc. Pub.* **19**, 95–101.

Davis, G. L. 1977. The ages and uranium contents of zircons from kimberlites and associated rocks. *Carnegie Inst. Washington Yearb.* **76**, 631–635.

Dawson, J. B. 1970. The structural setting of African kimberlite magmatism. In Clifford, T. N., and Gass, I. G. (eds.). *African Magmatism and Tectonics*. Oliver and Boyd, Edinburgh, pp. 321–335.

Dawson, J. B. 1980. *Kimberlites and Their Xenoliths*. Springer-Verlag, New York.

Dawson, J. B. 1984. Contrasting types of upper mantle metasomatism. In Kornprobst (1984) q.v., Vol. 2, pp. 289–294.

Dawson, J. B. 1987a. The kimberlite clan: Relationship with olivine and leucite lamproites and inferences for upper mantle metasomatism. In Fitton and Upton (1987) q.v., pp. 95–107.

Dawson, J. B. 1987b. The MARID suite of xenoliths in kimberlite: Relationships to veined and metasomatised peridotite xenoliths. In Nixon (1987) q.v., pp. 465–473.

Dawson, J. B., Smith, J. V. 1977. The MARID (mica-amphibole-rutile-ilmenite-diopside) suite of xenoliths in kimberlite. *Geochim. Cosmochim. Acta* **41**, 309–323.

Dawson, J. B., Stephens, W. E. 1975. Statistical analysis of garnets from kimberlites and their xenoliths. *J. Geol.* **83**, 589–607.

Deans, T., Powell, J. L. 1968. Trace elements and strontium isotopes in carbonatites, fluorites and limestones from India and Pakistan. *Nature* **218**, 750–752.

Debelmas, J., Lemoine, M. 1970. The western alps, palaeogeography and structure. *Earth Sci. Rev.* **6**, 221–256.

DeMarco, L. 1959. Su alcuni filoni lamprofirici radioattivi del complesso Sesia-Lanzo. *Stud. Richerche Div. Geomin. Cnen* **1**, 1–30.

DeMarco, L. 1958. Su alcuni filoni lamprofirici radioattivi del complesso Sesia-Lanzo. Comitato Nazionale per le richerch Nucleari, Divisione Geomineraria Studie richerche, Rome, Italy, Vol. 1, pp. 499–526.

DePaolo, D. J., Wasserberg, G. J. 1976. Nd isotopic variations and petrogenetic models. *Geophys. Res. Lett.* **3**, 249–252.

Derrick, G. M., Gellatly, D. C. 1972. New leucite lamproites from the west Kimberley, Western Australia. *Aust. Bur. Mineral Res. Bull.* **125**, 103–119.

Derrick, G. M., Playford, P. E. 1973. Lennard River, Western Australia. Aust. Bur. Mineral Res. 1:250,000 Geol. Map with Explan. Notes.

Deshpande, M. L. 1980. Diamond-bearing kimberlites. *Indian Minerals* **34**, 1–9.

Dewey, J. F. 1988. Extensional collapse of orogens. *Tectonics* **7**, 1123–1140.

De Yarza, R. A. 1895. Roca eruptiva de Fortuna (Provincia de Murcia) *Boll. Comm. Mapa Geol. España* **20** (1893), 349–353.

Di Battistini, G., Toscani, L., Iaccarino, S., Villa, I. M. 1987. K–Ar ages and the geological setting of calc-alkaline volcanic rocks from Sierra de Gata, S. E. Spain. *N. J. Mineral. Monats.* **1987**, 369–383.

Dickinson, W. R. 1975. Potash-depth (K-h) relations in continental margin and intraoceanic magmatic arcs. *Geology* **3**, 53–56.

Dort, W. 1972. Late Cenozoic volcanism in Antarctica. In, Adie, R. J. (ed) *Antarctic Geology and Geophysics*. Universitetsforlaget, Oslo, pp. 645–652.

Drew, G. J. 1986. A geophysical case history of the AK1 lamproite. 4th. Internat. Kimberlite Conf. Perth, Western Australia, Extended Abstr. pp. 454–456.

Droop, G. T. R. 1987. A general equation for estimating Fe^{3+} concentrations in ferromagnesian silicates and oxides from microprobe analyses using stoichiometric criteria. *Mineral. Mag.* **51**, 431–435.

Drummond, B. J. 1988. A review of crust/upper mantle structure in the Precambrian areas of Australia and implications for Precambrian crustal evolution. *Precambrian Res.* **40/41**, 101–116.

Drygalski, E. 1912. Der Gaussberg, seine kartierung und seine formen. Deutsche Sudpolar Expedition 1901–1903. *Geogr. Geol.* **II**(1), 1–46.

Dubeau, M. I., Edgar, A. D. 1985. The stability of priderite in the system $K_2MgTi_7O_{16}$–$BaMgTi_7O_{16}$. *Mineral. Mag.* **49**, 603–606.

Dubey, V. S., Mehrs, S. 1949. Diamondiferous plug of Majhgawan in central India. *Q. J. Geol. Mining Metall. Soc. India* **21**, 1–5.

Duda, A., Schminke, H. U. 1985. Polybaric differentiation of alkali basaltic magmas: Evidence from green-core clinopyroxenes (Eifel, FRG). *Contrib. Mineral. Petrol.* **91**, 340–353.

Edgar, A. D. 1979. Mineral chemistry and petrogenesis of an ultrapotassic-ultramafic volcanic rock. *Contrib. Mineral. Petrol.* **71**, 171–175.

Edgar, A. D. 1983. Relationship of ultrapotassic magmatism in the western U.S.A. to the Yellowstone plume. *N. J. Mineral. Abh.* **147**, 35–46.

Edgar, A. D. 1987. The genesis of alkaline magmas with emphasis on their source regions: Inferences from experimental studies. In Fitton and Upton (1987) q.v., pp. 29–52.

Edgar, A. D. 1989. Barium- and strontium-enriched apatites in lamproites from West Kimberley, Western Australia. *Am. Mineral.* **74**, 889–895.

Edgar, A. D., Condliffe, E., Barnett, R. L., Shirran, R. J. 1979. An experimental study of an olivine ugandite magma and mechanisms for the formation of its K-enriched derivatives. *J. Petrol.* **21**, 475–497.

REFERENCES

Edgar, A. D., Green, D. H., Hibberson, W. O. 1976. Experimental petrology of a highly potassic magma. *J. Petrol.* **17**, 339–356.

Eggler, D. H. 1987. Geochemistry of upper mantle and lower crust beneath Colorado and Wyoming. *Geol. Soc. Am. Ann. Mtg. Abstr. with Progr.* **19**, 272–273.

Eggler, D. H. 1987. Discussion of recent papers on carbonated peridotite, bearing on mantle metasomatism and magmatism: an alternative. *Earth Planet. Sci. Lett.* **82**, 398–400.

Eggler, D. H., Baker, D. R. 1982. Reduced volatiles in the system C-O-H: Implications to mantle melting, fluid formation and diamond genesis. In Akimoto, S., and Manghnani, M. H. (eds.), *High Pressure Research in Geophysics.* Center for Academic Publications, Tokyo, Japan, pp. 237–250.

Eggler, D. H., Meen, J. K., Welt, F., Dudas, F. O. Furlong, K. P., McCallum, M. E., Carlson, R. W. 1989. Tectonomagmatism of the Wyoming province. In Drexler, J. and Larson, E. E. (eds.), *Colorado Volcanism.* Colorado School Mines. Quarterly, in press.

Ehrenberg, S. N. 1978. Petrology of potassic volcanic rocks and ultramafic xenoliths from the Navajo volcanic field, New Mexico and Arizona, Ph.D. thesis. Univ. California, Los Angeles.

El Goresy, A. 1976. Oxide minerals in lunar rocks. *Oxide Minerals.* Mineral Soc. Am. Rev. Mineral. Vol. 3, pp. EG1–EG46.

El-Hinnawi, E. E. 1965. Petrochemical characters of African volcanic rocks, Part III: Central Africa. *N. J. Mineral. Abh.* **103**, 126–146.

Ellam, R. M., Hawkesworth, C. J., Menzies, M. A., Rogers, N. W. 1989. The volcanism of southern Italy: Role of subduction and the relationship between potassic and sodic alkaline volcanism. *J. Geophys. Res.* **94**, 4589–4601.

Emeleus, C. H., Upton, B. G. J. 1976. The Gardar period in southern Greenland. In, Escher and Watt (1976) q.v., pp. 152–181.

Emmons, S. F. 1877. Descriptive Geology. In King, C. (ed.), *Report of the Geological Exploration of the 40th Parallel.* U.S. Army Engineer Dept. Prof. Papers 2, pp. 236–238.

Eremeyev, N. V., Kononova, V. A., Makhotkin, I. L., Dmitrieva, M. T., Aleshii, V. G., Vashchenko, A. N. 1989. Native metals in lamproites of the central Aldan. *Dokl. Akad. Nauk SSSR* **303**, 1464–1467 (Russian).

Erlank, A. J. 1973. Kimberlite potassium richterite and the distribution of potassium in the upper mantle. Internat. Kimberlite Conf. Cape Town, Extended Abstr. pp. 103–106.

Erlank, A. J., Waters, F. G., Hawkesworth, C. J., Haggerty, S. E., Allsopp, H. L., Rickard, R. S., Menzies, M. A. 1987. Evidence for mantle metasomatism in peridotite nodules from the Kimberley pipes, South Africa. In Menzies and Hawkesworth (1987), q.v., pp. 221–311.

Ernst, W. G. 1960. Stability relations of magnesioriebeckite. *Geochim. Cosmochim. Acta* **19**, 10–40.

Ernst, W. G. 1962. Synthesis, stability relations and occurrence of riebeckite-arfvedsonite solid solutions. *J. Geol.* **70**, 689–736.

Ernst, W. G. 1988. Metamorphic terranes, isotopic provinces, and implications for crustal growth of the western United States. *J. Geophys. Res.* **93**, 7634–7642.

Ervin, C. P., McGinnis, L. D. 1975. Reelfoot rift: Reactivated precursor to the Mississippi embayment. *Geol. Soc. Am. Bull.* **86**, 1287–1295.

Escher, A., Watt, W. S. (eds.) 1976. *Geology of Greenland.* Geol. Surv. Greenland, Copenhagen.

Escher, A., Sörensen, K., Zeck, H. P. 1976. Nagssugtoqidian mobile belt in West Greenland. In Escher and Watt (1976) q.v., pp. 77–95.

Esperanca, S., Holloway, J. R. 1987. On the origin of some mica lamprophyres: experimental evidence from a mafic minette. *Contrib. Mineral. Petrol.* **95**, 207–216.

Ewart, A., LeMaitre, R. W. 1980. Some regional compositional differences within Tertiary-to-Recent orogenic magmas. *Chem. Geol.* **30**, 257–283.

Ewing, T. E., Caran, S. C. 1982. Late Cretaceous volcanism in south and central Texas—Stratigraphic, structural and seismic models. *Trans. Gulf. Coast Assoc. Geol. Soc.* **32**, 137–145.

Fallot, P., Jeremine, E. 1932. Rémarque sur une variété nouvelle de jumillite et sur l'extension des laves de ce groupe. *C. R. Congr. Natl. Soc. Savanates* **1929**, 1–13.

Farmer, G. L., Boettcher, A. L. 1981. Petrologic and crystal-chemical significance of some deep-seated phlogopites. *Am. Mineral.* **66**, 1154–1163.

Farquharson, R. A. 1920. Petrologic work. Ann. Rep. Geol. Surv. W. Australia for 1919, p. 42.

Farquharson, R. A. 1922. Petrologic work. Ann. Rep. Geol. Surv. W. Australia for 1921, p. 56.

Feraud, J., Fornari, M., Geffroy, J., Lenck, P. 1977. Minéralisations arseniées et ophiolites: le filon a realgar et stibine de Matra et sa place dans le district a Sb–As–Hg de la Corse alpine. *Bull. B.R.G.M.* **2**(II), 91–112.

Ferguson, A. K., Cundari, A., 1982. Feldspar crystallization trends in leucite-bearing and related assemblages. *Contrib. Mineral. Petrol.* **18**, 212–218.

Fermoso, M. L. 1967a. Composicion quimica de las sanidinas de las rocas lamproiticas espanolas. *Estud. Geol.* **23**, 29–30.

Fermoso, M. L. 1967b. El diopsido de las rocas volcanicas de Jumilla. *Estud. Geol.* **23**, 31–33.

Fernandez, S., Hernandez-Pacheco, A. 1972. Lamproitic rocks of Cabezo Negro, Zeneta, Murcia. *Estud. Geol.* **28**, 267–276.

Fesq, H. W., Kable, E. J. D., Gurney, J. J. 1975. Aspects of the geochemistry of kimberlites from the Premier Mine and other South African occurrences, with particular reference to the rare earth elements. *Phys. Chem. Earth* **9**, 686–707.

Fielding, D. C., Jaques, A. L. 1989. Geology, petrology and geochemistry of the Bow Hill lamprophyre dikes, Western Australia. In Ross *et al.* (1989) q.v., Vol. 1, pp. 206–219.

Fisher, R. V., Schminke, H. U. 1984. *Pyroclastic Rocks.* Springer-Verlag, New York.

Fitton, J. G., Upton, B. G. J. 1987. *Alkaline Igneous Rocks.* Geol. Soc. London Spec. Publ. No. 30.

Fitzgerald, W. V. 1907. Reports on portions of the Kimberleys (1905–6) W. Aust. Parlimentary Papers No. 19 of 1907.

Flower, M. F. J. 1971. Evidence for the role of phlogopite in the genesis of alkaline basalts. *Contrib. Mineral. Petrol.* **32**, 126–137.

Foden, J. D. 1983. The petrology of the calc-alkaline lavas of Rindjani Volcano, East Sunda Arc: A model for island arc petrogenesis. *J. Petrol.* **24**, 98–130.

Foden, J. D., Varne, R. 1980. The petrology and tectonic setting of Quaternary-Recent volcanic centers of the Lombok and Sumbawa, Sunda Arc. *Chem. Geol.* **30**, 201–226.

Foley, S. F. 1985. The oxidation state of lamproitic magmas. *Tschermaks Mineral. Petrol. Mitteil.* **34**, 217–238.

Foley, S. F. 1986. The genesis of lamproitic magmas in a reduced fluorine-rich mantle. 4th. Internat. Kimberlite Conf., Perth, Western Australia, Extended Abst. pp. 173–175.

Foley, S. F. 1989. The genesis of lamproitic magmas in a reduced fluorine-rich mantle. In Ross *et al.* (1989) q.v., Vol. 1, pp. 616–630.

Foley, S. F., Taylor, W. R., Green, D. H. 1986a. The effect of fluorine on phase relationships in the system $KAlSiO_4$–Mg_2SiO_4–SiO_2 at 28 kbar and the solution mechanism of fluorine in silicate melts. *Contrib. Mineral. Petrol.* **93**, 46–55.

Foley, S. F., Taylor, W. R., Green, D. H. 1986b. The role of fluorine and oxygen fugacity in the genesis of the ultrapotassic rocks. *Contrib. Mineral. Petrol.* **94**, 183–192.

Foley, S. F., Venturelli, G., Green, D. H., Toscani, L. 1987. The ultrapotassic rocks: Characteristics, classification and constraints for petrogenetic models. *Earth Sci. Rev.* **24**, 81–134.

Forbes, W. C., Flower, M. J. F. 1974. Phase relations of titanphlogopite $K_2Mg_4TiAl_2Si_6O_{20}(OH)_4$: A refractory phase in the upper mantle? *Earth Planet. Sci. Lett.* **22**, 60–66.

Fourie, G. P. 1958. Die diamantvoorkomste in die omgewing van Swartruggens, Transvaal. *Geol. Surv. S. Africa Bull.* **26**, 1–16.

Fox, C. S. 1930. Intrusive igneous rocks. In *The Jharia Coalfield.* Mem. Geol. Surv. India vol. 56, pp. 113–127.

Fragomeni, P. R. Z. 1976. Tectonic control of Paranatinga kimberlitic province. *Boll. Nucl. Centro-Oeste Socied. Brazil Geol.* **5**, 3–10 (Portuguese).

Franks, P. C. 1959. Pectolite in mica peridotite, Woodson County, Kansas. *Am. Mineral.* **44**, 1082–1086.

Franks, P. C. 1966. Ozark Precambrian-Paleozoic relations: Discussion of igneous rocks exposed in E. Kansas. *Am. Assoc. Petrol. Geol. Bull.* **50**, 1035–1042.

Franks, P. C., Bickford, M. E., Wagner, H. C. 1971. Metamorphism of Precambrian granitic xenoliths in a mica peridotite at Rose Dome, Woodson County, Kansas. Part 2, Petrologic and mineralogic studies. *Geol. Soc. Am. Bull.* **82**, 2869–2890.

Frantsesson, E. V. 1968. *The Petrology of Kimberlites.* (Translated from the Russian by D. A. Brown) Dept. Geology Pub. No. 150, A.N.U. Canberra.

Fraser, K. J. (1987) Petrogenesis of kimberlites from South Africa and lamproites from Western Australia and North America. Ph.D. thesis, The Open Univ., Milton Keynes, U.K.

Fraser, K. J., Hawkesworth, C. J., Erlank, A. J., Mitchell, R. H., Scott Smith, B. H. 1985. Sr, Nd, and Pb isotope and minor element geochemistry of lamproites and kimberlites. *Earth Planet. Sci. Lett.* **76**, 57–70.

Frey, F. A., Ferguson, J., Chappell, B. W. 1977. Petrogenesis of South African and Australian kimberlite suites. 2nd. Internat. Kimberlite Conf. Santa Fe, Extended Abstr. (unpaginated).

REFERENCES

Fuster, J. M. 1956. Las erupciones delleniticas del Terciario superior de la Fosa de Vera (Provincia de Almeria). *Boll. R. Soc. Esp. Hist. Nat.* **54**, 53–58.

Fuster, J. M., de Pedro, F., 1953. Estudio petrologico de las rocas lamproiticas de Cabezo Maria (Almeria). *Estud. Geol.* **9**, 477–508.

Fuster, J. M., Gastesi, P. 1964. Estudio petrologico de las rocas lamproiticas de Barqueros (Provincia de Murcia). *Estud. Geol.* **20**, 299–314.

Fuster, J. M., Gastesi, P., Sagredo, J., Fermoso, M. L. 1967. Las rocas lamproiticas del Sureste de España. *Estud. Geol.* **23**, 35–69.

Fuster, J. M., Ibarrola, E., Lobato, M. P. 1954. Analisis quimicos de rocas españolas publicados hosta 1952. C.S.I.C. Monogr. del Inst. Lucas Mallada Invest. Geol. No. 14.

Fyfe, W. S., McBirney, A. 1975. Subduction and the structure of andesite volcanic belts. *Am. J. Sci.* **275A**, 285–297.

Gallo, F., Giametti, F., Venturelli, G., Vernia, L. 1984. The kamafugitic rocks of San Venanzo and Cupaello, Central Italy. *N. J. Mineral. Monats.* **1984**, 198–210.

Gaspar, J. C., Wyllie, P. J. 1984. The alleged kimberlite–carbonatite relationship: Evidence from ilmenite and spinel from Premier and Wesselton Mines and the Benfontein Sill, South Africa. *Contrib. Mineral. Petrol.* **85**, 133–140.

Gee, B. R. 1932. The geology and coal reserves of Raniganj coalfield. *Mem. Geol. Surv. India* **61**.

George, G. R. 1979. Leucite Hills. In Bradley, W. A., *et al.* (eds.) *Wyoming Oil and Gas Fields Symposium Greater Green River Basin*, Wyoming Geologists Association, Casper, Wyoming, Vol. 2, pp. 114–115.

Gerryts, E. 1970. Diamond prospecting by geophysical methods: A review of current practice. In Morley, L. W. (Ed.), *Mining and Groundwater Geophysics*. Geol. Surv. Can. Econ. Geol. Rep. Vol. 26, pp. 439–446.

Getreuer, F., Rehkopff, A. 1980. Grundfjeldsgeologien i Marranguit-Kangilerssua området. Nordre Strømfjord den centrale del af de Vesgrønlandske Nagssugtoqider. Del 1. Petrografi og strukturer. Thesis, Aarhus Univ. Denmark.

Ghiara, M. R., Lirer, L. 1977. Mineralogy and geochemistry of the "low potassium" series of the Roccamonfina volcanic suite (Campania, South Italy). *Bull Volcanol.* **40**, 39–56.

Ghose, C. 1949. A petrochemical study of lamprophyres and associated intrusive rocks of the Jharia coalfield. *Q. J. Geol. Mining Metall. Soc. India* **21**, 133–147.

Giardini, A. A., Melton, C. E. 1975a. The nature of cloud-like inclusions in two Arkansas diamonds. *Am. Mineral.* **60**, 931–933.

Giardini, A. A., Melton, C. E. 1975b. Gases released from natural and synthetic diamonds by crushing under high vacuum at 200°C and their significance to diamond genesis. *Fortschr. Miner. Spec. Issue IMA Papers 9th Mtg.* **52**, 455–464.

Giardini, A. A., Melton, C. E. 1975c. Chemical data on a colorless Arkansas diamond and its black amorphous C–Fe–Ni–S inclusions. *Am. Mineral.* **60**, 934–936.

Giardini, A. A., Hurst, V. J., Melton, C. E., Stormer, J. C. 1974. Biotite as a primary inclusion in diamond: Its nature and significance. *Am. Mineral.* **59**, 783–789.

Gilbert, M. C., Briggs, D. F. 1974. A comparison of the stabilities of OH and F-potassic richterites—A preliminary report. *Trans. Am. Geophys. Union* **55**, 480–481.

Gilbert, M. C., Helz, R. T., Popp, P. K., Spear, F. S. 1982. Experimental studies of amphibole stability. In, *Amphiboles: Petrology and Experimental Phase Relations*, Mineral. Soc. Am. Rev., Mineral. Vol. 9B, 229–353.

Giles, C. W. 1981. A comparative study of Archean and Proterozoic felsic associations in southern Australia. Ph.D. thesis, Univ. Adelaide.

Gill, J. B. 1981. *Orogenic Andesites and Plate Tectonics*. Springer Verlag, New York.

Gilluly, J. 1927. Analcite diabase and related alkaline syenite from Utah. *Am. J. Sci.* **14**, 199–211.

Gittins, J. 1979. The feldspathoidal alkaline rocks. In Yoder, H. S. (ed.), *The Evolution of the Igneous Rocks, Fiftieth Anniversary Perspectives*. Princeton Univ. Press, pp. 351–390.

Gittins, J., Fawcett, J. J., Brooks, C. K., Rucklidge, J. C. 1980. Intergrowths of nepheline-K-feldspar and kalsilite-K-feldspar: A reexamination of the pseudo-leucite problem. *Contrib. Mineral. Petrol.* **73**, 119–126.

Gittins, J., MacIntyre, R. M., York, D. 1967. The ages of carbonatite complexes in eastern Canada. *Can J. Earth Sci.* **4**, 651–655.

Glover, J. E., Harris, P. G. (Eds.) 1985. *Kimberlite Occurrence and Origin: A Basis for Conceptual Models in Exploration*. 2nd. Edtn. Geol. Dept. and Univ. Extension Univ. Western Australia Publ. No. 8.

Glover, J. E., Phillips, C. 1980. Kimberlites and diamonds, a selected bibliography for exploration geologists. In, Glover, J. E., and Groves, D. J. (eds.). *Kimberlites and Diamonds.* Geol. Dept. of Univ. Extension Univ. Western Australia Publ. 5, pp. 96–123.

Gogineni, S. V., Melton, C. E., Giardini, A. A. 1978. Some petrological aspects of the Praire Creek diamond-bearing kimberlite diatreme, Arkansas. *Contrib. Mineral. Petrol.* **66**, 251–261.

Grantham, D. R. 1964. The diamond deposits of Panna, central India. *Industrial Diamond Rev.* **24**, 30–35.

Grantham, D. R. 1969. The age of the diamond-bearing rocks at Panna, Madhya Pradesh. *Current Sci.* **16**, 1–3.

Green, D. H., Fallon, T. J., Taylor, W. R. 1987. Mantle-derived magmas—Roles of variable source peridotite and variable C–H–O fluid compositions. In Mysen (1987) q.v., pp. 139–154.

Green, T. H. 1980. Island arc and continent building magmatism: A review of petrogenetic models based on experimental petrology and geochemistry. *Tectonophys.* **63**, 367–385.

Green, T. H. 1981. Experimental evidence for the role of accessory phases in magma genesis. *J. Volcan. Geothem. Res.* **10**, 405–422.

Green, T. H., Pearson, N. J. 1986. Ti-rich accessory phase saturation in hydrous mafic-felsic compositions at high P. T. *Chem. Geol.* **54**, 185–201.

Gregory, G. P. 1969. Geochemical dispersion patterns related to kimberlite intrusives in North America. Ph.D. thesis, Royal School of Mines, Imperial College, London.

Gregory, G. P. 1984. Exploration for primary diamond deposits with special emphasis on the Lennard shelf, Western Australia. In Purcell, P. G. (ed.), *The Canning Basin.* Proc. Symp. Geol. Soc. Aust./Petroleum Expl. Soc. Aust. Perth, pp. 475–484.

Gregory, G. P., Tooms, J. S. 1969. Geochemical prospecting for kimberlites. *Q. J. Colorado School Mines* **64**, 265–305.

Griffin, W. L., Jaques, A. L., Sie, B. H., Ryan, C. G., Cousens, D. R., Suter, G. F. 1988. Conditions of diamond growth: A proton microprobe study of inclusions in Western Australian diamonds. *Contrib. Mineral. Petrol.* **99**, 143–158.

Gunter, W. D., Pajani, G. E., Hoinkes, G., Trembath, L. T. 1983. Mineral flow layering in the Leucite Hills volcanics. *Geol. Assoc. Can. Mineral. Assoc. Can. Ann. Mtg. Progr. Abstr.* **8**, A29.

Gupta, A. K., Fyfe, W. S. 1975. Leucite survival: Alteration to analcime. *Can. Mineral.* **13**, 361–363.

Gupta, A. K., Yagi, K. 1980. *Petrology and Genesis of Leucite-bearing Rocks.* Springer Verlag, New York.

Gupta, A. K., Yagi, K., Lovering, J., Jaques, A. L. 1986. Geochemical and microprobe studies of the diamond-bearing ultramafic rocks from central and south India. 4th. Internat. Kimberlite Conf. Perth, Western Australia, Extended Abstr., pp. 27–29.

Gupta, S. P., Phukan, S. 1971. Mineralogy of the altered diamondiferous rock at Panna, M. P. *Geol. Surv. India Spec. Publ.* **19**, 114–119.

Haebig, A. E., Jackson, D. G. 1986. Geochemical expression of some west Australian kimberlites and lamproites. 4th. Internat. Kimberlite Conf. Perth, Western Australia, Extended Abstr., pp. 466–468.

Haggerty, S. E. 1976. Opaque mineral oxides in terrestrial igneous rocks. In *Oxide Minerals.* Mineral. Soc. Am. Rev. Mineral. Vol. 3, pp. HG 101–HG 300.

Haggerty, S. E. 1982. Kimberlites in western Liberia: An overview of the geological setting in a plate tectonic framework. *J. Geophys. Res.* **81**, 10811–10826.

Haggerty, S. E. 1988. Redox heterogeneities in the upper mantle. Geol. Soc. Am. Ann. Mtg. Abstr. with Progr. p. A31.

Haggerty, S. E., Grey, I. E., Madsen, I. C., Criddle, A. J., Stanley, C. J., Erlank, A. J. 1989. Hawthorneite Ba[Ti$_3$Cr$_4$Fe$_4$Mg]O$_{19}$: A new metasomatic magnetoplumbite-type mineral from the upper mantle. *Am. Mineral.* **74**, 668–675.

Haggerty, S. E., Nagieb, M. S. 1989. Diamonds in non-kimberlitic, non-lamproitic diatremes from northwest Syria. 28th. Internat. Geol. Congr. Washington Extended abstr. Vol. 2, pp. 6–7.

Haggerty, S. E., Erlank, A. J., Grey, I. E. 1986. Metasomatic mineral titanate complexing in the upper mantle. *Nature* **319**, 761–763.

Haggerty, S. E., Hardie, R. E., McMahon, R. M. 1979. The mineral chemistry of ilmenite nodule associations from the Monastery diatreme. In Boyd and Meyer (1979) q.v., Vol. 2, pp. 249–256.

Haggerty, S. E., Tompkins, L. A. 1983. Redox state of Earth's upper mantle from kimberlitic ilmenites. *Nature* **303**, 295–300.

Halder, D., Ghose, D. B. 1974. Tectonics of kimberlites around Majhgawan, Madyha Pradesh, India. In Murthy, M. V. N. (ed.), *International Seminar on Tectonics and Metallogeny of Southeast Asia and Far East.* Geol. Surv. India, Vol. 47–48, 12 pp.

Hall, A. 1982. The Pendennis peralkaline minette. *Mineral. Mag.* **45**, 257–266.

REFERENCES

Hall, A. 1986. *Igneous Petrology*. Longman Sci. & Tech. Publ., Harlow U. K.
Hall, A. E., Smith, C. B. 1985. Lamproite diamonds: Are they different? In Glover and Harris (1985) q.v., pp. 167–212.
Hancock, S. L., Rutland, R. W. R. 1984. Tectonics of an early Proterozoic geosuture: The Halls Creek orogenic subprovince, northern Australia. *J. Geodynam.* **1**, 387–432.
Hardman, E. T. 1884. Report of the geology of the Kimberley District, Western Australia. West. Aust. Parliament. Papers 31.
Hargraves, R. B., Onstott, T. C. 1980. Paleomagnetic results from some southern African kimberlites, and their tectonic significance. *J. Geophys. Res.* **85**, 3587–3597.
Harris, J. W. 1987. Recent physical, chemical and isotopic research of diamond. In Nixon (1987) q.v., pp. 477–500.
Harris, J. W., Collins, A. T. 1985. Studies of Argyle diamonds. *Ind. Diamond Rev.* **3**, 128–130.
Harris, J. W., Gurney, J. J. 1979. Inclusions in diamond. In Field, J. E. (ed.) *The Properties of Diamond*. Academic Press, London, pp. 555–594.
Harte, B. 1983. Mantle peridotites and processes—The kimberlite sample. In Hawkesworth, C. J., and Norry, M. J. (eds.) *Continental Basalts and Mantle Xenoliths*. Shiva Publ. Ltd. Nantwich, Cheshire U.K., pp. 46–91.
Hartman, P. 1969. Can Ti^{4+} replace Si^{4+} in silicates? *Mineral. Mag.* **37**, 366–369.
Hasui, Y., Cordani, U. G. 1968. Idades K–Ar de rochas eruptivas mesozoica do oeste mineiro e sul de Goias. 22nd. Cong. Brasil. Geol. Belo Horizonte, pp. 139–143.
Hatherton, T., Dickinson, W. R. 1969. The relationship between andesitic volcanism and seismicity in Indonesia, the Lesser Antilles and other island arcs. *J. Geophys. Res.* **74**, 5301–5310.
Hatch, F. H., Wells, A. K., Wells, M. K. 1961. *Petrology of the Igneous Rocks*. Thomas Murby & Co. Ltd., London.
Hausel, W. D., McCallum, M. E., Woodzick, T. L. 1979. Exploration for diamond-bearing kimberlite in Colorado and Wyoming: An evaluation of exploration techniques. *Geol. Surv. Wyoming Rep. Invest.* **19**, 1–26.
Hawkesworth, C. J., van Calsteren, P. W. C. 1974. Radiogenic isotopes, some geological applications. In Henderson, P. (ed.), *Rare Earth Element Geochemistry*. Elsevier, New York, pp. 375–421.
Hawkesworth, C. J., Vollmer, R. 1979. Crustal contamination versus enriched mantle, $^{143}Nd/^{144}Nd$ and $^{87}Sr/^{86}Sr$ evidence from the Italian volcanics. *Contrib. Mineral. Petrol.* **69**, 151–165.
Hawkesworth, C. J., Fraser, K. J., Rogers, N. W. 1985. Kimberlites and lamproites: Extreme products of mantle enrichment processes. *Trans. Geol. Soc. S. Africa* **88**, 439–447.
Hay, R. 1883. The igneous rocks of Kansas. *Trans. Kansas Acad. Sci.* **8**, 14–18.
Heald, K. C. 1926. Geology of the Ingomar anticline, Treasure and Rosebud counties, Montana. *U.S. Geol. Surv. Bull.* **786**, 1–37.
Heaman, L. M. 1989. The nature of the subcontinental mantle from Sr–Nd–Pb isotopic studies on kimberlitic perovskite. *Earth Planet Sci. Lett.* **92**, 323–334.
Hearn, B. C. Jr. 1968. Diatremes with kimberlitic affinities in north-central Montana. *Science* **159**, 622–625.
Helmstaedt, H., Carmichael, D. M., Percival, J. A. 1979. Grosspydite xenoliths from the Zagodochnaya kimberlite pipe, Yakutia: High grade metamorphic rodingites. *Geol. Soc. Am. Ann. Mtg. Abstr. with Progr.* **11**, 442.
Henage, L. F. 1972. A definitive study of the origin of lamproites. M.Sc. thesis, Univ. Oregon.
Henage, L. F., Best, M. G. 1968. Association of potassic silica-poor igneous rocks with post-Laramide andesites in northeastern Utah. *Geol. Soc. Am. Spec. Pap.* **115**, 423–424.
Henderson, C. M. B. 1965. Minor element chemistry of leucite and pseudoleucite. *Mineral. Mag.* **35**, 596–603.
Henshaw, D. E. 1955. The structure of wadeite. *Mineral. Mag.* **30**, 585–595.
Hernandez, J. 1975. Sur la caractère shoshonitique des andesites du Gourongou, Rif oriental (Maroc). *C.R. Acad. Sci. Paris*, **280**, 233–236.
Hernandez-Pacheco, A., 1965. Una richterita potasica de rocas volcanicas alcalinas, Sierra de las Cabras (Albacete). *Estud. Geol.* **20**, 265–270.
Hernandez-Pacheco, F. 1935. Estudio fisiographico y geologico del territorio comprendido entre Hellin y Cieza. *Anales Univ. Madrid (Ciencias)* **4**, 1–38.
Hernandez, J., Larouziere, F. D. de, Boize, J., Bordet, P. 1987. Le magmatisme néogène bético-rifain et le couloir de decrochément trans-Alboran, *Bull. Soc. Geol. France* **8**, 257–267.
Hertogen, J., Lopez-Ruiz, J., Rodriguez-Badiola, E., Demaiffe, D., Weis, D. 1985. A mantle sediment mixing model for the petrogenesis of an ultrapotassic lamproite from S. E. Spain. EOS, 46, Abstr. V12B-08, p. 114.
Hervieu, H., Germain, P., Desgardin, G., Raveau, B. 1979. Non-stoichiometric titanates $A_{2-2x}Ba_{3x}Ti_{6-x}O_{13}$ with a tunnel structure. *Materials Res. Bull.* **14**, 267–272.
Herz, N. 1977. Timing of spreading in the South Atlantic: information from Brazilian alkalic rocks. *Geol. Soc. Ann. Bull.* **88**, 101–112.

Hickey, R. L., Frey, F. A. 1982. Geochemical characteristics of boninite series volcanics: Implications for their source. *Geochim. Cosmochim. Acta* **46**, 2099–2115.

Higazy, R. A. 1954. Trace elements of volcanic ultrabasic potassic rocks of southwestern Uganda and adjoining part of the Belgian Congo. *Geol. Soc. Am. Bull.* **65**, 39–70.

Hills, E. S. 1956. A contribution to the morphotectonics of Australia. *J. Geol. Soc. Australia* **3**, 1–15.

Hills, F. A., Houston, R. S. 1979. Early Proterozoic tectonics of the central Rocky Mountains, North America. *Contrib. Geol. Univ. Wyoming* **17**, 89–109.

Hintze, L. F. 1980. Geological map of Utah (1:500,000 scale, 2 sheets). Utah Geol. Mineral Surv. Salt Lake City.

Hintze, L. F. 1988. Geologic history of Utah. Brigham Young Univ. Geol. Studies Spec. Publ. 7.

Hoffman, P. F. 1988. United plates of America, the birth of a craton: Early Proterozoic assembly and growth of Laurentia. *Ann. Rev. Earth Planet. Sci.* **16**, 543–603.

Hogarth, D. D., Brown, F. F., Pritchard, A. M. 1970. Bi-absorption, Mössbauer spectra, and chemical investigation of five phlogopite samples from Quebec. *Can. Mineral.* **19**, 710–722.

Hogg, N. C. 1972. Shoshonitic lavas in west-central Utah. *Brigham Young Univ. Geol. Stud.* **19**, 133–184.

Holland, T. H., Saise, W. 1985. On the igneous rocks of the Giridih coalfield. *Rec. Geol. Surv. India.* **28**, 121–137.

Holm, P. M. 1982. Mineral chemistry of perpotassic lavas of the Vulsinian district, the Roman province, Italy. *Mineral. Mag.* **46**, 379–386.

Holm, P. M., Munksgaard, N. C. 1982. Evidence for mantle metasomatism: An oxygen and strontium isotope study of the Vulsinian district, central Italy. *Earth Planet. Sci. Lett.* **60**, 376–388.

Holm, P. M., Lou, S., Nielsen, A. 1982. The geochemistry and petrogenesis of the lavas of the Vulsinian district, Roman province, central Italy. *Contrib. Mineral. Petrol.* **80**, 367–378.

Holmes, A. 1937. The petrology of katungite. *Geol. Mag.* **74**, 200–219.

Holmes, A. 1950. Petrogenesis of katungite and its associates. *Am. Mineral.* **35**, 772–792.

Holmes, A., Harwood, H. F. 1932. Petrology of the volcanic fields east and southeast of Ruwenzori, Uganda. *Q. J. Geol. Soc. (London)* **88**, 370–442.

Honda, M., Reynolds, J. H. Roedder, E., Epstein, S. 1987. Noble gases in diamonds: Occurrence of solar-like helium and neon. *J. Geophys. Res.* **92**, 12507–12521.

Horvarth, F., Berckhemer, H. 1982. Mediterranean back arc basins, In Berckhemer, H. (ed.), *Alpine–Mediterranean Geodynamics,* Am. Geophys. Union Geodynamics Ser. 7, pp. 141–173.

Houston, R. S., Karstrom, K. E., Hills, F. A., Smithson, S. B. 1979. The Cheyenne belt: A major Precambrian crustal boundary in the western United States. *Geol. Soc. Am. Ann. Mtg. Abstr. with Progr.* **11**, 446.

Hughes, C. J. 1982. *Igneous Petrology.* Elsevier, New York.

Hughes, T. W. H. 1870. Karharbari coalfield—Deoghan coalfield. *Mem. Geol. Surv. India,* **7**, 2.

Hunter, D. R., Reid, D. L. 1987. Mafic dyke swarms in southern Africa. In Halls, H. F. C., and Fahrig, W. F. (eds.) *Mafic Dyke Swarms.* Geol. Assoc. Can. Spec. Pap. 34, pp. 445–456.

Hunter, D. R., Pretorius, D. A. 1981. Southern Africa structural framework, In Hunter, D. R. (ed.), *Precambrian of the Southern Hemisphere.* Elsevier, Amsterdam, pp. 397–422.

Hunziker, J. C. 1974. Rb–Sr age determination and the alpine tectonic history of the western Alps. *Mem. Inst. Geol. Min. Univ. Padova* **31**, 1–54.

Iddings, J. P. 1985. Absarokite–shoshonite–banakite series. *J. Geol.* **3**, 935–959.

Jakes, P., White, A. J. R. 1972. Major and trace element abundances in volcanic rocks of orogenic areas. *Geol. Soc. Am. Bull.* **83**, 29–40.

Janse, A. J. A. 1985. Kimberlites—Where and when. In Glover and Harris (1985) q.v., pp. 19–61.

Janse, A. J. A., Sheahan, P. A. 1987. Bibliochrony of igneous rocks of Arkansas with particular emphasis on diamonds. In Morris and Pasteris (1987) q.v., pp. 249–292.

Jaques, A. L. 1989. Lamproite diamonds and their inclusions: New insights from the West Australian deposits. Workshop on Diamonds. 28th. Internat. Geol. Congr. Washington, Extended Abstr. pp. 36–39.

Jaques, A. L., Foley, S. F. 1985. The origin of Al-rich spinel inclusions in leucite from the leucite lamproites of Western Australia. *Am. Mineral.* **70**, 1143–1150.

Jaques, A. L., Creaser, R. A., Ferguson, J., Smith, C. B. 1985. A review of the alkaline rocks of Australia. *Trans. Geol. Soc. S. Africa* **88**, 311–334.

Jaques, A. L., Gregory, G. P., Lewis, J. D., Ferguson, J. 1982. The ultra-potassic rocks of the West Kimberley region, Western Australia and a new class of diamondiferous kimberlite. *Terra Cognita* **2**, 251–252.

Jaques, A. L., Lewis, J. D., Gregory, G. P., Ferguson, J., Smith, C. B., Chappell, B. W., McCulloch, M. T. 1983. The ultrapotassic diamond-bearing rocks of the West Kimberley region, *Western Australia. Proc. 6th Geol. Soc. Aust. Conv.* **9**, 286–287.

REFERENCES

Jaques, A. L., Lewis, J. D., Gregory, G. P., Ferguson, J., Smith, C. B., Chappell B. W., McCulloch, M. T. 1984. The diamond-bearing ultrapotassic (lamproitic) rocks of the West Kimberley region, Western Australia. In Kornprobst (1984) q.v., Vol. 1, pp. 225–254.

Jaques, A. L., Lewis, J. D., Smith, C. B. 1986. The kimberlites and lamproites of Western Australia. *Geol. Surv. Western Aust. Bull.* **132**. 268 pp.

Jaques, A. L., Haggerty, S. E., Lucas, H., Boxer, G. L. 1989a. Mineralogy and petrology of the Argyle (AK1) lamproite pipe, Western Australia. In Ross *et al.* (1989) q.v., Vol. 1, pp. 153–169.

Jaques, A. L., Hall, A. E., Sheraton, J. W., Smith, C. B., Sun, S. S., Drew, R. M., Foudoulis, C., Ellingsen, K. 1989b. Composition of crystalline inclusions and C-isotopic composition of Argyle and Ellendale diamonds. In Ross *et al.* (1989) q.v., Vol. 2, 966–989.

Jaques, A. L., Sun, S. S., Chappell, B. W. 1989c. Geochemistry of the Argyle (AK1) lamproite pipe, Western Australia. In Ross *et al.* (1989) q.v., Vol. 1, pp. 170–188.

Jaques, A. L., Kerr, I. D., Lucas, H., Sun, S. S., Chappell, B. W. 1989d. Mineralogy and petrology of monchiquites from Wandagee, Carnarvon Basin, Western Australia. In Ross *et al.* (1989) q.v., Vol. 1, pp. 120–139.

Jenke, G. 1983. The role of geophysics in the discovery of the Ellendale and Fitzroy kimberlites. 3rd. Bienn. Conf. Aust. Soc. Expl. Geophys. Brisbane, Extended Abstr. pp. 66–72.

Jeremine, E., Fallot, P., 1929. Sur la présence d'une varieté de jumillite aux environs de Calasparra (Murcia). *C. R. Acad. Sci.* **188**, 800.

Jewett, J. M. 1964. Geologic map of Kansas (1:500,000 scale). State Geol. Surv. Kansas.

Johannsen, A. 1932. Die quantitative mineralogische klassifikation der eruptivgesteine. *Centralb. Mineral. Abt. A.* **1932**, 146–150.

Johanssen, A. 1939. *A Descriptive Petrography of the Igneous Rocks.* (4 vols.) 2nd. Ed. Univ. Chicago Press.

Johnson, C. M., O'Neill, J. R. 1984. Triple junction magmatism: A geochemical study of Neogene volcanic rocks in western California. *Earth Planet. Sci. Lett.* **71**, 241–262.

Johnson, W. D., Smith, H. R. 1964. Geology of the Winnett–Mosby area, Petroleum, Garfield, Rosebud and Fergus Counties, Montana. *U.S. Geol. Surv. Bull.* **1149**, 1–91.

Johnston, R. H. 1959. Geology of the Northern Leucite Hills, Sweetwater County, Wyoming. M.Sc. thesis, Univ. Wyoming, Laramie.

Jones, A. P., Smith, J. V. 1983. Petrological significance of mineral chemistry in the Agathla Peak and the Thumb minettes, Navajo volcanic field. *J. Geol.* **91**, 643–656.

Jones, A. P., Smith, J. V. 1985. Phlogopite and associated minerals from Permian minettes in Devon, South England. *Bull. Geol. Soc. Finland* **57**, 89–102.

Jones, A. P., Smith, J. V., Dawson, J. B. 1982. Mantle metasomatism in 14 veined peridotites from the Bultfontein mine, South Africa. *J. Geol.* **90**, 435–453.

Joplin, G. A. 1965. The problem of the potash-rich basaltic rocks. *Mineral. Mag.* **34**, 266–275.

Joplin, G. A. 1968. The shoshonite association: A review. *J. Geol. Soc. Aust.* **15**, 275–294.

Jordan, T. H. 1985. The continental tectosphere. *Rev. Geophys. Space Phys.* **13**, 1–12.

Juckes, L. M. 1972. The geology of northeastern Heimefrontfjella, Dronning Maud Land. *Brit. Antarct. Surv. Sci. Rep.* **65**.

Kaila, K. L., Roy Chowdhury, K., Reddy, P. R., Krishna, V. G., Hari, N., Subbotin, S. I., Sollogub, V. B., Chekunov, A. V., Kharetchko, G. E., Lazarenko, M. A., Ilchenko, T. V. 1979. Crustal structure along the Kavali–Udipi profile in the Indian peninsular shield from deep seismic sounding. *J. Geol. Soc. India* **20**, 307–333.

Kaila, K. L., Tewari, H. C., Roy Chowdhury, K., Rao, V. K., Sridhar, A. R., Mall, D. M. 1987. Crustal structure of the northern part of the Proterozoic Cuddapah basin of India from deep seismic soundings and gravity data. *Tectonophys.* **140**, 1–12.

Kaminskii, F. V. 1984. *The Diamond Content of Non-kimberlitic Rocks.* Nedra, Moscow (Russian).

Kaplan, G., Faure, G., Elloy, R., Heilammer, R. 1967. Contribution a l'etude de l'origine des lamproites. *Bull. Centre Rech. Pau, Soc. Nat. Petroles d'Aquitaine* **1**, 153–159.

Kapustin, Y. L. 1963. On the new occurrence of wadeite in the Soviet Union. *Dokl. Akad. Nauk SSSR* **151**, 1410–1412 (Russian).

Kapustin, Y. L. 1980. *Mineralogy of Carbonatites.* Amerind Publ. Co. Pvt. Ltd, New Delhi.

Katz, M. B. 1979. Tectonic model of the Halls Creek mobile zone, northwest Australia, and its comparison with the Athapascan aulacogen of the northwest Canadian shield. 1978 Basement Tectonic Contrib. No. 40, pp. 255–267.

Kay, R. W., Gast, P. W. 1973. The rare earth content and origin of alkali-rich basalts. *J. Geol.* **81**, 653–682.

Kay, S. M., Kay, R. W., Hangas, J., Snedden, T., 1978. Crustal xenoliths from potassic lavas, Leucite hills, Wyoming. *Geol. Soc. Am. Ann. Mtg. Abstr. with Progr.* **7**, 432.

Keller, J. 1983. Potassic lavas in the orogenic volcanism of the Mediterranean area. *J. Volcan. Geotherm. Res.* **18**, 321–335.

Kemp, J. F. 1897. The Leucite Hills of Wyoming. *Geol. Soc. Ann. Bull.* **8**, 169–182.

Kemp, J. F., Knight, W. C. 1903. Leucite Hills of Wyoming. *Geol. Soc. Am. Bull.* **14**, 305–336.

Kesson, S. E., White, T. J. 1986. $[Ba_xCs_y][(Ti,Al)^{3+}_{2x+y}Ti^{4+}_{8-2x-y}]O_{16}$ synroc-type hollandites: I. Phase chemistry. *Proc. R. Soc. London* **405A**, 73–101.

Kidwell, A. L. 1951. Mesozoic igneous activity in the northern Gulf Coast plain. *Trans. Gulf. Coast Assoc. Geol. Soc.* **1**, 182–199.

Kilinc, A. I., Carmichael, I. S. E., Rivers, M. L., Sack, R. O. 1983. The ferric–ferrous ratio of natural silicate liquids equilibrated with air. *Contrib. Mineral. Petrol.* **83**, 136–140.

King, W. 1872. Kadapa and Karnul formations of the Madras Presidency. *Mem. Geol. Surv. India* **8**, 1–346.

Kingston, M. J. 1989. Spectral reflectance features of kimberlites and carbonatites: Implications for remote sensing for exploration. In Ross *et al.* (1989) q.v., Vol. 2, pp. 1135–1145.

Kinsland, G. L. 1985. Transcontinental transform fault across North America. Internat. Basement. Tectonics Assoc. Publ. 5, pp. 267–284.

Kinsman, D. J. J. 1975. Rift valley and sedimentary history of trailing continental margins. In Fisher, A. G., and Judson, S. (eds.), *Petroleum and Continental Tectonics,* Princeton Univ. Press, Princeton, New Jersey, pp. 83–126.

Kirkley, M. B. 1987. Aspects of the geochemistry of kimberlite carbonates. Ph.D. thesis, Univ. Cape Town, South Africa.

Kirkley, M. B., Smith, H. S., Gurney, J. J. 1989. Kimberlite carbonates—A carbon and oxygen isotope study. In Ross *et al.* (1989) q.v., Vol. 1, pp. 264–281.

Knight, G. L., Landes, K. K. 1932. Kansas laccoliths. *J. Geol.* **40**, 1–15.

Knittel, U. 1987. The Cordon syenite complex: An undersaturated potassic igneous center in the Philippine Island arc. *N. J. Mineral. Abh.* **157**, 57–80.

Knopf, A. 1936. Igneous geology of the Spanish Peaks region. *Bull. Geol. Soc. Am.* **47**, 1728–1784.

Knopf, D., 1970. *Les Kimberlites et les Roches Apparentes de Cote d'Ivoire.* Sodemi, Abidjan.

Kornprobst, J. (Ed.) 1984. *Proceedings of the Third International Kimberlite Conference.* Vol. 1. *Kimberlites I: Kimberlites and Related Rocks.* Vol. 2. *Kimberlites II: The Mantle and Crust-Mantle Relationships.* Developments in Petrology 11A and 11B, Elsevier, New York.

Korstgård, J. A. 1979. Metamorphism of the Kangamiut dykes and the metamorphism and structural evolution of the southern Nagssugtoqidian boundary in the Itivdleq–Ikertoq region, West Greenland. *Rapp. Grøn. Geol. Undersøk.* **89**, 63–75.

Kostyuk, V. P. 1983. The potassic alkalic magmatism of the Baikal Aldan Belt. *Sov. Geol. Geophys.* **24**, 31–38.

Kresten, P., Paul, D. K. 1976. Mineralogy of Indian kimberlites—A thermal and X-ray study. *Can. Mineral.* **14**, 487–490.

Krummenacher, D., Evernden, J. F. 1960. Determination d'age isotopique fautes sur quelques roches des Alpes par le methode K-Ar. *Schweiz. Mineral. Petrog. Mitteil.* **40**, 267–277.

Kuehner, S. M. 1980. Petrogenesis of ultrapotassic rocks, Leucite Hills, Wyoming. M.Sc. thesis, Univ. Western Ontario, London, Ontario.

Kuehner, S. M., Edgar, A. D., Arima, M., 1981. Petrogenesis of the ultrapotassic rocks from the Leucite Hills, Wyoming. *Am. Mineral.* **66**, 663–677.

Kumar, S. 1971. Mining for diamonds at Majhgawan, Panna. *Geol. Surv. India Misc. Pub.* **19**, 163–169.

Kumarapelli, P. S., Saull, V. A. 1966. The St. Lawrence rift system: A North American equivalent of the East African rift valley system. *Can. J. Earth Sci.* **3**, 639–658.

Kunz, G. F. 1908. The occurrence of diamond in North America. *Geol. Soc. Am. Bull.* **17**, 692–694.

Kunz, G. F., Washington, H. S. 1903. On the peridotite of Pike County, Arkansas, and the occurrence of diamond therein. *N. Y. Acad. Sci. Ann.* **18**, 350.

Kutina, J., Carter, W. D. 1977. Landsat contributions to studies of plate tectonics, In Woll, P. W., and Fischer, W. A. (eds.), Proceedings of the first annual W. T. Pecora memorial symposium, October 1975, Sioux Falls, South Dakota, U.S. Geological Survey Prof. Paper 1015, pp. 75–82.

Lacroix, A. 1926. La systematique des roches leucitiques: Les types de la famille syenitique. *C.R. Acad. Sci. Paris* **182**, 597–601.

Lacroix, A. 1933. Contribution a la connaissance de la composition chimique et mineralogique des roches eruptives de l'Indochine. *Bull. Serv. Geol. Indochine* **20**, 1–208.

Lambert, R. St. J. 1984. Earth tectonic and thermal history: Review and a hot spot model for the Archean. In Kroner, A. (ed.), *Precambrian Plate Tectonics.* Elsevier, Amsterdam, pp. 453–467.

Lange, R. A., Stebbins, J. F., Carmichael, I. S. E. 1984. Phase transitions in leucite and kaliophilite. *Geol. Soc. Am. Ann. Mtg. Abstr. with Progr.* **16**, 568–569.

Langford, R. E. 1974. A study of the origin of Arkansas diamonds by mass spectrometry. Ph.D. thesis, Univ. Georgia, Athens GA.

Langworthy, A. P., Black, L. P. 1978. The Mordor complex: A highly differentiated potassic intrusion with kimberlitic affinities in central Australia. *Contrib. Mineral. Petrol.* **67**, 51–62.

Larouzière, F. D. de 1985. Etude tectono-sedimentaire et magmatique des bassins néogènes d'Hinojar et de Mazarron (Cordilleres bétiques internés, Espagne). Theses Univ. Paris VI, 316 pp.

Larouzière, F. D., Bordet, P. 1983. Sur la genése de certains types de lamproites du bassin de Mazarron (Espagne). *C. R. Acad. Sci. Paris* **296**, 1071–1074.

Larsen, E. S. 1940. Petrographic province of central Montana. *Geol. Soc. Am. Bull.* **51**, 887–948.

Larsen, L. M. 1980. Lamprophyric and kimberlitic dykes associated with the Sarfartoq carbonatite complex, southern West Greenland. *Rapp. Gron. Geol. Undersok.* **100**, 65–69.

Larsen, L. M., Rex, D. C., Secher, K. 1983. Age of carbonatites, kimberlites and lamprophyres from southern Greenland: recurrent alkaline magmatism during 2500 million years. *Lithos* **16**, 215–221.

Larsen, W. N. 1954. Precambrian geology of the W. Uinta Mountains, Utah. M.Sc. Thesis, Univ. Utah, Salt Lake City.

Lathram, E. M., Raynolds, R. G. H. 1977. Tectonic deductions from Alaskan space imagery, In Woll, P. W., Fischer, W. A. (eds.), Proceedings of the first annual W. T. Pecora memorial symposium, October 1975, Sioux Falls, South Dakota, *U.S. Geological Survey Prof. Pap.* 1015, pp. 179–192.

Laughlin, A. W., Charles, R. W., Aldrich, M. J. 1989. Heteromorphism and crystallization paths of katungites, Navajo volcanic field, Arizona, U.S.A. In Ross et al. (1989) q.v., Vol. 1, pp. 582–591.

Lazebnik, K. A., Lazebnik, Y. D., Makhotko, V. F. 1984. Davanite, $K_2TiSi_6O_{15}$, a new alkali titanosilicate. *Zap. Vses. Mineral. Obshch.* **118**, 95–97 (Russian).

Lazebnik, K. A., Makhotko, V. F., Lazebnik, Y. D. 1985. The first find of priderite in eastern Siberia. *Mineral. Zh.* **7**, 81–83 (Russian).

Leake, B. E. 1978. Nomenclature of amphiboles. *Am. Mineral.* **63**, 1023–1052.

Le Bas, M. J. 1971. Peralkaline volcanism, crustal swelling and rifting. *Nature* **230**, 85–87.

Le Bas, M. J. 1987. Ultra-alkaline magmatism with and without rifting. *Tectonophysics* **143**, 75–84.

Le Bas, M. J., Le Maitre, R. W., Steckeisen, A., Zanetti, B. 1986. A chemical classification of volcanic rocks based on the total alkali-silica diagram. *J. Petrol.* **27**, 745–750.

Leelanandam, C. 1981. Some observations on the alkaline province in Andhra Pradesh. *Current Sci.* **50**, 799–802.

Le Maitre, R. W. 1982. *Numerical Petrology*. Developments in Petrology, Vol. 8, Elsevier, New York.

Leonardos, O. H., Ulbrich, M. N. C. 1987. Lamproitos de Presidente Oligario. Soc. Brasil. Para. Prog. di Ciencias, 39th. Ann. Mtg. Abstr. 634.

Le Roex, A. P. 1986. Geochemical correlation between southern African kimberlites and south Atlantic hotspots. *Nature* **324**, 243–245.

Leudke, R. G., Smith, R. L. 1978. Map showing the distribution, composition and age of Late Cenozoic volcanic centers in Colorado, Utah and S. W. Wyoming. *U.S. Geol. Surv. Map.* I-1091-B.

Leudke, R. G., Smith, R. L. 1983. Map showing the distribution, composition and age of Late Cenozoic volcanic centers in Idaho, W. Montana, W-central S. Dakota and N.W. Wyoming. *U.S. Geol. Surv. Map* I-1091-E.

Lewis, J. D. 1987. The geology and geochemistry of the West Kimberley lamproite province, Western Australia. M.Sc. thesis, Univ. Western Australia, Perth.

Lewis, R. D. 1977. Mineralogy, petrology and geophysical aspects of Prairie Creek kimberlite, near Murfreesboro, Arkansas. M.Sc. thesis, Purdue Univ., Indiana.

Lewis, R. D., Meyer, H. O. A. 1977. Diamond-bearing kimberlites of Prairie Creek, Murfreesboro, Arkansas. 2nd. Internat. Kimberlite Conf. Santa Fe, Extended Abstr. (unpaginated).

Lewis, R. D., Meyer, H. O. A., Bolivar, S. L., Brookins, D. B. 1976. Mineralogy of the diamond-bearing "kimberlite," Murfreesboro, Arkansas. EOS 57, 761 (abstract).

Lindsley, D. H., Kesson, S. E., Hartzmann, M. J., Cushman, M. W. 1974. The stability of armalcolite: Experimental studies in the system Mg-Fe-Ti-O. *Proc. Lunar Sci. Conf.* **5**, 521–534.

Linthout, K. 1984. Alkali zirconosilicate in peralkaline rocks. *Contrib. Mineral. Petrol.* **86**, 155–158.

Linthout, K., Nobel, F. A., Lustenhouwer, W. J. 1988. First occurrence of dalyite in extrusive rock. *Mineral. Mag.* **52**, 705–708.

Liou, J. G. 1971. Analcime equilibria. *Lithos* **4**, 389–402.

Lipman, P. W., Prosta, H. J., Christiansen, R. L. 1971. Evolving subduction zone in the western United States, as interpreted from igneous rocks. *Science* **174**, 821–825.

Lisenbee, A., Karner, F., Fashbaugh, E., Halvorseon, D., O'Toole, F., White, S., Wilkinson, M., Kirchener, J. 1981. Geology of the Tertiary intrusive province of the northern Black Hills, S. Dakota and Wyoming. In Rich, F. J. (ed.), *Geology of the Black Hills, S. Dakota and Wyoming*. Am. Geol. Inst. Field Trip Guidebk., pp. 33–105.

Lloyd, F. E. 1987. Characterization of mantle metasomatic fluids in spinel lherzolites and alkali clinopyroxenites from the West Eifel and South West Uganda. In Menzies and Hawkesworth (1987) q.v., pp. 91–124.

Lopez Ruiz, J., Rodriguez Badiola, E. 1980. La region volcanica neogena del sureste de España. *Estud. Geol.* **36**, 5–63.

Lorenz, V., McBirney, A. R., Williams, H. 1970. An investigation of volanic depressions III: Maars, tuff-rings, tuff-cones and diatremes. NASA Progr. Rept. NGR-38-003-012, Houston Texas.

Lorenz, V. 1984. Explosive volcanism of the West Eifel volcanic field, Germany. In Kornprobst (1984) q.v., Vol. 1, pp. 299–307.

Lorenz, V. 1985. Maars and diatremes of phreatomagmatic origin, a review. *Trans. Geol. Soc. S. Africa* **88**, 459–470.

Lorenz, V. 1986. On the growth of maars and diatremes and its relevance to the formation of tuff-rings. *Bull. Volcanol.* **48**, 265–274.

Love, J. D., Weitz, J. L., Hose, R. K. 1955. Geological map of Wyoming (scale = 1/500,000), U.S. Geological Survey, Washington.

Lucas, H., Ramsay, R. R., Hall, A. E., Smith, C. B., Sobolev, N. V. 1989. Garnets from Western Australian kimberlites and related rocks. In Ross *et al.* (1989) q.v., Vol. 2, pp. 809–819.

Luhr, J. F., Carmichael, I. S. E. 1981. The Colima volcanic complex, Mexico: Part II. Late Quaternary cinder cones. *Contrib. Mineral. Petrol.* **76**, 127–147.

Luth, W. C. 1967. Studies in the system $KAlSiO_4–Mg_2SiO_4–SiO_2–H_2O$ I. inferred phase relations and petrologic application. *J. Petrol.* **8**, 372–416.

Macdonald, R., Thorpe, R. S., Gaskarth, J. W., Grindrod, A. R. 1985. Multi-component origin of Caledonian lamprophyres of northern England. *Mineral. Mag.* **49**, 485–494.

MacIntyre, R. M., Dawson, J. B. 1976. Age and significance of some South African kimberlites. 4th. Eur. Colloq. Geochron. Cosmochron. Isotope Geol. Amsterdam, Abstr. 66.

MacKenzie, W. S., Richardson, D. M., Wood, B. J. 1974. Solid solution of SiO_2 in leucite. *Soc. Fr. Mineral. Cristallogr. Bull.* **97**, 257–260.

MacNae, J. C. 1979. Kimberlites and exploration geophysics. *Geophysics* **44**, 1395–1416.

Mansker, W. L., Ewing, R. C., Keil, K. 1979. Barian titanian biotites in nephelinites from Oahu, Hawaii. *Am. Mineraol.* **64**, 156–159.

Marinelli, G., Mittempergher, M. 1966. On the genesis of some magmas of typical Meditteranean (potassic) suite. *Bull. Volcanol.* **29**, 113–140.

Marsh, J. S. 1973. Relationships between transform directions and alkaline igneous lineaments in Africa and South America. *Earth Planet. Sci. Lett.* **18**, 317–323.

Marvin, R. F., Hearn, B. C. Jr., Mehnert, H. H., Naeser, C. W., Zartman, R. E., Lindsey, D. A. 1980. Late Cretaceous–Paleocene igneous activity in north-central Montana. *Isochron West* **29**, 5–25.

Mason, B. 1977. Elemental distribution in minerals from the Wolgidee Hills intrusion, Western Australia. *New Zealand D. S. I. R. Bull.* **218**, 114–120.

Mathur, S. M. 1951. Diamond cutting and polishing in Panna, Vindhya Pradesh. *Ind. Minerals* **5**, 125–127.

Mathur, S. M. 1953. Diamond mining and recovery at the Majhgawan mine in Panna, Vindhya Pradesh. *Ind. Minerals* **7**, 34–42.

Mathur, S. M. 1955. Some aspects of diamond mining and milling in Panna. *Ind. Minerals* **9**, 222–228.

Mathur, S. M. 1962. Geology of the Panna diamond deposits. *Rec. Geol. Surv. India* **87**, 787–818.

Mathur, S. M. 1981. The diamond deposits of India—Presidential Address. *Proc. Indian Sci. Cong.* **68**, 1–30.

Mathur, S. M. 1986. Panna revisited. *Indiaqua* **44**, 23–27.

Mathur, S. M., Singh, H. N. 1963. Geology and sampling of the Majhgawan diamond deposit, Panna District, Madhya Pradesh. *Bull. Geol. Surv. India. Ser. A* **21**, 1–58.

Mathur, S. M., Singh, H. N. 1971. Petrology of the Majhgawan pipe rock. *Geol. Surv. India Misc. Publ.* **19**, 78–85.

Matson, R. E. 1960. Petrography and petrology of Smoky Butte Intrusives, Garfield County, Montana. M.Sc. thesis, Montana State Univ. Missoula.

Maughan, E. K. 1983. Tectonic setting of the Rocky Mountain region during the late Paleozoic and the early Mesozoic. *Symposium on the Genesis of the Rocky Mountain Ore Deposits: Changes with Time and Tectonics*. Proc. Denver Reg. Expl. Geol Soc. pp. 39–50.

REFERENCES

McBirney, A. R. 1984. *Igneous Petrology*. Freeman Cooper & Co., San Francisco.
McBirney, A. R., Murase, T. 1984. Physical properties of magmas. *Rev. Earth Planet. Sci.* **12**, 337–357.
McCallum, M. E., Eggler, D. H., Burns, L. K. 1975. Kimberlite diatremes in northern Colorado and southern Wyoming. *Phys. Chem. Earth* **9**, 149–162.
McCandless, T. E. 1982. The mineralogy, morphology and chemistry of detrital minerals of a kimberlitic and eclogitic nature, Green River Basin, Wyoming. M.Sc. thesis, Univ. of Utah, Salt Lake City.
McCandless, T. E. 1984. Detrital minerals of mantle origin in the Green River Basin, Wyoming. Soc. Min. Eng. A.I.M.E. Preprint No. 84-395.
McCulloch, M. T., Jaques, A. L., Nelson, D. R., Lewis, J. D., 1983. Nd and Sr isotopes in kimberlites and lamproites from Western Australia: An enriched mantle origin. *Nature* **302**, 400–403.
McDowell, F. W. 1966. Potassium–argon dating of cordilleran intrusives. Ph.D. thesis, Columbia Univ., New York.
McDowell, F. W. 1971. K–Ar ages of igneous rocks from the western United States. *Isochron West* **2**, 1–16.
Meen, J. K., Eggler, D. H. 1987. Petrology and geochemistry of the Cretaceous Independence volcanic suite, Absaroka Mountains, Montana: Clues to the composition of the Archean sub-Montana mantle. *Geol. Soc. Am. Bull.* **98**, 238–247.
Megartsi, M. 1985. Le volcanisme mio-plio-quaternaire de l'Oranie nord-occidentale. Theses Sci., Univ. Alger, 296 pp.
Mehr, S. 1952. Further study of the Majhgawan diamond mine, Panna, central India. *Q. J. Geol. Mining Met. Soc. India* **24**, 125–132.
Meijer, A. 1980. Primitive arc volcanism and a boninite series: Examples from the western Pacific island arcs. In Hayes, D. E. (ed.) *The Tectonic and Geologic Evolution of Southeast Asian Seas and Islands*. Am. Geophys. Union Mongr. No. 23, pp. 269–282.
Meijerink, A. M. J., Rao, D. P., Rupke, J. 1984. Stratigraphic and structural development of the Precambrian Cuddapah basin, S. E. India. *Precambrian Res.* **26**, 57–104.
Melton, C. E., Giardini, A. A. 1975. Experimental results and theoretical interpretation of gaseous inclusions found in Arkansas natural diamonds. *Am. Mineral.* **60**, 413–417.
Melton, C. E., Giardini, A. A. 1976. Experimental evidence that oxygen is the principal impurity in natural diamond. *Nature* **263**, 309–310.
Melton, C. E., Giardini, A. A. 1980. The isotopic composition of Ar included in an Arkansas diamond and its significance. *Geophys. Res. Lett.* **7**, 461–464.
Melton, C. E., Salotti, C. A., Giardini, A. A. 1972. The observation of nitrogen, water, carbon dioxide, methane and argon as impurities in natural diamonds. *Am. Mineral.* **57**, 1518–1523.
Menzies, M. A., Hawkesworth, C. J. (eds.) 1987. *Mantle Metasomatism*. Academic Press, London.
Merrill, R. B., Bickford, M. E., Irving, A. J. 1977. The Hills Pond peridotite, Woodson County, Kansas: A richterite-bearing Cretaceous intrusive with kimberlitic affinities. 2nd Internat. Kimberlite Conf. Santa Fe, Extended Abstr. (unpaginated).
Meseguer Pardo, J. 1924. Estudio de los yacimientos de azufre de las provincias de Murcia y Albacete. *Boll. Inst. Geol. España* **45**, 131–214.
Meyer, H. O. A. 1976. Kimberlites of the continental United States: A review. *J. Geol.* **84**, 377–403.
Meyer, H. O. A. 1987. Inclusions in diamonds. In Nixon (1987) q.v., pp. 501–522.
Meyer, H. O. A., Lewis, R. D., Bolivar, S., Brookins, D. G. 1977. Prairie Creek Kimberlite, Murfreesboro, Pike County, Arkansas. 2nd Internat. Kimberlite Conf. Santa Fe, N.M. Extended Abstr. (unpaginated).
Meyer, H. O. A., Mitchell, R. H., Svisero, D. P. 1988. Lamproite in Brazil. *Geol. Soc. Am. Ann. Mtg. Abstr. with Progr.* **20**, A74–A75.
Middlemost, E. A. K. 1975. The basalt clan. *Earth Sci. Rev.* **11**, 337–364.
Middlemost, E. A. K. 1985. *Magmas and Magmatic Rocks*. Longmans, New York.
Middlemost, E. A. K., Paul, D. K. 1984. Indian kimberlites and the genesis of kimberlites. *Chem. Geol.* **47**, 249–260.
Middlemost, E. A. K., Paul, D. K., Fletcher, I. R. 1988. Geochemistry of the minette–lamproite association from the Indian Gondwanas. *Lithos* **22**, 31–42.
Milanovsky, E. E. 1981. Aulacogens of ancient platforms: Problems of their origin and tectonic development. *Tectonophys.* **73**, 213–248.
Miller, T. P. 1972. Potassium-rich alkaline intrusive rocks of western Alaska. *Geol. Soc. Am. Bull.* **83**, 2111–2128.
Miser, H. D. 1914. New areas of diamond-bearing peridotite in Arkansas. *U.S. Geol. Surv. Bull. Contrib. Econ. Geol.* **1912**(1), 534–546.
Miser, H. D., Purdue, A. H. 1929. Geology of the Dequeen and Caddo Gap quadrangles, Arkansas. *U.S. Geol. Surv. Bull.* **808**, 99–115.

Miser, H. D., Ross, C. S. 1922. Diamond-bearing peridotite in Pike county, Arkansas. *Econ. Geol.* **17**, 662–674.
Miser, H. D., Ross, C. S. 1923a. Diamond-bearing peridotite in Pike County, Arkansas. *U.S. Geol. Surv. Bull.* **735-I**, 279–322.
Miser, H. D., Ross, C. S. 1923b. Peridotite dikes in Scott County, Arkansas. *U.S. Geol. Surv. Bull.* **735-H**, 271–278.
Miser, H. D., Ross, C. S. 1923c. Volcanic rocks in the Upper Cretaceous of southwestern Arkansas and southeast Oklahoma. *Am. J. Sci.* **9**, 113–126.
Mitchell, A. H. G., Garson, M. S. 1981. *Mineral Deposits and Global Tectonic Settings.* Academic Press, New York.
Mitchell, R. H. 1979. The alleged kimberlite–carbonatite relationship: Additional contrary mineralogical evidence. *Am. J. Sci.* **279**, 570–589.
Mitchell, R. H. 1981. Titaniferous phlogopites from the leucite lamproites of the West Kimberley area, Western Australia. *Contrib. Mineral. Petrol.* **76**, 243–251.
Mitchell, R. H. 1983. Lamproites: petrography and mineralogy. Symp. Mantle metasomatism and the origin of ultrapotassic and related rocks. Univ. Western Ontario, London (extended abstract).
Mitchell, R. H. 1985. A review of the mineralogy of lamproites. *Trans. Geol. Soc. S. Africa* **88**, 411–437.
Mitchell, R. H. 1986. *Kimberlites: Mineralogy, Geochemistry and Petrology.* Plenum Press, New York.
Mitchell, R. H. 1988. The lamproite clan of potassic rocks. *Zap. Vses. Mineral. Obshch.* **117**, 575–586. (Russian)
Mitchell, R. H. 1989a. Aspects of the petrology of kimberlites and lamproites: some definitions and distinctions. In Ross et al. (1989) q.v., Vol. 1, pp. 7–45.
Mitchell, R. H. 1989b. Compositional variation of micas from the Leucite Hills lamproites. 28th. Internat. Geol. Congr. Washington, Extended Abstr. 2, pp. 446–447.
Mitchell, R. H., Bell, K. 1976. Rare earth geochemistry of potassic lavas from the Birunga and Toro-Ankole regions of Uganda. *Contrib. Mineral. Petrol.* **58**, 293–303.
Mitchell, R. H., Brunfelt, A. O. 1975. Rare earth geochemistry of kimberlite. *Phys. Chem. Earth* **9**, 671–686.
Mitchell, R. H., Haggerty, S. E. 1986. A new K–V–Ba titanate related to priderite from the New Elands kimberlite, South Africa. *N.J. Mineral. Monats.* **1986**, 376–384.
Mitchell, R. H., Hawkesworth, C. J. 1984. Geochemistry of potassic lavas from Smoky Butte, Montana. *Geol. Soc. Am. Ann. Mtg. Abstr. with Progr.* **16**, 597.
Mitchell, R. H., Lewis, R. D. 1983. Priderite-bearing xenoliths from the Prairie Creek mica peridotite, Arkansas. *Can. Mineral.* **21**, 59–64.
Mitchell, R. H., Meyer, H. O. A. 1980. Mineralogy of micaceous kimberlite from the Jos dike, Somerset Island, N.W.T., Canada. *Can. Mineral.* **18**, 241–250.
Mitchell, R. H., Meyer, H. O. A. 1989a. Mineralogy of micaceous kimberlites from the New Elands and Star Mines, Orange Free State, South Africa. In Ross et al. (1989) q.v., Vol. 1, pp. 83–96.
Mitchell, R. H., Meyer, H. O. A. 1989b. Niobian K–Ba–V titanates from micaceous kimberlite, Star Mine, Orange Free State, South Africa. *Mineral. Mag.* **53**, 451–456.
Mitchell, R. H., Platt, R. G. 1978. Mafic mineralogy of ferroaugite syenite from the Coldwell alkaline complex, Ontario, Canada. *J. Petrol.* **19**, 627–651.
Mitchell, R. H., Platt, R. G. 1979. Nepheline-bearing rocks from the Poohbah Lake complex, Ontario: Malignites and malignites. *Contrib. Mineral. Petrol.* **69**, 255–264.
Mitchell, R. H., Platt, R. G., Downey, M. 1987. Petrology of lamproites from Smoky Butte, Montana. *J. Petrol.* **28**, 645–677.
Mitchell, R. H., Reed, S. J. B. 1988. Ion microprobe determination of rare earth elements in perovskites from kimberlites and alnöites. *Mineral. Mag.* **52**, 331–339.
Mitra, B. 1953. Some aspects of the petrology of the Rajmahal traps. *Q. J. Geol. Min. Met. Soc. India* **25**, 73–83.
Miyashiro, A. 1986. Hot regions and the origin of marginal basins in the western Pacific. *Tectonophys.* **122**, 195–216.
Miyashiro, A., Aki, K., Sengor, A. M. C. 1982. *Orogeny*, John Wiley, Chichester.
Modreski, P. J., Boettcher, A. L. 1972. The stability of phlogopite and enstatite at high pressures: A model for micas in the interior of the Earth. *Am. J. Sci.* **272**, 852–869.
Modreski, P. J., Boettcher, A. L. 1973. Phase relationships of phlogopite in the system K_2O-MgO-CaO-Al_2O_3-SiO_2-H_2O to 35 kilobars: A better model for micas in the interior of the Earth. *Am. J. Sci.* **273**, 385–414.
Molin, D. 1980. Les caracteres du volcanisme néogène des provinces de Murcia et Almeria (Espagne): Implications pour l'evolution de la Mediterranee occidentale. *Reun. Ann. Sci. Terre. Progr. Abstr.* **2**, 292.
Montenat, C. 1973. Les formations néogènes et quaternaires du Levant espagnol (Provinces d'Alicante et de Murcia). Thesis, Univ. Paris Sud.

Montenat, C. 1975. Le néogène des Cordilleres beltiques. Essai de synthese stratigraphique et paleogeographique. Rapport Beicip Inedit, pp. 1–187.
Montenat, C. 1977. Les bassin néogènes du Levant d'Alicante et de Murcia (Cordilleres betiques orientales) stratigraphie, paleogeographie et evolution dynamique. Doc. Lab. Geol. Fac. Sci. Lyon No. 69, 1–345.
Moody, C. L. 1949. Mesozoic igneous rocks of northern Gulf Coast plain. *Am. Assoc. Petrol. Geol. Bull.* **33**, 1410–1428.
Mooney, S. 1984. Geochemistry of micas from some Australian lamproites and MARID-suite xenoliths, Bultfontein, South Africa. H.B.Sc. Thesis, Lakehead Univ. Thunder Bay, Ontario.
Morgan, W. J. 1981. Hotspot tracks and the opening of the Atlantic and Indian oceans. In Emiliani, C. (ed.), *The Sea*, Vol. 7, Wiley, New York.
Morgan, W. J. 1968. Rises, trenches, great faults and crustal blocks. *J. Geophys. Res.* **73**, 1959–1982.
Morgan, W. J. 1972. Deep mantle convection plumes and plate motions. *Am. Assoc. Petrol. Geol. Bull.* **56**, 203–213.
Morris, E. C. 1953. Geology of the Big Piney area, Summit County, Utah. M.Sc. thesis, Univ. Utah, Salt Lake City.
Morris, E. M. 1987. The Cretaceous Arkansas alkalic province: A summary of petrology and geochemistry. In Morris and Pasteris (1987) q.v., pp. 217–233.
Morris, E. M., Pasteris, J. D. (eds.) 1987. *Mantle Metasomatism and Alkaline Magmatism*. Geol. Soc. Am. Spec. Pap. 215.
Morrison, G. W. 1980. Characteristics and tectonic setting of the shoshonite rock association. *Lithos* **13**, 97–108.
Mossman, D. J. 1976. Dykes, their relationship to rifting, basement, orogenic trends and metallogenesis in the Luangwa Valley of the southern rift zone, northeastern Zambia. *Geol. Assoc. Can. Spec. Pap.* **14**, pp. 321–338.
Mueller, P. A., Wooden, J. L., Odom, A. L., Bowes, D. R. 1982. Geochemistry of the Archean rocks of the Quad Creek and Hellroaring Plateau area of the eastern Beartooth Mountains. In Mueller, P. A., and Wooden, J. L. (eds.) *Precambrian Geology of the Beartooth Mountains, Montana and Wyoming*. Montana Bur. Mines Geol. Spec. Publ. 84, pp. 69–82.
Mueller, P. A., Wooden, J. L., Schulz, K., Bowes, D. R. 1983. Incompatible element-rich andesitic amphibolites from the Archean of Montana and Wyoming: evidence for mantle metasomatism. *Geology* **11**, 203–206.
Mukherjee, K. K. 1961. Petrology of lamprophyres of the Bokaro coalfield, Bihar. *Q. J. Geol. Mining Metall. Soc. India* **33**, 69–87.
Murayama, J. K., Nakai, S., Kato, M., Kumazawa, M. 1986. A dense polymorph of $Ca_3(PO_4)_2$: A high pressure phase of apatite decomposition and its geochemical significance. *Phys. Earth Planet. Inter.* **44**, 293–303.
Murta, R. L. L. 1965. Nota sobre a roche leucitica de Sacramento, Minas Gerais, Cien. Cult., S. Paulo 17, p. 135.
Murthy, Y. G. K., Babu Rao, V., Guptasarma, D., Rao, J. M., Rao, M. N., Bhattacharji, S. 1987. Tectonic, petrochemical and geophysical studies of mafic dyke swarms around the Proterozoic Cuddapah basin, S. India. In Halls, H. C., and Fahrig, W. F. (eds.) *Mafic Dyke Swarms*. Geol. Assoc. Can. Spec. Pap. 34, pp. 303–316.
Murthy, Y. G. K., Rao, M. G., Misra, R. C., Reddy, T. A. K. 1980. Kimberlite diatremes of Andhra Pradesh—Their assessment and search for concealed bodies. Seminar on Diamonds. Geol. Surv. India, Bombay (unpaginated).
Myrha, S., White, T. J., Kesson, S. E., Riviere, C. J. 1988. X-ray photoelectron spectroscopy for direct identification of Ti valence in $[Ba_xCs_y][(Ti,Al)^{3+}_{2x+y}Ti^{4+}_{8-2x-y}]O_{16}$ hollandites. *Am. Mineral.* **73**, 161–167.
Mysen, B. O. (Ed.) 1987. *Magmatic Process: Physicochemical Principles*. Geochem Soc. Spec. Publ. No 1. Pennsylvania State Univ.
Mysen, B. O., Boettcher, A. L. 1975a. Melting of a hydrous mantle: I. Phase relations of natural peridotite at high pressures and temperatures with controlled activities of water, carbon dioxide, and hydrogen. *J. Petrol.* **16**, 520–548.
Mysen, B. O., Boettcher, A. L. 1975b. Melting of a hydrous mantle: II. Geochemistry of crystals and liquids formed by anatexis of mantle peridotite at high pressures and high temperatures as a function of controlled activities of water, hydrogen, and carbon dioxide. *J. Petrol.* **16**, 549–593.
Mysen, B. O., Virgo, D. 1978. Influence of pressure, temperature and bulk composition on melt structure in the system $NaAlSi_2O_6$-$NaFe^{3+}Si_2O_6$. *Am. J. Sci.* **278**, 1307–1322.
Mysen, B. O., Virgo, D., Seifert, F. A. 1985. Relationships between properties and structure of aluminosilicate melts. *Am. Mineral.* **70**, 88–105.
Nag, S. 1983. The alkaline rocks of the Eastern Ghats orogenic belt. India. *N. J. Mineral. Abh.* **148**, 97–112.
Nakamura, K. 1974. Preliminary estimate of global volcanic production rate. In Furumento, S., and Yuhara, K. (eds.), *The Utilization of Volcanic Energy*. Sandia Lab, Albuquerque, NM, pp. 273–284.
Nane, S. G. 1971. The Angore ultramafic intrusive and its potential with respect to diamonds: A review. *Geol. Surv. India Spec. Publ.* **19**, 109–113.

Naqvi, S. M., Rogers, J. J. W. 1987. *Precambrian Geology of India*. Clarendon Press, Oxford.
Nayak, S. S., Kasivisiwanathan, C. V., Reddy, T. A. K., Nagaraja Rao, B. K. 1988. New find of kimberlitic rocks in Andhra Pradesh, near Maddur, Mahaboobnagar Distrct. *J. Geol. Soc. India* **31**, 343–346.
Nelson, D. R. 1989. Isotopic characteristics and petrogenesis of the lamproites and kimberlites of central West Greenland. *Lithos* **22**, 265–274.
Nelson, D. R., McCulloch, M. T. 1989. Enriched mantle components and mantle recycling of sediments. In Ross *et al.* (1989) q.v., Vol. 1, pp. 560–570.
Nelson, D. R., McCulloch, M. T., Sun, S. S. 1986. The origins of ultrapotassic rocks as inferred from Sr, Nd, and Pb isotopes. *Geochim. Cosmochim. Acta* **50**, 231–245.
Němec, D. 1972. Micas of the lamprophyres of the Bohemian Massif. *N. J. Mineral. Abh.* **117**, 196–216.
Němec, D. 1973. Amphibole der lamprophyrischen und lamproiden Gestine in Ostfeil der Böhmischen Masse. *Cas. Mineral. Geol.* **18**, 31–46.
Němec, D. 1974. Petrochemistry of the dyke rocks of the Central Bohemian Pluton. *N.J. Mineral. Monats.* **1974**, 193–209.
Němec, D. 1975. Petrochemie und Genese der lamprophyrischen und lamproiden Ganggesteine im Nordostteil der Böhmischen Masse (CSSR). *Z. Geol. Wissen. Berlin* **3**, 37–52.
Němec, D. 1985. Ba and its carriers in dyke rocks of the minette series. Geol. Assoc. Can. Mineral. Assoc. Can. Ann. Mtg. Progr. with Abstr. **10**, A43.
Němec, D. 1987. Barium in dyke rocks of the minette series. *Chem. Erde* **47**, 117–124.
Němec, D. 1988. The amphiboles of potassium-rich dyke rocks of the southeastern border of the Bohemian massif. *Can. Mineral.* **26**, 89–95.
Newton, M. G., Melton, C. E., Giardini, A. A. 1977. Mineral inclusion in an Arkansas diamond. *Am. Mineral.* **62**, 583–586.
Nickel, E. H., Grey, I. E., Madsen, I. C. 1987. Lucasite-Ce, $CeTi_2(O,OH)_6$, a new mineral from Western Australia: Its description and structure. *Am. Mineral.* **72**, 1006–1010.
Nicolaysen, L. O. 1985. On the physical basis for the extended Wilson cycle, in which most continents coalesce and then disperse again. *Trans. Geol. Soc. S. Africa* **88**, 562–580.
Nicoll, R. S. 1981. Conodont colour alteration adjacent to a volcanic plug, Canning basin, Western Australia. *Aust. Bur. Min. Res. J. Geol. Geophys.* **6**, 265–267.
Nielsen, B. L. 1976. Economic minerals, diamonds, olivine, garnets in Greenland. In Escher and Watt (1976) q.v., pp. 481–486.
Niggli, P. 1920. *Lehrbuch der Mineralogie*. Verlag Gebrüder Borntraeger, Berlin. 476 pp.
Niggli, P. 1923. *Gesteins- und Mineralprovinzen*. Verlag Gebrüder Borntraeger, Berlin. 586 pp.
Niggli, P. 1931. Die quantitative mineralogische Klassifikation der Eruptivgesteine. *Schweiz. Mineral. Petrol. Mitteil.* **11**, 296–364.
Nixon, P. H. 1980. Regional diamond exploration—Theory and practice. In Glover, J. E., and Groves D. I. (eds.) *Kimberlites and Diamonds*. Geol. Dept. Univ. Extension Univ. Western Australia, Perth, Pub. 5, pp. 65–80.
Nixon, P. H. (Ed.) 1987. *Mantle Xenoliths*. J. Wiley & Sons, New York.
Nixon, P. H., Bergman, S. C. 1987. Anomalous occurrences of diamonds. *Indiaqua* **47**, 21–27.
Nixon, P. H., Boyd, F. R., Lee, D. C. 1987. Western Australia—Xenoliths from kimberlites and lamproites. In Nixon (1987) q.v., pp. 281–286.
Nixon, P. H., Davies, G. R. 1987. Mantle xenolith perspectives. In Nixon (1987) q.v., pp. 741–756.
Nixon, P. H., Davies, G. R., Condliffe, E., Baker, R., Baxter Brown, R. 1989. Discovery of ancient source rocks of Venezuela diamonds. Workshop On Diamonds. 28th. Internat. Geol. Congr. Washington Extended Abstr. pp. 73–75.
Nixon, P. H., Mitchell, R. H., Rogers, N. W. 1980. Petrogenesis of alnöitic rocks from Malaita, Solomon Islands, Melanesia. *Mineral. Mag.* **43**, 587–596.
Nixon, P. H., Thirwall, M. F., Buckley, F. 1982. Kimberlite-lamproite consanguinity. *Terra Cognita* **2**, 252–254.
Nixon, P. H., Thirwall, M. F., Buckley, F., Davies, C. J. 1984. Spanish and Western Australian lamproites: Aspects of whole rock chemistry. In Kornprobst (1984)q.v., Vol. 1, pp. 285–296.
Nobel, F. A., Andriessen, P. A. M., Hebeda, E. H., Priem, H. N. A., Rondeel, H. E. 1981. Isotopic dating of the post-Alpine Neogene volcanism in the Betic Cordilleras, southern Spain. *Geol. Minjb.* **60**, 209–214.
Nockolds, S. R. 1940. Petrology of rocks from Queen Mary Land. Aust. Antarct. Exped. 1911–1914. Rept. Ser. A, IV, Part 2, pp. 15–86.
Noe-Nygaard, A., Ramberg, H. 1961. Geological reconnaissance map of the country between latitudes 69°N and 63°45N, West Greenland. *Medd. Grønland* **123**, 3.

REFERENCES

Norrish, K., 1951. Priderite, a new mineral from the leucite lamproites of the West Kimberley area, Western Australia. *Mineral. Mag.* **24**, 496–501.
Novgorodova, M. I., Galuskin, Y. V., Boyarskaya, R. V., Mokhov, A. V. 1987. Accessory minerals in lamproite-like rocks from central Asia. *Internat. Geol. Rev.* **29**, 295–306.
O'Brien, H. E., Irving, A. J., McCallum, I. S. 1988. Complex zoning and resorption of phenocrysts in mixed potassic mafic magmas of the Highwood Mountains, Montana. *Am. Mineral.* **73**, 1007–1024.
O'Driscoll, E. S. T. 1986. Observations of the lineament-ore relation. *Phil. Trans. R. Soc. London* **A317**, 195–218.
Ogden, P. R. 1978. Evidence for contamination in the petrogenesis of madupite. Leucite Hills, Wyoming. *Geol. Soc. Am. Ann. Mtg. Abstr. with Progr.* **10**, 465.
Ogden, P. R. 1979. The geology, major element geochemistry and petrogenesis of the Leucite Hills volcanic rocks, Wyoming. Ph.D. thesis, Univ. Wyoming, Laramie.
Ogden, P. R., Gunter, W. D., Fundry, C. B. 1977. A new occurrence of madupite: Leucite Hills, Wyoming. *Geol. Soc. Am. Ann. Mtg. Abstr. with Progr.* **9**, 754.
Ogden, P. R., Sperr, J. T., Gunter, W. D., Pajari, G. E. 1978. Morphology of a recent ultrapotassic volcanic field, Leucite Hills, S. W. Wyoming. *Geol. Soc. Am. Ann. Mtg. Abstr. with Progr.* **10**, 140.
O'Hara, M. J. 1968. The bearing of phase equilibria studies in synthetic and natural systems on the origin and evolution of basic and ultrabasic rocks. *Earth Sci. Rev.* **4**, 69–133.
O'Hara, M. J., Yoder, H. S. 1967. Formation and fractionation of basic magmas at high pressure. *Scottish J. Geol.* **3**, 67–117.
Ollier, C. D. 1967. Maars, their characteristics, varieties and definitions. *Bull. Volcanol.* **31**, 45–73.
O'Neill, H. St. C., Jaques, A. L., Smith, C. B., Moon, J. 1986. Diamond-bearing peridotite xenoliths from the Argyle (AK1) pipe. 4th Internat. Kimbelite Conf. Perth, Western Australia, Extended Abstr. 300–302.
O'Nions, R. K., Carter, S. R., Evensen, N. H., Hamilton, P. J. 1979. Geochemical and cosmochemical applications of Nd isotope analysis. *Ann. Rev. Earth Planet. Sci.* **7**, 11–38.
Onuoha, K. M. 1985. Rheological properties of parts of the African lithosphere and their geodynamical significance. *J. African Earth Sci.* **3**, 437–442.
Orlova, M. P. 1987. New information on the geology of the Malo-Murun alkaline massif. *Sov. Geol.* **1987**, 83–92 (Russian).
Orlova, M. P. 1988. Petrochemistry of the Malo-Murun alkaline massif, Yakutia. *Izv. Akad. Nauk SSR Ser. Geol.* **1988**, 15–27 (Russian).
Osann, A. 1889. Beitrage Zur Kentniss der Eruptivegesteine des Cabo de Gata I. *Z. Deutsch Geol. Gesellschaft* **41**, 297–311.
Osann, A. 1906. *Über einige Alkaligestein aus Spanien*. Festchr. Rosenbusch, Stuttgart, pp. 283–301.
Ott d'Estévou, P., Montenat, C. 1985. Evolution structurale de la zone bétique orientale (Espagne) du Tortonien a l'Holocene. *C.R. Acad. Sci. Paris II* **300**, 363–368.
Page, R. W. 1988. Geochronology of early-to-middle Proterozoic fold belts in northern Australia: Review. *Precambrian Res.* **40/41**, 1–19.
Page, R. W., Hannock, S. L. 1988. Geochronology of a rapid 1.85–1.86 Ga tectonic transition: Halls Creek orogen, northern Australia. *Precambrian Res.* **40/41**, 447–467.
Pantaleo, N. S., Newton, G. S., Gogineni, S. V., Melton, C. E. 1979. Mineral inclusions in four Arkansas diamonds: Their nature and significance. *Am. Mineral.* **64**, 1059–1062.
Parga Pondal, I. 1935. Quimismo de las manifestaciones magmaticas Cenozoicas de la Peninsula Iberica. Trab. Museo Nac. Cien. Nat. Ser. Geol. 39, 174 pp.
Patchett, P. J. 1978. Rb/Sr ages of Precambrian dolerites and syenites in southern and central Sweden. *Sver. Geol. Unders.* **C747**, 1–63.
Paterson, N. R., MacFadyen, D. A., Turkei, A. 1977. Geophysical exploration for kimberlites, with special reference to Lesotho. *Geophysics* **42**, 1531.
Paul, D. K. 1979. Isotopic composition of strontium in Indian kimberlites. *Geochim. Cosmochim. Acta* **43**, 389–394.
Paul, D. K. 1980. Indian diamonds and kimberlite. In Glover, J. E., and Groves D. I. (eds.) *Kimberlites and Diamonds*. Geol. Dept., Univ. Extension, Univ. Western Australia, Pub. 5, pp. 15–23.
Paul, D. K., Crocket, J. H., Nixon, P. H. 1979. Abundance of palladium, iridium and gold in kimberlites and associated nodules. In Boyd and Meyer (1976) q.v., Vol. 1, pp. 272–279.
Paul, D. K., Gale, N. H., Harris, P. G. 1977. Uranium and thorium abundances in Indian kimberlites. *Geochim. Cosmochim. Acta* **41**, 335–339.
Paul, D. K., Potts, P. J. 1981. Rare earth abundances and origin of some Indian lamprophyres. *Geol. Mag.* **118**, 393–399.

Paul, D. K., Rex, D. C., Harris, P. G. 1975. Chemical characteristics and K–Ar ages of Indian kimberlites. *Geol. Soc. Am. Bull.* **86**, 364–366.

Paul, D. K., Sarkar, A. 1986. Petrogenesis of some Indian lamprophyres. *Geol. Surv. India Spec. Publ.* **12**, 45–54.

Peacor, D. R. 1968. A high temperature single crystal diffraction study of leucite $(K,Na)AlSi_2O_6$. *Z. Krist.* **127**, 213–224.

Pearce, T. H. 1968. A contribution to the theory of variation diagrams. *Contrib. Mineral. Petrol.* **19**, 142–157.

Peccerillo, A., Manetti, P. 1985. The potassic alkaline volcanism of central southern Italy; a review of the data relevant to petrogenesis and geodynamic significance. *Trans. Geol. Soc. S. Africa* **88**, 379–394.

Peccerillo, A., Poli, G., Serri, G. 1988. Petrogenesis of orenditic and kamafugitic rocks from central Italy. *Can. Mineral.* **26**, 45–65.

Pellicer, M. J. 1973. Estudio petrologico y geoquimico de un nuevo yacimiento de rocas lamproiticas situado en las proximades de Aljorra (Murcia). *Estud. Geol.* **29**, 99–106.

Peterman, Z. E. 1979. Geochronology and the Archean of the United States. *Econ. Geol.* **74**, 1544–1562.

Phillipi, E. 1912. Geologische Beschreibung des Gaussbergs. Deutsche Sudpolar Expedition 1901–1903. *Geogr. Geol.* **11**, 47–71.

Phipps, S. P. 1988. Deep rifts as sources for alkaline intraplate magmatism in eastern North America. *Nature* **334**, 27–31.

Pidgeon, R. T., Smith, C. B., Fanning, C. M. 1989. Kimberlite and lamproite emplacement ages in Western Australia. In Ross *et al.* (1989) q.v., Vol. 1, pp. 369–381.

Plumb, K. A. 1979. The tectonic evolution of Australia. *Earth Sci. Rev.* **19**, 205–249.

Plumb, K. A., Gemuts, I, 1976. Precambrian geology of the Kimberley region, Western Australia. 25th. Internat. Geol. Congr. Excursion Guide 44C, p. 1–69.

Plumb, K. A., Derrick, G. M., Needham, R. S., Shaw, R. D. 1981. The Proterozoic of northern Australia. In Hunter, D. R. (ed.), *The Precambrian Geology of the Southern Hemisphere*. Elsevier, Amsterdam, pp. 2045–2307.

Post, J. E., von Dreele, R. B., Buseck, P. R. 1982. Symmetry and cation displacements in hollandites: structure refinements of hollandites, cryptomelane and priderite. *Acta Cryst.* **B38**, 1056–1065.

Powell, J. L., Bell, K. 1970. Strontium isotopic studies of alkalic rocks. Localities from Australia, Spain, and the western United States. *Contrib. Mineral. Petrol.* **27**, 1–10.

Powell, W. B. 1842. Geological report upon the Fourche Cove and its immediate vicinity. Private Publ. by Author, Little Rock, AR, 22 pp.

Prider, R. T. 1939. Some minerals from the leucite-rich rocks of the West Kimberley area, Western Australia. *Mineral. Mag.* **25**, 373–387.

Prider, R. T. 1960. The leucite lamproites of the Fitzroy Basin, Western Australia. *J. Geol. Soc. Aust.* **6**, 71–118.

Prider, R. T. 1965. Noonkanbahite, a potassic batisite from the lamproites of Western Australia. *Mineral. Mag.* **34**, 403–405.

Prider, R. T. 1969. Phanerozoic volcanism in Western Australia. *Spec. Publ. Geol. Soc. Aust.* **2**, 123–125.

Prider, R. T. 1982. A glassy lamproite from the West Kimberley area, Western Australia. *Mineral. Mag.* **45**, 279–282.

Prider, R. T., Cole, W. F. 1942. The alteration products of olivine and leucite in the leucite-lamproites from the West Kimberley area, Western Australia. *Am. Mineral.* **27**, 373–384.

Primel, L. 1963. Etude geologique et metallogenique de la partie meridionale du cap Corse. Ph.D. thesis, Univ. Paris.

Pring, A., Jefferson, D. A. 1983. Incommensurate superlattice ordering in priderite. *Mineral. Mag.* **47**, 65–68.

Pryce, M. W., Hodge, L. C., Criddle, A. 1984. Jeppeite, a new K–Ba–Fe titanate from the Walgidee Hills, Western Australia. *Mineral. Mag.* **48**, 263–266.

Puri, S. N. 1972. A note on the Angor kimberlite. *Ind. Mineral.* **26**, 133–134.

Radde, G. 1987. Davinite, $K_2TiSi_6O_{15}$, in the Smoky Butte (Montana) lamproites: Discussion of X-ray powder data. *Am. Mineral.* **72**, 1014–1015.

Radnakrishna, B. P., Naqvi, M. 1986. Precambrian continental crust of India and its evolution. *J. Geol.* **94**, 145–166.

Rahmann, M. A. A., Wright, R. L. 1976. Air photo lineations, joints and bedding in part of southeast Spain. *Proc. Internat. Conf. New Basement Tectonics* Vol. 2, pp. 225–234.

Ramberg, I. B., Morgan, P. 1984. Physical characteristics and evolutionary trends of continental rifts. *Proc. 27th Internat. Geol. Congress*, Vol. 7, pp. 165–216, VNU Press, Utrecht, The Netherlands.

Ravich, M. G., Krylov, A. H. 1964. Absolute ages of rocks from Eastern Antarctica. In Adie, R. J. (ed.) *Antarctic Geology*. North-Holland, Amsterdam, pp. 578–589.

Reddy, T. A. K. 1987. Kimberlite and lamproite rocks of the Vajrakarur area, Andhra Pradesh. *J. Geol. Soc. India* **30**, 1–12.
Reichenbach, I., Parrish, R. R. 1988. Age of crystalline basement in British Columbia and Northwest Territories, Canada, and Colorado and Arkansas, U.S.A., as inferred from U–Pb zircon geochronology of diatremes. *Geol. Soc. Am. Ann. Mtg. Abstr. with Progr.* **20**, A110.
Reid, D. L., Barton, E. S. 1983. Geochemical characterization of granitoids in the Namaqualand geotraverse. *Geol. Soc. S. Africa Spec. Publ.* **10**, 67–82.
Reinisch, R. 1912. Petrographische Beschreibungder Gaussberg-Gesteins. Deutsch Sudpolar-Expedition 1901–1903. *Geogr. Geol.* **II**, 73–78.
Rex, D. C. 1972. K–Ar age determinations on volcanic and associated rocks from the Antarctic Peninsula and Dronning Maud Land. In Adie, R. J. (ed.) *Antarctic Geology and Geophysics.* Universitetsforlaget, Oslo, pp. 133–136.
Rhodes, R. C. 1971. Structural geometry of subvolcanic ring complexes as related to pre-Cenozoic motions of continental plates. *Tectonophys.* **12**, 111–117.
Richardson, S. W. 1986. Latter-day origins of diamonds of eclogitic paragenesis. *Nature* **322**, 623–626.
Ringwood, A. E. 1989. Constitution and evolution of the mantle. In Ross *et al.* (1989) q.v., Vol. 1, pp. 457–485.
Rittman, A. 1973. *Stable Mineral Assemblages of Igneous Rocks.* Springer-Verlag, Berlin.
Robert, J. L. 1976. Titanium solubility in synthetic phlogopite solid solutions. *Chem. Geol.* **17**, 213–227.
Robert, J. L. 1981. Etudes cristallochimiques sur les micas et les amphiboles. Applications à la petrographie et à la geochemie. Thèse de Doctorat d'Etat. Univ. Paris-Sud, Orsay.
Robertson, D., Fox, J. J., Martin, A. E. 1934. Two types of diamond. *Phil. Trans. R. Soc. London* **A232**, 463–538.
Robey, J. V. A., Bristow, J. W., Marx, M. R., Joyce, J., Danchin, R. V., Arnot, F. 1986. Alkalic ultrabasic dykes of the south-east Yilgarn margin, West Australia. 4th. Internat. Kimberlite Conf. Perth, Western Australia, Extended Abstr. pp. 142–144.
Rock, N. M. S. 1977. The nature and origin of lamprophyres: Some definitions, distinctions and derivations. *Earth Sci. Rev.* **13**, 123–169.
Rock, N. M. S. 1982. The Late Cretaceous alkaline igneous province in the Iberian Peninsula and its tectonic significance. *Lithos* **15**, 111–131.
Rock, N. M. S. 1984. Nature and origin of calc-alkaline lamprophyres: Minettes, vogesites, kersantites, spessartites. *Trans. R. Soc. Edinburgh* **74**, 193–227.
Rock, N. M. S. 1986. The nature and origin of lamprophyres: Alnöites and allied rocks (ultramafic lamprophyres). *J. Petrol.* **27**, 155–196.
Rock, N. M. S. 1987. The nature and origin of lamprophyres: An overview. In Fitton and Upton (1987) q.v., pp. 191–226.
Rock, N. M. S. 1989. Kimberlites as varieties of lamprophyres: Implications for geological mapping, petrological research and mineral exploration. In Ross *et al.* (1989) q.v., Vol. 1, pp. 46–59.
Roden, M. F. 1981. Origin of coexisting minette and ultramafic breccia, Navajo volcanic field. *Contrib. Mineral. Petrol.* **77**, 195–206.
Roedder, E. 1984. *Fluid Inclusions.* Rev. Mineral. 12, Mineral. Soc. Am., Washington, D.C.
Rogers, N. W., Bachinski, S. W., Henderson, P., Parry, S. J. 1982. Origin of potash-rich basic lamprophyres: Trace element data from Arizona minettes. *Earth Planet. Sci. Lett.* **57**, 305–312.
Rosenbusch, H. 1877. *Mikroskopische Physiographie der Mineralien und Gesteine. II. Mikroskopische Physiographie der Massigen Gesteine.* E. Schweizerbartsche Verlagshandlung, Stuttgart, 1st Ed.
Rosenbusch, H. 1899. Über Euktolith, ein neues Glied der Theralithischen Effusivmagmen. *Sitzb. Akad. Wiss. Berlin* **1889**, 110–112.
Ross, C. P., Andrews, D. A., Witkind, I. J. 1955. Geological map of Montana (1:500,000 scale) Montana Bur. Mines. Geol.
Ross, C. S. 1926a. Nephelite-hauynite alnöite from Winnett, Montana. *Am. J. Sci.* **11**, 218–227.
Ross, C. S. 1926b. A Colorado lamprophyre of the verite type. *Am. J. Sci.* **12**, 217–229.
Ross, C. S., Miser, H. O., Stephenson, L. W. 1929. Water-laid volcanic rocks of early Upper Cretaceous age in southwestern Arkansas, southern Oklahoma and northeastern Texas. *U.S. Geol. Surv. Prof. Pap.* 154F, pp. 175–202.
Ross, J., Jaques, A. L., Ferguson, J., Green, D. H., O'Reilly, S. Y., Danchin, R. V., Janse, A. J. A. (eds.) 1989. *Proceedings of the Fourth International Kimberlite Conference. Kimberlites and Related Rocks.* Vol. 1. *Their Composition, Occurrence, Origin and Emplacement;* Vol. 2. *Their Mantle/Crust Setting, Diamonds and Diamond Exploration.* Geol. Soc. Aust. Spec. Publ. No. 14.

Rowell, W. F., Edgar, A. D. 1983. Cenozoic potassium-rich mafic volcanism in the western U.S.A.: Its relationship to deep subduction. *J. Geol.* **91**, 338–341.

Ruddock, D. I., Hamilton, D. L. 1978b. The system $KAlSi_2O_6-CaMgSi_2O_6-SiO_2$ at 4 kb. *Progr. Exp. Petrol.* N.E.R.C. Publ. Ser. D11, pp. 25–27.

Rutland, R. W. R. 1973. Tectonic evolution of the continental crust of Australia. In Tarling, D. H., and Runcorn, S. K. (eds.). *Continental Drift, Sea Floor Spreading and Plate Tectonics.* Academic Press, London, pp. 1011–1033.

Ryabchikov, I. D., Green, D. H. 1978. The role of CO_2 in the petrogenesis of highly potassic magmas. *Trudy Inst. Geol. Geof. Akad. Nauk SSSR, Novosibirsk* **403**, 49–64 (Russian).

Sabatini, V. 1899. I vulcani di S. Venanzo. *Boll. R. Comm. Geol. Italia* **30**, 60.

Sabatini, V. 1903. La pirossenite melilitica di Coppaeli. *Boll. R. Comm. Geol. Italia* **35**, 376–378.

Saether, E. 1950. On the genesis of peralkaline rock provinces. 18th Internat. Geol. Congr. London, Part II, pp. 123–130.

Sahama, Th. G. 1974. Potassium-rich alkaline rocks. In Sörensen, H. (ed.), *The Alkaline Rocks.* Wiley, New York, pp. 96–109.

Sakuntala, S., Krishna Brahmann, N. 1984. Diamond mines near Raichur. *J. Geol. Soc. India* **25**, 780–786.

Salpas, P. A., Taylor, L. A., Shervais, J. W. 1985. The Blue Ball, Arkansas kimberlite: Mineralogy, petrology and geochemistry. *J. Geol.* **94**, 891–901.

Salter, V. J. M., Barton, M. 1985. The geochemistry of ultrapotassic lavas from the Leucite Hills, Wyoming. *EOS* **66**, 1109 (abstract).

Sandiford, M., Wilson, C. J. L. 1983. The geology of the Fyfe Hills–Khmara Bay region, Enderby Land. In Oliver, R. L., James, P. R., and Jago, J. B. (eds.), *Antarctic Earth Science.* Aust. Acad. Sci., Canberra, pp. 16–19.

San Miguel, M. 1935. Una erupcion dc jumillita en la Sierra de las Cabras (Albacete). *Boll. R. Soc. Esp. Hist. Nat.* **35**, 147–154.

San Miguel, M. 1936. Estudio de las rocas eruptivas de España. *Mem. Acad. Cienc. Exact. Fisic. Natur.* 1–660.

San Miguel, M., Almela, A., Fuster, J. M. 1951. Sobre un volcan de veritas recientemente descubierto en el Miocene de Barqueros (Murcia). *Estud. Geol.* **7**, 411–429.

San Miguel, M. De Pedro, F. 1945. Afloramientas en Puebla de Mula (Murcia). *Notas Comm.* **33**, 9–24.

Sanyal, S. P. 1964. Petrology of certain lamprophyres from the Jharia Coalfield, Bihar, with a discussion on the differentiation of the Sudamdih sill. *Geol. Surv. India Misc. Publ.* **8**, 27–44.

Sarkar, A., Paul, D. K., Balasubrahmanyan, M. N., Sengupta, N. R. 1980. Lamprophyres from Indian Gondwanas: K–Ar ages and chemistry. *J. Geol. Soc. India* **21**, 188–193.

Sarma, B. S. P. 1983. A report on ground magnetic survey over Chelima dyke (Cuddapah basin). Nat. Geophys. Res. Inst. Hyderabad, India, Tech. Report 83-210.

Sathe, R. V., Oka, S. S. 1975. Petrogenesis of lamprophyres of Mt. Girnar, Saurahtra. *Q. J. Geol. Mining Metall. Soc. India* **46**, 61–67.

Satian, M. A., Khanzatyan, G. A. 1987. Rocks of the lamproite series in the ophiolite section of the Vedi ophiolite zone of the Lesser Caucasus. *Izvt. Akad. Nauk Armyanskoy SSR* **40**, 64–67 (Russian).

Scambos, T. A. 1987. Sr and Nd isotope ratios for the Missouri Breaks diatremes, Central Montana. *Geol. Soc. Am. Abstr. Progr.* **19**, 830.

Scarfe, C. M., Luth, W. C., Tuttle, O. F. 1966. An experimental study bearing on the absence of leucite in plutonic rocks. *Am. Mineral.* **51**, 726–735.

Scatena-Watchel, D. E., Jones, A. P. 1984. Primary baddeleyite (ZrO_2) in kimberlite from Benfontein, South Africa. *Mineral. Mag.* **48**, 257–261.

Scheuring, B., Ahrendt, H., Hunziker, J. C., Zing, A. 1974. Paleobotanical and geochronological evidence for the Alpine age of the metamorphism in the Sesia Zone. *Geol. Rundsch.* **63**, 305–326.

Schmahl, W. W., Tillmans, E. 1987. Isomorphic substitutions, straight Si–O–Si geometry and disorder of tetrahedral tilting in batisite, $(Ba,K)(K,Na)Na(Ti,Fe,Nb,Zr)_2Si_4O_{14}$. *N. J. Mineral. Monats.* **1987**, 107–118.

Schreyer, W., Massone, H. J., Chopin, C. 1987. Continental crust subducted to depths near 100 km: Implications for magma and fluid genesis in collision zones. In Mysen (1987) q.v., pp. 155–163.

Schultz, A. R., Cross, W. 1912. Potash-bearing rocks of the Leucite Hills, Sweetwater County, Wyoming. *U.S. Geol. Surv. Bull.* **512**, 1–39.

Schulze, D. J., Smith, J. V., Němec, D. 1985. Mica chemistry of lamprophyres from the Bohemian massif, Czechoslovakia. *N. J. Mineral. Abh.* **152**, 321–334.

Scott, B. H. 1977. Petrogenesis of kimberlites and associated potassic lamprophyres from central West Greenland. Ph.D. thesis, Univ. Edinburgh, U.K.

Scott, B. H. 1979. Petrogenesis of kimberlites and associated potassic lamprophyres from central West Greenland In Boyd and Meyer (1979) q.v., Vol. 1, pp. 190–205.
Scott, B. H. 1981. Kimberlite and lamproite dykes from Holsteinsborg, W. Greenland. *Medd. Grønland Geosci.* **4**, 3–23.
Scott Smith, B. H. 1989. Lamproites and kimberlites in India. *N. J. Mineral.* **161**, 193–225.
Scott Smith, B. H., Skinner, E. M. W. 1982. A new look at Prairie Creek, Arkansas. *Terra Cognita* **2**, 210.
Scott Smith, B. H., Skinner, E. M. W. 1984a. Diamondiferous lamproites. *J. Geol.* **92**, 433–438.
Scott Smith, B. H., Skinner, E. M. W. 1984b. A new look at Prairie Creek, Arkansas. In Kornprobst (1984), q.v., Vol. 1, pp. 255–284.
Scott Smith, B. H., Skinner, E. M. W. 1984c. Kimberlite and American Mines, near Prairie Creek, Arkansas. *Ann. Sci. Univ. Clermont-Ferrand II* **74**, 27–36.
Scott Smith, B. H., Danchin, R. V., Harris, J. W., Stracke, K. J. 1984. Kimberlites near Orroroo, South Australia. In Kornprobst (1984) q.v., Vol. 1, pp. 121–142.
Scott Smith, B. H., Skinner, E. M. W., Loney, P. E. 1989. The Kapamba lamproites of the Luangwa valley, Eastern Zambia. In Ross et al. (1989) q.v., Vol. 1, pp. 189–205.
Sekine, T., Wyllie, P. J. 1982. Phase relationships in the system $KAlSiO_4$–Mg_2SiO_4–SiO_2–H_2O as a model for hybridization between hydrous siliceous melts and peridotite. *Contrib. Mineral. Petrol.* **79**, 368–374.
Sen, S. N., Narashima Rao, C. 1967. Igneous activity in the Cuddapah basin and adjacent area and suggestions on the paleogeography of the basin and adjacent areas. Proc. Symp. Upper Mantle Proj. GRB & NGRI Publ. 8, Hyderabad, pp. 261–285.
Sen, S. N., Narashima Rao, C. 1970. Chelima dykes. Proc. 2nd. Symp. Upper Mantle Sess. 5, Combined Geosurvey in Dharwar and Cuddapah Basin, Hyderabad, pp. 435–439.
Sen, S. N., Narashima Rao, C. 1971. Chelima dykes—A source for diamonds in Kurnool district, Andhra Pradesh. *Geol. Surv. India Misc. Publ.* **19**, 92–94.
Seyfert, C. K. 1987. Mantle plumes and hot spots. In Seyfert, C. K. (ed.), *The Encyclopedia of Structural Geology and Plate Tectonics*, Encyclopedia of Earth Sciences, Van Nostrand Reinhold Co., New York, Vol. 10, pp. 412–430.
Shand, S. J. 1922. The problem of the alkaline rocks. *Proc. Geol. Soc. S. Africa* **25**, xix–xxxii.
Shand, S. J. 1943. *Eruptive Rocks*. 2nd Ed., J. Wiley & Sons, New York.
Sharp, W. E. 1974. A plate tectonic origin for diamond-bearing kimberlites. *Earth Planet. Sci. Lett.* **21**, 351–354.
Shaw, H. R. 1972. Viscosities of magmatic silicate liquids: An empirical method of prediction. *Am. J. Sci.* **272**, 870–893.
Sheraton, J. W. 1983. Geochemistry of mafic igneous rocks of the northern Prince Charles Mountains, Antarctica. *J. Geol. Soc. Aust.* **30**, 295–300.
Sheraton, J. W., Cundari, A. 1980. Leucites from Gaussberg, Antarctica. *Contrib. Mineral. Petrol.* **71**, 417–427.
Sheraton, J. W., England, R. N. 1980. Highly potassic mafic dykes from Antarctica. *J. Geol. Soc. Aust.* **27**, 129–135.
Sheraton, J. W., Thomson, J. W., Collerson, K. D. 1987a. Mafic dyke swarms of Antarctica. In Halls, H. C., and Fahrig, W. F. (eds.) *Mafic Dyke Swarms*. Geol. Assoc. Can. Spec. Pap. 34, pp. 419–432.
Sheraton, J. W., Tingey, R. J., Black, L. R., Offe, L. A., Ellis, D. J. (1987b). Geology of Enderby Land and western Kemp Land, Antarctia. *Aust. Bur. Min. Res. Bull.* **223**, 1–51.
Sheriff, S. D., Shive, P. N. 1980. Paleomagnetism of the Leucite Hills volcanic field, southwestern Wyoming. *Geophys. Res. Lett.* **7**, 1025–1028.
Sheriff, S. D., Shive, P. N., Ogden, P. R. 1979. Paleomagnetism of the Leucite Hills of southwestern Wyoming. *EOS* **60**, 244–245.
Sigvaldson, G. E., Oskarson, N. 1986. Fluorine in basalts from Iceland. *Contrib. Mineral. Petrol.* **94**, 263–271.
Simpson, E. S. 1925. Contributions to the mineralogy of Western Australia. *J. R. Soc. W. Aust. Ser. I* **12**, 57–66.
Singh, G. D. 1971. Recovery of diamonds from Majhgawan tuff. *Geol. Surv. India Spec. Pub.* **19**, 169–175.
Sinor, K. P. 1930. *The Diamond Mines of Panna State in Central India*. The Times of India Press, Bombay, 189 pp.
Skeats, E. W., Richards, H. C. 1926. Report of the alkaline rocks of Australia and New Zealand Committee. *Aust. Assoc. Adv. Sci.* **18**, 36–45.
Skinner, E. M. W. 1989. Contrasting group I and group II kimberlite petrology: Towards a genetic model for kimberlites. In Ross et al. (1989) q.v., Vol. 1, pp. 528–544.
Skinner, E. M. W., Clement, C. R. 1979. Mineralogical classification of Southern African kimberlites. In Boyd and Meyer (179) q.v., Vol. 1, pp. 129–139.

Skinner, E. M. W., Scott, B. H. 1979. Petrography, mineralogy, and geochemistry of kimberlite and associated lamprophyre dykes near Swartruggens, W. Transvaal, R.S.A. Kimberlite Symposium II Cambridge (Extended Abstract).

Skinner, E. M. W., Smith, C. B., Bristow, J. W., Scott Smith, B. H., Dawson, J. B. 1985. Proterozoic kimberlites and lamproites and a preliminary age for the Argyle lamproite pipe, Western Australia. *Trans. Geol. Soc. S. Africa* **88**, 335–340.

Slingerland, R. L. 1977. The effects of entrainment on the hydraulic equivalence relations of light and heavy minerals in sands. *J. Sed. Petrol.* **47**, 753–770.

Smith, Christopher B. 1984. The genesis of the diamond deposits of the West Kimberley, W. A. In Purcell, P. G. (ed.), *The Canning Basin*. Proc. Geol. Soc. Aust./Petroleum Expl. Soc. Aust. Symp. Perth, pp. 463–474.

Smith, Christopher B., Lorenz, V. 1989. Volcanology of the Ellendale lamproite pipes, Western Australia. In Ross *et al.* (1989) q.v., Vol. 1, pp. 505–519.

Smith, Craig B. 1977. Kimberlite and mantle-derived xenoliths at Iron Mountain, Wyoming. M.Sc. thesis. Colorado State Univ. Fort Collins.

Smith, Craig B. 1983. Pb, Sr, and Nd isotopic evidence for sources of African Cretaceous kimberlites. *Nature* **304**, 51–54.

Smith, Craig B., Allsopp, H. L., Kramers, J. D., Hutchinson, G., Roddick, J. C. 1985. Emplacement ages of Jurassic-Cretaceous South African kimberlites by the Rb–Sr method on phlogopite and whole rock samples. *Trans. Geol. Soc. S. Africa* **88**, 249–266.

Smith, Craig B., Gurney, J. J., Skinner, E. M. W., Clement, C. R., Ebrahim, N. 1985. Geochemical character of southern Africa kimberlites: A new approach based upon isotopic constraints. *Trans. Geol. Soc. S. Africa* **88**, 267–280.

Smith, J. V., Breenesholtz, R., Dawson, J. B. 1978. Chemistry of micas from kimberlites and xenoliths. I. Micaceous kimberlites. *Geochim. Cosmochim Acta* **42**, 959–971.

Smithson, S. B. 1959. The geology of the southeastern Leucite Hills, Sweetwater County, Wyoming. M.Sc. thesis Univ. Wyoming, Laramie.

Smyslov, S. A. 1986. Kalsilite-bearing rocks of the Malo-Murun massif. *Geol. Geof.* **27**, 33–38 (Russian).

Snelling, N. J. 1965. Age determinations on three African carbonatites. *Nature* **205**, 491–492.

Snyder, F. G., Gerdemann, P. E. 1965. Explosive igneous activity along an Illinois–Missouri–Kansas axis. *Am. J. Sci.* **263**, 465–493.

Sobolev, A. V., Sobolev, N. V., Smith, C. B., Dubessy, J. 1989. Fluid and melt compositions in lamproites and kimberlites based on the study of inclusions in olivine. In Ross *et al.* (1989) q.v., Vol. 1, pp. 220–241.

Sobolev, A. V., Sobolev, N. V., Smith, C. B., Kononkova, N. N. 1985. New data on the petrology of olivine lamproites of Western Australia revealed by a study of magmatic inclusions in olivine. *Dokl. Akad. Nauk SSSR* **284**, 196–201.

Sobolev, V. S., Bazarova, T. Y., Yagi, K. 1975. Crystallization temperature of wyomingite from Leucite Hills. *Contrib. Mineral. Petrol.* **49**, 301–308.

Soloviev, D. A. 1972. Platform magmatic formations of East Antarctica. In Adie, R. J. (ed.), *Antarctica Geology and Geophysics*. Universitetsforlaget, Oslo, pp. 531–538.

Sörensen, H. 1974. *The Alkaline Rocks*. J. Wiley & Sons, New York.

Sörensen, K. 1983. Growth and dynamics of the Nordre Strømfjord shear zone. *J. Geophys. Res.* **88**, 3419–3437.

Spera, F. J. 1980. Aspects of magma transport. In Hargraves, R. B. (ed.) *Physics of Magmatic Processes*. Princeton Univ., Press, Princeton, New Jersey, pp. 265–3232.

Stacey, J. S., Kramers, J. D. 1975. Approximation of terrestrial lead isotope evolution by a two stage model. *Earth Planet. Sci. Lett.* **26**, 207–221.

Stearn, N. H. 1932. Practical geomagnetic exploration with the Hotchkiss superdip. *Trans. A.I.M.E. Geophys. Pros.*, 169.

Stecher, O., Thy, P. 1982. Kimberlite and lamproite dykes, West Greenland. Implications for melting of richterite, phlogopite and clinopyroxene in LIL-enriched mantle. *Terra Cognita* **2**, 212.

Stecher, O., Thy, P., Carlson, R. W. 1987. Sub-crustal metasomatism below West Greenland: Isotopic and geochemical evidence from lamproite and kimberlite dykes. Conf. on Oceanic and Continental Lithosphere: Similarities and Differences. Loudon, Edinburgh (abstract).

Steele, K. F., Wagner, G. H. 1979. Relationship of the Murfreesboro kimberlite and other igneous rocks of Arkansas. In Boyd and Meyer (1979) q.v., Vol. 1, pp. 293–399.

Stefanova, M. 1966. Petrochemical peculiarities of the Svidnya potassium alkaline rocks. *Bulgar. Acad. Sci. Bull.* **40**, 191–203 (Bulgarian).

Stefanova, M., Boyadzhieva, R. 1974. Genetic and chemical characterization of apatite from the biotite-kataphorite lamproites near the village of Svidnya, District of Sofia. *Bulgar. Acad. Sci. Geol Inst. Mineral Genesis* **1974**, 277–285 (Russian).
Stefanova, M., Boyadzhieva, R. 1975. On the geochemistry of Nb and Ta in K-alkaline lamproitic rocks from the village of Svidnya, district of Sofia. *Geochem. Mineral. Petrol.* **1**, 16–30 (Russian).
Stefanova, M., Pavlova, M., Amov, B. 1974. Geochemistry and isotopic composition of lead from potassium alkaline rocks of lamproite character. *Bulgar. Acad. Sci. Geol. Inst. Mineral Genesis* **1974**, 333–347 (Russian).
Stern, R. J., Bloomer, S. H., Lin, P.-N., Ito, E., Morris, J. 1988. Shoshonitic magmas in nascent arcs: New evidence from submarine volcanoes in the northern Marianas. *Geology* **16**, 426–430.
Stewart, D. B., Wright, T. L. 1974. Al/Si order and symmetry of natural alkali feldspars and the relationship of strained cell parameters to bulk composition. *Bull. Soc. Mineral. Cristall.* **97**, 356–377.
Stewart, J. W. 1970. Precambrian alkaline ultramafic/carbonatite volcanism at Qagssiarssuk, South Greenland. *Grønl. Geol. Unders. Bull.* **84**, 1–70.
Stokes, W. L. 1976. What is the Wasatch Line? *Rocky Mtn. Assoc. Geol. Symp.*, 11–25.
Stolz, A. J., Varne, R., Wheller, G. E., Foden, J. D., Abbott, M. J. 1988. The geochemistry and petrogenesis of K-rich alkaline volcanics from Batu Tara volcano, eastern Sunda arc. *Contrib. Mineral. Petrol.* **98**, 374–389.
Stone, C. G., Sterling, P. J. 1964. Relationship of igneous activity to mineral deposits in Arkansas. Arkan. Geol. Comm. Misc. Publ., Little Rock, Ak. 55 pp.
Stracke, K. J., Ferguson, J., Black, L. P. 1979. Structural setting of kimberlites in southeastern Australia. In Boyd and Meyer (1979) q.v., Vol. 1, pp. 71–91.
Streckeisen, A. 1978. Classification and nomenclature of volcanic rocks, carbonatites and melilititic rocks. I.U.G.S. Subcommission on the systematics of igneous rocks. *N. J. Mineral. Abh.* **134**, 1–14.
Suhadolc, P., Panza, G. F. 1988. The European–African collision and its effects on the lithosphere–asthenosphere system. *Tectonophys.* **146**, 59–66.
Sun, S. S., Jaques, A. L., McCulloch, M. T. 1986. Isotopic evolution of the Kimberley block, Western Australia. 4th Internat. Kimberlite Conf. Perth, Western Australia, Extended Abstract, pp. 346–348.
Sundeen, D. A., Cook, P. L. 1971. K–Ar dates from the upper Cretaceous volcanic rocks in the subsurface of west-central Mississippi. *Geol. Soc. Am. Bull.* **88**, 1144–1146.
Svisero, D. P., Hasui, Y., Drumond, D. 1979a. Geologia de kimberlitos de Alto Paranaíba, Minas Gerais. *Min. Metal.* **42**, 34–38.
Svisero, D. P., Meyer, H. O. A., Tsai, H. 1979b. Kimberlites in Brazil: An initial report. In Boyd and Meyer (1987) q.v., Vol. 1, pp. 92–100.
Svisero, D., Meyer, H. O. A., Haralyi, N. L. E., Hasui, Y. 1984. A note on the geology of some Brazilian kimberlites. *J. Geol.* **92**, 331–338.
Sykes, L. R. 1978. Intraplate seismicity, reactivation of pre-existing zones of weakness, alkaline magmatism and other tectonism post-dating continental fragmentation. *Rev. Geophys. Space Phys.* **16**, 621–688.
Tainton, K. M. 1987. A petrographic and geochemical investigation of the 81 Mile Vent lamproite, Western Australia. H.B.Sc. thesis. Univ. Natal Pietermaritzburg.
Tateyama, H., Shimoda, S., Sudo, T. 1974. The crystal structure of synthetic Mg_{iv} mica. *Z. Kristall.* **139**, 196–206.
Taylor, H. P. 1968. The oxygen isotope geochemistry of igneous rocks. *Contrib. Mineral. Petrol.* **19** 1–71.
Taylor, H. P., Turi, B. 1976. High ^{18}O igneous rocks from the Tuscan magmatic province, Italy. *Contrib. Mineral. Petrol.* **55**, 34–54.
Taylor, H. P., Gianetti, B., Turi, B. 1979. Oxygen isotope geochemistry of the potassic igneous rocks from the Roccamonfina volcano, Roman comagmatic region. *Earth Planet. Sci. Lett.* **46**, 81–106.
Taylor, L. A. 1984. Kimberlitic magmatism in the eastern United States: relationships to mid-Atlantic tectonism. In Kornprobst (1984) q.v., Vol. 1, pp. 417–424.
Taylor, W. R., Green, D. H. 1989. The role of C–O–H fluids in mantle partial melting. In Ross *et al.* (1989) q.v., Vol. 1, pp. 592–602.
Thiessen, R., Burke, K., Kidd, W. S. F. 1979. African hotspots and their relation to the underlying mantle. *Geology* **7**, 263–266.
Thoenen, R. S., Hill, R. S., Howe, E. G., Runke, S. M. 1949. Investigation of the Prairie Creek diamond area, Pike County, Arkansas. U.S. Bur. Mines Rept. Invest. 4549, 24 pp.
Thomas, G. A. 1958. Noonkanbah, WA-4 mile geological series, sheet SE51-12, Bur. Min. Resour. Australia explan. notes.

Thomas, W. A. 1978. Basement faults along the Appalachian–Ouachita continental margin. Basement Tectonic Contrib. No. 56, pp. 347–356.

Thompson, R. N. 1977. Primary basalt and magma genesis III. Alban Hills, Roman comagmatic province, central Italy. *Contrib. Mineral. Petrol.* **60**, 91–108.

Thompson, R. N., Fowler, M. B. 1986. Subduction-related shoshonitic and ultrapotassic magmatism: A study of Siluro-Ordovician syenites from the Scottish Caledonides. *Contrib. Mineral. Petrol.* **94**. 507–522.

Thornber, C. R., Roeder, P. L., Foster, J. R. 1980. The effect of composition on the ferric–ferrous ratio in basaltic liquids at atmospheric pressure. *Geochim. Cosmochim. Acta* **44**, 525–532.

Thy, P. 1982. Richterite–arfvedsonite–riebeckite–actinolite assemblage from MARID dykes associated with ultrapotassic magmatic activity in central West Greenland. *Terra Cognita* **2**, 247–249.

Thy, P. 1985. Contrasting crystallization trends in ultrapotassic lamproites from central W. Greenland. *Geol. Assoc. Can. Mineral. Assoc. Can. Ann. Mtg. Progr. with Abstr.* **10**, A63.

Thy, P. Stecher, O., Korstgard, J. A. 1987. Mineral chemistry and crystallization sequences in kimberlite and lamproite dikes from the Sisimut area, central west Greenland. *Lithos* **20**, 391–417.

Tikhonenkov, I. P., Kukharchik, M. V., Pyatenko, Y. A. 1960. Wadeite from the Khibiny massif and the conditions of its formation. *Dokl. Akad. Nauk. SSSR* **134**, 920–923 (Russian).

Tingey, R. J., 1981. Geological investigations in Antarctica 1968–1969: The Pryce Bay–Amery Ice Shelf–Prince Charles Mountains area. *Aust. Bur. Miner. Res. Geol. Geophys. Rec.* **1981**, 34.

Tingey, R. J., McDougall, I., Gleadow, A. J. W. 1983. The age and mode of formation of Gaussberg, Antarctica. *J. Geol. Soc. Aust.* **30**, 241–246.

Tombs, G. A., Sechos, B. 1986. Examination of surface features of Argyle diamonds from Western Australia. *Proc. 20th. Internat. Gemm. Conf. Aust. Gemmol.* **16**, 41–44.

Tompkins, L. A., Haggerty, S. E. 1984. The Koidu kimberlite complex, Sierre Leone: Geological setting, petrology and mineral chemistry. In Kornprobst (1984) q.v., Vol. 1, pp. 83–105.

Tompkins, L. A., Haggerty, S. E. 1985. Groundmass oxide minerals in the Koidu kimberlite dikes, Sierre Leone, West Africa. *Contrib. Mineral. Petrol.* **91**, 245–263.

Trail, D. S. 1963. The 1961 geological reconnaissance in the Southern Prince Charles Mountains, Antarctica. *Aust. Bur. Miner. Res. Geol. Geophys. Rec.* 1963, p. 155.

Tröger, W. E. 1935. *Spezielle Petrographie der Eruptivgesteine. Ein Nomenklatur Kompendium*. Verlag der Deutschen Mineralogischen Gesellschaft, Berlin.

Tronnes, R. G., Edgar, A. D., Arima, M. 1985. A high temperature study of TiO_2 solubility in Mg-rich phlogopite: Implications to phlogopite chemistry. *Geochim. Cosmochim. Acta* **49**, 2323–2329.

Tronnes, R. G., Takahashi, E., Scarfe, C. M. 1989. Stability and phase relations of K-richterite and phlogopite to 15 GPa. *Geol. Assoc. Canad. Mineral. Assoc. Can. Ann. Mtg. Progr. with Abstr.* **14**, A93.

Turner, F. J., Verhoogen, J. 1960. *Igneous and Metamorphic Petrology*. McGraw-Hill, New York.

Turpin, L., Velde, D., Pinte, G. 1988. Geochemical comparison between minettes and kersantites from the western European Hercynian orogen: Trace element and Pb–Sr–Nd isotope constraints on their origin. *Earth Planet. Sci. Lett.* **87**, 73–86.

Twenhofel, W. H., Bremer, B. H. 1928. An extension of the Rose Dome intrusives Kansas. *Am. Assoc. Petroleum. Geol. Bull.* **12**, 757–762.

Twenhofel, W. H., Edwards, E. C. 1921. The metamorphic rocks of Woodson County, Kansas. *Am. Assoc. Petroleum Geol. Bull.* **5**, 63–74.

Ulbrich H. H. G. J., Gomes, C. B. 1981. Alkaline rocks from continental Brazil. *Earth Sci. Rev.* **17**, 135–154.

Umbgrove, J. H. F. 1947. *The Pulse of the Earth* (second edition), Martinus Nijhoff, The Hague.

Upton, B. G. J., Stephenson, D., Martin, A. R. 1985. The Tugtutoq Older Giant Dyke Complex: Mineralogy and geochemistry of an alkali gabbro-augite syenite-foyaite association in the Gardar Province of South Greenland. *Mineral. Mag.* **49**, 623–642.

Ussing, N. V. 1912. Geology of the country around Julianhaab, Greenland. *Medd. Grønland* **38**, 1–146.

Van Bergen, H. J., Ghezzo, C., Ricci, C. A. 1983. Minette inclusions in the rhyodacite lavas of Mt. Amiata (Central Italy): Mineralogical and chemical evidence of mixing between Tuscan and Roman type magmas. *J. Volcanol. Geotherm. Res.* **19**, 1–35.

Van Kooten, G. K. 1980. Mineralogy, petrology, and geochemistry of an ultrapotassic basaltic suite, Central Sierra Nevada, California, U.S.A. *J. Petrol.* **21**, 651–684.

Van Kooten, G. K. 1981. Pb and Sr systematics of ultrapotassic and basaltic rocks from the Central Sierra Nevada, California. *Contrib. Mineral. Petrol.* **76**, 378–385.

Van Schmus, W. R., Hinze, W. J. 1985. The midcontinent rift system. *Ann. Rev. Earth Planet. Sci.* **13**, 345–383.

REFERENCES

Van Schmus, W. R., Bickford, M. E., Zietz, I. 1987. Early and Middle Proterozoic provinces in the central United States. In, Kroner, A. (ed.) *Proterozoic Lithospheric Evolution*. Am. Geophys. Union Geodynam. Ser. 17, pp. 43–68.

Veevers, J. J. 1958. Lennard River—4 mile geological series. Bur. Min. Resour. Australia Explan. Notes 11.

Veevers, J. J. 1984. *Phanerozoic Earth History of Australia*. Clarendon Press, Oxford. U.K.

Velde, D. 1967. Sur un lamprophyre hyperalcalin potassique: La minette de Sisco (Ile de Corse). *Bull. Soc. France Mineral. Crist.* **90**, 214–223.

Velde, D. 1968. A new occurrence of priderite. *Mineral. Mag.* **36**, 867–870.

Velde, D. 1969a. Micas from lamprophyres: Kersantites, minettes, and lamproites. *Bull. Soc. France. Mineral. Cristall.* **92**, 203–223.

Velde, D. 1969b. Minettes et kersantites. Une contribution a l'etude des lamprophyres. Thèse de Doctorat d'Etat, Fac. Sci. Univ. Paris.

Velde, D. 1975. Armalcolite-Ti-phlogopite-diopside-analcite-bearing lamproites from Smoky Butte, Garfield County, Montana. *Am. Mineral.* **60**, 566–573.

Velde, D. 1979. Trioctahedral mica in melilite-bearing eruptive rocks. *Carnegie Inst. Washington Yearbk.* **78**, 468–475.

Vemban, N. A. 1946. Chemical and petrological study of some dyke rocks in the Precambrian (Cuddapah traps). *Proc. Indian Acad. Sci. Sec. A* **22**, 347–378.

Venkatakrishnan, R., Dotiwalla, F. E. 1987. The Cuddapah salient: A tectonic model for the Cuddapah basin, India, based on Landsat image interpretation. *Tectonophysics* **136**, 237–253.

Venkataraman, K. 1960. Petrology of the Majhgawan agglomeritic tuff and the associated rocks. *Q. J. Geol. Mining Metall. Soc. India* **32**, 1–10.

Venturelli, G., Capedri, S., Di Battistini, G., Crawford, A., Kogarko, L. N., Celestini, S. 1984a. The ultrapotassic rocks from south eastern Spain. *Lithos* **17**, 37–54.

Venturelli, G., Thorpe, R. S., Dal Piaz, G. V., Del Moro, A., Potts, P. J. 1984b. Petrogenesis of calc-alkaline, shoshonitic and associated ultrapotassic Oligocene volcanic rocks from the northwestern Alps, Italy. *Contrib. Mineral. Petrol.* **86**, 209–220.

Venturelli, G., Mariani, E. S., Foley, S. F., Capredi, S., Crawford, A. J. 1988. Petrogenesis and conditions of crystallization of Spanish lamproitic rocks. *Can. Mineral.* **26**, 67–79.

Verkamp, J. C., Kalamarides, R. I. 1989. Hybridization processes in leucite tephrites from Vulsini, Italy and the evolution of the Italian potassic suite. *J. Geophys. Res.* **94**, 4603–4618.

Vila, J. M., Hernandez, J., Velde, D. 1974. Sur la présence d'un filon deroche lamproitique (trachyte potassique à olivine) recoupant le flysch de type Guerrouch entre Azzaba (ex-Jemmapes) et Hammam-Meskoutine dans l'Est du Constantinois (Algerie). *C. R. Acad. Sci. Paris Ser. D* **278**, 2589–2591.

Vink, G. E., Morgan, W. J., Vogt, P. R. 1985. The Earth's hot spots. *Sci. Am.* **252**(4), 50–57.

Vishnevskii, S. A., Dolgov, Y. A., Sobolev, N. N. 1986. Lamproites of the Talakhtakh diatreme on the eastern slope of the Anabar Shield. *Geol. Geof.* **27**, 171–27 (Russian).

Vladikin, N. V. 1985. First find of lamproites in the USSR. *Dokl. Akad. Nauk SSSR* **280**, 718–722 (Russian).

Vollmer, R. 1976. Rb–Sr and U–Th–Pb systematics of alkaline rocks: The alkaline rocks of Italy. *Geochim. Cosmochim. Acta* **40**, 283–295.

Vollmer, R. 1977. Isotopic evidence for genetic relations between acid and alkaline rocks in Italy. *Contrib. Mineral. Petrol.* **60**, 109–118.

Vollmer, R. 1989. On the origin of the Italian potassic magmas. *Chem. Geol.* **74**, 229–239.

Vollmer, R., Hawkesworth, C. J. 1980. Lead isotopic composition of the potassic rocks from Roccamonfina (South Italy). *Earth Planet. Sci. Lett.* **47**, 91–101.

Vollmer, R., Norry, M. J. 1983. Possible origin of K-rich volcanic rocks from Virunga, East Africa, by metasomatism of continental crustal material: Pb, Nd, and Sr isotopic evidence. *Earth Planet. Sci. Lett.* **64**, 374–386.

Vollmer, R., Ogden, P. R., Schilling, J. G., Kingsley, R. H., Waggoner, D. G. 1984. Nd and Sr isotopes in ultrapotassic volcanic rocks from the Leucite Hills, Wyoming. *Contrib. Mineral. Petrol.* **87**, 359–368.

Volynets, O. N., Anoshin, G. N., Puzankov, Y. M., Perepelov, A. B., Antipin, V. S. 1987. Potassium basaltoid of western Kamchatka: Appearance of rocks of the lamproite series in an island arc system. *Sov. Geol. Geophys.* **28**, 36–44.

Von Gümbel, C. W. 1874. *Die Paleolithischen Eruptivgesteine des Fichtelgebirges*. Verlag Franz, Munich.

Von Knorring, O., Cox, K. G. 1961. Kennedyite, a new mineral of the pseudobrookite series. *Mineral. Mag.* **32**, 676–682.

Vredenburg, E. W. 1906. Geology of the State of Panna. *Rec. Geol. Surv. India* **33**, 261–314.

Vyalov, O. S., Sobolev, V. S. 1959. Gaussberg, Antarctica. *Internat. Geol. Rev.* **1**, 30–40.

Wade, A. 1924. Petroleum prospects: Kimberley district of Western Australia and Northern Territory. Australia Commonwealth Parliament. Pap. No. 142.
Wade, A. 1937. The geological succession in the West Kimberley district of Western Australia. *Rep. Aust. Assoc. Adv. Sci.* **73**, 93.
Wade, A., Prider, R. T. 1940. The leucite-bearing rocks of the West Kimberley area, Western Australia. *Q. J. Geol. Soc. London* **96**, 39–98.
Wagner, C. 1986. Mineralogy of the type kajanite from Kalimantan: similarities and differences with typical lamproites. *Bull. Mineral.* **109**, 589–598.
Wagner, C., Velde, D. 1986a. The mineralogy of K-richterite-bearing lamproites. *Am. Mineral.* **71**, 17–37.
Wagner, C., Velde, D. 1986b. Davanite, $K_2TiSi_6O_{15}$ in the Smoky Butte (Montana) lamproites. *Am. Mineral.* **71**, 1473–1475.
Wagner, C., Velde, D. 1986c. Lamproites in North Vietnam? A re-examination of cocites. *J. Geol.* **94**, 770–776.
Wagner, C., Velde, D. 1987. Aluminous spinels in lamproites: Occurrence and probable significance. *Am. Mineral.* **72**, 689–696.
Wagner, C., Velde, D., Mokhtari, A. 1987. Sector zoned phlogopites in igneous rocks. *Contrib. Mineral. Petrol.* **96**, 186–191.
Wagner, H. C. 1954. Geology of the Fredonia Quadrangle, Kansas. *U.S. Geol. Surv. Map* GQ49.
Wagner, P. A. 1914. The diamond fields of Southern Africa. Transvaal Leader, Johannesburg.
Waldman, M. A., McCandless, T. E., Dummett, H. T. 1985. Geology and petrography of the Twin Knobs #1 lamproite, Pike County, Arkansas. In Morris and Pasteris (1987) q.v., pp. 205–216.
Walker, E. C., Edgar, A. D. 1989. High pressure–high temperature melting experiments on a diamondiferous olivine lamproite from Prairie Creek, Arkansas. *Geol. Assoc. Can. Mineral. Assoc. Can Ann. Mtg Progr. with Abstr.* **14**, A23.
Walker, K. R., Mond, A. 1971. Mica lamprophyre (alnöite) from Radok Lake, Prince Charles Mountains Antarctica. Aust. Bur. Miner. Res. Geol. Geophys. Rec. 1971/108. (unpublished)
Washington, H. S. 1906. *The Roman Comagmatic Region.* Carnegie Inst. Washington Publ. 57, 140 pp.
Waters, F. G. 1987. A suggested origin of MARID xenoliths in kimberlites by high pressure crystallization of an ultrapotassic rock such as lamproite. *Contrib. Mineral. Petrol.* **95**, 523–533.
Watson, J. V. 1980. Flaws in the continental crust. *Mercian Geologist* **8**, 1–10.
Watson, K. D. 1967. Kimberlites of eastern North America. In Wyllie P. J. (ed.) *Ultramafic and Related Rocks.* J. Wiley & Sons, New York, pp. 312–323.
Wellman, P. 1973. Early Miocene potassium-argon age for the Fitzroy lamproites of Western Australia. *J. Geol. Soc. Aust.* **19**, 471–474.
Wellman, P. 1983. Hotspot volcanism in Australia and New Zealand: Cainozoic and mid-Mesozoic. *Tectonophys.* **96**, 225–243.
Wellman, P. 1988. Development of the Australian Proterozoic crust as inferred from gravity and magnetic anomalies. Precambrian Res. 40/41, 89–100.
Wellman, P., McDougall, I. 1974. Cainozoic igneous activity in eastern Australia, *Tectonophysics* **23**, 49–65.
Wellman, P., Cundari, A., McDougall, J. 1970. Potassium-argon ages for leucite-bearing rocks from New South Wales, Australia. *J. Proc. R. Soc. New South Wales* **103**, 103–107.
Wendlandt, R. F. 1977. Barium phlogopite from Haystack Butte, Highwood Mountains, Montana. *Carnegie Inst. Washington Yearbk.* **76**, 534–539.
Wendlandt, R. F., Eggler, D. H. 1980a. The origins of potassic magmas 1: Melting relations in the systems $KAlSiO_4$–Mg_2SiO_4–SiO_2 and $KAlSiO_4$–MgO–SiO_2–CO_2 to 30 kb. *Am. J. Sci.* **280**, 385–420.
Wendlandt, R. F., Eggler, D. H. 1980b. The origin of potassic magmas 2. Stability of phlogopite in natural spinel lherzolite and in the system $KAlSiO_4$–MgO-SiO_2–H_2O–CO_2 at high pressures and temperatures. *Am. J. Sci.* **280**, 421–458.
Wezel, F. C. 1982. The Tyrrhenian Sea: A rifted krirkogenic-swell basin. *Mem. Geol. Soc. Italy* **24**, 531–568.
Wheller, G. E., Varne, R., Foden, J. D., Abbott, M. J. 1987. Geochemistry of Quaternary volcanism in the Sunda-Banda Arc and three component genesis of island arc basaltic magmas. *J. Volcanol. Geotherm. Res.* **32**, 137–160.
White, R. S., Spence, G. D., Fowler, S. R., MacKenzie, D. P., Westbrook, G. K., Bowen, A. N. 1987. Magmatism at rifted continental margins. *Nature (London)* **330**, 439–444.
White, S. H., Bretan, P. G., Rutter, E. H. 1986. Fault zone reactivation: Kinematics and mechanisms, *Phil. Trans. R. Soc. London* **A317**, 81–97.
Williams, A. F. 1932. *The Genesis of Diamond* (2 Vols.). Ernest Benn Ltd., London.
Williams, H., McBirney, A. R. 1979. *Volcanology.* Freeman, Cooper & Co., San Francisco.

REFERENCES

Williams, H. R., Williams, R. A. 1977. Kimberlites and plate tectonics in West Africa. *Nature* **270**, 507–508.
Williams, J. F. 1891a. Distribution and petrographic character of the igneous rocks from Pike County. In Williams (1891b) q.v., pp. 376–391.
Williams, J. F. (ed.) 1891b. *The Igneous Rocks of Arkansas*. Arkansas Geol. Surv. Rept. for 1890, 432 pp.
Wilson, J. T. 1963. Evidence from islands on the spreading of ocean floors. *Nature* **197**, 536–538.
Wilson, J. T., 1968. Static or mobile earth: The current scientific revolution. *J. Am. Philosoph. Soc.* **112**, 309–320.
Wilson, J. T. 1973. Mantle plumes and plate motions. *Tectonophys.* **19**, 149–164.
Wimmenauer, W. 1974. The alkaline province of central Europe and France. In Sørensen H (1974) q.v., pp. 238–271.
Winter, J. 1971. Geologisk beskrivelse af Tungarnit nunatak området, Vestgrønland, samt en kemisk undersøgelse af en bjergart fra området. Thesis, Univ. Aarhus, Denmark.
Winter, J. 1974. Precambrian geology of Tungarnit nunatak area, outer Nordre Strømfjord, central West Greenland. *Rapp. Grøn. Geol. Undersk.* **61**, 17 pp.
Wohletz, K. H., Sheridan, M. F. 1983. Hydrovolcanic explosions II. Evolution of basaltic tuff rings and tuff cones. *Am. J. Sci.* **283**, 385–413.
Wood, D. A. 1979. A variably veined suboceanic upper mantle. Genetic significance for mid-ocean ridge basalts from geochemical evidence. *Geology* **7**, 499–503.
Wooden, J. L., Mueller, P. A. 1988. Pb, Sr, and Nd isotopic compositions of a suite of Late Archean, igneous rocks, eastern Beartooth Mountains: Implications for crust-mantle evolution. *Earth Planet. Sci. Lett.* **87**, 59–72.
Wright, J. B. 1985. *Geology and Mineral Resources of West Africa*, George Allen and Unwin, London.
Wyborn, L. A. I., Page, R. W., Parker, A. J. 1987. Geochemical and geochronological signatures in Australian Proterozoic igneous rocks. In Pharoah, T. C., Beckinsale, R. D., and Rickards, D. (eds.) *Geochemistry and Mineralization of Proterozoic Volcanic Suites*. Geol. Soc. London Spec. Publ. 33, pp. 377–394.
Wyllie, P. J. 1980. The origin of kimberlite. *J. Geophys. Res.* **85**, 6902–6910.
Wyllie, P. J. 1988. Magma genesis, plate tectonics and chemical differentiation of the earth. *Rev. Geophys.* **26**, 370–404.
Yagi, K., Matsumoto, H. 1966. Note on the leucite-bearing rocks from the Leucite Hills, Wyoming, U.S.A. *J. Fac. Sci. Hokkaido Univ. Ser. IV Geol. Mineral.* **13**, 301–311.
Yoder, H. S. 1986. Potassium-rich rocks: Phase analysis and heteromorphic relations. *J. Petrol.* **27**, 1215–1228.
Yoder, H. S., Kushiro, I. 1969. Melting of a hydrous phase: Phlogopite. *Am. J. Sci.* **267A**, 558–582.
Zartmann, R. E. 1977. Geochronology of some alkalic rock provinces in eastern and central United States. *Ann. Rev. Earth Planet. Sci.* **5**, 257–286.
Zartman, R. E., Brock, M. R., Heyle, A. V., Thomas, H. H. 1967. K–Ar and Rb–Sr ages of some alkalic intrusive rocks from the central and eastern United States. *Am. J. Sci.* **265**, 848–870.
Zhuravleva, L. N., Yukina, N. V., Ryabeva, Y. G. 1978. Priderite, first find in the USSR. *Dokl. Akad. Nauk. SSSR* **239**, 141–143.
Zingg, A., Hunziker, J. C., Frey, M., Ahrendt, H. 1976. Age and degree of metamorphism of the Canavese Zone and the sedimentary cover of the Sesia Zone. *Schweiz. Mineral. Petrogr. Mitteil.* **56**, 361–375.
Zirkel, F. 1876. Microscopical petrography. U.S. Geol. Expl. 40th Parallel Rep. 6, pp. 259–261.
Zirkel, F. 1893. *Lehrbuch der Petrographie*. Bonn.
Zoback, M. L., Zoback, M. D. 1980. State of stress in the conterminous United States. *J. Geophys. Res.* **85**, 6113–6156.
Zwart, H. J., Dornslepen, U. F. 1978. The tectonic framework of central and western Europe. *Geol. Mijnbouw* **57**, 627–654.
Zyryanov, V. N. 1986. Experimental investigation of the lamproite formation. 4th. Internat. Kimberlite Conf. Perth, Western Australia, Extended Abstr., pp. 217–218.
Zyryanov, V. N., Zharikov, V. A. 1985. Experimental study of lamproite formation. *Dokl. Akad. Nauk SSSR* **283**, 116–119.

Index

Alkalinity
   definitions
      agpaitic, 10
      alkaline, 9–10
      miascitic, 10
      peralkalinity index, 10
      perpotassic, 10
      potassic, 10, 27
      sodic, 10
      ultrapotassic, 10, 27, 30
Alnöite, 11, 15, 35, 38, 61, 78, 82, 88, 119, 264–265, 286, 406
Amphibole
   classification, 217–218
   compositional trends and variation, 219–229, 232
   conditions of crystallization, 232–233
   in leucitites and leucite-bearing rocks, 232
   in mantle xenoliths, 230
   in minettes, 232
   paragenesis, 218–219, 354–356
   trace elements, 229–230
Analcite, 253–254
Anatase: *see* Titanium dioxides
Antimony, 342
Apatite
   composition, 283–286
   paragenesis, 282
Armalcolite, 274–276, 359
Arsenic, 341

Barite, 294
Barium, 121, 319–323, 341, 342, 351, 382, 399
Beryllium, 341
Biotite, xenocrystal, 194–195
Boninites, 27, 406
Boron, 341

Calcite, 59, 291

Carbonates, 291, 295, 365
Carbonatite, 45, 57, 59, 75, 93, 99–101, 118–119, 272, 276, 280, 393
Cesium, 340–341
Chlorine, 341
Chromium, 317, 343, 382
Clinopyroxene
   composition
      groundmass, 234, 238
      macrocrysts, 239–240
      phenocrystal, 224–229
   in group 2 kimberlites, 240
   in leucitites, 241
   in MARID-suite xenoliths, 240
   in minettes, 241
   paragenesis, 223–224, 354–356
Cobalt, 317, 382
Cocite, 12, 89–90
Cognate xenoliths, 178, 239, 356, 358, 361, 375–377, 388
Cohenite, 294
Composition
   isotopic
      group 1 kimberlites, 347, 350, 401
      group 2 kimberlites, 347, 350, 402
      lead, 348–350, 396–397, 400–401
      minettes, 347, 350
      neodymium, 345–348, 396–397, 399–400
      oxygen, 350–351
      potassic lavas, 347, 350–351
      strontium, 343, 345–347, 396–397, 399–400
   major elements
      alteration, 19, 295, 301, 309, 311–312
      averages, 295–298
      contamination, 296, 309, 387, 402
      differentiation trends, 301, 304–306, 308, 314, 386–397
      groundmass glasses, 311–313

443

Composition (*cont.*)
- group 2 kimberlites, 301
- inclusions of glass in minerals, 312
- inter-provincial, 296–297
- intra-provincial, 296–297, 301–311
- minettes, 301
- potassic lavas, 298–300
- trace elements
  - average abundances, 315, 320, 329–330
  - compatible, 314–318
  - group 2 kimberlites, 314, 316–318, 321–322, 326–327, 332, 334, 337, 339–340
  - incompatible, 319–342, 386, 388, 396
  - inter-element relationships, 342–343, 351
  - minettes, 314, 316–318, 322, 324, 326–327, 334, 337, 340–341
  - potassic lavas, 314, 316–318, 321–322, 326–327, 334, 336–337, 340–341
  - volatile, 341
  - *See also* individual elements

Copper, 317–318

Crater and pyroclastic facies
- base surge and pyroclastic flow deposits, 142–145, 147–148, 150
- epiclastic deposits, 146–147, 152
- examples of, 147–157
- pyroclastic fall deposits, 139–142, 148–157
- vent formation, 162–164
- vent morphology, 138–139, 148–151, 156–157, 162–166

Dalyite, 281–282
Davenite, 289–290
Diamond, 5–7, 46, 59, 72, 75, 79, 83, 86, 93, 95, 98, 107, 122, 127, 369, 371–374, 378–379, 383, 394, 400
- in lamproites
  - inclusion suites, 372–373
  - isotopic composition, 374
  - morphology, 372, 374

Diatremes, kimberlite, 129, 138, 165–166
Dolomite, 291

Experimental studies
- anhydrous melting, 353–354
- high pressure
  - olivine/madupitic lamproite, 360–362
  - phlogopite lamproite, 357–360
- oxygen fugacity, 367–368
- synthetic systems
  - garnet harzburgite—KOH, 367
  - $KAlSiO_4$—Mg_2SiO_4—SiO_2—CO_2, 364–365, 390–391, 390–391
  - $KAlSiO_4$—Mg_2SiO_4—SiO_2—F, 365, 393–394, 393–394
  - $KAlSiO_4$—Mg_2SiO_4—SiO_2—H_2O, 363–365, 393–394, 393–394

Experimental studies (*cont.*)
- $KAlSiO_4$—Mg_2SiO_4—SiO_2—H_2O—CO_2, 365, 391
- K_2O—MgO—Al_2O_3—SiO_2—H_2O, 395
- leucite—diopside—silica—H_2O, 362–363
- phlogopite-diopside, 366
- phlogopite-enstatite, 366
- water-saturated melting, 354–356

Exploration techniques
- area selection, 107–117, 119–120, 123–124, 378–379
- electrical, 380–381
- geochemical, 382–383
- gravity, 381
- indicator mineral, 6, 7, 86, 383–384
- magnetic, 380
- radiometric, 381–382
- remote sensing, 379
- seismic, 381

Fergusite, 17
Fluorine, 341, 365

Gallium, 341
Garnet, 11, 15, 38, 376–377, 383
Gas transfer, 386–387

Hafnium, 120, 323, 3351
Harmotome, 291
Harzburgite: *see* Mantle or Petrogenesis
Hollandite
- in carbonatites, 272
- in intrusive potassic rocks, 64, 273
- in kimberlites, 272–273
- in lamproites, 268–272
- *See also* Priderite

Hypabyssal facies
- characteristics, 157–158
- dikes, 158, 161
- examples of, 158–161
- sills, 158, 161
- volcanic necks, 157–159

Ilmenite, 276–278

Jeppeite, 273–274

Kajanite, 12, 90, 232
Kalsilite, 14, 17, 22–23, 33, 38, 55, 98, 251
Kamafugite, 14, 18, 21–22, 24, 30, 32–33, 38, 64, 73, 299–300, 317–318, 326–327, 336–337, 339–341, 343, 351, 365, 405–406
Katungite, 14
Khamrabayevite, 294
Kimberlite
- diatremes, 129, 138, 165–166
- group 1
  - isotopic composition, 347, 350

Kimberlite (cont.)
    mineralogy, 15, 212–213, 249, 272, 276, 286
    group 2
        isotopic composition, 347, 350
        major element composition, 10, 16, 18, 21, 26, 301
        mineralogy, 16, 38, 213, 240, 272
        trace element composition, 314, 317–318, 321, 326–327, 332–333, 337, 339–340, 351
    sensu lato, 5–8, 10–11, 39–40, 42, 33, 35, 38–39, 45, 57, 59, 103–106, 72, 75–76, 78–79, 82, 93, 95, 99–100, 103–106, 113, 118–119, 121–122, 127, 129, 138, 365, 374–375, 378–379, 381, 383–384, 386–387, 401–404

Lamproite
    ages, 39, 101, 104–107
    classification
        chemical, 2–3, 17–31
        mineralogical, 33–38
        modal, 4–5, 8, 31–34
    contemporaneous magmatism, 117–119
    definition, 36
    etymology, 2–3
    madupitic, 7, 34, 64, 178, 181–182, 189, 191, 194, 198, 216, 219, 221, 225, 233, 258, 261, 297–298, 300, 308–309, 353–354, 358, 361–362, 389–391, 400–401
    modern nomenclature, 35–36
    olivine 18, 24, 33, 179–180, 205, 244, 257–258, 261, 297–298, 300, 304–306, 312, 314–321, 323, 325–328, 338, 340–341, 343–348, 361–362, 365, 375, 377, 382–383, 394, 399–400, 402, 405
    otiose terminology, 11–12, 35
    recognition
        geochemical criteria, 37
        mineralogical criteria, 37–38
Lamprophyre, 3, 5, 15, 59, 61, 96–97, 99, 107, 121–122, 126, 264–265, 384, 405–406
Lava flow facies
    characteristics, 128–129
    examples, 130–136
    lava lakes, 128, 136, 156, 164–165
    lava viscosities, 137
Lead, 341
Leucite
    composition, 251–252, 267
    in leucitites, 252
    paragenesis, 249–251, 253–254, 354–356, 364
Leucite phonolite, 2, 10, 32
Leucite trachyte, 2, 32
Leucitite, 2, 5, 7, 10–13, 19, 21, 24, 32–33, 38, 79, 232, 241, 252, 257, 278

Lherzolite: *see* Mantle xenoliths
Lithium, 327–328
Lucasite, 292

Madupitic mica, 34, 170, 364
Mafurite, 10, 14, 18
Magnophorite, 217
Malignite, 17
Mantle
    metasomatism, 120–121, 124, 358, 388–389, 392–393, 397, 399
    oxygen fugacity, 369–370, 394
    plumes, 111–114, 118–124
    xenoliths
        harzburgite 375–377
        lherzolite, 127, 230, 375–376, 387
        MARID-suite, 108, 125, 230, 241, 403–404
Melilitite, 5, 14, 82, 365, 393, 406
Minette, 5, 10–11, 15, 18, 21, 24, 64–65, 90, 93, 95, 97, 126, 211–212, 232, 241, 266, 301, 314, 317–318, 321–322, 326–327, 334, 337, 339–340, 351, 384, 394, 405–406
Missourite, 17, 29, 90, 98
Moissanite, 294

Native metals, 294
Nephelinite, 4–5, 79, 79, 82
Nickel, 317, 343, 351, 382, 391
Niggli values, 3, 18–19
Niobium, 120, 325–327, 342–343, 351, 382
Noonkanbahite: *see* Shcherbakovite

Occurrence of lamproite
    descriptions of, 40–101
    fields, definition of, 40
    provinces, definition of, 40
    localities, definition of, 40
    volume, 39, 101, 103, 105
Olivine
    composition, 245–249
    in kimberlites, 249
    paragenesis, 244–245, 354–356, 358–359, 364, 375–376
Orangeite, 402–403
Orthopyroxene
    composition, 242–244
    paragenesis, 242, 358–359
Oxygen fugacity, 189, 216–217, 232–233, 362, 367–370, 400

Paleo-Benioff zone, 17, 47, 54, 116, 119–120, 124, 396, 398
Perovskite
    composition, 286
    paragenesis, 286

Petrogenesis
    contamination of kimberlites, 387, 401
    eclogite fractionation, 388
    fractional crystallization of kimberlites, 385–387
    fractional fusion, 390–391, 394
    incongruent melting of phlogopite, 389
    partial melting
        harzburgite, 358, 360, 365–366, 370, 390, 393–394, 398–400
        lherzolite, 388–390
        metasomatized mantle 392–393, 398–400
    within provinces, 399–401
    relationships to
        group 1 kimberlites 402–403
        group 2 kimberlites 402–403
        lamprophyres, 405–406
        MARID-suite xenoliths, 403–404
        potassic lavas, 404–405
    source mantle
        depletion of, 396
        metasomatism of 396–398
    subduction-related models, 395–396
    within provinces, 399–401
Phlogopite
    aluminous, 174–176, 178, 203, 217
    compositional trends, 178, 189–192, 195, 199–201, 206–207
    conditions of crystallization, 215–217
    groundmass composition, 174, 178, 180, 182, 187, 190–191, 194–196, 198
    in kimberlites, 212–213
    in leucitites, 214
    in mafurite, 215
    in minettes, 211–212
    paragenesis, 169–172, 354–356, 358–359
    phenocryst, composition, 172–173, 175, 179–181, 184–188, 190–205
    in Roman province type lavas, 214
    solid solutions, 207–211
    trace elements, 205–206
Phonolite, 21, 32–33
Platinum group elements, 318
Plutonic facies, 161–162
Potash nitre, 294
Potassic intrusive rocks, 17, 90, 98
Priderite
    Ba-titanites related to, 272
    composition, 268–271, 273
    paragenesis, 268, 359, 376
    See also Hollandite
Pseudoleucite, 250–251
Pyroclastic facies: see Crater facies

Rare earth elements, 327–337, 342
Rhabdophane, 293

Richterite: see Amphibole
Roedderite-like minerals, 291–292
Roman province type (RPT) lavas, 12, 17–18, 21, 23–25, 30, 214, 241, 252, 257, 266, 298–300, 314, 317–318, 321–322, 326–327, 334–335, 337, 339–341, 343, 351, 405
Rubidium, 338–340, 382
Rutile: see Titanium dioxides

Sanidine
    composition, 255–257
    in leucitites, 257
    paragenesis, 254–255, 354–356
Secondary phases, 294
Selenium, 341
Scandium, 314
Shcherbakovite, 287, 289
Shoshonite, 16–17, 55, 68, 71, 109, 119, 123, 387, 405
Silica activity, 22–23
Silicon dioxide, 293
Spinel
    in kimberlites, 264–265
    in lamprophyres, 265–266
    primary spinel
        composition, 258–264
        paragenesis, 257–258
    secondary aluminous spinels, 266–267
Strontianite, 291
Strontium, 319–323, 342–343, 351, 382
Sulfur, 341
Sylvite, 294

Tantalum, 325–326, 342–343, 351
Tectonic setting
    basement, 117
    continental rift zones, 56, 58–59, 74–75, 89, 114–115, 117, 123
    hot spots, 111–114, 122, 124
    lineaments, 45, 50, 56, 60, 63, 90, 92, 110–111, 117, 124
    mantle plumes, 111–114, 118, 124
    mobile belts, 58, 72, 75–76, 79, 82, 84, 89, 96, 99, 116, 120, 123–124, 378–379, 396–397
    orogenic zones, 65–66, 68, 70, 115–117, 123
    subduction, 16–17, 65, 108–109, 118–123, 343, 387, 395–397
    transform faults, 110, 124
Tetraferriphlogopite, 37, 174, 178, 182, 187–191, 194–196, 203, 206, 210–211, 213, 215–217
Thallium, 342
Thorium, 326–327, 343, 351, 381
Tin, 341–342
Titanate, Na-Fe, 294
Titanian potassium richterite: see Amphibole

INDEX

Titanium dioxides, 278–279
Titanosilicate K-Ba, unamed, 290
Tungsten, 342

Ugandite, 11, 14
Uranium, 327, 343, 351, 381

Vanadium, 314, 316

Wadeite, 279–282
Witherite, 291

Xenoliths: *see* Mantle

Yakutite, 17, 98
Yttrium, 327

Zeolite, 291
Zinc, 317–318
Zircon, 282
Zirconium, 120, 323, 325–326, 351, 382